Derivative

Securities

An Interactive
Dynamic Environment
with Maple V and Matlab

Pricing Derivative Securities

An Interactive Dynamic Environment with Maple V and Matlab

ELIEZER Z. PRISMAN
Schulich School of Business
York University

ACADEMIC PRESS
A Harcourt Science and Technology Company
San Diego San Francisco New York Boston London Sydney Tokyo

Front cover photographs: Examples of (bottom) PayoffCost, (top left) Function x sin (π/x), and (top right) arbitrage pricing. For more details, see Figures 6.16, 15.8, and 1.5, respectively.

Academic Press and the author have done their best to ensure the accuracy and completeness of the information presented in this book. The purpose of this book is to provide the reader with general information about pricing derivative securities. It is not intended as, and you should not consider it, financial or investment advice applicable to our specific situation. Academic Press and the author do not guarantee or warrant the content and do not endorse the purchase or sale of any security or investment or the use of any particular trading strategy. Academic Press and the author will not be liable to you or anyone else for damages or losses, including consequential damages, for investment decisions based on the information provided in this book.

This book is printed on acid-free paper. 

Academic Press
A Harcourt Science and Technology Company
525 B Street, Suite 1900, San Diego, California 92101-4495, U.S.A.
http://www.academicpress.com

Academic Press
Harcourt Place, 32 Jamestown Road, London NW1 7BY, UK
http://www.hbuk.co.uk/ap/

Library of Congress Catalog Card Number: 99-68356

International Standard Book Number: 0-12-564915-0

PRINTED IN THE UNITED STATES OF AMERICA
00 01 02 03 04 05 EB 9 8 7 6 5 4 3 2 1

This book is dedicated to

my father z"l
who taught me the foundations of logical thinking

my mother
who showed me how to expand beyond those bounds

my wife Esther,
whose support has been the inspiration for
so many other chapters of my life—without her
this book could not have been written

my children
Asaf, Eitan, Hanoch, and Ehud, for being there,
even though sometimes I was not

Contents

Preface **xv**
Software . . . xxii

1 Theory of Arbitrage **1**
1.1 A Basic One-Period Model . . . 1
1.2 Defining the No-Arbitrage Condition . . . 5
1.2.1 Identifying an Arbitrage Portfolio . . . 8
1.2.2 Law of One Price . . . 11
1.3 Pricing by Replication . . . 13
1.3.1 Three Special Contingent Cash Flows . . . 14
1.4 Stochastic Discount Factors (SDFs) . . . 18
1.4.1 SDFs and Risk-Neutral Probability . . . 22
1.5 Concluding Remarks . . . 28
1.6 Questions and Problems . . . 29
1.7 Appendix . . . 32
1.7.1 Complete Market . . . 32
1.7.2 Incomplete Market . . . 34
1.7.3 Incomplete Market and Arbitrage Bounds . . . 35
1.7.4 The No-Arbitrage Condition and Its Geometric Exposition . . . 42

2 Arbitrage Pricing: ***Equity Markets*** **47**
2.1 Market Structure and the Risk-Free Rate . . . 47
2.2 One-Period Binomial Model . . . 50
2.3 Valuing Two Propositions . . . 57
2.4 Forwards: A First Look . . . 61
2.4.1 Forward Contract on a Security . . . 62
2.4.2 Forward Contract on the Exchange Rate . . . 67
2.5 Swaps: A First Look . . . 72

2.5.1 Currency Swaps 72
2.5.2 Equity (Asset) Swap 74
2.6 General Valuation 77
2.6.1 The Risk-Free Rate of Interest Implicit in the Market 78
2.6.2 The Two Propositions 78
2.6.3 Forwards 79
2.6.4 Swaps 83
2.7 Concluding Remarks 86
2.8 Questions and Problems 87

3 Pricing by Arbitrage: *Debt Markets* **91**
3.1 Setting the Framework 91
3.2 Arbitrage in the Debt Market 94
3.2.1 Distinct Features of the Debt Market 99
3.2.2 Defining the No-Arbitrage Condition 101
3.3 Discount Factors 104
3.4 Discount Factors and Continuous Compounding 108
3.4.1 Continuous Compounding 108
3.5 Concluding Remarks 110
3.6 Questions and Problems 111
3.7 Appendix 113
3.7.1 No-Arbitrage Condition in the Bond Market 113

4 Fundamentals of Options **115**
4.1 Extending the Simple Model 115
4.2 Two Types of Options 116
4.3 Trading Strategies 125
4.3.1 Portfolios of Calls and Puts with the Same Maturity Date 127
4.4 Payoff Diagrams and Relative Pricing 141
4.4.1 Pricing Bounds Obtained by Relative Pricing Results 143
4.4.2 Put–Call Parity 148
4.5 From Payoffs to Portfolios 154
4.6 Concluding Remarks 164
4.7 Questions and Problems 165
4.8 Appendix 168
4.8.1 Explanation of Stripay 168
4.8.2 Procedural Issues 169

5 Risk-Neutral Probability and the SDF **183**
5.1 Infinite vs. Finite States of Nature 184
5.2 SDF for an Infinite Ω 187
5.3 Risk-Neutral Probability and the SDF 191
5.4 A First Look at Stock Prices 193
5.5 The Distribution of the Rate of Return 196
5.6 Paths of the Price Process 204
5.7 Specifying a Risk-Neutral Probability 208
5.8 Lognormal Distributions and the SDF 213
5.9 The Stochastic Discount Factor Function 215
5.10 Concluding Remarks 220
5.11 Questions and Problems 221

6 Valuation of European Options **223**
6.1 Valuing a Call Option 224
6.2 Valuing a Put Option 230
6.3 Combinations across Time 234
6.4 Dividends and Option Pricing 255
6.5 Volatility and Implied Volatility 259
6.5.1 Estimating Volatility from Historical Data 259
6.5.2 Implied Volatility 261
6.6 Concluding Remarks 265
6.7 Questions and Problems 266
6.8 Appendix 268
6.8.1 Estimating Implied Volatility Using Trial and Error . 268

7 Sensitivity Measures **271**
7.1 The Theta Measure 272
7.2 The Delta Measure 281
7.3 The Gamma Measure 288
7.4 The Vega Measure 293
7.5 The Rho Measure 298
7.6 Concluding Remarks 302
7.7 Questions and Problems 304
7.8 Appendix 307
7.8.1 Derivation of Sensitivity Measures 307
7.8.2 Sensitivities of Other Options 312
7.8.3 Signs of the Sensitivities 317

8 Hedging with the Greeks **323**
8.1 Hedging: The General Philosophy 323
8.2 Delta Hedging 326
8.2.1 Solving for a Delta Neutral Portfolio 326
8.3 Delta Neutral Portfolios 341
8.4 General Hedging 347
8.5 Optimizing Hedged Portfolios 364
8.6 Concluding Remarks 370
8.7 Questions and Problems 371

9 The Term Structure and Its Estimation **373**
9.1 The Term Structure of Interest Rates 374
9.1.1 Zero-Coupon, Spot, and Yield Curves 377
9.2 Smoothing of the Term Structure 383
9.2.1 Smoothing and Continuous Compounding 389
9.3 Forward Rate 393
9.3.1 Forward Rate: A Classical Approach 393
9.3.2 Forward Rate: A Practical Approach 396
9.4 A Variable Rate Bond 399
9.5 Concluding Remarks 402
9.6 Questions and Problems 404
9.7 Appendix 408
9.7.1 Theories of the Shape of the Term Structure 408
9.7.2 Approximating Functions 411

10 Forwards, Eurodollars, and Futures **413**
10.1 Forward Contracts: A Second Look 414
10.2 Valuation of Forward Contracts 415
10.3 Forward Price of Assets 423
10.3.1 Forward Contracts, Prior to Maturity, of Assets That Pay Known Cash Flows 427
10.3.2 Forward Price of a Dividend-Paying Stock 430
10.4 Eurodollar Contracts 432
10.4.1 Forward Rate Agreements (FRAs) 432
10.5 Futures Contracts: A Second Look 435
10.6 Deterministic Term Structure (DTS) 439
10.7 Futures Contracts in a DTS Environment 441
10.8 Concluding Remarks 448
10.9 Questions and Problems 449

11 Swaps: *A Second Look* **453**
11.1 A Fixed-for-Float Swap 453
11.1.1 Valuing an Existing Swap 458
11.2 Currency Swaps 461
11.3 Commodity and Equity Swaps 472
11.3.1 Equity Swaps 475
11.4 Forwards and Swaps: A Visualization 478
11.5 Concluding Remarks 479
11.6 Questions and Problems 481

12 American Options **485**
12.1 American Call Option 486
12.1.1 Arbitrage Bounds 486
12.1.2 Early Exercise Decision 487
12.2 American Put Options 488
12.2.1 Arbitrage Bounds 488
12.2.2 Early Exercise Decision 490
12.3 Put–Call Parity 492
12.4 The Effect of Dividends 495
12.4.1 A Call Option 495
12.4.2 A Put Option 501
12.5 Concluding Remarks 502
12.6 Questions and Problems 502

13 Binomial Models I **505**
13.1 Setting the Premises 505
13.2 No-Arbitrage and SDFs 511
13.2.1 No-Arbitrage 511
13.2.2 SDF 512
13.3 Valuation 521
13.3.1 Valuation with SDFs 521
13.3.2 Valuation by Replication 522
13.4 Numerical Valuation 529
13.4.1 Price Evolution 529
13.4.2 European Call 530
13.4.3 European Put 539
13.4.4 American Options 546
13.5 Concluding Remarks 554
13.6 Questions and Problems 555

14 Binomial Models II **557**
14.1 Binomial Model and Black–Scholes Formula 558
14.1.1 Binomial vs. Lognormal 558
14.1.2 Numerical Implementations 562
14.1.3 The Effect of Dividends 568
14.2 Risk-Neutral Probabilities 571
14.3 Futures and Forwards: A Symbolic Example 579
14.4 Brownian Motion 585
14.5 Concluding Remarks 590
14.6 Questions and Problems 592
14.7 Appendix 593
14.7.1 The Black–Scholes Formula as a Limit of the Binomial Formula 593

15 The Black–Scholes Formula **599**
15.1 An Overview 599
15.2 The Price Process: A Second Look 602
15.2.1 Stochastic Evolution: The Discrete Case 605
15.3 Simulation of Stochastic Evolution 608
15.4 Stochastic Evolution 615
15.5 Ito's Lemma 621
15.5.1 Heuristic Proofs of Ito's Lemma 623
15.5.2 Examples Utilizing Ito's Lemma 628
15.6 The Black–Scholes Differential Equation 632
15.6.1 A Second Derivation 640
15.7 Reconciliation with Risk-Neutral Valuation 642
15.8 American vs. European 644
15.9 Concluding Remarks 649
15.10 Questions and Problems 651
15.11 Appendix 652
15.11.1 A Change over an Instant 652
15.11.2 The Limit of a Random Variable 656
15.11.3 A More Rigorous Insight into Ito's Lemma 666

16 Other Types of Options **673**
16.1 Early Exercise, Dividends and Binomial Models 674
16.2 Indexes, Foreign Currency, and Futures 677
16.2.1 Stock Index Options 677
16.2.2 Currency Options 679
16.2.3 Options on Futures Contracts 682

16.3 Examples of Exotic Options 688
16.3.1 Binary (Digital) Options 689
16.3.2 Combinations of Binary and Plain Vanilla Options . . 694
16.3.3 Gap Options 695
16.3.4 Paylater (Cash on Delivery) Options 700
16.4 Interest Rate Derivatives 704
16.4.1 Black's Model 705
16.4.2 The Black, Derman, and Toy Model 714
16.5 Concluding Remarks 729
16.6 Questions and Problems 731

17 The End or the Beginning? 735

Index 743

Preface

It is an indisputable fact that derivative securities form part of the essential landscape of managerial financial education. An understanding of options is becoming a must for anyone whose career touches financial markets. Indeed, it is difficult to find a business school with a curriculum that does not contain as part of its core requirements some exposure to the theory of derivative securities.

When writing such a book, one is faced with numerous decisions, some of which include:

- What should be assumed about readers' prior knowledge of financial economics?
- What level of mathematical maturity should be assumed?
- How should balance be achieved (or equilibrium reached) between the institutional features of the subject matter and its core theory?

This book is directed at an audience of MBA or advanced BBA students. Since almost all business schools require the students to have taken basic courses in calculus, statistics, and probability, the book assumes this knowledge has been obtained. In the area of Financial Economics, the book counts on students having already been exposed to a basic course in Finance: concepts such as time-value-of-money and risk premium are assumed to be familiar to the readers.

However, as will be explained below, the uniqueness of the book lies in its way of overcoming mathematical complexity by explaining the intuition behind the scenes. Therefore, mastery of the contents of this book and comprehension of its methodology should not be an impossible task for students who are "rusty" in calculus and statistics or probability.

Regarding the third item of debate above, the book approaches the material to be learned from the viewpoint that picking up the institutional details is easier than learning the theory. Indeed, there are many excellent publications by the various financial exchanges that one can choose from in order to become familiar with these details (some of which are referenced in the book). Therefore, this book concentrates on the understanding of the core concepts of Option Pricing. It ensures that the students obtain knowledge enabling them to extend these concepts to situations not addressed in books or in the classroom. In my professional teaching experience, lack of such fundamental knowledge has proven to be an evident shortcoming for students.

Philosophy

I have chosen this equilibrium, one which leans heavily toward mastery of core theoretical concepts, based on my experience of teaching both graduate and undergraduate students. The academic community has a variety of different opinions on this subject. I have witnessed many occasions in which students are familiar with the jargon of derivative securities and are readily able to carry on a "cocktail party" conversation, but fail to appreciate the basic foundation.

It is for this reason that the book approaches the subject at hand from a no-arbitrage perspective, in a one-period state-preference model. The core foundation of modern valuation techniques is the no-arbitrage principle. Establishing intuitive insight and a solid understanding of the one-period model is the key to teaching the philosophy espoused in this book.

Equipped with this insight, even the reader who is not familiar with the intricacies of non-Euclidean spaces has a frame of reference with which to investigate and comprehend financial innovations never before tackled. The reader is able to translate the innovation from the continuous-time case (real world) to the equivalent one-period model, and verify intuition based on the principles of the simpler model. In almost all situations, understanding the phenomena in a one-period model and proceeding with care to the real-world case leads to the right conclusion. Thus, a thorough understanding of the one-period model serves as the frame of reference for those readers who are not familiar with, or who do not wish to spend time on, mathematical details. This understanding allows the reader to cope with (perhaps not the derivation) even the very complex issues which arise in considering this subject.

In keeping with this philosophy, the first chapter of the book concerns itself with a discrete-time model in which the major concepts of valuation are introduced. This chapter builds intuition and explains the basic concepts. Valuation techniques, stochastic discount factors, and risk-neutral probabilities, in the simplest of models are only some examples discussed. When a reader is faced with the more complex and realistic continuous-time version of this model, he or she has a solid foundation and intuition to which to refer.

Some Unique Features

One of the principal obstacles in teaching and learning the financial economic theory of derivative securities is the intricate, yet crucial, mathematical foundation of contingent claim pricing. A unique and attractive feature of this book is its use of symbolic computation. As the name of the book indicates, MATLAB and MAPLE V are used, making the analytical complexities of derivative securities' formulas attainable for the mathematically unsophisticated. MAPLE V, a symbolic computational language, can manipulate, solve, and evaluate mathematical expressions, both symbolically and numerically. MATLAB allows access to a kernel of MAPLE for symbolic computation and is a powerful numerical computational language. MAPLE also facilitates a direct linkage to MATLAB for numerical computations.

The book comes in an on-line version that presents an interactive and dynamic friendly environment allowing readers to learn through hands-on experience. The on-line version can be read with the MAPLE student version 5.0, that comes with the book. The book comes with a software package that presents an **I**nteractive **D**ynamic Environment for **A**dvanced **L**earning, called **IDEAL**. Pricing Derivative Securities, with the **IDEAL** software harnesses the power of MATLAB and MAPLE V, using modules written for the teaching and practice of derivative securities.

These packages can manipulate, solve, and evaluate mathematical expressions, analytically and numerically. For example, the user can input a system of equations into the software package and then prompt the machine for an analytic solution as separate algebraic structures. This is quite distinct from a computer package that solves equations numerically. Symbolic computation can leave the underlying variables unevaluated throughout the solution procedure and immediately produce comparative static results. Thus, sensitivity analysis and sensitivity measures or "stress testing" are easily done with either MAPLE and/or MATLAB.

The process of computing integrals, derivatives, transforms, convolutions, many other indispensables (for the purposes of option pricing) and mathematical operations can now be performed with a few keystrokes. An additional luxury of both packages is their excellent graphical capabilities.

Animation can be used to understand live price movements, and two- and three-dimensional on-screen images generated by the reader allow instant visual verification of the intuition behind the mathematical analysis.

This book comes with a collection of software modules, written in MATLAB and MAPLE V, which utilize the power of a symbolic (MAPLE) and a numeric (MATLAB) computational language. These modules will explain, illustrate, and, most importantly, bypass the computational maturity needed to understand the fundamentals of derivative securities.

The instructor will be able to demonstrate the payoff structure of a portfolio of options, even if the students have never taken a course in linear algebra. Students will be able to experiment with diffusion and Ito processes, probability distributions, and random variables, matrixes, and linear systems just as a child learns about electromagnetic theory by playing with an electronics set.

These modules are fully portable and a professional version of the educational package can serve the students long after they have graduated. Indeed, graduates of such a course should be able to take this set of tools for use for professional, practical applications. The ease of use is analogous to a collection of recipes that can be used repeatedly. Financial managers will be able to explore and tailor payoff structures of specific derivative securities by simply plugging them into the modules.

An interactive technology creates a friendly environment allowing readers to learn through immediate application of theory and concepts. Readers can use prepared MAPLE files, follow the text on-screen, and explore different numerical examples with no prior programming knowledge. These files exist on the attached CD as well as in an HTML format, allowing the reader to browse through them on any platform, needing only a browser on their machines. The prepared **G**raphic **U**ser **I**nterface (**GUI**) of MATLAB (Runtime) allows execution of all the commands in the HTML files.

In fact, readers can keep generating their own examples, verifying and investigating different situations not addressed in the book. The book exploits the concept that options are building blocks. Understanding options is akin to mastering the child's game of "Lego." Putting different blocks together, the financial architect constructs the desired payoff. Using MAPLE V and MATLAB as an exceptional powerful "calculator," option pricing becomes child's play.

Notwithstanding the above, the book can be used for advanced courses and by professional practitioners. Readers interested in Financial Engineering will be able to build on knowledge acquired here, and use the supplied modules or add their own and connect to C++ or MATLAB for more efficient numerical valuations. For this type of use it is recommended that readers familiarize themselves with the programming language used by MATLAB and/or MAPLE V. This book, therefore, may be very useful for practitioners who might build onto the supplied modules and modify them to suit their own purposes.

Structure of the Book: *Use in the Classroom*

Professors can use the on-line version of the book and project it in the class from their own notebook using the MAPLE package provided with the book. In this way, the on-line version can serve as lecture notes that can be executed on–the–fly by the professor. It saves the preparation of lecture notes and provides the students with visualization on–the–fly demonstrating the material. The examples and MAPLE commands are imbedded in the text. Hence, during the lecture, professors can, with a key stroke, recalculate the examples for different parameters, and animations can be run resampling the variables, demonstrating a true random dynamic environment. The solution manual that comes with the book will also be available in a MAPLE worksheet that contains the Questions and the Answers. Professors who prefer to teach with a hands-on approach using case-like methods will be able to use these files to construct a lab.

One possible way which has been successfully tested in my class is to conduct the lectures in a PC lab and split the time between lectures on theoretical material and hands-on experience. The hands-on experience can be conducted by assigning the exercises to the students in the MAPLE worksheet and asking them to solve it with MAPLE or the MATLAB GUI. Professors can alter and/or modify these files to tailor it to their test, and thereby emphasize the aspects they would like to concentrate on further. As I mentioned above, I would be interested in receiving such files and, pending your permission, post it on the web for interested readers and/or other professors.

The progression through the book is suited for use as an introductory course aimed at a less technically oriented audience. In particular, many business schools now offer shorter courses for derivatives, e.g., half a semester. These courses aim to equip students with basic tools and a certain moderate level of understanding, without going into technicalities. Such a course would benefit from Chapters 1, 2, 4, 5, and 6, which achieve this goal. For this reason, the book avoids technical issues in these chapters. Rather, Chapters 1, 2, and 4 concentrate on building an intuitive understanding of derivative securities. A full term course in Derivative Securities, that wants to avoid some technical issues, can be constructed based on Chapters 1 and 2, certain issues from Chapter 3, Chapters 5 through 8 (omitting the last section of Chapter 8), Chapter 12, a selection of issues from Chapters 13 and 14, and Chapter 16.

Other courses aimed at a more mathematically mature student, e.g., courses in financial engineering, will find a more technical explanation and computational exploration of the subject matter in subsequent chapters. Such a course can be built on Chapters 1 through 16. The instructor can skip some issues in Chapters 3 and 9 regarding the estimation of the term structure, as well as some topics in Chapter 14, and decide what appendixes to use in Chapter 15. The book can be used as a first course for Ph.D. students by covering the appendix and adding supplementary reading of relevant academic papers from the finance literature.

The progression of material in this book is largely dictated by the progression of the models used, rather than by the list of topics to be covered. This book emphasizes the role of the no-arbitrage condition as the modern financial tool for valuation. Therefore, the book starts by introducing the simplest model of an equity market. The no-arbitrage condition is defined and, in a subsequent chapter, is used to value simplified financial assets. Gradually, the book extends the simple model to a more realistic situation, permitting the valuation of more complicated securities. The result of this structure is that it may take more than one look at a certain topic to fully develop that topic, the first pass being more simple than later exploration. In some sense, though, this multifaceted investigation of topics serves as a review and provides a more thorough comprehension.

Acknowledgments

Many people have been involved, directly and indirectly, throughout the progress of this book, and their contributions were invaluable to me. Simultaneously while writing this book, I was writing a perpetual American call option (that I hope will never be exercised) to the members of my family, without whose support this project would never have been completed.

First, Mr. Seymour Schulich, the benefactor of the Schulich School of Business (SSB), who established in memory of his late friend, Nigel Martin, the chair I am holding. The secretarial support and time release provided by his endowment were crucial to a project of this size. York University and SSB granted me educational leave. Sandy Bell has worked closely and efficiently with me, well beyond her duty, for almost a decade. My research assistants, Aron Gottesman and Nadeem Siddiqi, and my assistants, Susana Graca and Mary Stearns, have assisted me during part of the period I was writing the book. The students in the Collaborative Program in Financial Engineering and the Ph.D. program I taught at York University provided many constructive comments that improved the book. The people at Waterloo Maple, Inc. and in particular Sean Curry at the MathWorks Inc., and my editors at Academic Press, Scott Bentley and Jim Mulliner (the copy editor), have been of great assistance throughout. The entire book is typeset by Scientific Work Place 2.5 and I am grateful for the LaTeX assistance and patience of George Pearson from MacKichan Software, Inc. (http://www.mackichan.com).

My thanks to the Ph.D. students I supervised: Kavous Ardalan, Andy Aziz, Narat Charupat, Ioulia Ioffe, Alexandra MacKay, Moshe Arye Milevsky, and Yisong Tian. Each one has taught me some aspect of life in general and contributed in their own way to my intuition and the methodological approach used in the book.

Ioulia Ioffe built the MATLAB GUI and was the first student to learn about the Pricing of Derivative Securities from this book. Her curiosity and professional integrity contributed profoundly to the order and mathematical presentation throughout the book. Alexandra MacKay read, commented, and significantly improved the exposition and flow of almost all the chapters of the book. Their "magic touches" are evident throughout. Finally, but by no means least, I would like to thank two of my very good friends. My colleague, Eliakim Katz for fruitful discussions, and Ruth Rosen who was always willing to help and dedicated many hours of her time. Ruth was the last person to edit the complete book ensuring that the copy editor's comments were incorporated correctly and suggested improvements to the text and help files of the software.

The book has, of course, benefited from the research achievements and papers of many people in the profession. In writing this textbook, I have tried to present the material in an intuitively appealing manner. References to papers in the literature are thus very limited. They serve only as pointers for readers who wish to explore the literature. Any errors remaining in the book are, of course, mine and I welcome any comments regarding this or other aspects of the book. (See www.academicpress.com/sbe/authors, for links to communication with me.)

Eliezer Z. Prisman

Software

The book comes in on-line versions that present an interactive and dynamic friendly environment allowing students to learn through hands-on experience.

- The MAPLE student version 5.0, that comes with the book (on the CD) allows the user to read the complete book on-line and execute all the commands that are embedded in the text.
- To read the book on-line in the HTML format with your browser, open the file Dersim.html in the ideal-html subfolder on the CD and follow the links. The HTML files on the CD facilitate reading the complete book on-line without installation of any software package, provided the user has a browser. Both animation and three-dimensional graphs can be viewed but not executed.
- The MATLAB GUI on the CD can be used to execute all the commands in the book. Simply read the hardcopy and type the required commands, or read the HTML files and cut and paste the commands into the appropriate window of the MATLAB GUI to be executed.

The IDEAL software is embedded in the on-line version of the book in MAPLE and in the MATLAB GUI. *Detailed instructions of how to read the on-line versions of the book are provided in the readme file on the CD.*

To read the on-line book in MAPLE you need first to install the MAPLE that you have received with this package. After MAPLE is installed double click on the shortcut to IDEAL MAPLE, which is located in the 'ideal-maple' subfolder on the CD.[1] Double click on the shortcut and open MAPLE. Choose File from the upper bar, then Open, and click on the file IDEAL.mws. Follow the on-line instructions, see Figure 0.1, to start your journey.

To execute the commands with the MATLAB GUI, double click on the shortcut to IDEAL MATLAB found in the 'ideal-matlab' subfolder on the CD. Click the HELP button in the main window of the 'IDEAL GUI' window and follow the on-line instructions to start your journey. This GUI, see Figure 0.2, opens two fields on your screen: an input field and an output field. Using the same syntax as for MAPLE (as exemplified in the

[1] If you have modified the standard installation of MAPLE (or if your hard drive is not designated by the letter C and/or your CD is not designated by the letter D) consult the readme file on the CD.

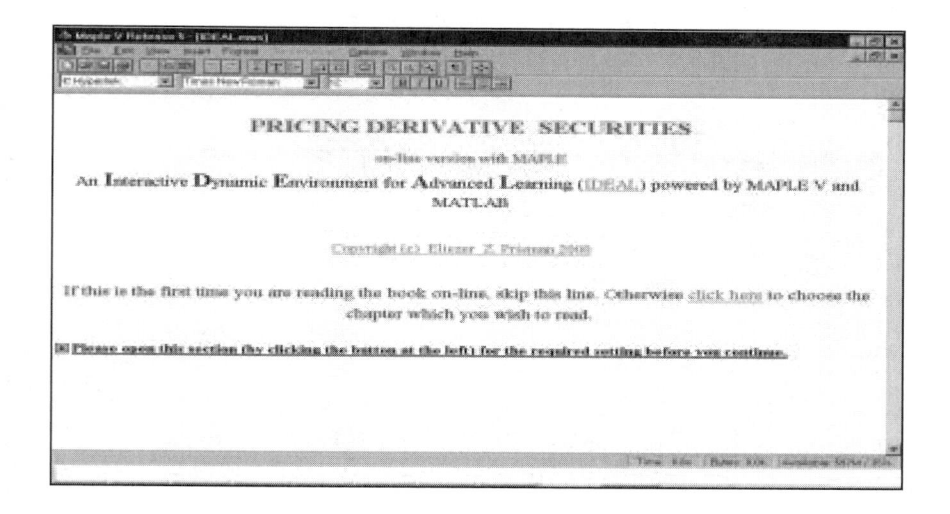

Figure 0.1: Reading the On-Line Book in MAPLE

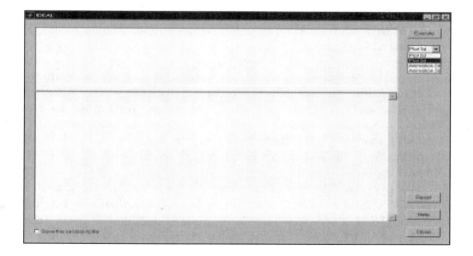

Figure 0.2: Executing the Commands via MATLAB

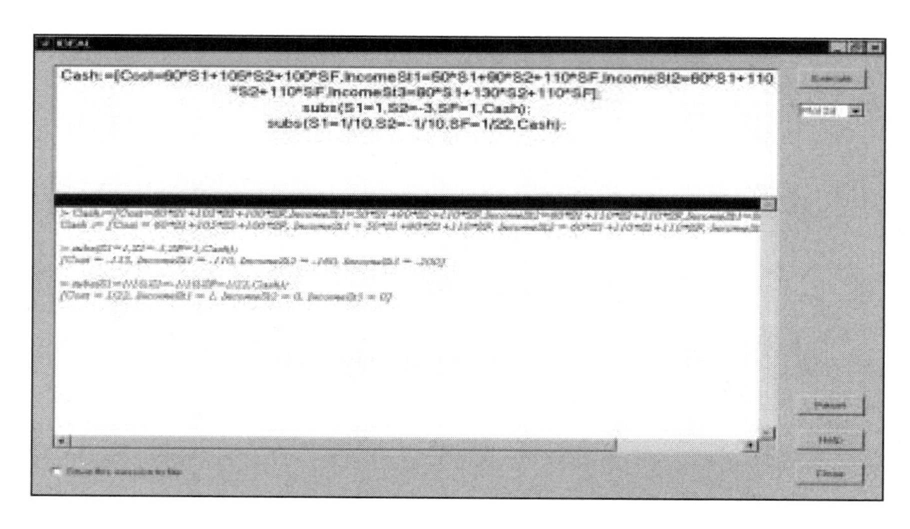

Figure 0.3: Input and Output in the MATLAB GUI

book) you can type in the input field (or in the plot window, see Figure 0.4) any command that appears in the book, keeping the same syntax as in the book. Once the input is inserted, you have to click on the Execute button to have it executed. The output is printed into the output file as demonstrated in Figure 0.3. Given the size of the output field after a few lines of output, you will have to scroll down in the output field to see the new output. Alternatively, you can click Reset to clean both the input and the output fields. You can save to file the diary of your session. However, this file must be created before you start your session. You can use any text editor (e.g., Notepad) to create an empty file and enter its name and path in the appropriate place. The output file, however, is better read with Word. Check the "Save this session to a file" box to have the input window of the file name appear.

The MATLAB GUI was built to keep the uniformity of the syntax between MAPLE and MATLAB. Since the book was first written with MAPLE, the MATLAB GUI was built to conform to the syntax required by MAPLE. ***Furthermore, when the text refers to submission of commands to MAPLE it should be interpreted to MAPLE and/or to the MATLAB GUI****. The IDEAL software whether read in the MATLAB GUI or with MAPLE comes with extensive help that can be read on-line by following the instructions in the packages.*

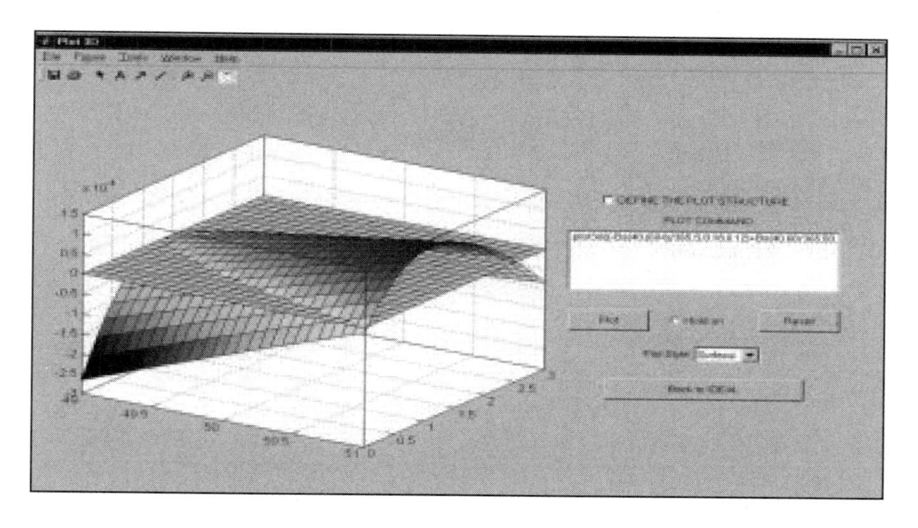

Figure 0.4: The Plot3d Window of the MATLAB GUI

Readers are introduced to the syntax as the material progresses, but only to the bare necessities of MAPLE and MATLAB which are essential for the best use of the book (not to detailed programming in MATLAB and/or MAPLE). The on-line version of the book contains a hyperlink to files that cover only a friendly set of commands. Working through these files should not be difficult, and the reader will have reviewed sufficient commands of MAPLE to be able to proceed through the remainder of the book. Readers who are already familiar with MAPLE and who do not need a refresher should proceed directly to Chapter 1.

MAPLE sessions are incorporated throughout the book and appear as part of the text. The entire book can be read on-line as HTML files or MAPLE files called Maple worksheets: identified as files with the ***.mws** extension. Having started MAPLE, the reader chooses the file dersim.mws from the subfolder MAPLE, which is in the IDEAL folder on the provided CD, and then "clicks" on the hyperlink to the file which she or he would like to load. These worksheet files (*.mws) are self-contained. The reader can (and should) experiment with running the different modules, using different examples, and verifying the valuation of various assets.

The commands in these files can be operated either via the MATLAB GUI or with MAPLE.

The on-screen MAPLE prompt is “>”. A command is issued by typing in the command immediately following such a prompt. A MAPLE command is printed in the book in the following font (*to execute the command place the cursor at the “ ; ” and hit return, or with the mouse click the “ ! ” in the top bar)*:

```
> Cash:={Cost=60*S1+105*S2+100*SF,IncomeSt1=50 *S1+90*S2
> +110*SF,IncomeSt2=60*S1+110*S2+110*SF,IncomeSt3=80*S1+
> 130*S2+110*SF};
```

This command defines a set in MAPLE called **Cash**. It has four elements: *Cost, IncomeSt1, IncomeSt2,* and *IncomeSt3*.

Every command in MAPLE must end with either “ :” or “ ;”. The latter is used when a reply from MAPLE is warranted and the former when the user desires only that MAPLE execute the command. A MAPLE reply will be printed in the following font (this is the MAPLE response to the above command):

$$Cash := \{Cost = 60\,S1 + 105\,S2 + 100\,SF,\ IncomeSt1 = 50\,S1 + 90\,S2 \\ +110\,SF,\ IncomeSt2 = 60\,S1 + 110\,S2 + 110\,SF,\ IncomeSt3 = 80\,S1 \\ +130\,S2 + 110\,SF\}$$

A reader who follows the book on-screen, see Figure 0.5, will find that the commands are already typed in the appropriate files. The reader can merely reexecute the printed commands. Learning is enhanced by altering the commands, varying them at will, in order to experiment with applications of the concepts and different (reader-generated) examples, in addition to the ones already in the prepared files. **It is this interaction and experimentation, making use of MAPLE and MATLAB together with the ability to bring to life on the screen the theoretical material of the chapter, which provide the unique, powerful, and entertaining way to learn about options.**

You can consult the **web page** of the book,

www.academicpress.com/sbe/authors,

for updates of the software and for links to communication with me on any aspect of the book or the software. If you like, you may contribute MAPLE worksheets that build on the book or suggest exercises, lab sessions or cases. Such worksheets, if allowed by the author, will be made accessible on the web.

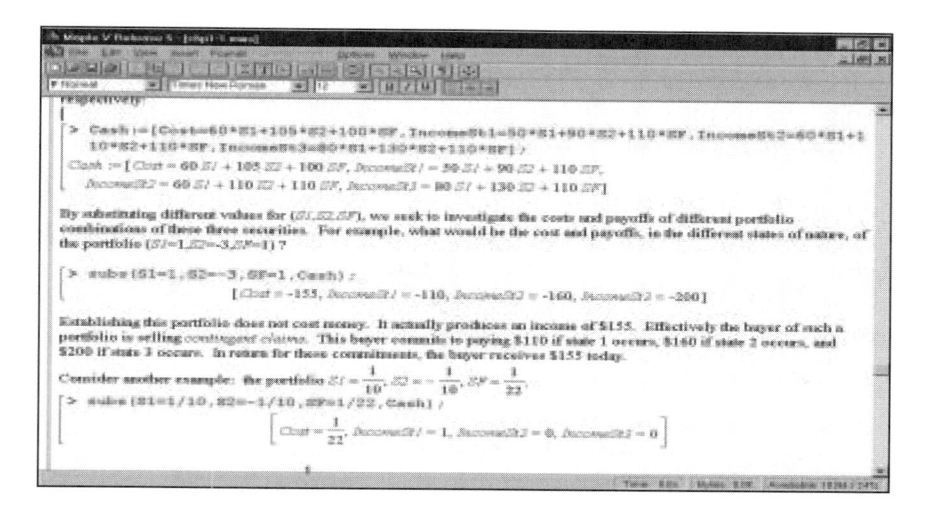

Figure 0.5: Input and Output in MAPLE

Disclaimer

Academic Press and the author have done their best to ensure the accuracy and completeness of the information presented in this book. The purpose of this book is to provide the reader with general information about pricing derivative securities. It is not intended as, and you should not consider it, financial or investment advice applicable to your specific situation. Academic Press, and the author do not guarantee or warrant the content and do not endorse the purchase or sale of any security or investment, or the use of any particular trading strategy. Academic Press and the author will not be liable to you or anyone else for damages or losses, including consequential damages, for investment decisions based on the information provided in this book.

Chapter 1

Theory of Arbitrage

1.1 A Basic One-Period Model

One of the key concepts in modern financial theory is "arbitrage", or rather the lack of it, in financial markets. The no-arbitrage (NA) condition, frequently referred to as the "law of one price", or by the idiom "no free lunch", is an essential tool in developing pricing methodologies. The consequences of this condition will be demonstrated by examining a simple model, (see [1]).

The real-world-conscious reader may be dismayed at the simplicity of the model, and may fear for its inability to address concepts of real-world significance, believing it so simple a model that it cannot possibly capture anything of use. However, this model is, in fact, extremely powerful. Though it seems, and indeed is, unrealistic, it provides the foundation for solid intuitive understanding, and can readily be augmented to capture the complexity of the illusive real-world case. Before a formal definition of the no-arbitrage condition is given, an example to motivate the definition is investigated. The example will be followed throughout this Chapter.

Consider a "simple" market with three securities and only three possible outcomes in the next time period. We refer to these possible outcomes as states of nature. Security 1 has a current price of \$60. Its price will decrease to \$50 if state 1 occurs, stay at \$60 if state 2 occurs, and will increase to \$80 if state 3 occurs. Similarly, security 2 has a current price of \$105 and its price will be \$90, \$110, and \$130 in states 1, 2, and 3, respectively.

One can think about the three states of nature as collections of exogenous circumstances which will cause the prices of the securities to be as above. At the present time, it is not known which state will be realized in

Price	State 1	State 2	State 3	Security
\$60	\$50	\$60	\$80	$S1$
\$105	\$90	\$110	\$130	$S2$
\$100	\$110	\$110	\$110	SF (bond)

Table 1.1: A One-Period Model

the next period. It is known only that one (and only one) of these states will occur. There is no assumption made about the probability of each state's occurrence, either subjective or objective, except that each state has a positive probability of occurrence. States, in this model, capture the uncertainty about the prices of the securities in the next time period.

The value of the third security in the next period does not depend on which of the three states of nature will occur. To put it differently, its value is not *contingent* on the state of nature in the next time period. It is a risk-free security, or a bond. In a later Chapter we will see that such a security is a 10% coupon bond. Its value will be \$110 in the next time period regardless of which state of nature materializes, and its present value (current price) is \$100. Hence, the interest rate in this simple model is 10%. Table 1.1 summarizes the prices and payoffs.

Our simple market is assumed to be a "perfect market" or frictionless market. It is a stylized market in which there are no transaction costs, no margin requirements, no taxes, and no limit on short sales. It is a market in which borrowing and lending are done at the same rate of interest, the risk-free rate, which is 10%. All investors in this market agree that each of the three states is a possible outcome in the next time period, and that one (and only one) of the states will actually occur. The investors all agree that each of the states has a positive probability of occurrence; the investors may, however, disagree about the actual value of the probability. We shall keep these assumptions throughout the book.

Consider buying a portfolio in this market of $S1$ units of security 1, $S2$ of security 2, and SF of the bond. We interpret positive values of $S1$ and $S2$ as buying the securities, or taking a long position, and negative values as short positions. Similarly, a positive value of SF means a long position in the bond, lending, and a negative value for SF means shorting the bond, which is equivalent to borrowing.

Taking a short position in a security generates positive proceeds. The short seller receives the price of the security being shorted in exchange for the commitment to pay back the value of that security in the next time period, whatever that value may end up being. Thus, taking a short position in security 1, say, for 2 units ($S1 = -2$), would cost $-2 \times \$60 = -\$120$ (i.e., it produces an income of \$120). A long position of two units of security 2, $S2 = 2$, would cost $2 \times \$105 = \$210$. The total **net cost** of such a portfolio is $-120 + 210 = \$90$. In general, the net cost of buying a portfolio of $S1$, $S2$, and SF units of each of the three securities is

$$\$60S1 + \$105S2 + \$100SF.$$

If this last quantity is positive, then establishing this position does indeed cost money; if it is zero, then establishing the position costs nothing; and if it is negative, then establishing the position actually produces income: **proceeds of the buy**. Portfolios for which $\$60S1 + \$105S2 + \$100SF \leq \$0$ (in our example), or more generally, for which the net cost is nonpositive, are referred to as self-financing portfolios. We also use the terminology of a negative cost portfolio if $\$60S1 + \$105S2 + \$100SF$ is actually negative. Self-financing portfolios require no out-of-pocket cost to establish the position; the buyer, however, may be committed for some payments in the future. These commitments are *contingent* on the state of nature in the next time period.

We will denote a portfolio by an *n-tuple* where n is the number of securities in the market. The first component of the *n-tuple* is the number of units of security 1 in the portfolio, the second component is the number of units of security 2 in the portfolio, and so on. Similarly, cash flows will be denoted by an *s-tuple*, where s is the number of states. Thus the cash flow of security 1 is $(50, 60, 80)$ and is interpreted as \$50 in state 1, \$60 in state 2, and \$80 in state 3. A cash flow should actually be thought of as a random variable taking the values specified in the *s-tuple*. The cash flow $(110, 110, 110)$ is a sure amount since it is not contingent on the state of nature. Whenever there is a possibility of confusion (as in the case of the example where $n = s = 3$), the tuple will be referred to specifically as either a cash flow or as a portfolio. The general form of the cash flow resulting from the transaction of buying the portfolio$(S1, S2, SF)$ is specified below.

In the current period, the cost of the portfolio is

$$\$60 \times S1 + \$105 \times S2 + \$100 \times SF,$$

and in the next time period, the portfolio will pay

$$\$50 \times S1 + \$90 \times S2 + \$110 \times SF$$

if state 1 is realized,

$$\$60 \times S1 + \$110 \times S2 + \$110 \times SF$$

if state 2 is realized, and

$$\$80 \times S1 + \$130 \times S2 + \$110 \times SF$$

if state 3 is realized.

Let us look at a few numerical examples. The array **Cash** is defined in MAPLE as having four components: **Cost,** the cost of the portfolio, and **IncomeSt1**, **IncomeSt2**, **IncomeSt3,** the payoffs from the portfolio in states $1, 2$, and 3, respectively.

```
> Cash:=[Cost=60*S1+105*S2+100*SF,\
> IncomeSt1=50*S1+90*S2+110*SF, \
> IncomeSt2=60*S1+110*S2+110*SF,\
> IncomeSt3=80*S1+130*S2+110*SF];
```

$$\begin{aligned} Cash := [Cost &= 60\,S1 + 105\,S2 + 100\,SF, \\ IncomeSt1 &= 50\,S1 + 90\,S2 + 110\,SF, \\ IncomeSt2 &= 60\,S1 + 110\,S2 + 110\,SF, \\ IncomeSt3 &= 80\,S1 + 130\,S2 + 110\,SF] \end{aligned}$$

By substituting different values for $(S1, S2, SF)$, we seek to investigate the costs and payoffs of different portfolio combinations of these three securities. For example, what would be the cost and payoffs, in the different states of nature, of the portfolio $(S1 = 1,\ S2 = -3,\ SF = 1)$?

```
> subs(S1=1,S2=-3,SF=1,Cash);
```

$$\begin{aligned} &[Cost = -155,\ IncomeSt1 = -110, \\ &\quad IncomeSt2 = -160,\ IncomeSt3 = -200] \end{aligned}$$

Establishing this portfolio does not cost money. It actually produces an income of \$155. Effectively the buyer of such a portfolio is selling *contingent claim.* This buyer commits to paying \$110 if state 1 occurs, \$160 if state 2 occurs, and \$200 if state 3 occurs. In return for these commitments, the buyer receives \$155 today.

Consider another example, the portfolio $S1 = \frac{1}{10}$, $S2 = -\frac{1}{10}$, $SF = \frac{1}{22}$.

```
> subs(S1=1/10,S2=-1/10,SF=1/22,Cash);
```

$$[Cost = \frac{1}{22},\ IncomeSt1 = 1,\ IncomeSt2 = 0,\ IncomeSt3 = 0]$$

Establishing this position will cost $\$\frac{1}{22}$. This position is actually a contingent claim on state 1. If state 1 occurs, the portfolio will pay $1; in any other state, the portfolio will pay nothing. This means that, in this market, the present value of $1 to be obtained contingent on state 1 occurring in the next time period is $\$\frac{1}{22}$. It would seem, therefore, that the price of $1, contingent on state 1, is implicit in the structure of the market. We shall return to this important key point in the pricing of derivative securities.

1.2 Defining the No-Arbitrage Condition

The no-arbitrage condition[1] is a statement about the nonexistence of certain self-financing portfolios. A self-financing portfolio must have short and long positions, since effectively the short positions finance the cost of the long positions. Such a portfolio can thus be decomposed into two subportfolios: a long portfolio and a short portfolio. Our definition of the no-arbitrage condition is given in terms of these portfolios.

Definition 1 (the No-Arbitrage Condition) *Consider all the portfolios for which*

- *(i) the income from the long part of the portfolio is* ***at least*** *equal to the payoff required for the short part of the portfolio, for every possible state of nature in the next time period, and for which*
- *(ii) the cost of the long part of the portfolio* ***does not exceed*** *the proceeds from the sale of the short part of the portfolio.*

The no-arbitrage condition is satisfied if, for all such portfolios, conditions (i) and (ii) are satisfied in such a way that

- *(iii) the income from the long part of the portfolio is* ***equal*** *to the payoff required for the short part of the portfolio, for every possible state of nature in next time period, and*
- *(iv) the cost of the long part of the portfolio* ***equals*** *the proceeds from the sale from the short part of the portfolio.*

[1] See [41] and [39] for a discussion of the no-arbitrage condition in a more complete setting.

To better illustrate this definition, let us first formulate more precisely the process of finding a portfolio satisfying conditions (i) and (ii) in the above definition. We formulate a problem whose solution will provide a portfolio satisfying conditions (i) and (ii), and, in case there is more than one such portfolio, we further specify that we are searching for the one which is the least expensive. We do this hoping to identify a negative cost portfolio, a portfolio which generates proceeds up front.

Consider a model of a market in which there are n securities and s possible states of nature, each of which could occur (and one of which will occur) in the next time period. The payoff from security i in state j is denoted by a_{ij}, x_i is the position in security i (either long or short, positive or negative), and P_i is the price of security i. To find the minimum cost portfolio satisfying conditions (i) and (ii), the problem to be solved[2] is

$$\begin{aligned} \min_{x_1,\ldots,x_n} \quad & \sum_{i=1}^{n} x_i P_i \\ \textit{such that} \quad & \sum_{i=1}^{n} x_i a_{ij} \geq 0, \; j = 1, \ldots, s. \end{aligned} \tag{1.1}$$

In the context of our example, $n = 3$, $s = 3$, we know the components of a_{ij} and we know each of the three P_i. We seek to solve

$$\begin{aligned} \min_{S1,\, S2,\, SF} \quad & 60S1 + 105S1 + 100SF \\ \textit{such that} \quad & 50S1 + 90S2 + 110SF \geq 0 \\ & 60S1 + 110S2 + 110SF \geq 0 \\ & 80S1 + 130S2 + 110SF \geq 0. \end{aligned} \tag{1.2}$$

Here is the formulation of this problem in MAPLE, and its solution:

```
> simplex[minimize](60*S1+105*S2+100*SF,\
> {50*S1+90*S2+110*SF>=0, 60*S1+110*S2+110*SF>=0,\
> 80*S1+130*S2+110*SF>=0});
```

$$\{S1 = 0,\ S2 = 0,\ SF = 0\}$$

The solution to the minimization problem (1.2) is ($S1 = 0$, $S2 = 0$, $SF = 0$). This means that the only way of satisfying conditions (i) and (ii) is by not buying any portfolio. It should be clear that, for the portfolio ($S1 = 0$, $S2 = 0$, $SF = 0$), conditions (i) and (ii) are satisfied as conditions (iii) and

[2] $\sum_{i=1}^{n} x_i$ stands for $x_1 + x_2 + \cdots + x_n$.

(iv). (The inequalities of (i) and (ii) are actually satisfied as equalities, as $0 = 0$.) Consequently, the no-arbitrage condition is satisfied.

It should be noted that having a value zero for the objective function in problem (1.1) does not guarantee that the no-arbitrage condition is satisfied.[3] If the value of the objective function in (1.1) is zero (i.e., $\sum_{i=1}^{n} x_i P_i = 0$), but for one state j, $\sum_{j=1}^{s} x_i a_{ij} > 0$ rather than $\sum_{j=1}^{s} x_i a_{ij} = 0$, x is an arbitrage portfolio. It costs nothing and it has a positive probability of generating income.

The procedure **CheckNA** was defined in MAPLE to check whether the no-arbitrage condition is satisfied. The inputs to the procedure are the set of prices and the value of the securities in each state of nature. If the no-arbitrage condition is satisfied, the procedure will print that out; if it is not satisfied, an arbitrage portfolio and the resultant arbitrage profit will be identified and printed out.[4] Here are a few examples:

First let us operate the procedures using our example. Note that the input to the procedure **CheckNA** is actually the rows of Table 1.1, e.g., [[row 1],[row 2],[row 3]], and the column of the security prices, e.g., [price of $S1$, price of $S2$, price of SF]:

```
> CheckNA([[50,60,80],[90,110,130],\
> [110,110,110 ]],[60,105,100]);
```

The no − arbitrage condition is satisfied.

Let us now examine what the output will be when the price of security 2 is changed to $100. (If you are working in a Windows environment, simply copy the above, paste it and change the number 105 to 100.)

```
> CheckNA([[50,60,80],[90,110,130],\
> [110,110,110 ]],[60,100,100]);
```

The no − arbitrage condition is not satisfied

An arbitrage portfolio is :

Short, 1, *of security*, 1

Buy, $\frac{7}{10}$, *of security*, 2

Short, $\frac{1}{10}$, *of security*, 3

---

[3] In the example followed in this Chapter, if the optimal value of (1.1) is zero the no-arbitrage condition is satisfied. This is due to the structure of the market in our example. It is a *complete* market (see explanation in the Appendix) and thus the possibility of having profit in the future can be transferred back to a value in the current period. The Appendix will explain this point further.

[4] While MAPLE is primarily a symbolic computation software, this procedure cannot accept a parameter as input unless that parameter is a numerical value.

The cost of this portfolio is zero

This portfolio produces income of, 2, *in state*, 1

This portfolio produces income of, 6, *in state*, 2

This portfolio produces income of, 0, *in state*, 3

The reader is invited to try a few market structures to see if they satisfy the no-arbitrage condition. The next section will look into the process of identifying an arbitrage portfolio.

1.2.1 Identifying an Arbitrage Portfolio

Now consider the change in the price of security 2 from \$105 to \$100. Is the no-arbitrage condition still satisfied? We know from the output of **CheckNA** that it is not. Let us take a deeper look at why arbitrage is possible. We solve the same minimization problem as we did before, adjusting the price of security 2 appropriately.

There are only two possible optimal values to the problem of the form (1.1), given the way in which it has been structured. One possible value, the optimal value of zero, we have already encountered above. The optimal value cannot be negative since $x = 0$, i.e., do not buy any portfolio, is a feasible solution and the objective function value at zero is zero. The other possibility is that the problem is unbounded. That is, there exists a portfolio for which the minimization resulted in a negative cost. However, this portfolio can be scaled so that the minimization results in larger and larger negative costs (positive up front proceeds of the buy). As this portfolio becomes larger and larger, the minimization value approaches infinity. Such an outcome would mean that there is an opportunity to make an infinite amount of money: an arbitrage portfolio has been identified.

```
>  simplex[minimize](60*S1+100*S2+100*SF,\
>  {50*S1+90*S2+110*SF>=0,60*S1+110*S2+110*SF>=0,\
>  80*S1+130*S2+110*SF>=0});
```

A NULL is returned by MAPLE, meaning that the problem is not bounded. To better understand this phenomenon, let us actually find a particular arbitrage portfolio in this market.

Since the problem had an unbounded solution, some sort of restriction is now added to the problem to bound the solution. In other words, we modify the problem so that the procedure can identify a particular arbitrage portfolio (rather than the procedure returning a NULL value). The restriction will be that the short position in the stocks cannot exceed 2 units. (Borrowing is still allowed in unlimited amounts.) This will result in a finite value for the problem, identifying an optimal portfolio which is an arbitrage portfolio.

```
> simplex[minimize](60*S1+100*S2+100*SF,\
> {50*S1+90*S2+110*SF>=0,60*S1+110*S2+110*SF>=0,\
> 80*S1+130*S2+110*SF>=0,S1>=-2, S2>=-2});
```

$$\{S1 = -2,\ SF = \frac{-7}{22},\ S2 = \frac{3}{2}\}$$

The optimal solution is now the portfolio $\left(S1 = -2,\ S2 = \frac{3}{2},\ SF = -\frac{7}{22}\right)$. We have identified how many units of each of the three securities to purchase (sell). How much arbitrage profit will this portfolio generate?

The array **Cash** is now redefined to reflect the change in the price of security 2. We proceed to substitute the optimal solution (the number of units of each security in the arbitrage portfolio, identified above) into the array **Cash** in order to find the cash flow (income) from this arbitrage portfolio:

```
> Cash:=[Cost=60*S1+100*S2+100*SF,\
> IncomeSt1=50*S1+90*S2+110*SF,\
> IncomeSt2=60*S1+110*S2+110*SF,\
> IncomeSt3=80*S1+130*S2+110*SF];
```

$$\begin{aligned} Cash := [Cost &= 60\,S1 + 100\,S2 + 100\,SF, \\ IncomeSt1 &= 50\,S1 + 90\,S2 + 110\,SF, \\ IncomeSt2 &= 60\,S1 + 110\,S2 + 110\,SF, \\ IncomeSt3 &= 80\,S1 + 130\,S2 + 110\,SF] \end{aligned}$$

```
> subs(S2=3/2,S1=-2,SF=-7/22,Cash);
```

$$[Cost = \frac{-20}{11},\ IncomeSt1 = 0,\ IncomeSt2 = 10,\ IncomeSt3 = 0]$$

The arbitrage portfolio identified above produces money with no risk and no initial investment: it is self-financing. Actually, buying the portfolio produces income now (the current time period) — it is a negative cost portfolio. Better yet, there is still the chance that state 2 will be realized, generating not only income up front, but income in the next time period (since $IncomeSt2 = 10$). It should be obvious why such a portfolio produces a "free lunch". In a well-functioning economic market, however, such prices for these three securities could not hold for very long. Why not?

Note that the procedure **CheckNA** identified an arbitrage portfolio, but not the one we identified above. **CheckNA** identified a self-financed portfolio which produces income next period in states 1 and 2. The arbitrage portfolio just identified produces income in the current period. It is a negative cost portfolio. If arbitrage opportunities exist, as here, there may be more than one way to go about exploiting them.

Suppose instead of the portfolio identified above, one were to buy that same portfolio, x, five times, i.e., $\left(S1 = -2 \times 5,\ S2 = \frac{3}{2} \times 5,\ SF = -\frac{7}{22} \times 5\right)$. (Remember, there are no transaction costs.) How much arbitrage profit would be generated by such a purchase?

In fact, having identified an arbitrage portfolio, albeit with finite arbitrage possibilities (recall that we bounded the problem so that we were not confronted with an infinite solution), an investor is not limited to buying such a portfolio only once, or even only five times. The investor could buy it any number of times, consequently multiplying his or her arbitrage profit by that number. Consider the portfolio

$$\left(S2 = \alpha \times \frac{3}{2},\ S1 = \alpha \times (-2),\ SF = \alpha \times \left(-\frac{7}{22}\right)\right),$$

that is, multiplying the holding of each security by α, where α is any positive number:

```
> subs(S2=alpha*3/2,S1=alpha*(-2),SF=alpha*(-7/22),Cash);
```

$$\left[Cost = -\frac{20}{11}\alpha,\ IncomeSt1 = 0,\ IncomeSt2 = 10\,\alpha,\ IncomeSt3 = 0\right]$$

It is evident that the larger α is, the more negative the cost of the portfolio, and hence the larger the arbitrage profit. The investor cannot generate infinite arbitrage profit, but can generate a very large finite amount of arbitrage profit.

```
> alpha:=2;
```

$$\alpha := 2$$

```
> subs(S2=alpha*3/2,S1=alpha*(-2),SF=alpha*(-7/22),Cash);
```

$$\left[Cost = \frac{-40}{11},\ IncomeSt1 = 0,\ IncomeSt2 = 20,\ IncomeSt3 = 0\right]$$

```
> alpha:=200;
```

$$\alpha := 200$$

```
> subs(S2=alpha*3/2,S1=alpha*(-2),SF=alpha*(-7/22),Cash);
```

$$\left[Cost = \frac{-4000}{11},\ IncomeSt1 = 0,\ IncomeSt2 = 2000,\ IncomeSt3 = 0\right]$$

Any or all investors who have identified this arbitrage portfolio will want to buy as much of it as possible. Thus, investors' demand for this portfolio will increase and prices will react accordingly. Demand for a short position in a security will decrease its price, while demand for a long position in a security will increase that price. Even if an arbitrage portfolio can be identified, market forces will not allow it to remain in existence. Prices that do not satisfy the no-arbitrage condition will not hold in the market for very long.

1.2.2 Law of One Price

Before reverting to a discussion of prices satisfying the no-arbitrage condition, a justification for the name the "law of one price" is given. The negative cost arbitrage portfolio identified above had a current price of $-\$\frac{20}{11}$. The portfolio

$$x = \left(S1 = -2,\ S2 = \frac{3}{2},\ SF = -\frac{7}{22} \right)$$

is decomposed into its long part

$$xl = \left(S1 = 0,\ S2 = \frac{3}{2},\ SF = 0 \right)$$

and into its short part

$$xs = \left(S1 = -2,\ S2 = 0,\ SF = -\frac{7}{22} \right)$$

such that[5] $x = xl + xs$. The cash flow from each subportfolio is given below. The long part produces

```
> subs(S2=3/2,S1=0,SF=0, Cash);
```

$$[Cost = 150,\ IncomeSt1 = 135,\ IncomeSt2 = 165,\ IncomeSt3 = 195]$$

and the short part produces

```
> subs(S2=0,S1=-2,SF=-7/22, Cash);
```

$$[Cost = \frac{-1670}{11},\ IncomeSt1 = -135,\ IncomeSt2 = -155,\ IncomeSt3 = -195]\ .$$

[5] $xl + xs$ should be interpreted as the portfolio whose $S1$ component is the sum of $S1$ in xl and $S1$ in xs, i.e., $0 + (-2) = -2$.

Reverse the short portfolio to be a long one instead — that is, the portfolio $-xs$. The resulting cash flow from $-xs$ is

```
> subs(S2=0,S1=2,SF=7/22, Cash);
```

$$\left[Cost = \frac{1670}{11},\ IncomeSt1 = 135,\ IncomeSt2 = 155,\ IncomeSt3 = 195\right]$$

The portfolio $-xs$ is more expensive than xl since it costs $\$\frac{1670}{11} = \$151\frac{18}{22}$, whereas xl costs only \$150. It produces the same cash flow as xl in states 1 and 3, and \$10 less than xl in state 2. Herein lies the core of the arbitrage opportunities.

An investor shorts the dearer portfolio and longs the cheaper, collects the difference in their prices, and assumes no risk. Since, in every state of nature, the long portfolio produces at least the same amount of income as should be paid out to the short position, no risk is assumed. Something is wrong in the prices of the securities which leads to an absurdity.

On one hand the cost of $-xs$ was just calculated to be $\$\frac{1670}{11}$. However, if we solve for the least-cost portfolio that generates a cash flow of at least $-xs$, we find the following portfolio:

```
> simplex[minimize](60*S1+100*S2+100*SF,\
> {50*S1+90*S2+110*SF>=135, 60*S1+110*S2+110*SF>=155,\
> 80*S1+130*S2+110*SF>=195,S1>=-2,S2>=-2});
```

$$\left\{S1 = -2,\ SF = \frac{-7}{22},\ S2 = 3\right\}$$

Substituting the solution into the array **Cash** yields

```
> subs(S1=-2,S2=3,SF=-7/22, Cash);
```

$$\left[Cost = \frac{1630}{11},\ IncomeSt1 = 135,\ IncomeSt2 = 175,\ IncomeSt3 = 195\right]$$

This portfolio produces \$20 more than $-xs$ in state 2 but it costs only $\$\frac{1630}{11}$; thus $-xs$ should cost even less. However, the price of $-xs$ was calculated above to be $\$\frac{1670}{11}$. In this market, the same portfolio xs has two prices. *It violates the law of one price.* The law of one price and the no-arbitrage condition are not equivalent conditions. A question at the end of this Chapter elaborates on this issue.

The no-arbitrage condition can also be defined in an equivalent way.

Definition 2 (Equivalent Definition of the No-Arbitrage Condition) *There exists no portfolio $x = (x_1, ..., x_n)$ such that*

$$
\begin{array}{lcl}
x_1 P_1 + x_2 P_2 +, \ldots, +x_n P_n & \leq & 0 \\
x_1 a_{11} + x_2 a_{21} +, \ldots, +x_n a_{n1} & \geq & 0 \\
x_1 a_{12} + x_2 a_{22} +, \ldots, +x_n a_{n2} & \geq & 0 \\
\cdot & \cdot & \cdot \\
\cdot & \cdot & \cdot \\
x_1 a_{1s} + x_2 a_{2s} +, \ldots, +x_n a_{ns} & \geq & 0
\end{array}
$$

and at least one inequality is strict.

A portfolio satisfying the above inequalities is a self-financing portfolio which imposes no commitments on its holder in the future time period. The left-hand side of the first inequality is the cost of the portfolio. The left-hand side of the second inequality is the income from the portfolio in state 1, and so on. If the first inequality is strict, the cost of the portfolio is negative. If only one of the other inequalities is strict, the cost of the portfolio is zero, but it grants its holder a positive probability of making money with no risk. If the state of nature in the future time period corresponding to the strict inequality occurs, the holder will receive income.

In a market where arbitrage opportunities do not exist, two portfolios that produce the same cash flow (even if the compositions of the portfolios are different) must have the same price. Thus, given a cash flow, its price should be equal to the price of a portfolio replicating this cash flow. There may be markets, however, where not every cash flow can be generated by the securities in the market. Such cash flows are said "not to be spanned" by the securities in the market. Such markets are called incomplete. In an incomplete market (where the no-arbitrage condition holds), only bounds on the prices, rather than one price, of such cash flows can be determined.[6] The discussion in this book will concentrate on complete markets in which the law of one price holds and thus allows the pricing of cash flows by replication.

1.3 Pricing by Replication

In a market where there are no arbitrage opportunities, the following question can be posed: Given the profile of a certain cash flow in the future, what should its price be? Only in markets with no arbitrage opportunities can such a question make sense. As noted above, if arbitrage opportunities exist there may be more than one price assigned to a portfolio, as the law

[6]The Appendix offers a more elaborate explanation of this concept. It also explains how to generate the bounds on the prices of these cash flows.

of one price does not hold. Moreover, one would not bother buying any security if one can make infinite amounts of money with no risk and no initial investment.

The question posed, however, may not have a unique solution. The absence of a unique solution may occur in markets termed *incomplete markets*. An incomplete market is a market where, given a cash flow profile, there might not be a portfolio that generates this cash flow. The Appendix elaborates more on this point. The market in our example is complete.

We thus return the price of security 2 to its original value of \$105 so that the no-arbitrage condition holds, and accordingly redefine the set **Cash**.

```
> Cash:=[Cost=60*S1+105*S2+100*SF,\
> IncomeSt1=50*S1+90*S2+110*SF,\
> IncomeSt2=60*S1+110*S2+110*SF,\
> IncomeSt3=80*S1+130*S2+110*SF];
```

$$\begin{aligned} Cash := [Cost &= 60\,S1 + 105\,S2 + 100\,SF, \\ IncomeSt1 &= 50\,S1 + 90\,S2 + 110\,SF, \\ IncomeSt2 &= 60\,S1 + 110\,S2 + 110\,SF, \\ IncomeSt3 &= 80\,S1 + 130\,S2 + 110\,SF] \end{aligned}$$

1.3.1 Three Special Contingent Cash Flows

There are three "special" cash flows in this market, sometimes also referred to as elementary cash flows. These can be priced now. The meaning of these cash flows is the topic of the next section. The first cash flow is $(1, 0, 0)$, i.e., in the next time period it promises \$1 in state 1, and \$0 in states 2 and 3. In order to price this cash flow, we use our minimization problem, as before:

```
> simplex[minimize](60*S1+105*S2+100*SF,\
> {50*S1+90*S2+110*SF>=1, 60*S1+110*S2+110*SF>=0,\
> 80*S1+130*S2+110*SF>=0});
```

$$\{SF = \frac{1}{22},\ S2 = \frac{-1}{10},\ S1 = \frac{1}{10}\}$$

The optimal solution of the minimization problem,

$$\left(S1 = \frac{1}{10},\ S2 = -\frac{1}{10},\ SF = \frac{1}{22}\right),$$

is the portfolio we seek. It stipulates the combination of the three securities to be purchased now in order to generate the first elementary cash flow in the next time period. Substituting the optimal solution of the minimization

Price	State 1	State 2	State 3
$\$\frac{1}{22}$	\$1	\$0	\$0
$\$\frac{25}{44}$	\$0	\$1	\$0
$\$\frac{13}{44}$	\$0	\$0	\$1

Table 1.2: Three Elementary Cash Flows

problem into **Cash** (as always) verifies the solution and allows the cost to be calculated.

```
>   subs(S1=1/10,S2=-1/10,SF=1/22,Cash);
```

$$[Cost = \frac{1}{22},\ IncomeSt1 = 1,\ IncomeSt2 = 0,\ IncomeSt3 = 0]$$

If $\frac{1}{10}$ units of security 1 are purchased, $\frac{1}{10}$ units of security 2 are shorted, and $\frac{1}{22}$ units of the bond are purchased, the resultant portfolio will generate the first elementary cash flow. This portfolio will cost $\$\frac{1}{22}$ now. In the same manner, the prices of the other two elementary cash flows, $(0, 1, 0)$ and then $(0, 0, 1)$, are calculated and substituted in **Cash**:

```
>   simplex[minimize](60*S1+105*S2+100*SF,\
>   {50*S1+90*S2+110*SF>=0, 60*S1+110*S2+110*SF>=1,\
>   80*S1+130*S2+110*SF>=0});
```

$$\{S2 = \frac{3}{20},\ S1 = \frac{-1}{5},\ SF = \frac{-7}{220}\}$$

```
>   subs(S1=-1/5,S2=3/20,SF=-7/220,Cash);
```

$$[Cost = \frac{25}{44},\ IncomeSt1 = 0,\ IncomeSt2 = 1,\ IncomeSt3 = 0]$$

```
>   simplex[minimize](60*S1+105*S2+100*SF,\
>   {50*S1+90*S2+110*SF>=0, 60*S1+110*S2+110*SF>=0,\
>   80*S1+130*S2+110*SF>=1});
```

$$\{S1 = \frac{1}{10},\ SF = \frac{-1}{220},\ S2 = \frac{-1}{20}\}$$

```
>   subs(S1=1/10,S2=-1/20,SF=-1/220,Cash);
```

$$[Cost = \frac{13}{44},\ IncomeSt1 = 0,\ IncomeSt2 = 0,\ IncomeSt3 = 1]$$

$EPor_1$	$EPor_2$	$EPor_3$	Security
$\frac{1}{10}$	$\frac{-1}{5}$	$\frac{1}{10}$	$S1$
$\frac{-1}{10}$	$\frac{3}{20}$	$\frac{-1}{20}$	$S2$
$\frac{1}{22}$	$\frac{-7}{220}$	$\frac{-1}{220}$	SF

Table 1.3: The Portfolios Generating the Elementary Cash Flows

We now know the portfolio combinations of securities 1 and 2, and of the bond, to purchase in the current time period in order to generate each of the three elementary cash flows in the next time period. We also know how much these portfolios cost. These results are summarized in Table 1.2.

In fact, now that we have the prices of these elementary cash flows, it is possible to value any cash flow in this market in a very simple way. Consider the cash flow (c_1, c_2, c_3) and suppose we can find three portfolios: portfolio j denoted by Por_j with a price of $PPor_j$ that pays $\$c_j$ if state j occurs and zero otherwise, for $j = 1, 2, 3$. In this case the price of the cash flow (c_1, c_2, c_3) must equal (by the replication arguments) $\Sigma_{j=1}^3 PPor_j$. However, the cash flow from Por_j is actually c_j times the cash flow from the jth elementary cash flow. Consequently, portfolio Por_j that produces $\$c_j$ contingent on state j is equivalent to buying c_j times the portfolio producing $1 contingent on state j. Portfolios producing the elementary cash flows will be referred to as elementary portfolios and will be denoted by $EPor_j$, $j = 1, 2, 3$. In our example the compositions of the elementary portfolios are summarized in Table 1.3.

Indeed, we already know the prices of the elementary portfolios. In our example the portfolio producing $1 contingent on state 1 costs $\$\frac{1}{22}$. Thus the portfolio producing $\$c_1$ contingent on state 1 will cost $\$\frac{1}{22}c_1$ and the cost of the cash flow (c_1, c_2, c_3) is simply

$$\frac{1}{22}c_1 + \frac{25}{44}c_2 + \frac{13}{44}c_3. \tag{1.3}$$

The above argument shows that the cash flow (c_1, c_2, c_3) is generated by buying c_1 times the portfolio $EPor_1$, c_2 times the portfolio $EPor_2$, and c_3 times the portfolio $EPor_3$; i.e., the required portfolio is thus

$$\sum_{j=1}^{3} c_j EPor_j. \tag{1.4}$$

Let us now apply this valuation technique to an example and compare it to our first valuation method. Consider the cash flow $(100, -20, 45)$. Based on equation (1.3) its price should be

```
>  (1/22)*100 -(25/44)*20 +(13/44)*45;
```

$$\frac{285}{44}$$

This result can be confirmed by solving the optimization problem that finds the least-cost portfolio producing this cash flow:

```
>  simplex[minimize](60*S1+105*S2+100*SF,\
>  {50*S1+90*S2+110*SF>=100, 60*S1+110*S2+110*SF>=-20,\
>  80*S1+130*S2+110*SF>=45}) ;
```

$$\{SF = \frac{219}{44},\ S2 = \frac{-61}{4},\ S1 = \frac{37}{2}\}$$

Substitution of the solution (the portfolio) in the array **Cash** results in the cost and the cash flows of the portfolio.

```
>  subs(S1=37/2,S2=-61/4,SF=219/44,Cash);
```

$$[Cost = \frac{285}{44},\ IncomeSt1 = 100,\ IncomeSt2 = -20,\ IncomeSt3 = 45]$$

Alternatively, the confirmation can be carried out without the need to solve an optimization problem. Since we already know the units of securities 1 and 2 and the bond in each of the elementary portfolios, we can find the portfolio producing this cash flow utilizing equation (1.4). This is done by summing the units of security 1 in $EPor_j$ (follow the first row of Table 1.3) and multiplying it by c_j for $j = 1, 2, 3$.

```
>  100*(1/10) -20*(-1/5) +45*(1/10);
```

$$\frac{37}{2}$$

The result of the above calculation is the number of units of security 1 in the portfolio producing the cash flow (c_1, c_2, c_3). We then repeat this for security 2

```
>  100*(-1/10)-20*(3/20) +45*(-1/20);
```

$$\frac{-61}{4}$$

and finally for the bond

```
>  100*(1/22) -20*(-7/220) +45*(-1/220);
```

$$\frac{219}{44}$$

We see that the two methods produce the same result. The prices of the portfolios producing the elementary cash flows can be interpreted as discount factors. This is elaborated on in the next section from a different perspective.

1.4 Stochastic Discount Factors (SDFs)

There is another way to calculate prices in markets which do not allow arbitrage opportunities. Consider the structure of a security from the first example of this Chapter. Take security 1: it can actually be "stripped" into three securities[7]:

- a security which pays \$50 contingent on state 1 occurring, and \$0 in both states 2 and 3;
- a security which pays \$60 contingent on state 2 occurring, and \$0 in either states 1 or 3; and
- a third security which pays \$80 contingent on state 3 occurring, and \$0 in the first two states.

In fact, these three securities can be further decomposed into even more elementary securities. The security that pays \$50 in state 1 and \$0 otherwise can actually be thought of as 50 units of a security that pays \$1 contingent on state 1 and \$0 otherwise, or 50 units of the first "special cash flow" (elementary cash flow), as described above.

In the same manner,[8] the other two securities into which security 1 was stripped can also be decomposed into the corresponding elementary cash flows. In a market where the no-arbitrage condition is satisfied, the law of one price holds. Thus, if the price of security 1 is \$60, it must equal the sum of the prices of the elementary cash flows which make it up. If the price of \$1 contingent on state j is d_j, $j = 1, 2, 3$, then the price of security 1 should be equal to the sum of the prices of the securities from which it is composed, namely,

$$50d_1 + 60d_2 + 80d_3 = 60. \qquad (1.5)$$

[7]This terminology is borrowed from the fixed income market where bonds are decomposed ("stripped") into their component parts. A bond promises payment of a coupon (the interest) every half a year (usually) and, at maturity, a payment of the last coupon and return of the principal value. Consider a bond maturing one year from today with a principal value of \$1000 and which pays 8% interest. It pays \$40 in half a year and \$40 plus \$1000 in a year. It can be stripped into a bond (zero coupon) which pays \$40 and matures in half a year, another zero coupon bond paying \$40 and maturing in one year, and a third zero coupon bond paying \$1000 when it matures also in one year. The price of the original bond must be the sum of the prices of its component parts, the parts into which it can be stripped. The states of nature in our example play the role of time in the bond example.

[8]The reader who is familiar with linear algebra will realize that this is nothing but representing the vectors of cash flows in terms of the natural basis of R^3 and requiring the prices of the securities to satisfy the same linear relation.

The other two securities in the market can be stripped in the same manner. Actually these securities are all composed of the same elementary cash flows, just in different quantities. It is analogous to different recipes which each use the same ingredients, but of different amounts, resulting in treats of different taste. Consequently, an equation like (1.5) can be written for each security in the market. Since each state has a positive probability of occurrence, the d_j, $j = 1, 2, 3$ must each be positive. The chance of getting \$1 in the next time period (albeit, not for certain) must be worth something today. The following system of equations is obtained:

$$\begin{aligned} 50d_1 + 60d_2 + 80d_3 &= 60 \\ 90d_1 + 110d_2 + 130d_3 &= 105 \\ 110d_1 + 110d_2 + 110d_3 &= 100 \\ d_j &> 0, \quad j = 1, 2, 3. \end{aligned} \tag{1.6}$$

The reasoning above suggests that if the law of one price holds, there must be a solution to this system.[9] The solution should equal the value calculated for these elementary cash flows by the replication method in the previous section.

The system of equations (1.6) includes the constraints $d_j > 0$, which cannot be handled numerically, but which MAPLE, being a symbolic computation package, can solve.[10] If a set of three d's satisfying the system of equations (1.6) can be found, then MAPLE will report the solution. Of course, the output reported by MAPLE would also indicate to us if there were no possible set of three d's which were able to satisfy the constraints. In this instance, MAPLE would report an output to the optimization problem of the empty set $\{\quad\}$.

The solution to the system of equations is

[9] In fact, the no-arbitrage condition is equivalent to this system of equations being consistent. In incomplete markets, there will be no unique solution for the d_j, $j = 1, 2, 3$. See the Appendix for more on this issue.

[10] In an older version of MAPLE (version 3), to handle the nonnegativity constraints, we have to proceed in a round-about way. Constraints of the type $d_j > 0$ could not be handled at all. To ensure the satisfaction of the nonnegativity constraintst an optimization problem with a "dummy" objective function must be set up. The artificial objective function is $0d_1 + 0d_2 + 0d_3$. We ask MAPLE to maximize this objective function subject to the system of equations (1.6). MAPLE will provide a solution to this optimization problem, which is a set of three d's, satisfying, among other things, the nonnegativity constraints. The optimization problem is actually the (linear programming) dual problem of (1.2). If one knows for sure that there exists a unique solution to the system and that the solution is positive, one can simply solve the system of equations leaving out the nonnegativity constraints.

```
> solve({50*d1+60*d2+80*d3=60,90*d1+110*d2+130*d3=105,\
> 110*d1+110*d2+110*d3=100,d1>0,d2>0,d3>0},{d1,d2,d3});
```

$$\{d1 = \frac{1}{22},\ d2 = \frac{25}{44},\ d3 = \frac{13}{44}\}$$

If, however, the no-arbitrage condition is not satisfied, there will exist no solution to this system. To demonstrate, the price of security 2 is changed to \$100 and the problem is submitted to MAPLE.

```
> solve({50*d1+60*d2+80*d3=60,90*d1+110*d2+130*d3=110,\
> 110*d1+110*d2+110*d3=100,d1>0,d2>0,d3>0},{d1,d2,d3});
```

It can be seen now that the solutions to (1.6), $d_1 = \frac{1}{22}$, $d_2 = \frac{25}{44}$, $d_3 = \frac{13}{44}$, have the same values as the prices for the three elementary cash flows in Table 1.2. The meaning of the d_j is confirmed: each d_j is the present value of a \$1 contingent claim on the state of nature j in the next period. The present value of the cash flow $(1, 0, 0)$ is $d_1 = \$\frac{1}{22}$.

Every security, or contingent claim, in this market can be "stripped" into its basic building blocks of \$1 contingent on each of the states of nature j with respective prices of d_j. Thus, the price of every security in this market can be calculated as the sum of the prices of the elementary cash flows composing the security. It is useful, then, to define a function in MAPLE which calculates the prices of securities in the market using the elementary cash flows. The arguments of the function are the component pieces of the security to be priced (in terms of the elementary cash flows), and the function will calculate the price of the security. The function

$$Vdis\,(c_1, c_2, c_3) = \sum_{j=1}^{3} c_j d_j \tag{1.7}$$

is defined in MAPLE:

```
> Vdis:=(c1,c2,c3)->c1*1/22+c2*25/44+c3*13/44;
```

$$Vdis := (c1,\ c2,\ c3) \to \frac{1}{22}\,c1 + \frac{25}{44}\,c2 + \frac{13}{44}\,c3$$

We are now in a position of being able to value every type of cash flow in this market. Let us start by valuing two cash flows $(0, 10, 30)$ and $(10, 20, 40)$. The values of these cash flows are:

```
> Vdis(0,10,30);
```

$$\frac{160}{11}$$

```
> Vdis(10,20,40);
```

$$\frac{260}{11}$$

Consider a somewhat more complicated cash flow. This type of a cash flow may seem bizarre now. Shortly, however, we will see that it is associated with a certain contingent claim that is very common in the market. Consider a cash flow that is contingent on the value of security 1 in the following way. If the value of security 1 in the next time period is below \$50 it offers nothing. If the value of security 1, in the next time period, is above \$50 it offers the difference between 50 and the value of the security. The reader may verify that the cash flow $(0, 10, 30)$ is a result of such a contingent cash flow, since security 1's payoff, depending on the realized state of nature, will be \$50, \$60, or \$80. This type of a cash flow can be written as

$$(\max(0, 50 - 50), \max(0, 60 - 50), \max(0, 80 - 50)),$$

where $\max(x, y)$ is x if $x > y$ and y if $y \geq x$.

We can further generalize this form of cash flow by replacing the \$50 above with some number K, and investigate the value of the cash flow

$$(\max(0, 50 - K), \max(0, 60 - K), \max(0, 80 - K)).$$

For the moment K is left unassigned. Thus, in general, the value of such a cash flow will be

```
> Vdis(max(0,50-K),max(0,60-K),max(0,80-K));
```

$$\frac{1}{22}\max(0,\ 50 - K) + \frac{25}{44}\max(0,\ 60 - K) + \frac{13}{44}\max(0,\ 80 - K)$$

Here we have made use of the symbolic computation power of MAPLE. Rather than always specifying numerical values, in most cases MAPLE allows the user to input some variables as symbols. This allows the user to see a more general form of the output formula: the resultant expression in terms of the variable specified only as a symbol. Here, the symbol is K. We wish to understand the role of K in the pricing of these two propositions. K is different for each proposition, but the general form of the propositions is the same. We can also use MAPLE to get a visual image of what the resultant session for the value of the proposition looks like in terms of the variable K.

```
> plot(Vdis(max(0,50-K),max(0,60-K),max(0,80-K)),\
> K=0..100,title=‘Figure 1.1:The Value of the\
> Proposition as a Function of K‘,\
> titlefont=[TIMES,BOLD,10]);
```

Can you explain the shape of the graph in Figure 1.1? What is the value of the proposition when $K = 0$? Why is there a kink at $K = 60$? Why is

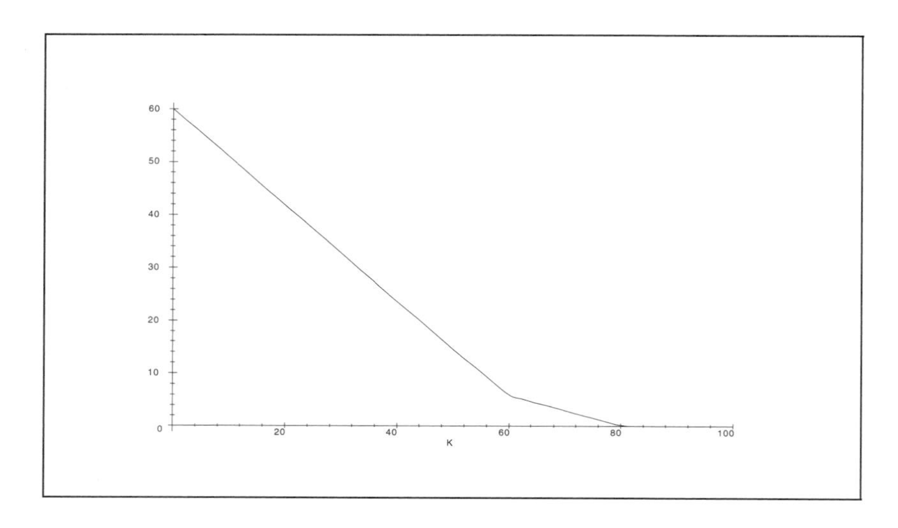

Figure 1.1: The Value of the Proposition as a Function of K

the value 0 when $K \geq 80$? How will the value of a proposition change as a function of the price of security[11] 1 in state 3 when the payoff in the third state of nature in the next time period, c_3, is allowed to vary between 0 and 160?

```
> plot3d(Vdis(max(0,50-K),max(0,60-K),max(0,c3-K)),\
> K=0..180,c3=0..160, title=`Figure 1.2:  The Value\
> of the Proposition as a Function of K and c3`,\
> titlefont=[TIMES,BOLD,8], axes=normal,\
> labels=[K,c3,Value], orientation=[-51,57]);
```

On the screen you can actually look at Figure 1.2 from different perspectives and you can see it in color. Can you describe the region where the value of the proposition is zero? This region is in the (c_3, K) plane.

1.4.1 SDFs and Risk-Neutral Probability

The d's should be thought of as discount factors — each state of nature in the next time period has its own discount factor. The d's are therefore also

[11] The value of the discount factor is assumed to be unaffected by this change. So we actually are examining the value of such a proposition on a new security that is added to the market.

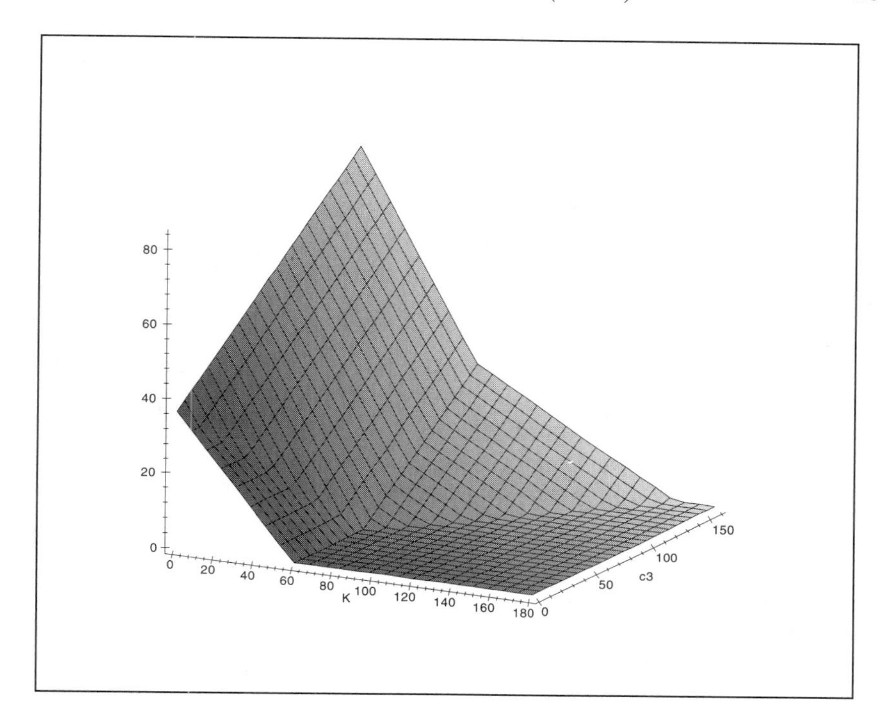

Figure 1.2: The Value of the Proposition as a Function of K and c_3

referred to as the *states' prices*. Since the states are random or stochastic (it is not known for sure which state will be realized), the discount factors are termed[12] *stochastic discount factors*. The discount factor of the first state of nature in the next time period is $\frac{1}{22}$: the present value of \$1 contingent on state 1.

The d's are actually *risk-adjusted* discount factors. They take care of two effects:

- time value of money: the amount to be earned will be received in the next time period, not now, and
- risk: it is not known for sure which state of nature will be realized next period. The amount to be received in the next time period from

[12] A more precise explanation is based on the fact that the discount factors are outcomes of a random variable taking on the value d_j if state j occurs.

a given security is a risky amount and thus cannot be discounted by the risk-free rate of interest.

Intuition suggests that there should be a connection between these risk-adjusted discount factors and the risk-free rate of interest. Since the *d*'s transform risky cash flows from the next time period into their present value today, perhaps the risk-free rate of interest is implicit in the value of the *d*'s. This would provide a second way of determining the risk-free rate of interest. (The reader will recall that the first was via a risk-free security (or bond) in the market.)

Consider the portfolio of the following three contingent claims in a market for which there are only three states of nature possible in the next time period. Claim 1 is for $1 contingent on state 1 ($0 if any other states are realized), claim 2 is for $1 contingent on state 2, and claim 3 is for $1 contingent on state 3. The price of this portfolio should be[13]

$$d_1 + d_2 + d_3.$$

On the other hand, buying this portfolio guarantees receiving $1 in the next time period, regardless of the state of nature. Hence, the price of the portfolio should also be equal to the present value of a (sure) $1 in the next time period. We conclude that r_f, the risk-free interest rate, must satisfy the relation

$$d_1 + d_2 + d_3 = \frac{1}{1 + r_f}, \tag{1.8}$$

otherwise arbitrage opportunities will exist. Can you show why? Indeed this is nothing but the third[14] equation in (1.6). The relation (1.8) holds even in the absence of a risk-free security (bond) in the market. Even if the *primary securities* (securities 1 and 2 and the bond from the very first example in this Chapter) in the market do not include a bond, the risk-free interest rate is implicit in the prices of the risky securities. Section 2.1 in Chapter 2 elaborates on this point.

Let us reexamine the present value of a cash flow (c_1, c_2, c_3) to be received in the next time period, contingent on which state of nature will occur. The present value of this cash flow is

$$d_1 c_1 + d_2 c_2 + d_3 c_3. \tag{1.9}$$

[13] In our example (see Table 1.2), if you sum the rows of the table the number under the column price will be the value of the cash flow (1,1,1).

[14] The equation is $110d_1 + 110d_2 + 110d_3 = 100$. Dividing it by 110 yields $d_1 + d_2 + d_3 = \frac{100}{110} = \frac{1}{1+\frac{10}{100}}$, which is equation (1.8) with $r_f = \frac{10}{100}$.

If one owns this contingent cash flow today one can transform it, in the market place, to a noncontingent cash flow (i.e., a sure amount) to be received next period. If you know what something is worth today (its value or current price), then you can calculate what it will be worth (or ought to pay you) in the next time period. This is true provided you know the relevant way to move cash flows through time. The noncontingent cash flow to be received in the next time period, given its present value, will simply be the future value of (1.9), i.e.,

$$\begin{aligned} & (1+r_f)(d_1c_1+d_2c_2+d_3c_3) \\ = \; & (1+r_f)\,d_1c_1+(1+r_f)\,d_2c_2+(1+r_f)\,d_3c_3. \end{aligned} \tag{1.10}$$

By equation (1.8)

$$1+r_f=\frac{1}{d_1+d_2+d_3}$$

and thus the future value is

$$\frac{d_1}{d_1+d_2+d_3}c_1+\frac{d_2}{d_1+d_2+d_3}c_2+\frac{d_3}{d_1+d_2+d_3}c_3. \tag{1.11}$$

The contingent cash flow (c_1,c_2,c_3), to be received in the next time period, can be replaced in the market by the sure amount (1.11), also to be received in the next time period. This last expression, (1.11), is in a sense, the *certainty equivalent* of the contingent cash flow (c_1,c_2,c_3).

Since d_j, $j=1,2,3$, are all positive numbers,

$$q_j=\frac{d_j}{d_1+d_2+d_3} \tag{1.12}$$

must be a positive number smaller than one and greater than zero. Furthermore,

$$q_1+q_2+q_3=1.$$

The q_j therefore have the interpretation of a probability distribution. Substituting the q_j in (1.11) yields

$$q_1c_1+q_2c_2+q_3c_3. \tag{1.13}$$

Thus, in this market, the contingent claim (c_1,c_2,c_3) has the same value as the sure amount (the certainty equivalent) in equation (1.13). Using the interpretation of the q_j's as probabilities, the certainty equivalent is the expected value of the contingent claim.

It is important to note that this probability distribution is an "artificial" probability distribution. It is neither the subjective nor the objective probability distribution which investors assign to the likelihood of each of the future states of nature in the market. Rather, it is simply a consequence of the no-arbitrage condition: the no-arbitrage condition is equivalent to the existence of the d's, and the probability distribution is simply a normalization[15] of the d's.

This probability distribution is termed *the risk-neutral probability* distribution for the following reason. The cash flow (c_1, c_2, c_3) is a contingent cash flow to be obtained in the next time period. The discussion above has shown that its certainty equivalent is equal to its expected value, under the probability distribution defined above. Investors who are risk-averse[16] demand some compensation for taking a risk. Thus such investors would value a risky cash flow at less than its expected value. On the other hand, investors who are risk lovers value contingent cash flows (risky cash flows) above their expected values — risk constitutes positive value for those types of investors. Investors who are risk-neutral value contingent cash flows at their expected values.

It is essential to emphasize here that the no-arbitrage condition allows us to value contingent cash flows using their expected value under a certain probability. The no-arbitrage condition does not claim that investors in the market are risk-neutral, nor does it state that the probability distribution obtained above is the true probability of the possible states of nature in the next time period.

Denote the contingent cash flow (c_1, c_2, c_3) by c, and interpret c as a random variable that takes the value c_j in state j with the probability q_j. Thus $q_1c_1+q_2c_2+q_3c_3$ becomes the expected value of c under this probability denoted by

$$E_q(c) = \sum_{j=1}^{3} q_j c_j, \tag{1.14}$$

[15] Normalization is used here to mean calibration. Dividing the d's by the sum of the d's conveys a sense of the relative value of \$1 in state j, with respect to its value in other states.

[16] The classical definition of risk attitude uses the concept of utility function. This function ranks investments based on the utility of the investor (the higher the number assigned by the function to the investment, the better the investment). Uncertain prospects are ranked based on the expected utility. An investor with a linear utility function is risk-neutral, and an investor with concave (convex) utility function is risk-averse (lover). In the first case ranking can be done based on the expected value of the investment and thus the name above.

where q stands for the probability (q_1, q_2, q_3).

Consequently, the present value of the contingent cash flow (c_1, c_2, c_3) is[17]

$$\frac{1}{1+r_f}E_q(c). \tag{1.15}$$

The above discussion can be summarized in the following way.

Theorem 3 *There are no arbitrage opportunities in the market if and only if there exists a probability distribution over the states of nature (i.e., q_j is the probability of state j) such that for each security with a price P and a cash flow c*

$$P = \frac{1}{1+r_f}E_q(c).$$

We conclude this section with a slightly different way of calculating the certainty equivalent of the cash flow (c_1, c_2, c_3). This will enhance our understanding of stochastic discount factors.

The present value of $1 contingent on state 1 is d_1. Thus, the value of this contingent claim equals the sure amount $d_1\,(1+r_f)$ to be received in the next period. Let us check this in our example when we substitute $\frac{11}{10}$ for r_f and $\frac{1}{22}$ for d_1 and utilize the function **Vdis**, defined by (1.7) to calculate the value of the cash flow $(d_1\,(1+r_f), d_1\,(1+r_f), d_1\,(1+r_f))$.

```
> Vdis((11/10)*1/22,(11/10)*1/22,(11/10)*1/22);
```

$$\frac{1}{22}$$

Indeed the result is the value of d_1. In general, therefore, receiving the value of c_1 contingent on state 1 in the next period is equal to the sure value of

```
> c1*d1*(1+rf);
```

$$c1\ d1\ (1+rf)$$

to be received in the next period. Continuing in this way, the value of the contingent claim (c_1, c_2, c_3) to be received in the next period is like the sure amount

```
> c1*d1*(1+rf)+c2*d2*(1+rf)+c3*d3*(1+rf);
```

$$c1\ d1\ (1+rf) + c2\ d2\ (1+rf) + c3\ d3\ (1+rf)$$

---

[17] Mathematically, this result is simply the consequence of multiplying and dividing an equation by the same factor. Namely, the equation $d_1c_1 + d_2c_2 + d_3c_3$ is simply multiplied and divided by $1 + r_f$, but the multiplication is with $\frac{1}{d_1+d_2+d_3}$, which, as proven above, is equal to $1 + r_f$.

to be received in the next period. Substituting $\frac{1}{d_1+d_2+d_3}$ for $1+r_f$ in the above expression (% in MAPLE stands for the "above expression"), yields

```
> subs((1+rf)=1/(d1+d2+d3),%);
```

$$\frac{c1\ d1}{d1+d2+d3}+\frac{c2\ d2}{d1+d2+d3}+\frac{c3\ d3}{d1+d2+d3}$$

The last expression simply confirms that the value of a contingent claim c is the present value of its expected value (the certainty equivalent under the risk-neutral probability) in the next period.

The risk-neutral probabilities also induce a relation between the expected return (under that probability) of the securities in the market. Applying equation (1.15) to security i with a payoff $(a_{i1},\ldots,a_{is})$ yields

$$\frac{1}{1+r_f}E_q(a_i)=P_i. \tag{1.16}$$

One plus the rate of return on security i in state j is $\frac{a_{ij}}{P_i}$ and thus $\frac{a_{ij}}{P_i}-1$ is the rate of return on security i in state j. Dividing equation (1.16) by P_i, multiplying it by $1+r_f$, and subtracting one from each side implies that

$$E_q\left(\frac{a_i}{P_i}-1\right)=r_f. \tag{1.17}$$

In words, the expected rate of return (under the risk-neutral probability) of each security in the market is equal to the risk-free rate of return. Indeed, this is an intuitively appealing property. A risk-neutral investor judges uncertain prospects based on expected values. If equation (1.17) had been violated in a market populated with risk-neutral investors, equilibrium could not have prevailed. The investors would long securities with an expected rate of return greater than r_f and short the bond. Effectively the investors would finance buying securities with an expected rate of return greater than r_f with short positions in the bond. This activity would continue without impediment; thus equilibrium could not have been obtained unless the expected rates of return for all securities in the market would be equal.

1.5 Concluding Remarks

This Chapter describes three methods of pricing:

- by replication,
- utilizing the stochastic discount factors, and

- via risk-neutral probabilities.

The last two are virtually the same. There is an advantage to the second method over the first. Once the stochastic discount factors have been found, they can be used to price any contingent cash flow in the market. Valuing thus becomes simply evaluating a linear function. On the other hand, valuing by replication requires the solution to an optimization problem for each valuation.[18]

Although the model described in this Chapter seems naive, it possesses many of the features of the real world. Specifically, the example followed throughout the Chapter looks at three possible states of nature. Everything in the Chapter would still hold if the number of possible future states of nature were allowed to be the more general value n. In theory, the price of a stock can take on any value between zero and infinity. Nevertheless, the guidelines of the model presented in this Chapter still hold true. There is still a set of stochastic discount factors, but in the real-world case there will be a discount factor for each possible state in the next time period. So if the stock can take any value s in $[0, \infty)$, the stochastic discount factor will be a function $d(s)$. Summation will be replaced by an integral, but the *core concepts will stay unchanged.* This extension of the simple model will be discussed in Chapter 4.

Another possible extension (modification) of the simple model facilitates the study of the bond market. In Chapter 3 the states of nature are replaced (interpreted as) by time periods and only static portfolio strategies are employed. Consequently an equivalent model to the one investigated here is obtained. Such a model is utilized to analyze and value, by arbitrage, many debt instruments, as well as to estimate the interest rates prevailing in the market. The next Chapter presents real-world applications of the methods studied here.

1.6 Questions and Problems

Problem 1. Consider the equity market, in the table below, with two states of nature and two securities:

	State I	*State II*	*Prices*
Security 1	*\$108*	*\$115*	*\$92*
Security 2	*\$115*	*\$108*	*\$113*

[18] See [36].

1. Use the MATLAB GUI or MAPLE to solve the system of equations that the SDF should satisfy in this market (see equation (1.6)) and confirm that the no-arbitrage condition does not hold in this market.
2. Use the MATLAB GUI or MAPLE to set up the optimization problem (equation (1.1)) and confirm that the problem is unbounded. Add the constraints that a short position cannot exceed one unit in each security and resubmit your problem. Identify an arbitrage portfolio.
3. Use the MATLAB GUI or MAPLE to run the **CheckNA** procedure and record the arbitrage portfolio suggested by it. Compare the two arbitrage portfolios you identified.
4. Correct the price of one of the securities in the market so as to satisfy the NA condition and solve for the SDF in this new market.
5. Depict the NA condition for this market graphically, as in Figure 1.6 in Section 1.7, and spot the two arbitrage portfolios you identified in this market. Can you identify the set of all arbitrage portfolios?
6. Is this market complete?

Problem 2. Consider the equity market, in the table below, with three states of nature and three securities:

	State I	*State II*	*State III*	*Prices*
Security 1	*$20*	*$90*	*$80*	*$65*
Security 2	*$60*	*$70*	*$100*	*$80*
Security 3	*$130*	*$130*	*$130*	*$120*

1. Utilize the MATLAB GUI or MAPLE to run the **CheckNA** procedure and determine if an arbitrage opportunity exists in the above market.
2. Solve for the SDF in the above market.
3. Solve for the risk-neutral probabilities in this market.
4. Value a state-contingent cash flow with payoffs of $(20, -23, 100)$ using the three methods.

5. What is the expected value of the cash flow $(c1, c2, c3)$?
6. Assume that you short a portfolio that produces the cash flow $(c1, c2, c3)$ and invest the proceeds in security 3 (the risk-free security). Compare the results of this part and part 5 of this question with respect to the concept of certainty equivalent.

Problem 3. In certain markets there exists a type of financial instrument called an index fund (a certain combination of securities in the market). This type of fund usually pays the investor the return on a certain index, but guarantees that in any event (state of nature) the principal will be returned.

1. Consider an index, 2 in the market of Problem 2, half of which is represented by investment in security 1 and the other half is represented by investment in security 2 in the market of Problem 2. The index fund is based on principal units of $100. What is the no-arbitrage value of such an index fund?
2. Consider another fund that pays the investor the lowest return available in the market of Problem 2, in each state. The principal units are $100. What is the no-arbitrage value of such a fund?

Problem 4. Consider the equity market, in the table below, with three states of nature and three securities:

	State I	*State II*	*State III*	*Prices*
Security 1	*$120*	*$75*	*$165*	*$125*
Security 2	*$60*	*$100*	*$200*	*$115*
Security 3	*$65*	*$65*	*$145*	*$55*

1. Use equation (1.6) to show that the no-arbitrage condition is not satisfied in the above market.
2. Design a portfolio that costs nothing at initiation and has positive payoffs in all states.
3. Design a portfolio that costs nothing at initiation and has a positive payoff in only one state and no payoff in other states.
4. Design a portfolio that pays at initiation but pays nothing at the end of the period.

Problem 5. Consider the equity market, in the table below, with three states of nature and two securities:

	State I	*State II*	*State III*	*Prices*
Security 1	*\$100*	*\$0*	*\$0*	*\$80*
Security 2	*\$100*	*\$0*	*\$50*	*\$80*

1. Identify an obvious arbitrage position in this market.
2. Run the optimization problem (as in equation (1.1)) and explain the results.
3. Run the **CheckNA** procedure to confirm your answer.
4. Can you find an arbitrage portfolio such that the arbitrage profit is realized in the first period?
5. Discuss the difference between the no-arbitrage condition and the law of one price based on your findings.

Problem 6. Consider the equity market given in Table 1.2.

1. What is the value of a cash flow defined by $\max_j a_{i,j}$, where a_{ij} is the payoff from the elementary cash flow i in state j?
2. Can you explain the shape of the graph in Figure 1.1?
3. What is the value of the proposition (on page 21) when $K = 0$?
4. Why is there a kink at $K = 60$?
5. Why is the value of the proposition 0 when $K \geq 80$?
6. How will the value of the proposition change as a function of the price of security 1 in state 3, when the its payoff in the third state of nature in the next time period, c_3, is allowed to vary between 0 and 160?
7. Can you describe the region where the value of the proposition is zero? This region is in the (c_3, K) plane.

1.7 Appendix

1.7.1 Complete Market

A market is said to be complete if, for any cash flow c_i across states of nature i, there exists a portfolio that pays the same cash flow. The existence of such a portfolio for any cash flow c occurs if and only if the payoff vectors associated with each primary security are linearly independent and the number of securities is at least as large as the number of states of nature. In other words, the rank of the payment matrix must be equal to the number of states of nature. Thus if there are n states of nature the row of the payment matrix must contain a basis for R^n.

Consider a market with two securities and two states of nature as follows:

	State 1	State 2
Security 1	110	110
Security 2	90	120

Using MAPLE it is easy to verify if a market is complete or not. The following procedure asks MAPLE to evaluate the rank of the payment matrix. First, designate the matrix **Paymat:**

```
> Paymat:=matrix([[110,110],[90,120]]);
```

$$Paymat := \begin{bmatrix} 110 & 110 \\ 90 & 120 \end{bmatrix}$$

Rank is then calculated by the following procedure:

```
> linalg[rank](Paymat);
```

$$2$$

Since both the number of states and the rank of the matrix are two, this matrix defines a complete market. Thus, for any given cash flow $c = (c_1, c_2)$ a portfolio x exists that solves the matrix equation $x\ Paymat = c$. Let us define the vector c

```
> c:=array(1..2,[]);
```

$$c := \text{array}(1..2, [])$$

and ask MAPLE to solve the matrix equation $x\ Paymat = c$.

```
> x:=linalg[linsolve](Paymat, c);
```

$$x := \left[\frac{2}{55}\,c_1 - \frac{1}{30}\,c_2,\ -\frac{3}{110}\,c_1 + \frac{1}{30}\,c_2\right]$$

Hence, to obtain the cash flow (c_1, c_2) one should buy

```
> 'x[1]'=x[1];
```

$$x_1 = \frac{2}{55}\,c_1 - \frac{1}{30}\,c_2$$

units of security 1 and

```
> 'x[2]'=x[2];
```

$$x_2 = -\frac{3}{110}\, c_1 + \frac{1}{30}\, c_2$$

of security 2.

1.7.2 Incomplete Market

Consider a market with two securities and two states of nature, as specified below:

	State 1	State 2	price
Security 1	1	2	1.5
Security 2	3	6	4.5

Using the procedure discussed earlier we can define the payoff matrix for the market to be

```
> Paymat:=matrix([[1,2],[3,6]]);
```

$$Paymat := \begin{bmatrix} 1 & 2 \\ 3 & 6 \end{bmatrix}$$

and we check if the rank of the matrix is equal to the number of states of nature:

```
> linalg[rank](Paymat);
```

$$1$$

Since the rank is one, but there are two states of nature, the payment matrix defines an incomplete market. In the case of an incomplete market, certain state-contingent cash flows cannot be generated by existing securities. Indeed, in this market one can only generate cash flows with the characteristic such that the cash flow in the first period, c_1, is equal to half the cash flow in the second period c_2. This limitation can be verified by multiplying the vector (x_1, x_2) by the payment matrix as below,

```
> evalm([x1,x2]&*Paymat);
```

$$[x1 + 3\,x2,\; 2\,x1 + 6\,x2]$$

Here $x_1 + 3x_2$ is the payoff in state 1 and $2x_1 + 6x_2$ is the payoff in state 2.

We find therefore that a contingent cash flow, (c_1, c_2), that can be generated in this market, must satisfy both of the following equations:

$$\begin{aligned} c_1 &= x_1 + 3x_2 \\ c_2 &= 2x_1 + 6x_2. \end{aligned} \tag{1.18}$$

Eliminating x_1 and x_2 from both equations allows us to simplify the requirement as follows:

$$\frac{c_1}{c_2} = \frac{1}{2}. \tag{1.19}$$

Consequently, cash flows that are not of this form cannot be generated in this market.

Another way of looking at this concept is as follows. The row vectors of the payment matrix span only a subspace of R^2. MAPLE determines the basis of the subspace by issuing the command

```
> linalg[basis](Paymat,'rowspace');
                              [[1, 2]]
```

Those cash flows that do not fall in the subspace of the payment matrix cannot be valued uniquely using arbitrage arguments. Instead, contingent cash flows that are not spanned by the securities in the market can only be assigned upper and lower bounds on their value as will be discussed in the following subsection.

1.7.3 Incomplete Market and Arbitrage Bounds

The procedure **Valucash** verifies whether the no-arbitrage condition is satisfied in a given market. If the no-arbitrage condition *is* satisfied, **Valucash** determines whether the market is complete or incomplete. As well, the procedure determines the value of a state-contingent cash flow. If the market is incomplete and the cash flow is not spanned by the securities in the market, the upper and lower bounds for the value of the cash flow are determined. For example, let's consider the market discussed earlier:

	State 1	State 2	price
Security 1	1	2	1.5
Security 2	3	6	4.5

Is this market complete? What is the value of a state-contingent security paying cash flows of $(4, 8)$ in the next time period?

The inputs to the procedure **Valucash** are similar to those of **CheckNA**. The payment matrix is inputted first, followed by the vector of prices and then by a vector representing the cash flow to be valued.

```
> Valucash([[1,2],[3,6]],[1.5,4.5],[4,8]);
```

this market is incomplete

the cashflow is spanned by the primary assets

the value of the cashflow is 6.000000000

An exact value for the contingent securities is found, even though the market is incomplete. This is possible in this case because the cash flow $(4, 8)$ is spanned by the primary securities in the market. Indeed the cash flow $(4, 8)$ satisfies equation (1.19). Let us see what is the value of a state-contingent security paying a cash flow of $(4, 9)$ in the next time period.

```
> Valucash([[1,2],[3,6]],[1.5,4.5],[4,9]);
```

this market is incomplete

the cashflow is not spanned by the primary assets

only bounds on the price can be calculated

the minimal value of the cashflow is, 6.000000000

the min value is induced by [1.500000000, 0]

the maximal value of the cashflow is 6.750000000

the max value is induced by, [0, .7500000000]

Since the cash flow of this contingent security is not spanned by the primary securities in the market, no exact value can be determined. Instead, upper and lower bounds on its value are determined. The bounds can be determined as follows. We first hold c_1 constant at 4 and find a c_2 to fit the requirement $\frac{c_1}{c_2} = \frac{1}{2}$. Since $c_1 = 4$, c_2 must equal 8 and we know the value of this cash flow is 6. We are able to value it since it is spanned by the primary securities in the market. Furthermore, the value of $(4, 8)$ must be less than the value of $(4, 9)$ since it offers equal or smaller payments in every state of nature. (If the values did not satisfy this hierarchy, arbitrage opportunities would exist.)

We next hold c_2 constant at 9 and find a c_1 to fit the requirement $\frac{c_1}{c_2} = \frac{1}{2}$. Since $c_2 = 9$, c_1 must have a value of 4.5. The cash flow $(4.5, 9)$ can also be valued uniquely since it is spanned by the primary securities in the market. The value of $(4.5, 9)$ must be larger than the value of $(4, 9)$ for the same reasons we outlined above regarding the lower bound. Let us value it with the **Valucash** procedure.

```
> Valucash([[1,2],[3,6]],[1.5,4.5],[4.5,9]);
```

this market is incomplete

the cashflow is spanned by the primary assets

the value of the cashflow is, 6.750000000

We thus confirmed the result of our first run of **Valucash**: the upper bound is 6.75 and the lower bound is 6.

Valuation in incomplete markets can also be analyzed from a dual point of view. Such a view is in the spirit of valuation with SDF while the above analysis was induced by valuation via the replication method. The name "valuation operators" also refers to stochastic discount factors and we will use one or the other interchangeably. In a complete market there exists a unique valuation operator with which any contingent cash flow (security) in the market can be valued uniquely. However, in an incomplete market there is a set of valuation operators. In our example the set of valuation operators is the solution set to the following system of inequalities and equalities:

```
> solve( {1*d1+2*d2=15/10,3*d1+6*d2=45/10,d1>0,d2>0},\
> {d1,d2});
```

$$\{0 < d2,\ d1 = -2\,d2 + \frac{3}{2},\ d2 < \frac{3}{4}\}$$

The fact that no unique solution exists is just a confirmation of what we already knew: we are dealing with an incomplete market. Many valuation operators can be selected. Each induces a different value for a contingent cash flow that is not spanned by the primary securities in the market. The set of the different values forms the range of possible (no-arbitrage) values of a security for which its cash flows are not spanned by the primary securities in the market. The minimal and maximal values in this set correspond to the lower and upper bound, respectively, on the no-arbitrage value of the security in question.

Let us consider some of the possible valuation operators implied by the above example and the pricing induced by them. Since $d_1 = -2d_2 + \frac{3}{2}$, it is easy to verify that $d_1 = \frac{1}{2}$ and $d_2 = \frac{1}{2}$ is a valuation operator. Thus if we graph in the (c_1, c_2) plane the function $\frac{c_1}{2} + \frac{c_2}{2}$, we can visualize (Figure 1.3) the no-arbitrage prices based on this valuation operator.

```
> plot3d(c1/2+1*c2/2,c1=-10..10,c2=-10..10,\
> title='Figure 1.3:  Arbitrage Valuation of the\
```

```
> Cash Flow (c1,c2) by d1=3 and d2=1`,\
> axes=box, style=patchnogrid, titlefont=[TIMES,BOLD,8]);
```

Another valuation operator is $d_1 = \frac{1}{4}$ and $d_2 = \frac{5}{8}$ as can be verified by the general solution $d_1 = -2d_2 + \frac{3}{4}$. We can plot this valuation operator, Figure 1.4, in the same plane as the previous valuation operator.

```
> plot3d({c1/2+c2/2,c1/4+5*c2/8},c1=-10..10,c2=-10..10,\
> title=`Figure 1.4:  Arbitrage Bounds for the Cash Flow\
> (c1,c2), with Two Valuation Operators`,\
> orientation=[45,45], labels=[C1,C2, Value],\
> axes=box, titlefont=[TIMES,BOLD,8]);
```

Yet, another valuation operator is $d_2 = \frac{3}{4}$ and $d_1 = \frac{3}{8}$. (Again verifiable by substitution of these values into the general solution.) We can now plot, Figure 1.5, all three valuation operators in the same plane.

```
> plot3d({c1/2+c2/2,c1/4+5*c2/8,3*c1/4+3*c2/8},\
> c1=-10..10,c2=-10..10, title=`Figure 1.5:  Arbitrage\
> Bounds for the Cash Flow (c1,c2) with Three Valuation\
> Operators`, orientation=[45,45],labels=[C1,C2,Value],\
> titlefont=[TIMES,BOLD,8], axes= box);
```

We observe that all the valuation operators assign the same value to cash flows that are spanned by the primary securities in the market, i.e., those cash flows satisfying $2c_1 = c_2$. The projection of the intersection of the graphs of all the valuations operators onto the (c_1, c_2) plane is the set $\{(c_1, c_2) \mid 2c_1 = c_2\}$. This is a consequence of the fact that every cash flow that satisfies $2c_1 = c_2$ is assigned the same value by each of the SDFs.

Let us consider the other method of calculating the bounds on the value of a cash flow to which we alluded earlier. Namely, let us calculate the maximal and minimal value assigned to a cash flow by the valuation operators. Let us consider again the cash flow $(c_1 = 4, c_2 = 9)$. Using linear programming, as below, we can determine the SDF that minimizes the value of this cash flow.

```
> simplex[minimize](4*d1+9*d2,{1*d1+2*d2=15/10,\
> 3*d1+6*d2=45/10},NONNEGATIVE);
```

$$\{d1 = \frac{3}{2},\ d2 = 0\}$$

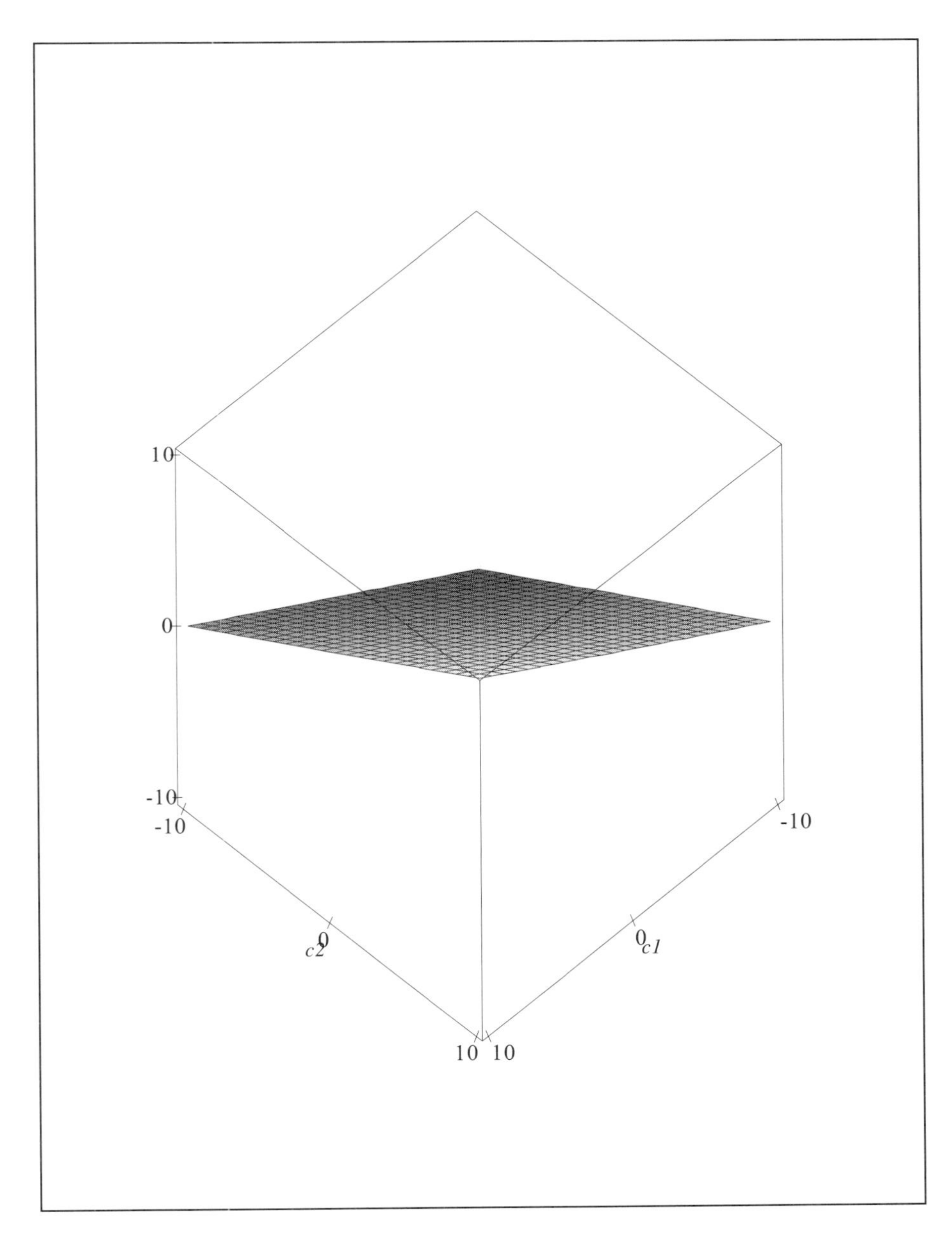

Figure 1.3: Arbitrage Valuation of the Cash Flow $(c1, c2)$ by $d_1 = 3$ and $d_2 = 1$

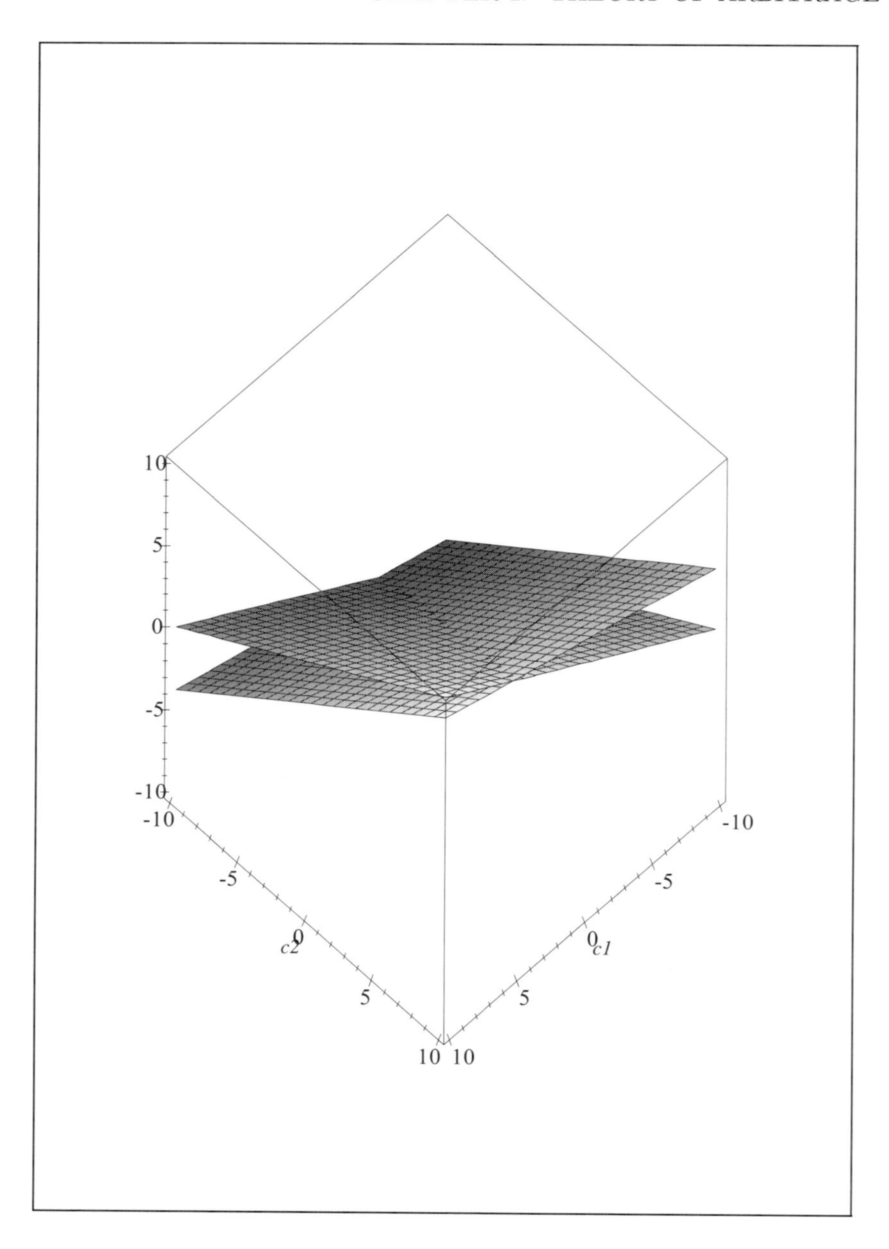

Figure 1.4: Arbitrage Bounds for the Cash Flow $(c1, c2)$ with Two Valuation Operators

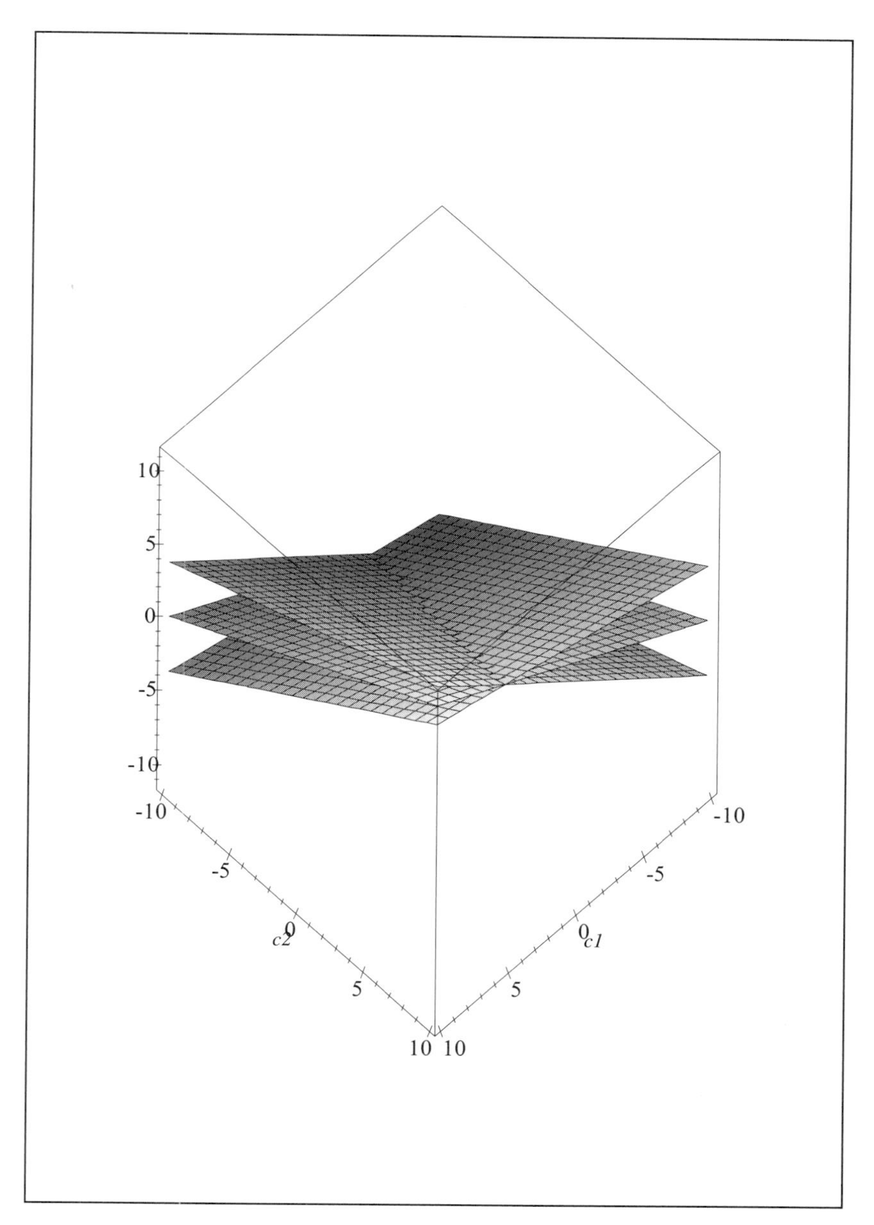

Figure 1.5: Arbitrage Pricing for the Cash Flow $(c1, c2)$, with Three Valuation Operators

Thus the minimal value (to be accurate, it is the infimum rather than the minimum since d_1 is constrained to be strictly positive and thus the minimum is not attained) of the cash flow will be

```
> 4*3/2;
```

$$6$$

Similarly, we can determine the valuation operator which maximizes the value of this cash flow:

```
> simplex[maximize](4*d1+9*d2, {1*d1+2*d2=15/10,\
> 3*d1+6*d2=45/10},NONNEGATIVE);
```

$$\{d1 = 0, \; d2 = \frac{3}{4}\}$$

Thus the maximum value (the reservation voiced before applies here as well, it is actually the suprimum, not the maximum) of the cash flow will be:

```
> evalf(3/4*9);
```

$$6.750000000$$

We have now confirmed that the two methods produce the same lower and upper bounds. We also hope that the reader, courtesy of the example, has gained some more insight into the connections among the upper and lower bounds, the valuation by SDF, and the replication method. A slightly different exposition of this issue is offered in [42].

1.7.4 The No-Arbitrage Condition and Its Geometric Exposition

As explained in the text, there are two methods of determining whether or not the no-arbitrage condition is satisfied in a given market. It is possible to solve for a portfolio that maximizes arbitrage profit and to determine if an arbitrage portfolio exists. Alternatively, it is possible to solve for a set of stochastic discount factors that equate prices to their discounted cash flows. Such discount factors exist if and only if the no-arbitrage condition is satisfied. Consider a market with two securities and a payment matrix given by

```
> Paymat:=matrix([[110,110],[90,120]] );
```

$$Paymat := \begin{bmatrix} 110 & 110 \\ 90 & 120 \end{bmatrix}$$

where the prices of the securities are $p_1 = 100$, $p_2 = 110$. We can determine if the no-arbitrage condition is satisfied by solving for a set of discount factors in this market.

```
> solve({110*d1+110*d2=100,90*d1+120*d2=110,d1>0,d2>0},\
> {d1,d2});
```

MAPLE returns a NULL, meaning no solution exists, and thus we know that the no-arbitrage condition is not satisfied. Let us see what happens if the prices of the two securities had instead been $p_1 = 100$, $p_2 = 95$.

```
> solve( {110*d1+110*d2=100,90*d1+120*d2=95,d1>0,d2>0},\
> {d1,d2});
```

$$\{d1 = \frac{31}{66}, d2 = \frac{29}{66}\}$$

For this second set of prices of the two securities, the no-arbitrage condition is satisfied.

We will now examine the geometric interpretation of the no-arbitrage condition. Consider a market with two securities and two states of nature for which the payoff matrix is

```
> Paymat:=matrix([[110,110],[90,120]] );
```

$$Paymat := \begin{bmatrix} 110 & 110 \\ 90 & 120 \end{bmatrix}$$

The set of portfolios (x_1, x_2) with **positive** cash flows for each state of nature is the solution to the system of inequalities

$$0 \leq 110x_1 + 90x_2 \text{ and } 0 \leq 110x_1 + 120x_2.$$

Graphically, this set is plotted in Figure 1.6 (where the horizontal axis is x_1 and the vertical axis is x_2):

```
> plots[inequal]({110*x1+90*x2>=0,110*x1+120*x2>=0} ,\
> x1=-100..100,x2=-100..100,optionsfeasible=(color=pink),\
> optionsopen=(color=purple,thickness=1),\
> optionsclosed=(color=red, thickness=2), \
> optionsexcluded=(color=yellow), title=\
> `Figure 1.6:  Geometric Exposition of the No-Arbitrage\
> Condition` , titlefont=[TIMES,BOLD,8]);
```

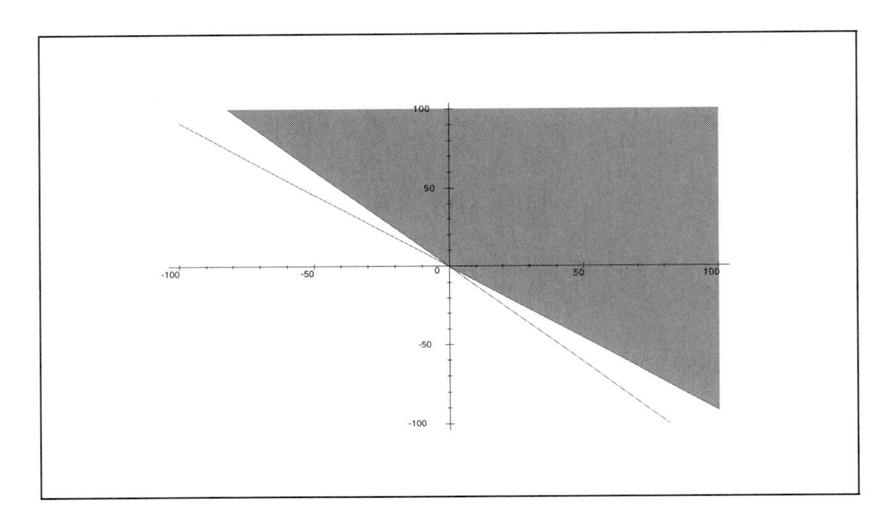

Figure 1.6: Geometric Exposition of the No-Arbitrage Condition

The red[19] lines are the graphs of $110x_1 + 90x2 = 0$ and $110x_1 + 120x_2 = 0$. The shaded pink region is the set of portfolios that produce positive cash flows for each state of nature. The shaded pink region is thus the intersection of

$$0 \leq 110x_1 + 90x_2 \text{ and } 0 \leq 110x_1 + 120x_2.$$

The no-arbitrage condition essentially says that every portfolio in the pink region must have a positive price. Since the cost of the portfolio (x_1, x_2) is $x_1p_1 + x_2p_2$, this last expression must be *positive for every* (x_1, x_2) *in the shaded pink region*, i.e., the shaded pink region must be completely contained in the region $\{(x_1, x_2)|\ 0 \leq x_1p_1 + x_2p_2\}$. This is the concept of "no-free lunch". A portfolio producing a positive cash flow in at least one state of nature, and non-negative cash flows in every other state of nature in the future, must cost something to purchase today. We can verify that the vector of prices (p_1, p_2) satisfies the no-arbitrage condition by adding the equality

$$x_1p_1 + x_2p_2 = 0 \tag{1.20}$$

to the set of inequalities in the MAPLE command. If the new line intersects the feasible (pink) set in the first orthant the no-arbitrage condition is not

[19] Throughout the book, colors refer to the on-line versions of the figures only.

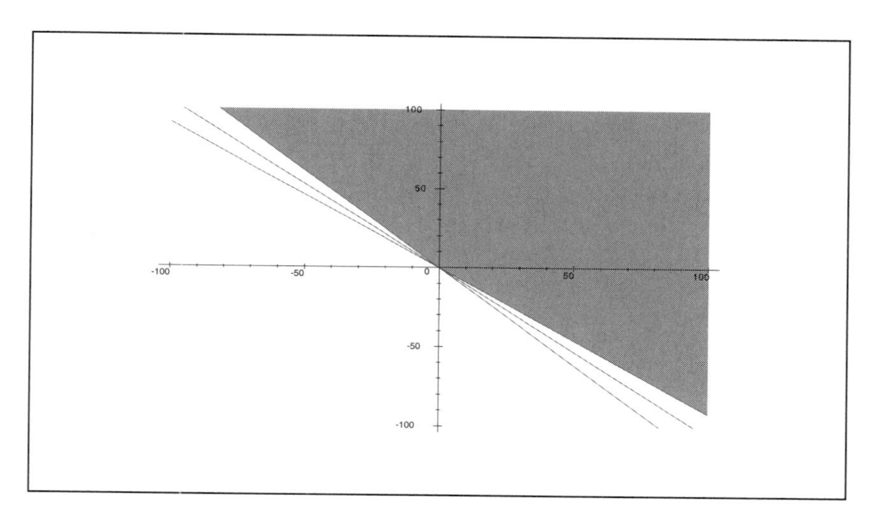

Figure 1.7: The Price Vector and the No-Arbitrage Condition

satisfied. For example, to determine if $p_1 = 100$ and $p_2 = 95$ satisfy the no-arbitrage condition we plot the graph in Figure 1.7:

```
> plots[inequal]({110*x1+90*x2>=0,110*x1+120*x2>=0,\
> 100*x1+95*x2=0},x1=-100..100,x2=-100..100,\
> optionsfeasible=(color=pink), \
> optionsopen=(color=red,thickness=1),\
> optionsclosed=(color=red, thickness=2),\
> optionsexcluded=(color=white), \
> title=`The Price Vector and the No-Arbitrage Condition`);
```

Since the new line does not intersect the pink region, $p_1 = 110$ and $p_1 = 95$ satisfy the no-arbitrage condition. The text discusses the equivalency of the no-arbitrage condition to the existence of a valuation operator. Mathematically we can prove this equivalency by applying to separation theorems in the spirit of the above geometric exposition.[20]

[20]The no-arbitrage condition in a market with taxes and transaction costs are explored in [36] and [19].

Chapter 2

Arbitrage Pricing: *Equity Markets*

The purpose of this Chapter is twofold. It practices the valuation methods studied in the former Chapter and thereby introduces the reader to a variety of financial instruments that are common in the marketplace. We also hope that at the end of this Chapter, the reader will be convinced that indeed the simple model of Chapter 1 is useful, even in the setting of real markets ([14] and [26]).

The Chapter begins with an analysis of a market in which there is no risk-free asset. It continues with the introduction of the binomial model. The binomial model is a special case of the model presented in the previous Chapter. Alas, that model has only two securities and two states of nature. The binomial model is used extensively for numerical valuation when analytical valuation is not possible (does not exist). In the process of introducing the binomial model, the three valuation methods are reviewed and practiced. The Chapter then proceeds to analyze a variety of financial securities. The last section of the Chapter demonstrates that while the valuation of the variety of securities may seem to utilize different approaches, this impression is misleading. A unifying framework for valuations by arbitrage does exist.

2.1 Market Structure and the Risk-Free Rate

We start this section with a few examples using the procedure **Narbit**. The procedure **Narbit** checks whether or not the no-arbitrage condition is satisfied. If it is satisfied, **Narbit** defines a function **Vdis** which values cash flows in this market based on the stochastic discount factors. The procedure

Narbit also reports the value of the stochastic discount factors, the risk-neutral probabilities, and the value of a risk-free interest rate implicit in the prices of the securities. After all, even if a bond is not included among the primary securities, the risk-free rate of interest is implicit in the structure of the market. Why is this the case? The examples below demonstrate.

The reader can (and is encouraged to) experiment with this procedure further. Let us look at the example we followed throughout the former Chapter. The input to the **Narbit** procedure is exactly like that of **CheckNA**:

```
> Narbit([[50,60,80],[90,110,130],[110,110,110]],\
> [60,105,100]);
```

$$\textit{The no - arbitrage condition is satisfied.}$$

$$\textit{The stochastic discount factors are,}\ \left[\frac{1}{22}, \frac{25}{44}, \frac{13}{44}\right]$$

$$\textit{The risk - neutral probability is,}\ \left[\frac{1}{20}, \frac{5}{8}, \frac{13}{40}\right]$$

$$\textit{A risk - free rate in this market is,}\ .1000$$

$$\textit{The function Vdis}([c1, c2, ..]),\ \textit{values the cashflow}\ [c1, c2, ..]$$

Since the no-arbitrage condition is satisfied, the procedure **Narbit** also defines the function **Vdis**. What should the market price of the contingent cash flow $(120, 30, 200)$ be?

```
> Vdis([120,30,200]);
```

$$\frac{1795}{22}$$

```
> evalf(%);
```

$$81.59090909$$

Let us see what happens if we add a security with this cash flow to the market, but at a price of \$81. Can you guess?

```
> Narbit([[50,60,80],[90,110,130],[110,110,110],\
> [120,30,200]],[60,105,100,81]);
```

$$\textit{The no - arbitrage condition is not satisfied}$$

$$\textit{An arbitrage portfolio is :}$$

$$\textit{Short, 1, of security, 1}$$

$$\textit{Buy,}\ \frac{162}{247},\ \textit{of security, 2}$$

$$\textit{Short,}\ \frac{30}{247},\ \textit{of security, 3}$$

$$\textit{Buy,}\ \frac{10}{247},\ \textit{of security, 4}$$

$$\textit{The cost of this portfolio is zero}$$

$$\textit{This portfolio produces income of}, \frac{10}{19}, \textit{in state}, 1$$

Can you explain why this happened? We added a security to the market, but the price which we assigned to it allows for the existence of arbitrage opportunities in the market. It is possible to construct a portfolio of primary securities which will produce the same cash flow as that of the new security, $(120, 30, 200)$, but which has a price greater than \$81. Confronted with such a possible portfolio, investors would short this portfolio and take a long position in the new (added) security, thereby generating arbitrage profit. The procedure **Narbit** reports yet another portfolio which generates arbitrage profit. The portfolio costs nothing but generates positive income in state 1.

Let us conclude this discussion by replacing the bond in our market with the cash flow we just valued as a primary security, but at the correct price.

```
> Narbit([[50,60,80],[90,110,130],[120,30,200]],\
> [60,105,1795/22]);
```

$$\textit{The no} - \textit{arbitrage condition is satisfied.}$$

$$\textit{The stochastic discount factors are}, \left[\frac{1}{22}, \frac{25}{44}, \frac{13}{44}\right]$$

$$\textit{The risk} - \textit{neutral probability is}, \left[\frac{1}{20}, \frac{5}{8}, \frac{13}{40}\right]$$

$$\textit{A risk} - \textit{free rate in this market is}, .1000$$

$$\textit{The function } Vdis([c1, c2, ..]), \textit{ values the cashflow } [c1, c2, ..]$$

Nothing has changed in the market, not even the risk-free rate implicit in the security prices. Let us use the function **Vdis** to value the cash flow $(110, 110, 110)$ just to verify that indeed the risk-free rate is the same.

```
> Vdis([110,110,110]);
```

$$100$$

We have valued, in the current time period, a security which provides a sure cash flow of \$110 in the next time period. This is the bond from the very beginning of the previous Chapter. We are able to confirm that the risk-free rate of interest is 10 percent.

Indeed, the risk-free rate of interest is implicit in the prices of the securities of the market. To recover the risk-free rate in a given market we simply have to value a sure dollar in the next time period. In a framework such as the one we have just explored (three states of nature), a sure dollar in the next time period is a cash flow of the form $(1, 1, 1)$. Even in a market where a bond is not one of the primary securities, we can still infer the risk-free rate of interest. Consider yet another example. In this example there is no bond among the primary securities.

```
> Narbit([[50,60,80],[0,10,30],[120,30,200]],\
> [60,160/11,1795/22]);
```

$$The\ no-arbitrage\ condition\ is\ satisfied.$$

$$The\ stochastic\ discount\ factors\ are,\ \left[\frac{1}{22}, \frac{25}{44}, \frac{13}{44}\right]$$

$$The\ risk-neutral\ probability\ is,\ \left[\frac{1}{20}, \frac{5}{8}, \frac{13}{40}\right]$$

$$A\ \ risk-free\ rate\ in\ this\ market\ is,\ .1000$$

$$The\ function\ Vdis([c1, c2, ..]),\ values\ the\ cashflow\ [c1, c2, ..]$$

Solving now for the implicit risk-free rate of interest, we recognize that this interest rate should be the solution to the equation $\frac{1}{1+r} = Vdis([1, 1, 1])$. That is, the risk-free rate is r that equates the value of the cash flow $(1, 1, 1)$ to $\frac{1}{1+r}$. We submit this expression to MAPLE and request a solution. The solution confirms that the risk-free rate implicit in this market is indeed still 10 percent.

```
> solve(1/(1+r)=Vdis([1,1,1]));
```

$$\frac{1}{10}$$

2.2 One-Period Binomial Model

Consider now the following situation.[1] The market consists of only two securities: a bond and a stock. There are only two possible states of nature in the next time period: "Up" and "Down". Let us denote by Up one plus the rate of return if state Up occurs and by Do one plus the rate of return if state Down occurs. The bond costs $\$B$ today and will be worth $\$B(1+r_f)$ in the next time period, regardless of which state of nature occurs, where r_f is the risk-free rate of interest. (We have encountered this concept before in the three-security, three-state example earlier in the Chapter. We were able to determine the risk-free rate from the risk-free security, the bond, in the model.) Denote $1+r_f$ by R_f so that the bond will have value $\$BR_f$ in either state of nature in the next time period. The stock costs $\$S$ today, and will be worth $\$S\,Up$ in the next time period if state Up occurs, and $\$S\,Do$ if state

[1] See [38] for an example.

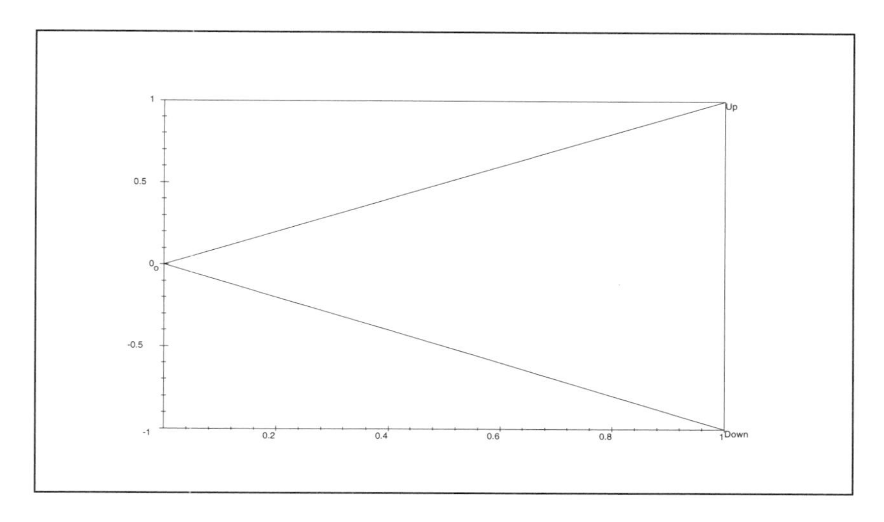

Figure 2.1: One-Period Binomial Model

Down occurs. Figure 2.1 presents a one-period binomial model and Chapter 13 contains a MAPLE procedure that generates an n-nodes binomial model.

```
> geometry[point](Up,1,1);
```

$$Up$$

```
> geometry[point](O,0,0);
```

$$O$$

```
> geometry[point](Down,1,-1);
```

$$Down$$

```
> geometry[segment](OU, [O,Up] );
```

$$OU$$

```
> geometry[segment](OD, [O,Down] );
```

$$OD$$

```
> geometry[draw]([OU,OD],title=`Figure 2.1:\
> One-Period Binomial Model`);
```

It should be clear that Up is a number greater than 1 and Do is a positive number smaller than 1. To avoid obvious arbitrage[2] opportunities, it must

[2]Suppose $R_f \geq Up$. Consider the following portfolio: invest \$1 in the bond and short \$1 of the risky asset. Obviously, it costs nothing to obtain such a portfolio. At time $t = 1$

be the case that $Do < R_f$ and $R_f < Up$. If this inequality relationship does not hold, either the bond dominates the stock (pays off more in at least one state of nature and the same in other states, if $Up \leq R_f$) or vice versa (if $R_f \leq Do$). We let MAPLE know about the assumptions regarding Do by the following set of commands:

```
> assume(0<Do);
> additionally(Do<1);
```

We can check now that MAPLE indeed has received this information (given that we typed the assumptions in correctly). We check the assumptions regarding Do by typing the command

```
> about(Do);

Originally Do, renamed Do~:
  is assumed to be: RealRange(Open(0),Open(1))
```

Similarly, we tell MAPLE about our other required assumptions.

```
> assume(Rf>1);
> about(Rf);

Originally Rf, renamed Rf~:
  is assumed to be: RealRange(Open(1),infinity)

> additionally(Rf<Up);
> about(Up);

Originally Up, renamed Up~:
  Involved in the following expressions with properties
    -Up+Rf assumed RealRange(-infinity,Open(0))
  is assumed to be: real
  also used in the following assumed objects
  [-Up+Rf] assumed RealRange(-infinity,Open(0))
```

We wish to demonstrate in a symbolic manner, and in the setting of the current model, the equivalency of the three pricing methods discussed above. To this end we need to solve for the discount factors which move cash flows through time and are required by two of the pricing methods. The price of a security today, in a well-functioning financial market, should be equal to the present value of its future cash flows. We arrive at the current market

---

the contingent cash flow will be $(R_f - Up, R_f - Do)$, which is nonnegative (with $R_f - Do$ always positive), regardless of the realized state of nature. Hence, we have a self-financed portfolio (a zero-cost portfolio) with a nonnegative cash flow in the future. In other words, there is an arbitrage opportunity, and therefore we must have $R_f < Up$. Similarly it can be shown that the assumption of $R_f \leq Do$ will lead to an arbitrage opprtunity. This is left to the reader as an exercise.

price by discounting (moving) the future cash flows to the present time using discount factors.

We know the value of the bond now and in each of the two possible states of nature in the next time period. We have the same information about the stock. We wish to find the discount factors which will equate the possible future cash flows of each of these two securities to their current prices: thus, we have two equations and two unknowns.

$$\begin{array}{lcccc} B & = & BR_f d_u & + & BR_f d_d \\ S & = & SUpd_u & + & SDod_d, \end{array} \tag{2.1}$$

where d_u and d_d are the discount factors for states Up and Down, respectively. As we can see, the first equation in (2.1) can be divided by B, and thus our result will be independent of B. This simply means that we can decide that the current price of the bond is \$1. MAPLE can find expressions for the discount factors, not as numerical values (though indeed MAPLE is capable of doing this also), but as algebraic expressions. This symbolic capability of MAPLE allows us to arrive at expressions for the discount factors in terms of the variables in the equations. To this end we have to solve the system of equations in (2.1). We arrive at a solution for the two discount factors, written in terms of R_f, Up, and Do.

```
> Sols1:=solve({B=du*B*Rf+dd*B*Rf,\
> S=du*S*Up+dd*S*Do},{du,dd});
```

$$Sols1 := \{du = \frac{-Rf\tilde{} + Do\tilde{}}{Rf\tilde{}\,(-Up\tilde{} + Do\tilde{})},\ dd = \frac{-Up\tilde{} + Rf\tilde{}}{Rf\tilde{}\,(-Up\tilde{} + Do\tilde{})}\}$$

These are the algebraic expressions for the two discount factors, done symbolically in MAPLE.

```
> assign(%);
> dd;
```

$$\frac{-Up\tilde{} + Rf\tilde{}}{Rf\tilde{}\,(-Up\tilde{} + Do\tilde{})}$$

```
> is(dd,positive);
```

true

```
> du;
```

$$\frac{-Rf\tilde{} + Do\tilde{}}{Rf\tilde{}\,(-Up\tilde{} + Do\tilde{})}$$

```
> is(du,positive);
```

true

Each discount factor is verified to be positive. We see now that by Theorem 3 there are no arbitrage opportunities in this market. The existence of these discount factors satisfying the relation in **Sols1** guarantees the satisfaction of the no-arbitrage condition.

```
> dd+du;
```

$$\frac{-Up\tilde{} + Rf\tilde{}}{Rf\tilde{}\,(-Up\tilde{} + Do\tilde{})} + \frac{-Rf\tilde{} + Do\tilde{}}{Rf\tilde{}\,(-Up\tilde{} + Do\tilde{})}$$

```
> simplify(%);
```

$$\frac{1}{Rf\tilde{}}$$

The preceding MAPLE commands demonstrate that the sum of the two (stochastic) discount factors, du and dd, is the reciprocal of $R_f = 1 + r_f$ (see equation (1.7)). This is actually the first equation in **Sols1**. (Recall the first section for an explanation of the risk-free discount factor in a market without a risk-free asset.) Armed with du, dd, and the risk-free discount factor, $\frac{1}{R_f}$, the price of every security (any security) in this market can be calculated. We will review the three methods: stochastic discount factors, replication, and risk-neutral probability.

Consider a general security, C, paying $\$C_u$ in the Up state and $\$C_d$ in the Down state and its basic component parts. The first component is a security that pays $\$C_u$ in the Up state and $0 in the Down state, and the price of this component is $\$duC_u$. The second component is a security that pays $\$C_d$ in the Down state and $0 in the Up state, the price of which is $\$ddC_d$. The price, P_c, of the original security, C, is calculated below. It will be denoted FM as the first method of calculating the price. This pricing method is via the stochastic discount factors method.

```
> FM:=(du*Cu)+ (dd*Cd);
```

$$FM := \frac{(-Rf\tilde{} + Do\tilde{})\,Cu}{Rf\tilde{}\,(-Up\tilde{} + Do\tilde{})} + \frac{(-Up\tilde{} + Rf\tilde{})\,Cd}{Rf\tilde{}\,(-Up\tilde{} + Do\tilde{})}$$

```
> simplify(%);
```

$$\frac{-Cu\,Rf\tilde{} + Cu\,Do\tilde{} - Cd\,Up\tilde{} + Cd\,Rf\tilde{}}{Rf\tilde{}\,(-Up\tilde{} + Do\tilde{})}$$

The preceding expression gives the price of security C in the current time period. It is not a numerical value; it is a general algebraic expression.

As discussed in an earlier section, a second way to price the security C is via replication. Find a portfolio that mimics, or replicates, the cash flows of the security C and see what the price of such a portfolio should be. Denote the holdings of the bond (the portfolio position in the bond) by X_B, and the holdings of the stock by X_S. We seek the values of X_B and X_S that produce income of C_u and C_d in the Up and Down states, respectively. In order to do so we have to solve an already familiar system of equations:

$$\begin{array}{rcl} Cu &=& X_B B Rf + X_S S Up \\ Cd &=& X_B B Rf + X_S S Do \end{array} \tag{2.2}$$

```
> Sols2:=solve({Cu=XB*B*Rf+XS*S*Up,\
> Cd=XB*B*Rf+XS*S*Do},{XB,XS});
```

$$Sols2 := \{XS = -\frac{-Cd + Cu}{S\,(-Up\tilde{} + Do\tilde{})},\ XB = \frac{Cu\,Do\tilde{} - Cd\,Up\tilde{}}{(-Up\tilde{} + Do\tilde{})\,B\,Rf\tilde{}}\}$$

```
> assign(%);
```

The portfolio composition is given by the expressions for X_B and X_S in the **Sols**2 set. The price of this portfolio is thus $X_S S + X_B B$. The price of the security C calculated by the second method will be denoted SM. It must equal the price of the replicating portfolio, otherwise an arbitrage opportunity exists.

This price, of course, should be equal to the price calculated by our first pricing method, the stochastic discount factors method.

```
> SM:=(XS*S+XB*B);
```

$$SM := -\frac{-Cd + Cu}{-Up\tilde{} + Do\tilde{}} + \frac{Cu\,Do\tilde{} - Cd\,Up\tilde{}}{(-Up\tilde{} + Do\tilde{})\,Rf\tilde{}}$$

```
> simplify(%);
```

$$\frac{-Cu\,Rf\tilde{} + Cu\,Do\tilde{} - Cd\,Up\tilde{} + Cd\,Rf\tilde{}}{Rf\tilde{}\,(-Up\tilde{} + Do\tilde{})}$$

We would now like to confirm that the two methods produced the same price for the security C. Examining the expressions that were obtained for FM and SM demonstrates that indeed they are the same. We can also utilize MAPLE's capabilities to do so. We can ask MAPLE whether $FM = SM$, and see if the answer is true. This is done below.

```
> is(FM=SM);
```

$$true$$

We have now confirmed that calculating the price of a security by discounting its future cash flows using stochastic discount factors and by pricing a portfolio which spans the same future cash flows are equivalent pricing methods (of course, in a complete market with no arbitrage opportunity).

The third method of valuing a security is done via what is known as the "risk-neutral probability". In the next time period, either state Up or state Down can occur, and one of these states will occur. The sum of the probabilities of each state occurring must, therefore, equal 1. Both d_u and d_d are positive (confirmed earlier). We can, then, divide each of d_u and d_d by their sum, and arrive at probabilities for states Up and Down, denoted q_u and q_d, respectively.

```
> qd:=simplify(dd/(dd+du));
```

$$qd := \frac{-Up\tilde{} + Rf\tilde{}}{-Up\tilde{} + Do\tilde{}}$$

```
> qu:=simplify(du/(dd+du));
```

$$qu := \frac{-Rf\tilde{} + Do\tilde{}}{-Up\tilde{} + Do\tilde{}}$$

```
> qd+qu;
```

$$\frac{-Up\tilde{} + Rf\tilde{}}{-Up\tilde{} + Do\tilde{}} + \frac{-Rf\tilde{} + Do\tilde{}}{-Up\tilde{} + Do\tilde{}}$$

```
> simplify(%);
```

$$1$$

```
> is(qd,positive);
```

true

```
> is(qu,positive);
```

true

```
> is(qd<1);
```

true

```
> is(qu<1);
```

true

These MAPLE statements confirm that (a) each of the two probabilities is positive (recall from Theorem 3 that we require this), (b) each of the two probabilities is less than 1, and (c) their sum is 1. Remember, though, that this is an artificial probability: it does not represent the subjective probabilities of investors in the market. It does, however, as we saw in Theorem 3, allow valuation of securities in a risk-neutral way, and hence its name.

We are now in a position to price the general security C using the risk-neutral valuation method. We can also examine the relationship among that valuation method, valuation using stochastic discount factors, and valuation using replicating portfolios (spanning). Valuing a security via the risk-neutral valuation method requires two steps:

- calculating the expected value (under the risk-neutral probability) of the cash flows, and
- discounting it at the risk-free rate.

Applying these steps to our case yields equation (2.3) for the price of the security.

$$P_c = \frac{q_u C_u + q_d C_d}{R_f}. \tag{2.3}$$

We can verify that this expression is equal to the price determined by the first pricing method, stochastic discount factors, FM.

```
> is((1/Rf)*(qu*Cu +qd*Cd)=FM);
                              true
```

The expression for the price of the security C from the risk-neutral valuation method is also equal to the price determined via replicating portfolios, SM.

```
> is((1/Rf)*(qu*Cu +qd*Cd)=SM);
                              true
```

The discussion above is a general proof for the case where there are only two possible states of nature. (It is a general proof since it does not involve any numerical values, but demonstrates the desired result in terms of general algebraic expressions.) It is however true even when there are multiple (but finitely many) states of nature. As we shall see later, it can be extended to a more realistic case, under certain assumptions, where the possible states of nature are a continuum: a case where the price of the stock could take any value between zero and infinity. As promised, the intuition being developed in these simple models has power and applicability in the more complex real-world environment.

2.3 Valuing Two Propositions

This section uses the setting of the simple model of Chapter 1 in order to value two contingent claims. The reader will later see that these claims actually have special names which will be formally defined. Recall that the current price of security 1 is \$60 and that its price will decrease to \$50 if state 1 occurs, stay at \$60 if state 2 occurs, and increase to \$80 if state 3 occurs. Consider the following proposition.

> *Pay \$14 now and in return you have the right,* ***not the obligation,*** *to purchase security 1 in the next time period for \$50* ***regardless of its actual market price in the next period.***

This type of a contingent claim is termed a *call option* and it is defined formally in Chapter 4. Let us analyze the next time-period cash flows that are a consequence of this proposition. The price of security 1 in state 1 is \$50. Hence, if state 1 is realized, an investor that entered into the terms of the proposition would be able to buy security 1 for \$50 without paying \$14 today. Thus in state 1 the proposition's cash flow is zero. If state 2 is realized, the price of security 1 will be \$60, and thus the right to buy it for \$50 is equivalent to an income of \$10, contingent on state 2 occurring. Such an investor would be able to buy the security for \$50 and immediately sell it in the market for \$60. In a similar manner, if state 3 is realized, the right to buy security 1 for \$50 is equivalent to \$30 of income, contingent on state 3 occurring. In summary, the proposition could be rephrased as pay \$14 now in exchange for a cash flow in the next time period of $(0, 10, 30)$ in states 1, 2, and 3 respectively.

The contingent cash flows can also be written as

$$(0,10,30) = (\max(0,50-50), \max(0,60-50), \max(0,80-50)), \quad (2.4)$$

where $\max(a, b)$ stands for the maximum of a or b. The presentation of the cash flows as in equation (2.4) will prove useful in generalizing the proposition.

Based on the *law of one price*, the value of this proposition should be equal to the price of a portfolio generating a cash flow of \$0 in state 1, \$10 in state 2, and \$30 in state 3. Is it possible[3] to identify portfolios which yield at least the above cash flow? If such a portfolio can be found, and if there are many such portfolios, choose from among them using a criterion of minimum cost.

```
> simplex[minimize](60*S1+105*S2+100*SF,\
> {50*S1+90*S2+110*SF>=0,60*S1+110*S2+110*SF>=10,\
> 80*S1+130*S2+110*SF>=30});
```

$$\{S2 = 0,\ S1 = 1,\ SF = \frac{-5}{11}\}$$

[3]If it is known that the market is complete, one need simply solve a system of three equations with three unknowns to find the replicating portfolio. Such a system is identical to the constraints of the minimization problem posed earlier in this Chapter, but for which $\geq$ is replaced with $=$. This point will be revisited at the end of this Chapter using the procedure **Narbit.**

We have identified the desired portfolio and we substitute it in **Cash.** However, first the object **Cash** is redefined.

```
> Cash:=[Cost=60*S1+105*S2+100*SF,\
> IncomeSt1=50*S1+90*S2+110*SF,\
> IncomeSt2=60*S1+110*S2+110*SF,\
> IncomeSt3=80*S1+130*S2+110*SF];
```

$$\begin{aligned} \mathit{Cash} := [\mathit{Cost} &= 60\,\mathit{S1} + 105\,\mathit{S2} + 100\,\mathit{SF}, \\ \mathit{IncomeSt1} &= 50\,\mathit{S1} + 90\,\mathit{S2} + 110\,\mathit{SF}, \\ \mathit{IncomeSt2} &= 60\,\mathit{S1} + 110\,\mathit{S2} + 110\,\mathit{SF}, \\ \mathit{IncomeSt3} &= 80\,\mathit{S1} + 130\,\mathit{S2} + 110\,\mathit{SF}] \end{aligned}$$

```
> subs(S1=1,S2=0,SF=-5/11,Cash);
```

$$[\mathit{Cost} = \frac{160}{11},\ \mathit{IncomeSt1} = 0,\ \mathit{IncomeSt2} = 10,\ \mathit{IncomeSt3} = 30]$$

The solution to the minimization problem was substituted in **Cash** to verify two things:

- Does the portfolio produce exactly the cash flow $(0, 10, 30)$, or does it perhaps produce more in some states of nature? After all, the minimization problem had inequality constraints (of the form "greater than or equal to") and we are actually looking for a portfolio satisfying it as equality constraints (of the form "equal to").
- What is the price of the portfolio? It should be the amount an investor would be willing to pay for the proposition.

Indeed the portfolio $H = \left(S1 = 1,\ S2 = 0,\ SF = -\frac{5}{11}\right)$ generates exactly the cash flow of the proposition and not more. To generate this cash flow, the investor needs to buy 1 unit of security 1 and short (or borrow) $\frac{5}{11}$ of the bond, resulting in a total net cost of $\frac{160}{11} = \$14\frac{6}{11}$. It follows that such a proposition should be accepted, if it were offered to an investor.

In fact, such a proposition will not be found in the market: it generates arbitrage opportunities. An investor can short the portfolio H, i.e., acquire the portfolio $-H = \left(S1 = -1,\ S2 = 0,\ SF = \frac{5}{11}\right)$ and buy the proposition. Shorting the portfolio imposes some risk: the commitments to pay claims contingent on the resultant states of nature, undertaken when shorting the portfolio. The cash flows in the next time period from the short portfolio and from the proposition are exactly the same, but with opposite signs. Thus they offset each other and no risk is assumed by the combined position of buying the proposition and shorting the portfolio. (This combined position is risk-free provided no default is possible by the party offering the proposition.) The activity of eliminating risk from a position is called *hedging*.

An investor can thus *hedge* the risk implicit in the short portfolio $(-H)$ by buying the proposition. The price difference of $\$\frac{6}{11}$, between the portfolio and the proposition, is therefore pure arbitrage profit. The price of the cash flow is denoted Val(0, 10, 30) and is calculated, in decimal form, below:

```
> Val(0,10,30):=evalf(160/11);
```

$$\text{Val}(0, 10, 30) := 14.54545455$$

In the same manner, the value of a different proposition, similar to that above, can be calculated. The offer now is to buy security 1 for $40 (rather than for $50):

```
> simplex[minimize](60*S1+105*S2+100*SF,\
> {50*S1+90*S2+110*SF>=10,60*S1+110*S2+110*SF>=20,\
> 80*S1+130*S2+110*SF>=40});
```

$$\{SF = \frac{-4}{11}, S1 = 1, S2 = 0\}$$

```
> subs(S1=1,S2=0,SF=-4/11,Cash);
```

$$[Cost = \frac{260}{11}, IncomeSt1 = 10, IncomeSt2 = 20, IncomeSt3 = 40]$$

In decimal form, the price of the cash flow from this second proposition is

```
> Val(10,20,40):=evalf(260/11);
```

$$\text{Val}(10, 20, 40) := 23.63636364$$

The difference in the future cash flows of the two propositions is $10 in each state. That is, the difference is the cash flow $(10, 10, 10)$ for states 1, 2, and 3, respectively, i.e., a nonrisky $10. This $10 to be received in the next time period will not be contingent on the state of nature which actually occurs. If the difference in value of the two propositions is equal to the present value of $10 received in the next time period, then the valuation method is consistent. This can be calculated directly since the risk-free rate of interest is known (from the third security, a risk-free security, the bond). Alternatively, it can be valued by replication.

```
> simplex[minimize](60*S1+105*S2+100*SF,\
> {50*S1+90*S2+110*SF>=10,60*S1+110*S2+110*SF>=10,\
> 80*S1+130*S2+110*SF>=10});
```

$$\{SF = \frac{1}{11}, S1 = 0, S2 = 0\}$$

```
> subs(S1=0,S2=0,SF=1/11,Cash);
```

$$[Cost = \frac{100}{11}, IncomeSt1 = 10, IncomeSt2 = 10, IncomeSt3 = 10]$$

At last, we are able to confirm that indeed the difference in the value of the two propositions is the present value of $10.

```
> evalf(100/11);
                              9.09090909
> Val(10,20,40)-Val(0,10,30);
                              9.09090909
```

2.4 Forwards: A First Look

In the former section we dealt with a proposition that gave an *investor the option* to buy a certain security in the future for a certain price. The price is agreed upon now but will be paid, if the option is exercised, in the future. Hence, in such a transaction, money is exchanged at the current time and may be exchanged in the future. This section deals with a similar idea, but which is still slightly different.

In contrast to the case of an option, the agreement which we explore in this section is **binding**. The parties to this agreement are **obligated** to a transaction which will take place in the future for a price agreed upon now **but no transfer of money** occurs in the present time. For this reason these types of transactions fall into a category of financial assets called *price fixing*.

The seller and the buyer agree on the price at which a particular transaction involving a specified good or a financial security will take place in the future. The price at which the transaction will take place is fixed now but will be paid in the future. Not surprisingly, these types of contracts are called *futures* or *forward contracts.* There is a distinction between a futures contract and a forward contract. However, in the setting of the one-period model (where only two points in time are considered, now and the future) this distinction is moot. We shall soon revisit these issues in a multiperiod context and at that time will clarify the difference between futures and forward contracts. The agreed-upon price for future delivery of the specified asset is called the *futures* or *forward price.*

Forward agreements are mostly transacted between two parties, sometimes with the help of a financial institution. Futures contracts are standardized and traded on exchanges such as the Chicago Board of Trades (CBOT). Both contracts are not default free, forward contracts are subject to a much higher risk. This is due to the fact that futures contracts are standardized and guaranteed by the CBOT. The analysis below will ignore this risk.

Futures and forwards are examples of contingent claims. We shall see shortly that the profit or loss from the transaction described by these contracts depends on the price of the good or security which prevails in the future. Hence, the profit or loss from the contract is contingent on the price of an underlying asset. The underlying asset could be a good, a financial security, or a foreign currency. This is a nonexhaustive list of possible underlying assets for forward and futures contracts. The possibilities are limited only by the imagination of the parties involved in the transaction and by their willingness to agree to exchange the asset on the agreed-upon terms. Forward and futures contracts are agreements which introduce another market for goods and securities. The investor has at least two markets in which to trade these assets or commodities: the *spot market* and the *forward market*. The spot market is the usual and familiar market. Assets are exchanged on the spot and delivered immediately.[4] In the forward market, assets or commodities are bought and sold for future delivery. Two examples are provided in this subsection to cement these ideas. They make use of the procedure **Narbit**.

2.4.1 Forward Contract on a Security

Assume that at time $t = 0$ you wish to ensure that you will receive security 2 at time $t = 1$ at a fixed price. Thus you are interested in a contract **obligating** your counterparty to deliver to you security 2, denoted $S2$, at $t = 1$ for a fixed amount, say, F, that you will pay at $t = 1$. No cash is exchanged at the initial time period $t = 0$. Note that both parties to this agreement have **obligated** themselves to a transaction that will take place in the future for a price agreed upon in the current time period. What fixed amount, F, should be paid at $t = 1$?

Let us consider the above question in the framework of the example introduced at the beginning of this Chapter. We rerun the **Narbit** procedure so that the function **Vdis** is defined to confirm with this example.

```
> Narbit([[50,60,80],[90,110,130],[110,110,110]],\
> [60,105,100]);
```

$$\textit{The no} - \textit{arbitrage condition is satisfied.}$$

$$\textit{The stochastic discount factors are,}\ \left[\frac{1}{22}, \frac{25}{44}, \frac{13}{44}\right]$$

$$\textit{The risk} - \textit{neutral probability is,}\ \left[\frac{1}{20}, \frac{5}{8}, \frac{13}{40}\right]$$

[4] In fact it usually takes two or three days for delivery.

$$A\ \ risk - free\ rate\ in\ this\ market\ is,\ .1000$$
$$The\ function\ Vdis([c1, c2, ..]),\ values\ the\ cashflow\ [c1, c2, ..]$$

What is the payoff from the above arrangement? The payoff is the difference between the fixed amount F paid at $t = 1$ and the actual security price at $t = 1$. Since the value of the security in each of the three states is

$$(90, 110, 130) \tag{2.5}$$

the cash flow will be one of the following:

$$(90 - F,\ 110 - F,\ 130 - F)\,. \tag{2.6}$$

We can value this cash flow, as a function of F, by applying to it the function **Vdis**.

```
>  Vdis([90-F,110-F,130-F]);
```

$$105 - \frac{10}{11}F$$

Since obtaining this payoff costs nothing, the present value of this payoff must be zero. Hence, we can find the value of F by equating the present value of the payoff to zero.

```
>  solve(Vdis([90-F,110-F,130-F])=0);
```

$$\frac{231}{2}$$

Recalling that the risk-free interest rate in this market is 10 percent, the future value factor is $1+0.1 = \frac{11}{10}$. The price of security 2 is \$105 at $t = 0$. If the \$105 had been invested at the risk-free rate of interest at $t = 0$ it would grow to

```
>  105*11/10;
```

$$\frac{231}{2}$$

in the next time period. Can you provide an explanation for this result?

We can also use the pricing by replication to solve for the value of F. Since no cash is exchanged at $t = 0$, the replicating portfolio must be a self-financing portfolio. In other words it must satisfy

$$60\,S1{+}105S2{+}100\,SF{=}\ 0. \tag{2.7}$$

At time $t = 1$ the buyer of the forward contract (the party with the long position in the contract) pays $\$F$ and in return receives security 2. Hence at time $t = 1$ the cash flow to the party with the long position in the forward contract is the value of security 2 less $\$F$. In order to solve for the replicating portfolio we need to solve the following system of equations:

$$\begin{array}{cccccc} -60S1 & -105S2 & -100SF & = & 0 \\ 50SI & +90S2 & +110S3 & = & 90-F \\ 60SI & +110S2 & +110S3 & = & 110-F \\ 80SI & +130S2 & +110S3 & = & 130-F \end{array} \tag{2.8}$$

Solving this system using MAPLE we get the following:

```
> solve({-60*S1-105*S2-100*SF=0,\
> 50*S1+90*S2+110*SF=90-F,\
> 60*S1+110*S2+110*SF=110-F,\
> 80*S1+130*S2+110*SF=130-F});
```

$$\{S2 = 1,\ SF = \frac{-21}{20},\ S1 = 0,\ F = \frac{231}{2}\}$$

The replicating portfolio is composed of a long position in security 2 and a short position in $-\frac{21}{20}$ units of the bond (risk-free security). Stated differently, the replicating portfolio is composed by taking a loan (a short position in the bond) of

```
> 100*(21/20);
```

$$105$$

and using the proceeds of the loan to buy security 2, whose price at $t = 0$ is, conveniently,[5] \$105.

The writer of the forward contract (the one who has a short position in it) can thus buy this portfolio at time $t = 0$. It costs nothing. At time $t = 1$ the short position is closed (the loan is repaid). Since the interest rate is 10 percent, closing the short position costs

```
> 105*1.1;
```

$$115.5$$

The replicating portfolio also had a long position in security 2 which is now delivered to the buyer (the one who holds the long position) in the forward contract. Thus, at $t = 1$, the writer of the contract delivers security 2, which was held long in the portfolio, receives $\$F$ from the buyer of the contract, and closes (repays) the loan. Hence the cash flow to the writer of the forward contract at time $t = 1$ is $\$F$ less the amount of cash which is used up in repayment of the loan. This cash flow must be zero in total, since it costs nothing to initiate the transaction. Therefore $\$F$ must be the value of the loan repayment and its value is given by

[5] This is, of course, more than just a convenience — it is by design.

```
> evalf(231/2);
```

$$115.5000000$$

In summary, the replicating portfolio for the forward contract involves buying security 2 at $t = 0$ and holding it until $t = 1$, at which time it is delivered in exchange for receipt of a payment of $\$F$. At $t = 0$ no payment is made, since the long position in security 2 is financed by a short position in the bond (a loan). At $t = 1$, the receipt of $\$F$ is sufficient to (repay) close the short position in the bond (repay the loan). Since the short position in the bond was initiated to cover the cost of purchasing security 2, the amount required to close the short position is the future value of the spot price of security 2. Since calculating the value of F in this way involves buying the security and holding it until delivery, it is known as the *cost-of-carry model.*

We can show that in general the futures price is equal to the future value of the security, i.e., $1+r_f$ times the spot price of the security. This is done by replacing the numerical value with symbols and asking MAPLE to solve for the replicating portfolio. As before, we denote the spot price of the ith security by P_i and the price of security i in state j by P_{ij}. We thus need to solve the system

$$\begin{array}{llllll} -P1S1 & -P2S2 & -SF & = & 0 \\ P11S1 & +P21S2 & +(1+r)SF & = & P21-F \\ P12S1 & +P22S2 & +(1+r)SF & = & P22-F \\ P13S1 & +P23S2 & +(1+r)SF & = & P23-F, \end{array} \tag{2.9}$$

which is submitted to MAPLE:

```
> solve({-P1*S1-P2*S2-1*SF=0,\
> P11*S1+P21*S2+(1+r)*SF=P21-F,\
> P12*S1+P22*S2+(1+r)*SF=P22-F,\
> P13*S1+P23*S2+(1+r)*SF=P23-F},{S1,S2,SF,F});
```

$$\{S2 = 1,\ S1 = 0,\ SF = -P2,\ F = r\,P2 + P2\}$$

We can therefore see that the futures price F is equal to the current spot price $P2$ plus the incurred cost of borrowing the amount $P2$. Hence $F = P2+rP2$ or $F=P2(1+r)$.

In certain instances there may be some costs associated with holding the security which is to be delivered in the future. This situation affects the cash flows from the replicating portfolio and the futures price (forward price) F. The payoff from the replicating portfolio should be reduced to reflect that cost. In other words, the payoff from the portfolio in state one will be

$$P11S1 + P21\,S2 + (1+r)\,SF - StoCos, \tag{2.10}$$

where **StoCos** is the cost deducted from the payoff to carry the position. Thus, when solving for the replicating portfolio in the presence of such "costs of carry", we should solve the following system instead:

$$\begin{array}{lllllll} -P1S1 & -P2S2 & -SF & & = & 0 \\ P11S1 & +P21S2 & +(1+r)SF & -StoCos & = & P21-F \\ P12S1 & +P22S2 & +(1+r)SF & -StoCos & = & P22-F \\ P13S1 & +P23S2 & +(1+r)SF & -StoCos & = & P23-F. \end{array} \tag{2.11}$$

The request to solve the system (2.11) is submitted to MAPLE below.

```
>  solve({-P1*S1-P2*S2-1*SF=0,\
>  P11*S1+P21*S2+(1+r)*SF-StoCos=P21-F,\
>  P12*S1+P22*S2+(1+r)*SF-StoCos=P22-F,\
>  P13*S1+P23*S2+(1+r)*SF-StoCos=P23-F},\
>  {S1,S2,SF,F});
```

$$\{S2 = 1,\ S1 = 0,\ SF = -P2,\ F = r\,P2 + P2 + StoCos\}$$

As we can see in this case the futures price is equal to the spot price of the commodity plus the carrying (storage) cost. However, note that the carrying cost includes both the interest parameter, in $rP2$, and the storage cost parameter **StoCos**.

We can see that the futures price of the security F is greater than the spot price since both **StoCos** and r are positive. This set of circumstances is called *contango*. In certain real market situations, these circumstances are reversed. It can be observed that the futures price is smaller than the spot price. This situation is called *backwardation* or is referred to as an *inverted market*.

The theoretical explanation for the existence of such a situation relies on the concept of a convenience yield. If F is smaller than the spot price then there must be some added benefit for holding the portfolio over holding the forward contract. It can be assumed that holding the portfolio incurs some storage cost but also accrues some benefit to the holder. We then have to reformulate the cash flow from the portfolio.

We have to solve a system of equations in which **ConvY** denotes the benefit (in dollars received in the next time period) which accrues from holding the actual good or security rather than the futures contract. The system to be solved is thus

$$\begin{aligned}
-P1S1 - P2S2 \quad & -SF & = \quad & 0 \\
P11S1 + P21S2 \quad & +(1+r)SF - StoCos + ConvY & = \quad & P21 - F \\
P12S1 + P22S2 \quad & +(1+r)SF - StoCos + ConvY & = \quad & P22 - F \\
P13S1 + P23S2 \quad & +(1+r)SF - StoCos + ConvY & = \quad & P23 - F,
\end{aligned} \tag{2.12}$$

and is submitted to MAPLE below.

```
> solve({-P1*S1-P2*S2-1*SF=0,\
> P11*S1+P21*S2+(1+r)*SF-StoCos+ConvY=P21-F,\
> P12*S1+P22*S2+(1+r)*SF-StoCos+ConvY=P22-F,\
> P13*S1+P23*S2+(1+r)*SF-StoCos+ConvY=P23-F},\
> {S1,S2,SF,F});
```

$$\{F = r\,P2 + P2 + StoCos - ConvY,\ SF = -P2,\ S1 = 0,\ S2 = 1\}$$

It should now be apparent from the solution to F that if **ConvY** is large enough, we can arrive at the relation $F < P2$. **ConvY** is sometimes expressed in terms of the yield of the spot price which produces a value of **ConvY** at the expiration of the contract. Hence it is often referred to as a *convenience yield*, though it is only the difference between $(1+r)P2+StoCos$ and the futures price.

One possible explanation for the convenience yield may be related to the benefits which accrue to farmers who hold the actual commodity (grain, for example) rather than the futures contract. Holding the commodity allows the possibility of consuming the commodity before the expiration of the futures contract. Holding the futures contract allows consumption only at the expiration of the contract. For a firm which requires a certain raw material, for example, there may be additional benefits derived from holding the actual commodity.

2.4.2 Forward Contract on the Exchange Rate

Different countries use different currencies as their local denomination. Naturally a market has developed for the exchange of currencies. Exchange rates are quoted in terms of the amount of domestic currency needed to buy one unit of the foreign currency. Thus, in Canada, for example, an exchange rate of 1.38/USD means that each U.S. dollar costs 1.38 Canadian dollars. Notice that if one unit of a foreign currency costs $F0$ units of the domestic currency, then one unit of the domestic currency costs $\frac{1}{F0}$ units of the foreign currency.

There are two types of markets which deal in foreign exchange: the *spot market* and the *forward market*, as explained above. In the forward market currency is bought and sold for future delivery. Maturities of the forward contracts range between 1 and 12 months. We can analyze a similar situation to the forward contract, where the underlying asset is a foreign currency. In doing so we think about the opportunity to invest in a foreign currency just as another financial asset.

Assume that at time $t = 1$ the exchange rate between the foreign and domestic currencies is F_0 of the domestic currency per one unit of the foreign currency. Assume further that at time $t = 1$ the exchange rate will be F_1, F_2, and F_3 in states 1, 2, and 3, respectively. Let us first consider the cash flow obtained when 1 unit of the domestic currency is invested in a foreign market. Consider an investment of 1 unit of the domestic currency in the foreign market at the foreign interest rate (R_F) from time $t = 0$ to time $t = 1$. In terms of the domestic currency, at time $t = 1$ we will have the following cash flows:

$$\left(\frac{F1}{F0} (1 + R_F), \frac{F2}{F0} (1 + R_F), \frac{F3}{F0} (1 + R_F) \right). \tag{2.13}$$

The process of investing domestic currency in a foreign market encompasses three steps:

- Exchange 1 unit of the domestic currency for a foreign currency at time $t = 0$. This will result in $\frac{1}{F0}$ units of the foreign currency.
- Invest the foreign currency at the foreign risk-free rate of interest, R_F, applicable from time $t = 0$ to time $t = 1$. This will result in having $\frac{1+R_F}{F0}$ units of the foreign currency at time $t = 1$.
- Exchange the foreign currency back to the domestic currency at time $t = 1$. The resulting cash flow in terms of the domestic currency is stipulated in equation (2.13).

Hence, the opportunity to invest in a foreign market adds another asset (security) to the model. The price of the added asset is \$1 of the local currency, and the cash flow from this asset is given, in denomination of local currency, by equation (2.13). Note that, in spite of this added asset being an investment in a foreign market, we denominate its price and its payoff in the local currency.

Consider now two parties that enter into an agreement whereby one party is obligated to deliver to the other, in the next period, some units of the foreign currency. In exchange, at time $t = 1$, the other party pays a fixed amount, FF, of units of the domestic currency per one unit of the foreign currency. The party which has committed itself to sell the foreign currency is referred to as the one with the short position in the contract, or the *writer* of the contract. The other party is referred to as the one with a long position in the contract.

The writer of the contract will, regardless of the realization of the state of nature at time $t = 1$, sell one unit of the foreign currency for FF units of the local currency. What fixed exchange rate, FF, should be paid at time $t = 1$? This agreement is virtually the same agreement discussed in the previous example but here the exchange rate is the underlying asset.

Consider a contract, written at time $t = 0$, which obligates two counterparties to exchange, at time $t = 1$, $\frac{1+R_F}{F0}$ units of the foreign currency for FF units of the domestic currency per unit of the foreign currency (payable at time $t = 1$). Let us also consider a portfolio composed of the following:

- Invest 1 unit of the local currency at the foreign interest rate of R_F, and
- take a short position in a contract which obligates you to sell $\frac{1+R_F}{F0}$ units of the foreign currency at time $t = 1$ for FF local currency (per one unit of the foreign currency).

The cost of such a portfolio is of course 1 unit of the local currency. One unit of the local currency buys $\frac{1}{F0}$ units of the foreign currency which, if invested at the foreign risk-free rate, grows to $\frac{1+R_F}{F0}$ at time $t = 1$. Now make use of the short position in the contract to generate the sure amount $\frac{FF}{F0}(1 + R_F)$ of the local currency. Regardless of the state of nature, the short contract guarantees an exchange rate of one unit of foreign currency for FF units of the local currency. Hence $\frac{1}{F0}(1 + R_F)$ will be exchanged for $FF\frac{1}{F0}(1 + R_F)$ regardless of the state of nature. The present value of the cash flows from this portfolio is then

$$\frac{\frac{FF}{F0}(1 + R_F)}{1 + r_f}, \tag{2.14}$$

which must equal the cost of the portfolio, since otherwise arbitrage opportunities would exist. We therefore arrive at the equation

$$\frac{FF}{F0}\frac{1 + R_F}{1 + r_f} = 1, \tag{2.15}$$

or the equivalent relation

$$\frac{FF}{F0} = \frac{1 + r_f}{1 + R_F}. \tag{2.16}$$

Equation (2.16) is named the *interest rate parity condition*: it connects the risk-free rates of interest in the two countries. The interest rate parity condition determines the value of FF given the current exchange rate and the risk-free rates of interest in each country. The forward exchange rate is given by

$$FF = F0 \frac{1 + r_f}{1 + R_F}. \tag{2.17}$$

The arguments just presented follow the cost-of-carry model. If an investor is obligated to deliver a particular asset in the next time period, the investor buys it now and stores it until that future time period. Thus the investor ensures having it on hand in order to make delivery. When the underlying asset (or deliverable good) is money, the storage costs involved are negative. An investor can make money while storing money via interest earned. Thus there is a need to have on hand, now, an amount which is smaller than the amount to be delivered in the future time period. The investor buys the present value of the deliverable amount and invests it at the risk-free rate of interest in the currency to be delivered. Thus the investor ensures the availability of the amount to be delivered at the delivery time in the future.

An identical result can be derived in an equivalent way starting from the point of view of the writer of a forward contract to deliver one unit of the foreign currency at time $t = 1$. We repeat this argument from this point of view, just for clarity, and perhaps to emphasize that, although we cast the situation in a simple framework, the result is general. We also make use of MAPLE for some of the substitution commands.

Assume the general structure as before, i.e., P_i is the current price of security i, the price of security i in state j is P_{ij}, r_f is the risk-free rate of interest in the local market, R_F is the risk-free rate of interest in the foreign market, $F0$ is the current exchange rate, and Fi is the exchange rate in state i. The value of one unit of foreign currency at time $t = 1$ denominated in the local currency will be $F1$, $F2$, or $F3$, respectively, in states of nature 1, 2, or 3. In order to value FF by replication, we need to know what the replicating portfolio is. The system of equations is virtually identical to that formulated for the example above. However, here there is one additional asset: the investment in the foreign currency. We denote by FE the dollar amount, denominated in the domestic currency, invested in the foreign currency. From the point of view of the writer of such a contract, the cash flow obtained in the next time period is

$$(F1 - FF,\ F2 - FF,\ F3 - FF)\,. \tag{2.18}$$

The writer of the contract delivers Fi in state i and receives FF. A portfolio replicating this cash flow is the solution to the following system of equations:

$$\begin{array}{lllll} P11S1 & +P21S2 & +(1+r)SF + \frac{FE(1+R_F)F1}{F0} & = & F1 - FF \\ P12S1 & +P22S2 & +(1+r)SF + \frac{FE(1+R_F)F2}{F0} & = & F2 - FF \\ P13S1 & +P23S2 & +(1+r)SF + \frac{FE(1+R_F)F3}{F0} & = & F3 - FF. \end{array} \tag{2.19}$$

We submit this system to MAPLE, but for simplicity we constrain $S1$ and $S2$ to zero as we know such a solution exists.

```
> solve({S1=0,S2=0,\
> P11*S1+P21*S2+(1+r[f])*SF+(FE/F0)*(1+R[F])*F1=F1-FF,\
> P12*S1+P22*S2+(1+r[f])*SF+(FE/F0)*(1+R[F])*F2=F2-FF,\
> P13*S1+P23*S2+(1+r[f])*SF+(FE/F0)*(1+R[F])*F3=F3-FF},\
> {S1,S2,SF,FE});
```

$$\{S1 = 0,\ S2 = 0,\ SF = -\frac{FF}{1 + r_f},\ FE = \frac{F0}{1 + R_F}\}$$

The cost of the identified portfolio is given by

```
> subs(FE=F0/(1+R[F]),SF=-FF/(1+r[f]),S1=0,S2=0,\
> -P1*S1-P2*S2-SF-FE);
```

$$\frac{FF}{1 + r_f} - \frac{F0}{1 + R_F}$$

Since entering this contract costs nothing, the cost of the portfolio must be zero, otherwise arbitrage opportunities exist. Hence, we can solve for FF by equating the cost to zero. Thus we obtain a value of FF given by

```
> solve({%=0},{FF});
```

$$\{FF = \frac{F0\,(1 + r_f)}{1 + R_F}\}$$

This is, of course, the same result obtained before, i.e., equation (2.17). This expression is induced by the interest rate parity condition of equation (2.16). The most recent derivation emphasizes that our result is independent of the number of states of nature assumed to be possible in the future time period. The result is quite general.

The latter equation and equation (2.16) were derived by taking into account only the investment universe of the risk-free instruments available in the local and foreign markets. Given the structure of the investment universe in the domestic market, there are further constraints possible on the values of R_F and r_f which can coexist without the possibility of arbitrage opportunities. We shall examine this notion later in this Chapter.

2.5 Swaps: A First Look

A swap contract is an agreement between two parties to exchange securities, currencies, or some payments at some point in the future. There are a variety of swap contracts possible and they will be visited shortly. Here we make use of our simple framework in order to introduce the reader to the valuation of swaps. We will start with a currency swap and will cast it in our simple model. Currency swap is the term used to describe a contract between two parties. The parties agree to exchange interest and principal payments in one currency for interest and principal payments in another currency.

2.5.1 Currency Swaps

Consider a domestic firm which is required to pay an amount of foreign currency $N1(1+R1)$ at time $t=1$. This required payment could be a loan repayment for a loan which the firm obtained in the foreign market at a rate $R1$. $N1$ would then be the principal amount of the one-period loan. Depending on the actual exchange rate which prevails at time $t=1$, the cost of the required payment in domestic currency will be either $N1(1+R1)F1$, $N1(1+R1)F2$, or $N1(1+R1)F3$ in states of nature 1, 2, or 3, respectively.

Consider now a foreign firm which has to pay an amount of domestic currency $N0(1+r0)$ at time $t=1$. Again, this could be a loan payment on a loan obtained in the local market at a rate $r0$, where $N0$ is the principal amount. Depending on the realization of the exchange rate at time $t=1$, the foreign firm will have to pay $\frac{N0\,(1+r0)}{F1}$, $\frac{N0\,(1+r0)}{F2}$, or $\frac{N1\,(1+r0)}{F3}$, in states of nature 1, 2, or 3, respectively, in terms of the foreign[6] currency.

[6]It is possible that the domestic firm is not recognized in the foreign market, and as

These two firms may enter into a currency swap, in which the domestic firm will pay the foreign firm the amount $N0(1+r0)$ in local currency at time $t=1$, and the foreign firm will pay the domestic firm the amount $N1(1+R1)$ in foreign currency at time $t=1$. The domestic firm will therefore use the foreign currency obtained at time $t=1$ to pay off its debt, and the foreign firm will use the domestic currency obtained to pay off its foreign debt. Both firms are thus not subjected to the risk induced by the fluctuating exchange rates. We would like to investigate the value of such a swap agreement to the domestic firm. The cash flow at time 1, from the point of view of the domestic firm, in terms of the domestic currency, is

$$\begin{array}{ll} \textit{State } 1 & N1(1+R1)F1 - N0(1+r0), \\ \textit{State } 2 & N1(1+R1)F2 - N0(1+r0), \\ \textit{State } 3 & N1\,(1+R1)F3 - N0(1+r0). \end{array} \tag{2.20}$$

The value can be calculated in the following manner. The value of the cash inflow as of time zero, in terms of the foreign currency, is $\frac{N1\,(1+R1)}{1+R_F}$, where R_F is the risk-free rate in the foreign market. The value of this cash flow in terms of the domestic currency is $\frac{N1\,(1+R1)\,F0}{1+R_F}$. From the point of view of the domestic firm, this is a cash inflow. The domestic firm will pay $N0(1+r0)$ at time 1 regardless of which state of nature occurs. The present value of this sure payment is $\frac{N0\,(1+r0)}{1+r_f}$, where r_f is the risk-free rate in the domestic market. Therefore, the net present value of the swap agreement, from the point of view of the domestic firm in terms of the domestic currency, is

$$\frac{N1(1+R1)F0}{1+R_F} - \frac{N0(1+r0)}{1+r_f}. \tag{2.21}$$

The swap will thus have a value of zero if $N1$ and $N0$ are related via

```
> solve({N1*(1+R1)/(1+R[F])*F0-N0*(1+r0)/(1+r[ f])=0},N1);
```

$$\{N1 = \frac{N0\,(1+r0)\,(1+R_F)}{(1+r_f)\,(1+R1)\,F0}\}$$

If $N1$ is larger than

$$\frac{N0(1+r0)(1+R_f)}{(1+r_f)(1+r1)\,F0} \tag{2.22}$$

the swap will have a positive value, and if $N1$ is smaller than the expression in equation (2.22) the swap will have a negative value.

such might be required to pay a higher rate on a loan than if they had taken the loan in the domestic market. Note also that our analysis does not consider the riskiness of the cash flows involved in the swap. We are assuming that there is no possible default by either of the counterparties in the agreement.

2.5.2 Equity (Asset) Swap

An equity swap allows the parties to exchange the return on an index (a variable cash flow) or on a specific security with, for example, a fixed rate of interest. The variable cash flow is a random variable which depends on the realized state of nature in the future time period.

For this type of swap the two swapped cash flows depend on some principal amount called the *notional principal* as only the return is swapped. The fixed cash flow is Nr, where r is the relevant rate of interest. The variable cash flow depends on the return on an agreed-upon index or security. As before, we can cast this swap agreement and analyze it in the context of our simple model. The variable cash flow is given by $r_i N$, where r_i is the rate of return on the ith security. The rate of return[7] is a random variable and is defined as $r_i = \frac{P_{ij}-P_i}{P_i}$, where P_i is the current price of security i and P_{ij} is the price of security i in state of nature j.

Let us consider valuation from the point of view of the party who pays the fixed rate of interest. We assume that the variable cash flow is the return on security 2,

$$\left(\frac{N(P21-P2)}{P2}, \frac{N(P22-P2)}{P2}, \frac{N(P23-P2)}{P2}\right), \tag{2.23}$$

and that the fixed rate of interest is r. We begin by valuing the variable cash flow by replication. Hence, we first solve for the replicating portfolio by submitting to MAPLE the request to solve the system of equations (2.24).

$$\begin{array}{llll} P11S1 & +P21S2 & +(1+r)SF & = \frac{N(P21-P2)}{P2} \\ P12S1 & +P22S2 & +(1+r)SF & = \frac{N(P22-P2)}{P2} \\ P13S1 & +P23S2 & +(1+r)SF & = \frac{N(P23-P2)}{P2}. \end{array} \tag{2.24}$$

```
> solve({P11*S1+P21*S2+(1+r)*SF=N*(P21-P2)/P2,\
> P12*S1+P22*S2+(1+r)*SF=N*(P22-P2)/P2,\
> P13*S1+P23*S2+(1+r)*SF=N*(P23-P2)/P2},{S1,S2,SF});
```

$$\{S2 = \frac{N}{P2}, SF = -\frac{N}{r+1}, S1 = 0\}$$

[7] In reality, some dividends might be paid on the security or on the index, and these cash flows should be taken into account when calculating the return. We will return to this subject later.

The MAPLE results tell us that the replicating portfolio is formed by borrowing $\$\frac{N}{1+r}$ (having a short position in the bond) and investing $\$N$ in security 2. The cost of the replicating portfolio is thus

$$\frac{Nr}{1+r} \tag{2.25}$$

as calculated below.

```
> simplify(subs(S2 = N/P2, SF = -N/(r+1),\
> S1 =0,S1*P1+S2*P2+SF));
```

$$\frac{N\,r}{r+1}$$

We can easily verify that this is the solution. Borrowing $\$\frac{N}{1+r}$ at time $t = 1$ and investing $\$N$ in security 2 makes the out-of-pocket cost of this portfolio $\frac{Nr}{1+r}$. The $\$N$ invested in security 2 buys $\frac{N}{P2}$ units of security 2. The $\frac{N}{P2}$ units of security 2 will be worth $\frac{N\,P2j}{P2}$ in state j. However, the loan should be paid back and therefore the value of the portfolio in state j will be $\frac{N\,P2j}{P2} - N$ or

```
> simplify((N*P2j)/P2-N);
```

$$-\frac{N\,(P2 - P2j)}{P2}$$

as required. This replication strategy is quite simple. We elaborate now to demonstrate that indeed its results are general and that it holds not only for this simple setting, but for a more complicated and realistic one.

To generate security 2 return on a principal amount of $\$N$ (or the return from any other security), the investor needs to invest $\$N$ in security 2 (or the particular security) at time zero and pay back $\$N$ at time one. The value, at time one, of the $\$N$ invested in security 2, minus the investment N, is simply the return on $\$N$ invested in security 2. In order to generate the payout of N at time one, the investor needs to borrow $\frac{N}{1+r}$ at time zero. In other words, the investor needs to have a short position in the bond for $\$\frac{N}{1+r}$. The second step is to invest N in security 2 at time zero. However, since shorting the bond generates the amount $\$\frac{N}{1+r}$, the investor needs only add an out-of-pocket amount equal to $N - \frac{N}{1+r}$ to be able to invest N in security 2. The out-of-pocket cost of the strategy is

```
> simplify(N-N/(1+r));
```

$$\frac{N\,r}{r+1}$$

In order to avoid arbitrage possibilities, this cost must also be the value of the cash flow

$$\left(\frac{N(P21-P2)}{P2}, \frac{N(P22-P2)}{P2}, \frac{N(P23-P2)}{P2}\right). \tag{2.26}$$

From the point of view of the party who pays the fixed rate r, the present value of the cash flow (2.26) obtained at time one is $\frac{Nr}{1+r}$. This party has to pay rN at time one. The present value of this payment is $\frac{Nr}{1+r}$. The value of the swap agreement is therefore zero: the value of the cash inflow is equal to the value of the cash outflow.

This result may seem puzzling at first glance. Moreover, it does not matter what is being swapped: the return on N will always be equal to $\frac{Nr}{1+r}$. This is verified by noting that the value is independent of the designation "2". (Alternatively, the reader could replace "2" by "1", and repeat the argument with respect to security 1.) Continuing in this line of reasoning, if the return on N is invested in security 1, it has a current value of $\frac{Nr}{1+r}$. This is the same as the present value of the return on security 2. This means that the values of the two cash flows (2.26) and

$$\left(\frac{N(P11-P1)}{P1}, \frac{N(P12-P1)}{P1}, \frac{N(P13-P1)}{P1}\right) \tag{2.27}$$

are the same. Hence, swapping these two cash flows will be a fair deal.

A second way of thinking about these swapped cash flows may convince you that there is no puzzle here. If you are given \$100 today, you can turn the \$100 into $100(1+r)$ in the next time period. Alternatively, you can turn the \$100 into $100(1+r_i)$, where r_i is the random return on some risky security. In both cases you require exactly the same amount now to generate either $100(1+r)$ or $100(1+r_i)$. Hence, the value of the cash flow received in the second period is like the sure amount $100(1+r)$ received in the second period. The value of $100r_i$ must be the same as the value of $100(1+r_i) - 100$ in the second period. This, though, yields the result that the value of $100r_1, 100r_2$, and $100r_3$ is the same as the value of $100r$, which is our result. Obviously, \$100 can be replaced by a general amount N.

Indeed, keeping the above argument in mind, the question of which part of the swap provides greater return on \$100 — the \$100 earning the risk-free rate of interest or the \$100 invested in another security, — reminds one of the old question, *what weighs more, a kilogram of feathers or a kilogram of iron? The value of an asset is its cost regardless of the type of the asset, risk-free or risky. It is much as the weight of a matter is its mass in kilograms regardless of the type of the substance.*

2.6 General Valuation

In the preceding examples we demonstrated how a selection of financial contracts work and how they can be valued. In this section we want to emphasize yet again that while it may seem that the valuation technique for each contract is particular to that contract, this appearance is misleading. This impression might have been obtained since we used pricing by replication methods which may not have seemed uniform. This is not actually the case. In fact, each example can be analyzed using a generic method of valuation. The value of the financial contract (or asset) is obtained merely by discounting, at the risk-free rate of interest, the expected value, based on the risk-neutral probability, of the future cash flows from that contract. Equivalently, the value is the discounted value of the cash flows based on stochastic discount factors. The valuation method is the one which is induced by equation (1.8).

To demonstrate this generic valuation, we take advantage of the procedure **Valucash,** which was also used in the appendix of the previous Chapter. The reader is also reminded that in the on-line version of the book one can use the Help option to search for **Valucash** and obtain instructions on how to use this procedure (or any other procedure mentioned in the book). The input to the procedure is similar to the input to the **Narbit** procedure, with the addition of one parameter: the cash flow to be valued. Thus, the procedure checks if the no-arbitrage condition is satisfied, as did **Narbit**. If the no-arbitrage condition is satisfied, **Valucash** values the cash flow. If the no-arbitrage condition is not satisfied, the procedure reports that this is the case and stops there. In the case for which the market specified by the user is not complete, the procedure reports this also. In such an instance the price of the cash flow valued by arbitrage may not be unique. Readers who are interested in pursuing pricing in incomplete markets are referred to the appendix of the previous Chapter. In what follows, we repeat the valuation of the examples above utilizing the **Valucash** procedure. In each case we submit the structure of the market to the procedure: the prices of the securities and the cash flow to be valued. The reader is encouraged to try and value different cash flows and think about different contracts that can be valued with this approach. Furthermore, the reader is encouraged to try and change the structure of the market and see its effect on the valuation.

2.6.1 The Risk-Free Rate of Interest Implicit in the Market

We begin with the first example in which we were given a market which does not have a bond as one of the primary securities. Even in this market, the risk-free rate of interest is implicit in the prices of the securities. The risk-free rate is recovered by valuing in the current time period the cash flow $(1,1,1)$ to be received at a future time period. The current value of this cash flow is $\frac{1}{1+r}$, where r is the risk-free rate of interest implicit in the market. We submit the market structure to MAPLE (the payoff from each security, their prices, and the cash flow to be valued).

```
> Valucash([[50,60,80],[0,10,30],[120,30,200]],\
> [60,160/11,1795/22],[1,1,1]);
```

this market is complete

$$\textit{the valuation is induced by, } \left[\frac{1}{22}, \frac{25}{44}, \frac{13}{44}\right]$$

$$\textit{the portfolio generating this cashflow is , } \left[\begin{array}{ccc} \frac{1}{50} & \frac{-1}{50} & 0 \end{array}\right]$$

$$\textit{the value of the cashflow is, } \frac{10}{11}$$

We confirm that the value of the $(1,1,1)$ cash flow is $\frac{10}{11}$, which is equal to $\frac{1}{1+r}$, and so r is equal to 10 percent. The procedure **Valucash** also reports the set of stochastic discount factors under which valuation is induced, and reports the portfolios which span this cash flow. The reader may wish to experiment with other examples of this sort. If the market specified by the reader induces arbitrage opportunities, the procedure **Valucash** will report that this is the case and will not value the cash flow in question.

2.6.2 The Two Propositions

In the second example from Section 2.3, we valued two propositions. The details of the proposition can be translated into the problem of valuing two cash flows. The first proposition was the option to buy security 1 at time one for \$50, regardless of its market price at time one. Hence, if we hold this option, we actually own the cash flow $(50-50, 60-50, 80-50)$. This can be submitted to MAPLE and valued.

```
> Valucash([[50,60,80],[90,110,130],[110,110,110]],\
> [60,105,100],[0,10,30]);
```

this market is complete

$$\textit{the valuation is induced by } \left[\frac{1}{22}, \frac{25}{44}, \frac{13}{44}\right]$$

$$\textit{the portfolio generating this cashflow is } \left[\begin{array}{ccc} 1 & 0 & \frac{-5}{11} \end{array}\right]$$

$$\textit{the value of the cashflow is } \frac{160}{11}.$$

The second proposition was the ability to buy the security for \$40 regardless of its market price. To value this proposition, we ask MAPLE to value the cash flow $(50-40, 60-40, 80-40)$.

```
> Valucash([[50,60,80],[90,110,130],[110,110,110]],\
> [60,105,100],[10,20,40]);
```

this market is complete

$$\textit{the valuation is induced by } \left[\frac{1}{22}, \frac{25}{44}, \frac{13}{44}\right]$$

$$\textit{the portfolio generating this cashflow is } \left[\begin{array}{ccc} 1 & 0 & \frac{-4}{11} \end{array}\right]$$

$$\textit{the value of the cashflow is } \frac{260}{11}.$$

2.6.3 Forwards

Forward Contract on a Security

Next we review an example in which the price of a security is fixed. Assume that the investor has entered into an agreement in which he or she is obligated to buy security 2 for a fixed price F. We are interested in valuing the cash flow

$$(90-F,\ 110-F,\ 130-F)\,. \tag{2.28}$$

```
> Valucash([[50,60,80],[90,110,130],[110,110,110]],\
> [60,105,100], [90-F,110-F,130-F]);
```

this market is complete

$$\textit{the valuation is induced by } \left[\frac{1}{22}, \frac{25}{44}, \frac{13}{44}\right]$$

$$\textit{the portfolio generating this cashflow is } \left[\begin{array}{ccc} 0 & 1 & -\frac{1}{110}F \end{array}\right]$$

$$\textit{the value of the cashflow is } 105-\frac{10}{11}F.$$

The value of this cash flow is clearly a function of F. The reader has likely already recognized that $\frac{10}{11}$ above is simply the more general term $\frac{1}{1+r}$. If entering this agreement costs the investor nothing, then the value of the reproduced cash flow at time zero must also be zero. In other words, F must satisfy $105 - \frac{11F}{10} = 0$. As we have already seen, this means that the futures price F is $1+r_f$ times the current price of the security.

Foreign Exchange

We consider now an example which involves a foreign market. We saw in equation (2.16) that, in order to avoid arbitrage opportunities, the following relationship should always hold: $FF= F0\frac{1+r_f}{1+R_F}$. We can derive this relationship again by valuing the cash flows which are a consequence of the forward agreement to buy $\frac{1+R_F}{F0}$ units of foreign currency at a price of FF units of local currency per unit of foreign currency. Investing one unit of the domestic currency in the foreign market and entering such an agreement will result in the cash flow

$$\left(\frac{FF(1+R_F)}{F0}, \frac{FF(1+R_F)}{F0}, \frac{FF\,(1+R_F)}{F0}\right) \tag{2.29}$$

denominated in the domestic currency. This cash flow can be valued using the function **Vdis**.

```
> Vdis([(1*FF/F0)*(1+R[F]),(1*FF/F0)*(1+R[F]),\
> (1*FF/F0)*(1+R[F])]);
```

$$\frac{10}{11}\,\frac{FF\,(1+R_F)}{F0}$$

The reader will not be surprised to discover that the result shows that the value is the discounted value of the amount

$$\frac{FF\,(1+R_F)}{F0}. \tag{2.30}$$

The cost of the writer's portfolio is \$1. (One dollar was invested in the foreign market and the cost of entering the contract was zero.) Thus, the value calculated above must also be equal to \$1 or it would be the case that arbitrage opportunities exist. We can solve for a relationship between $1+R_F$ and FF by equating the above value to 1. We ask MAPLE to solve it for us.

```
> solve({Vdis([(1*FF/F0)*(1+R[F]),(1*FF/F0)*(1+R[F]),\
> (1*FF/F0)*(1+R[F])])=1}, {R[F]+1});
```

$$\{1+R_F = \frac{11}{10}\,\frac{F0}{FF}\}$$

The $\frac{11}{10}$ is indeed one plus the interest rate. We therefore obtain the following general result when we make the substitution of $1+r_f$ for $\frac{11}{10}$.

```
> subs(11/10=1+r[f],%);
```

$$\{1 + R_F = \frac{F0\,(1 + r_f)}{FF}\}$$

We can also solve for the value of FF

```
> solve(%,{FF});
```

$$\{FF = \frac{F0\,(1 + r_f)}{1 + R_F}\}$$

This ought to seem remarkably familiar: it is indeed equation (2.17) which was obtained above by the replication method. If we replace FF in the above equation with the expected value of the exchange rate in the next time period, based on the risk-neutral probability, the expression above would still hold true. This is proved by considering the cash flow

$$\left(\frac{F1}{F0}(1+R_F),\ \frac{F1}{F0}(1+R_F),\ \frac{F3}{F0}(1+R_F)\right), \tag{2.31}$$

which results from investing 1 unit of the domestic currency in the risk-free rate of interest that prevails in the foreign market. Recall that the value of this cash flow is its expected value (equation (1.15)), based on the risk-neutral probability, discounted at the domestic risk-free rate of interest. Hence, the value of this cash flow is

$$\frac{E\left(\frac{Fi}{F0}\,(1 + R_F)\right)}{1 + r_f}. \tag{2.32}$$

Since $F0$, R_F, and r_f are deterministic constants, equation (2.32) can be rewritten as

$$\frac{1 + R_F}{1 + r_f}\frac{E(Fi)}{F0}. \tag{2.33}$$

The result of this valuation must be 1 unit of the local currency; otherwise an arbitrage opportunity exists. Therefore we have

$$\frac{1 + R_F}{1 + r_f}\frac{E(Fi)}{F0} = 1 \tag{2.34}$$

and we can solve for $F0$ in terms of the other parameters. This is done with the next MAPLE statement.

```
> solve({(1+R[F])*E(Fi)/((1+r[f])*F0)=1},{(F0)});
```

$$\{F0 = \frac{(1 + R_F)\,\mathrm{E}(Fi)}{1 + r_f}\}$$

Hence we get the relation

$$F0 = \frac{1 + R_F}{1 + r_f} E(Fi). \tag{2.35}$$

The arbitrage argument which induces this relation involves investment in only two riskless assets in both the domestic and foreign markets. We now investigate the relation in the richer framework which includes three securities and their prices in the three possible future states of nature. As mentioned, expanding the environment to include investment in the foreign market is akin to adding another asset to the market. We can define an asset with a price of 1 and a cash flow equal to

$$\left(\frac{F1}{F0}(1 + R_F), \frac{F2}{F0}(1 + R_F), \frac{F3}{F0}(1 + R_F)\right). \tag{2.36}$$

Including such an asset will not introduce arbitrage opportunities provided the value of the asset added based on **Vdis** is indeed equal to 1. In other words, there must exist $F1$, $F2$, $F3$, $F0$, and R_F, each non negative, which satisfy the equation

$$d1\frac{F1}{F0}(1 + R_F) + d2\frac{F2}{F0}(1 + R_F) + d3\frac{F3}{F0}(1 + R_F) = 1, \tag{2.37}$$

where $d1 = \frac{1}{22}$, $d2 = \frac{25}{44}$, and $d3 = \frac{13}{44}$.

```
> Vdis((F1/F0)*(1+R[F]),(F2/F0)*(1+R[F]),\
> (F3/F0)*(1+R[F]))=1;
```

$$\frac{1}{22}\frac{F1\,(1 + R_F)}{F0} + \frac{25}{44}\frac{F2\,(1 + R_F)}{F0} + \frac{13}{44}\frac{F3\,(1 + R_F)}{F0} = 1$$

In order to investigate the existence of such numerical parameters, the reader may make use of the procedure **CheckNA**. This procedure checks to see if the no-arbitrage condition is satisfied for a given market. Its output is similar to that of **Narbit**. We can introduce a new asset with a price of 1, first by introducing the values of $F1$, $F2$, $F3$, $F0$, and R_F. Consider the following situation. Since **CheckNA** is not a (fully) symbolic procedure, we take two steps.

```
> subs(F0=1,F1=1.2,F2=0.9,F3=1.4,R[F]=.03,\
> 'CheckNA([[50,60,80],[90,110,130],[110,110,110],\
> [(F1/F0)*(1+R[F]),(F2/F0)*(1+R[F]),(F3/F0)*(1+R[F])]],\
> [60,105,100,1])');
```

CheckNA([[50, 60, 80], [90, 110, 130], [110, 110, 110], [1.236, .927, 1.442]], [60, 105, 100, 1])

```
> %;
```

$$\begin{gathered}
\textit{The no} - \textit{arbitrage condition is not satisfied} \\
\textit{An arbitrage portfolio is}: \\
\textit{Short}\ 1.000000000\ \textit{of security}\ 1 \\
\textit{Buy}\ .5896414343\ \textit{of security}\ 2 \\
\textit{Short}\ .1784860558\ \textit{of security}\ 3 \\
\textit{Buy}\ 15.93625498\ \textit{of security}\ 4 \\
\textit{The cost of this portfolio is zero} \\
\textit{This portfolio produces income of}\ 3.131474104,\ \textit{in state}\ 1 \\
\textit{This portfolio produces income of}\ 0\ \textit{in state}\ 2 \\
\textit{This portfolio produces income of}\ 0\ \textit{in state}\ 3
\end{gathered}$$

These parameter values do not satisfy the no-arbitrage condition. The reader may wish to substitute different values for the parameters and rerun the procedure **CheckNA**. Of course, it should be clear that setting $F0 = Fi = 1$ for all i, and setting R_F equal to $\frac{1}{10}$, will satisfy the no-arbitrage condition.

```
> subs(F0=1,F1=1,F2=1,F3=1,R[F]=1/10,
> 'CheckNA([[50,60,80],[90,110,130],[110,110,110],\
> [(F1/F0)*(1+R[F]),(F2/F0)*(1+R[F]),(F3/F0)*(1+R[F])]],\
> [60,105,100,1])');
```

$$\mathrm{CheckNA}([[50, 60, 80], [90, 110, 130], [110, 110, 110], [\frac{11}{10}, \frac{11}{10}, \frac{11}{10}]], [60, 105, 100, 1])$$

```
> %;
```

$$\textit{The no} - \textit{arbitrage condition is satisfied.}$$

2.6.4 Swaps

Currency Swaps

Consider the currency swap now from the point of view of the domestic firm. The cash flows which are being swapped are

$$(N1(1 + R1)F1,\ N1(1 + R1)F2,\ N1(1 + R1)F3) \tag{2.38}$$

as cash inflows to the domestic firm and $N0(1+r0)$ as cash outflow to the domestic firm. As before, our aim is to show that the valuation methods already described are generic. We can use **Vdis** to value this currency swap.

We begin by valuing the cash flows which are to be received by the domestic firm. Their value is[8]

```
> Vdis([N1*(1+R1)*F1,N1*(1+R1)*F2,N1*(1+R1)*F3]);
```

$$\frac{1}{22}\,N1\,(1+R1)\,F1+\frac{25}{44}\,N1\,(1+R1)\,F2+\frac{13}{44}\,N1\,(1+R1)\,F3$$

where, as before, Fj is the exchange rate at time 1 in state j. Fj is the value of one unit of foreign currency in the domestic currency. Collecting terms from the previous expression, we obtain the following.

```
> collect(%,[N1,R1],distributed);
```

$$(\frac{1}{22}\,F1+\frac{25}{44}\,F2+\frac{13}{44}\,F3)\,N1\,R1+(\frac{1}{22}\,F1+\frac{25}{44}\,F2+\frac{13}{44}\,F3)\,N1$$

Recognizing that the expression in parentheses is the expected value of Fj based on the risk-neutral probability, we substitute its value and collect terms with respect to $N1$.

```
> subs(1/22*F1+25/44*F2+13/44*F3=1/(1+r[f])*E(Fj),%);
```

$$\frac{\mathrm{E}(Fj)\,N1\,R1}{1+r_f}+\frac{\mathrm{E}(Fj)\,N1}{1+r_f}$$

```
> collect(%,N1);
```

$$(\frac{\mathrm{E}(Fj)\,R1}{1+r_f}+\frac{\mathrm{E}(Fj)}{1+r_f})\,N1$$

The value of the currency swap will be the above less the amount the domestic firm is required to pay out. This cash outflow is valued next.

```
> %-Vdis([N0*(1+r0),N0*(1+r0),N0*(1+r0)]);
```

$$(\frac{\mathrm{E}(Fj)\,R1}{1+r_f}+\frac{\mathrm{E}(Fj)}{1+r_f})\,N1-\frac{10}{11}\,N0\,(1+r0)$$

In general, $\frac{10}{11}$ can be replaced with $\frac{1}{1+r_f}$. To arrive at a more general expression, we make this substitution.

```
> subs(-10/11=-1/(1+r[f]),%);
```

$$(\frac{\mathrm{E}(Fj)\,R1}{1+r_f}+\frac{\mathrm{E}(Fj)}{1+r_f})\,N1-\frac{N0\,(1+r0)}{1+r_f}$$

---

[8]The reader is reminded that some of the MAPLE commands must be executed in order. If the reader begins reading in the middle, it may be necessary to redefine the function **Vdis**. This could be done simply by reexecuting the procedure **Narbit**, i.e., by issuing the MAPLE command

```
> Narbit([[50,60,80],[90,110,130],[110,110,110] ],[60,105,100]):
```

Furthermore, to demonstrate that this expression is equivalent to the value which we obtained in equation (2.21), we rewrite $E(Fj)$ in terms of R_F and r_f, the risk-free rates of interest in the foreign and domestic markets, respectively. Converting 1 unit of domestic currency to foreign currency will produce $\frac{1}{F0}$ units of the foreign currency. Investing this at the risk-free rate of interest in the foreign market will produce $(1 + R_F)\frac{1}{F0}$ in terms of the domestic currency at time 1. This amount will be worth $(1+R_F)\frac{1}{F0}Fj$ units of domestic currency in state i. Hence, its value is the expected value, based on the risk-neutral probability, discounted by the risk-free rate of interest. However, the value of the cash flow must be equal to 1, and so we can solve for $E(Fj)$ as below.

```
> solve((1+R[F])*E(Fj)/((1+r[f])*F0)=1,(E(Fj)));
```

$$\frac{(1 + r_f)\,F0}{1 + R_F}$$

Substituting for $E(Fj)$, the solution obtained by MAPLE is executed next. (Recall that %% stands for the expression obtained two lines above.)

```
> subs(E(Fi)=%,%%);
```

$$\left(\frac{F0\;R1}{1 + R_F} + \frac{F0}{1 + R_F}\right) N1 - \frac{N0\,(1 + r0)}{1 + r_f}$$

Collecting terms with respect to $F0$, we obtain

```
> collect(%,F0);
```

$$\left(\frac{R1}{1 + R_F} + \frac{1}{1 + R_F}\right) N1\;F0 - \frac{N0\,(1 + r0)}{1 + r_f}$$

which can be further simplified to

$$(1 + R1)N1\frac{F0}{1 + R_F} - N0\frac{1 + r0}{1 + r_f}. \tag{2.39}$$

This is the same result as that obtained when the valuation was done by replication.

Equity (Asset) Swaps

To value an equity swap using stochastic discount factors, we input the cash flows of the swap to **Vdis**. As before, looking at it from the view point of the party who pays the fixed rate cash flows, the amount paid out is Nr_f and the cash flow received is $\left(\frac{N\,P_{i1}}{P_i} - N, \frac{N\,P_{i2}}{P_i} - N, \frac{N\,P_{i3}}{P_i} - N\right)$. Hence, the net cash flow at time 1 is

$$\left(N\left(1 + r_f - \frac{P_{i1}}{P_i}\right), N\left(1 + r_f - \frac{P_{i2}}{P_i}\right), N\left(1 + r_f - \frac{P_{i3}}{P_i}\right)\right). \tag{2.40}$$

Define R_i to be the rate of return of security i. In state 1, the return on security i is denoted by R_{i1} and has the value $\frac{P_i - P_{i1}}{P_i}$. Let us rewrite $\frac{P_{i1}}{P_i}$ as $\frac{P_{i1} - P_i + P_i}{P_i}$ and thus, in state 1,

$$1 + r_f - \frac{P_{i1}}{P_i} = 1 + r_f - (1 + R_{i1}) = r_f - R_{i1}.$$

Thus to value the equity swap, we need only value the cash flow

$$(r_f - R_{i1}, r_f - R_{i2}, r_f - R_{i3}) \tag{2.41}$$

and recognize that the value of the swap will be N times the value of the above cash flow.

Consider now the result for security 2 of our model. We input the cash flow in question to **Vdis**.

```
> Vdis([(1/10)-(90-105)/105,(1/10)-(110-105)/105,\
> (1/10)-(130-105)/105]);
                              0
```

This is the anticipated result. The reader may wish to recall equation (1.17) of the previous Chapter. According to that equation, the expected rate of return of every security under the risk-neutral probability is r_f. The cash flow of equation (2.41) can also be valued as the discounted value of the expected value of the cash flow. That is, as $\frac{E(r_f - R_i)}{1+r_f}$, but since $E(r_f - R_i) = 0$ by equation (1.17), the result is confirmed in a general way.

2.7 Concluding Remarks

As a final note we would like to again point out to the reader that our analysis of the foreign exchange and the forward contract is more general than it may appear at first glance. Although the analysis took place in the setting of a simple model with only three possible states of nature in the next time period, the analysis can be extended to a more general model in a straightforward way. The conclusion summarized by equation (2.16), which holds true even if the foreign exchange or the price of the asset on which the forward contract is written is a random variable that takes on many more possible values than the three in that example. The reader could review the analysis and would then realize that indeed nothing is lost when this assumption of three states of nature is relaxed.

In the setting of the one-period model, swaps and forwards are essentially the same. A forward contract on the foreign exchange, for example, can be reinterpreted as swapping local currency for foreign currency. Under such interpretation the forward price determines the amount of local currency that is swapped for the foreign currency. We will review again swaps and forwards in a multiperiod setting and will comment at that time on the intimate relation that exists between them.

2.8 Questions and Problems

Problem 1. The text states that in the binomial model $Do < R_f$ and $R_f < Up$; otherwise arbitrage opportunities exist. Assume that $Up \leq R_f$. Identify an arbitrage opportunity in such a market.

Problem 2. Consider a market with the following securities when answering the following questions:

	State 1	*State 2*	*State 3*	*Prices*
Security 1	30	65	95	75
Security 2	90	110	145	120
Security 3	100	100	100	93

1. Determine if the no-arbitrage condition is satisfied in the above market. If the no-arbitrage condition is satisfied, determine the stochastic discount factors, the risk-neutral probability, and the risk-free rate.
2. Determine the forward price associated with a forward contract written on each of the three securities by discounting the resultant cash flow from each contract.
3. Repeat the above utilizing the replication method.
4. Verify that the results of the above two parts are the same and equal the future value of the current price of each security.
5. What should the cost be, if any, of a forward contract on each security with a forward price of $100? Explain your results and confirm your answer utilizing the replication and the SDF methods.

Problem 3. Consider the market discussed in Problem 2 and assume that it describes a local market with a currency denoted by Lc. Assume that an investment in a foreign currency, denoted by Fr, with a risk-free rate denoted by RFr, becomes available.

1. What constraints are imposed on the $\frac{Lc}{Fr}$ exchange rate in this market? Why?
2. Consider the following alternatives for the exchange rates and the foreign interest rates.

	Spot	*State 1*	*State 2*	*State 3*	*RFr*
Alternative 1	1.085484	$\frac{49}{50}$	1	$\frac{26}{25}$	$\frac{1}{20}$
Alternative 2	$\frac{25011}{25000}$	$\frac{9}{10}$	1	$\frac{21}{20}$	$\frac{1}{20}$

   Which alternatives, if any, satisfy the no-arbitrage condition?
3. If an alternative satisfies the no-arbitrage condition, determine a portfolio composed of only local securities that is equivalent to an investment of Lc 100 in the risk-free asset in the foreign market.
4. Determine the spot exchange rate if the exchange rate in each of the three states is $\frac{83}{100}$, $\frac{87}{100}$, $\frac{88}{100}$, and the foreign risk-free rate is $\frac{2}{100}$. What will the forward price be of Fr in the local market in such a market.

Problem 4. Consider the market discussed in Problem 2 and an investor who holds 100 shares of security 2 and would like to hedge the risk implied by this portfolio.

1. How can this be accomplished by adding a forward contract to the portfolio? What are the details of this contract? Should the investor assume a long or a short position in it?
2. What would be the forward price in the contract above?
3. What will the value be of the new portfolio in each state of nature?
4. Can you recommend another strategy that will accomplish the same goal?

Problem 5. Consider a mutual fund company that manages an index fund in the market discussed in Problem 2. The index is composed of $400,000 invested in the first security, $500,000 in the second security, and $200,000 in the risk-free rate. The fund manager would like to reshuffle the index so that the investment in security 2 is moved to the risk-free rate, without operating in the spot market. (The manager might be concerned with acting in the spot market as it may adversely affect the security price.) Can you offer an alternative strategy?

Chapter 3

Pricing by Arbitrage: *Debt Markets*

The Chapter 1 model in which there were only a finite number of states of nature seems unrealistic. However, as we mentioned already, this is not in fact the case. With a slight modification of the interpretation of that model, a very realistic model describing the debt market is obtained.

If the states of nature are interpreted as time periods, a realistic model describing the debt market is obtained. The counterpart of the stochastic discount factors is deterministic factors that value cash flows across time periods. The use of these discount factors is virtually the same as that of the stochastic discount factors. The only difference is that here the discounting is due only to the time value of money. Most of the cash flows dealt with in this Chapter involve no risk; these are dollars to be obtained for certain in the future. This Chapter derives the discount factors and their intimate relation to the no-arbitrage condition. It will demonstrate the use of the derived factors for valuing certain assets.

3.1 Setting the Framework

The *debt market*, as its name implies, is a market in which debts are bought and sold. This market is also referred to as the *bond market.* A bond is a security issued by a particular entity which promises to pay the holder of the bond a fixed amount of money at fixed times in the future, $t_1, t_2, \ldots, t_k$. At each of the payment times except the last one, t_k, the bond pays the same amount of money which we will denote by c. On the last payment date, referred to as the *maturity* of the bond, the bond pays an amount equal to

$c + FC$. The amount FC is referred to as the *principal* or the *face value* of the bond and the payments of the amount c are called *coupon payments.* In Chapter 1 we were introduced to a bond that matures in the next time period and paid $FC = 100$ and $c = 10$.

The value of c is a certain percentage of the face value, FC, and is called the *coupon rate*. The coupon rate is specified when the bond is issued and remains fixed for the life of the bond. Thus, for example, a bond may pay \$5 every year for the next three years and at the end of the three years pay \$(100 + 5). Such a bond has a maturity of three years, and a face value of \$100, and the annual coupon payments are \$5 each. One immediately recognizes that the cash flow from a bond is like a repayment of a loan that was taken for three years at an interest rate of 5 percent, paid annually. Indeed, buying a bond is giving a loan to the issuer. Such a bond, as in our example, is called a 5 percent bond since it is like a loan taken at 5 percent. In most countries though, the payments from a bond are made semiannually: a 5 percent bond will pay 2.5 percent of the face value every six months.

The coupon rate at which the bond is issued depends, of course, on the interest rate that prevails at the time of issue in the market. In order to induce investors to buy the bond (lend their money), the bond must offer a competitive interest rate. Similarly, after the bond has been issued, it can also be bought and sold in the market (called the *secondary market*). The price in the secondary market will reflect current market conditions with respect to the interest rate prevailing at that time.

Consider the bond in our example that was issued with a coupon rate of 5 percent. An investor holding the bond for the first six months and selling it then in the secondary market may get more or less than \$100 when it is sold. Buying a bond six months after the bond was issued is like giving a loan of \$100 to the issuer for 2.5 years. If at that time the interest rate prevailing in the market for loans over 2.5 years is, for example, 4 percent, the bond will not be sold for \$100. If the bond did sell for \$100, it would constitute a loan at an interest rate higher than the one prevailing in the market. An investor therefore would prefer not to buy such a bond. Hence it must be sold at a price, say P, such that \$5 is 4 percent of P: that is, $P = \left(\frac{5}{4}\right)(100)$. In such an environment, the bond will sell for more than its face value. Such a bond is called a *premium bond.*

Suppose instead that interest rates rise to 6 percent. The competitive forces in the market will alter the price of the bond in such a way that \$5 will be 6 percent of its new price P; hence $P = \left(\frac{5}{6}\right)(100)$. Thus, the bond will be sold at less then its face value. Such a bond is called a *discount bond.*

We thus see that there is an inverse relationship between the price of a bond and the level of interest rates. Therefore, implicit in the prices of bonds in the market is some information about the interest rates in the market. This information can be uncovered using a technique that we have seen before.

Moreover, recovering information about interest rates implicit in bond prices is intimately related to the no-arbitrage condition — a condition with which we have already familiarized ourselves. The fuzzy term we just used ("competitive forces in the market") will soon be seen to be the force of investors seeking arbitrage opportunities. Consequently, these investors affect the market so that prevailing prices eliminate such opportunities. Stating it differently, prices in the market satisfy the no-arbitrage condition. Furthermore, a slight modification of the way the condition was formulated in the former Chapter will make it adaptable to the bond market. Consequently, the stochastic discount factors of Chapter 1 will soon be seen to be the structure of interest rates in the market.

We limit our focus, almost throughout this Chapter, to national government bonds. These securities are regarded as risk-free securities, since governments do not usually default on their obligations.[1] Bonds, as we see, represent fixed payment amounts which are paid at fixed, deterministic times. For this reason, bonds are also referred to as fixed income securities. Thus, if an investor holds a bond to its maturity, the amount of the payments and their timing are certain, provided that the issuer does not default on some payments. Hence, national government bonds are considered nonrisky securities.

A bond which is issued by a less creditworthy issuer must offer a higher interest rate, in comparison to a government bond, in order to compensate the investor for taking the risk of the issuer defaulting on the bond. The lower the creditworthiness of the issuer, the higher the interest rate the bond must offer. Indeed, we observe this in the market for corporate bonds (issued by corporations) which offer higher interest rates than government-issued bonds. Agencies exist in the market which engage in rating the creditworthiness of different issuers. The lower the rating is, the higher the interest rate they must offer on their bonds. There are other factors that may affect the interest rate at which the bond is issued.

Certain bonds have features that affect the interest rate. For example, some bonds, called *callable bonds*, allow the issuer to call the bond back

[1] Note, however, that the government can act in a number of different ways which essentially reduce the value of its obligations. Examples include printing too much money and opting for strategies which increase inflation. This however is beyond the scope of our analysis. We will assume that the inflation rate is zero.

prior to its maturity. The issuer can pay the holder the principal plus a certain amount and so buy back the bond, at certain times subject to certain conditions. If interest rates decrease it might be advantageous to the issuer to "call" the bond. If it is advantageous to the issuer, it is disadvantageous to the holder of the bond. The investor holding such a bond would require compensation for this callable feature. The compensation is in terms of a higher interest rate offered on the bond. Bonds with no extra features are sometimes referred to as *straight bonds*.

Studying the interest rate structure is conducted in the market for national government bonds and includes only straight bonds. In this market, the interest rate implicit in the prices of these bonds does not include compensation for risk. It reflects only the economic competitive conditions in the market. For this reason we limit our attention to the government bond market and, for the time being, to straight bonds. As we will proceed we will discover that there are a few rates in the market and that what we referred to as the "interest rate" in the market is a more complicated structure of interest rates.

3.2 Arbitrage in the Debt Market

Consider a government bond market with n bonds where P_i denotes the price[2] of bond $i = 1, ..., n$. These bonds have been issued at various times in the past and thus offer different coupon rates and pay coupons on different days. For example a bond that was issued three years ago with a maturity of six years on February 2 will pay on the second of February and the second of August of each year until it matures on the second of February three years from now. At that last date it will pay the final coupon and will repay the principal (face value) to the holder of the bond. The payment dates of a bond that was issued two years ago on the second of February, with a maturity of two years, will coincide with the dates of the earlier bond, but will mature earlier (almost immediately).

Assume that the collection of outstanding bonds in the market pays on N distinct days and define a_{ij} to be the payment from bond i on date j,

[2] The prices quoted in the newspaper should be corrected in order to serve as our prices. If a bond changes ownership in-between coupon payments, the last coupon should actually be divided between the last owner and the new owner. The new owner thus pays the price that is written in the newspaper and pays the person from whom he or she bought the bond a portion of the next coupon. The portion is prorated based on the portion of the time each owner held the bond during the time between the last two coupon payments. This payment is referred to as accrued interest.

Price/Time	1	2	3	Security
$94.5	$105	$0	$0	$B1$
$97	$10	$110	$0	$B2$
$89	$8	$8	$108	$B3$

Table 3.1: A Simple Bond Market Specification

$j = 1, ..., N$. Note that since N is the collection of the dates on which the bonds make payments, for a given bond i_o, there might be many dates j for which a_{ij} is zero; i.e., bond i_o does not pay on some of the dates in $\{1, 2, ..., N\}$. The reader may already sense that this type of model is familiar. In the model of the former Chapter where an equity market was investigated we looked at n securities which paid uncertain amounts (or have uncertain value) in the next time period. Security i paid a_{ij} if in the next time period the state of nature was j. Thus the payment next period was not certain.

In the current model, an investor pays the price P_i for bond i now and in so doing purchases a certain sequence of cash flows to be paid in the future at specified times, the amount a_{ij} to be paid at the future times j, $j = 1, ..., N$. If the states of nature of the last model are replaced with the time periods of this model, it would seem that we obtain a model describing the bond market. Indeed, this is nearly the case and we will shortly point out a characteristic of the current model that does not exist in the former model.

Let us look at an example in which we try to form an arbitrage portfolio. Assume we have a market with three bonds that pay on three distinct payment dates, and for simplicity we assume these dates are equally spaced in time.[3] The prices and payments from the bond are summarized in Table 3.1.

[3]Indeed in a realistic market this is not the case, and typically the number of payment dates is about two or three times the number of bonds. Also the payment dates are not necessarily equally spaced. The user may decide to adopt a smaller time unit to accommodate other structures of payments. One may choose the smallest time period between two consecutive payment dates, from any outstanding bonds, as the time unit. They will accommodate any structure of payments at the expense of having more variables (the d's) although the cash flow from the bonds would include many zeros (be very sparse). We shall come back to these assumptions and either relax them or examine how to treat such markets.

In order to form an arbitrage portfolio we try to set up a zero cost portfolio, or a portfolio which results in positive income as proceeds of the transaction (negative cost portfolio). Let us denote by B_1, B_2, and B_3 the number of units of bonds one, two, and three, respectively, in the portfolio. At time zero, the proceeds from the transaction are

$$-94.5B_1 - 97B_2 - 89B_3. \tag{3.1}$$

The short part of the portfolio produces income at the time the transaction takes place. A short position in a bond is much the same as a short position in a stock. The short seller can be thought of as the issuer of the bond. These proceeds may be enough to finance the long part of the portfolio, or may even produce positive net cash inflow at time zero. Note that in formulating an arbitrage portfolio in the equity market, equation (1.1), we minimized the cost of the portfolio, while here we will maximize the proceeds of the transaction. Since a short position, for example, in bond one, is designated by a negative value of B_1, we have a minus sign next to the variables in order to obtain the *proceeds* of the sale rather than the cost of the sale. (Of course maximizing a function or minimizing its negative produces the same solution.) A negative or zero cost portfolio would be identified by maximizing the proceeds of the sale.

We are seeking an arbitrage portfolio that imposes no future liability on the investor. We will solve a maximization problem subject to some constraints. At each future time period, the long part of the portfolio should be at least enough to cover the commitment resulting from the short part of the portfolio. Hence we must ensure that at each future payment time, the payoff from the portfolio is nonnegative: the portfolio must satisfy the following constraints. At the first-time payment,

$$105B_1 + 10B_2 + 8B_3 \geq 0. \tag{3.2}$$

This constraint ensures that at time one the payoff from the portfolio is nonnegative. We have a similar constraint for each time payment. At time two we must make sure that

$$110B_2 + 8B_3 \geq 0, \tag{3.3}$$

and at time three that

$$108B_3 \geq 0. \tag{3.4}$$

We thus arrive at the optimization problem in which we maximize (3.1) subject to (3.2), (3.3), and (3.4). The general form of this problem is

$$\max_{x_1,\ldots,x_n} \sum_{i=1}^{n} -x_i P_i \tag{3.5}$$
$$\text{such that} \quad \sum_{i=1}^{n} x_i a_{ij} \geq 0, \; j = 1, \ldots, N.$$

The reader might recognize that this type of argument and the optimization problem (3.5) are not new. Indeed Table 3.1 and the optimization problem (3.5) are just the bond market counterparts of Table 1.1 and Problem 1.1. It is mathematically the same model, but time periods are replaced with the states of nature. Indeed the definition of complete market and elementary securities extend to this model as well. (Albeit, here an elementary security is one that pays \$1 at time j and zero otherwise.) Certain bonds pay only one payment at the maturity time of the bond. These bonds are referred to as *zero-coupon bonds*, since they make no coupon payments. The market is thus complete if it contains zero-coupon bonds for every maturity. Let us submit this problem (3.5) to MAPLE and see if an arbitrage portfolio can be identified.

```
> simplex[maximize](-94.5*B1-97*B2-89.2*B3,\
> {105*B1+10*B2+8*B3>=0, 110*B2+8*B3>=0,108*B3>=0});
```

$$\{B1 = 0, \; B3 = 0, \; B2 = 0\}$$

We see that no arbitrage portfolio was found as the only solution was $B_1 = B_2 = B_3 = 0$.

The above example may give the impression that the no-arbitrage condition in the bond market is defined as was its counterpart in the case of the equity market, Definition 1[4]. Definition 1 should be read now interpreting the states of nature as time periods and the a_{ij} payoffs as payments from bond i at time j. Even so, there is a distinct feature of the debt market that does not exist in the equity market.

[4] It should be noted though that in the equity market we were examining a one-period model. Here we are dealing with a multiperiod model. The strategies which we are investigating in either context are static. Such strategies are often referred to as buy-and-hold strategies. Once a portfolio has been purchased, no change in the portfolio is made in subsequent time periods. This fact allows us to use the one-period equity model to investigate the multiperiod bond market. We will investigate some dynamic strategies in later chapters. Meanwhile, we continue to see what can be said in the current context, assuming that no buy-and-hold strategies exist which generate arbitrage profit.

Before we will explore this difference we would like to introduce the procedure **NarbitB**. This procedure is the counterpart of **Narbit**. Its input is exactly the input to the **Narbit** procedure. One enters the payoff from the bonds — a "0" must be entered if the bond does not make a payment in that time period. Then the prices of the bonds are entered. The reader should recall that if reading the book on-line, the "help function" can be consulted by entering the name of the procedure. Instructions followed by examples will appear on the screen. In order to determine whether an arbitrage portfolio can be identified in our example, one can submit the following to MAPLE:

```
> NarbitB([[105,0,0],[10,110,0],[8,8,108]],[94.5,97,89]);
```

The no – arbitrage condition is satisfied.

The discount factor for time, 1, *is given by*, .9000000000

The interest rate spanning the time intreval, [0, 1], *is given by*, .111

The discount factor for time, 2, *is given by*, .8000000000

The interest rate spanning the time intreval, [0, 2], *is given by*, .250

The discount factor for time, 3, *is given by*, .6981481481

The interest rate spanning the time intreval, [0, 3], *is given by*, .432

The function Vdis([*c1* , *c2*, ..]), *values the cashflow* [*c1* , *c2*, ..]

The procedure identifies an arbitrage portfolio if it exists, and reports its composition. If an arbitrage portfolio does not exist the procedure states this and produces further output that will be discussed shortly. The reader may already anticipate the meaning of this output given the similarities to the **Narbit** procedure. You, the reader, may wish to explore this procedure further, changing the prices or the payment structure, before reading on. For example, let us examine the effect of making bond two mature at time three, instead of at time two, while leaving its price unchanged.

```
> NarbitB([[105,0,0],[10,10,110],[8,8,108]],[94.5,97,89]);
```

The no – arbitrage condition is not satisfied

An arbitrage portfolio is :

Buy, .02910762160, *of Bond*, 1

Short, 1., *of Bond*, 2

Buy, 1.058981233, *of Bond*, 3

Buying this portfolio produces income of, $.2\,10^{-7}$, *at time 0*

This portfolio produces income of, 1.528150132, *at time*, 1

This portfolio produces income of, −1.528150136, *at time*, 2

This portfolio produces income of, 4.3699732, *at time*, 3

3.2.1 Distinct Features of the Debt Market

In the equity market, the states of nature are to be realized in the future and *one and only one* of the states is actually realized. In this model the time periods are subsequent to one another and so a dollar obtained at time zero can be carried over to time one. A dollar obtained at time one can also be carried over to time two. This possibility was not accounted for by the search for an arbitrage portfolio in the former optimization problem. Let us refine this distinction by exploring another example.

Let us consider a bond market with two bonds: a 6 percent semiannual coupon bond maturing in one year which costs \$90 now and a 22 percent semiannual coupon bond maturing in 1.5 years which costs \$130 now. The first bond will pay \$3 in six months and \$103 in a year. The second bond will pay \$11 in half a year, \$11 in one year, and \$111 in 1.5 years.[5] We search now for a possible arbitrage portfolio in this bond market by submitting to MAPLE the following optimization problem.

$$\max_{B_1,B_2} -90B_1 - 130B_2 \tag{3.6}$$

$$\begin{aligned} \text{such that} \quad 3B_1 + 11B_2 &\geq 0 \\ 103B_1 + 11B_2 &\geq 0 \\ 111B_2 &\geq 0. \end{aligned} \tag{3.7}$$

```
> simplex[maximize](-90*B1-130*B2,{3*B1+11*B2>=0,\
> 103*B1+11*B2>=0,111*B2>=0});
```

$$\{B1 = 0,\ B2 = 0\}$$

Again no arbitrage portfolio could be identified. Let us try to check the no-arbitrage condition in this market with the aid of **NarbitB**.

```
> NarbitB([[3,103,0],[11,11,111]],[90,130]);
```

This is an incomplete market.

The no − arbitrage condition is not satisfied

An arbitrage portfolio is :

$$Buy,\ \frac{119}{87},\ of\ Bond,\ 1$$

$$Short,\ 1,\ of\ Bond,\ 2$$

[5] Twenty-two percent is a high coupon rate. It is used to exaggerate the case and make the point clearer. For the reader who is uncomfortable with such a high coupon rate, consider instead the bond as an 11 percent bond paying every year instead of every half a year. Under this interpretation the first bond matures in two years and the second in three years. It is only a change of the time unit, which does not affect our analysis.

$$\textit{Buying this portfolio produces income of}, \frac{200}{29}, \textit{ at time 0}$$
$$\textit{This portfolio produces income of}, \frac{-200}{29}, \textit{ at time}, 1$$
$$\textit{This portfolio produces income of}, \frac{11300}{87}, \textit{ at time}, 2$$
$$\textit{This portfolio produces income of}, -111, \textit{ at time}, 3$$

When the same market is treated as a debt market, the no-arbitrage condition is not satisfied. Perhaps the fact that we omitted the possibility of carrying money from one period to the next has something to do with what appears to be a contradiction. Examine the following portfolio: $B_1 = 119$ and $B_2 = -87$. (We simply multiplied the portfolio identified by **NarbitB** by a factor of 87.) We submit to MAPLE the following substitution to find the proceeds from the transaction of buying this portfolio.

```
>  subs(B1 = 119, B2 =-87 ,-90*B1-130*B2);
```

$$600$$

We see that as a result of buying this portfolio we pocket \$600 since the proceeds from the short part of the portfolio exceed the value of the long part of the portfolio by that amount. Next we calculate the cash flow obtained from this portfolio at times one, two, and three. This is given below where the first element is the cash flow at time one, the second, is that at time two, and the third is that at time three.

```
>  subs(B2 = 119,B3 =-87,[3*B2+11*B3,103*B2+11*B3,111*B3]);
```

$$[-600, 11300, -9657]$$

Of course each component of the cash flow has been multiplied by 87 in comparison to the cash flow reported by **NarbitB**. We confirm that this portfolio is an arbitrage portfolio. The cash flow obtained at time zero can be carried over to the next period and be used to pay the \$600 obligation generated by the short part of the portfolio due at time one. The amount received from the portfolio at time two is more than sufficient to cover the liability due at time three. Thus this portfolio generates an arbitrage profit of at least $\$11,300 - \$9,657 = \$1,643$. Why was the arbitrage portfolio not identified by the optimization problem in equation (3.6)? It is due to the fact that our current formulation does not allow us to carry money from one period to another period in the future. This can be corrected.

Assuming that no inflation exists in the market we know that a dollar today is at least a dollar in a future period. It might be worth more depending on the interest rate that prevails in the market at that time. That is, the portfolio above which generates a cash inflow of \$600 at time zero and a cash outflow of \$600 at time one does not impose a future liability on the investor. Moreover, the investor might end up with a net cash inflow. At time zero the investor received \$600 and can invest it at the risk-free rate of interest prevailing in the market at that time, for one period. However, if the cash outflow at time one is even \$600.5 we are not sure whether this portfolio will result in a net cash inflow or outflow. The \$600 received at time zero and invested for one period may or may not be sufficient to cover the \$600.5 payable at time one.

At time two the investor receives \$11,300 which can be invested for one period until time three. Let r be the interest rate which will prevail in the market for investment from time two to time three. At time three the investor will thus have \$11,300$(1 + r)$. Consequently, if the cash outflow from the portfolio at time three is \$11,300 or less, the investor will certainly have enough to cover the liability since $0 < r$. However, if the cash outflow is even slightly more than \$11,300, \$11,300$(1 + r)$ may or may not be enough to cover the liability. At time zero we cannot know for sure the magnitude of \$11,300$(1 + r)$ because the value of r as of time zero is uncertain.

3.2.2 Defining the No-Arbitrage Condition

In order to construct a portfolio completely free of risk we have no other alternative but to assume that a dollar in period one can cover at most a dollar liability in period two. If we allow for more than a dollar liability at time two, some risk will be involved in the position. If even an "infinitesimal" risk is involved in the portfolio, it is no longer an arbitrage portfolio. Thus, as in the above example, we can take the excess cash obtained at time j and use it to cover a shortage, if such exists, at time $j+1$. If we reformulate the optimization problem equation (3.6) to allow for this possibility we see that indeed the portfolio above is an arbitrage portfolio.

This is done by introducing the additional variables z_0, z_1, and z_2, which are constrained to be nonnegative. The variable z_0 represents the cash we move from period zero to period one. Thus it can be used to cover any shortage that might arise in period one. In the same manner z_1 is the amount we might want to transfer from period one to period two. Hence we subtract it from the cash flow obtained at time one and add it to the cash flow obtained at time two. The z's are constrained to be positive, since negative values mean we allow cash to be moved backward through time. (In other words, negative values of z mean borrowing at the current period against excess cash which will be obtained in the next time period.) However, as of time zero, if we would like to have a portfolio free of risk, we have no way of knowing for sure what a dollar obtained in some future period $1 < j$ might be worth at time $j-1$. We reformulate problem (3.6) along these lines and resubmit to MAPLE the revised version, problem (3.8).

$$\max_{B_1, B_2, z_0 \geq 0, z_1 \geq 0, z_2 \geq 0} \quad -90B_1 - 130B_2 - z_0 \tag{3.8}$$

$$\begin{aligned} \text{such that} \quad 3B_1 + 11B_2 + z_0 - z_1 &\geq 0 \\ 130B_1 + 11B_2 + z_1 - z_2 &\geq 0 \\ 111B_2 + z_2 &\geq 0. \end{aligned} \tag{3.9}$$

```
> simplex[maximize](-90*B1-130*B2-z0,\
> {3*B1+11*B2+z0-z1>=0,103*B1+11*B2+z1-z2>=0,\
> 111*B2+z2>=0,z0>=0,z1>=0,z2>=0});
```

This time we see that the problem is unbounded. A **NULL** is returned meaning an infinite arbitrage profit can be made. Thus the prices of the bonds allow for arbitrage opportunities. The no-arbitrage condition thus needs to be reformulated for the debt market. Here is the general formulation of this condition.

Definition 4 (The No-Arbitrage Condition (Debt Market)) *Consider all the portfolios for which*

- *(i) the income from the long part of the portfolio is at least equal to the payoff required for the short part of the portfolio, for every time period, where a dollar obtained at time j may be used to cover a dollar shortage at time $j+1$, and for which*
- *(ii) the cost of the long part of the portfolio does not exceed the proceeds from the sale from the short part of the portfolio.*

The no-arbitrage condition definition is satisfied if, for all such portfolios, conditions (i) and (ii) are satisfied in such a way that

- *(iii) the income from the long part of the portfolio is equal to the payoff required for the short part of the portfolio, for every time period, and*
- *(iv) the cost of the long part of the portfolio equals the proceeds from the sale from the short part of the portfolio.*

Alternatively it can be stated in a more formal way as follows:

Definition 5 (The No-Arbitrage Condition (Debt Market)) *The optimal value of the optimization problem below is zero and, for the optimal solution, all the inequalities are satisfied with equality.*

$$\max_{x_1,\ldots,x_n,z_0\geq 0,\ldots,z_k\geq 0} \sum_{i=1}^{n} -x_i\,P_i - z_0$$

$$\text{such that} \quad \sum_{i=1}^{n} x_i\,a_{ij} + z_{j-1} - z_j \geq 0,\ j = 1, ..., N-1 \tag{3.10}$$

$$\sum_{i=1}^{n} x_i a_{ik} + z_k \geq 0$$

We conclude this subsection by changing the prices of the bonds in the last example and running **NarbitB** again. We see that indeed there are bond prices which preclude arbitrage opportunities. However, this example is of an incomplete market and there are a multiplicity of discount factors as seen in the output of **NarbitB** below.

```
> NarbitB([[3,103,0],[11,11,111]],[100,122]);
```

This is an incomplete market.

The no − arbitrage condition is satisfied.

The set of discount factors is,

$$\{d_1 = -\frac{103}{3}\,d_2 + \frac{100}{3},\ \frac{97}{103} < d_2,\ d_2 < \frac{50}{53},\ d_3 = \frac{1100}{333}\,d_2 - \frac{734}{333}\}$$

The reader is encouraged to experiment with other combinations of bond prices and bond payoffs to arrive at an appreciation of the distinct features of the debt market. For example, the reader may like to investigate the result of **Narbit** and **NarbitB** applied to the same input. Can you identify an example in which the no-arbitrage condition of the debt market is not satisfied but that of the equity market is? Remember the results of **NarbitB** when applied to two bonds, a 6 percent bond maturing at time two, and an 22 percent bond maturing at time three? If not you can rerun it below.

```
> NarbitB([[3,103,0],[11,11,111]],[90,130]):
```

This is an incomplete market.

The no − arbitrage condition is not satisfied

An arbitrage portfolio is :

$$Buy,\ \frac{119}{87},\ of\ Bond,\ 1$$

$$Short,\ 1,\ of\ Bond,\ 2$$
$$Buying\ this\ portfolio\ produces\ income\ of,\ \frac{200}{29},\ at\ time\ 0$$
$$This\ portfolio\ produces\ income\ of,\ \frac{-200}{29},\ \ at\ time,\ 1$$
$$This\ portfolio\ produces\ income\ of,\ \frac{11300}{87},\ \ at\ time,\ 2$$
$$This\ portfolio\ produces\ income\ of,\ -111,\ \ at\ time,\ 3$$

One of the exercises at the end of this Chapter asks the reader to explain why **Narbit** reports that the no-arbitrage condition is satisfied when the above market is viewed as an equity market.

```
> Narbit([[3,103,0],[11,11,111]],[90,130]):
```

$$This\ is\ an\ incomplete\ market.$$
$$The\ no-arbitrage\ condition\ is\ satisfied.$$
$$The\ stochastic\ discount\ factors\ are,\ \left[\frac{45}{53},\frac{45}{53},\frac{5900}{5883}\right]$$
$$The\ risk-neutral\ probability\ is,\ \left[\frac{999}{3178},\frac{999}{3178},\frac{590}{1589}\right]$$
$$A\ \ risk-free\ rate\ in\ this\ market\ is,\ -.6298$$
$$The\ function\ Vdis([c1,c2,..]),\ values\ the\ cashflow\ [c1,c2,..]$$

3.3 Discount Factors

In the equity model, the price of a security reflects the uncertainty of the cash flow that it promises the investor. We saw that the no-arbitrage condition was equivalent to the existence of a price of \$1 contingent on state j, $0 < d_j$, for each j so that for every security

$$\sum_{j=1}^{s} a_{ij} d_j = P_i. \tag{3.11}$$

As we explained in the first Chapter, d_j discounts the dollar of state j in two ways: First, the dollar is not a certain dollar; it is received only if state j occurs. (The discounting accounts for the riskiness of the cash flow.) Second, if state j occurs, the dollar is received in the future. Thus d_j takes care of both the time value of money and the uncertainty. We can therefore expect that the no-arbitrage condition in the debt market is equivalent to the existence of d_j satisfying the same equation. After all, our interpretation of the states of nature as time periods does not conceptually affect the mathematical formulation of the no-arbitrage condition. Indeed, our formulation of the maximization problem equation (3.5) and Definition 5 confirm the above analysis.

Thus, in a similar manner to the equity model, the d_j here can be interpreted as the price of a dollar to be received in time j. The discount factors here, however, take care of the time value of money alone since the amount is to be received for sure. The similarities extend further. In the same manner in which we interpreted the basic building blocks of the equity securities as a dollar contingent on state j (state-contingent cash flows), here we can identify them as a dollar to be received in time j. Indeed, an examination of the bonds will verify that they are composed of the same basic units: a dollar to be received at time j.

Consider a bond maturing at time one and paying \$105. It is composed of 105 units of the time one basic building block: a dollar to be received in time one. Suppose the same bond were to mature at time two, i.e., it pays \$5 at time one and \$105 at time two. Such a bond is composed out of 5 units of the time one building block and 105 units of the time two building block — a dollar received at time two. Hence, in a parallel manner to the equity market, we should expect that the market assigns a price to a dollar to be received at time j. Let us denote these prices by d_j; then these d's would satisfy an equation for every bond i such as the following:

$$\sum_{j=1}^{k} a_{ij}\, d_j = P_i. \tag{3.12}$$

There are some characteristics of the current market that do not exist in the equity market. We alluded to these a few paragraphs back. In the debt market, the d_j are the prices of one dollar at time j. We expect that in spite of the dollar being a sure amount obtainable at some future time j, its value (price) should be less than a dollar today. Having a dollar now is worth more than having it later, simply because by having it now one can invest it in a risk-free asset to generate more than a dollar in the future. Hence the d_j reflects the discounting of one dollar due to the time value of money.

The distinct feature of the debt market is a consequence of the fact that the time periods occur sequentially and only one state of nature occurs. A dollar from time one can be carried forward to time two, and at that time must be equal to at least a dollar. (If the dollar were invested at a positive rate of interest, it would be worth more.) We should expect therefore that if

$m < n$ then $d_n < d_m$. This is one property which the d_j's do not enjoy in the model of the equity market. This property is referred to as the *monotonicity condition of the discount factors.* We should also imagine that $d_1 < 1$. After all, a dollar at time one should have less value than a dollar now. This is simply a time value of money argument.[6] Indeed we see that the solution to our example satisfies this condition.

The interpretation for equation (3.12) is identical to that of equation (1.7) of the equity market. Furthermore, the use of the discount factors in the context of the debt market is analogous to the use of the stochastic discount factors in the equity market. The d_j's in the debt market facilitate valuation of cash flows across time periods exactly in the same manner as the stochastic discount factors allow the valuation of cash flows across states of nature.

Consider again the example which is summarized in Table 3.1. The **NarbitB** procedure confirmed that the no-arbitrage condition was satisfied for that market. Let us reexecute **NarbitB** applying it to this example in order to explain another part of the output from the procedure.

```
> NarbitB([[105,0,0],[10,110,0],[8,8,108]],
> [945/10,97,89]);
```

$$\textit{The no}-\textit{arbitrage condition is satisfied.}$$

$$\textit{The discount factor for time, } 1, \textit{ is given by, } \frac{9}{10}$$

$$\textit{The interest rate spanning the time intreval, } [0,\,1], \textit{ is given by, } .1111$$

$$\textit{The discount factor for time, } 2, \textit{ is given by, } \frac{4}{5}$$

$$\textit{The interest rate spanning the time intreval, } [0,\,2], \textit{ is given by, } .2500$$

$$\textit{The discount factor for time, } 3, \textit{ is given by, } \frac{377}{540}$$

$$\textit{The interest rate spanning the time intreval, } [0,\,3], \textit{ is given by, } .4324$$

$$\textit{The function } Vdis([c1\,,c2\,,..]), \textit{ values the cashflow } [c1\,,c2\,,..]$$

Since the no-arbitrage condition is satisfied, we expect to find a solution to the system of equations below.

$$94.5 = 105\, d_1 \tag{3.13}$$

$$97 = 10\, d_1 + 110\, d_2 \tag{3.14}$$

$$89 = 8\, d_1 + 8\, d_2 + 108\, d_3 \tag{3.15}$$

[6] In an exercise at the end of this Chapter, the reader is asked to explain why this must be the case using arbitrage arguments.

Furthermore, since the d's are prices, and given the time value of money arguments, we also expect that the d's will satisfy

$$1 > d_1, \quad d_j > d_{j+1}, \quad j = 1, ..., k, \quad \textit{and} \quad d_k > 0. \tag{3.16}$$

Let us submit thus to MAPLE the request to solve the system of equations (3.13) and (3.16).

```
> solve({d1*105=945/10,d1*10+d2*110=97,d1*8+d2*8+d3*108=89\
> ,d1<1,d2<d1,d3<d2,d3>0},{d1,d2,d3});
```

$$\{d1 = \frac{9}{10},\ d2 = \frac{4}{5},\ d3 = \frac{377}{540}\}$$

MAPLE confirmed that there is a solution to this system of equalities and inequalities. The reader may now compare this solution to the discount factors reported from the **NarbitB** procedure. They are the same (of course) and have the same meaning as the stochastic discount factors of the equity market. The meaning of the d_j's is as in the equity market. It is the price of \$1 obtainable at time j. Furthermore, as in the equity market, the no-arbitrage condition is satisfied if and only if the following system of equalities and inequalities is consistent (i.e., a solution to it exists):

$$a_{i1}\,d_1 + a_{i2}\,d_2 + \cdots + a_{ik}\,d_k = P_i, \quad i = 1, ..., n \tag{3.17}$$

$$d_{j+1} < d_j, \quad j = 1, ..., k-1 \tag{3.18}$$

$$d_1 < 1 \tag{3.19}$$

$$d_k > 0. \tag{3.20}$$

In fact, the existence of a solution to (3.17) can be used as a dual definition of the no-arbitrage condition. The Appendix will explore this point further. In the same manner as the d's are used in the equity market they are used in the debt market. We utilize the d's to value different cash flows. The function **Vdis,** which is an output defined by the procedure **NarbitB,** allows us to value a given cash flow. Thus in our example, given the cash flow $(c1, c2, c3)$, we may value it by issuing the following command:

```
> Vdis([c1,c2,c3]);
```

$$.9000000000\, c1 + .8000000000\, c2 + .6981481481\, c3$$

Indeed, the **Vdis** procedure defined here acts in the same way as its equity market counterpart, albeit, here we are valuing certain cash flows which will be obtained at different times in the future. In the equity market, we valued state-contingent cash flows. For example, we can ask, what should the value of $c3$ in $(0, -1, c3)$ be if the present value of $(0, -1, c3)$ is zero? This will be the solution to $Vdis([0, -1, c3]) = 0$.

The reader may now try to value different cash flows across time periods using the **Vdis** function. Alternatively, it is also possible to define a new market using the **NarbitB** procedure and thus a new **Vdis** function which will be defined in such a way to allow the reader to value cash flows in the new market.

Note that for now the **Vdis** function allows the valuation of cash flows as long as the time of payments coincides with the payments of existing bonds in the market. Of course, in real-world markets, there may be a need to value a cash flow whose time of payment does not coincide with the cash flow from an existing bond. Hence, there is a need to extract, somehow, the discount factors for times which do not correspond to existing bond payments, i.e., for some time period t where t belongs to a continuous interval $[0, T]$. We shall see soon how this is done.

3.4 Rates, Discount Factors, and Continuous Compounding

There is, of course, a link between the interest rates and the discount factors. This is a part of the output of **NarbitB** that we have not explicitly addressed. The discounting is a result of the opportunity to invest (the opportunity cost) at the risk-free rate. Hence, if the discount factor from time zero to some time t is d then

$$d = \frac{1}{1+r}. \tag{3.21}$$

The risk-free rate spanning the interval $[0, t]$ is thus recovered from d by

$$r = \frac{1}{d} - 1 \tag{3.22}$$

and is reported by **NarbitB**. We shall return to the structure of interest rate and its connection to the discount factors in Chapter 9.

3.4.1 Continuous Compounding

The state-of-the-art models of financial markets assume continuous trading, and in the same manner, assume that interest is paid at every instant in time. The result is that the familiar risk-free discount factor is replaced by another discount factor. We next review the notion of continuous compounding.

It will be useful in the subsequent analysis when nonrisky cash flows obtained at different periods in time are to be compared. Continuous compounding will also come in handy in forthcoming chapters when the model is generalized even further to deal with continuous trading. The reader who is familiar with this issue can bypass this subsection.

There are two possible ways to explain continuous compounding and we will concisely mention both. Assume that the risk-free rate of interest for one year is r and interest is being paid only once at the end of the year. One dollar invested for one year will thus grow to $1 + r$ at the end of the year. If interest is being paid semiannually instead, then after six months $1 + \frac{r}{2}$ can be reinvested and will grow to $(1 + \frac{r}{2})^2$ at the end of the year. When interest is paid n times a year, at the end of the year \$1 will become $(1 + \frac{r}{n})^n$. The case of continuous payments of interest rate will correspond to the limit of $(1 + \frac{r}{n})^n$ as n approaches infinity. We can use MAPLE to calculate this limit. Assuming we invest $\$V_0$ for t years at a continuously compounded interest rate, what will be its value at the end of the year?

```
> limit(V0*(1+r/n)^(t*n),n=infinity);
```

$$e^{(r\,t)}\;V0$$

The discount factor from time zero to time t will be the reciprocal of the future factor generating $\frac{1}{e^{rt}} = e^{-rt}$. Thus, if the annual risk-free rate of interest is r and we would like to use continuous compounding, we should use the rate r_f such that $e^{r_f} = 1 + r$; hence

$$r_f = \ln(1 + r), \tag{3.23}$$

as confirmed by MAPLE:

```
> solve(exp(r[f])=1+r,{r[f]});
```

$$\{r_f = \ln(1 + r)\}$$

There is another way of producing the same result. If interest is being paid every instant of time, then the value of an amount invested at the risk-free rate of interest grows every instant in time. Let us denote the value V of the amount invested as a function of time t, by $V(t)$. Thus, every instant of time adds $V(t)r_f$ to the amount $V(t)$. Or to put it more precisely, for t_1, t_2, and $\Delta t = t_2 - t_1$, the difference between $V(t_1)$ and $V(t_2) = V(t_1 + \Delta t)$, divided by Δt,

$$\frac{V(t_1 + \Delta t) - V(t_1)}{\Delta t}, \tag{3.24}$$

is the average growth rate of $V(t_1)$ per unit time. The instantaneous growth rate at time t_1 is the limit of (3.24) as Δt approaches zero, or as t_2 approaches

t_1. Therefore, the instantaneous growth rate is $\frac{\partial V(t)}{\partial t}$, the derivative (as defined in calculus) of $V(t)$ with respect to t.

Assume that at time t_0 the amount we invested was V_0. At time t the value of the invested amount is $V(t)$ and it grows by $V(t)r_f$. We are therefore in a situation where we know the derivative of a function, with respect to t, and we would like to know the function. Thus, we have what is called a *differential equation* as specified below:

$$\frac{\partial V(t)}{\partial t} = V(t)r_f. \tag{3.25}$$

Furthermore, we know that at time t_0 the value of the function is V_0. This is called an *initial condition*, since it specifies the value of the function at the initial time. (The initial condition allows determination of the constant of integration.) The following differential equation with the initial condition that at time zero, $V(0) = V_0$, is submitted to MAPLE, and the solution produced by MAPLE is displayed below:

```
> dsolve({diff(V(t),t)=V(t)*rf, V(0)=V0}, V(t));
```

$$\mathrm{V}(t) = e^{(rf\,t)}\,V0$$

We see that V_0 invested for the period 0 to t grows to $V_0\,e^{r_f t}$. The future value coefficient is $e^{r_f t}$, and consequently the discount factor is its reciprocal $e^{-r_f t}$. If the risk-free rate is not the same for each time period, but rather a function of time, say, $r_f(t)$, this can also be solved to generate our discount factor:

```
> dsolve({diff(V(t),t)=V(t)*rf(t), V(0)=V0}, V(t));
```

$$\mathrm{V}(t) = e^{\left(\int_0^t \mathrm{rf}(u)\,du\right)}\,V0$$

We can check this solution by differentiating $V(t)$ (as defined by the MAPLE solution) with respect to t. This will verify that the instantaneous growth rate at time t is indeed $r_f(t)$. Executing the next line does exactly that, where $rhs(\%)$ refers to the right-hand side of the solution, i.e., to $e^{\int_0^t r_f(u)du}V0$.

```
> is(diff(rhs(%),t)=rhs(%)*rf(t));
```

$$true$$

3.5 Concluding Remarks

We will return to the topic of valuation in the bond market, taking advantage of the discount factors implied by the prices of bonds. However, this will be done in a multiperiod setting in Chapters 10 and 11 where a second look at swaps, forwards, and futures will be taken. The order of subjects covered in this book is governed by the generalization of the simple model of Chapter 1. We prefer to stay in a one-period model before moving to a multiperiod model, and extend first the one-period model in a different direction. This is done in the next Chapter to facilitate the investigation of equity options over the next few chapters.

3.6 Questions and Problems

Problem 1. Show that assuming that the interest rate in the market is positive implies that the discount function should be monotonically decreasing. (Hint: Assume that $d_{j+1} > d_j$, where d_j is the discount factor for time j, and devise an arbitrage strategy.)

Problem 2. Assume that the payoff matrix of the bonds available in the market is $A = \begin{bmatrix} 110 & 0 \\ 5 & 105 \end{bmatrix}$ and that the price of the 10 percent bond is \$100.

1. Determine all possible prices for the 5 percent bond that would make the market arbitrage free.
2. What is the set of prices (P_1, P_2) that if assigned to the two bonds will keep the market arbitrage free?

Problem 3. Suppose that the bond market is represented by the payoff matrix A and a price vector P, where $A = \begin{bmatrix} 7 & 107 \\ 111 & 0 \end{bmatrix}$ and $P = \begin{bmatrix} 93 \\ 101 \end{bmatrix}$.

1. Is this market arbitrage free?
2. Is it complete?
3. In this market you can price cash flows by the problem

$$\min\{c'd| \quad Ad = P, \quad d_1 > d_2 > 0\},$$

where A and P are as above and c and d are assumed to be column vectors of cash flows and discount factors, respectively.

4. Explain the meaning of d's.
5. Calculate the problem dual to the minimization problem above.
6. Interpret the dual problem and explain the meaning of the dual variables.

Problem 4. Suppose that the bond market is represented by the payoff matrix $A = \begin{bmatrix} 105 & 0 \\ 8 & 108 \end{bmatrix}$ and a price vector $P = \begin{bmatrix} 98 \\ 95 \end{bmatrix}$.

1. Is this market arbitrage free?
2. Is it complete?
3. Consider a cash flow $c = [1, 1]$. What can be said about the value of this cash flow?
4. Answer parts 1, 2, and 3 above using the discount factors and replication arguments.
5. Give an example of a security in this market such that only lower and upper bounds (*distinct!!!*) can be assigned to its value.
6. What can be said about the value of cash flow $[0, 0]$?

Problem 5. Suppose that the bond market is represented by the payoff matrix $A = \begin{bmatrix} 6 & 106 & 0 \\ 10 & 10 & 110 \end{bmatrix}$ and a price vector $P = \begin{bmatrix} 95.5 \\ 100 \end{bmatrix}$.

1. Is this market arbitrage free?
2. Is it complete?
3. Formulate the optimization problems that determine the lower and upper bound for the price of a cash flow c.
4. Calculate problems dual to the ones you have formulated above.
5. Identify the set of all cash flows that can be priced uniquely.

Problem 6. Assume the market defined in the previous problem and consider the following strategy:

- *buy one unit of the 6 percent bond now*

- *sell it immediately after the first coupon payment*

- *invest all the proceeds in the second bond and hold until it matures.*

  (a) Is this a riskless strategy?

  1. If you think that this strategy is riskless explain why and give an example of one that would involve risk.
  2. If you think that this strategy involves risk explain why and give an example of a riskless strategy.

Problem 7. Investigate the result of **Narbit** and **NarbitB** applied to the same input on page 103. Can you identify another example in which the no-arbitrage condition of the debt market is not satisfied but that of the equity market is?

3.7 Appendix

3.7.1 No-Arbitrage Condition in the Bond Market

This Chapter defined the no-arbitrage condition in terms of the optimization problem below.

$$\max_{x_1,\ldots,x_n,z_0\geq 0,\ldots,z_k\geq 0} \sum_{i=1}^{n} -x_iP_i - z_0 \tag{3.26}$$

$$\text{such that} \quad \sum_{i=1}^{n} x_ia_{ij} + z_{j-1} - z_j \geq 0, \; j = 1, \ldots, k-1 \tag{3.27}$$

$$\sum_{i=1}^{n} x_ia_{ik} + z_k \geq 0. \tag{3.28}$$

An alternative definition can be given in terms of the system of inequalities below:

$$\sum_{i=1}^{n} -x_iP_i - z_0 \geq 0 \tag{3.29}$$

$$\sum_{i=1}^{n} x_ia_{ij} + z_{j-1} - z_j \geq 0, \; j = 1, \ldots, k-1 \tag{3.30}$$

$$\sum_{i=1}^{n} x_ia_{ik} + z_k \geq 0. \tag{3.31}$$

Definition 6 (Equivalent Definition of the No-Arbitrage Condition) *The no-arbitrage condition in the bond market is satisfied if for every solution of the system (3.29) — (3.31) all the inequalities are satisfied as equalities.*

As in the equity case, if there are no arbitrage opportunities, the existence of the discount factors is proven by application of separation theorems or by applying a set of results known as the theorem of alternatives. Application of a theorem of alternatives known as Farkas' lemma (actually a slight modification of it [30]) yields that the condition in the definition is satisfied if and only if the system of equations and inequalities below is consistent.

$$a_{i1}\, d_1 + a_{i2}\, d_2 + \cdots + a_{ik}\, d_k = P_i, \quad i = 1, ..., n \tag{3.32}$$

$$d_{j+1} < d_j, \quad j = 1, ..., k-1 \tag{3.33}$$

$$d_1 < 1 \tag{3.34}$$

$$d_k > 0. \tag{3.35}$$

Chapter 4

Fundamentals of Options

4.1 Extending the Simple Model

Former chapters concerned themselves only with a "simple" market. This Chapter gets a few steps closer to reality. In real financial markets prices of securities can take on many values, not only two or three corresponding to the states of nature of the simple model. In fact, the price of a stock can take an infinite number of possible values.

The model of the former chapters possessed another simplifying feature. It described a market for which there were only two relevant points in time: the current time (the beginning of a period) and a future time. The reality is that prices of financial assets fluctuate continually.[1] The time dimension is a continuous interval from the current time t_0 to some time in the future, T. Traded financial assets may have different values at any instant of time t. In contrast, for the model in the previous Chapter, t_0 was the current time and T the future time. Using that model, we were only able to investigate trading strategies of the form "buy today, sell tomorrow." No trading was allowed to take place at any intermediate time t such that $t_0 < t < T$. In reality, trading can take place at any time t.

The model we used in Chapter 1 simplified reality via discretization. The continuous time interval $[t_0, T]$ was replaced by two discrete points in time: t_0 and T. The *a priori* set of possible values that a stock price can assume is actually the continuous interval $[0, \infty)$. That interval was replaced with a finite number of possible prices that depends on the realized states of nature.

[1] Prices fluctuate continually provided those markets are open. When making the transition from a "simple" model to a "real" model, most studies ignore the issue of markets being open or closed and allow trading at any instant of time, denoted t.

The set of states of nature had n elements, and a finite number of discrete states, a set $\Omega = \{\omega_1, \omega_2, \ldots, \omega_n\}$. In the example of Chapter 1, the set[2] Ω had three elements: ω_1-state 1, ω_2-state 2, and ω_3-state 3. The price of the first (risky) security, P_1, had three possible outcomes, depending on the realized state of nature (as a function of the state of nature): $P_1(\omega_1) = 50$, $P_1(\omega_2) = 60$, and $P_1(\omega_3) = 80$.

This Chapter introduces the two most simple types of options. It also relaxes, at least to a certain extent, some of the assumptions governing the previous Chapter. Rather than the set Ω being restricted to a finite number of states of nature, Ω becomes the more real case of having an infinite number of elements, allowing the price of the stock to take on values in the interval $[0, \infty)$. We will maintain, at least for the moment, the assumption that there are only two time periods, and that there is no intermediate date which would allow trading between the current and the next time period.

4.2 Two Types of Options

Options are nothing more than contingent claims. Their value is contingent on the state of nature which is realized in the future time period, exactly in the same way as we have already seen a contingent claim in Chapter 2. The reader will recall the discussion of a claim of \$1 contingent on state 1 occurring in the next time period. As a matter of fact, we have already "met" an option. Remember the first proposition, on page 57, of Chapter 2?

> "Pay \$14 now and in return you have the right, not the obligation, to purchase security *1* in the next time period for \$50 regardless of its actual market price in the next period."

We were able to translate the structure of the proposition into a contingent cash flow of the type $c = (c_1, c_2, c_3)$, i.e., receive c_j if state $j = 1, 2, 3$, occurs.

The value of the proposition was calculated utilizing stochastic discount factors as implied by the no-arbitrage condition. Thus, the price of the contingent cash flow was calculated by discounting it, using stochastic discount factors as in equation (1.9), or as the present value of the expected cash

[2] It is customary to use the notation Ω for the set of possible states. This notation is "borrowed" from probability theory where the set Ω is the sample space of a random variable, i.e., the collection of all possible realizations. Here, each realization is a state of nature. Thus these states are referred to as $\{\omega_1, \omega_2, \ldots, \omega_n\}$.

flow, as in equation (1.15). We will later see that this guideline for valuation stays intact even after some assumptions are relaxed. We begin first with a definition.

Definition 7 (Call Option (European)) *The holder of a call option (long position) has the right,* **not the obligation,** *to buy a certain stock for a certain price (called the strike price or exercise price) on a certain date, the expiration date (or maturity date).*

The name European (as opposed to American) is given to an option that gives the holder the right to buy a certain stock on a certain date. By contrast, an American option grants the right to buy a certain stock at a certain price at any time up to and including a certain date. In this Chapter we will deal only with **European options on a stock that does not pay dividends.**

A European call option is exactly like the type of proposition we encountered in Chapter 2. The strike price, or exercise price, was \$50. The option would have expired worthless had the stock price in the next time period been less than or equal to \$50, as was the case in state 1. In state 2, where the stock price was \$60, the option created a cash flow of \$60 − \$50, that is, the price of the stock minus the strike price. Similarly in state 3 the value of the option was \$80 − \$50. The price paid for the proposition in Chapter 2 was \$14. This is referred to as the *premium.*[3]

Options are also called *derivative* securities since their values are derived from another security — the stock in our case. The payoff from the option is a function of the value of a certain stock, referred to as the *underlying security* (or the underlying, for short). The only information needed to determine the value of the option at time T, the maturity time, is the value S_T of the underlying at T. At the expiration date, T, the holder of the call option has an asset worth zero if at time T the price of the stock S_T is smaller than the exercise price, K. In this case the call option expires worthless.

If at time T, $S_T > K$, the holder of the call option pays the exercise price to the writer of the option and receives the stock. The stock can then be sold immediately in the market, generating proceeds of $S_T - K$. Thus the value of the call option at time T depends on the sign of the quantity $S_T - K$. If it is positive, the value of the call option is $S_T - K$, while if it is

[3] This Chapter concentrates on the payoff from the option, or from a portfolio of options, rather than on the profit or loss, or the value of the option. We will tackle the issue of the value of the option in the next Chapter.

negative, the value is zero. The payoff[4] from a call option is thus a function of the value of the stock. It is the maximum of 0 and $S_T - K$, and it is denoted by

$$\max(0, S_T - K). \tag{4.1}$$

We can associate the concept of states of nature with the value of the stock. In the simple model of Chapter 1, the payoff from an option (like the proposition) depends solely on the state of nature which determines the price of the underlying security. In precisely the same manner, here the price of an option at time T depends on the price of the stock. The state of nature, in the simple model, thus plays the role of the price of the stock in this model. We can therefore think of each possible value for the price of the stock as a possible state of nature.

Since the state of nature is the price of the stock, which can be any nonnegative number, it is no longer possible to represent the cash flow from a security in terms of a table such as Table 1.1. The number of columns in such a table will be like the number of points in the interval $[0, \infty)$. Instead, we will use functions to represent the cash flow from an option in each state of nature.

First, let us look at the graph of the cash flow from the stock plotted against the states of nature. The state of nature, the x-axis, is the price of the stock. The y-axis is the value of the stock as a function of the state of nature — simply, the state of nature which equals the value of the stock. Therefore, the function is $f(s) = s$, and if we graph the price of the stock against the state of nature we indeed see a line which begins at the origin and extends from the origin at a 45° angle[5] as in Figure 4.1.

```
> plot(s,s=0..80,title='Figure 4.1:  A Payoff from a Long
> Position in a Stock', titlefont=[TIMES,BOLD,8]);
```

The call option has no value if the state of nature (the stock price) is less than the strike price of the option. The value of the call increases by one dollar for each dollar increase in the state of nature, provided the state of nature is above the strike price. The graph is zero for the region where the

[4]Some authors like to speak about the profit or loss from holding an option. If this is done the time value of money should be taken into account since the premium (the price paid to buy the option) is paid at time t_0, the current period, while the payoff from the option is received at time T. When calculating the value of an option, the payoff is what matters. Thus, we will (most of the time) be looking at payoffs rather than losses or profits from the position.

[5]For simplicity, when using MAPLE, we will omit the subscript T from S_T. One should think about S_T as a random variable: the price of the stock at time T. Think of s as a possible realization of S_T.

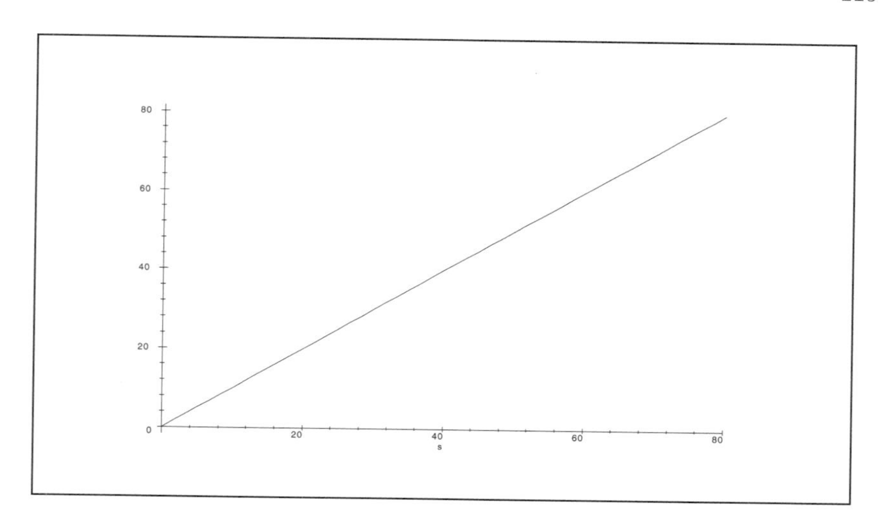

Figure 4.1: A Payoff from a Long Position in a Stock

state of nature is below the strike price. When the state of nature (stock price) is greater than the strike price, the graph shows a line which emanates from the strike price at a 45° angle to the x-axis.

The function **Call**$(\cdot,\cdot)$ has been defined in MAPLE to have two arguments. The first is the price of the stock, s, and the second, K, is the exercise price. The function **Call(s,K)** calculates the cash flow at maturity from a call option with a strike price K, if the state of nature is s. For a given s and K the value of **Call(s, K)** is given by equation (4.1). Thus the payoff from a call option with a strike price of \$50 when the stock price is \$55 is calculated by **Call**(55,50). We can ask MAPLE to evaluate it

```
> Call(55,50);
```

$$5$$

and graph the payoff from the call option with an exercise price of \$50 for states of nature ranging from 0 to 80. This is done by

```
> plot(Call(s,50),s=0..80, title=`Figure 4.2:  A Payoff \
> from a Long Position in a Call`);
```

and the graph is displayed in Figure 4.2.

Looking at the x-axis, from the strike price on, the graph of the value of the call option is parallel to the line which describes the payoff from the

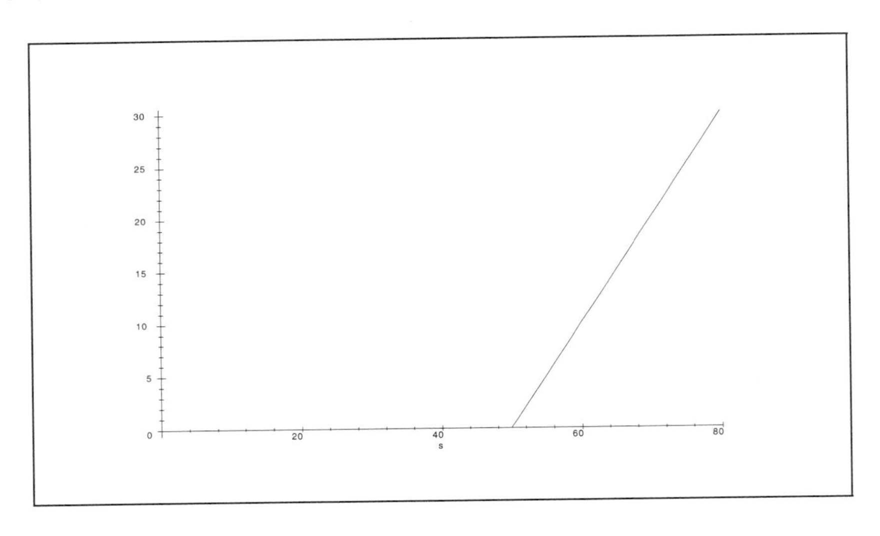

Figure 4.2: A Payoff from a Long Position in a Call

stock. Figure 4.3 displays the payoffs from a call and from a stock in the same plane.

```
> plot({Call(s,50),s},s=0..80, title=`Figure 4.3:  Payoffs\
> from a Long Position in a Call and a Long Position in\
> a Stock`, titlefont=[TIMES,BOLD,5] );
```

An option can be thought of as a side bet[6] between two investors: the holder of the option, the one who has the right to buy the stock and is said to have a *long position in the option*, and the writer of the option, who is said to have a *short position in the option*, and is committed to selling the stock at a certain price (the exercise price). The cash flow from holding the option long is the negative of the cash flow from holding the option short: the two positions have exactly offsetting cash flows, necessarily.

The writer of a call option, the one who has a short position in the option, faces the negative of the cash flow of the holder of the option. If the price of the stock moves above the strike price of the call option, the writer

[6] In reality the clearing house serves as intermediary between the two sides: the buyer and the seller of the option. Each side does not know the identity of the other (the counterparty to the buy or sell transaction). The over-the-counter marketplaces and organized trading exchanges produce publications which explain these arrangements, as well as margin requirement details.

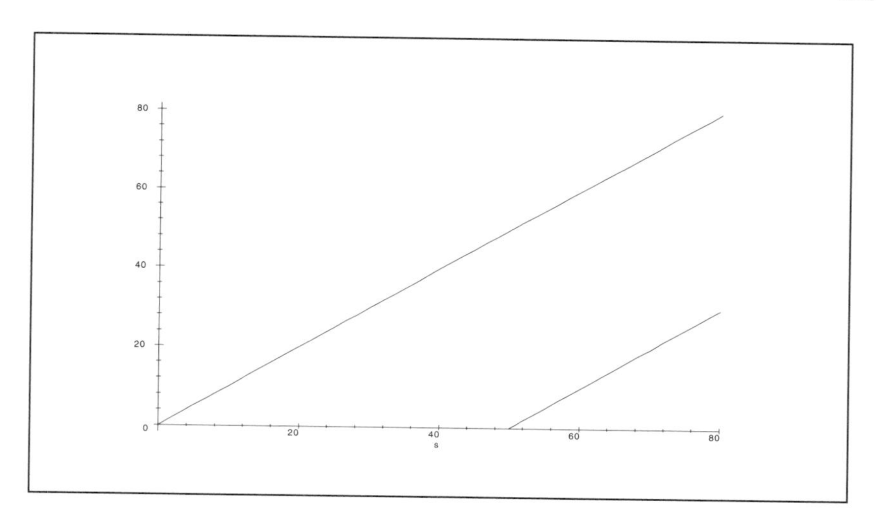

Figure 4.3: Payoffs from a Long Position in a Call and a Long Position in a Stock

of the call option must deliver the stock for the strike price. The writer thus buys[7] the stock in the market for S_T and sells it for the strike price K. Since $S_T > K$, the writer has to pay out $S_T - K$. If, on the other hand, $S_T \leq K$, the holder of the call option will not choose to exercise the right he or she has purchased, and the writer has to pay nothing. Figure 4.4 shows the graph of the cash flows from a short position in a call option (for the writer of the call).

```
> plot(-Call(s,50),s=0..80,title='Figure 4.4:  A Payoff\
> from a Short Position in a Call');
```

There is another type of option called a put option. This option gives the holder the right to sell a stock, in contrast to the call option which gives the holder the right to buy a stock. Here is the definition.

Definition 8 (A Put Option (European)) *The holder of a put option (long position) has the right,* ***not the obligation,*** *to sell a certain stock for*

[7]If the writer owns the stock, he or she can just deliver it to the party holding the long position in the call option. Whatever the case, the resultant cash flows are the same. The position of writing an option and holding the stock is called writing *a covered option.* In certain countries, the tax treatment of writing a covered option may be different than that of writing a *naked option*, i.e., writing the option and not holding the stock.

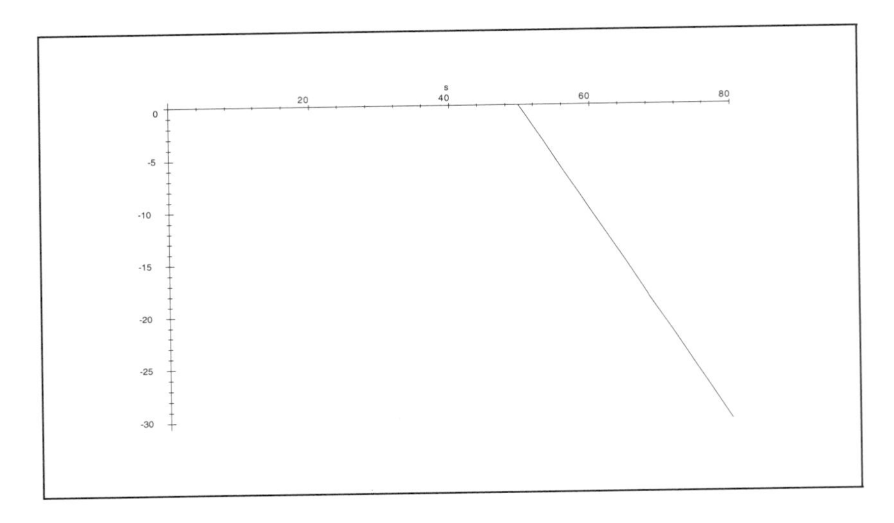

Figure 4.4: A Payoff from a Short Position in a Call

a certain price (called the strike price or exercise price) on a certain date, the ***expiration date*** *(or maturity date).*

Let us look at the payoff at maturity from holding a put option with a strike price of K, given various values of the stock. The holder of a put option has a valuable asset when the exercise price is greater than the stock price. In this case the holder of the option can buy the stock for its market value and sell it for more — the strike price. Thus the cash flow from holding a put option will be zero if the stock price is greater than the strike price. For each dollar that the price of the stock is less than the strike price, the value of the put option increases by one dollar. The payoff is given by

$$\max(0, K - S_T). \tag{4.2}$$

We define a MAPLE function **Put**$(\cdot, \cdot)$ which has two arguments, like the **Call** function, the first being s, the stock price, and the second, K, the exercise price of the put option. The payoff from a put option with a strike price of \$50 when the price of the stock is \$60 is given by **Put**(60,50). MAPLE can evaluate it

```
> Put(60,50);
```

$$0$$

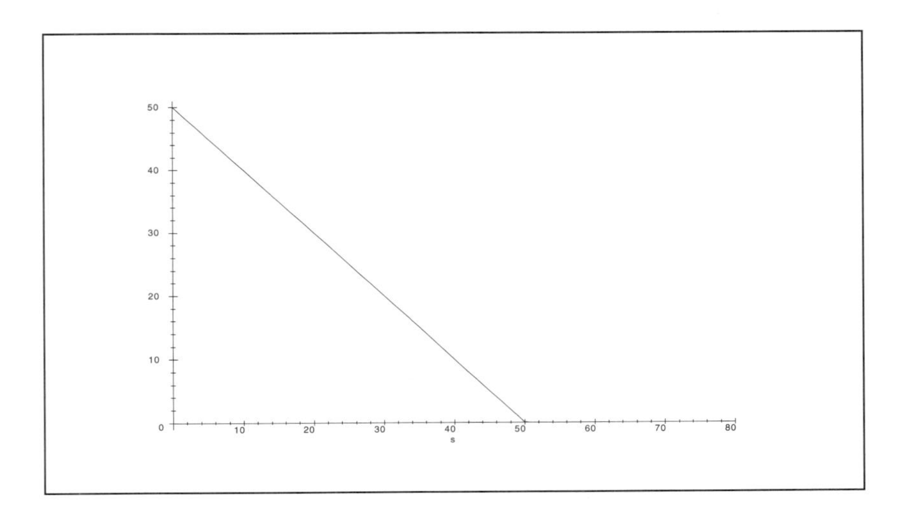

Figure 4.5: A Payoff from a Long Position in a Put Option

and graph it. The payoff is described for a stock price range of between \$0 and \$80 in Figure 4.5:

```
> plot(Put(s,50),s=0..80,title=‘Figure 4.5:  A Payoff from a
> Long Position in a Put‘,titlefont= [TIMES,BOLD,9]);
```

Similar to the case of a call option, the graph of a put option, in a certain region, parallels the graph of a short position in the stock. From the origin to the strike price, the graph of the put is parallel to the line which describes the payoff from a short position in the stock. Figure 4.6 displays the payoffs from a call and a short position in the stock in the same plane.

```
> plot({Put(s,50),-s},s=0..80, title=‘Figure 4.6:  Payoffs\
> from a Short Position in a Stock and a Long Position in\
> a Put‘, titlefont=[TIMES,BOLD,5]);
```

In an analogous way to the discussion of the call option, the writer of a put option has a short position in the put. The writer is obligated to buy the underlying stock for a certain price if the holder of the put option decides to exercise the right he or she has purchased. If at time T, the relation $S_T < K$ holds, the holder of the put option would buy the stock in the market for S_T and exercise the right to sell it for K to the writer of the option. The writer in turn can only sell this stock in the market for S_T and thus incurs

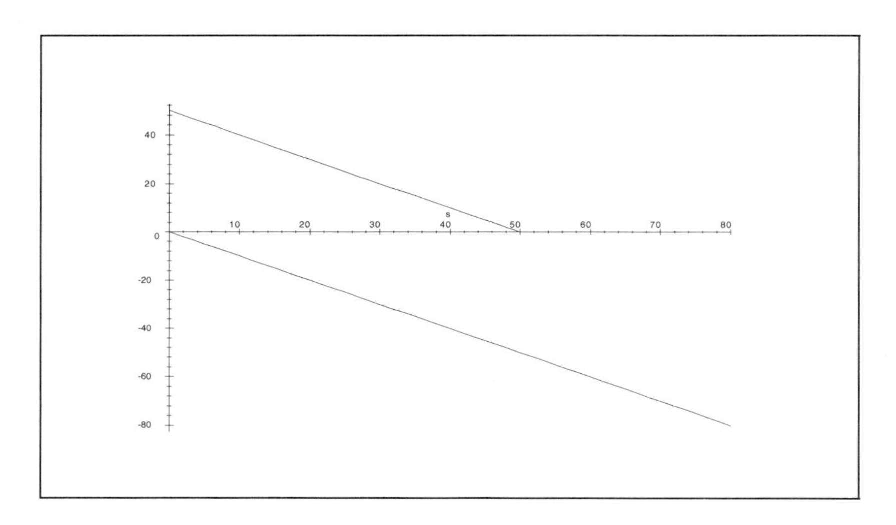

Figure 4.6: Payoffs from a Long Position in a Put and a Short Position in a Stock

a negative cash flow of $S_T - K$. This position will generate the cash flow (to the writer) as in Figure 4.7.

```
> plot(-Put(s,50),s=0..80,title= 'Figure 4.7:  A Payoff\
> from a Short Position in a Put', titlefont=
> [TIMES,BOLD,9] );
```

The graph of a short position in a put is parallel in a certain region to the graph of a long position in a stock. Can you verify it with the help of MAPLE?

Options are now traded on various exchanges around the world and their prices are reported in the press. An explanation of the institutional features of option markets and the details of the availability of strike prices and maturity dates are easily obtained from the various exchanges. The appendix to this Chapter includes some explanations of how to read the quotations in the newspaper and a glossary of terminology reproduced with the permission of the Montreal Exchange from their publication *Guide to the Stock Option Market*.

The reader is encouraged to experiment at this stage with different strike prices of call and put options and see what the visual images of the resultant cash flows look like. Perhaps even combinations of options (portfolios) are

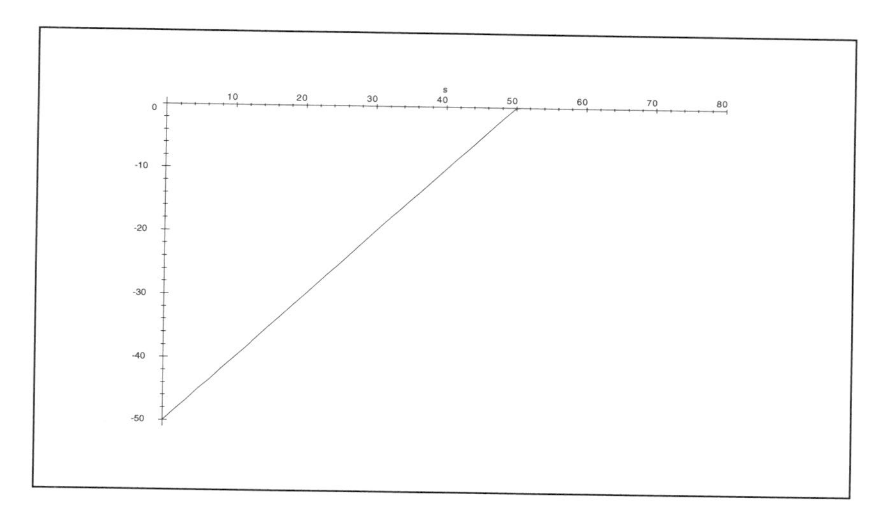

Figure 4.7: A Payoff from a Short Position in a Put

worth trying. We will point out a few combinations later in the text and the exercises at the end of the Chapter will cover a few more. Some combinations are so common that they have their own names. The next section will look at portfolios which combine puts and calls of different strike prices and see what types of portfolios can be composed out of these securities.

4.3 Trading Strategies

This section investigates the different trading strategies, *combinations*, that can be composed with put and call options of different strike prices and a stock. The reader can thus acquire an appreciation of different profiles of cash flows that can be generated. This section will visualize and exemplify some purposes of trading with options, e.g., elimination of risk (*hedging activities*), reduction of risk, speculation, or the creation of custom-made profiles of cash flows required by the investor. In this Chapter we will be concerned with trading strategies which involve puts and calls with the same maturity dates. In the next Chapter we will investigate strategies which involve different maturity dates, referred to as *calendar spreads*.

The example of Chapter 1 dealt with three securities (referred to as *primary securities*). It examined the feasibility of generating different cash flows from portfolios of various combinations of holdings of these primary securities. In the current context, the puts, the calls, and the stock play the role of primary securities. The various profiles of cash flows generated from combinations of these primary securities form the topic of this section.

In Chapter 1 we defined the set **Cash** in order to investigate the cash flows resulting in the different states of nature from various portfolios. In the same manner we now make use of combinations of our defined MAPLE functions **Call** and **Put,** which will calculate and graph the cash flows in every state of nature (for every possible stock price) from different portfolios of these options.

The set **Cash** calculated the cash flow in each state of nature for a given portfolio. Substitution of a portfolio position such as $(S1 = 1, S2 = 2, SF = -1)$ in **Cash** resulted in the cash flow from the portfolio of longing one unit in security 1, longing two units in security 2, and shorting the bond. The counterpart in this Chapter of the set **Cash** from the former chapters is a graph showing the cash flow from the portfolios in each state of nature in the region $[0, \infty)$.

In the simple model, to calculate the cash flow from a portfolio composed of the first and second security, we summed the cash flows from each security in each state. In this Chapter we have an infinite number of states (possible values of the stock price). The mathematical concept stays the same. To calculate the cash flow from a combination of a call and a put, both at exercise price \$50, when the stock price is \$40, we simply take the sum of two functions:

$$Call(40, 50) + Put(40, 50).$$

We can ask MAPLE to evaluate it by

```
>  Call(40,50)+ Put(40,50);
                              10
```

One should think about generating different cash flows with put options, call options, and the underlying stock as building with Lego building blocks. The put is simply one piece, which can come in a variety of colors (maturity dates) and sizes (strike price); likewise the call. Other blocks are the short call, the short put, and the short and long stock. We now try to see what structures can be created from these basic blocks.

We think of these blocks as elementary cash flows and try to construct different profiles of desired cash flows. It will be useful to visualize these positions as mechanically building a shape from the available Lego pieces. In what follows we will enhance this interpretation using the graphic capability of MAPLE. We will start with portfolios composed of a stock, and of calls and puts with the same maturity date.

4.3.1 Portfolios of Calls and Puts with the Same Maturity Date

The payoffs from puts and calls can be interpreted as payoffs that were generated from the stock by stripping it to its basic[8] components: cash and a put and a call with the same strike price. Let us look at the pattern of the payoff if we put together a portfolio of a short put and a long call with the same strike price (try to think about it as already suggested – mechanically putting together the pieces of the graph of the long call and the short put).

Assume that we create a portfolio of a call with a strike price of $60 and a short put with the same strike price. Thus, we are looking at a portfolio whose payoff in state s will be

$$Call(s, 60) - Put(s, 60).$$

For each s in the region $[0, \infty)$, we sum the value of the put and the call. MAPLE is capable of graphing the result. The resultant cash flow is shown in Figure 4.8.

```
> plot({Call(s,60)-Put(s,60)},s=0..100,title= 'Figure 4.8:\
> Payoff from a Long Position in a Call and a Short\
> Position in a Put',titlefont =[TIMES,BOLD,5]);
```

To highlight the building blocks of the above position, as is Figure 4.9, one can issue the following command (note that instead of plotting the sum of the **Call** and the negative of the **Put**, we simply plot them together; thus the "," instead of the "–" in Figure 4.8).

```
> plot({Call(s,60),Put(s,60)},s=0..100,title=
> Figure 4.9:  Payoff from a Long Position in a Call
> and a Short Position in a Put',titlefont = [TIMES,BOLD,5]);
```

Remembering that the graph of the payoff from a stock is a line emanating from the origin at a 45° angle, we see that we just need to "lift up" the line in Figure 4.8 by $60 to get the payoff from the underlying stock. This can be accomplished by investing an amount of money in a risk-free asset that

[8] In Chapter 5 we will see that it is possible to strip the stock to even smaller building blocks, like the $1 state-contingent claims of Chapter 1. However, for the present purpose, the above is sufficient. We will use the more fundamental stripping to generate the counterpart of the stochastic discount factors in a more realistic model.

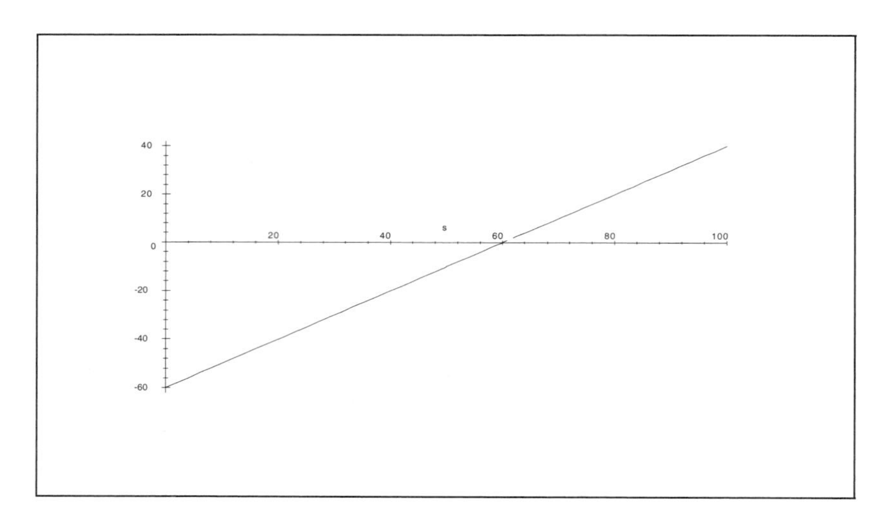

Figure 4.8: Payoff from a Long Position in a Call and a Short Position in a Put

will grow to \$60 by the maturity date of the call and the put. Thus, the payoff of the call and the put at maturity will be as above, plus \$60. The reader can graph this payoff by executing the command (Figure 4.10)

```
> plot(Call(s,60)-Put(s,60)+60,s=0..100,\
> title='Figure 4.10:  A Payoff from a Long Call \
> and a Short Put + $60',titlefont=[TIMES,BOLD,7] );
```

This is indeed the payoff from the stock as confirmed by MAPLE.[9]

```
> assume(s,positive);

> about(s);

Originally s, renamed s~:
  is assumed to be: RealRange(Open(0),infinity)

> is(Call(s,60)-Put(s,60)+60=s);
```

$$true$$

[9]We shall return later to further discussion of some implications of this relation, such as *put — call parity*.

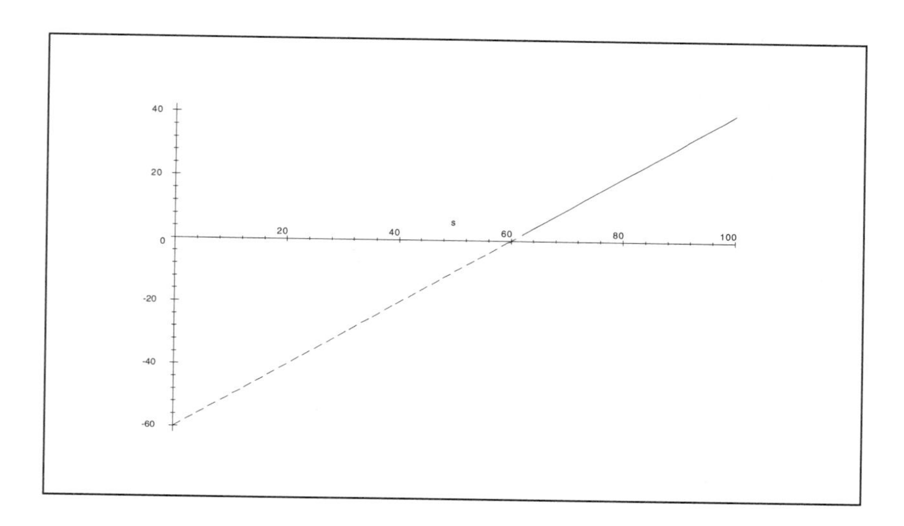

Figure 4.9: Payoff from a Long Position in a Call and a Short Position in a Put.

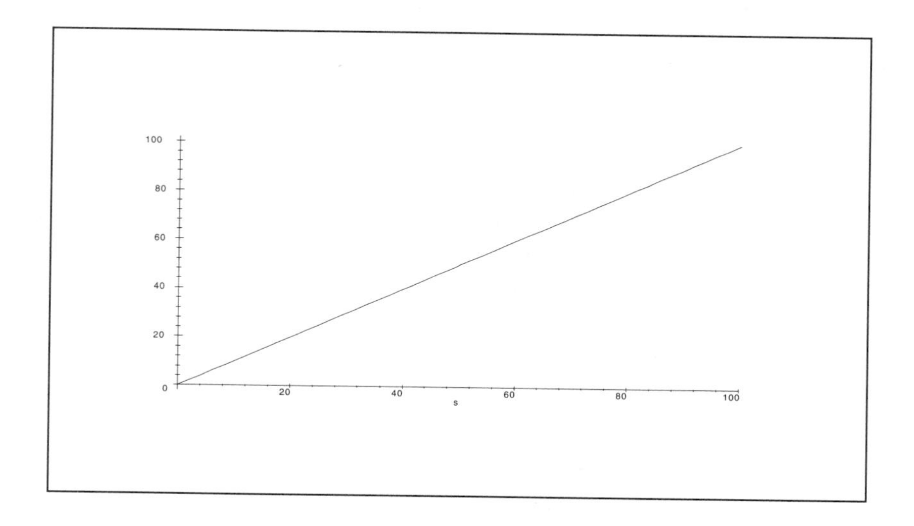

Figure 4.10: A Payoff from a Long Call and a Short Put (with an exercise price of \$60)+ \$60

It is not surprising that combining the stock with options on the same stock can either reinforce or reduce the risk inherent in the stock. The payoff from the stock in certain states of nature can be neutralized by taking an offsetting short position, and can be enhanced in other states of nature by taking a long position. Options allow us to give up future possible payoff from the stock in exchange for a decrease in the current price of the portfolio. They can facilitate risk reduction at the cost of an increase in the price of the portfolio. They also allow creation of a position for which the investor is paid today for buying risk out of somebody else's hands.

The analogy of holding options as a side bet between two investors on the outcome of the stock value should be clearer now. It is like a zero sum game between the writer and the holder of an option. If the writer profits, the holder loses and vice versa. The availability of options in the market allow buying and selling "parts of the risk" inherent in the stock. It introduces a market for risk. Options can be used to reduce risks assumed by certain positions.

Consider a portfolio which is only a long stock. There is a risk inherent in such a portfolio from the possibility of a decrease in the price of the stock. An investor can guarantee that the payoff from the portfolio will never fall below a certain value. This can be done by including a put in the portfolio. Let us first look at the graph, Figure 4.11, of the cash flow from a portfolio composed of a long stock and a long put with an exercise price of \$60.

```
> plot(s+Put(s,60),s=0..100,y=0..100, title='Figure 4.9:\
> A Payoff from a Long Stock and a Long Put', titlefont
> =[TIMES,BOLD,6]);
```

Figure 4.11 clearly demonstrates that holding a stock long and a put long ensures a payoff above the strike price of the put. Such a position is called a *protective put.* The larger the strike price is, the larger is the minimum payoff. This is demonstrated via three-dimensional illustration in Figure 4.12, where the minimum payoff increases as the exercise price K increases.

```
> plot3d(s+Put(s,K),s=0..100,K=20..90, title='Fig.\
> 4.12: The Payoff from a Protective Put',titlefont=\
> [TIMES,BOLD,8]);
```

Effectively, holding the (long) put protects the investor from price decreases of the stock. When the stock price decreases below the strike price of the put, the put becomes valuable and has a payoff equal to the difference between the strike price and the price of the stock. Thus, the put introduces a "floor" — a lower bound on the payoff from the portfolio. The investor

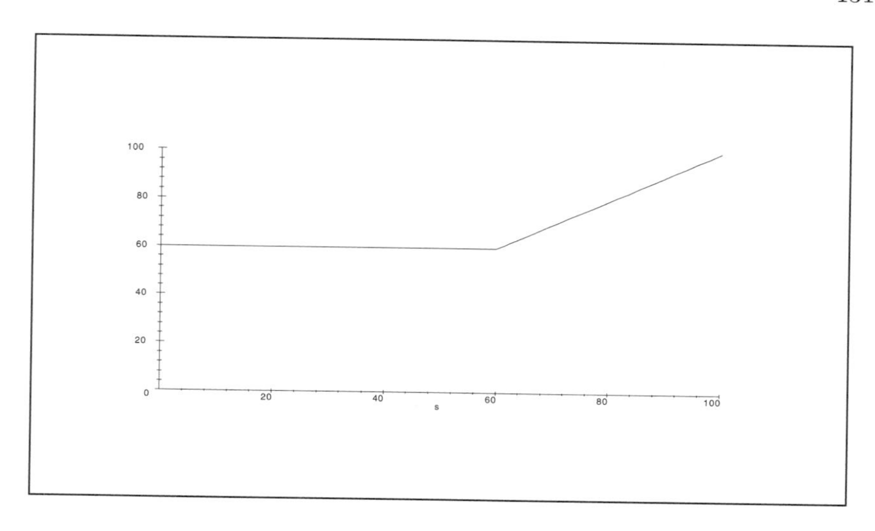

Figure 4.11: A Payoff from a Long Stock and a Long Put

purchased the put as an "insurance policy" which would guarantee that the payoff from the portfolio would not fall below the strike price. The writer of the put took on this risk from the hands of the holder of the stock (for a price — the premium). The writer was willing to receive a certain amount of money today (the price of the put, or the premium) and in return guaranteed payment in the future if the price of the stock fell below the strike price of the put. This insurance-type argument can also work for a short position.

Consider a short position in the stock. The potential payoff from such a position can be very negative — it can potentially be any negative number. There is no limit to the loss one can suffer from such a position. If the price of the stock increases to a huge number, the holder of the short position will have to pay it when closing the position. However, it is possible to truncate this potentially very negative payoff by taking a long position in a call. Let us first graph, Figure 4.13, the payoff from shorting a stock and longing a call at an exercise price of $60 on the same plane.

```
> plot({-s,Call(s,60)},s=0..199, title=`Figure 4.13:  A\
> Payoff from Shorting a Stock and Longing a Call`,\
> titlefont=[TIMES,BOLD,6] );
```

It is evident that the higher the stock price, the larger the amount the

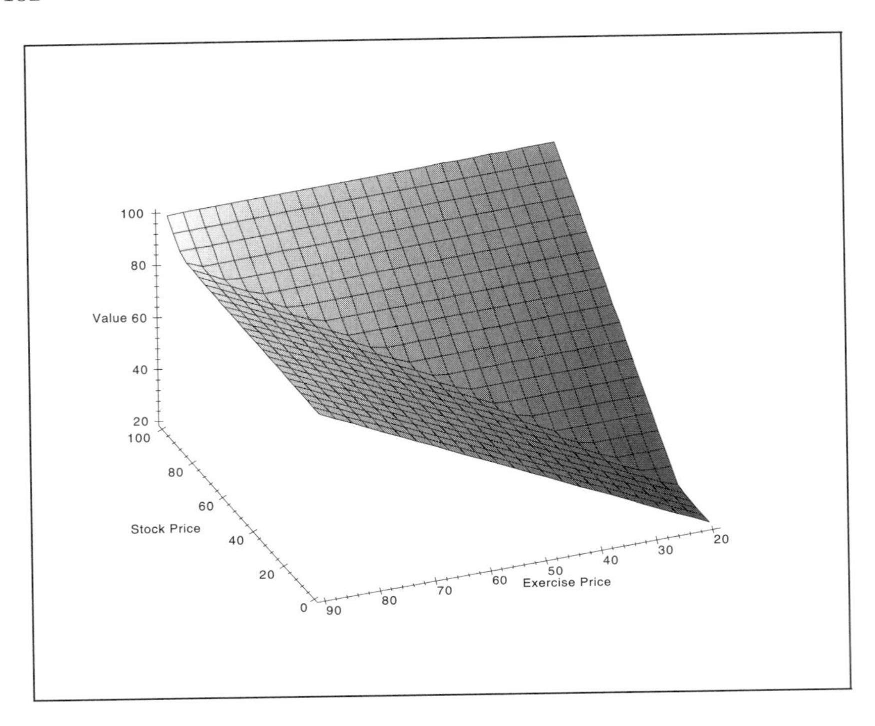

Figure 4.12: The Payoff from a Protective Put

investor who shorts the stock must pay. At the same time, it can be seen that a position in a call can offset the decrease in the stock price. Figure 4.14 demonstrates what happens when the short stock and the call are combined in a portfolio.

```
> plot(-s+Call(s,60),s=0..199,title=`Figure 4.14:  A Payoff\
> from a Short Position in a Stock and a Long Position in a\
> Call`, titlefont =[TIMES,BOLD,5]);
```

The loss from the short position is bounded by the negative of the strike price. Whenever the stock price goes above the strike price, the call becomes valuable and produces a payoff equal to the difference between the stock price and the call. Thus, the payoff from the portfolio of a short call and a stock cannot be more negative than the strike price.

Writing a call also exposes the investor to a potentially unbounded loss, in the same way that shorting a stock does. However, in the case of shorting

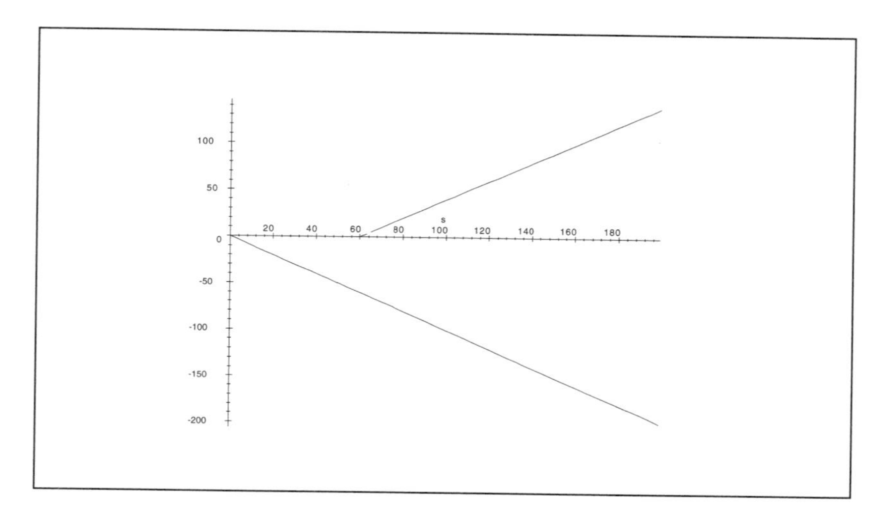

Figure 4.13: Building Blocks of Payoff from Shorting a Stock and Longing a Call

a call, the payoff can be negative only if the price of the stock increases above the strike price. This can also be offset by holding a long position in the stock. Such a position is called writing a *covered call.* Let us first graph the payoff from a short call position and a long stock in the same plane in Figure 4.15.

```
> plot({s,-Call(s,60)},s=0..199,title=`Figure 4.15:  A\
> Payoff from a Long Position in a Stock and a Short\
> Call Position`, titlefont=[TIMES,BOLD,5]);
```

Again it is seen that if the stock and the short call are combined in a portfolio, the potential payoff is bounded below. Figure 4.16 demonstrates that the payoff from such a position is positive for every state of nature.

```
> plot(s-Call(s,60),s=0..199, title=`Figure 4.16:  The
> Payoff from a Long Position in the Stock and a Short
> Position in the Call`,titlefont =[TIMES,BOLD,5] );
```

We can also investigate the effect of changes in the strike price of the option on such a position. Figure 4.17 demonstrates the three-dimensional graph of the payoff of this portfolio as a function of the state of nature, s, and the strike price, K. The reader should be convinced, even before seeing

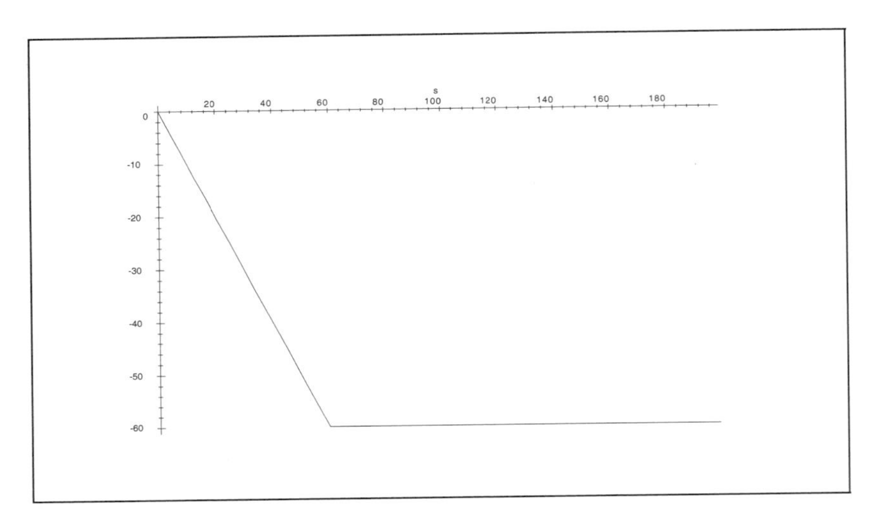

Figure 4.14: A Payoff from a Short Position in a Stock and a Long Position in a Call

the graph, that the higher the strike price, the larger the minimum payoff from such a portfolio.

```
> plot3d(s-Call(s,K),s=0..199,K=20..80,title=
> 'Figure 4.17:  The Payoff from a Long Position in a
> Stock Plus a Short Position in a Call as a Function of s
> and K',titlefont=[TIMES,BOLD,4]);
```

In writing a covered call, a stock is held long against every short call position. It is possible to cover the call, but not "fully". The ratio of the number of written calls to the number of long positions in the stock, referred to as a *hedge ratio*, is not one to one. In such cases, the payoff can be negative (indeed it is unbounded). It is, though, for every state of nature, less negative than the payoff from a simple short position in the call.

Shorting (writing) a call without holding the stock is called a *naked position*. The best way to illustrate it is to graph an example of such a payoff. Consider the payoff from shorting three calls and holding two units long in the stock. Figure 4.18 illustrates these two graphs in the same plane.

```
> plot({-3*Call(s,60),2*s-3*Call(s,60)},s=0..  300\,
> title='Figure 4.18:  A Payoff from Writing Three Calls,\
> and a Payoff from Two Long Positions in the Stock Plus\
```

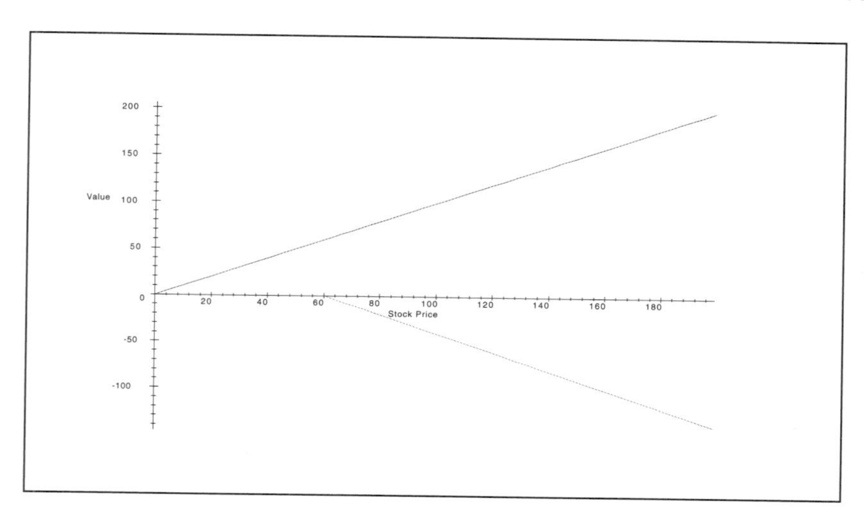

Figure 4.15: A Payoff from a Long Position in a Stock and a Short Call Position

```
>  Writing Three Calls`, titlefont =[TIMES,BOLD,4] );
```

While the payoff can be potentially very negative when the call is not fully covered, it is still less negative, for every state of nature, than that of a naked position. The payoff decrease per dollar increase in the stock price in the partially covered position is less than in the case of the naked call. This is evident from the slopes of the graph: shorting the three calls has a steeper slope (for a stock price greater than \$60) than that of the portfolio. The reader is encouraged to investigate how the payoff changes with changes in the ratio of written calls to holdings of the stock.

Consider a portfolio composed of a call with a strike price of \$50 and a put with a strike price of \$50. The cash flow in state s will be the sum of the function **Call**(s,50) and the function **Put**(s,50). (Try to visualize the two pieces put together — the call like a hockey stick with a handle from zero to the strike price and the put a hockey stick with an infinite handle starting at the strike price and continuing to infinity.)

This portfolio will increase in value if the price of the stock moves away from \$50. If the stock price is \$50 the value of this portfolio is zero (both the put and the call are worthless at this value). At any other value of the stock the portfolio is worth something. If the price of the stock either increases from \$50 by \$10 or decreases from \$50 by \$10 it will have the same effect

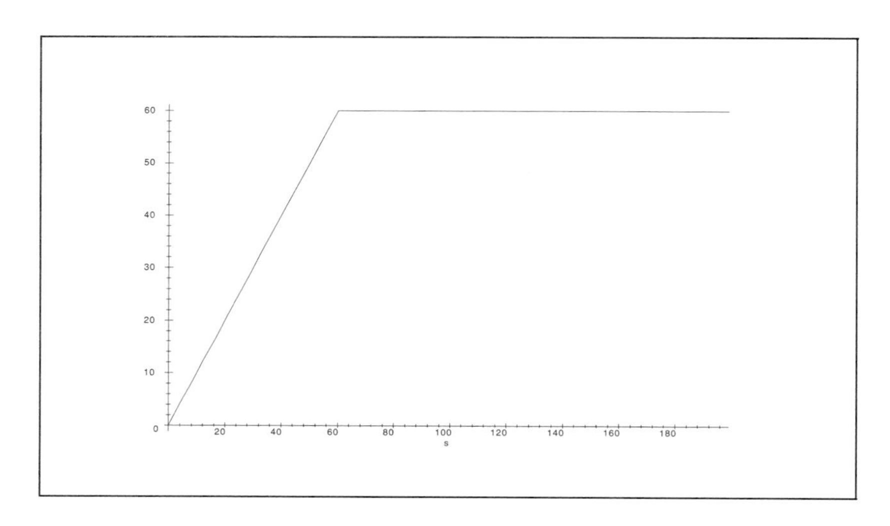

Figure 4.16: The Payoff from a Long Position in the Stock and a Short Position in the Call

on the cash flow. This position is called a *purchased straddle*. The result is depicted in Figure 4.19.

```
> plot(Call(s,50)+Put(s,50),s=0..100,title='Figure 4.19:\
> Straddle - Call Plus Put with the Same K', titlefont=\
> [TIMES,BOLD,8]);
```

To highlight the building blocks of the above position one can replace the "," with the "+" in Figure 4.19, as shown in the on-line version of the book. Can you graph the payoff from a written straddle (i.e., instead of holding the put and call long, hold them short)?

Hedging

Imagine a situation in which we would like to be able to fix today, in spite of any uncertainty, the price of a commodity or a stock which we know we will need in the future. Suppose we entered into an agreement in which we **obligated ourselves** to buy a stock (or more commonly, a certain commodity) at some time T in the future for a certain price. Suppose we had committed to buy a stock for \$100 at time T. Note that indeed the price of the transaction which will take place in the future has been fixed now, and we are obliged to buy the stock for \$100. However, no cash changes hands

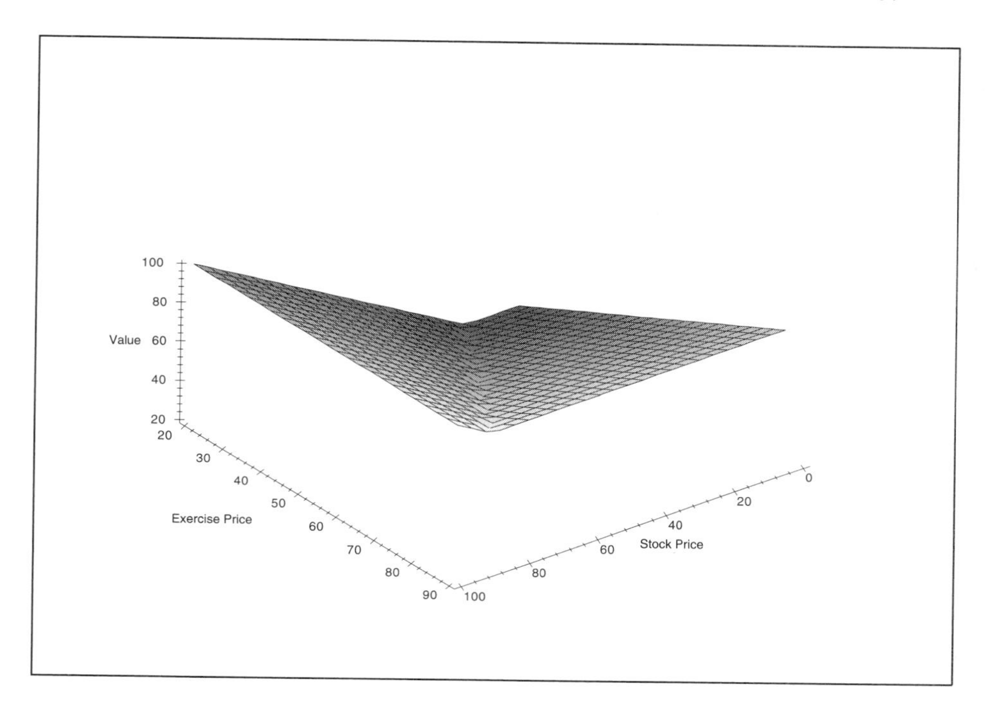

Figure 4.17: The Payoff from a Long Position in a Stock Plus a Short Position in a Call as a Function of s and K

now. This type of an agreement is of course a *forward contract*[10] which has been introduced in Section 2.4.1, albeit here we have a continuum of states of nature.

At the current time it is not known what the price of the stock will be at time T. If the price is above \$100, we achieve a positive cash flow which is the difference between the price in the future and \$100 (buy the stock for \$100 and sell it in the market for its market price). If the price of the stock at time T is less than \$100, we achieve a negative cash flow of the price difference. Thus the cash flow at time T, as a result of this commitment, is stochastic (or random): it is a function of the state of nature s as shown in Figure 4.20.

```
> plot(s-100, s=0..200, title='Figure 4.20:  Time T Cash\
```

[10]Later, we shall return to discuss further details of these types of agreements.

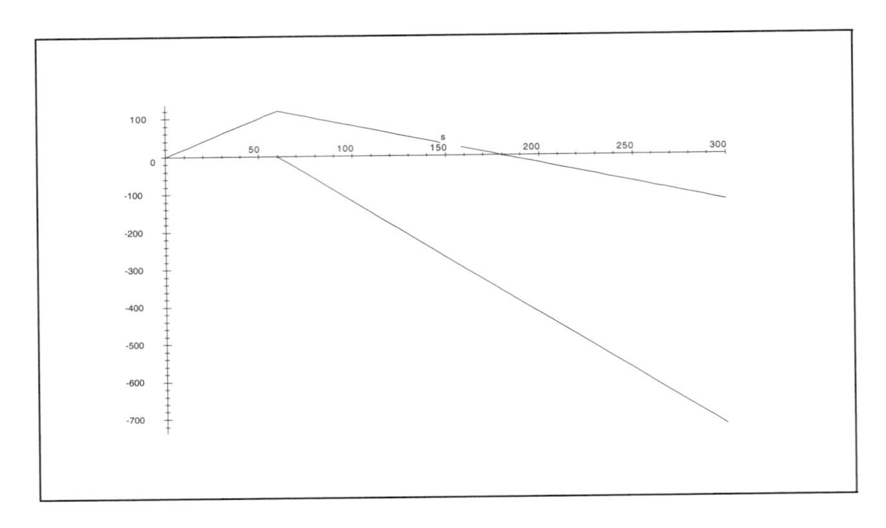

Figure 4.18: A Payoff from Writing Three Calls, and a Payoff from Two Long Positions in the Stock plus Writing Three Calls

```
> Flow Resultant From Commitment to Buy a Stock for $100`);
```

There is a way to make sure that regardless of the state of nature at time T, our commitment will be for exactly \$100. The graph above, Figure 4.20, nearly spells out the answer. We need to devise a portfolio which will have exactly the opposite cash flow to the above commitment. This portfolio should be worth \$1 if the price of the stock is \$99 and should be worth $-\$1$ if the price of the stock is \$101. If it is possible to devise such a position, it would lock in the value of the commitment for \$100 regardless of the price of the stock (state of nature) in the next period. Graphically, we are attempting to generate a line intersecting the x-axis at \$100 which makes a 90° angle with the line in Figure 4.20. Can we generate such a profile from puts, calls, and the underlying stock? It is actually easy to see what type of pieces we will need from our "Lego" box. Here it is in Figure 4.21:

```
> plot(Put(s,100)-Call(s,100),s=0..180,title=`Figure\
> 4.21:Put Minus Call at Strike of $100`,titlefont\
> =[TIMES,BOLD,9]);
```

If you would like to be convinced that indeed these two positions exactly offset each other, you can ask MAPLE to graph it. This is done by issuing

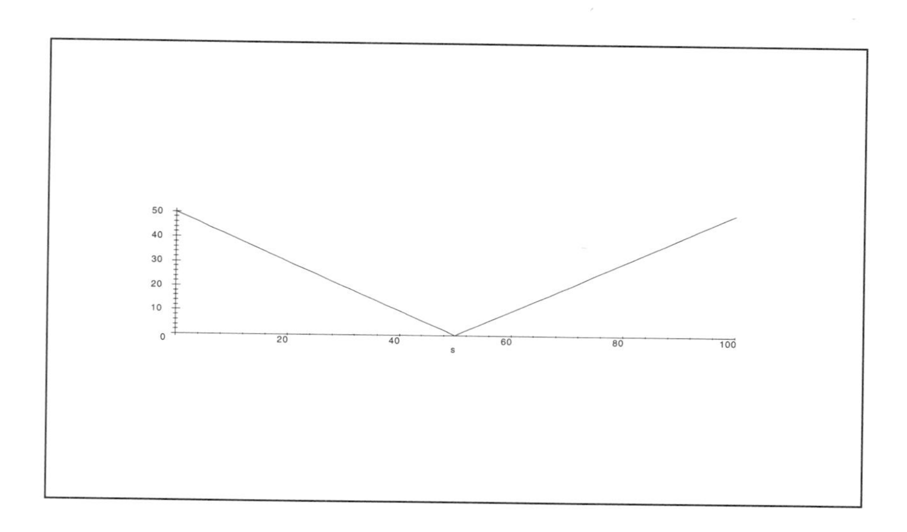

Figure 4.19: Straddle — Call Plus Put with the Same Exercise Price K

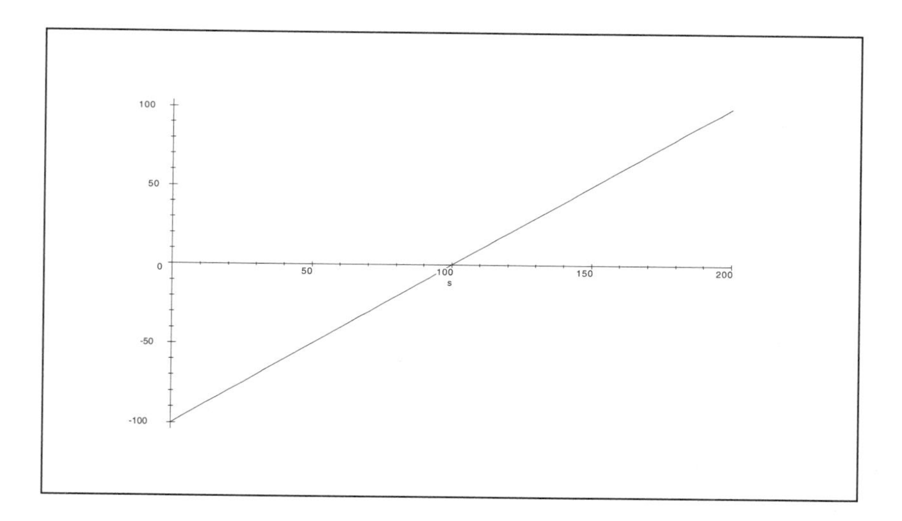

Figure 4.20: Time T Cash Flow Resultant From Commitment to Buy a Stock for \$100

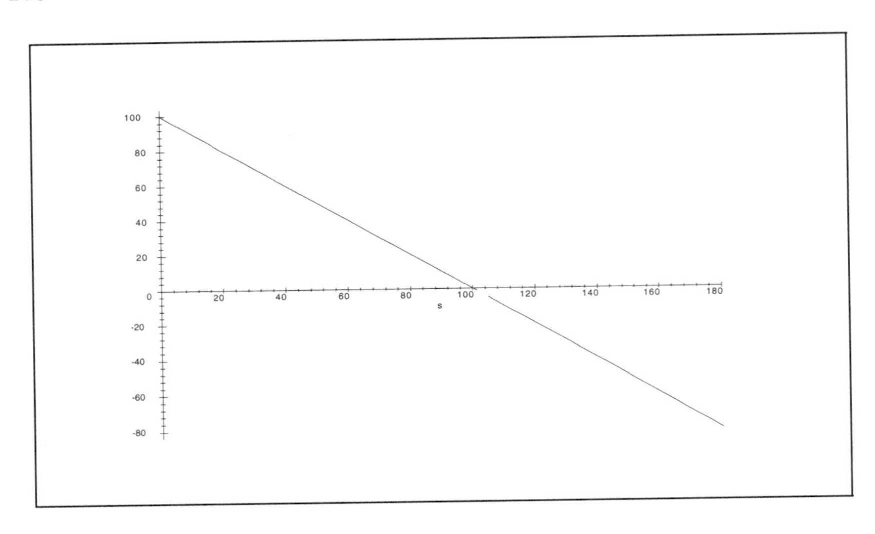

Figure 4.21: Put Minus Call with the Strike of $100

the command which generates Figure 4.22

```
> plot(Put(s,100)-Call(s,100)\
> +(s-100),s=0..180,title='Figure 4.22:\
> Sum of the Two Offsetting Positions);
```

or by evaluating it:

```
> assume(s,positive);
> is(Put(s,100)-Call(s,100)+(s-100)=0);
```

true

The reader is encouraged to look at some combinations defined in the appendix to this Chapter and to graph the corresponding payoff diagrams. Many other payoffs are possible, which can be tailored to your needs. This is done by combining different assets. Taking a look at our "Lego" pieces and at those trading strategies we have managed to devise, the reader should now be convinced that we can generate any payoff which can be represented as a piecewise linear function. This is indeed the case and the last section of this Chapter pertains to this issue.

Before we leave this section, we provide a "practical" demonstration by summing together many puts and calls whose exercise prices are close to one another so that we can approximate a smooth curve. Two examples

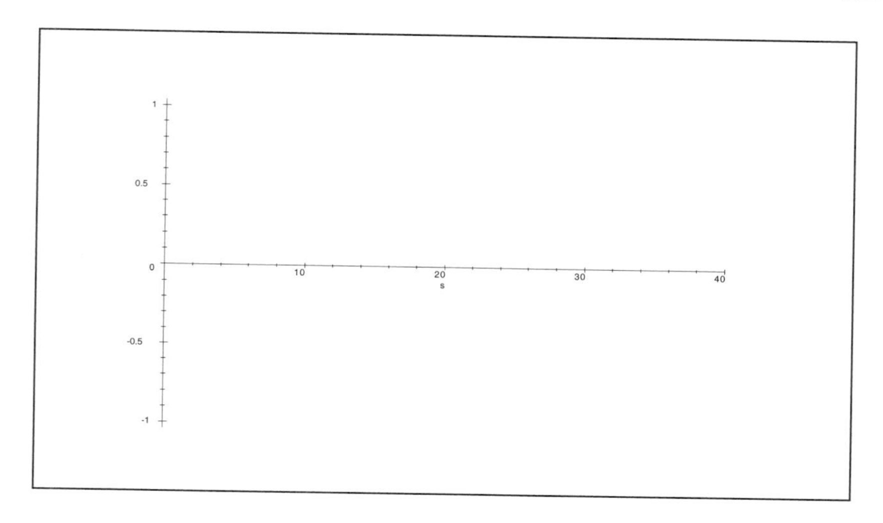

Figure 4.22: Sum of the Two Offsetting Positions

are shown here and are termed the "smile" (Figure 4.23), and the "frown" (Figure 4.24):

```
> plot(sum(Put(s,i)+Call(s,i),i=1..50),s=0..70, title =\
> 'Figure 4.22:  The Smile', titlefont = [TIMES,BOLD,10]);
```

```
> plot(sum(-Put(x,i)-Call(x,i),i=1..50),x=\
> 0..70,title='Figure 4.24:  The Frown',titlefont\
> = [TIMES,BOLD,10]);
```

Can you explain the shape of these graphs?

4.4 Payoff Diagrams and Relative Pricing

It is possible to deduce certain conclusions about the relative pricing of puts and calls by looking at the payoff diagrams of certain combinations, and by utilizing arbitrage arguments. The core of these arguments is based on a comparison of the payoffs, at maturity, of two portfolios, A and B. Namely, if one holds portfolios A and B to maturity, and at that time the payoff from A is larger than the payoff from B, A must be dearer than B.

Probably the best known such result is the *put-call parity*. It will be explored here after we investigate some simpler relationships. The conclusions

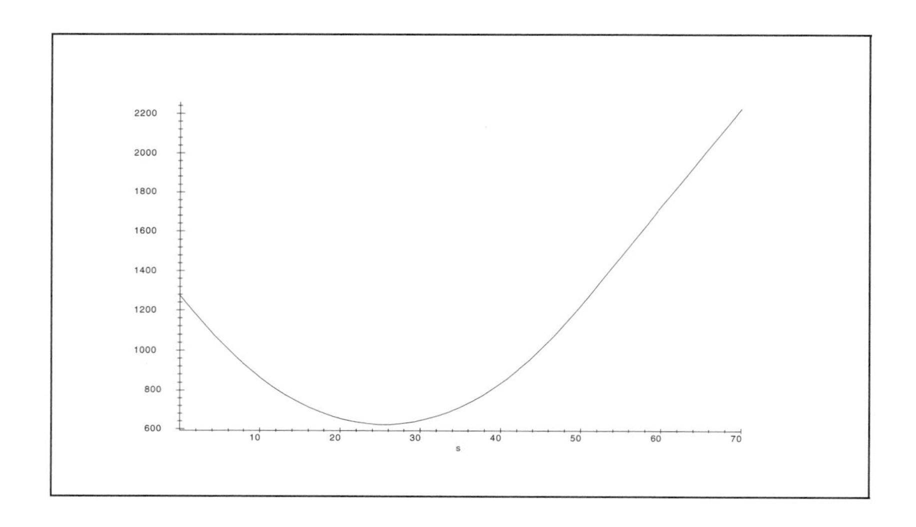

Figure 4.23: The Smile

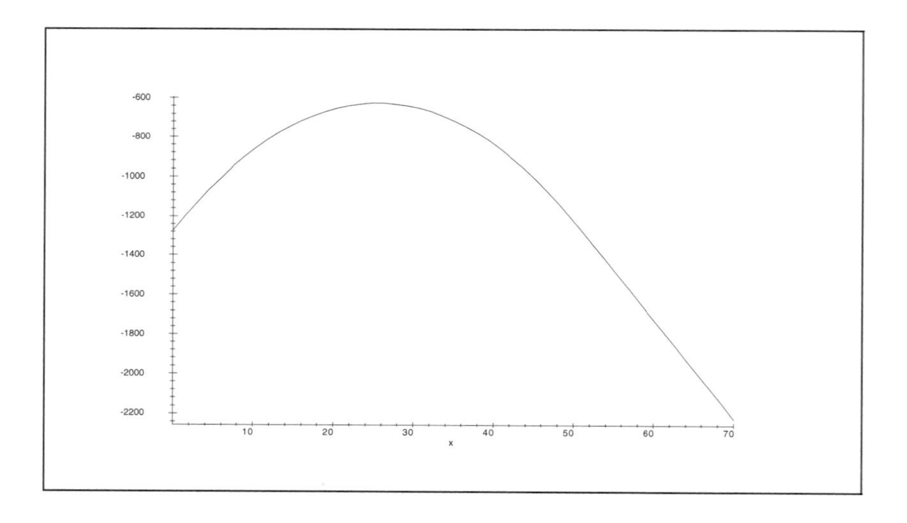

Figure 4.24: The Frown

reached in this section are free from assumptions about the price behavior[11] of the underlying securities. They may, however, depend on the rate of interest when present values are being calculated.

4.4.1 Pricing Bounds Obtained by Relative Pricing Results

The relations among the payoffs of puts, calls, and the stock facilitate the derivation of bounds on the prices of puts and calls.[12] These bounds are free from assumptions regarding the price behavior of the underlying security. This section investigates a few such relations. In Chapter 6 we will take a first look at pricing of derivative securities when an assumption about the distribution of the price of the underlying asset is made. We will then contrast the results from the two pricing methods.

Relative Pricing of European Calls

Consider the following two portfolios: (a) a call with a strike price of $50 and (b) the underlying stock. Figure 4.3 graphed the payoffs from these two portfolios at the maturity of the call option in the same plane.

Clearly the payoff from the stock dominates that of the call, and thus the price of the stock must always be greater than the price of the call. We can continue by investigating the difference in payoffs (i.e., $s - \mathbf{Call}(s, 50)$) at maturity in Figure 4.25. Such a payoff was displayed in Figure 4.16, only there it was demonstrated for $K = 60$.

```
> plot(s-Call(s,50),s=0..100,title=`Figure 4.25:The Payoff\
> from a Stock Minus a Call`,titlefont = [TIMES,BOLD,10]);
```

The graph in Figure 4.25 demonstrates that at maturity the payoff of the stock minus the price of the call is smaller than or equal to K, in our case $50. Intuition suggests that at any time prior to maturity a similar relation must hold; i.e., if the time until maturity is t and the price of the stock is S, then

$$S - P_{Call(S,50)} \leq 50e^{-r_f t}. \tag{4.3}$$

Or in words, if the length of the period to maturity is t, the price difference of the stock and the call is less than the present value of $50 to be paid at maturity. Thus, rearranging produces

$$S - 50\, e^{-r_f t} \leq P_{Call(S,50)}. \tag{4.4}$$

[11] We will compare the bounds obtained to the pricing obtained when an assumption about the behavior of the price process is made.

[12] See [32], [33], and [45].

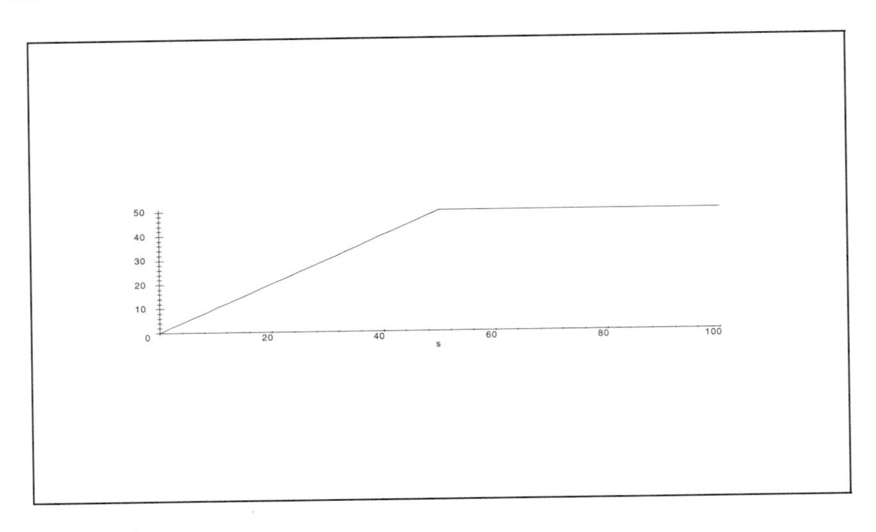

Figure 4.25: The Payoff from a Stock Minus a Call

We would like to conclude, therefore, that the price of the call should be greater than or equal to, at any time prior to maturity, the price of the stock minus the present value of the strike price. Such a conclusion is indeed obtainable by arbitrage arguments.

Assume that this pricing bound is violated at some time, say t units of time, prior to maturity. That is, assume that the relation (4.4) is violated and

$$S - 50\, e^{-r_f t} \geq P_{Call(S,50)}.$$

Thus,

$$S - P_{Call(S,50)} \geq 50\, e^{-r_f t}.$$

This gives rise to the following arbitrage strategy. Short the portfolio composed of long stock and a short call (that is, have a short position in the stock and a long position in the call) and invest $\$50\, e^{-r_f t}$ at the risk-free rate of interest until maturity. Since, by assumption, the proceeds of the short sale are larger than $\$50\, e^{-r_f t}$, the investor pockets the difference at that time. At maturity, the risk-free investment will be worth \$50 and the short portfolio will have a commitment of at most \$50. This was presented in Figure 4.14, for $K = 60$, and is replotted here in Figure 4.26.

```
> plot(-s+Call(s,50),,title='Figure 4.26:\
> The Payoff from a Short Position in \
```

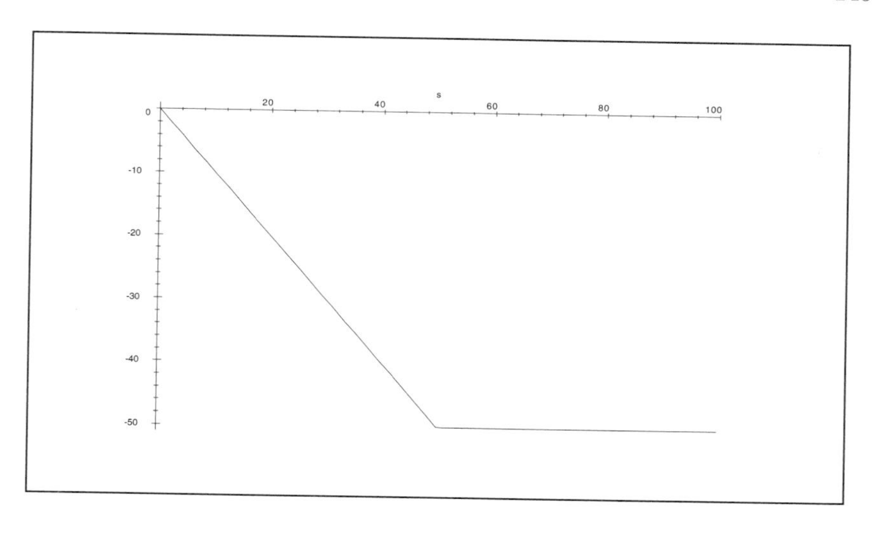

Figure 4.26: A Payoff from a Short Position in the Stock and a Long Position in the Call

```
>   the Stock and a Long Position in the Call'\
>   ,s=0..100);
```

In any state of nature the risk-free investment will be sufficient to cover the liability induced by the short portfolio. An arbitrage profit can be realized t units of time prior to maturity. An arbitrage profit may also be obtained at maturity, time T, if $S_T < 50$. In this case, the liability of the short position is $-S_T$ and the risk-free investment is valued at \$50. This is because the value of the portfolio, short stock and long call, as a function of S_T is as below:

$$-S_T + \max\{0, S_T - 50\} = \begin{cases} -S_T & if\ S_T < 50 \\ -50 & if\ S_T \geq 50. \end{cases} \tag{4.5}$$

The lower bound we obtained on the price of the call, relation (4.4), may be refined.[13] Since in every state of nature the payoff from the call is nonnegative, the price of the call must be nonnegative. Therefore we can conclude the following lower and upper pricing bound whenever the stock price is S and the time to maturity is t:

[13] In those states of nature where the price of the stock is less then the present value of the strike price, the lower bound is negative.

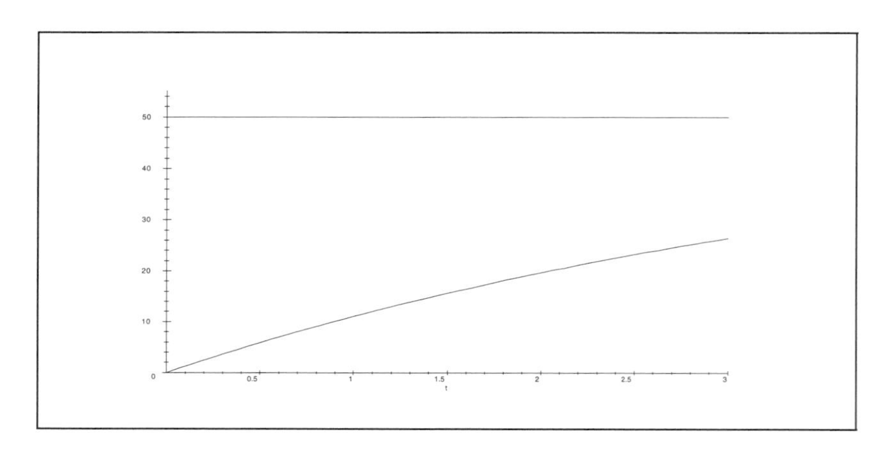

Figure 4.27: A Lower Bound on the Price of a Call when $S_T = 50$ and $r_f = .25$

$$\max(0, S - \$50\, e^{-r_f t}) \leq P_{Call(S,50)} \leq S_t. \tag{4.6}$$

Figure 4.27 shows the pricing bounds obtained above, for a fixed stock price (\$50) at different times prior to maturity (where we have displayed only \$0 through \$55 on the y-axis).

```
> plot({50,max(0,50-(50*exp(-.25*t)))},t=0..3 ,y=0..55,\
> title=`Figure 4.27:  A Lower Bound on the Price of a\
> Call when S_T =50 and r_f =.25`,titlefont=\
> [TIMES,BOLD,5]);
```

We can also graph, Figure 4.28, the pricing bounds at a fixed time, $t = 0.5$ prior to maturity, as a function of the price of the stock.

```
> plot({s,max(0,s-(50*exp(-.10*.5)))},s=0..10 0,y=0..55,\
> title=`Figure 4.28:  Pricing Bounds for a Call, at a\
> Fixed Time, t=0.5 Prior to Maturity as a Function of\
> S_t`,titlefont=[TIMES,BOLD,4] );
```

Finally we can investigate the pricing bounds as a function of t, the time to maturity, and of s, in the three-dimensional Figure 4.29.

```
> plot3d({s,max(0,s-(50*exp(-.10*t)))},s=0..100,t=0..1\
> ,title=`Figure 4.29:  Bounds on the Price of a Call as\
> a Function of s and t`, titlefont = [TIMES,BOLD,7]);
```

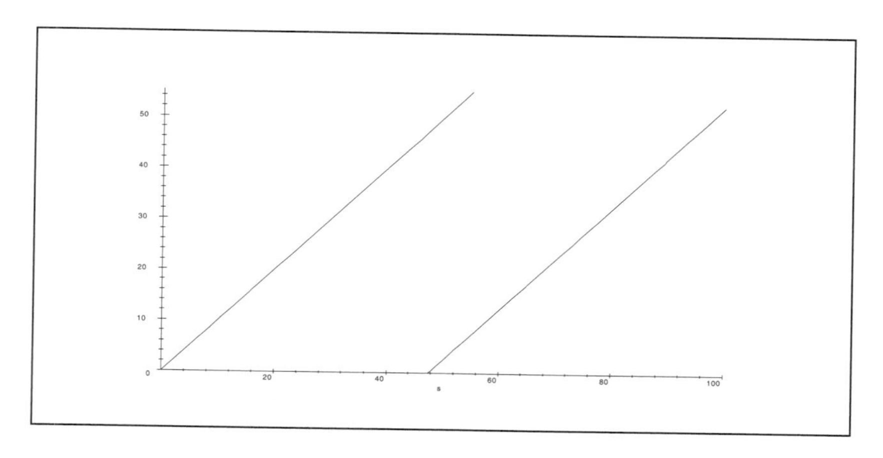

Figure 4.28: Pricing Bounds for a Call, at a Fixed Time, $t = 0.5$ Prior to Maturity as a Function of S

Relative Pricing of European Put

Consider the portfolio composed of a stock and a put with an exercise price of \$60 on the same stock. The payoff of this portfolio at maturity was presented in Figure 4.11. The payoff from this portfolio is greater than \$60, regardless of the state of nature. Using the same argument as above, we can deduce that if the time until maturity of the put is t, then the price of the put plus the price of the stock must be greater than or equal to the present value of \$60. That is greater than $\$60e^{-r_f t}$. Thus

$$S_t + P_{Put(S_t,60)} \geq 60e^{-r_f t} \tag{4.7}$$

and therefore

$$P_{Put(S_t,60)} \geq 60e^{-r_f t} - S_t. \tag{4.8}$$

The lower bound we obtained on the price of the put in equation (4.8) may be refined.[14] Since in every state of nature the payoff from the put is nonnegative, the price of the put must be nonnegative. Therefore we can conclude the following pricing bound:

$$P_{Put(S_{T-t},60)} \geq \max(0, 60\, e^{-r_f t} - S_{T-t}). \tag{4.9}$$

[14] In those states of nature where the price of the stock is greater then the present value of the strike price, the lower bound is negative. Hence the bound can be refined since the value of the position can never be negative.

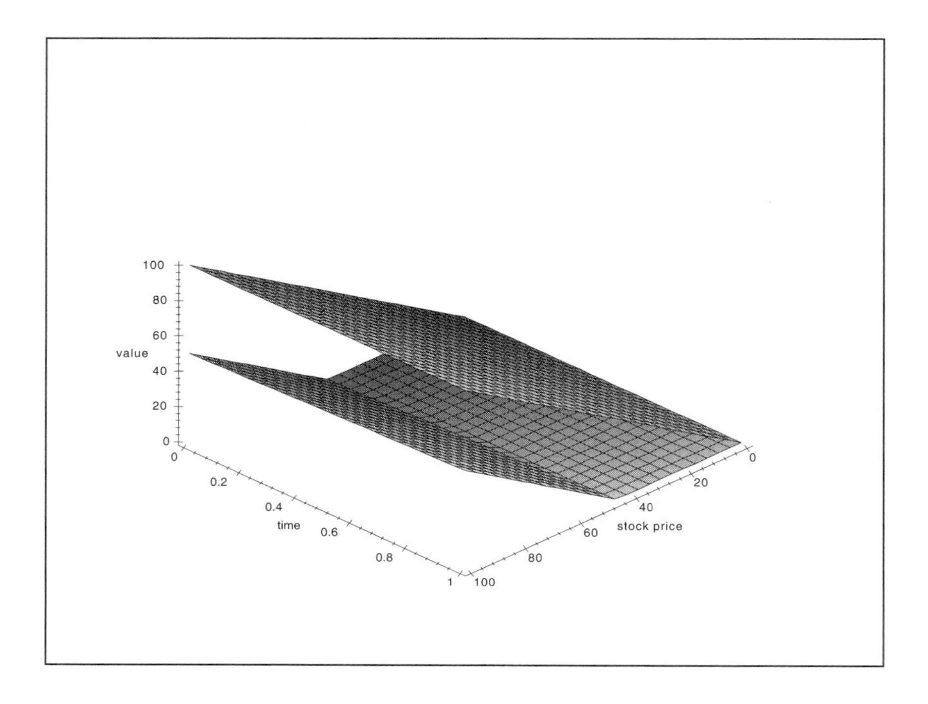

Figure 4.29: Bounds on the Price of a Call as a Function of s and t

4.4.2 Put–Call Parity

In the simple model of Chapter 1, we were able to generate a risk-free security from the elementary contingent claims. We simply bought a contingent claim of \$1 for each state of nature. Is it possible to generate a risk-free security from the Lego blocks we have here? The graphic representation of the payoff from a risk-free security (now that we have an infinite number of possible states of nature) will simply be a horizontal line. Is it possible to construct a portfolio with this resultant payoff? If so, is there more than one way of accomplishing this? Let us first try to solve the puzzle of combining together some basic graphs in order to generate a horizontal line. Later in the Chapter, we will introduce a procedure we programmed in MAPLE that will do this for us.

The idea leading to a way of generating a risk-free portfolio from puts and calls is fairly simple. We need to generate two portfolios such that the difference in their values does not depend on the state of nature (price of the

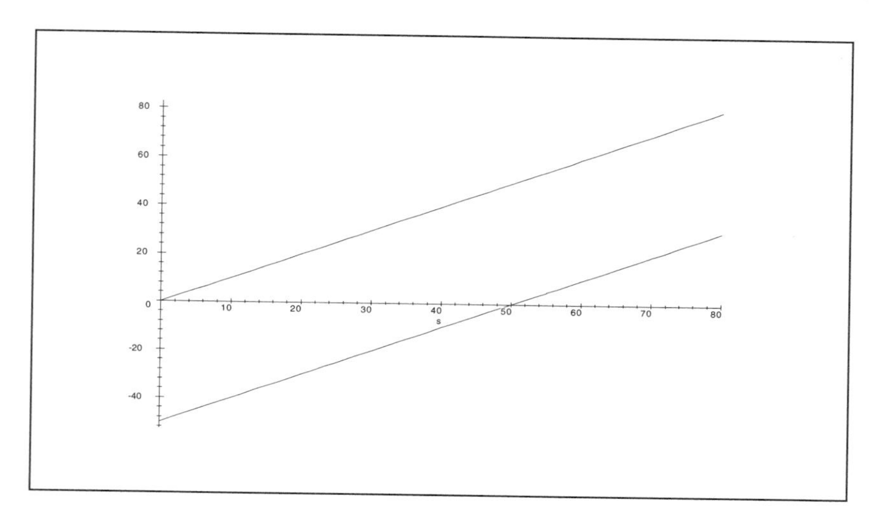

Figure 4.30: A Payoff Parallel to the Stock Payoff

stock): one minus the other will generate the same amount for each state of nature. Graphically, this means the two lines must be parallel to each other. Since the payoff from the stock is just a straight line emanating from the origin at a 45° angle, first generate a line parallel to this. The graph of a call (when its value is not zero) is indeed a 45° line starting from the K as in Figure 4.2. To make this line parallel to that of the stock, we need to add a piece so that it will continue below the horizontal axis in the same way and will not be broken for those states of nature smaller than K. This additional piece is a short position in a put with the same exercise price. Let us try this by graphing the payoff from a call minus a put together with a payoff from the stock in the same plane. Figure 4.30 demonstrates that whatever the price of the stock, the difference between these two payoffs, the forward (long call, short put) and the stock, is always the same.

```
> plot({s,Call(s,50)-Put(s,50)},s=0..80,title=\
> `Figure 4.30:  A Payoff Parallel to the Stock Payoff`\
> ,titlefont=[TIMES,BOLD,9]);
```

Indeed, this is confirmed below:

```
> assume(s,positive);
```

```
> about(s);
```

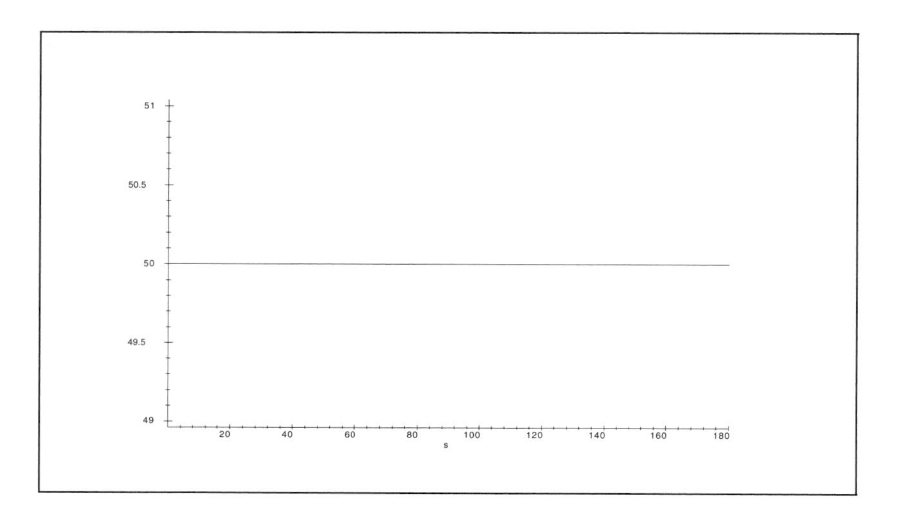

Figure 4.31: The Payoff Difference between a Stock and a Forward

```
Originally s, renamed s~:
  is assumed to be: RealRange(Open(0),infinity)
```

```
> is(s-(Call(s,50)-Put(s,50))=50);
```

true

```
> plot(s-(Call(s,50)-Put(s,50)),title=\
> 'Figure 4.31:The Payoff Difference\
> Between a Stock and a Forward',s=0..80);\
```

Whatever the price of the stock, the difference between these two payoffs is always the same: \$50. Since the graph of the parallel line starts to rise from the state of nature of \$50, it has the same slope as the line of the stock. There will always be 50 units under it. We obtained a risk-free portfolio. Its value is \$50 at time T regardless of the state of nature. It is now easy to generate[15] a risk-free asset paying any amount at time T by changing K.

We can modify the relation above in order to generate a payoff which is *between* two values, regardless of the price of the stock. Suppose we would like to guarantee a payoff between \$45 and \$50. Consider the portfolio in

[15]There are other ways of generating a risk-free asset without using the stock in the portfolio. We will see this later.

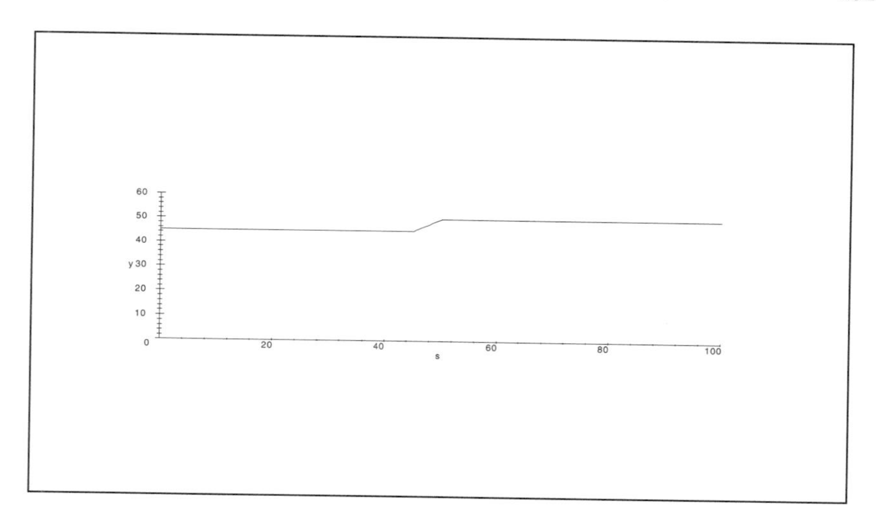

Figure 4.32: A payoff between \$45 and \$50

Figure 4.32 of a long stock, a long put with a strike price of \$45, and a short call with a strike price of \$50.

```
> plot(s-(Call(s,50)-Put(s,45)),s=0..100,y=0..6 0,title\
> ='Figure 4.32:  A Payoff Between $45 and $50',titlefont=\
> [TIMES,BOLD,10]);
```

The ability to generate a risk-free portfolio with a put, a call, and a stock induces a relation between the prices of these three securities. This is analogous to the methodology visited in Chapter 1. There we were able to reach equations (1.8) by constructing a portfolio paying \$1 in every state of nature.

Let P_S denote the current price of a stock and $P_{c(50)}$ and $P_{p(50)}$ denote the current price of a call and of a put, each with an exercise price of \$50, maturing in t units of time. Assume further that the risk-free rate spanning the time interval $[0, t]$ is r_f. Thus, the price of the portfolio consisting of a long position in a stock, a put option, and a short position in a call option (with the same maturity and the same exercise price) must be equal to the present value of \$50. Otherwise arbitrage opportunities would exist. Can you specify the arbitrage portfolios which would be available had this condition not been satisfied? The following relation, called the *put–call parity,* is thus obtained:

$$P_S + P_{p(50)} - P_{c(50)} = \frac{50}{1 + r_f}. \tag{4.10}$$

The same equation can be obtained in a general way for any value of the exercise price, i.e.,

$$P_S + P_{p(K)} - P_{c(K)} = \frac{K}{1 + r_f}. \tag{4.11}$$

With this in mind it is easy to see the put–call parity in a general form: a strike price K, a maturity in t units of time, and using continuous compounding[16] implies

$$P_{S_t} + P_{p(K)} - P_{c(K)} = K\, e^{-r_f t}. \tag{4.12}$$

Synthetic Loan

Utilizing the put–call parity, it is possible to generate a synthetic loan without tapping the market for risk-free securities. Let us see how this would work. This time we are going to use another way of generating a risk-free asset. Consider a long position in a call with an exercise price of \$50 and a short position in a put with the same exercise price; i.e., the cash flow in state s is

$$Call(s, 50) - Put(s, 50). \tag{4.13}$$

We have already seen that this will generate a line crossing the x-axis at \$50 which forms a 45° angle with the x-axis. In the same way, holding a long position in a call and a short position in a put both with an exercise price of \$60 will form a line parallel to the above line crossing the x-axis at \$60. The cash flow in state s of this position will be

$$Call(s, 60) - Put(s, 60). \tag{4.14}$$

The cash flow resulting from the portfolios in equation (4.13) is \$10 more, in every state of nature s, than the cash flow from the portfolio in equation (4.14). Thus, the payoff from holding short the portfolio in equation (4.13) and long the portfolio in equation (4.14) will generate a payoff of $-$\$10 regardless of the state of nature. This is verified in Figure 4.33 by asking MAPLE to evaluate and graph the payoff from this portfolio.

```
> is(Call(s,60)-Put(s,60)-(Call(s,50)-Put(s,50) )= -10);
                              true
```

[16] Here r_f is defined in terms of continuous compounding; the one dollar so invested will grow to $e^{r_f t}$ when r_f is the interest rate for one unit of time.

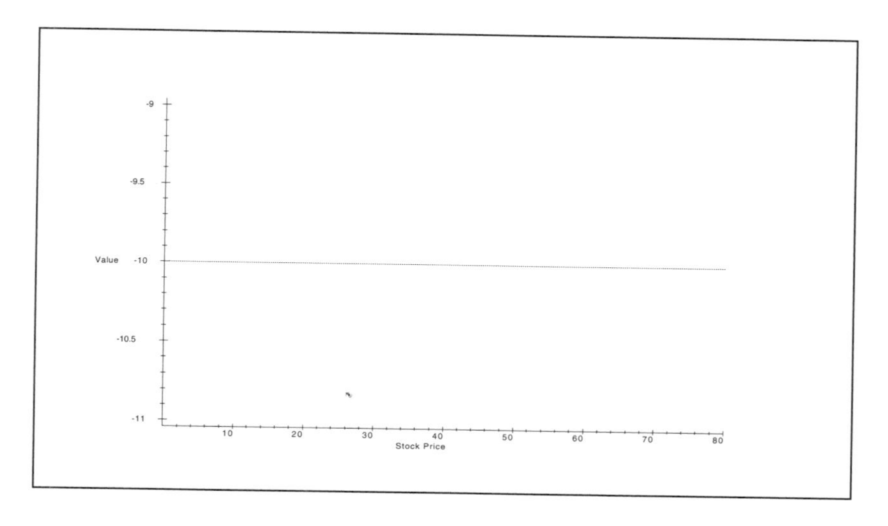

Figure 4.33: Synthetic Loan

```
> plot(Call(s,60)-Put(s,60)-(Call(s,50)-Put(s,5 0)),
> s=0..80, title='Figure 4.33:  Synthetic Loan', titlefont
> =[TIMES,BOLD,10]);
```

Thus, the holder of this portfolio is committed to paying \$10 at maturity (in every state of nature). When purchasing this portfolio, the proceeds from the short part over the long part must be the present value of \$10, that is, $10\,e^{-r_f T}$. If this were not the case, an arbitrage opportunity would exist. If buying this portfolio produces less than $10\,e^{-r_f T}$, then short this portfolio for a cost of less than $10e^{-r_f T}$ and borrow $10e^{-r_f T}$. The total proceeds of this transaction are positive. At maturity, the \$10 loan is paid back, but this is exactly the payoff from the portfolio. Can you prove why the proceeds of buying the portfolio cannot exceed $\$10e^{-r_f T}$?

This portfolio is therefore equivalent to borrowing money at the risk-free rate of interest until the maturity of the options. In exactly the same manner, one can invest money at the risk-free rate of interest even though dealing only in the equity market. This will be done by reversing the above portfolio; i.e., long the portfolio in equation (4.13) and short the portfolio in equation (4.14). This portfolio[17] must cost $10\,e^{r_f T}$ or else an arbitrage

[17] This may result in some creative risk-free investments (see [29]) that will be taxed at the capital gains rate rather than at the ordinary income tax rate. The latter is usually

opportunity exists.

4.5 From Payoffs to Portfolios

In the preceding subsection we investigated different trading strategies resulting from different combinations of puts, calls, and the underlying stock. This subsection is devoted to the inverse [12] question: Given a payoff, as a function of the state of nature (the price of the stock), what portfolio, or portfolios, generates it?

The graphs of puts, of calls, and of the stock are piecewise-linear and continuous. Thus, a combination of these graphs will also result in a continuous piecewise-linear function. The procedure **Stripay** that we defined in MAPLE takes as input the knot points of a piecewise-linear function and solves for the portfolio(s) that generates this payoff.

To demonstrate the use of **Stripay** let us start with a simple example. Consider an investor who is looking at a particular stock with a current price of \$100. The investor believes that in one month the price of that stock will be between \$60 and \$80, and would like to speculate and profit from this belief. Shorting the stock might accomplish this goal. However, shorting the stock will expose the investor to "more risk" than is desirable, e.g., the price of the stock rising to \$120. Shorting the stock will also offer more return in the case where the price of the stock decreases below \$60. The investor does not wish to take these extra risks or to profit from a state of nature in which the stock price is below \$50. Given the investor's held beliefs on the future stock price s, the investor seeks a portfolio with the following payoff as a function of S_T:

$$\text{Payoff}(S_T) = \begin{cases} 0 & \text{if } S_T < 50 \\ 4(S_T - 50) & \text{if } 50 \leq s \leq 60 \\ 100 - S_T & \text{if } 60 \leq s \leq 80 \\ 20 - 2(S_T - 80) & \text{if } 80 \leq s \leq 90 \\ 0 & \text{if } S_T > 90. \end{cases} \tag{4.15}$$

The graph of this payoff is presented in Figure 4.34.

Note that in the interval $[60, 80]$ the resultant profit is exactly like the one from shorting a stock. However, in the intervals $[50, 60]$ and $[80, 90]$ the payoff has the same sign as a payoff from shorting the stock, but it decreases gradually to zero when the price of the stock is above \$90 or below \$50. We

higher. However, careful study of the tax code in the country where such a strategy is planned must be done.

shall come back to this point. The fact that the investor seeks zero payoff when the stock price is below \$50 and above \$90 should make this transaction cheaper relative to a transaction in which, all other things being equal, a positive payoff is desired when the stock price is less than \$60.

The procedure **Stripay** solves for a portfolio of puts and calls and the stock that generates a payoff as in equation (4.15). The input to the procedure is the list of coordinates of the knot points of the graph. It must start with the origin and continue in an increasing manner (of the x-coordinate, from left to right along the x-axis). The last point must be the value of the function at an x-coordinate that is larger than the last knot point. This is how the knot points are inserted: we define the list (an object in MAPLE) Knotp as below,

```
> Knotp:=[[0,0],[50,0],[60,40],[80,20],[90,0],[ 95,0]];
```

$$Knotp := [[0, 0], [50, 0], [60, 40], [80, 20], [90, 0], [95, 0]]$$

To execute the procedure **Stripay** we submit to MAPLE the following command:

```
> Stripay(Knotp,Payoff);
```

Buy, 4, *call at strike price*, 50

Short, 5, *call at strike price*, 60

Short, 1, *call at strike price*, 80

Buy, 2, *call at strike price*, 90

The payoff as a function of the stock price, s, *is*, *Payoff*, (.)

The procedure solves for a portfolio[18] that will generate the required payoff. It also defines the function **Payoff** that we plot, Figure 4.34, in order to verify that indeed the required payoff was obtained.

```
> plot(Payoff(s),s=40..100,title=`Figure 4.34:  The Graph\
> of the Function Payoff`, titlefont=[TIMES,BOLD,10]);
```

For now we assume that we are dealing with payoffs that are always nonnegative. Such a piecewise-linear function is completely specified by the value of the function at a finite number of points. The input to this procedure is thus the list of points $x_0, ..., x_n$, and the value of the function at each point, $y_0, ..., y_n$. This input to the procedure is specified as

[18] It is also possible to input the list of coordinates directly in **Stripay** without defining it beforehand, i.e., as in

>Stripay([[0,0],[50,0],[60,40],[80,20],[90,0],[95,0]],Payoff);

See the Help description of the software to find other solutions when there are many solutions. Note also that **Stripay** defines some global variables.

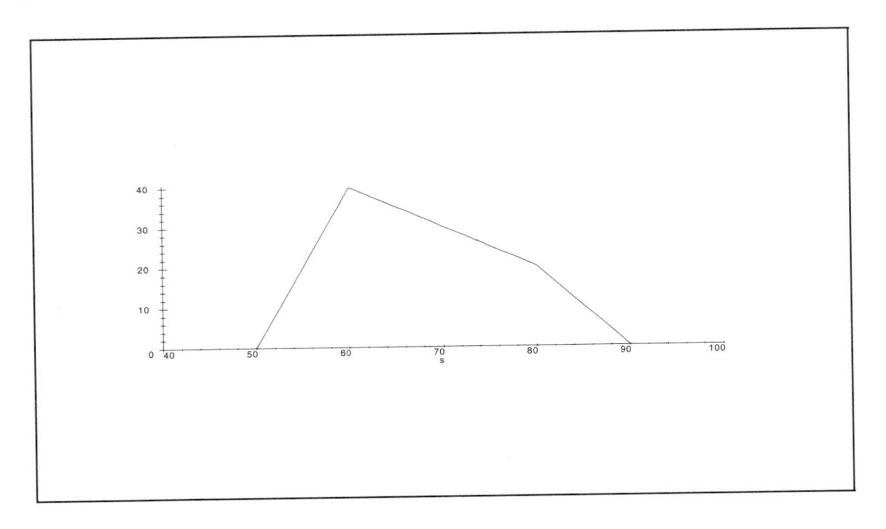

Figure 4.34: The Graph of the Function Payoff

$[[x_0, y_0], [x_1, y_1], ..., [x_n, y_n]]$. Each square bracket contains a knot point specified by the stock price and the payoff specific to the knot point. The first knot point must have an x-coordinate of zero that is $[x_0, y_0] = [0, y_0]$, where y_0 is the payoff if the stock price is zero. The last point must be a coordinate of a point with an x-coordinate greater than the x-coordinate of the right-most knot point of the graph. Please see the Appendix (4.8.1) for a description of the methodology by which the required portfolios are determined when the knot points are given.

Consider another example. The current price of a stock is \$100. An investor believes that in two months the price of the stock will be above its current value. The investor is seeking a position that, if the price of the stock decreases, will generate only half the return that would have been received had the stock been "longed", and twice the return when compared to a long position in the stock, should the price of the stock increase. The investor is thus looking for a position which pays $2(s - 100)$ if the price of the stock is above \$100 and $0.5(s - 100)$ should the price of the stock be smaller than \$100. Thus, the coordinates to be passed on to **Stripay** are [[0, −50], [100, 0], [110, 20], [120, 40]]. This time we ask MAPLE to define the payoff function and call it Pay2.

```
> Stripay([[0,-50],[100,0],[110,20],[120,40]],Pay2);
```

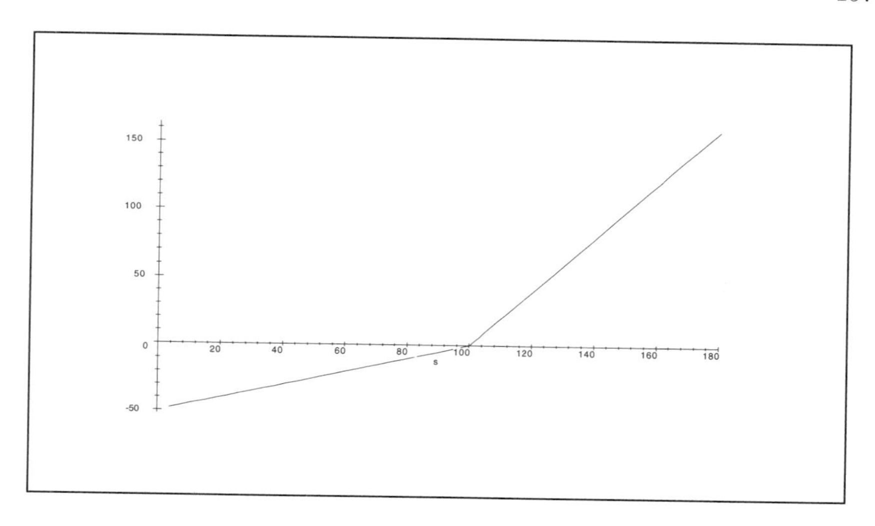

Figure 4.35: The Graph of the Function Pay2

$$Short, \frac{1}{2}, put\ at\ strike\ price, 100$$

$$Buy, 2, call\ at\ strike\ price, 100$$

$$The\ payoff\ as\ a\ function\ of\ the\ stock\ price,\ s,\ is,\ Pay2, (.)$$

The graph of the payoff **Pay2** is in Figure 4.35.

```
> plot(Pay2(s),s=0..180,title='Figure 4.35:  The Graph
> of the function Pay2',titlefont=[TIMES,BOLD,10]);
```

What can we say about the price of the above portfolio? Assume that the continuously compounded annual rate of interest is 10%. If one borrows the present value of $100 to be received in two months, i.e., the amount $100e^{-0.10\frac{2}{12}}$

```
> 100*exp((-0.10)*(2/12));
```

$$98.34714538$$

buys the stock and sells it at the end of the two months, one's payoff will be $s - 100$. We plot the two payoffs in the same graph in Figure 4.36.

```
> plot({s-100,Pay2(s)},s=0..160,title='Figure 4.36:  The
> Two Payoffs',titlefont=[TIMES,BOLD,10]);
```

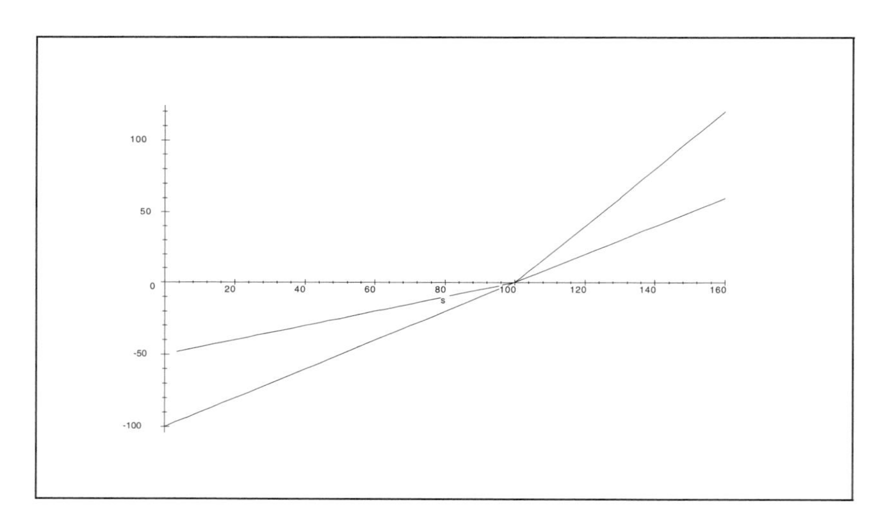

Figure 4.36: The Two Payoffs

It is evident that the payoff from borrowing the money and buying the stock is dominated by the combination of a short position of $\frac{1}{2}$ put with an exercise price of \$100 and a long position in two calls with an exercise price of \$100. The net cost of the first strategy (buy the stock and borrow the \$100) is

```
>   100-100*exp(-0.10)(2/12);
                                1.65285462
```

Thus the combinations of the option portfolio must cost more than that.

This example demonstrates again that puts and calls allow stripping the stock into some smaller components. Thereby one can engineer a position which, in certain states of nature, pays more than the stock and in certain states of nature pays less. The puts and calls allow us to alter the payoff from the stock and to generate a much richer profile of payoffs. This is the contribution of the derivative securities.[19]

Options facilitate the creation of positions that are "riskier" or those that assume less risk than the positions that could have been created without an option market. The example above investigated a position that is less risky

[19] We will see later that our pricing methodology made use of the assumption that these payoffs could be generated by a combination of stocks and bonds. However, it would have cost more to generate them in such a way.

than buying a stock and financing it with a loan. It was created with the aid of certain combinations of puts and calls.

The last example can be modified to create a position that is riskier than borrowing money and buying the stock. Suppose we are looking for a position that pays $\frac{1}{2}(s-100)$ if s, the price of the stock, is above \$100 and $2(s-100)$ if the price of the stock is smaller than \$100. Thus, the coordinates to be passed to **Stripay** are as defined in Knotp below.

```
> Knotp:=[[0,-200],[100,0],[105,10],[110,20]];
```

$$Knotp := [[0, -200], [100, 0], [110, 10], [120, 20]]$$

Such a position will pay less than the portfolio (borrow \$100 and buy the stock) in the preceding example if the price of the stock increases, but will be more negative if the price of the stock decreases. Essentially, in this position there is a trade-off between the current cost of the position and the risk taken. Can you solve for the portfolio generating this position and graph its payoff? Its cost should be less than \$1.65285462. Can you explain why?

There are some payoffs that we cannot generate from our building blocks of the put, the calls, and the stock. For example, we cannot create a non-continuous payoff. Recently, a new derivative security called a *binary (bet) option* was introduced into the market. Essentially, this security can be viewed as a bet on the price of a stock. By way of example, consider a binary option which pays \$30 if the price of the stock is greater than or equal to say \$80, and zero otherwise. The graph of this payoff, even though piecewise-linear, is not continuous. For the state of nature (price of the stock) \$80, the payoff is \$30, while at every state of nature less than \$80 (be it as close as possible to \$80), the payoff is zero. We can approximate such a payoff from a payoff which is zero for states of nature from zero to some number less than \$80, and is \$30 for states of nature \$80 or above. Let us try to generate a payoff that is zero from state of nature zero to state of nature \$75, and is \$30 for every state of nature above \$80. Passing the appropriate coordinates to **Stripay** produces:

```
> Stripay([[0,0],[75,0],[80,30],[120,30]],Approx1);
```

Buy, 6, *call at strike price*, 75

Short, 6, *call at strike price*, 80

The payoff as a function of the stock price, s, is, Approx1, (.)

The graph of the first approximation, Figure 4.37, exemplifies why we cannot create the exact payoff.

```
> plot(Approx1(s),s=0..130,title=`Figure 4.37:  The Graph\
```

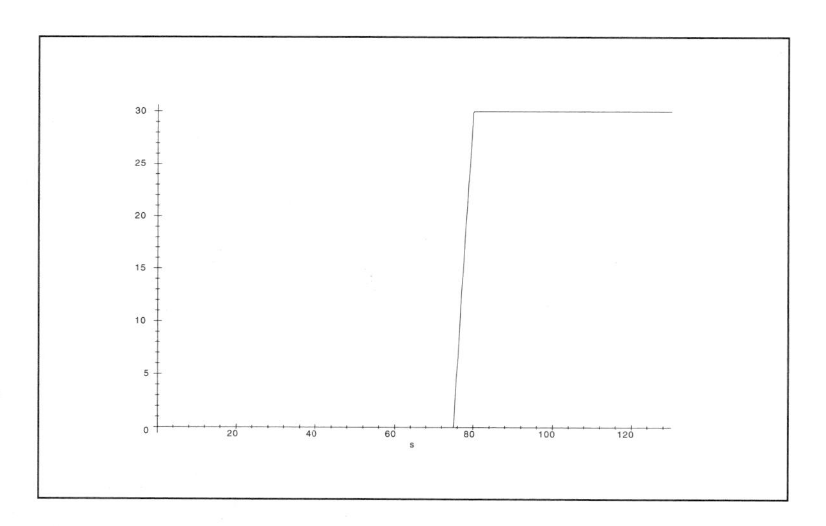

Figure 4.37: The Graph of the First Approximation

```
> of, Payoff1, The graph of the First Approximation.\
> ',titlefont=[TIMES,BOLD,10]);
```

We cannot create a payoff that will jump to \$30 at \$80 but will be zero for every state of nature that is strictly less than \$80. We can try to have a "better" approximation to the required payoff by choosing a point closer to \$80 than \$75, say \$79, and define the set of knot points as below.

```
> Knotp:=[[0,0],[79,0],[80,30],[120,30]];
```

$$Knotp := [[0, 0], [79, 0], [80, 30], [120, 30]]$$

We can solve for a portfolio that creates this payoff by calling **Stripay**.

```
> Stripay(Knotp,Approx2);
```

Buy, 30, *call at strike price*, 79

Short, 30, *call at strike price*, 80

The payoff as a function of the stock price, *s*, *is*, *Approx2*, (.)

The graph of the payoff Approx2, presented in Figure 4.38, will be

```
> plot(Approx2(s),s=0..130,title='Figure 4.38:  The Graph\
> of the Second Approximation');
```

Of course, the feasibility of actually creating such a portfolio depends on the availability of puts and calls at the exercise prices of \$79 and \$80.

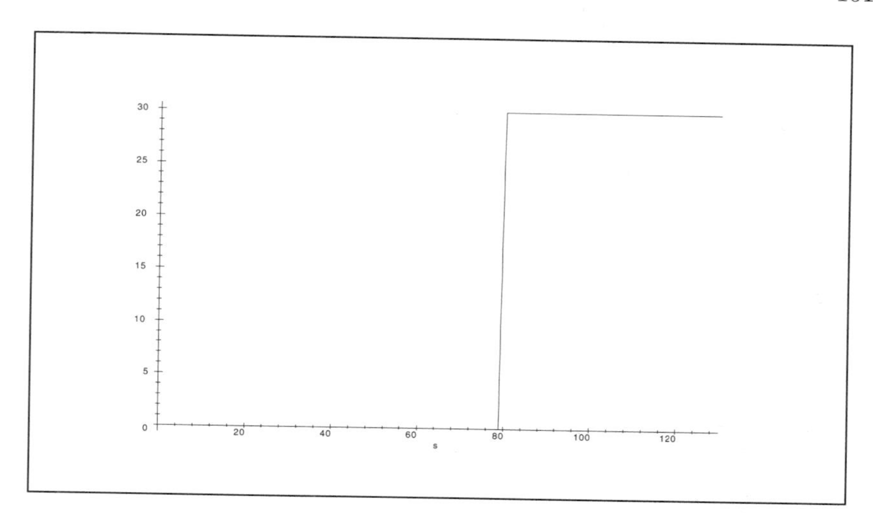

Figure 4.38: The Graph of the Second Approximation

Exercise prices are not available at any price: usually they are available in jumps of \$5. So the best we can do is to choose the knot point \$75 rather than \$79. The price of this cash flow must be greater than or equal to the price of the binary option. However, we can also find a lower bound on the price of the binary option. To this end we will create a payoff that is less than or equal, for every state of nature, to the payoff from the binary option. Consider a payoff of zero up to state of nature \$80, and \$30 for every state of nature larger than \$85. Redefining Knotp to reflect this payoff produces

```
> Knotp:=[[0,0],[80,0],[85,30],[120,30]];
```

$$Knotp := [[0, 0], [80, 0], [85, 30], [120, 30]]$$

and executing **Stripay** produces the portfolio.

```
> Stripay(Knotp,Approx3);
```

Buy, 6, *call at strike price*, 80

Short, 6, *call at strike price*, 85

The payoff as a function of the stock price, s, is, Approx3, (.)

We graph the payoff in Figure 4.37 together with the payoff in Figure 4.39. The payoff from the binary option is indeed between the two payoffs, Figure 4.39, and thus its price must be between the prices of the above two portfolios. So while we can create many profiles of payoffs, we remain

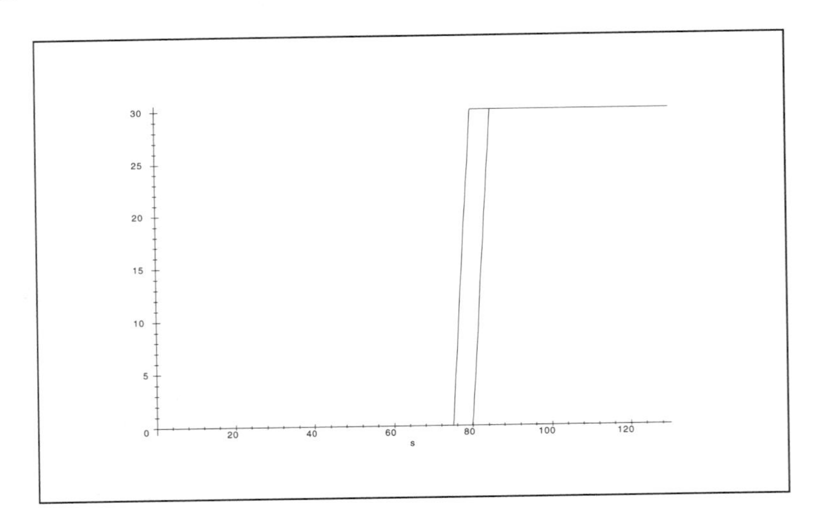

Figure 4.39: Approximating a Binary Option

constrained by the fact that payoffs created from the building blocks of stocks, puts, and calls must be continuous and piecewise-linear.

```
> plot({Approx1(s),Approx3(s)},s=0..130,title ='Figure 4.39:
> Approximating a Binary Option',titlefont
> =[TIMES,BOLD,10]);
```

We conclude this section with the following example. Consider an investor that has a liability to pay at time T. The liability is contingent on the price of a certain stock and is given by a piecewise-linear function. The knot points of the function and the values of the function at these knot points are given below:

```
> Lib(s)=piecewise(s<=5,0,s=6,27,s=8,35,s=10,45 ,s=12,50,\
> s=15,50,s=18,30,s=20,15,s=22,-20,s=25,-10,\
> s=28,-5,s>=30,0);
```

$$\mathrm{Lib}(s) = \begin{cases} 0 & s \leq 5 \\ 27 & s = 6 \\ 35 & s = 8 \\ 45 & s = 10 \\ 50 & s = 12 \\ 50 & s = 15 \\ 30 & s = 18 \\ 15 & s = 20 \\ -20 & s = 22 \\ -10 & s = 25 \\ -5 & s = 28 \\ 0 & 30 \leq s. \end{cases}$$

The investor would like to find a portfolio whose value will be equal to the liability value at time T. If such is found, the investor can buy the portfolio and walk away from the liability. Indeed, we are actually looking for the present value of the liability and whether or not there is a portfolio that produces a cash flow equal to the liability. We proceed by redefining the list Knotp according to the coordinates of the function.

```
> Knotp:=[[0,0],[5,0],[6,27],[8,35],[10,45],[12 ,50],
> [15,50],[18,30],[20,15],[22,-20],[25,-10],[28,-5],
> [30,0],[35,0]];
```

$$Knotp := [[0, 0], [5, 0], [6, 27], [8, 35], [10, 45], [12, 50], [15, 50], [18, 30], [20, 15], [22, -20], [25, -10], [28, -5], [30, 0], [35, 0]]$$

To solve for the portfolio we execute

```
> Stripay(Knotp,Liability);
```

$$Buy,\ 167,\ put\ at\ strike\ price,\ 5$$
$$Short,\ 140,\ call\ at\ strike\ price,\ 5$$
$$Short,\ 173,\ put\ at\ strike\ price,\ 6$$
$$Buy,\ 150,\ call\ at\ strike\ price,\ 6$$
$$Buy,\ 1,\ put\ at\ strike\ price,\ 8$$
$$Short,\ \frac{5}{2},\ put\ at\ strike\ price,\ 10$$
$$Short,\ \frac{5}{2},\ put\ at\ strike\ price,\ 12$$
$$Short,\ \frac{20}{3},\ put\ at\ strike\ price,\ 15$$
$$Short,\ \frac{5}{6},\ put\ at\ strike\ price,\ 18$$

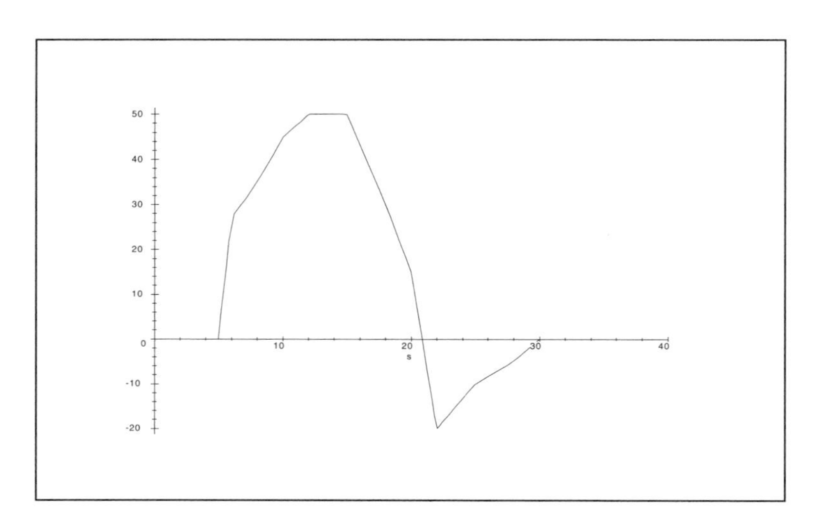

Figure 4.40: The Graph of the Liability

$$\begin{gathered} \textit{Short},\ 10,\ \textit{call at strike price},\ 20 \\ \textit{Buy},\ \frac{125}{6},\ \textit{put at strike price},\ 22 \\ \textit{Short},\ \frac{5}{3},\ \textit{put at strike price},\ 25 \\ \textit{Buy},\ \frac{5}{6},\ \textit{put at strike price},\ 28 \\ \textit{Short},\ \frac{5}{2},\ \textit{put at strike price},\ 30 \end{gathered}$$

The payoff as a function of the stock price, s, is, Liability, (.)

Finally we plot the function "Liability", Figure 4.40, to verify that indeed we created a portfolio producing the required payoff.

```
> plot(Liability(s),s=0..40,title='Figure 4.40: The Graph
> of the Liability',titlefont=[TIMES,BOLD,10]);
```

4.6 Concluding Remarks

Option markets facilitate the creation of positions that were not possible (or were difficult) to create without options. A market for risk is introduced.

Investors are thus able to buy or sell "risks" which would not have been possible without an options market. It is not surprising, therefore, that options can be used as a vehicle for speculation as well as for hedging.

There are limitations to the types of payoffs that can be engineered in this market. We cannot produce a "smooth" function for a nonlinear payoff, but we can approximate such a function by piecewise-linear functions. The smaller we are able to make the linear pieces, the better the approximation. For example, an interesting payoff that we cannot produce is $1 if the state of nature is s for a certain price of the stock. This is not feasible. We can only approximate such a payoff given the availability of options at different strike prices.

Such a payoff is the counterpart, in the discrete model of Chapter 1, of $1 contingent on state ω_j, where ω_j is a possible state in $\Omega = \{\omega_1, \ldots, \omega_n\}$. It was investigated in Chapter 1 and its price was defined to be the stochastic discount factor for state j. The payoff of $1 contingent on state s will play a key role in the discussion of Chapter 5. In real markets we can only approximate such a payoff, e.g., a payoff of $1 contingent on state $s = 80$, by a payoff that is zero up to $75, $1 from $75 to $80, and zero from $85 on.

The goodness-of-the-approximation depends on the "size of the pieces", which in turn depend on the availability of options at different strike prices in the market. Usually the availability is in discrete units of $5. However, given the availability of strike prices and the constraint imposed by the continuity, not much more can be done. *This is one reason that, when we move to a continuous time model, we cannot solve for the stochastic discount factors. Rather, we must make some assumptions about the factors in order to derive pricing relations.*

The reader has now acquired, through presentation here and through personal experimentation with the MAPLE procedures, a fundamental understanding of what can be done with derivative securities. We now move on to some evaluation issues wherein we investigate the stochastic discount factors in realistic markets.

4.7 Questions and Problems

Problem 1. Suggest a combination composed of put(s) and call(s) written on the same underlying asset, with the same maturity date, such that at maturity the payoff is the function

$$\begin{cases} 0 & if\ S_T < 0 \\ \text{negative} & if\ 0 < S_T < K_1 \\ \text{positive} & if\ K_1 < S_T < K_2 \\ 0 & if\ S_T > K_2, \end{cases}$$

where $0 < K_1 < K_2$, S_T is the price of the underlying asset at time T, and the exercise prices of the puts and the calls that are available in the market are K_1, K_2, and K, where $K_1 < K < K_2$.

Problem 2. Suggest a combination that guarantees a payoff of C $(C > 0)$ if the price of the underlying asset at maturity, S_T, is less than K_1 $(K_1 > 0)$ and is zero if the price of the underlying asset exceeds K_2 $(K_2 > K_1)$.

Problem 3. You are holding a portfolio that is contingent on the value of a stock on a certain date. Your portfolio value will be zero if the value of the stock is zero or if it is between K_1 and K_2, where $0 < K_1 < K_2$. Between zero and K_1, the slope of the payoff function takes two values, $\frac{1}{5}$ and $-\frac{1}{3}$, and it only reverts once from one slope to another. If the value of the stock is above K_2, the slope of the payoff function is $\frac{1}{6}$. You are interested in hedging this position. Suggest a portfolio of puts and calls that will satisfy your goal and that has a continuous payoff.

Problem 4. Suggest a combination that consists of puts only, written on the same underlying asset with the same maturity dates, such that if every put in the combination is replaced by a call the cash profile does not change.

Problem 5. Assume a market with an underlying asset and with puts and calls that are written on this asset with the same maturity dates and exercise prices of $1, 2, 3$, and 4. How would you approximate the price of a dollar contingent on the value of the underlying asset being between 2 and 3.

Problem 6. Specify the arbitrage portfolio asked about in the text, just prior to equation (4.10), which proves the put – call parity relation.

Problem 7. Consider the combination of

$$Call(s, 60) - Put(s, 60) - Call(s, 50) + Put(s, 50),$$

where the puts and the calls mature in t units of time and are all written on the same underlying asset. Prove that buying this portfolio produces a cash inflow and that the proceeds of buying this portfolio cannot exceed $\$10e^{-r_f t}$.

Problem 8. Consider a portfolio of short position of $\frac{1}{2}$ put with an exercise price \$100, and two long positions in a call with exercise prices of \$100. The payoff from this portfolio at maturity is displayed by Figure 4.36. Develop the arbitrage bounds on the price of this portfolio, assuming that the continuously compounded risk-free rate spanning the time to the maturity of the options is 10%.

Problem 9. Consider two call options that both mature in t units of time with exercise prices of K_1 and K_2 ($K_2 > K_1$), respectively. Let r_f be the continuously compounded risk-free rate per unit of time. Prove that

$$-(K_2 - K_1)e^{-r_f t} < P_{c(K_2)} - P_{c(K_1)} \leq 0.$$

Problem 10. Assume a market where the borrowing rate is r_b and the lending rate is $r_l < r_b$. How will the put-call parity relation be affected in such a market? Prove your result utilizing arbitrage arguments.

Problem 11. Suggest a combination that will have a positive payoff if the price of the underlying asset is in the interval $[K - e, K + e]$ and zero otherwise, where $K > 0$, $e > 0$, and, $K - e > 0$. Assume that in the market there exist puts and calls at strike prices of K, $K - e$, and $K + e$.

Problem 12. Consider a market with a stock and puts and calls, written on the stock, with the same exercise price. Let the price of the stock be P, the price of the call be $C(t, K)$, and the price of the put be $P(t, K)$, where K is the exercise price and t the time to maturity.

1. What can you say about the risk-free rate r, spanning the time interval $[0, t]$?
2. What can you say about the relation between $P(t_1, K) - P(t_2, K)$ and $C(t_1, K) - C(t_2, K)$, where $t_2 > t_1$?

4.8 Appendix

4.8.1 Explanation of Stripay

The payoff graphs of puts and calls, as a function of the value of the stock, are piecewise-linear and continuous while the graph of the stock's value against itself is obviously linear. Thus, a combination of these graphs will also result in a continuous piecewise-linear function. Hence, puts, calls, and a stock can be used as building blocks of a portfolio generating piecewise-linear payoffs.

In order to replicate a piecewise-linear payoff, the puts and calls must be chosen to include those with the exercise prices equal to the x-coordinate of the knot points of the payoff function. Then the payoff from a portfolio consisting of these puts, calls, and a stock, as a function of S_T, will be

$$xS_T + \sum_{i=2}^{n} p_i Put(S_T, K_i) + \sum_{i=2}^{n} c_i Call(S_T, K_i), \tag{4.16}$$

where n is the number of knot points, K_i is the x-coordinate of the ith knot point, x is the number of units of the stock in the portfolio, and p_i and c_i are the number of units of puts and calls in the portfolio with exercise price K_i.

From here on we will assume that $K_1 = 0$ and $K_i < K_{i+1}$ for all i's. These assumptions cause no loss of generality. However, since we do not use puts and calls with $K_1 = 0$, the summations in expression (4.16) will start with $i = 2$.

Two linear functions coincide if and only if they have two points in common. Applying this fact to our situation, we conclude that given a piecewise-linear payoff a portfolio will generate the desired payoff if and only if its payoff coincides with the given payoff at all knot points and one other point with an x-coordinate larger than the right-most knot point.

Hence, x, p_i, and c_i should satisfy

$$xS_1 + \sum_{i=2}^{n} p_i Put(S_1, K_i) = Payoff_1$$

$$xS_j + \sum_{i=j}^{n} p_i Put(S_j, K_i) + \sum_{i=2}^{j} c_i Call(S_j, K_i) = Payoff_j, \quad for\ j = 2, ..., n$$

$$xS_{n+1} + \sum_{i=2}^{n} c_i Call(S_{n+1}, K_i) = Payoff_{n+1}, \tag{4.17}$$

where $S_1 = 0$, $S_j = K_j$ for $j = 1, ..., n$, S_{n+1} is an arbitrarily chosen point such that $K_n < S_{n+1}$, and $Payoff_j$ is the desired payoff if the stock price is S_j for $j = 1, ..., n+1$.

By solving this system of equations for x, p_i's and c_i's we get the composition of a portfolio generating the required payoff. As usual, negative values of x (c_i or p_i) refer to a short position.

The procedure **Stripay** that we defined in MAPLE takes as input the knot points of a piecewise-linear function and solves for the portfolio(s) that generates the desired payoff.

It is worthwhile mentioning that the system of equations in (4.17) may have a unique solution, a multiple solution, or no solution at all. The latter is not possible in our case since the system has more variables than equations and is always consistent. If the system has multiple solutions the values of the free parameters are arbitrarily set to zero. See the HELP file for **Stripay** for an explanation of how to extract all the solutions.

4.8.2 Procedural Issues

In this Appendix we provide an explanation of how to read the quotations in the newspaper[20] and definitions of some glossary items[21]. End of day prices for exchange traded options are reported daily in business newspapers. A generic table that explains how to read the codes and symbols is detailed below:

STOCK SERIES		BID	ASK	LAST	VOL	OP Int[1].
ABC Inc.[2]	$40.85[3]				770[4]	22217[5]
August	$40	$1.85	$2.10	$1.95	10	35
August	$40 p	$0.95	$1.10	$1.00	19	82
October	$42.50	$1.70	$1.95	$1.60	34	299
XYZ Ltd.	$28.50					
October	$30	$1.90	$2.15	$1.80	5	133
January[6]	$29[7]	$3.30	$3.55	$3.25	5	5
January	$32.50 p[8]	$5.50[9]	$5.75[10]	$5.65[11]	5[12]	55[13]

Table 4.1: A Typical Newspaper Option Table

[20] We thank the Toronto Stock Exchange for permission to replicate the included table and the explanation taken from their web site.

[21] We thank the Montreal Exchange for the permission to replicate the explanations taken from '*Options Guide and Investment Strategies - The Monteral Exchange*'.

1. The open interest represents the number of option contracts written on a particular stock that have not yet been exercised.
2. Stock's name
3. Closing Price
4. Total volume for all ABC Inc. option contracts.
5. Total open interest for all ABC Inc. option contracts.
6. Expiration Month
7. Strike Price
8. "p" indicates a put option
9. Bid for the Jan $32.50 put
10. Ask for the Jan $32.50 put
11. Last price for the Jan $32.50 price
12. Volume for the Jan $32.50 put
13. Open interest for the Jan $32.50 put

The Canadian Derivatives Clearing Corporation

The Canadian Derivatives Clearing Corporation (CDCC) is the official clearing house for options traded in Canada. Created in 1976 under the name TransCanada Options (TCO), this corporation acts as guarantor and is responsible for maintaining margins, issuing and clearing options traded at the Montreal Exchange. The CDCC acts as a guarantee agent for every option contract traded on Canadian exchanges. After ensuring that every option contract brings together a seller and a buyer, it takes the place of the parties and administers each parties obligations. After the transaction is executed, the original buyer and seller deal with the CDCC and no longer have any obligations towards each other.

For example, if a seller wants to liquidate his opening position, he must contact the CDCC, not the initial buyer of the option. In order to get out of his obligation (to deliver the stock at some point), he can buy an option contract belonging to the same series as the options he holds, which will enable him to liquidate his position.

Because only one organization, the CDCC, looks after clearing, it is not necessary to evaluate the risk related to whether the parties are solvent. Thanks to the standards established by the CDCC with regard to having sufficient equity and complying with the daily coverage requirements, the contracts traded are always based on solid guarantees. Furthermore, in terms of clearing and regulations, the CDCC ensures that the parties act in a disciplined manner when they trade and take a position.

Finally, the CDCC ensures that option holders can take and dispose of a position. Thus, everyone has the opportunity to trade options on a liquid, transparent market. If the options are kept until the expiry date, the CDCC ensures that the stocks are delivered in exchange for the final payment.

Glossary

The options market has its own terminology. When talking to your financial advisor, broker or agent, don't hesitate to ask for an explanation if they use terms or expressions you don't understand.

- American option - Option which the holder may exercise at any time up to and including the expiry date.
- Assignation - An assignation takes place when a holder exercises an option. The option writer receives an exercise notice that obliges him to sell (in the case of a call option) or buy (in the case of a put option) the shares at the stipulated exercise price.
- At-the-money option - When the option exercise price and share price are identical.
- Bear spread - This strategy can be used with both call and put options. In both cases, you buy an option with a higher exercise price and sell an option with a lower exercise price; the two options usually have the same expiry date.
- Bull spread - You buy an option with a lower exercise price and sell an option with a higher exercise price; the two options usually have the same expiry date. You can use call or put options.
- Call - Option contract which gives the holder the right to buy and obliges the writer to sell a specified number of shares at a specified exercise price, any time before the contract expiry date.
- Class of options - All option contracts (call and put options) on the same security.
- Combination - Or straddle. Purchase (or sale) of call and put options on the same stock, with different expiry months or different exercise prices.

- Covered write - The writer of a call option is covered if he holds an equivalent quantity of the underlying security for each option contract he sells or writes.
- European option - Option which the holder can only exercise on the expiry date.
- Exercise price - Or strike price. Specified price at which the holder of an option can buy (call option) or sell (put option) the underlying stock.
- Expiry cycle - Cycle that determines when the options expire. There are five expiry cycles:
  1. Cycle 1: January / April / July / October
  2. Cycle 2: February / May / August / November
  3. Cycle 3: March / June / September / December
  4. Cycle 4: Four months - three consecutive months and the next month from cycle 3 (e.g., December, January, February, March)
  5. Cycle 5: Four months - two consecutive months and the next two months from cycle 1, 2 or 3 (e.g., cycle 5 (3) - December, January, March, June)
- Expiry date - Date on which the option ceases to exist. Options expire at noon on the Saturday following the third Friday of the expiry month.
- In-the-money option - When the exercise price of a call (or put) option is lower (or higher) than the share price.
- Intrinsic value - Positive difference between the stock price and the exercise price of a call option or between the exercise price of a put option and the stock price. By definition, the intrinsic value cannot be negative.
- Last trading day - A business day, usually the third Friday of the expiry month, at 4:00 p.m. (EST/EDT).
- Liquidating operation - Sale of an option you originally bought or purchase of an option you originally sold.
- Margin requirement - Or coverage. The amount or the securities you must deposit with your broker to guarantee purchase or delivery of the shares when you sell an option.

- Naked write - Selling or writing a call (or put) option without holding an equivalent quantity of the underlying security.
- Open interest - Or open position. This is an important concept but it has no direct impact on the price of an option. It is the number of option contracts on a particular share that have not yet been exercised. An opening operation increases the open interest whereas a liquidating operation reduces it. The greater the open interest, the more liquidity is in the market, therefore the easier it is to take or dispose of a position.
- Out-of-the-money option - When the exercise price of a call (or put) option is higher (or lower) than the share price.
- Premium, option premium - Price of the option. Price the buyer pays the seller for the rights associated with the option contract.
- Put - Option contract which gives the holder the right to sell and obliges the writer to buy a specified number of shares at a specified exercise price, any time before the contract expiry date.
- Series of options - All options in the same class that have the same exercise price and same expiry date.
- Straddle - Simultaneous purchase or sale of a call and put option with the same characteristics (same expiry month and same exercise price).
- Time value - Or time premium. Portion of the premium that represents the remaining time until the expiry of the option contract and the fact that the factors which determine the value of the option premium can change during this period. The time value is equal to the difference between the option premium and the intrinsic value. The time value is usually positive and decreases with the passage of time.
- To exercise an option - To exercise an option is the process by which the holder exercises his right to buy (in the case of a call option) or sell (in the case of a put option) according to the terms specified in the contract.
- Volatility - Variability of a stock as measured by the standard deviation of its return. A measure of the tendency of the share price to rise or fall.
- Writer - Or seller. The person who sells or issues an option.

Sources of Exchanges, and Industry Information

Canadian Exchanges and Regulators

Montreal Exchange
URL: http://www.me.org/
Telephone: (514) 871-3582, 1-800-361-5353
Tour de la Bourse
P.O. Box 61
800 Victoria Square
Montreal, Quebec
H4Z 1A9
CANADA

Toronto Stock Exchange
URL: http://www.tse.com/
Telephone: 416-947-4700, 1.888.TSE.8392
Toronto Stock Exchange
The Exchange Tower
130 King Street West
Toronto, ON
M5X 1J2
CANADA

Vancouver Stock Exchange
URL: http://www.vse.com/
Telephone: (604) 689-3334
Stock Exchange Tower
P.O. Box 10333
609 Granville Street
Vancouver, British Columbia
V7Y 1H1
CANADA

Canadian Derivatives Clearing Corporation
URL: http://www.cdcc.ca/
Telephone: (416) 367-2463
120 Adelaide Street West
Suite 1016
Toronto, Ontario
M5H 1T1
CANADA

Alberta Stock Exchange

URL: http://www.ase.ca/
Telephone: 403-974-7400
10th Floor, 300 - 5th Avenue S.W.
Calgary, Alberta
T2P 3C4
CANADA

Winnipeg Commodity Exchange
URL: http://www.wce.mb.ca/
Telephone: 204-925-5008
500 Commodity Exchange Tower
360 Main Street
Winnipeg, Manitoba
R3C 3Z4
CANADA

Investment Funds Institute of Canada
URL: http://www.ific.ca/
Telephone: (416) 363-2158
151 Yonge Street
5th Floor
Toronto, Ontario
M5C 2W7
CANADA

Ontario Securities Commission
URL: http://www.osc.gov.on.ca/
Telephone: (416) 593-8314
E-Mail: inquiries@osc.gov.on.ca
Suite 800, Box 55
20 Queen Street West
Toronto Ontario
M5H 3S8
CANADA

Canadian Investor Protection Fund
URL: http://www.cipf.ca/
Telephone: (416) 866-8366
E-Mail: geoff.peers@cipf.ca
P.O. Box 192
200 Bay Street
Toronto, Ontario
M5J 2J4

CANADA

Canadian Securities Institute
URL: https://www.csi.ca/index.html
Telephone: (416) 364-9130
121 King Street West
Suite 1550
P.O. Box 113
Toronto, Ontario
M5H 3T9
CANADA

US Exchanges and Regulators

American Stock Exchange
URL: http://www.amex.com/
Telephone: 212-306-1000
86 Trinity Place
New York, NY 10006
USA

Arizona Stock Exchange
URL: http://www.azx.com/
Telephone: 602-222-5800
2800 N. Central Avenue
Suite 1575
Phoenix, AZ 85004
USA

Board of Trade Clearing Corporation
URL: http://www.botcc.com/
Telephone: (312) 786-5700
E-Mail: info@botcc.com
141 West Jackson Boulevard
Suite 1460
Chicago IL 60604
USA

Chicago Board of Trade
URL: http://www.cbot.com/
Telephone: 312-435-3500
141 W. Jackson Blvd.
Chicago, IL. 60604-2994

USA

Chicago Board Options Exchange
URL: http://www.cboe.com/
Telephone: 1-800-OPTIONS
Investor Services Department
400 S. LaSalle Street
Chicago, IL 60605
USA

Chicago Mercantile Exchange
URL: http://www.cme.com/
Telephone: 312-930-1000
30 S. Wacker Drive
Chicago, IL 60606
USA

Coffee, Sugar, Cocoa Exchange
URL: http://www.csce.com/
Telephone: 212-742-6000
4 World Trade Center
New York, NY 10048
USA

Chicago Stock Exchange
URL: http://www.chicagostockex.com/
Telephone: 312-663-2980
Fax: 312-663-2396
The Chicago Stock Exchange
440 South LaSalle Street
Chicago, Illinois 60605
USA

GLOBEX
URL: http://www.cme.com/globex2/
Telephone: 312-456-6700
30 S. Wacker Drive, Suite 6N
Chicago, IL 60606
USA

Kansas City Board of Trade
URL: http://www.kcbt.com/
Telephone: 816-753-7500
4800 Main Street, Suite 303

Kansas City, MO 64112
USA

MidAmerican Commodity Exchange
URL: http://www.midam.com/
Telephone: 312-341-3000
141 W. Jackson Blvd.
Chicago, IL 60604
USA

Minneapolis Grain Exchange
URL: http://www.gromedia.com/mge/mge.html
Telephone: 612-338-6212
400 S. Fourth Street
Minneapolis, MN 55415
USA

NASD Regulation Inc.
URL: http://www.nasdr.com/1000.asp
Telephone: 202-728-6954
1735 K Street NW
Washington, DC 20006-1500
USA

NASDAQ
URL: http://www.nasdaq.com/welcome.htm
NASD MediaSource
P.O. Box 9403
Gaithersburg, MD 20898-9403
USA

New York Mercantile Exchange
URL: http://www.nymex.com/
Telephone: 212-299-20004
One North End Avenue
World Financial Center
New York, NY 10282-1101
USA

New York Stock Exchange
URL: http://www.nyse.com/
Telephone: 212-656-3000
11 Wall Street
New York, NY 10005

USA

New York Cotton Exchange
URL: http://www.nyce.com
Telephone: 212-742-5050
4 World Trade Center
New York, NY 10048
USA

Securities and Exchange Commision
URL: http://www.sec.gov/
Telephone: (202) 942-7040
Office of Investor Education and Assistance,
Securities and Exchange Commision
450 Fifth Street, N.W.
Washington, D.C. 20549
USA

Philadelphia Stock Exchange
URL: http://www.phlx.com/
Telephone: 1-800-843-7459
1900 Market Street
Philadelphia, PA 19103-3584
USA

Pacific Stock Exchange
URL: http://www.pacificex.com/index.html
Telephone: 415-393-4000
301 Pine Stret
San Francisco, CA 94104
USA

Commodity Futures Trading Commision
URL: http://www.cftc.gov/index.htm
Telephone: (202)418-5000
2033 K St., N.W.,
Washington, DC 20581
USA

Futures Industry Association
URL: http://www.fiafii.org/
2001 Pennsylvania Ave., N.W., Suite 600,
Washington, DC 20006
USA

International Swaps and Derivatives Association
URL: HTTP://www.isda.org/
Telephone: 212-332-1200
600 5th Avenue, 27th Fl
New York NY 10020
USA

National Futures Association
URL: http://www.nfa.futures.org/
Telephone: 312-781-1410
200 W. Madison St., Suite 1600
Chicago, IL 60606-3447
USA

International Exchanges

Athens Stock Exchange
URL: http://www.ase.gr/

Australian Stock Exchange
URL: http://www.asx.com.au/

Amsterdam Stock Exchange
URL: http://www.aex.nl/

Budapest Commodity Exchange
URL: http://www.bce-bat.com/indexen.html

Deutsche Terminbörse
URL: http://www.exchange.de/index.html

Hong Kong Futures Exchange
URL: http://www.hkfe.com/

Italian Stock Exchange
URL: http://robot1.texnet.it/finanza/

Kuala Lampur Stock Exchange
URL: http://www.klse.com.my/

Lisbon Stock Exchange
URL: http://www.bvl.pt/

London Metal Exchange
URL: http://www.lme.co.uk/

London Stock Exchange
URL: http://www.londonstockex.co.uk/

The London International Financial Futures and Options Exchange
URL: http://www.liffe.com/

Madrid Stock Exchange
URL: http://www.bolsamadrid.es/

MATIF (Marché à Terme International de France)
URL: http://www.matif.fr/

New Zealand Futures & Options Exchange Ltd.
URL: http://www.nzfoe.co.nz/

Singapore International Monetary Exchange Ltd.
URL: http://www.simex.com/

South African Futures Exchange
URL: http://www.safex.co.za/

MEFF Renta Fija (Spanish Financial Futures and Options Exchange)
URL: http://www.meff.es/

Sydney Futures Exchange
URL: http://www.sfe.com.au/Presentation/home/

Tel-Aviv
URL: http://www.tase.co.il/index.htm

Tokyo Grain Eschange
URL: http://www.tge.or.jp/

Tokyo International Financial Futures Exchange
URL: http://www.tiffe.or.jp/

Tokyo Stock Exchange
URL: http://www.tse.or.jp/

Warsaw Stock Exchange
URL: http://bimbo.ippt.gov.pl/gielda/gielda.html

International Organization of Securities Commissions
URL: http://www.iosco.org/iosco

Chapter 5

Risk-Neutral Probability and the SDF

This Chapter begins with an interpretation which will highlight the similarities of a realistic model to the simple[1] model of Chapter 1. The state space of the realistic model, $\Omega = [0, \infty)$, is infinite. The model of Chapter 1 was discrete: there were only n (a finite number) states of nature in $\Omega = \{\omega_1, \ldots, \omega_n\}$. In fact, in Chapter 1, n was equal to three. The interpretation presented here will prove to be key to the understanding of the option pricing mechanism.[2] In this Chapter we apply the intuition developed in the previous chapters. Parallels are drawn between the finite and the infinite case, and we sidestep some of the technical issues to be dealt with later.

Reality reminds us that security prices (and derivative prices) can take on more than just a finite number of possible values. Even so, the essential guidelines of the valuation method described in Chapter 1 remain intact for the realistic case. The no-arbitrage condition is equivalent to the existence of a set of stochastic discount factors. These factors facilitate the valuation of risky (and risk-free) cash flows, as in equation (1.9), or in terms of expected value,[3] as in Theorem 3 of Chapter 1. There is, though, a significant difference which we would like to highlight now.

In Chapter 1, Ω had only a finite number of states and there were a finite number of securities whose prices were known. Consequently, we could solve for the stochastic discount factors, as in equation (1.6). These factors were

[1] See [22] and [25].

[2] In later chapters we will take a closer look at this mechanism.

[3] One only needs to be comfortable with interchanging the Σ notation of the finite case with the $\int$ notation for the infinite case. This should be very natural for readers who are familiar with calculus or with an engineering background.

implicit in the market prices of the securities: the solution to equation (1.6) determines[4] the stochastic discount factors.

In the infinite case there are an infinite number of states in Ω and therefore it is harder to extract the stochastic discount factors from the prices of securities in the market. In this Chapter we will assume the form of the stochastic discount factors, and will investigate their properties. We will compare the properties of the stochastic discount factors from the realistic case (infinitely many possible states of nature) to the properties we already know of the stochastic discount factors of the simple model (finitely many possible states of nature). The comparison will serve to generate a pricing formula. Subsequent chapters will take a more in-depth[5] look at the assumption made here regarding the form of the stochastic discount factors in the infinite case.

5.1 Infinite vs. Finite States of Nature

In Chapter 4 we used the interpretation of building blocks. The payoffs from call options, put options, and the underlying stock were combined in various ways to design different desired payoff patterns. Chapter 1 investigated an example of three primary securities $S1$, $S2$, and SF in a market with three states of nature. The conclusion reached in Chapter 1 was that the payoffs from the primary securities were actually made from the same elementary building blocks. Each security was shown to be composed from the same elementary contingent cash flows, albeit in different quantities.

The building blocks of Chapter 1 were claims of \$1 contingent on each possible state of nature; e.g., $c_1 = (1, 0, 0)$ was a \$1 cash flow contingent on state 1 occurring. Furthermore, the no-arbitrage condition was shown to be equivalent to the existence of a set of (positive) prices, d_1, d_2, and d_3, for the contingent cash flows. These d's were the stochastic discount factors of

[4] In a complete market this equation *uniquely* determines the set of discount factors.

[5] The assumptions made here will be only partly justified in this chapter, but in Chapter 15 a model leading to these assumptions will be investigated in a more rigorous way. Recent trends in the finance literature (see [11] and [46]) explore the possibility of extracting the stochastic discount factors from market prices of securities. In some cases, this requires a discretization of reality, even though the resultant system of equations is underdetermined. Recall that in the finite case, each possible state of nature introduced a variable in the system of equations — the stochastic discount factor for that state of nature. The number of equations is equal to the number of securities whose payoff depends on these states. Consequently, a solution is not uniquely specified for the case where there are allowed to be infinitely many possible states of nature. One thus has to employ a selection criterion, for example, minimization of the relative entropy function.

Chapter 1. The discount factor d_1 was the present value (price in the current time period) of the elementary cash flow $c_1 = (1, 0, 0) - \$1$ contingent claim on state 1. The values of the stochastic discount factors were implicit in the prices of the primary securities as in equation (1.6). Once recovered, these factors could be used to value every cash flow in the market (see equation (1.9)).

In the same manner, in Figure 4.8, a stock could be stripped to its basic components: cash, a call, and a put on the same stock with the same strike price. The payoffs of call and put options are contingent on the value of the underlying stock. The value of the stock thus plays the role of the state of nature and every point in $\Omega = [0, \infty)$ is a possible state of nature. In an analogous manner to a case where Ω is finite, an infinite collection of elementary state-contingent claims can be defined. Each such claim pays \$1 if state s occurs in the next time period, and \$0 otherwise; i.e., it pays \$1 if the price of the stock is s in the next time period. Visually, a state s elementary contingent claim is depicted by a graph which has a value of \$1 at the point s, and 0 elsewhere. A contingent claim which pays $\$s$ if state s occurs can be generated by buying s units of \$1 contingent claim on state s. See Figure 5.1.

```
> plot( s,s=0..100,style=point,title='Figure 5.1:\
> A Contingent Claim Which Pays $s in State s',\
> titlefont=[TIMES,BOLD,10]);
```

Consider a portfolio composed of an infinite collection of these contingent claims. The portfolio includes s contingent claims for one dollar in state s, for every s in $[0, \infty)$. This generates an asset that pays $\$s$ if state s occurs, and, since the state of nature is the value of the stock, the asset is a reconstruction of the stock. This is depicted in Figure 5.1. It is much the same as we thought about security 1 in Chapter 1 that paid \$50 if state 1 occurred, \$60 if state 2 occurred, and \$80 if state 3 occurred. There, security 1 was composed of 50 units of the \$1 state 1 contingent claim, 60 units of the \$1 state 2 contingent claim, and 80 units of the \$1 state 3 contingent claim. Analogously, a call option with a strike price of K is also an infinite collection of elementary contingent claims, i.e., $s - K$ units of the \$1 state-s contingent claim for every state s for which it is the case that $s > K$. This is depicted in Figure 5.2 for $K = 60$.

```
> plot(Call(s,60),s=0..100,style=point,title='Figure 5.2:\
> A Contingent Claim Which Pays $s-60 in State s, s>60',\
> titlefont=[TIMES,BOLD,10]);
```

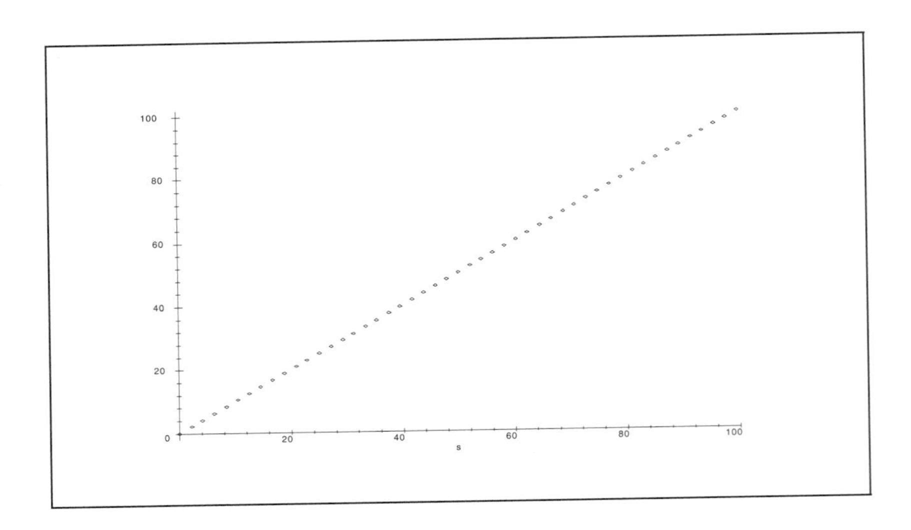

Figure 5.1: A Contingent Claim Which Pays $\$s$ in State s

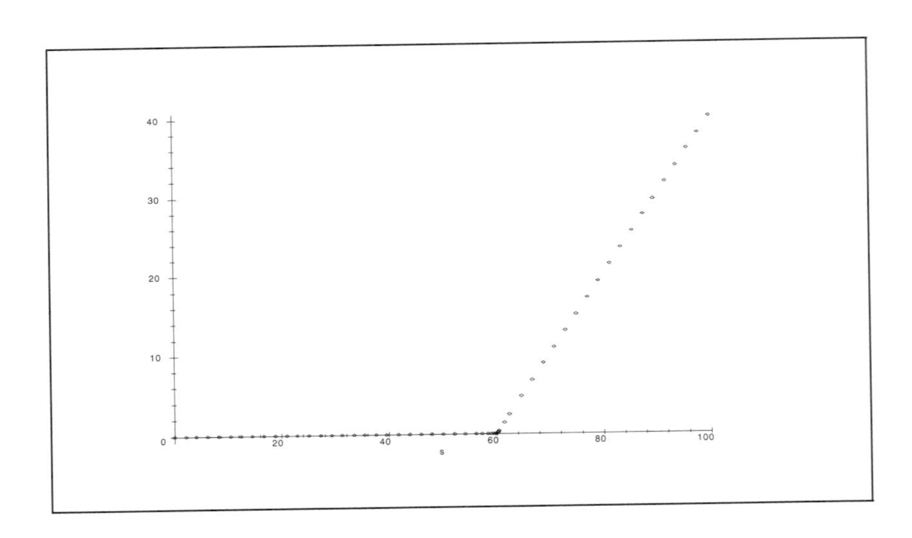

Figure 5.2: A Contingent Claim Which Pays $\$s - 60$ in State s, $s > 60$

5.2 SDF for an Infinite Ω

We might anticipate, as was true in Chapter 1, that the absence of arbitrage opportunities in the market implies (even for an infinite Ω) the existence of stochastic discount factors. Here we should have an infinite set of stochastic discount factors: for each s in $[0, \infty)$ there should be a discount factor. Thus the stochastic discount factors are represented as a function, much like the payoff was a function in the model of Chapter 4, where Ω was also infinite.

Let us denote by $\psi_t(s)$ the stochastic discount factor for state s occurring in t units of time. This function must depend on the length of the time period. As in the discrete case, these factors allow valuation of risky future cash flows. They account for both the time value of money and the riskiness of the cash flow which is being valued. In Chapter 1, the length of time was fixed and was thus suppressed from the discussion. Here, payoffs for securities which mature in any future time period t are investigated and hence $\psi_t(\cdot)$ depends on the length of the time period. Furthermore, we would like to add two assumptions (and fully justify them later[6]) about $\psi_t(\cdot)$:

- The stochastic discount factor for state s (recall that s is a future possible value of a certain stock) depends on the current conditions only through the current price of the stock. Thus the current price of the stock, in a sense, summarizes all relevant information for the future value of the stock. This assumption is related to the concept of "market efficiency," which is dealt with extensively in the literature, and the current price is actually a parameter[7] of $\psi_t(\cdot)$ much the same as t is.

- The dependence of $\psi_t(\cdot)$ on time is only through the length of the interval. Thus, if at time $t_1 \neq t_2$ market conditions are the same, i.e.,

[6] This will be further explored when we investigate, (in Sections 5.4 and 15.2), the price behavior of stocks in the market. See also [23] and [43].

[7] For an elaborate discussion of market efficiency see, for example, [23] and [24]. Indeed we should have used the notation $\psi_{t,P}(s)$, but to remain less cumbersome we have kept our notation $\psi_t(s)$. If indeed all relevant information is summarized in the current price, P, of the stock, then provided $P_{t_1} = P_{t_2}$, the future as seen from time t_1 should be indistinguishable from the future as seen from time t_2. Put differently, the risk-neutral density of the future price of the stock as of time t_1 should equal that of time t_2. Recall that the risk-neutral density is obtained by a normalization of the stochastic discount factors. Hence the discount factors as of time t_1 equal those as of time t_2 and this leads to the second assumption made here. This argument implicitly assumes that market efficiency is defined with respect to the risk-neutral density.

$P_{t_1} = P_{t_2}$, then the stochastic discount factors for time $t_1 + t$, as of time t_1, equal those of time $t_2 + t$ as of time t_2, namely, $\psi_{(t_1+t)-t_1}(s) = \psi_{(t_2+t)-t_2}(s)$.

For simplicity let us refer to the current time period t_0 as zero, so the future time $t_0 + t$ is t and the length of time until the realization of the state of nature in question is also t. Therefore, $\psi_t(s)$ is the current value of a $1 state -$s$ contingent claim (occurring at time t). It is the counterpart of d_j, the stochastic discount factor of state j in the model of Chapter 1.

Consider an asset that pays $1 at time t, regardless of which state of nature occurs at time t. Such an asset is risk-free. In the model of Chapter 1, the price of such an asset was given by equation (1.8), i.e., by $\sum_{j=1}^{n} \$1 \times d_j$. The price of such an asset must equal the present value of $1 to be received in the next time period; otherwise arbitrage opportunities exist. Thus we must have

$$\sum_{j=1}^{n} d_j = (1 + r_f)^{-1}, \tag{5.1}$$

where r_f is the risk-free rate spanning that period of time.

In the current model, to find the price of $1 to be received in t units of time, regardless of the state of nature, a similar operation is performed. The discount factor $\psi_t(s)$ for state s plays the same role as d_j for state j. In the discrete time case, d_j is strictly positive, being the present value of $1 to be received in the future. Also here $\psi_t(s)$ should be strictly positive for every s. However, to find the price of the contingent claim, $1 in state s for every s, we need to sum over all s in $[0, \infty)$ the product

$$1 \times \psi_t(s). \tag{5.2}$$

Infinite summation leads to the integral notation.[8] The integral $\int$ is the infinite dimensional counterpart of the summation Σ of the finite case.

[8] Recall that if the (Riemann) integral of the function $\psi_t(\cdot)$ exists, it is given in terms of the limit of a sum. Consider for a moment the $1 claim contingent on s being one of the states in $[0, M]$. Divide the interval $[0, M]$ into N equal subintervals, each of which has a width of $\frac{M}{N} = \Delta_N s$. (A generalized definition of an integral is obtained when $\Delta_N s$, the length of the interval, is replaced with some function of the length.) Each such interval represents a collection of states of nature, e.g., the first such interval is $[0, \frac{M}{N}]$, the second is $[\frac{M}{N}, 2\frac{M}{N}]$, etc. Let s_j be the left point of the jth interval. As N increases to infinity, $\Delta_N s \to 0$, each interval shrinks in width, and there will be just one state of nature in the limit. If $\psi_t(s)$ is the discount factor for state s, the discount factor for the jth subinterval can be approximated by $d_j(N) = \psi_t(s_j)\Delta_N s$ (by assuming that $\psi_t(s)$ is constant over the jth interval being equal to $\psi_t(s_j)$). The value of this claim can be approximated by $\sum_{j=1}^{N} d_j(N) \times 1$. When N increases to infinity, the limit of the last expression is the

Therefore, the price of an asset paying \$1 contingent on state s occurring, for every s in $[0, \infty)$, must be

$$\int_0^\infty 1 \times \psi_t(s)ds. \tag{5.3}$$

By the same argument, as in the finite case, the price of such a risk-free asset must be the present value of \$1 to be received at time t. Since we are using the continuously compounded rate of interest r_f, we obtain the following equation:

$$\int_0^\infty 1 \times \psi_t(s)ds = 1 \times e^{-r_f t}. \tag{5.4}$$

Following the same guidelines which were applied in the discrete time model of Chapter 1, the price of the stock should be the discounted value of its future cash flows. The stochastic discount factors take care of both time-value-of-money considerations and the riskiness of the cash flows. Since a stock is a risky asset, the discounting should be done using the stochastic discount factors. In Chapter 1, the equation for a price of a stock paying $\$a_j$ in state j was

$$\sum_{j=1}^{n} a_j d_j. \tag{5.5}$$

In this model the same relation should hold when the $\sum$ is replaced with $\int$ and the d_j are replaced with their infinite dimensional counterparts, the discount factors $\psi_t(s)$. Since the stock pays $\$s$ in state s, a_j is replaced with s and we thus obtain the following expression, which the price of the stock at time $t_0 = 0$, P, should satisfy:

$$P = \int_0^\infty s\psi_t(s)ds. \tag{5.6}$$

Extending the logic of equation (5.6) to the pricing of other assets yields that the price of an asset paying $h(s)$ in state s, for some (integrable) function $h(\cdot)$, should be

$$\int_0^\infty h(s)\psi_t(s)ds. \tag{5.7}$$

---

integral, namely, $\lim_{N\to\infty} \sum_{j=1}^{M} d_j(N) = \int_0^M \psi_t(s)ds$. (In a more general setting, if the length of the interval is replaced by a function of it, say α, the limit above is denoted by $\int_0^M \psi_t(s)d\alpha$.) Therefore, one may interpret $\Delta_N s$ as converging to ds, the "length" of a point, and thus $\psi_t(s_j)$ is replaced with $\psi_t(s)$ while Σ is replaced with $\int$. When the upper limit of the integral is ∞ it is calculated as $\lim_{M\to\infty} \int_0^M \psi_t(s)ds$.

Consequently, the price of a call option with a strike price K maturing in t units of time should be

$$\int_0^\infty \max(0, s - K)\, \psi_t(s) ds. \tag{5.8}$$

Since

$$\max(0, s-K) = \begin{cases} 0 & s < K \\ s - K & s \geq K, \end{cases}$$

we need to integrate only over the nonzero part of $\max(0, s - K)$. Equation (5.8) yields that $P_{Call_t(K)}$, the price of a call option with a strike price of K maturing in t units of time, satisfies

$$P_{Call_t(K)} = \int_K^\infty (s - K)\psi_t(s) ds. \tag{5.9}$$

One should note however that, although suppressed here, the discount factor $\psi_t(\cdot)$ is stock dependent. That is, each stock[9] induces its own stochastic discount factors. In the model of Chapter 1 the stochastic discount factors, the d_j's, were defined over the states of nature of the economy. Thus, they were applicable to valuing any asset in that economy. In contrast, in the present model, the states of nature are defined as the prices of the stock. Consequently, the stochastic discount factors in the current model could be applied only to valuation of claims which are contingent on the prices of this stock.

Consequently, the stochastic discount factors in the current model are applicable only to calculating the values of claims which are contingent on the prices of this stock.

While the analysis above pertains to stock options, it applies to other contingent claims as well. Equation (5.7) applies to any derivative security with payoff $h(s)$ contingent on the value of some underlying asset being s. Thus, for example, the derivative security might be an option to buy a foreign currency for a specific exercise price. In this case, $\psi_t(\cdot)$ will be the stochastic discount factor implied by the exchange rates, but the valuation equation (5.7) stays intact.

It is now apparent that the missing part needed to price derivative securities is the specification of $\psi_t(\cdot)$. However, before the specification is given we would like to explore some properties of $\psi_t(\cdot)$ and their connection to the risk-neutral probability.

[9] In more general cases a discount factor may be induced by each underlying process on which the claim is contingent, i.e., foreign exchange, interest rate, etc.

5.3 Risk-Neutral Probability and the SDF

In the model of Chapter 1, the stochastic discount factors were normalized to generate the risk-neutral probability. In Chapter 1 we normalized d_j, dividing it by $\sum_{j=1}^{n} d_j$ to generate q_j, the risk-neutral probability of state j. The counterparts of d_j and $\sum_{j=1}^{n} d_j$ are $\psi_t(s)$ and $\int_0^\infty \psi_t(s)ds$, respectively.

We have already mentioned that $\psi_t(s)$ is positive, since it is the present value of \$1 contingent on state s. Thus, both $\int_0^\infty \psi_t(s)ds$ and

$$\frac{\psi_t(s)}{\int_0^\infty \psi_t(s)ds} \tag{5.10}$$

must be positive. Consequently,

$$\int_0^\infty \frac{\psi_t(s)}{\int_0^\infty \psi_t(s)ds} ds = \frac{1}{\int_0^\infty \psi_t(s)ds} \int_0^\infty \psi_t(s)ds = 1, \tag{5.11}$$

exactly as

$$\sum_{j=1}^{n} \frac{d_j}{\sum_{j=1}^{n} d_j} = 1. \tag{5.12}$$

The counterpart of $q_j = \frac{d_j}{\sum_{j=1}^{n} d_j}$ is the function

$$f_t(s) = \frac{\psi_t(s)}{\int_0^\infty \psi_t(v)dv}. \tag{5.13}$$

Since

$$f_t(s) \geq 0 \tag{5.14}$$

and from equations (5.11) and (5.13) it follows that

$$\int_0^\infty f_t(s)ds = 1, \tag{5.15}$$

$f_t(s)$ can be interpreted as a density function. The counterpart of q_j, the risk-neutral probability for state j, is the value of the density function at a state s, $f_t(s)$.

By equation (5.4), $\int_0^\infty \psi_t(s)ds = e^{-r_f t}$ and we can therefore substitute $e^{-r_f t}$ for the integral in equation (5.13) to yield

$$f_t(s) = \frac{\psi_t(s)}{e^{-r_f t}}. \tag{5.16}$$

Indeed this is the same relation that held in Chapter 1, $q_j = \frac{d_j}{(1+r_f)^{-1}}$. Hence, multiplying equation (5.6) by $e^{r_f t}$ yields

$$e^{r_f t} P = \int_0^\infty s f_t(s) ds, \tag{5.17}$$

where s are the possible realizations in $[0, \infty)$ of the value of the stock at time t. Equivalently, equation (5.17) can be written as

$$P = e^{-r_f t} E_{f_t}(S_t), \tag{5.18}$$

where S_t, a random variable, the value of the stock at time t, and the expected value (the E in the above equation) is taken with respect to the risk-neutral density function $f_t(\cdot)$. The right-hand side of equation (5.18) is the counterpart of equation (1.15) of Chapter 1. Both these equations convey the same message: the price of a stock is its expected value, under the risk-neutral probability, discounted by the risk-free rate.

We repeat these calculations in a MAPLE session. Consider a function $\psi(\cdot)$ satisfying

```
> int(psi(s),s=0..infinity)=exp(-r*t);
```

$$\int_0^\infty \psi(s)\, ds = e^{(-rt)}$$

We know the stochastic discount factors should satisfy the above equation. Multiplying it by e^{rt}

```
> %*exp(r*t);
```

$$e^{rt} \int_0^\infty \psi(s)\, ds = e^{rt}\, e^{(-rt)}$$

and then simplifying the above expression we obtain

```
> simplify(%);
```

$$e^{rt} \int_0^\infty \psi(s)\, ds = 1$$

Substitution is used to define $f(\cdot)$,

```
> subs(psi(s)=f(s)*exp(-r*t),%);
```

$$e^{rt} \int_0^\infty \mathrm{f}(s)\, e^{(-rt)}\, ds = 1$$

which indeed shows that $\int_0^\infty f_t(s) ds = 1$, and thus that $f(\cdot)$ is a density function.

5.4 A First Look at Stock Prices

As was mentioned in the discussion after equation (5.9), only the specification of $\psi_t(\cdot)$ is needed to be applied to equation (5.7) in order to price an option. Alternatively, the specification of the risk-neutral probability, $f_t(\cdot)$, may be given, as one induces the other as in equation (5.13). In terms of $f_t(\cdot)$ the price of a derivative security paying $h(s)$, contingent on the underlying asset (or the stock in the case of an equity option) having the value s, in t units of time, is

$$e^{-r_f t} \int_0^\infty h(s) f_t(s) ds. \tag{5.19}$$

We have investigated the requirements $\psi_t(\cdot)$ should satisfy in order for it to behave like the stochastic discount factors (d_j's) of Chapter 1. This does not determine, though, either $\psi_t(\cdot)$ or $f_t(\cdot)$. Given the price of a stock P, it is possible to adopt any function, say $g(\cdot)$, satisfying equations (5.4) and (5.6) and scale it such that it satisfies equation (5.17). Some assumptions must therefore be made about the price behavior of stocks in the market.

To characterize the behavior of stock prices, as in [23], for example, we would like to see what can be said about the rate of return of stocks. Dividing equation (5.17) by P we obtain

$$e^{r_f t} = \int_0^\infty \frac{s}{P} f_t(s) ds, \tag{5.20}$$

or equivalently

$$e^{r_f t} = E_{f_t}\left(\frac{S_t}{P}\right). \tag{5.21}$$

Substituting $S_t = P + (S_t - P)$ in equation (5.21) yields

$$e^{r_f t} = 1 + E_{f_t}\left(\frac{S_t - P}{P}\right). \tag{5.22}$$

The random variable

$$\frac{S_t - P}{P} \tag{5.23}$$

is the rate of return of the stock over the period $[0, t]$. Equation (5.22) speaks about the expected value of this variable. It says that under the risk-neutral probability, the expected return of each stock in the market, over a period of time of length t, is the same. Moreover, it is equal to the risk-free rate return. Equation (5.21) is the equivalent of equation (1.17) in Chapter 1 and the intuitive explanation is as that offered in Chapter 1. In a market populated with risk-neutral investors, equilibrium can prevail only if the expected return of every security in the market is the same.

The caution voiced in Chapter 1 is worth repeating here. There is no claim made that the market is populated only with risk-neutral investors nor that the density function (probability distribution) of prices is f_t. Rather it is claimed[10] that the absence of arbitrage opportunities in the market is equivalent to the existence of a density function f_t, satisfying equation (5.18).

Our Chapter 3 discussion of continuous compounding concluded that (see Section (3.4.1)) if the rate of return over a unit of time is r, the instantaneous (continuously compounded) rate of return[11] is $\ln(1+r)$. The right-hand side of equation (5.21), $E_{f_t}\left(\frac{S_t}{P}\right)$, for $t = 1$, is the expected value of one plus the rate of return over a unit of time. Thus taking the natural logarithm of both sides of equation (5.21) yields that, under a risk-neutral probability, the expected (instantaneous) rate of return of the stock over a unit of time is r_f.

The instantaneous return on a stock, however, as opposed to the instantaneous risk-free rate, is not deterministic. Rather it evolves through time in a stochastic manner. For this reason S_t is a random variable and so $\frac{S_t}{P}$ will also be a random variable. Hence, if we omit the E in $E_{f_t}\left(\frac{S_t}{P}\right) = e^{r_f t}$, the right-hand side of this equation, $e^{r_f t}$, should also be a random variable.

We might therefore expect that in this case $r_f t$ might be replaced with a random variable. *This random variable is the continuously compounded rate of return* of the stock over the period $[0, t]$ and thus **depends** on t. If we denote this random variable by Y_t, replace $r_f t$ with Y_t, and omit the expectation operator (the E) in equation (5.22), we arrive at

$$e^{Y_t} = 1 + \frac{S_t - P}{P} = \frac{S_t}{P}. \tag{5.24}$$

In equation (5.24) the random variable Y_t is the counterpart of $r_f t$. The expression $r_f t$ is a linear function of t, while the dependence of Y_t on t is soon to be specified. Equation (5.24) also implies that $S_t = Pe^{Y_t}$, and hence that the possible realizations of the price of the stock at time t are the interval $[0, \infty)$. This is consistent with the fact that the price of the stock cannot be negative (if it were it would violate the limited liability assumed

[10] The claim is not proven here for the continuous case. See [25] for details.

[11] From now on the term *return* may mean either the (instantaneous) *continuously compounded return* or the simple return. It will be clear from the context which one is appropriate.

by stockholders) but has no upper bound. Indeed for every value of Y_t, e^{Y_t} is positive and approaches zero when Y_t approaches negative infinity.

Given a random variable Y, with expected value μ_Y and standard deviation σ_Y, a random variable Z with $\mu_Z = 0$ and $\sigma_Z = 1$ may be defined by $Z = \frac{Y-\mu_Y}{\sigma_Y}$. Thus, each random variable Y may be written as $Y = \mu_Y + \sigma_Y Z$, where Z is a random variable having the same distribution as Y but with[12] $\mu_Z = 0$ and $\sigma_Z = 1$. Equation (5.24) may therefore be written as

$$e^{\mu_{Y_t} + \sigma_{Y_t} Z} = 1 + \frac{S_t - P}{P}. \tag{5.25}$$

As explained above, the implications of the no-arbitrage conditions in this continuous setting do not uniquely determine a risk-neutral probability (or equivalently a stochastic discount factor, $\psi_t(\cdot)$). Some assumptions are therefore needed in order to choose a risk-neutral probability, which can be used for pricing as in equation (5.6). It is at this point that we make the following assumptions[13] on the "real-life" distribution of Y_t, from which the risk-neutral probability is determined.

Assumption I: Y_t has the familiar normal distribution (bell curve) for all t.

Assumption II: Equation (5.25) and Assumption I imply that the dependence of Y_t on t may be expressed through its variance and expected value. We thus assume that the expected value of Y_t is μt and its variance is $\sigma^2 t$, for some scalars μ and σ.

Assumption III: Y_t depends on t only via the length of the time interval $[0, t]$. Thus, the rates of return, over two disjoint time intervals having equal length, will be identically distributed independent random variables.

Y_t so defined is a stochastic process. A stochastic process is a family of random variables indexed by some parameter, commonly time. This is indeed the case in our situation: for each t we have a random variable Y_t — the continuously compounded rate of return from time 0 to time t.

---

[12] Applying the properties of the expected value and the variance yields $E(Z) = E\left(\frac{Y-\mu_Y}{\sigma_Y}\right) = E\left(\frac{Y}{\sigma_Y}\right) - E\left(\frac{\mu_Y}{\sigma_Y}\right) = \frac{\mu_Y}{\sigma_Y} - \frac{\mu_Y}{\sigma_Y} = 0$, and $Var\left(\frac{Y-\mu_Y}{\sigma_Y}\right) = Var\left(\frac{Y}{\sigma_Y}\right) = \frac{1}{\sigma_Y^2} Var(Y) = \frac{\sigma_Y^2}{\sigma_Y^2} = 1$, where Var stands for variance.

[13] These assumptions are consistent with those made for $\psi(\cdot)$ in the beginning of the preceding Section. In a later Capter we will review and justify these assumptions.

The stochastic process Y_t is called a Brownian motion or sometimes also a Wiener process. In fact, the classical definition of a standard Brownian motion also stipulates that $\sigma = 1$ and that $Y_0 = 0$ with probability one. The next two sections explore some properties of Brownian motions. We shall return to look at this process yet again in Chapter 15. Readers who are interested in further exploration of Brownian motions can consult other books, e.g., [40] and [28]. The rest of this Chapter investigates Y_t, the paths of the price process, and the $\psi_t(\cdot)$ and $f_t(\cdot)$ that are implied by the above assumptions. The next Chapter will utilize the determined $\psi_t(\cdot)$ and $f_t(\cdot)$ to price European options based on equation (5.6).

5.5 The Distribution of the Rate of Return

Let us utilize MAPLE to visualize some consequences of our assumptions about Y_t. The normal density function with expected value of μ and standard deviation of σ at a point x is given by

$$\Phi_{\mu,\sigma}(x) = \frac{e^{-\frac{(x-\mu)^2}{2\sigma^2}}}{\sqrt{2\pi\sigma^2}}. \tag{5.26}$$

We defined this function in MAPLE and called it **Normalpdf.**

```
>   Normalpdf:=(x,mu,sigma)->\
>   exp(-(x-mu)^2/2/(sigma)^2 )/sqrt(2*Pi*(sigma)^2);
```

$$\mathit{Normalpdf} := (x,\, \mu,\, \sigma) \to \frac{e^{(-\frac{1}{2}\frac{(x-\mu)^2}{\sigma^2})}}{\sqrt{2\pi\sigma^2}}$$

In our case, the random variable Y_t (we will omit the subscript t in the Maple calculation) is the continuously compounded rate of return for a period of length t. We assumed that Y_t is normally distributed such that its expected value and standard deviation are dependent on time, and are given by μt and $\sigma\sqrt{t}$ for some constant μ and σ . Thus the density function of Y_t at a point y is given by

```
>   Normalpdf(y,mu*t,sigma*sqrt(t));
```

$$\frac{1}{2}\frac{e^{(-\frac{1}{2}\frac{(y-\mu t)^2}{\sigma^2 t})}\sqrt{2}}{\sqrt{\pi\sigma^2 t}}$$

When $t = 2$, $\mu = 0.15$, and $\sigma = 0.23$ the value of **Normalpdf** at a point y can be calculated using the MAPLE command.

```
> Normalpdf(y,.15*2,.23*sqrt(2));
```

$$2.173913043 \frac{e^{(-4.725897920\,(y-.30)^2)}}{\sqrt{\pi}}$$

For a specific value of y, say $y = 0.15 \times 2$, y is simply replaced with a numerical value:

```
> Normalpdf(.15*2,.15*2,.23*sqrt(2));
```

$$\frac{2.173913043}{\sqrt{\pi}}$$

To get MAPLE to evaluate it in a decimal form we utilize the **evalf** command:

```
> evalf(Normalpdf(.15*2,.15*2,.23*sqrt(2)));
```

$$1.226499094$$

Let us look at Figure 5.3, the density function of Y_2 where $\mu = 0.15$ and $\sigma = 0.23$.

```
> plot(Normalpdf(y,.15*2,.23*sqrt(2)),y=-3..4,\
> title='Figure 5.3:  The Normal Density Function)',\
> titlefont=[TIMES, BOLD, 10]);
```

This is the famous bell-shaped graph of the normal density function.

In order to visualize the effect of time on Y_t we run the animation below. The animation can be played to demonstrate how this graph changes as the time interval t shrinks[14] from 6 to 0.1. To ensure that the animation is run in the right direction click the arrow (<−) in the animation bar. Figure 5.4 presents a static graph of the distribution for $t = 1$, $t = 3$, and $t = 5$ in the same plane.

```
> plots[animate](Normalpdf(Y,.15*t,.23*sqrt(t)),Y=-3..5,\
> t=0.1..6,color=green,title='Figure 5.4:  The Density \
> Function of Yt as t Approaches Zero');
```

[14]In this animation the t parameter, time, shrinks from 6 to 0.1. Later we will speak about the situation where t approches zero.

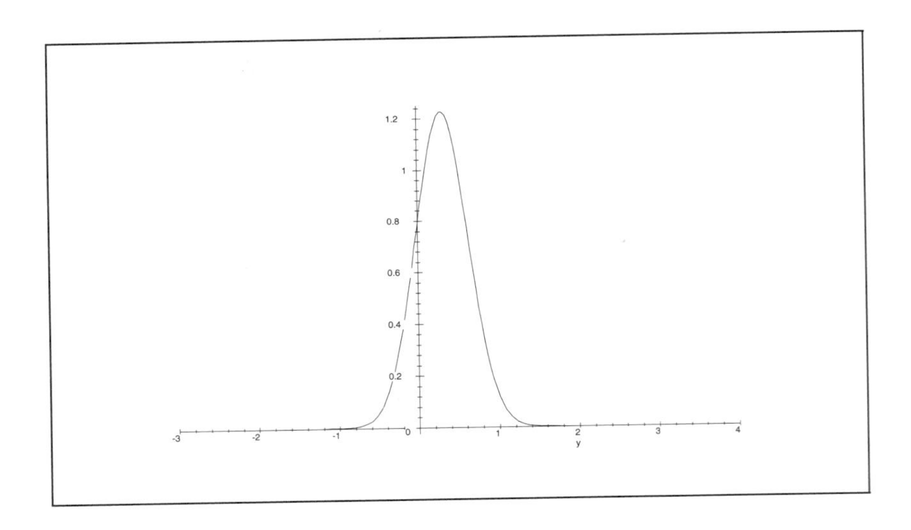

Figure 5.3: The Normal Density Function with $\mu = 0.15 \times 2$ and $\sigma = 0.23\sqrt{2}$

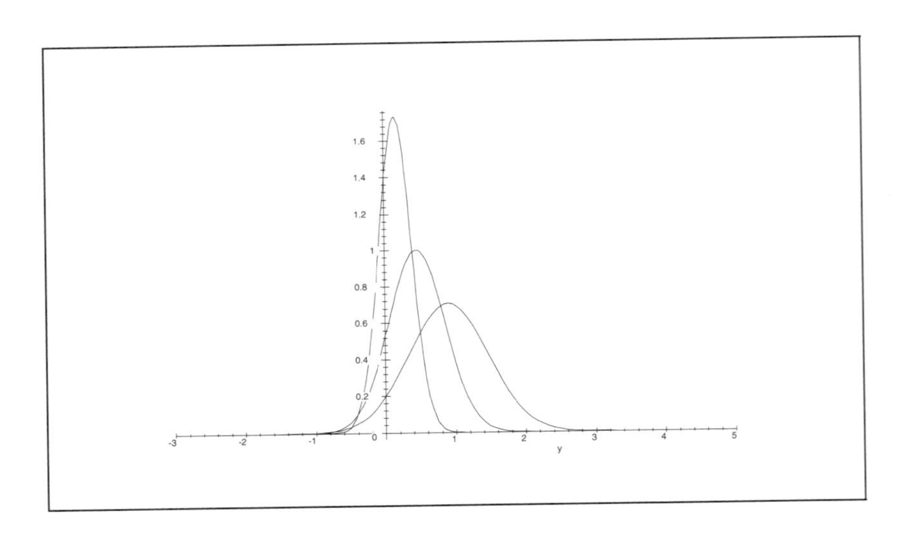

Figure 5.4: The Density Function of Y_t at $t = 1$, $t = 3$, and $t = 5$

Recall that the normal distribution has a peak at its expected value, at $y = \mu t$ in our case, and that the larger the standard deviation, $\sigma\sqrt{t}$ in our case, the flatter the graph and the riskier the situation. The area under the graph between two points y_0 and y_1 is the probability that $y_0 \leq Y_t \leq y_1$. Hence, as t decreases, $\sigma\sqrt{t}$ decreases, the graph gets steeper, and the probability of Y_t being close to its mean, μt, increases. Of course μt is also getting smaller as t is getting smaller; thus the peak of the graph is getting close to zero. Hence the area under the graph, the mass of the probability, is mostly concentrated close to zero. Thus, as t approaches zero the probability of Y_t being equal to zero approaches one.

The same point can be illustrated by looking at a three-dimensional picture, Figure 5.5. We observe the graph in a static version for each t between 0 and 6.

```
> plot3d(Normalpdf(y,10*t,15*sqrt(t)),y=-110..200,t=0..6,\
> title ='Figure 5.5:  The Density of Y[t]\
> as a Function of t and y');
```

It is now apparent that as t approaches zero, the mass of probability, the area under the graph, is concentrated around zero. Thus the probability that the instantaneous rate of return Y_t be close to zero as t approaches zero is close to one. Hence, as t approaches zero, e^{Y_t} is very likely to be one. The situation therefore becomes less and less risky. If at time $t = 0$ the price of the stock is P_0, its price at time t, P_t, will be the random variable $P_0\, e^{Y_t}$. For a small t, the probability of e^{Y_t} being one is close to one, and thus, the probability of P_t being close to P_0 is also close to one. Hence for short time intervals the realization of P_t is very likely to be close to P_0.

The normal cumulative distribution function of a random variable X at a point x is defined as the probability that X will have a value less than or equal to x. Thus, it is the area under the graph of **Normalpdf** from negative infinity to x. In other words it is

$$\int_{-\infty}^{x} \mathbf{Normalpdf}(v,\, \mu,\, \sigma)\, dv, \tag{5.27}$$

or in our notation, as in equation (5.26),

$$\int_{-\infty}^{x} \Phi_{\mu,\sigma}(v)\, dv = \int_{-\infty}^{x} \frac{e^{-\frac{(v-\mu)^2}{2\,\sigma^2}}}{\sqrt{2\,\pi\,\sigma^2}} dv. \tag{5.28}$$

We define this function in MAPLE and call it **Normalcdf**.

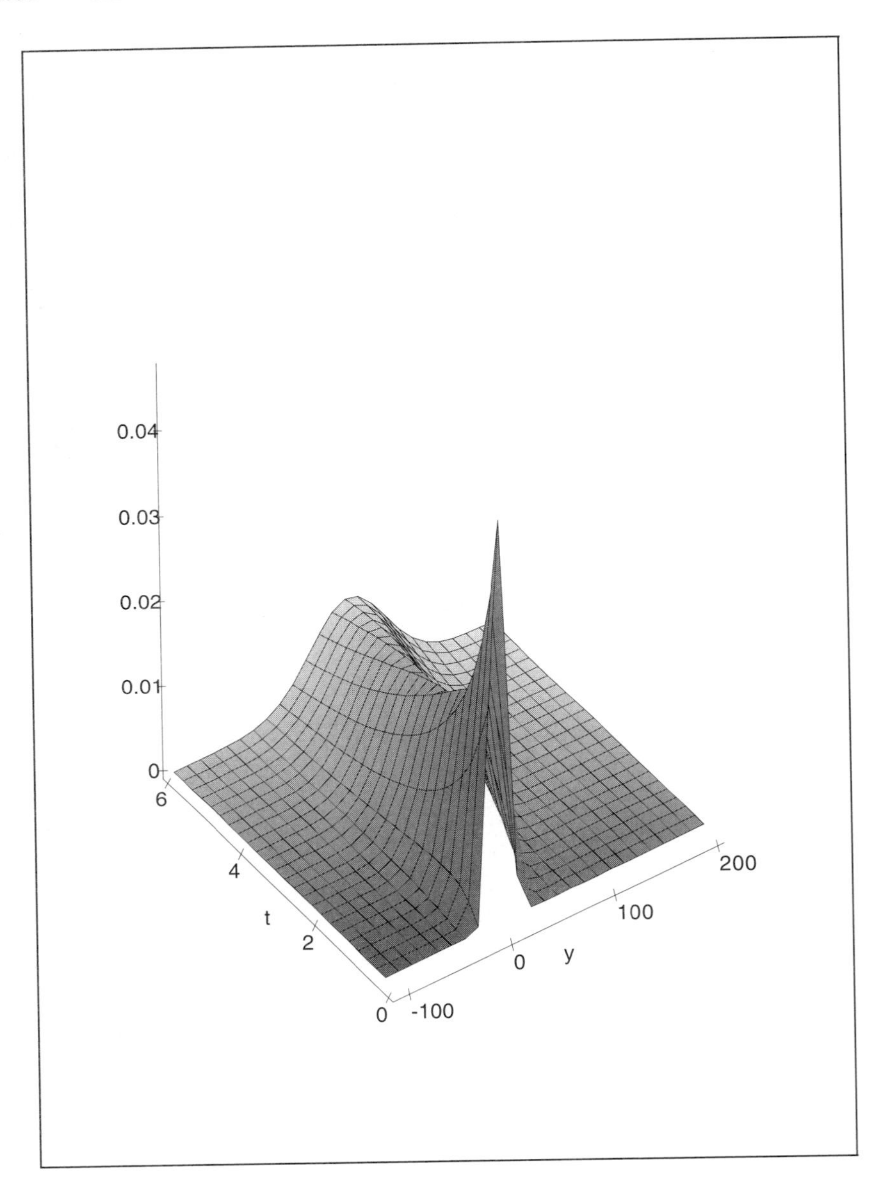

Figure 5.5: Three-Dimensional View of the Density of Y_t as a Function of t and y

```
> Normalcdf:=(x,mu,sigma)->\
> int(Normalpdf(v,mu,sigma),v=-infinity..x);
```

$$Normalcdf := (x, \mu, \sigma) \to \int_{-\infty}^{x} \mathrm{Normalpdf}(v, \mu, \sigma)\, dv$$

The effect of time on the distribution of Y_t can also be demonstrated with the cumulative function **Normalcdf**. This is performed by replacing the pdf with cdf as in the commands below. The graphs are shown only in the on-line version of the book.

```
> plot3d(Normalcdf(y,10*t,15*sqrt(t)),y=-110..200,t=0..6,\
> title ='The Probability Function of Y[t] as a Function\
> of t and y'):
```

```
> plots[animate](Normalcdf(Y,.15*t,.23*sqrt(t)),Y=-3..5,\
> t=0.1..6,color=green,title='Probability Function\
> of Y[t] as t Approaches Zero'):
```

The probability that the random variable Y_t will have a value between $\mu t - 2\sigma\sqrt{t}$ and $\mu t + 2\sigma\sqrt{t}$ is given by the area under the graph between these two points. We can calculate it in MAPLE with the aid of the function **Normalcdf**. However, we must let MAPLE know that σ and t are both positive.

```
> assume(sigma>0,t>0);
```

```
> evalf(Normalcdf(mu*t+2*sigma*sqrt(t),mu*t,sigma*sqrt(t))\
> -Normalcdf(mu*t-2*sigma*sqrt(t),mu*t,sigma*sqrt(t)));
```

$$.9544997361$$

It is well known that the probability of a normal random variable having a value between two standard deviations above or below the expected value is around 0.95. This is indeed confirmed above. As you can see, MAPLE was able to calculate the probability *without requiring us to specify the values* of t, σ, and μ. We see therefore that the probability does not depend on the actual values of these parameters.

However, consider the effect on the interval $[\mu t - 2\sigma\sqrt{t}, \mu t + 2\sigma\sqrt{t}]$, when t approaches zero. The interval clearly approaches zero, but the probability of the interval remains the same as it is independent of t. Hence the likelihood of Y_t being close to zero approaches one. Keep in mind that as the interval shrinks the tail ends of the probability shrink as well. In addition, most of the probability mass is now centered around zero. As t approaches zero, Y_t remains a random variable for every t, and hence the "limit of Y_t" as t approaches zero will still be a random variable.[15]

There are a few implications to these assumptions, including the instantaneous behavior of the price process. Some of these implications will be discussed shortly and some will be deferred to later chapters.[16] Our next task is to investigate the risk-neutral probability that is induced by the above assumptions. We would like to summarize and highlight a few issues demonstrated by the above graphic presentation.

- The larger the time interval t, the larger the variance and the expected value of Y_t are. Both the expected value and the variance of Y_t are linearly increasing functions of the length $[0, t]$, i.e., μt and $\sigma^2 t$, respectively. The *volatility* of the return is measured by the standard deviation of Y_t, $\sqrt{\sigma^2 t} = \sigma\sqrt{t}$, and thus depends on time via $\sqrt{t}$. Hence when t gets bigger, say, by a factor of 2, the expected return also increases by the same factor of 2. However, the volatility or the *"riskiness"* of the return increases only by a factor[17] of $\sqrt{2}$.

---

[15] The use of "limit of Y_t" should and could be made precise. However, it is used here in an intuitive way. See Appendix 15.11.2 for a further explanation and [13] for a more formal treatment.

For $\mu = 0.15$ and $\sigma = 0.23$ it is possible to ask MAPLE to generate the sequence of the intervals as t approaches zero from 1 to, say, $\frac{1}{10}$.

```
> seq(evalf([.15*(1/t)-2*.23*sqrt((1/t)),\
> .15*(1/t)+2*.23*sqrt(1/t)]),t=1..10);
```

[-.31, -.31], [-.2502691193, -.2502691193],
[-.2155811238, -.2155811238], [-.1925000000, -.1925000000],
[-.1757182540, -.1757182540], [-.1627942137, -.1627942137],
[-.1524350862, -.1524350862], [-.1438845596, -.1438845596],
[-.1366666666, -.1366666666], [-.1304647724, -.1304647724]

Alternatively it is possible to ask MAPLE to calculate the limit of the length of the interval as t approaches zero.

```
> limit((mu*t+2*sigma*sqrt(t))-(mu*t-2*sigma*sqrt(t)),t=0);
```

$$0$$

[16] These implications will be mainly dealt with in Chapter 15.

[17] Thus, when t shrinks to zero, μt will approach zero at a different speed than $\sigma\sqrt{t}$. This phenomenon can be illustrated by looking at the graph of $\sqrt{t}$ and t plotted on the same axis. Hence, if we refer to the per unit time return and volatility by $\frac{\mu t}{t}$ and $\frac{\sigma\sqrt{t}}{t}$, respectively, as t approaches zero the first expression is μ, while the second diverges to infinity. We will return to implications of this observation when we discuss the prices process more rigorously.

- As in the case of the risk-free rate, the stochastic return is additive across disjoint time intervals. If Y_t is the return over $[0, t]$, Y_{t_1} and Y_{t_2} are the returns over $[0, t_1]$ and $[t_1, t]$, respectively; then

$$Y_t = Y_{t_1} + Y_{t_2} \tag{5.29}$$

for every t. This equality, however, is between random variables, and indeed being held in the sense that the sum of two normal random variables is again a normal random variable. Moreover,

$$\mu t = E(Y_t) = E(Y_{t_1}) + E(Y_{t_2}) = \mu t_1 + \mu(t - t_1) = \mu t, \tag{5.30}$$

and since Y_{t_1} and Y_{t_2} are independent,

$$\begin{aligned} \sigma^2 t &= Var(Y_t) = Var(Y_{t_1}) + Var(Y_{t_2}) \\ &= \sigma^2 t_1 + \sigma^2 (t - t_1) = \sigma^2 t. \end{aligned} \tag{5.31}$$

- The reader must be alerted to the distinction between the risk-neutral probability and the real-life probability. The former is only an artificial consequence of the no-arbitrage condition, and the latter is the "real-life" probability of the continuously compounded return Y_t. Our assumptions above pertain to the real-life probability. We shall soon see that we actually assume that under both distributions Y_t is normally distributed with $E(Y_t) = \mu t$, and $\text{Var}(Y_t) = \sigma^2 t$, commonly denoted by $Y_t \sim N(\mu t, \sigma\sqrt{t})$. However, the value of the parameter μ is different under the risk-neutral probability and under the real-life probability.

- Assumption I (on page 195) and equation (5.25) imply that

$$S_t = Pe^{Y_t} = Pe^{\mu_{Y_t} + \sigma_{Y_t} Z} = Pe^{\mu t + \sigma\sqrt{t}Z} = Pe^{\mu t} e^{\sigma\sqrt{t}Z}, \tag{5.32}$$

where $Z \sim N(0, 1)$. The $e^{\mu t + \sigma\sqrt{t}Z}$ in equation (5.32) shows that the rate of return on a stock is the sum of the deterministic part μt and the stochastic part[18] $\sigma\sqrt{t}Z$. The deterministic part is a linear function of t, as is the risk-free return. The stochastic part $\sigma\sqrt{t}Z$, however, depends on the time t, via the square root of t. These last two points are visualized in the next section.

[18] Or putting it differently, one plus the rate of return on a stock, $e^{\mu t + \sigma\sqrt{t}Z} = e^{\mu t} e^{\sigma\sqrt{t}Z}$, has a deterministic part $e^{\mu t}$ which is multiplied by a random part $e^{\sigma\sqrt{t}Z}$.

5.6 Paths of the Price Process

We can visualize the above points in a manner that will let us have some insight into the path of the price process. One plus the rate of return on a stock, see footnote 18, has a deterministic part $e^{\mu t}$ which is multiplied by a random part $e^{\sigma\sqrt{t}Z}$. Consider, for example, a stock such that at $t = 0$, its price is $p = 100$, its expected return per unit of time is $\mu = 0.15$, and its standard deviation per unit of time is $\sigma = 0.23$. Its price at time $t = 1$ will be

$$100e^{0.15\times 1}e^{0.23\sqrt{1}Z}, \tag{5.33}$$

where Z is a random number, drawn from the normal distribution with $\mu = 0$ and $\sigma = 1$. The stock price at time $t = 1$ can also be expressed in terms of its price at time $t = 0.5$. This however will require two independent drawings; $Z_0 \sim N(0,1)$ drawn at time zero and $Z_{0.5} \sim N(0,1)$ drawn at time $t = 0.5$. At time $t = 0.5$ the stock price will be

$$S_{0.5} = 100e^{(0.15)(0.5)}e^{0.23\sqrt{0.5}Z_0} \tag{5.34}$$

for some realization of Z_0. At time $t = 1$ it will be

$$S_1 = S_{0.5}e^{(0.15)(0.5)}e^{0.23\sqrt{0.5}Z_{0.5}}$$

for yet another independent drawing of Z. Thus,

$$S_1 = 100e^{(0.15)(0.5)}e^{0.23\sqrt{0.5}Z_0}e^{(0.15)(0.5)}e^{0.23\sqrt{0.5}Z_{0.5}}, \tag{5.35}$$

where Z_0 and $Z_{0.5}$ are independent identically distributed, $N(0,1)$, random variables. In the absence of the stochastic part, that is, if Z is set equal to zero, we would have $S_1 = 100e^{(0.15)(0.5)+(0.15)(0.5)} = 100e^{(0.15)(1)}$.

In general we can divide the time interval to n parts and draw n numbers from the standard normal distribution. This will result in expressing S_1 as the product of n terms times the current price, that is, as

$$\begin{aligned} S_1 &= 100\prod_{k=0}^{n-1} e^{0.15\frac{1}{n}}e^{0.23\sqrt{\frac{1}{n}}Z_k} \\ &= 100\ e^{0.15\frac{1}{n}}e^{0.23\sqrt{\frac{1}{n}}Z_0} \cdot \ldots \cdot e^{0.15\frac{1}{n}}e^{0.23\sqrt{\frac{1}{n}}Z_{n-1}}, \end{aligned} \tag{5.36}$$

where $\prod_{k=1}^{n} a_k$ denotes the product of the a's, i.e., $a_1 \cdot a_2 \cdot \ldots \cdot a_n$.

The procedure **Gmbm** simulates this process. It takes the following parameters: the price of the stock at the initial time, the value of t_0 (which we assumed to be zero until now), the value of t_1, the value of n, and the

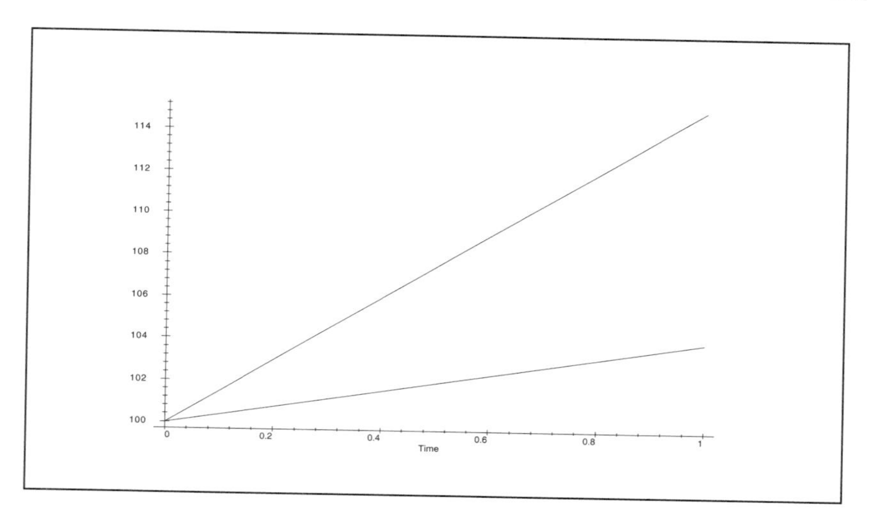

Figure 5.6: Expected Process vs. Stochastic Process for $n = 1$

values of μ and σ reported per the unit of time. The procedure animates the random price process against the expected price process by linear interpolation.

Thus if we choose, for example, $n = 1$ and μ and σ as in equation (5.33), the procedure simulates the price of the stock at time $t = 1$, as in Figure 5.6.

```
> Gmbm(100,0,1,1,.15,.23);
```

The graphs generated are thus linear. One graph connects the value of the stock at $t = 0$ to its expected value at $t = 1$, that is, to $100e^{0.15}e^{0.23}$. The other line connects it to a *realization* of the price of the stock at time $t = 1$. Every time you run this procedure a random number is drawn to generate the price of the stock at time $t = 1$. The expected price of course stays the same as long as the μ parameter stays the same. The reader is invited to run this procedure for different values of the parameters.

When $n = 2$ is chosen, the procedure also simulates the price of the stock at time $t = 0.5$. Thus the graph of the price of the stock is no longer linear. It has a knot point at $t = 0.5$ and its value there is based on equation (5.34). The graph of the expected price of the stock is of course the same. This is demonstrated in Figure 5.7 for $n = 2$.

```
> Gmbm(100,0,1,2,.15,.23,title=‘Figure 5.7:  Expected\
```

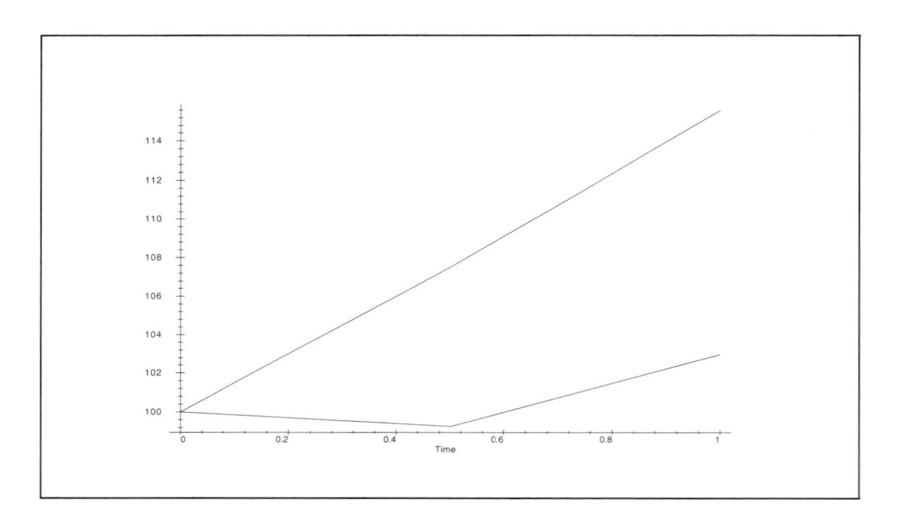

Figure 5.7: Expected Process vs. Stochastic Process for $n = 2$

```
> Process vs.  Stochastic Process for n=2',\
> titlefont=[TIMES, BOLD, 10]);
```

Running this procedure a few times, the reader can appreciate the random effect on the price of the stock by comparing it to the expected value of the stock price. The expected value of the stock price behaves like a risk-free asset with interest rate μ. Perhaps a better way of appreciating this risk is to run the procedure for a higher value of n with relatively low and high values of σ. For example, keep the value of μ at 0.15 and try two values for σ. Let us start with, say, $\sigma = 0.03$, as in Figure 5.8.

```
> Gmbm(100,0,1,50,.15,.03,title='Figure 5.8:  Expected\
> Process vs.  Stochastic Process for n=100',\
> titlefont=[TIMES, BOLD, 10]);
```

The reader should run this procedure a few times before moving to the next value of σ. Keep in mind that each time that it is run, the result is different. Each time random numbers are drawn from the normal distribution. Yet in most of the graphs the stochastic sample path is fairly close to the expected value graph. Let us now change the value of σ to be 0.63, as in Figure 5.9.

```
> Gmbm(100,0,1,50,.15,.63, title='Figure 5.9:  Expected\
> Process vs.  Stochastic Process for n=100 and a\
```

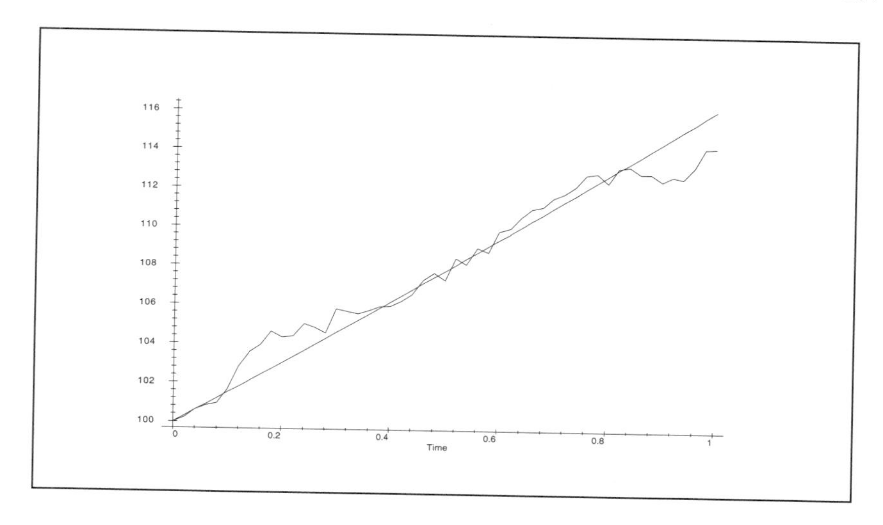

Figure 5.8: Expected Process vs. Stochastic Process for $n = 100$

```
>  'High' Value of Sigma`,titlefont=[TIMES, BOLD, 10]);
```

Figure 5.9 demonstrates how the volatility parameter measures the risk. For a low value of σ the simulated graph of the price runs very close to the expected value, while for the higher value it fluctuates considerably above and below the expected value. The reader is invited to run this procedure a few times so as to be convinced of the effect of the σ parameter, perhaps even trying a smaller value for σ to see how close the stochastic part is to the expected value. Keep in mind that what you see is a realization of random numbers, yet nearly every realization has the same characteristic: for a "low" value of σ the two graphs are very close. *Note however that regardless of the value of n, the value of the stock at time $t = 1$ follows the same distribution. It is just that for $n > 1$ we also get a peak at the values of the stock at some times prior*[19] *to time $t = 1$.*

[19]For the pricing of the European option the path that the stock price follows is irrelevant. However, for a more complex type option it is very relevant as we will see in the next chapters.

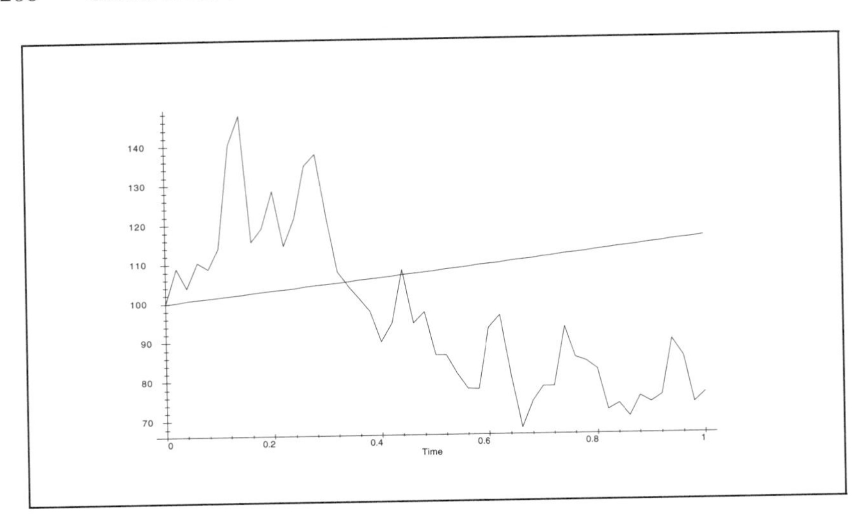

Figure 5.9: Expected Process vs. Stochastic Process for $n = 100$ and a "High" Value of σ

5.7 Specifying a Risk-Neutral Probability

The risk-neutral probability imposes certain conditions on the relation between μ and σ, the expected return and standard deviation per unit of time of Y_t. In this section, to avoid confusion, we denote the risk-neutral density of S_t by f_{S_t}. The risk-neutral probability of S_t, f_{S_t}, and the stochastic discount factor ψ_t must satisfy equations (5.18) and (5.17), i.e.,

$$P = e^{-r_f t} \int_0^\infty s f_{S_t}(s) ds = \int_0^\infty s \psi_t(s) ds. \tag{5.37}$$

The price of the stock at time t is $S_t = Pe^{Y_t}$, where P is the price of the stock at time 0 and Y_t is the continuously compounded **random** rate of return over the time period $[0, t]$. Therefore, the source of the uncertainty in S_t is Y_t as was demonstrated in the animation above. Hence, the probability distribution of Y_t and that of S_t are related and can be recovered from one another.[20]

[20] Denote by F_{S_t} and F_{Y_t} the cumulative distribution functions of S_t and Y_t, respectively. Then $F_{S_t}(s) = \Pr(S_t \leq s) = \Pr(Pe^{Y_t} \leq s) = \Pr\left(e^{Y_t} \leq \frac{s}{P}\right) = \Pr\left(Y_t \leq \ln\left(\frac{s}{P}\right)\right) = F_{Y_t}\left(\ln\left(\frac{s}{P}\right)\right)$. The relation between the density functions of S_t and Y_t is obtained by differentiating the relation $F_{S_t}(s) = F_{Y_t}\left(\ln\left(\frac{s}{P}\right)\right)$ with respect to s. Thus, we obtain

It will be more convenient to continue our investigation by looking at the probability distribution of Y_t and then deduce the probability distribution of S_t. Note that throughout this discussion we should be careful to distinguish between the risk-neutral probability and the real-life probability of both S_t and Y_t. We shall reserve the notation f_{Y_t} for the risk-neutral probability density function of Y_t, and f_{S_t} for that of S_t. Moreover, to avoid confusion in this section, whenever we calculate an expected value under the risk-neutral probability, either f_{S_t} or f_{Y_t}, we will use the notation $E_{risk-neutral}$.

If Y_t is a random variable such that $Y_t \sim N(\mu t, \sigma\sqrt{t})$ then the expected value of e^{Y_t} is given by[21]

$$E(e^{Y_t}) = e^{\mu t + \frac{1}{2}\sigma^2 t} = e^{\left(\mu + \frac{1}{2}\sigma^2\right)t}. \tag{5.38}$$

This can also be confirmed by MAPLE.

```
>   assume(t>0,sigma>0,mu>0);
>   int(exp(Y)*Normalpdf(Y,mu*t,sigma*sqrt(t)),\
>   Y=-infinity..infinity);
```

$$e^{\left(\frac{1}{2}t\,(\sigma^2 + 2\,\mu)\right)}$$

However, from equation (5.24), e^{Y_t} equals $\frac{S_t}{P}$, and equation (5.21) specifies that under the risk-neutral probability, f_t, the expected value of $\frac{S_t}{P}$, equals $e^{r_f t}$; thus

$$E_{risk-neutral}\left(\frac{S_t}{P}\right) = E_{risk-neutral}(e^{Y_t}) = e^{r_f t}. \tag{5.39}$$

The assumptions above and equations (5.24) and (5.25) imply that f_{S_t} and ψ_t satisfy (5.18) and (5.17) if, and only if, under the risk-neutral probability of Y_t, f_{Y_t},

$$E_{risk-neutral}\left(e^{Y_t}\right) = \int_{-\infty}^{\infty} e^{Y_t} f_{Y_t}(Y_t)dY_t = e^{r_f t}. \tag{5.40}$$

Therefore, if we maintain the assumption that f_{Y_t} is a normal distribution, like its counterpart the real-life distribution of Y_t, the parameters $\mu_{f_{Y_t}}$ and $\sigma_{f_{Y_t}}$, by equation (5.38) and (5.40), are constrained to satisfy

$$\left(\mu_{f_{Y_t}} + \frac{1}{2}\sigma^2_{f_{Y_t}}\right)t = r_f t. \tag{5.41}$$

$f_{S_t}(s) = \frac{f_{Y_t}\left(\ln\left(\frac{s}{P}\right)\right)}{s}$, where f_{S_t} and f_{Y_t} denote the density functions of S_t and Y_t, respectively. The expected values of Y_t and of S_t (or a function of Y_t or S_t) could be calculated based on either density function.

[21] Given a random variable V, its moment generating function is defined to be $M_V(x) = E(e^{vx})$. Thus, $E(e^{Y_t}) = M_{Y_t}(1)$, which is known to be equal to $e^{\mu_{Y_t} + \frac{1}{2}\sigma^2_{Y_t}} = e^{\mu t + \frac{1}{2}\sigma^2 t}$.

If we further assume that the value of $\sigma^2_{f_{Y_t}}$ is equal to the value of the real-life[22] σ^2 then $\mu_{f_{Y_t}}$ is determined by

$$\mu_{f_{Y_t}} = r_f - \frac{\sigma^2}{2}. \tag{5.42}$$

According to equation (5.42) the expected value of the risk-neutral probability f_{Y_t}, can be written[23] in terms of the *real-life* parameters σ and r_f. The common practice is to quote r_f, μ, and σ per annum, and this will be adopted here. The units of time, t, therefore, will be expressed in years. The above discussion is summarized in the following proposition.

Proposition 9 *Assume that the real-life distribution of Y_t is given by $Y_t \sim N(\mu t, \sigma\sqrt{t})$. Replacing the annual expected value μ with $r_f - \frac{1}{2}\sigma^2$ and keeping the volatility σ as in real-life produces a risk-neutral probability. Under the risk-neutral probability, therefore, $Y_t \sim N\left(\left(r_f - \frac{\sigma^2}{2}\right)t, \sigma\sqrt{t}\right)$.*

Let us utilize MAPLE to verify our derivation. The density function of Y_t will thus be given by

```
> Normalpdf(y,(r-(sigma^2)/2)*t,sigma*sqrt(t));
```

$$\frac{1}{2}\,\frac{e^{\left(-\frac{1}{2}\frac{(y-(r-\frac{1}{2}\sigma^2)\,t)^2}{\sigma^2\,t}\right)}\sqrt{2}}{\sigma\sqrt{\pi\,t}}$$

In order to calculate the expected value of e^{Y_t} we evaluate the following integral:

```
> int(exp(y)*Normalpdf(y,(r-(sigma^2)/2)*t,sigma*sqrt(t)),\
> y=-infinity..infinity);
```

$$e^{(t\,r)}$$

We thus confirm that indeed, under the risk-neutral probability, the expected value of the return on the stock is equivalent to that of the risk-free return. It is now apparent that under the risk-neutral probability the expected value of $S_t = Pe^{Y_t}$ will be P times the expected value of e^{Y_t} and hence, under the risk-neutral distribution, the expected value of S_t is $Pe^{r_f t}$.

[22] The issue of estimating σ from market data is discussed in the next Chapter.

[23] We assumed here that the risk-neutral probability of Y_t is the normal distribution and that the σ parameter of the risk-neutral probability is the same as that of the real-life probability. These assumptions can be deduced in a more rigorous way as being consequences of the assumption that the real-life probability distribution of Y_t is the normal distribution and utilize arbitrage arguments.

In other words, under the risk-neutral probability, the appreciation of the stock's price is the same as the appreciation of cash invested at the risk-free rate.

We are now in the position to determine the risk-neutral probability of S_t, which we denoted by f_{S_t}, and the stochastic discount factor ψ_t. The relationship between the cumulative distribution function of Y_t and of S_t is given (see footnote 20) by

$$F_{S_t}(s) = F_{Y_t}\left(\ln\left(\frac{s}{P}\right)\right). \tag{5.43}$$

The density function of S_t is obtained by differentiating $F_{Y_t}\left(\ln\left(\frac{s}{P}\right)\right)$ with respect to s. Thus, we obtain $f_{S_T}(s) = \frac{f_{Y_t}\left(\ln\left(\frac{s}{P}\right)\right)}{s}$, where f_{S_t} and f_{Y_t} denote the density functions of S_t and Y_t, respectively. Hence, to find the formula for f_{S_t}, based on equation (5.43) we should substitute $y = \ln\left(\frac{s}{P}\right)$ in the cumulative distribution of Y_t, and differentiate it with respect to s. Utilizing our MAPLE defined function **Normalcdf** we should substitute $y = \ln\left(\frac{s}{P}\right)$ in **Normalcdf** $\left(y, \left(r - \frac{\sigma^2}{2}\right) t, \sigma\sqrt{t}\right)$ and differentiate it with respect to s as follows:

```
> diff(subs(y=ln(s/P),Normalcdf(y,(r-(sigma^2)/2)*t,\
> sigma*sqrt(t))),s);
```

$$\frac{1}{2}\frac{e^{\left(-\frac{1}{8}\frac{(2\ln(\frac{s}{P})-2\,t\,r+\sigma^2\,t)^2}{\sigma^2\,t}\right)}\sqrt{2}}{s\,\sigma\,\sqrt{\pi\,t}}$$

The possible realizations of S are now all the nonnegative numbers, i.e., the interval $[0, \infty)$. Indeed $Y = \ln\left(\frac{s}{P}\right)$, or as we have seen before, $S = Pe^Y$, which can only take on positive values. Let us now verify that indeed the function in the MAPLE command above is a density function. It is easily verified that for every s in $[0, \infty)$ this function above is positive, and thus we need only to check that the integral of this function over $[0, \infty)$ is 1.

```
> simplify(int(%,s=0..infinity));
```

$$1$$

The above obtained distribution is called lognormal. In general, if Y follows a normal distribution ("is normally distributed") then e^Y (or a constant multiplication of e^Y) follows a lognormal distribution. For a review of the lognormal distribution and some of its properties see [16].

We define the lognormal density function in MAPLE under the name **Lnrn**. If MAPLE is asked to evaluate this function symbolically, i.e., not specifying numerical values for the parameters t, P, r_f, and σ, MAPLE responds with the actual formula:

```
> Lnrn(s,t,P,r,sigma);
```

$$\frac{1}{2}\frac{\sqrt{2}\,e^{\left(-1/2\,\frac{(\ln(s)-\ln(P)-(r-1/2\,\sigma^2)\,t)^2}{\sigma^2\,t}\right)}}{s\,\sigma\,\sqrt{\pi\,t}}$$

The following proposition summarizes the above discussion.

Proposition 10 *If the risk-neutral distribution of Y_t is given by*

$$N\left(\left(r_f-\frac{\sigma^2}{2}\right)t,\sigma\sqrt{t}\right)$$

then the distribution of both $\frac{S_t}{P}=e^{Y_t}$ and $S_t=Pe^{Y_t}$ is lognormal. If at time t_0 the price of a stock is P, the risk-neutral density function of $S_{t_0+t}=Pe^{Y_t}$ is given by

$$f_{S_t}(s)=\frac{1}{s\sqrt{2\pi\sigma^2 t}}e^{\left(-\frac{\left(\ln(s)-\ln(P)-\left(r_f-\frac{\sigma^2}{2}\right)t\right)^2}{2\sigma^2 t}\right)}. \tag{5.44}$$

Let us investigate with MAPLE some properties and conditions that should be satisfied by the risk-neutral probability **Lnrn** and the induced stochastic discount factor ψ. Given the current price of the stock P, we can confirm that the expected value of the stock in t units of time under the risk-neutral probability is indeed $Pe^{r_f t}$, as is claimed by equation (5.40). To this end we need to calculate the integral $\int_0^\infty sf_{S_t}(s)ds$, where $f_t(s)$ is as defined in equation (5.44).

In order to aid in the MAPLE calculation (these calculations will work in MAPLE version 5.1 and in the IDEAL MATLAB GUI; see the explanation below if you are working with MAPLE version 4), we assume the sign of some variables as in the next command.

```
> assume (P>0,t>0,r>0,sigma>0);
> int(s*Lnrn(s,t,P,r,sigma),s=0..infinity):
> simplify(%);
```

$$P^{\left(-\frac{1}{2}\,\frac{2\,r-\sigma^2}{\sigma^2}\right)}\,e^{\left(\frac{1}{2}\,\frac{2\,r\ln(P)+2\,\sigma^2\,t\,r+\ln(P)\,\sigma^2}{\sigma^2}\right)}$$

We can further simplify the result, using the expand command, to obtain the required confirmation.

```
> expand(%);
```

$$P\,e^{(t\,r)}$$

Due to a certain "feature" in MAPLE version 5 this integral cannot be calculated. However, the expected value of the price of the stock can be calculated based on the risk neutral distribution of the continuously compounded rate of return, i.e., as

```
> assume (P>0,t>0,r>0,sigma>0);
> int(P*exp(y)*Normalpdf(y,(r-sigma^2/2)*t,sigma*sqrt(t)) \
> ,y=-infinity..infinity);
```

$$e^{(t\,r)}\,P$$

If you are reading the MAPLE worksheets with MAPLE version 4 you should operate the sequence of commands as below (these commands also work in MAPLE version 5.1 and the IDEAL MATLAB GUI). In order to aid in the MAPLE calculation, we assume the sign of some variables and change the variable of integration as in the next command.

```
> assume (P>0,t>0,r>0,sigma>0);
> student[changevar](s=P*y,int(s*Lnrn(s,t,P,r,sigma),\
> s=0..infinity),y);
```

$$\frac{P\,e^{(\frac{1}{8}\,\frac{t\,(2\,r+\sigma^2)^2}{\sigma^2})}\,e^{(\frac{1}{2}\,t\,r)}\,\sqrt{t}\,\sqrt{\pi}}{\sqrt{\pi\,t}\,e^{(\frac{1}{2}\,\frac{t\,r^2}{\sigma^2})}\,(e^{(t\,\sigma^2)})^{\frac{1}{8}}}$$

We can further simplify the result to obtain the required confirmation:

```
> simplify(%);
```

$$P\,e^{(t\,r)}$$

5.8 Lognormal Distributions and the SDF

The risk-neutral distribution of S_t is thus the lognormal distribution. We would like now to visualize the effect of time on this distribution. This can be done in a similar manner to our animation of the distribution of Y_t.

Let us view the shape of this density function and also its dependence on time. To this end, we graph, Figure 5.10, the lognormal density function (equation (5.44)) in the same plane for two different values of t, $t = 3$ and $t = 1$, when $P = 70$, $r_f = 0.15$, and $\sigma = 0.22$.

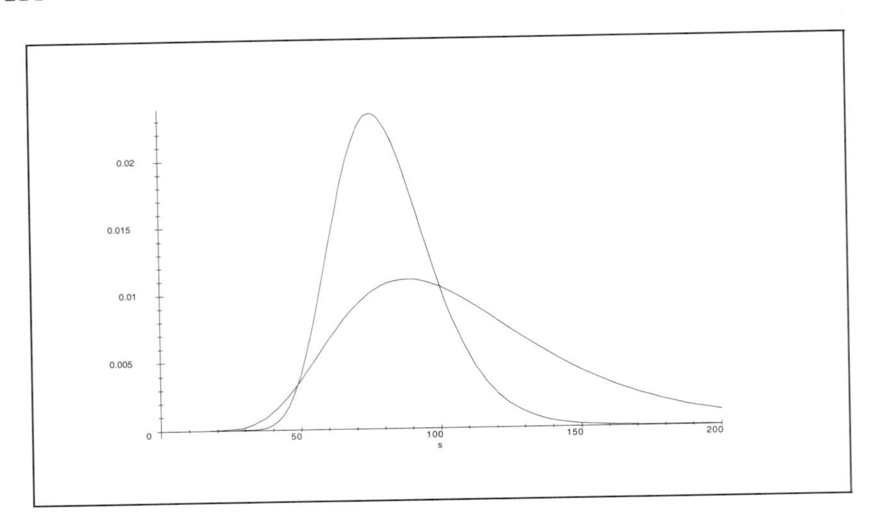

Figure 5.10: The Lognormal Distribution for $t = 1$ and $t = 3$

```
> plot({Lnrn(s,3,70,0.15,0.22),Lnrn(s,1,70,0.15,0.22)},\
> s=0..200,title='Figure 5.10:  The Lognormal\
> Distribution for t=1 and t=3',\
> titlefont=[TIMES,BOLD, 10]);
```

Note that the lognormal distribution is positive only for positive realizations and is skewed to the right. The spikier function is the one which corresponds to the shorter period of time. It represents less uncertainty as the probability mass is concentrated over a smaller region. Reading approximately from the graph, this is $[40, 140]$ vs. $[55, 90]$. The on-line version of the book demonstrates an animation to this effect. As t gets closer to zero, here also it is the case that most of the probability mass is concentrated around the current stock value ($60 in the example below).

```
> plots[animate](Lnrn(60,t,s,.25,.10),s=32..68,t=0..1,\
> title='The Lognormal Distribution as t Approaches\
> Zero', titlefont=[TIMES,BOLD, 10]);
```

We can also investigate the effect of the current stock price on the risk-neutral distribution. Figure 5.11 demonstrates the density function for two prices, $P = 70$ and $P = 40$, while the other parameters are as in Figure 5.10.

```
> plot({Lnrn(s,1,70,0.15,0.22),Lnrn(s,1,40,0.15,0.22)},\
```

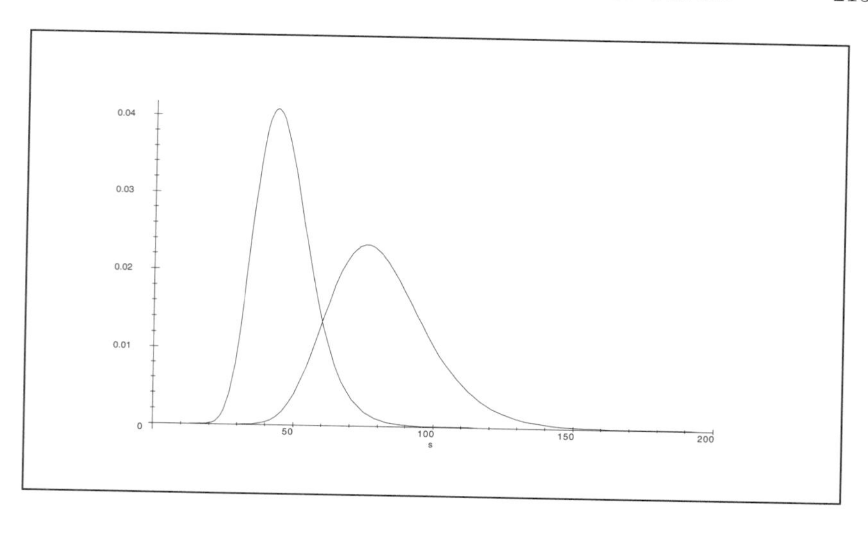

Figure 5.11: The Lognormal Distribution for Initial Stock Prices of $40 and $70

```
> s=0..200,title=`Figure 5.11:  The Lognormal\
> Distribution for Initial Stock Prices of $40 and $70`,\
> titlefont=[TIMES, BOLD, 10]);
```

In order to visualize the sensitivity of the curve to the current price of the stock and σ we produce a three-dimensional graph. Figure 5.12 displays the density function for initial stock prices in the range $[60, 100]$ as a function of the current price of the stock and σ.

```
> plot3d(Lnrn(s,1,P,.12,.22),s=0..120,P=60..100,\
> orientation=[45,45],axes=normal,labels=\
> [`Future realization`,`Current Stock Price`,Density],\
> title=`Figure 5.12:  Lognormal Density vs.  the Current\
> Stock Price`,titlefont=[TIMES,BOLD,10]);
```

5.9 The Stochastic Discount Factor Function

We proceed to generate the stochastic discount factor function. Based on equation (5.16), multiplying the risk-neutral density by $e^{-r_f t}$ yields the stochastic discount factor denoted by ψ. We define this function in MAPLE (the **unapply** command) with arguments listed at the end of the command.

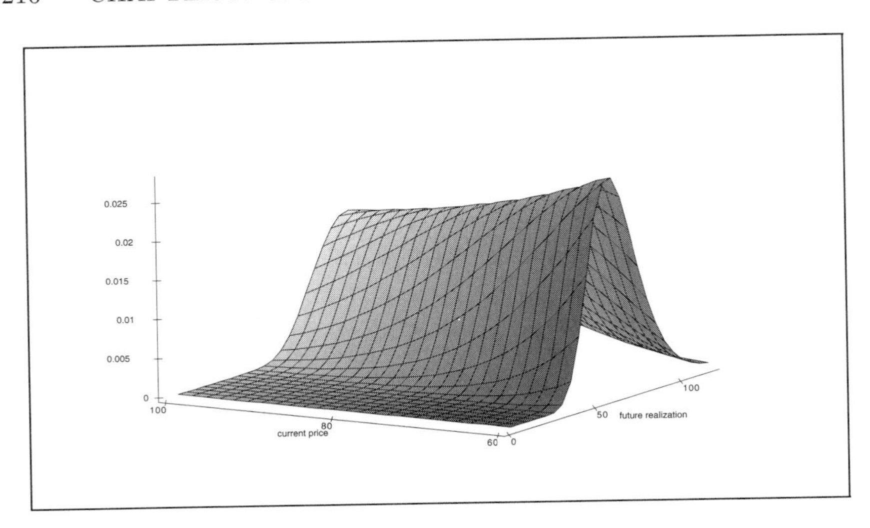

Figure 5.12: Lognormal Density vs. the Current Stock Price

```
> psi:=unapply(exp(-r*t)*Lnrn(s,t,P,r,sigma),\
> s, t,P,r,sigma);
```

$$\psi := (s,\, t,\, P,\, r,\, \sigma) \to \frac{1}{2}\,\frac{e^{(-r\,t)}\,\sqrt{2}\,e^{\left(-\frac{1}{2}\,\frac{(\ln(s)-\ln(P)-(r-\frac{1}{2}\,\sigma^2)\,t)^2}{\sigma^2\,t}\right)}}{s\,\sqrt{\pi\,\sigma^2\,t}}$$

To gain a visual image of the discount factor for points s in the range $[0, 300]$ we plot it together with the risk-neutral distribution. Figure 5.13 demonstrates the scaling of the lognormal distribution which produces the discount factors.

```
> plot({psi(s,1,100,.12,.22),Lnrn(s,1,100,.12,.22)},\
> s=0..300,title=`Figure 5.13:  Scaling the Lognormal\
> Distribution to Produce the Discount Factors`,\
> titlefont= [TIMES,BOLD,10]);
```

Indeed, since we have already verified that the function **Lnrn** satisfies equations (5.18) and (5.17), ψ defined as above must satisfy equations (5.4) and (5.6). Nevertheless we ask MAPLE to verify it. Equation (5.6) tells us what the price of the stock should be in terms of ψ. Let us see if indeed equation (5.6) is satisfied by this definition of ψ.

```
> assume(sigma>0,t>0,P>0,r>0);
> Int(psi(s,t,P,r,sigma),s=0..infinity)\
```

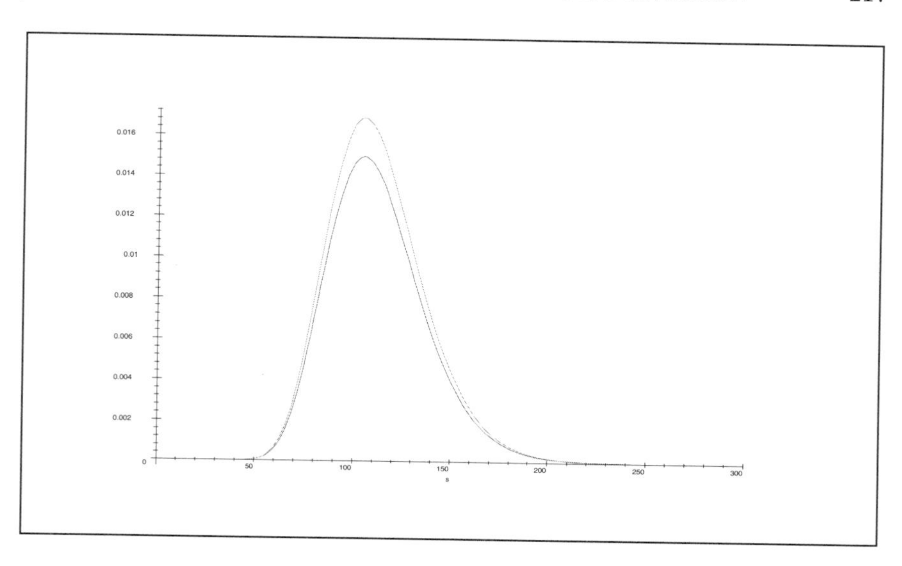

Figure 5.13: Scaling the Lognormal Distribution to Produce the Discount Factors

```
> =int(psi(s,t,P,r,sigma),s=0..infinity);
```

$$\int_0^\infty \frac{1}{2}\,\frac{e^{(-r\,t)}\,\sqrt{2}\,e^{\left(-\frac{1}{2}\,\frac{(\ln(s)-\ln(P)-(r-\frac{1}{2}\,\sigma^2)\,t)^2}{\sigma^2\,t}\right)}}{s\,\sqrt{\pi\,\sigma^2\,t}}\,ds = e^{(-r\,t)}$$

Recall that the risk-neutral distribution is defined with respect to the states of nature that are induced by a certain stock. The states of nature are the possible stock prices in t years. Thus, under this distribution the random variable S_t with possible realization s in $\Omega = [0, \infty)$ depends on the actual stock only via the current price of the stock P and its standard deviation. Hence, two different stocks with the same current price will induce different risk-neutral probabilities *if and only if* their standard deviations will be different.

This fact is in essence not surprising. We have seen before, in equation (5.4), that the expected value, under this distribution, of each stock should be the future value of its current price based on the risk-free rate. Moreover, given equation (5.8), the price of a call option cannot be dependent on the expected value of the stock, only on its standard deviation.

The pricing equation for an option[24] can be developed in two ways: applying equation (5.8) or applying the instantaneous properties of the price process and invoking directly some arbitrage arguments. Both ways are of course equivalent and rely on the absence of arbitrage opportunities in the market. Replacing $f_{S_t}(s)$ in

$$P_{Call_t(K)} = e^{-r_f t} \int_K^{\infty} (s - K) f_{S_t}(s) ds \tag{5.45}$$

with equation (5.44) produces the price of an option, as is demonstrated in the next Chapter.[25]

Equation (5.45) is a direct consequence of pricing with the SDF. In Chapter 15 we will take a second look at this formula and we will see that it can actually be derived utilizing replication arguments in a continuous time setting. Both ways are equivalent and rely on the absence of arbitrage opportunities in the market. Therefore, each could be viewed as being developed as if the market were populated with only risk-neutral investors. Such investors do not require compensation (in terms of expected return) for taking risk. Hence the expected return of every security in a market (in equilibrium) consisting of only risk-neutral investors, as we have mentioned, should be the risk-free rate.

Indeed this is the economic intuition behind Proposition 9 (on page 210) and equation (5.42) that establishes the substitution of $r_f - \frac{\sigma^2}{2}$ for the expected return of the real-life distribution to produce $f_t(s)$. For this reason it is not surprising that the expected return parameter of the real-life distribution drops out of the pricing equation. In fact it is consistent with our economic intuition that the pricing relation developed via arbitrage arguments should be agreed upon by all investors in the market regardless of their attitude toward risk. Hence the risk compensation, which is the real-life expected return parameter, cannot appear in the equation.

Investors with different degrees of risk aversion (or risk loving) would not be able to agree on the premium they require for taking risk and thus could not agree on the appropriate expected return a security should offer in order for them to hold it in their portfolios. Consequently, if the real-life expected return parameter had remained in the pricing equation, it could not have been agreed upon by all investors.

---

[24] The option pricing formula was developed by Black and Scholes [8] and by Merton [33].

[25] This is one way to derive the well-known Black–Scholes option pricing formula. Other approaches to this pricing will be explored in Chapter 15.

Investors with different appetites for risk can agree on equation (5.45) as long as they all agree on the risk of the security. Granted, these investors do require different compensation for the risk ultimately and hold a security only if its excepted return provides the appropriate risk compensation for their risk attitude. The pricing equation, though, does not depend on such compensation — the expected return. Rather than this result being surprising, it instead should be reassuring that it is consistent with our economic intuition. Perhaps even we should have anticipated that the pricing equation would not depend on risk-adjusted expected return. We shall encounter this phenomenon again, when the option pricing formula will be developed from other points of view in Chapter 15.

5.10 Concluding Remarks

This Chapter provides the bridge that facilitates application of pricing-by-arbitrage developed in the one-period model of Chapter 1 to a more realistic model investigated here. In dealing with contingent claims on a certain underlying asset, the relevant information is the value of the underlying asset. Hence, the space of discrete states of nature can be identified with the set, Ω, of possible realizations of the price of the underlying asset. Thus, when the underlying asset is a stock, the space of states of nature is $[0, \infty)$.

This Chapter approaches the goal at hand from an intuitive point of view. It encourages the reader to think about contingent claims in the current continuous setting as being the same "in spirit" as those of the one-period setting with its discrete states of nature. Consequently, an elementary contingent claim is defined in this current setting as one dollar contingent on the price of the underlying (in t units of time) being s, where s is a number in $[0, \infty)$. Interpreting the current model as being of the one-period type (the length of the period being t), albeit with a continuum of states of nature, and utilizing the above definition provides the desired linkage: The discrete stochastic discount factors d_j, $j = 1, ..., N$, are replaced with a continuous stochastic discount factor function $\psi(\cdot)$ defined over $[0, \infty)$. Replacing d_j with $\psi(s)$ and the summation sign with an integral sign, one obtains the pricing relations and the definition of a risk-neutral density in the current model.

However, in contrast to the discrete model where, given the prices of the security in each state of nature, the d_j were identified as a solution of a linear system of equations, the situation here is more complicated. Some assumptions must be made about the risk-neutral probability. This Chapter provides partial justification for the choice of the lognormal distribution as the risk-neutral probability and then solves for its parameters. It continues to investigate the path of the price process, thereby preparing the ground for a more rigorous investigation of Brownian motions and the stochastic process assumed to be followed by the price of the stock.

Chapter 15 will utilize the intuition built here to provide a more formal justification to the pricing relations motivated here simply as an extension of the one-period model. In spite of the absence of such a formal justification, the next few chapters exploit the implication of the pricing relation when the lognormal distribution is adopted as the risk-neutral distribution. This structure provides the opportunity for readers who are not interested in a formal derivation to appreciate and practice the pricing relation and its various implications for the hedging and pricing of derivative securities other than equity options. Such readers will be able to skip Chapter 15 without loss of continuity and read about other types of options in Chapter 16.

5.11 Questions and Problems

Problem 1. Assume that the price of a stock is \$100, the standard deviation of its continuously compounded rate of return is 0.23 a year, and the (annual) continuously compounded risk-free rate is 10%. (You can use the MATLAB GUI or MAPLE and some of the procedures mentioned in this Chapter to answer the questions below.)

1. What is the risk-neutral probability that the price of the stock will be greater than 95 but less than 110 in a day, in a month, in 6 months, and in a year?
2. What is the value of a dollar contingent on the stock price being in the above interval in a day, in a month, in 6 months, and in a year?
3. What is the risk-neutral probability that a European **call** option on the stock with an exercise price of \$150 and a maturity of 1 month will be exercised?
4. What is the risk-neutral probability that a European **put** option on the stock with an exercise price of \$150 and a maturity of 1 month will be exercised?
5. How would you answer the above if the standard deviation would have been 0.12?
6. How would you answer the above if the risk-free rate would have been 7%?
7. Reply to the above questions, calculating the probability under the real-life distribution, assuming that under this distribution the expected value of the continuously compounded rate of return (per annum) of the stock is 15%.

Problem 2. Assume that the price of a stock is \$100 and that the standard deviation of the continuously compounded rate of return is 0.23 a year. Given that the expected value of the stock price, under the risk-neutral distribution, in a year is \$120, what can you say about the (annual) continuously compound risk-free rate?

Problem 3. Assume that the price of a stock is \$100 and that the (annual) continuously compound risk-free rate is 10%. Given that the expected value of the price of the stock, under the risk-neutral distribution, in a year is \$120, what can you say about its standard deviation?

Problem 4. Consider a call and a put written on the same stock with the same exercise price and the same maturity. What can you say about the price of the stock if they have equal prices? Is your answer dependent on the price process of the stock?

Chapter 6

Valuation of European Options

The former Chapter, as was summarized in its Concluding Remarks, provided a bridge between the one-period discrete model and the one-period model with a continuum of states of nature. This latter type of model is sufficient to allow the valuation of European-type options. Utilizing the reader's intuition and extending the results of Chapter 1 to the continuous-time setting, enables the valuation of European-type contingent claims. Specifically if $h(s)$ is the payoff of the contingent claim, where s is the price of the underlying asset at maturity, the value of the derivative security is given by equation (5.7), i.e., by $\int_0^\infty h(s)\psi_t(s)ds$, where ψ is the SDF.

Chapter 5 was also, at least partially (see footnote 6 on page 187), intent on supporting the choice of the SDF and the risk-neutral probability as the lognormal distribution. We are, therefore, now at the point where calculating the value of a European-type option amounts to calculating the value of a certain integral. The current Chapter is dedicated to the valuation of European puts and calls, i.e., to European options where $h(S_T)$ is either given by $\max(0, S_T - K)$ or by $\max(0, K - S_T)$, where K is the exercise price and S_T the value of the underlying asset at maturity.

Having established the pricing formula, the investigation of combinations of options with different maturity dates follows. Prior to the derivation of this formula we lacked the "tool" to compare options with different maturity times and, of course, to value options prior to maturity. Consequently, we were limited to investigating only the payoffs of combinations at maturity. In this Chapter we have the tools and are able to explore combinations across time. The Chapter also looks into the effect of dividends on the price of

European-type options. The Chapter concludes with the issue of volatility vs. implied volatility.

6.1 Valuing a Call Option

We can now calculate the Black-Scholes formula (see [8]) for the price of a call option. As we have seen in equations (5.8) and (5.45), this formula can be derived in two ways. Equation (5.45) derives the formula for pricing call options as the expected value of the future payoff based on the risk-neutral probability, discounted by the risk-free rate of interest. Equation (5.9) expresses the price of a call option as the present value of the payoff based on the stochastic discount factor. In either approach, the formula is obtained by the evaluation of (essentially) the same integral.

If we approach the call option pricing formula from the perspective of the discounted expected value we must use the risk-neutral probability which we defined as the MAPLE function given below.

```
> Lnrn(s,t,P,r,sigma);
```

$$\frac{1}{2}\,\frac{\sqrt{2}\,e^{\left(-1/2\,\frac{(\ln(s)-\ln(P)-(r-1/2\,\sigma^2)\,t)^2}{\sigma^2\,t}\right)}}{s\,\sigma\,\sqrt{\pi\,t}}$$

The price of the call option based on equations (5.45) and (5.44) will thus be

```
> CallPrice:='exp(-r*t)*int(Call(s,K)*Lnrn(s,t,P,r,sigma)\
> ,s=0..infinity)';
```

$$CallPrice := e^{(-r\,t)}\int_0^\infty \mathrm{Call}(s,\,K)\,\mathrm{Lnrn}(s,\,t,\,P,\,r,\,\sigma)\,ds$$

We have seen in equation (5.16) the relationship between the risk-neutral probability and the stochastic discount factor ψ. The function ψ is defined below.

```
> psi:=unapply(exp(-r*t)*Lnrn(s,t,P,r,sigma),\
> s,t,P,r,sigma);
```

$$\psi := (s,\,t,\,P,\,r,\,\sigma\tilde{}) \to \frac{1}{2}\,\frac{e^{(-r\,t)}\,\sqrt{2}\,e^{\left(-1/2\,\frac{(\ln(s)-\ln(P)-(r-1/2\,\sigma\tilde{}^{2})\,t)^2}{\sigma\tilde{}^{2}\,t}\right)}}{s\,\sigma\tilde{}\,\sqrt{\pi\,t}}$$

The price of the call option can also be calculated, based on equation (5.45), as the expression **CallAlt**:

```
> CallAlt:=int((s -K)*psi(s,t,P,r,sigma),s=K..infinity);
```

$$CallAlt := \int_K^\infty \frac{1}{2}\,\frac{(s-K)\,e^{(-r\,t)}\,\sqrt{2}\,e^{\left(-1/2\,\frac{(\ln(s)-\ln(P)-(r-1/2\,\sigma^2)\,t)^2}{\sigma^2\,t}\right)}}{s\,\sigma\,\sqrt{\pi\,t}}\,ds$$

where $\max(s - K, 0)$ was substituted for $\text{Call}(s, K)$, and thus the integral is as above. Clearly, these two expressions are the same. We shall proceed now with the calculation of **CallPrice**. Some readers may wish to skip these calculations. The result is reported in equation (6.2) toward the end of this section. We begin by informing MAPLE that some of our variables must be positive:

```
> assume (P>0,t>0,r>0,K>0,sigma>0);
```

We then ask MAPLE to attempt an evaluation of the integral. To understand MAPLE's evaluation we need to introduce the definition of the function **erf**, a built-in MAPLE function. We previously defined the cumulative normal probability function in MAPLE as **Normalcdf**. **erf** is a built-in MAPLE function that is closely related to the function **Normalcdf**. **erf** is defined as

```
> erf(x) = 2/sqrt(Pi) * Int((exp(-v^2),v=0..x));
```

$$\text{erf}(x) = 2\,\frac{\int_0^x e^{(-v^2)}\,dv}{\sqrt{\pi}}$$

Similarly, the cumulative normal probability function is defined by

```
> Normalcdf:=(x,mu,sigma)->int(exp(-(v-mu)^2/2/sigma^2)/\
> sqrt(2*Pi*(sigma)^2), v=-infinity..x);
```

$$Normalcdf := (x,\, \mu,\, \sigma) \to \int_{-\infty}^{x} \frac{e^{(-1/2\,\frac{(v-\mu)^2}{\sigma^2})}}{\sqrt{2\,\pi\,\sigma^2}}\,dv$$

To demonstrate the relationship between **erf** and the normal probability function we evaluate **Normalcdf** at x for $\mu = 0$ and $\sigma = 1$.

```
> Normalcdf(x,0,1);
```

$$\frac{1}{2}\,\text{erf}(\frac{1}{2}\,x\,\sqrt{2}) + \frac{1}{2}$$

MAPLE recognizes that **Normalcdf** can be defined in terms of the MAPLE function **erf**. Conversely, we can express the function **erf** in terms of the function **Normalcdf**. Multiplying **Normalcdf** by two and evaluating it at the point $x\sqrt{2}$ is equivalent to the **erf** function plus one. This is demonstrated as follows:

```
> 2*Normalcdf(x*sqrt(2),0,1)-1;
```

$$\text{erf}(x)$$

To facilitate the MAPLE calculation the integral defining **CallPrice** is rewritten in terms of the density function of Y — the continuously compounded rate of return. As you recall, we assumed it to be the normal distribution with expected value of μt and a standard deviation of $\sigma\sqrt{t}$. This could be achieved by simply replacing the variable of integration in the above integral using the relation in equation (5.24), i.e., $S = P\,e^{Y}$.

Alternatively, one can apply the fact that given a random variable Y the expected value of $H(Y)$ may be calculated based on the density of Y as $\int_{-\infty}^{\infty} H(y)\,f(y)\,dy$. Thus in our case $H(Y) = P\,e^{Y}$ plays the role of S, and f is the normal density function. Consequently, we can equivalently define **CallPrice** as below.

```
> CallPrice:='exp(-r*t)*int(max(P*exp(y)-K,0)*Normalpdf\
> (y,(r-sigma^2/2)*t,sigma*sqrt(t)),y=0..infinity)';
```

$$CallPrice := e^{(-r\,t)} \int_0^{\infty} \max(P\,e^y - K,\,0)\,\mathrm{Normalpdf}(y,\,(r - \frac{1}{2}\,\sigma^2)\,t,\,\sigma\,\sqrt{t})\,dy$$

Since when $P\,e^y - K < 0$ the value of $\max(P\,e^y - K,\,0)$ vanishes, we can change the lower limit of the integral to be $\ln(\frac{K}{P})$ and replace $\max(P\,e^y - K,\,0)$ in the integrand with $P\,e^y - K$. Thus, the integral to be calculated is

```
> CallPrice:='exp(-r*t)*int((P*exp(y)-K)*Normalpdf\
> (y,(r-sigma^2/2)*t,sigma*sqrt(t)),y=ln(K/P)..infinity)';
```

$$CallPrice := e^{(-r\,t)} \int_{\ln(\frac{K}{P})}^{\infty} (P\,e^y - K)\,\mathrm{Normalpdf}(y,\,(r - \frac{1}{2}\,\sigma^2)\,t,\,\sigma\,\sqrt{t})\,dy$$

and as we can see below, the results of the calculation are written in terms of the MAPLE function **erf**.

```
> CallPrice:=expand(CallPrice);
```

$$CallPrice := \frac{1}{2}\,\frac{P\,\sqrt{\pi}\,\sqrt{t}}{\sqrt{\pi\,t}} - \frac{1}{2}\,\frac{K\,\sqrt{\pi}\,\sqrt{t}}{e^{(r\,t)}\,\sqrt{\pi\,t}} + \frac{1}{2}P\,\sqrt{\pi}\,\sqrt{t}\,\mathrm{erf}(-\frac{1}{2}\,\frac{\sqrt{2}\ln(K)}{\sigma\,\sqrt{t}} + \frac{1}{2}\,\frac{\sqrt{2}\ln(P)}{\sigma\,\sqrt{t}} + \frac{1}{2}\,\frac{\sqrt{2}\,\sqrt{t}\,r}{\sigma} + \frac{1}{4}\,\sqrt{2}\,\sigma\,\sqrt{t}) / \sqrt{\pi\,t} - \frac{1}{2}K\,\sqrt{\pi}\,\sqrt{t}\,\mathrm{erf}(-\frac{1}{2}\,\frac{\sqrt{2}\ln(K)}{\sigma\,\sqrt{t}} + \frac{1}{2}\,\frac{\sqrt{2}\ln(P)}{\sigma\,\sqrt{t}} + \frac{1}{2}\,\frac{\sqrt{2}\,\sqrt{t}\,r}{\sigma} - \frac{1}{4}\,\sqrt{2}\,\sigma\,\sqrt{t}) / (e^{(r\,t)}\,\sqrt{\pi\,t})$$

Let us ask MAPLE to simplify the above expression.

```
> CallPrice:=simplify(CallPrice);
```

$$CallPrice := \frac{1}{2}P - \frac{1}{2}K\,e^{(-r\,t)} + \frac{1}{2}P\,\mathrm{erf}(\frac{1}{4}\frac{\sqrt{2}\,(-2\ln(K)+2\ln(P)+2\,r\,t+t\,\sigma^2)}{\sigma\sqrt{t}}) - \frac{1}{2}K\,\mathrm{erf}(\frac{1}{4}\frac{\sqrt{2}\,(-2\ln(K)+2\ln(P)+2\,r\,t-t\,\sigma^2)}{\sigma\sqrt{t}})\,e^{(-r\,t)}$$

Now the resemblance between the **CallPrice** and the **erf** functions and consequently the **Normalcdf** function is more apparent.

Given the relationship between **Normalcdf** and **erf** we can convert this expression in terms of **Normalcdf**. Indeed, the classical way of stating this expression is not in terms of **erf**. However, at this time it is more efficient to use **erf** to perform calculations using MAPLE. We shall return to this point soon, but first we define a function in MAPLE that calculates the price of a call option. We name this function **Eqbs**. The following command in MAPLE defines the function **Eqbs** based on the result of our calculation:

```
> Eqbs:=unapply(CallPrice,K,t,P,sigma,r);
```

$$Eqbs := (K,\,t,\,P,\,\sigma,\,r) \to \frac{1}{2}P - \frac{1}{2}K\,e^{(-r\,t)} + \frac{1}{2}\,\mathrm{erf}(\frac{1}{4}\frac{\sqrt{2}\,(-2\ln(K)+2\ln(P)+2\,r\,t+\sigma^2\,t)}{\sigma\sqrt{t}})\,P - \frac{1}{2}\mathrm{erf}(\frac{1}{4}\frac{\sqrt{2}\,(-2\ln(K)+2\ln(P)+2\,r\,t-\sigma^2\,t)}{\sigma\sqrt{t}})\,K\,e^{(-r\,t)}$$

The reader can now calculate the price of a call option given specific parameter values. For example, an option that expires in one month ($\frac{1}{12}$ of a year) where σ and r are given per annum will have the following value:

```
> evalf(Eqbs(90,1/12,95,0.18,0.15));
```

$$6.34421922$$

Given the relation between **erf** and the normal distribution it is possible to write the pricing formula in terms of the normal distribution.

We can also ask MAPLE to express the pricing formula in terms of the cumulative standard normal distributional, i.e., in terms of **Normlacdf**$(x, 0, 1)$. For simplicity let us call this function **N**. Let us define a function called **newerf**; this function is simply the **erf** function in terms of the **N** function. As discussed earlier, this relationship is

```
> newerf:= t -> 2*N(t*sqrt(2))-1;
```

$$newerf := t \to 2\,\mathrm{N}(t\,\sqrt{2})$$

We can now substitute **newerf** for **erf** in the **CallPrice** expression. The pricing formula is now expressed in terms of the **N** function:

```
> NewCallPrice:=simplify(subs(erf=newerf,CallPrice));
```

$$NewCallPrice := P\,\mathrm{N}(\frac{1}{2}\,\frac{-2\ln(K)+2\ln(P)+t\,\sigma^2+2\,r\,t}{\sigma\,\sqrt{t}}) - K\,e^{(-r\,t)}\,\mathrm{N}(\frac{1}{2}\,\frac{-2\ln(K)+2\ln(P)+2\,r\,t-t\,\sigma^2}{\sigma\,\sqrt{t}})$$

which equals

$$PN\left(\frac{-\ln(K)+\ln(P)+rt+\frac{\sigma^2}{2}t}{\sigma\,\sqrt{t}}\right) - Ke^{-r\,t}\mathrm{N}\left(\frac{-\ln(K)+\ln(P)+rt-\frac{\sigma^2}{2}t}{\sigma\,\sqrt{t}}\right) \tag{6.1}$$

Recognizing that $\ln(P)-\ln(K)=\ln\left(\frac{P}{K}\right)$, we restate equation (6.1) as

$$PN\left(\frac{\ln\left(\frac{P}{K}\right)+\left(r+\frac{\sigma^2}{2}\right)t}{\sigma\sqrt{t}}\right) - Ke^{-rt}\mathrm{N}\left(\frac{\ln\left(\frac{P}{K}\right)+\left(r-\frac{\sigma^2}{2}\right)t}{\sigma\,\sqrt{t}}\right) \tag{6.2}$$

The above is the Black–Scholes option pricing formula. The formula is classically restated as

$$P\,\mathrm{N}(d_1) - K\,e^{-r\,t}\mathrm{N}(d_2), \tag{6.3}$$

where N is the cumulative normal distribution with $\mu=0$ and $\sigma=1$, i.e.,

$$\mathrm{N}(s) = \int_{-\infty}^{s}\frac{e^{-\frac{x^2}{2}}}{\sqrt{2\,\pi}}\,dx \tag{6.4}$$

$$d_1 = \frac{\ln\left(\frac{P}{K}\right)+\left(r+\frac{\sigma^2}{2}\right)t}{\sigma\,\sqrt{t}} \tag{6.5}$$

and

$$d_2 = \frac{\ln\left(\frac{P}{K}\right) + \left(r - \frac{\sigma^2}{2}\right) t}{\sigma \sqrt{t}} = d_1 - \sigma \sqrt{t}. \tag{6.6}$$

The procedure **BstCall** is a slight modification of the built-in MAPLE procedure which calculates the call price. Using the above example, the parameters for this procedure are as follows: exercise price ($95); time to expiration of the option (1 month); current price of the stock ($90); standard deviation of the continuous return on the stock per unit of time (0.18); and continuously compounded risk-free interest rate per unit of time (0.15).

Let us value an option with the above specifications. As discussed earlier it is a common practice to use a year as the basic unit of time. Hence a one month time period is inputted as $\frac{1}{12}$. We assume the interest rate and standard deviation are given in per annum values. We can now value such a call option as follows:

```
> BstCall(90,1/12,95,0.18,0.15);
                              6.34421923
```

Compare the results of the prices assigned by the built-in procedure **BstCall** to the function **Eqbs** defined above.[1] We see that the result from the procedure **BstCall** is the same as the calculation based on our function **Eqbs**.

Before we investigate some properties of the Black–Scholes formula, we would like to compare it to the bounds we obtained by relative pricing arguments. The bounds, in equation (4.6), specify that the value of a call option is always smaller than the value of the stock and larger than the value of the option at expiration. In Figure 6.1 we demonstrate these bounds in a three-dimensional graph. The value of a call option is graphed as a function of the current price of the stock and time to maturity. In the same plane we also graph the value of the stock (the upper bound) and the payoff at expiration (the lower bound). Looking at this picture the reader may gain insight as to how the value of the option approaches its value at expiration as time progresses and as the stock price changes. Recall that you may look at this graph from different perspectives by dragging the picture with the mouse. The insights gained by close examination of this picture will be revisited very shortly. Pay attention to the way the middle graph (the value of the call) gets closer to the value of the call at maturity as the time to maturity approaches zero.

```
> plot3d({'BstCall(20,t,P,0.25,0.10)',P,Call(P,20)},\
```

[1] The reader may have skipped this discussion as suggested.

```
> t=0..3,P=0..50,axes=NORMAL,orientation=[5,73],\
> labels=[time,'Stock Price','Call's Value'],\
> ambientlight=[0,0,0],style=patch,title='Figure 6.1:\
> The Value of a Call Option as a Function of S and t');
```

The on-line book shows an animation of the bounds. Run the animation and observe how as time approaches maturity the value of the call moves within the bounds toward its value at maturity.

6.2 Valuing a Put Option

Once we have derived the value of a call option we can extract the value of a put option based on the put–call parity relation. Equation (4.12) and the discussion preceding it invoked arbitrage arguments to prove that, by subtracting the current stock price and adding the present value of the exercise price from the price of a call, the price of the put is obtained. This is expressed in equation (4.12) and repeated here for convenience.

$$-P_S + Ke^{-r_f t} + P_{c(K)} = P_{p(K)} \tag{6.7}$$

Hence, we can use the defined procedure in MAPLE, **BstCall,** to obtain the price of a put option. Suppose we would like to price a put option given the parameters $(K,\ t,\ P,\ \sigma,\ r)$. Its price would therefore be given by the expression **BstPut,** as defined below.

$$\mathrm{BstPut}(K,\ t,\ P,\ \sigma,\ r) = \mathrm{BstCall}(K,\ t,\ P,\ \sigma,\ r) - P + K\,e^{-r\,t} \tag{6.8}$$

Let us verify this relationship using the example of a call option immediately preceding this section. The value of the call was given by

```
> BstCall(90,1/12,95,0.18,0.15);
```

$$6.34421923$$

The difference between the present value of the exercise price $90\,e^{-0.15\,\frac{1}{12}}$ and the price 95 is given by

```
> -95+(90*exp(-0.15*(1/12)));
```

$$-6.11799795$$

Hence the price of the put should be

```
> BstCall(90,1/12,95,0.18,0.15)-95+\
> (90*exp(-0.15*(1/12)));
```

$$.22622128$$

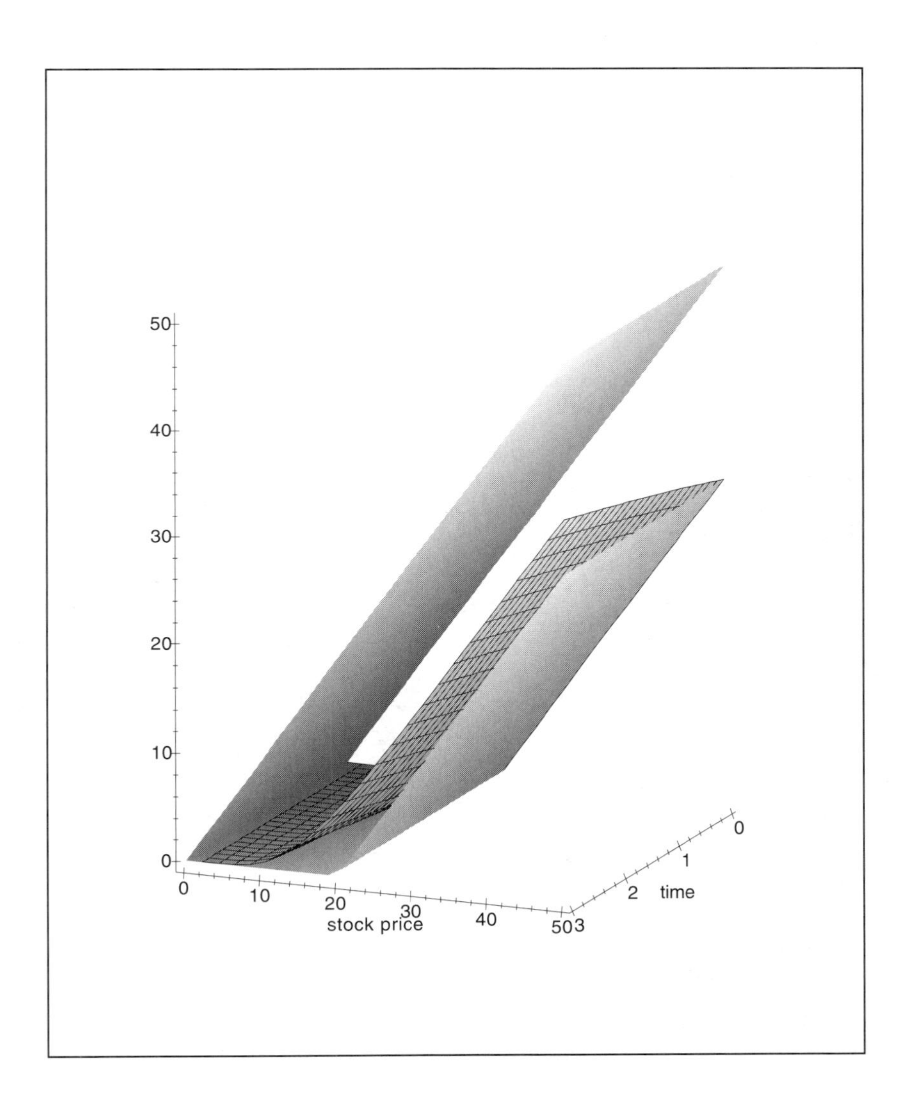

Figure 6.1: The Value of a Call Option and its Bounds as a Function of S and t

We can verify this by executing the **BstPut** procedure with the same parameters.

```
> BstPut(90,1/12,95,0.18,0.15);
```

$$.22622128$$

Alternatively, we can calculate the price of a put option from first principles as we did for the price of a call option. We thus only need to change the payoff function $h(s)$ in equation (5.7). The payoff from a put option is $\max(K - S, 0)$. We should thus replace, in equation (5.8), the payoff from a call option $\max(S - K, 0)$ with the former. Hence, we need to evaluate the following integral:

```
> PutAlt:=int((K-s)*psi(s,t,P,r,sigma),s=0..K);
```

$$PutAlt := \int_0^K \frac{1}{2} \frac{(K-s)\, e^{(-r\,t)} \sqrt{2}\, e^{(-1/2 \frac{(\ln(s)-\ln(P)-(r-1/2\,\sigma^2)\,t)^2}{\sigma^2\,t})}}{s\sqrt{\pi\,\sigma^2\,t}}\, ds$$

Those readers that skipped the derivation of the formula in the beginning of this section may want to skip the rest of this section. We now rewrite the integral in terms of the density of Y as we did in the former section. The resulting equation is in terms of the **erf** function as follows:

```
> PutAlt:=expand(exp(-r*t)*int((K-P*exp(y))*Normalpdf\
> (y,(r-sigma^2/2)*t,sigma*sqrt(t)),y=-infinity..ln(K/P)));
```

$$PutAlt := \frac{1}{2} \frac{K\sqrt{\pi}\sqrt{t}\operatorname{erf}(\frac{1}{2}\frac{\sqrt{2}\ln(K)}{\sigma\sqrt{t}} - \frac{1}{2}\frac{\sqrt{2}\ln(P)}{\sigma\sqrt{t}} - \frac{1}{2}\frac{\sqrt{2}\sqrt{t}\,r}{\sigma} + \frac{1}{4}\sqrt{2}\,\sigma\sqrt{t})}{e^{(r\,t)}\sqrt{\pi\,t}}$$

$$-\frac{1}{2} \frac{P\sqrt{\pi}\sqrt{t}\operatorname{erf}(\frac{1}{2}\frac{\sqrt{2}\ln(K)}{\sigma\sqrt{t}} - \frac{1}{2}\frac{\sqrt{2}\ln(P)}{\sigma\sqrt{t}} - \frac{1}{4}\sqrt{2}\,\sigma\sqrt{t} - \frac{1}{2}\frac{\sqrt{2}\sqrt{t}\,r}{\sigma})}{\sqrt{\pi\,t}}$$

$$+\frac{1}{2}\frac{K\sqrt{\pi}\sqrt{t}}{e^{(r\,t)}\sqrt{\pi\,t}} - \frac{1}{2}\frac{P\sqrt{\pi}\sqrt{t}}{\sqrt{\pi\,t}}$$

Simplifying and collecting terms this expression yields

```
> PutAlt:=collect(simplify(PutAlt),[P,K]);
```

$$PutAlt := (-\frac{1}{2}\operatorname{erf}(\frac{1}{4}\frac{\sqrt{2}\,(2\ln(K) - 2\ln(P) - t\,\sigma^2 - 2\,r\,t)}{\sigma\sqrt{t}}) - \frac{1}{2})\,P$$

$$+ (\frac{1}{2}\operatorname{erf}(\frac{1}{4}\frac{\sqrt{2}\,(2\ln(K) - 2\ln(P) - 2\,r\,t + t\,\sigma^2)}{\sigma\sqrt{t}})\,e^{(-r\,t)} + \frac{1}{2}e^{(-r\,t)})\,K$$

or in terms of the normal distribution function N, the expression below (which can be simplified further recalling that $N(z) = 1-N(-z)$) is obtained.

```
> NewPutAlt:=collect((simplify(subs(erf=newerf,PutAlt)),\
> [P,exp(-r*t),K]));
```

$$NewPutAlt := (\mathrm{N}(\frac{1}{2}\frac{-2\ln(K)+2\ln(P)+t\,\sigma^2+2\,r\,t}{\sigma\sqrt{t}})-1)\,P$$
$$+(-\mathrm{N}(\frac{1}{2}\frac{-2\ln(K)+2\ln(P)+2\,r\,t-t\,\sigma^2}{\sigma\sqrt{t}})+1)\,K\,e^{(-r\,t)}$$

We now define the function **PutAlt** in MAPLE as we did earlier with the function **Eqbs** for the call option.

```
> PutAlt:=unapply(PutAlt,K,t,P,sigma,r);
```

$$PutAlt := (K,\,t,\,P,\,\sigma,\,r) \to$$
$$(\frac{1}{2}\,\mathrm{erf}(\frac{1}{4}\frac{\sqrt{2}\,(-2\ln(K)+2\ln(P)+2\,r\,t+\sigma^2\,t)}{\sigma\sqrt{t}})-\frac{1}{2})$$
$$P+(-\frac{1}{2}$$
$$\mathrm{erf}(\frac{1}{4}\frac{\sqrt{2}\,(-2\ln(K)+2\ln(P)+2\,r\,t-\sigma^2\,t)}{\sigma\sqrt{t}})\,e^{(-r\,t)}$$
$$+\frac{1}{2}\,e^{(-r\,t)})\,K$$

Next we evaluate the price of a put option, with the parameters as above, to confirm that we receive the same value.

```
> evalf(PutAlt(90,1/12,95,0.18,0.15));
```

$$.22622128$$

```
> BstPut(90,1/12,95,0.18,0.15);
```

$$.22622128$$

The reader who followed this section on-line without skipping any commands now has the function **Eqbs** defined in the computer memory. Both **PutAlt** and **Eqbs** are not defined in an efficient way for valuation but they allow us to perform symbolic manipulation. We can thus check the integrity of the put-call parity relationship in a symbolic manner.

According to the put-call parity, the price difference between a put and a call (on the same stock with the same exercise price and time to expiration) should be equal to the difference between the current price of the stock and the present value of the exercise price. Recall that this relationship is actually a consequence of the fact that the payoff, at maturity, from a portfolio containing a long position in a call and a short position in a put is equivalent to the value of the stock at maturity, minus the exercise price.

Hence, at any other time prior to maturity, the value of the long-call, short-put portfolio is equivalent to the present value of the stock, at maturity, minus the present value of the exercise price. The present value of the stock at maturity, at a time prior to maturity, is obviously the current price of the stock.[2] The present value of the exercise price is obtained by multiplying K by $e^{-r_f t}$. This is verified below in a symbolic manner.

```
>  simplify(Eqbs(K,t,P,sigma,rf)-PutAlt(K,t,P,sigma,rf));
```

$$P - K\,e^{(-rf\,t)}$$

In the same manner as we visualize the bounds on a call option, we also visualize the lower bound on the value of a put option. The lower bound of the put option was developed in equation (4.9) as

$$\max\left(0, Ke^{-r_f t} - S\right) \leq P_{Put(S,K)}. \tag{6.9}$$

In this case, the bound is not static in time as it was in the case of the call option. Also, the animation (in the on-line version) displays the static value of the put option versus the bounds and the true value of the call at each time. In contrast to the call option, the value of a put option is not larger than its value at maturity for every stock price.

```
>  plots[animate]({Put(P,20),max(0,(-P+20*exp\
>  (-.10*(3-t)))),'BstPut(20,3-t,P,0.25,0.10)'},\
>  P=0..50,t=0..3,axes=NORMAL,color=red,thickness=2,\
>  labels=[`Stock Price`,`Call Value`]);
```

Can you provide the intuitive explanation for this animation?

6.3 Combinations across Time

Now that we have a model for pricing options, we can investigate trading strategies across time. We have already explored trading strategies in Chapter 4. These were limited to strategies involving options which all matured at the same time. Furthermore, because we did not yet have a pricing formula, our attention was limited to the payoffs at expiration.

In this section we will look at the value of portfolios of options across time, including portfolios composed of options which mature at different

[2]This is nearly a tautology. If you hold the stock from now until the maturity of the option, its value at maturity will be the value of the stock. Thus, its value now is its present value.

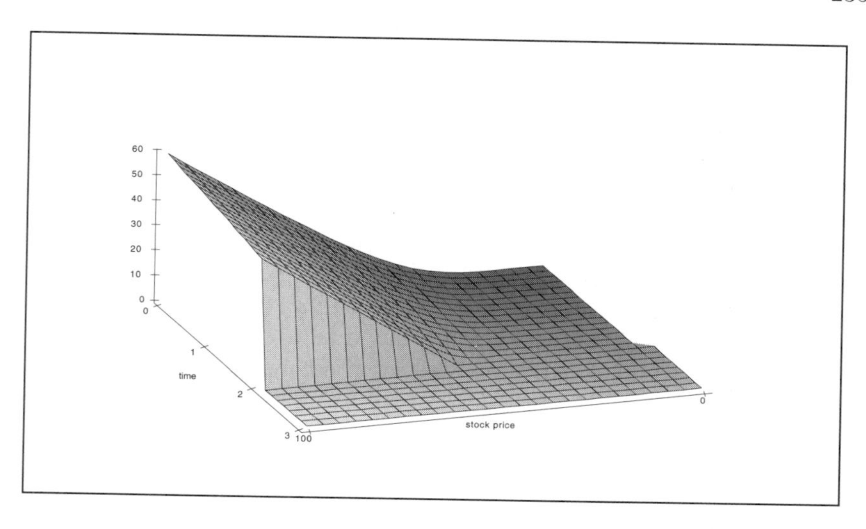

Figure 6.2: The Value of a Call Option before and after Maturity

times. In so doing, we assign a value of zero to an option whose maturity time has passed and will plot the graph in such a way that the time axis, the variable t, is calendar time, and $t = 0$ is the current time. Hence, at the maturity of the option, we will observe the graph of the payoff at maturity, and for t greater than the maturity of the option, we will observe a value of zero for each possible realization of the stock price.

Note that in order to generate this type of a graph, the parameter which usually denotes time to maturity is now simply labeled as *time*. This is illustrated below in Figure 6.2 where the value of the call option is plotted for a range of stock prices from \$0 to \$100 and for calendar time varying from zero to three. The call option matures at time two.

```
> plot3d('BstCall(50,2-t,s,.25,.10)',s=0..100,t=0..3,\
> labels=[`Stock Price`,`Time`,`Call Value`],\
> axes=normal,orientation=[70,49],title=`Figure 6.2:\
> Value of a Call Option before and after Maturity\
```

We commence by revisiting some of the option combinations whose payoffs at maturity were investigated in Chapter 4. Now, with the aid of the pricing formula, we can investigate the evolution of the value of these combinations across time. Most of the visualization in this Chapter will be done utilizing the three-dimensional capability of MAPLE. This is in sharp contrast to our previous examination of these combinations. Previously, when we investigated the payoffs of the option combinations at maturity, we were examining a function of one variable. This variable was the value of the combination at maturity for different values of the stock price (or, as they were then referred to, the states of nature).

Here, we are investigating the evolution of option combinations across time and, hence, we graph the value of the particular combination for a range of the stock prices at each time t. At the maturity of the option, $t = 2$ in Figure 6.2, the graph is the payoff at maturity. If we generate the payoff or value of the combination for a continuous range of calendar time t, the range $[0, 3]$ in Figure 6.2, we have an infinite collection of payoff or value functions, one for each t . In other words, we have a collection of functions, each of one variable, parameterized (or indexed) by t. We can therefore also demonstrate the evolution of the value across time by means of animation. This is a technique we have used in prior sections. This is done simply by graphing the value of the function for each t frame by frame, and running these frames one after the other.

We start with the simplest portfolio, one which includes only one call option. Let us look again at the value of this call option as time approaches the expiration of the option. Run the animation and watch how, as time progresses, the graph smoothly approaches the piecewise-linear payoff at maturity. Observe how as time approaches maturity, for a stock price smaller than the exercise price, the value of the call is nearly zero. As time approaches maturity, this range is closer and closer to the exercise price. For the situation where the stock price is smaller than the exercise price of the option, as time approaches maturity, the likelihood that this option will end up on or at the money decreases. This is reflected in the price of the call. Indeed, the price of the call is calculated as the expected value of the payoff. When the stock price is smaller than the exercise price, the risk-neutral probability is such that the expected value of the call[3] is smaller (Figure 6.3).

```
> plots[animate]('BstCall(50,1-t,P,.25,.10)',t=0..1,\
> P=20..70,frames=30,thickness=3,color=blue,\
> labels=['Stock Price', 'Call Value'],title='Figure 6.3:\
> Animation of the Value of the Call Option as Time\
> Approaches Maturity');
```

[3] We shall soon investigate the sensitivity of the price of the call to changes in the stock price. In this section, however, we are interested mainly in the visualization of different combinations across time.

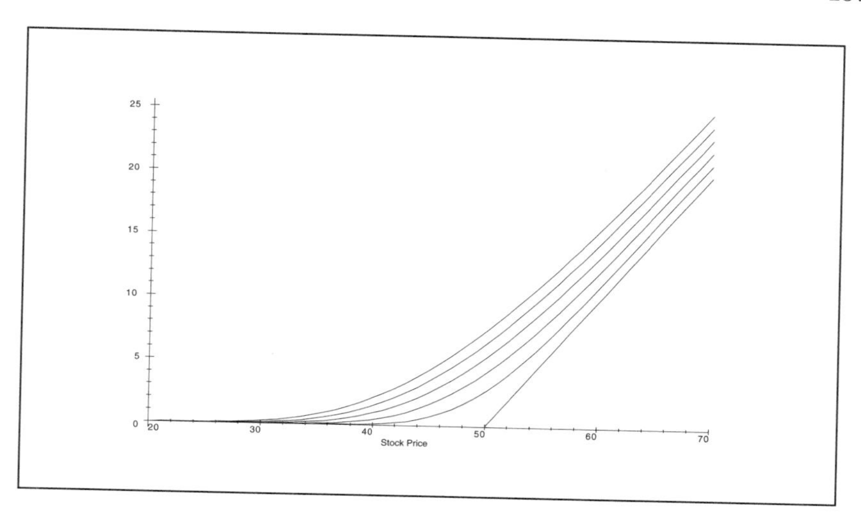

Figure 6.3: Animation (Static Version) of Call Option Value as Time Approaches Maturity

If we "put these graphs next to each other", a three-dimensional graph is generated. This is the graph illustrated in Figure 6.2. An intuitive way to think about the three-dimensional graph is as a collection of two-dimensional graphs, one for each t. The reader may like to take a second look at Figure 6.2 now viewed from a different perspective. If you run the command below, you will witness the t-axis in front as if the collection of one-dimensional graphs, indexed by t (the frames of animations), are placed next to each other.

```
> plot3d('BstCall(50,2-t,s,.25,.10)',s=0..100,t=0..3,\
> labels=[`Stock Price`,`Time`,`Call, Value`],\
> axes=normal,orientation=[-148,75],title=`Figure 6.4:\
> Call Value as Time and Stock Price Change`);
```

When examining the combination in Figure 6.4, it is convenient to recall that MAPLE permits the visualization of these graphs from different perspectives. Make use of this MAPLE feature and ensure that you fully appreciate all the angles of the combinations shown below. Of course, you can also view the equivalent image for a put option by executing the command below (Figure 6.5).

```
> plot3d('BstPut(50,2-t,s,.25,.10)',s=0..100,t=0..3,\
```

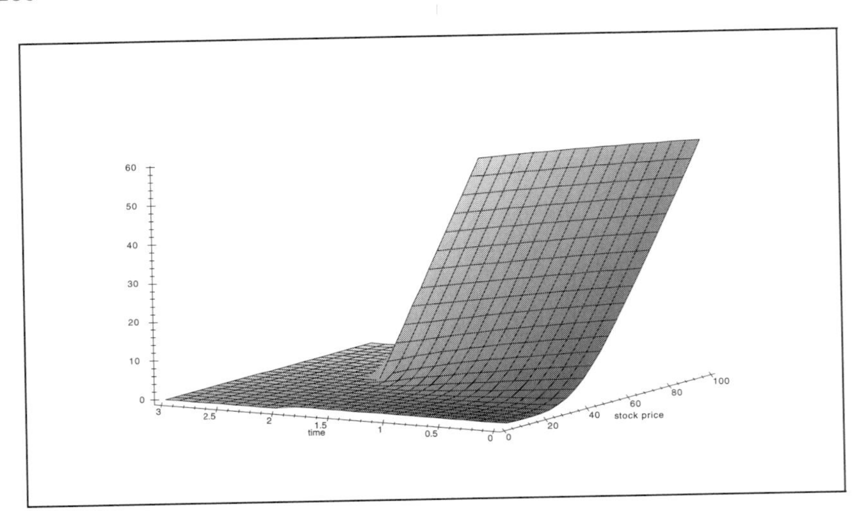

Figure 6.4: Call Value as Time and Stock Price Change

```
> labels=['Stock Price','Time','Call Value'],\
> axes=normal,orientation=[11,49],title='Figure 6.5:\
> Put Value as Time and Stock Price Change');
```

We now begin an examination of combinations which are less simple. We start with a portfolio that is not affected by the price of the stock. In Figure 6.6 we graph the put–call parity relationship for different calendar times. The portfolio is composed of a long call, a long stock, and a short put on the same stock, both with the same maturity time. We already know from equation (4.12) that the value of such a portfolio is the present value of the exercise price. Hence, the image is a nearly linear graph. Since the present value effect is not pronounced, what we see is the exercise price multiplied by $e^{-0.10t}$, and thus the graph depends only on calendar time. At the maturity time of the options $t = 2$, the value of the portfolio is the exercise price. At any time prior to maturity, the value of the portfolio is the present value of the exercise price, regardless of the price of the stock.

```
> plot3d(s-('BstCall(50,2-t,s,.18,.15)'\
> -'BstPut(50,2-t,s,.18,.15)'),s=0..100,t=0..2,\
> labels=['Stock Value','Time','Portfolio Value'],\
> axes=normal,orientation=[158,59],title='Figure 6.6:\
> Portfolio Consisting of a Call Option and a Put Option');
```

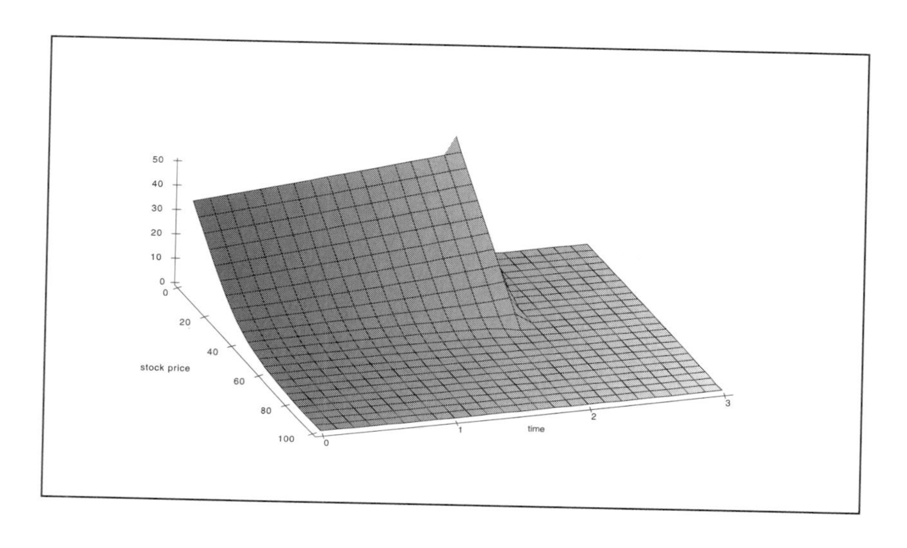

Figure 6.5: Put Value as Time and Stock Price Change

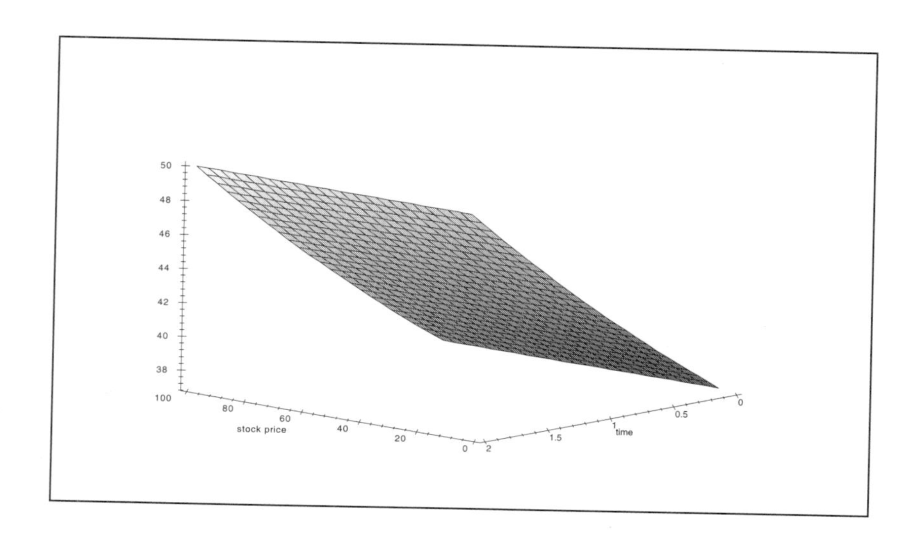

Figure 6.6: Portfolio Consisting of a Short Call a Stock and a Put Option

The reader may like to try and change the value of the interest rate from 0.15 to 0.45 (from 15 to 45 percent) and rerun the **plot3d** command just to see the curvature of the graph. (Note that such values are not at all realistic, assuming the absence of very strong inflation.)

The reader should approach the understanding of option combinations now exactly the same way as option combinations were approached earlier in the book. This is true even though previously we only examined the payoffs of option combinations at maturity, and here we are able to examine the combinations in more details. There are some basic building blocks. These can be put together in different ways to generate the desired payoffs or portfolio values. In order to refresh the reader's memory, we will examine the combinations just discussed from different perspectives and graph the combinations as well as their basic building blocks on the same set of axes. The recipe is the same, but the building blocks are three-dimensional graphs now[4] rather than two-dimensional.

Consider first combinations of options which have a common expiration date. These have been investigated in Chapter 4, but looking only at the payoffs from such combinations at their maturity date. Here, it can be seen how the values of these option combinations converge to the payoff at maturity from their values at times prior to maturity.

Consider the combination of a put option and a call option with the same exercise price (and the same maturity). This is a *straddle*, and its payoff diagram (representing payoff at maturity only) was shown in Figure 4.19. We begin by plotting the two basic building blocks on the same set of axes in Figure 6.7 — a put option which matures at time two, and a call option maturing at the same time, both with an exercise price of $50.

```
> plot3d({'BstPut(50,2-t,p,.28,.11),\
> BstCall(50,2-t,p,.28,.11)'},t=0..2,p=0..100,\
> labels=[`time`,`stock value`,``],\
> axes=normal,orientation =[-154,73],title=`Figure 6.7:\
> The Building Block of a Straddle`);
```

Note how the building blocks behave much like the payoff diagram at maturity which we saw in Chapter 4. Just prior to maturity, the graphs are smoother and have curvature rather than the two piecewise-linear functions as before. The reader might wish to run the next command in order to examine the same graphs from a different perspective.

---

[4]In the hard copy of the book, all of the images will not be reproduced, but the reader will see the command which generates the graph. The reader who is interested in seeing the image can run the on-line version.

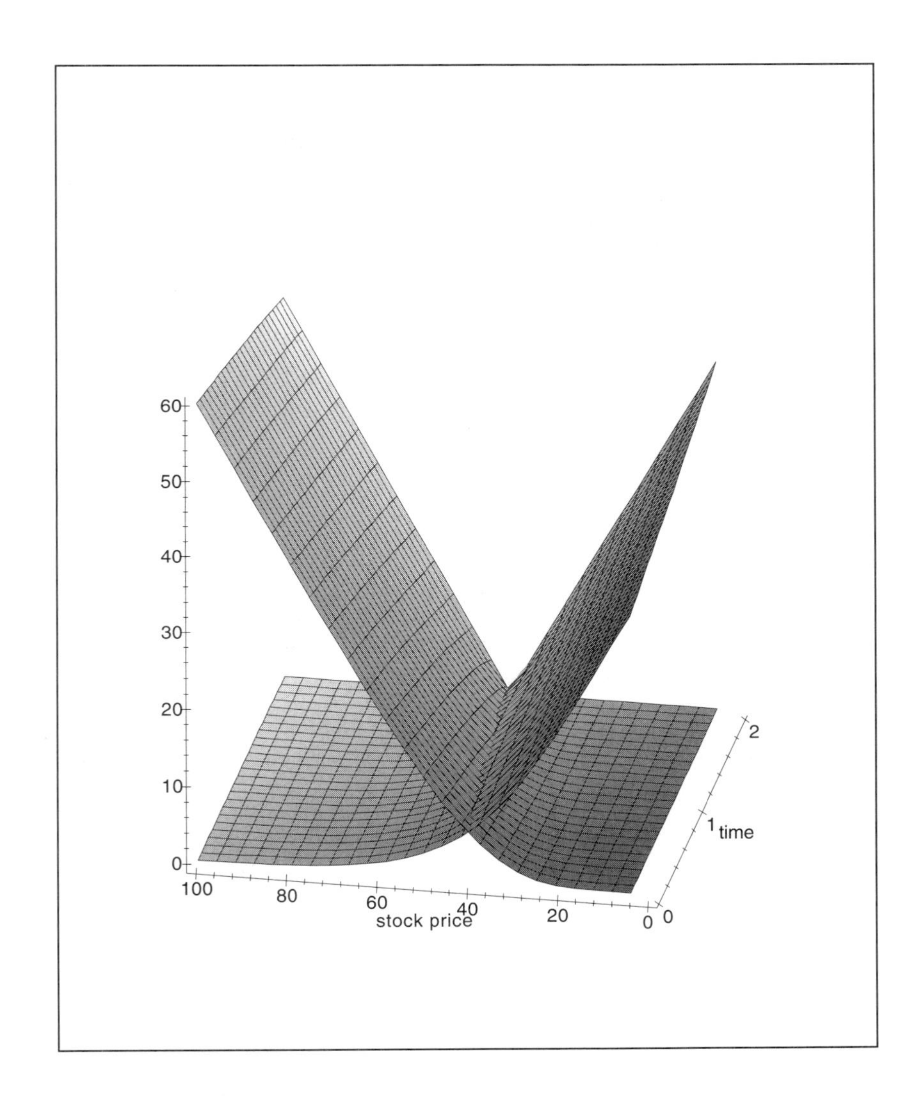

Figure 6.7: The Building Blocks of a Straddle

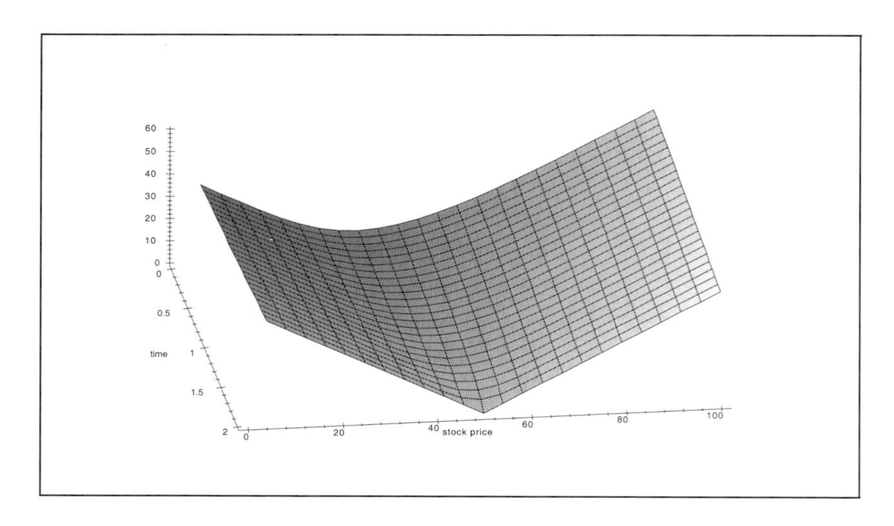

Figure 6.8: Straddle

```
> plot3d({'BstPut(50,2-t,p,.28,.11),\
> BstCall(50,2-t,p,.28,.11)'},t=0..2,p=0..100,\
> axes=normal,orientation=[-2,68],\
> labels=[`Time`,`Stock Value`,`Portfolio Value`]);
```

We can observe the portfolio value (rather than the building blocks) across time, as shown in Figure 6.8.

```
> plot3d('BstPut(50,2-t,p,.28,.11)+\
> BstCall(50,2-t,p,.28,.11)',t=0..2,p=0..100,axes=normal,\
> labels=[`time`,`stock price`,` `],\
> orientation=[-8,40],title=`Figure 6.8:  Straddle`);
```

Note that the slice of the graph which corresponds to the maturity date of the two options, at $t = 2$, is exactly the image which was viewed in Chapter 4. That image demonstrated the payoff of this combination at maturity.

Consider now portfolios of options which do not share their maturity date. The options making up these combinations differ only in their maturity date. Let us take a look at a portfolio composed of a long put and a long call, which mature at different times. The call option matures at time one, and the put option matures at time two. We begin our examination by plotting in Figure 6.9 the basic building blocks of the portfolio on the same set of axes.

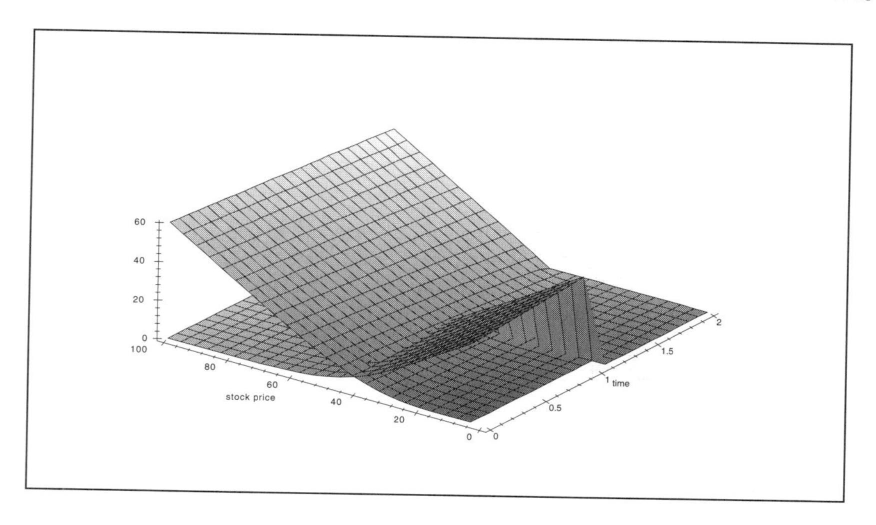

Figure 6.9: The Basic Building Blocks of a Portfolio Composed of Options with Different Maturities

```
> plot3d({'BstPut(50,1-t,p,.28,.11),\
> BstCall(50,2-t,p,.28,.11)'},t=0..2,p=0..100,\
> ,labels=[`time`,`stock price`,` `],\
> axes=normal,orientation=[145,39],title=`Figure 6.9:\
> The Basic Building Blocks of a Portfolio Composed\
> of Options with Different Maturities`);
```

The reader may wish to experiment with changing the value of the exercise price for one of the two options and so examine the effect on the basic building blocks. Consider the portfolio values in Figure 6.10 and imagine what the impact of changing the exercise price of one of the options would be on the value of the portfolio.

```
> plot3d('BstPut(50,1-t,p,.28,.11)+\
> BstCall(50,2-t,p,.28,.11)',t=0..2,p=0..100,\
> labels=[`time`,`stock Value`,``],\
> axes=normal,orientation=[19,45],title=`Figure 6.10:\
> Portfolio of Options with Different Maturities`);
```

Note again that at the maturity date of the option which matures later, we observe a piecewise-linear graph. This is the "slice" of the three-dimensional graph which corresponds to $t = 2$. Changing the exercise price of one of the

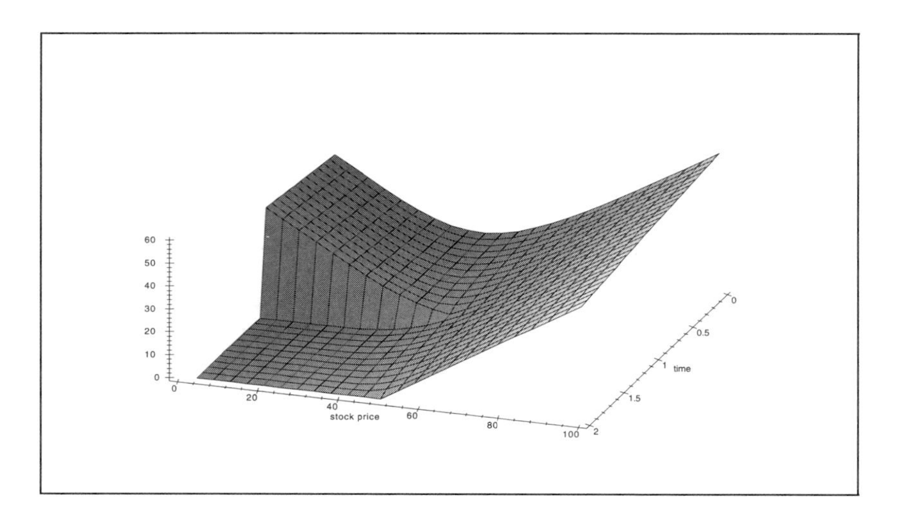

Figure 6.10: Portfolio of Options with Different Maturities

options will wedge a valley between the upward sloping parts of the graph. The reader may wish to rerun the above command to visualize these effects. This can be done by executing the MAPLE command below.

```
> plot3d('BstPut(50,1-t,p,0.28,0.11)'+\
> 'BstCall(90,2-t,p,0.28,0.11)',t=0..2,p=0..100,
> labels=[`Time`,`Stock Value`,`Portfolio Value`]\
> axes=normal,orientation=[-2,68]);
```

Without the luxury of three-dimensional viewing capabilities, the convention is to view the two-dimensional graph which corresponds to the maturity time of the shorter maturity option. This is illustrated in Figure 6.11. This graph is the slice of the three-dimensional image cut at $t = 1$. This slicing technique for extracting two-dimensional images from the three-dimensional ones should be becoming familiar.

```
> plot('BstPut(50,0,p,.28,.11)'+\
> 'BstCall(90,1,p,.28,.11)',p=0..100,axes=normal,labels=\
> [`stock price`,` `],title=Figure 6.11:\
> Slice of the Three-Dimensional Image Cut at t=1`);
```

Consider now the option portfolio combination which is known as a *calendar spread.* This is a portfolio composed of a short and long position in

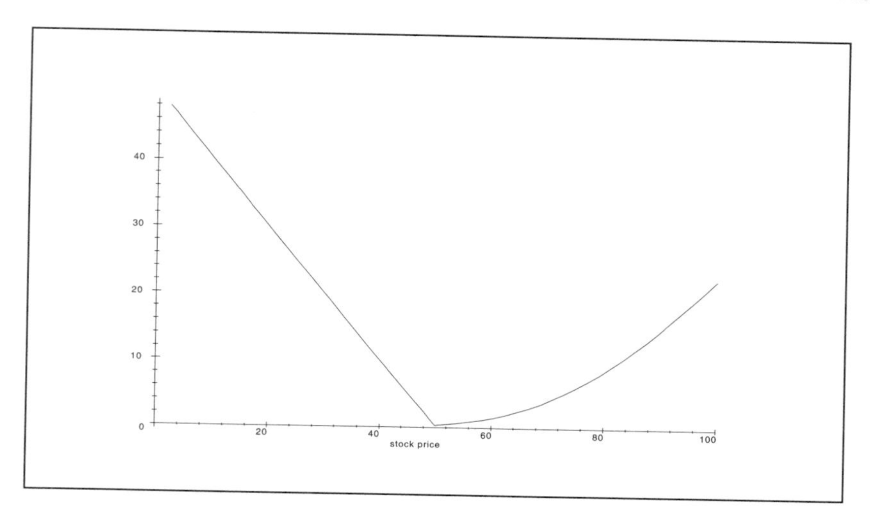

Figure 6.11: Slice of the Three-Dimensional Image Cut at $t = 1$

options (either both calls, or both puts) on the same stock, but with different maturity dates. We form our example of a calendar spread from a long call position with a relatively later maturity date, and a short call position with a relatively earlier maturity date. First, note the cost (value) of the portfolio at calendar time $t = 0$ when the price of the underlying stock is \$90.

```
> BstCall(90,2-0,40,.28,.11)-BstCall(90,1-0,40,.28,.11);
                         .59455998
```

This portfolio has positive value on its inception date. The call with the later maturity date is more valuable than the call with the sooner maturity date. This is a general property, and thus the value of such a portfolio is always positive. Can you explain why? If not, wait until Section 7.1 where this issue is discussed.

The three-dimensional visualization of the calendar spread's value is presented in Figure 6.12. Note that for the range of the stock prices shown, at $t = 0$, the graph is positive. The reasoning is identical to that discussed above.

```
> plot3d('BstCall(50,2-t,p,.28,.11)-\
> BstCall(50,1-t,p,.28,.11)',t=0..2,p=0..100,\
> axes=normal,orientation=[-8,40],\
```

```
> labels=[`time`,`stock price`,` `],\
> title=`Figure 6.12:  Calendar Spread`);
```

We make use of our (by now familiar) slicing technique to focus on the strip cut from the three-dimensional graph at $t = 1$. This corresponds to the maturity date of the option with the shorter maturity. In the absence of three-dimensional capabilities this slice is the most commonly shown. It is displayed in Figure 6.13.

```
> plot('BstCall(50,2-1,p,.28,.11)-\
> BstCall(50,1-1,p,.28,.11)',p=0..100,axes=normal,\
> labels=[`stock price`,` `],title=\
> `Figure 6.13:  Slice of the Previous 3D Graph at t=1`);
```

A calendar spread can also be created from a portfolio composed of two put options. The put option with the longer maturity date is held long, and the put option with the shorter maturity date is held short. This alternative version of a calendar spread is shown in Figure 6.14.

```
> plot3d('BstPut(50,2-t,p,.28,.11)-\
> BstPut(50,1-t,p,.28,.11)',t=0..2,p=0..100,\
> labels=[`time`,`stock price`,` `],\
> axes=normal,orientation=[151,44],title=`Figure 6.14:\
> A Calendar Spread Created from a Portfolio Composed\
> of Two Put Options`);
```

Note that for the calendar spread in Figure 6.14, the portfolio has negative value (for some values of the underlying stock price) during the time period from zero to one. Can you explain why this is so? One of the exercises of this Chapter asks the reader to prove it. Below, we again focus on the strip cut from Figure 6.14 at the maturity of the shorter put option. The two-dimensional graph in Figure 6.15 displays this strip — the value of the portfolio at the shorter maturity.

```
> plot('BstPut(50,2-1,p,.28,.11)-\
> BstPut(50,1-1,p,.28,.11)',p=0..100,axes=normal,\
> labels=[`stock price`,` `],title=\
> `Figure 6.15:  A Slice of the Previous 3D Plot`);
```

The figure showing the maturity value of the calendar spread created from call options, Figure 6.13, and that showing the maturity value of the calendar spread created from put options, Figure 6.15, look similar. They need not. The reader might be interested in investigating the building blocks

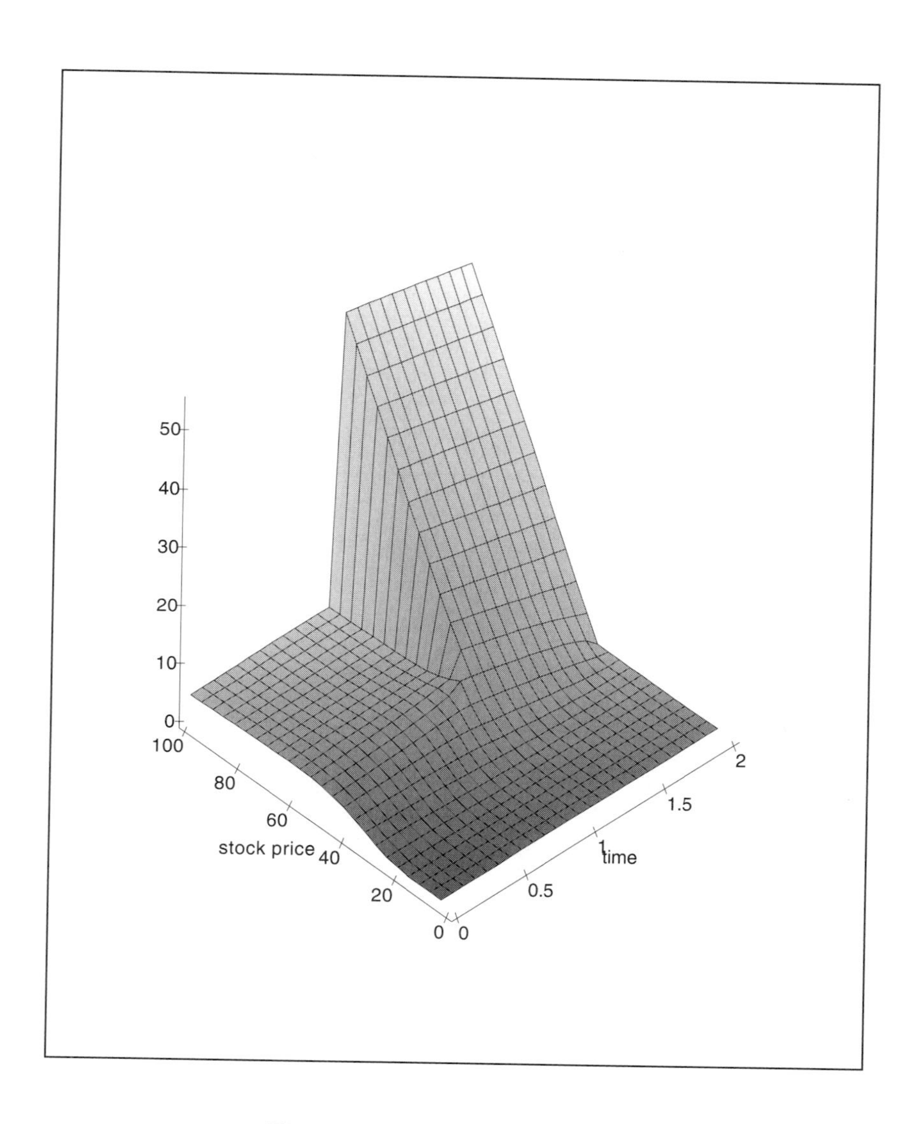

Figure 6.12: A Calendar Spread

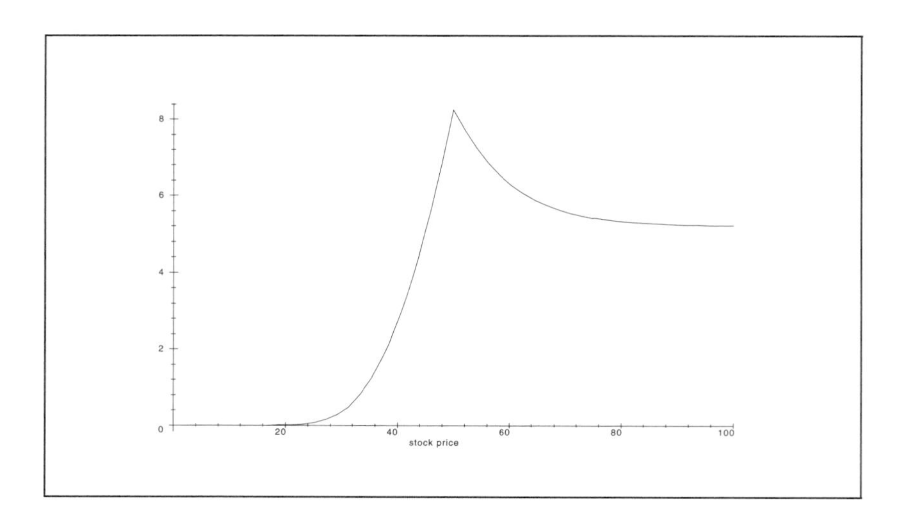

Figure 6.13: Slice of the Previous 3D Graph at $t = 1$

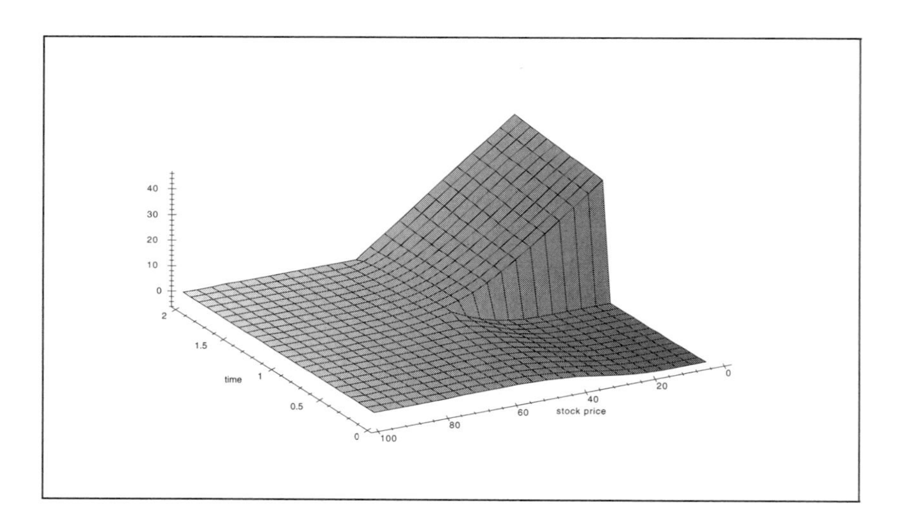

Figure 6.14: A Calendar Spread Created from a Portfolio Composed of Two Put Options

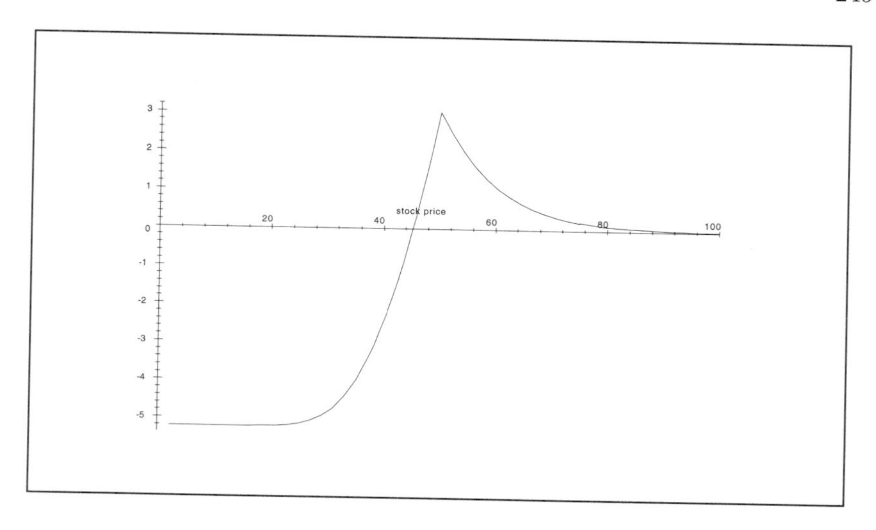

Figure 6.15: A Slice of the Previous 3D Plot

of the two positions. This can be done by executing the MAPLE command below.

```
> plot3d({'BstCall(50,2-t,p,.28,.11)',\
> -'BstCall(50,1-t,p,.28,.11)'},t=0..2,p=0..100,\
> labels=[`Time`,`Stock Value`,`Portfolio Value`],\
> axes=normal,orientation=[-156,54]);
```

The reader may want to run the animation below, in the on-line book, to see how different the two calendar spreads are. Here is an animation of the put calendar spread.

```
> plots[animate]('BstPut(50,2-t,p,.28,.11)-\
> BstPut(50,1-t,p,.28,.11)',p=0..100,t=0..2,\
> colour=red,title=`Put Calendar Spread Animation`);
```

The animation for the call calendar spread is displayed below (in the on-line book).

```
> plots[animate]('BstCall(50,2-t,p,.28,.11)-\
> BstCall(50,1-t,p,.28,.11)',p=0..100,t=0..2,\
> colour=blue,title=`Call Calendar Spread Animation`);
```

Recall the procedure **Stripay**. Given a contingent cash flow profile, which the investor desires at a given maturity date, the procedure determines the portfolio of call and put options which will create that profile. We are now in a position to investigate the evolution of such a portfolio through time. The procedure **PayoffCost** uses as its input the portfolio solution from **Stripay**, and solves for the cost of that portfolio given the usual parameters.

The reader is encouraged to experiment with different option combinations. It might be handy to make use of the procedure **Stripay** for such experimentation. **Stripay** defines certain variables so the payoff can be valued by **PayoffCost**. **PayoffCost** can be used to generate the three-dimensional images of the combination across time. Here is how this works.

Consider the following payoff solved for by **Stripay**. The function **Pay** and the needed variables for **PayoffCost** are defined by running **Stripay**.

```
> Stripay([[0,0],[5,0],[6,27],[8,35],[10,45],\
> [12,50],[15,50],[18,30],[20,15],[22,-20],\
> [25,-10],[28,-5],[30,0],[35,0]],Pay):
```

$$Buy, 312, put\ at\ strike\ price, 5$$

$$Short, 285, call\ at\ strike\ price, 5$$

$$Short, \frac{661}{2}, put\ at\ strike\ price, 6$$

$$Buy, \frac{615}{2}, call\ at\ strike\ price, 6$$

$$Buy, 1, put\ at\ strike\ price, 8$$

$$Short, \frac{5}{2}, put\ at\ strike\ price, 10$$

$$Short, \frac{5}{2}, call\ at\ strike\ price, 12$$

$$Short, \frac{20}{3}, call\ at\ strike\ price, 15$$

$$Short, \frac{5}{6}, call\ at\ strike\ price, 18$$

$$Short, 10, call\ at\ strike\ price, 20$$

$$Buy, \frac{125}{6}, put\ at\ strike\ price, 22$$

$$Short, \frac{5}{3}, put\ at\ strike\ price, 25$$

$$Buy, \frac{5}{6}, put\ at\ strike\ price, 28$$

$$Short,\ \frac{5}{2},\ call\ at\ strike\ price,\ 30$$

The payoff as a function of the stock price, s, is, Pay, (.)

The reader can plot the function **Pay** by issuing the command in order to see the payoff at maturity from the portfolio solved for above.

```
> plot(Pay(S),S=0..40);
```

In order to examine the evolution of this portfolio across time, the reader should run **PayoffCost** as below. The first argument is the set of knot points, the second argument is the time to maturity, the next argument is the current price of the stock, and the last two arguments are the standard deviation and the risk-free rate, respectively. The three-dimensional graph is displayed in Figure 6.16. The reader may solve for the value of the portfolio, at a certain time for a certain value of the stock, by simply substituting into **PayoffCost** the appropriate values instead of the parameters.

```
> plot3d(PayoffCost([[0,0],[5,0],[6,27],[8,35],[10,45],\
> [12,50],[15,50],[18,30],[20,15],[22,-20],[25,-10],\
> [28,-5],[30,0],[35,0]],t,S,.23,.09),t=0..2,S=2..45,\
> labels=[time,‘stock price‘, ‘‘],\
> axes=normal,orientation=[158,49],title=\
> ‘Figure 6.16:  PayoffCost‘);
```

Note how the slice of the graphs that corresponds to $t = 0$ is the same as the graph of the funciton **Pay** drawn above in the on-line version of the book.

Option combinations can also be created from options on different underlying securities. These can also be viewed. In order to do so, we will have to augment our states of nature. Recall that in Chapter 4 the states of nature for an option were defined as the possible prices of the underlying stock for that option. Once the price of the stock is determined, the payoff of the option in the future (at its maturity), and hence the value of the option (contingent claim) prior to maturity, is also known. If, however, we are dealing with options on different stocks, our set of possible states of nature is even "larger".

Every state of nature must now be defined by a couple (or a vector with two components). Each number corresponds to a possible stock price. In this way, the states of nature are all possible combinations[5] of two positive numbers.

[5] In more precise terminology, it is the Cartesian product of the nonnegative orthant of R with itself.

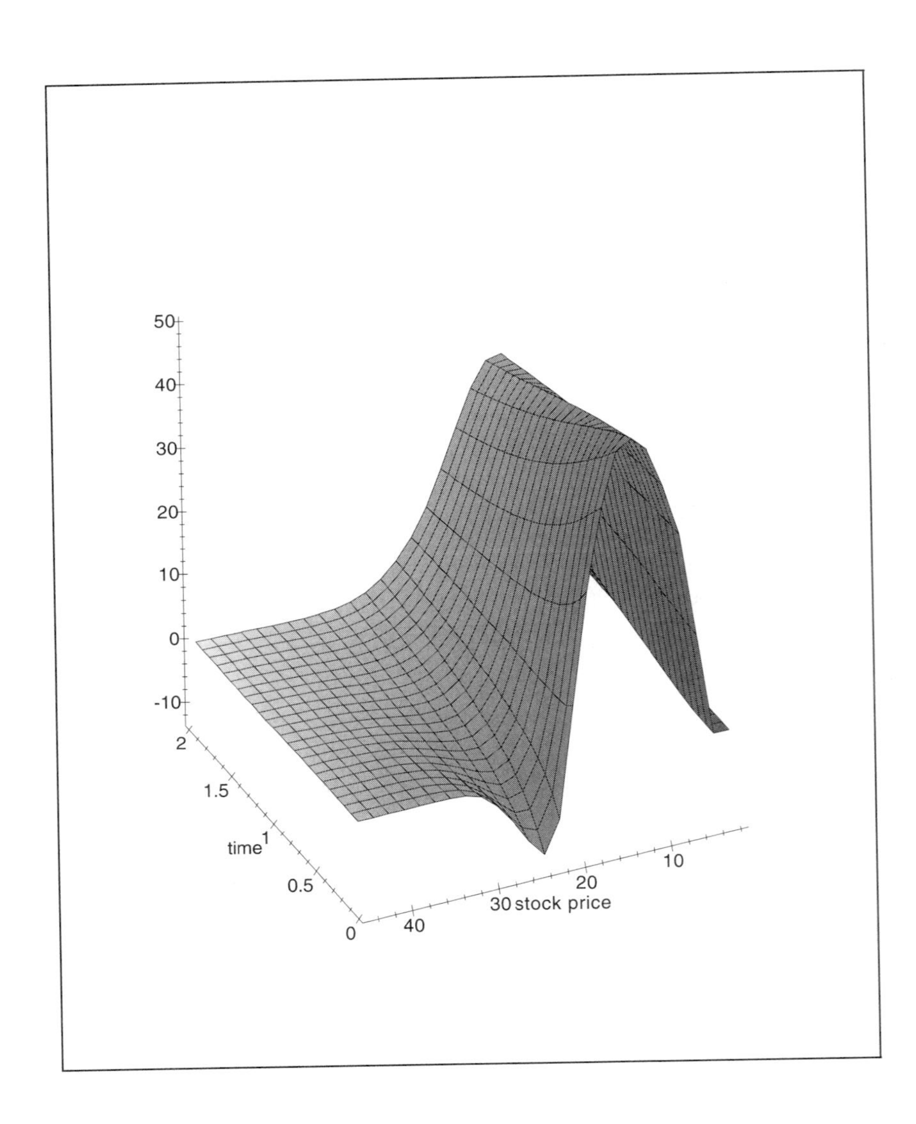

Figure 6.16: **PayoffCost**

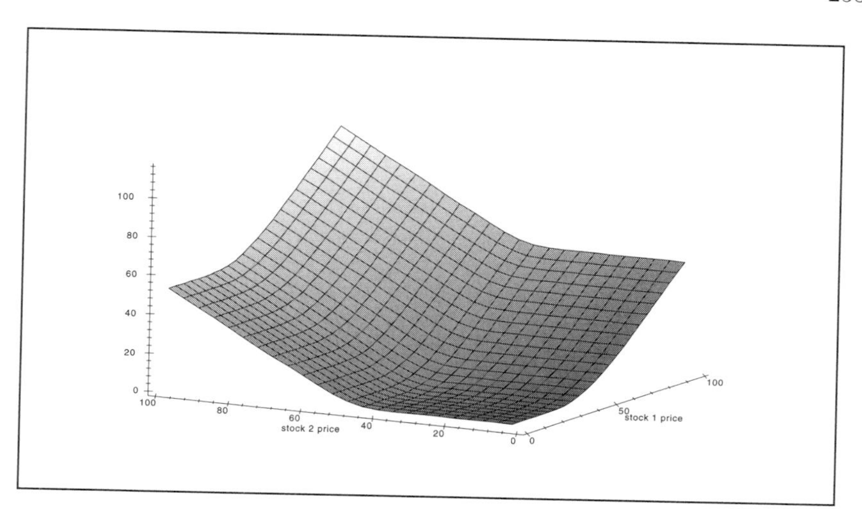

Figure 6.17: Two Call Options with the Same Maturity Written on Different Underlying Stocks

In order to visualize such positions, we require four dimensions: the states of nature occupy two dimensions, the value of the portfolio is the third, and time is the fourth. In order to show the value across time, we will use animation. We begin by examining the position already analyzed, composed of two call options with the same maturity but written on different underlying stocks. Therefore, there are two volatilities of which to take account: 0.28 and 0.18, respectively (Figure 6.17).

```
> plots[animate3d]('BstCall(50,2-t,s1,.28,.11)+\
> BstCall(50,2-t,s2,.18,.11)',s1=0..100,s2=0..100,\
> t=0..2,axes=normal,orientation =[-154,73],labels=\
> [`stock 2 price`,`stock 1 price`, ``],\
> title=`Figure 6.17:  Two Call Options With the Same\
> Maturity Written on Different Underlying Stocks`);
```

Such a graphing technique can also be used for other option combinations for which the times to maturity are different. For example, we might have different maturities for the options from the aforementioned portfolio combination (Figure 6.18).

```
> plots[animate3d]('BstCall(50,2-1,s1,.28,.11)+\
> BstCall(50,2-t,s2,.18,.11)',s1=0..100,s2=0..100,\
```

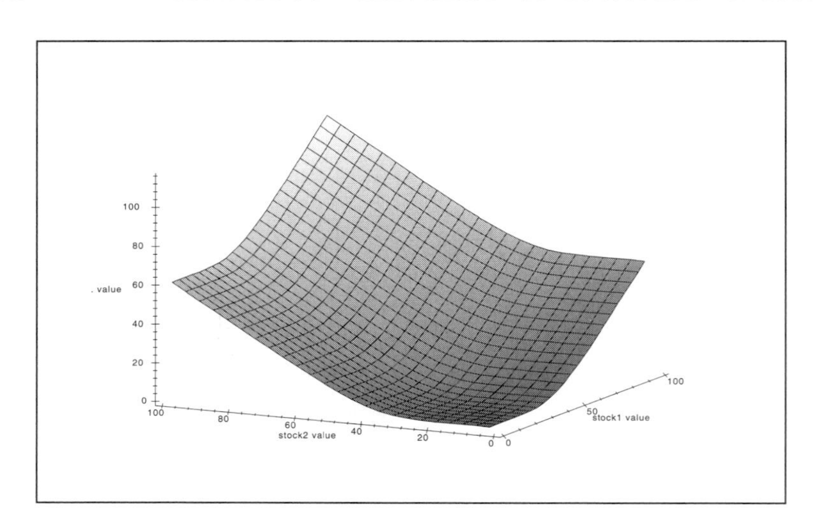

Figure 6.18: Two Call Options with Different Maturities Written on Different Underlying Stocks

```
> t=0..2,axes=normal, orientation =[-154,73],labels=\
> =[‘stock 1 price‘,‘ stock 2 price‘,‘ ‘],\
> title=‘Figure 6.18:  Two Call Options with Different\
> Maturities Written on Different Underlying Stocks‘);
```

The ability to visualize three-dimensional graphs of the position helps in judging the risk profile of different combinations across time. The reader is encouraged to try and visualize different positions using the methodology introduced here. With the aid of the animation, a three-dimensional animation can present changes with respect to four different parameters. This may be of interest to financial engineers exploring new positions and finding ways to visualize the risk to potential clients. The methods utilized here can also be explored by visualizing the sensitivity of positions to changes in certain parameters such as volatility. This will be further explored in the end-of Chapter exercises.

6.4 Dividends and Option Pricing

Until now, we assumed that the option is written on a stock that pays no dividend. Let us relax this assumption[6] and investigate its effect on the option pricing formula. One might initially believe that since the formula derived earlier represents the expected value of the option payoff at maturity, we can simply use this formula even when a dividend is paid. Such a perspective might even appear intuitive as the payoff from the option (at maturity) is not affected by the dividend on the stock. Hence, the expected value of the option payoff at maturity should also not be affected by dividends, and consequently the option pricing formula stays intact. Such logic is, however, faulty.

The pricing formula is indeed nothing but the expected value of the option payoff at maturity. It is, however, the expected value with respect to the risk-neutral probability. We already alerted the reader to the fact that this probability is specific to the stock in question. The dividend payment, as we shall soon see, affects the risk-neutral distribution of this stock and thereby the pricing formula for the option.

Let us assume that the stock is paying a known dividend yield. We cast this assumption in a continuous-time framework and assume that the stock produces a deterministic dividend (per each dollar invested in the stock) every instant. Thus, much like a risk-free investment is assumed to produce a continuous cash flow of r_f for every dollar invested in the account at time t, the dividend-paying stock produces a similar continuous cash flow. If we assume that the dividend yield is Di, the stock produces a continuous cash flow of Di for every dollar invested in the stock at time t.

This dividend yield thus has an effect on the continuously compounding rate of return of the stock and consequently affects the risk-neutral distribution of that stock. We maintain the assumption that the continuously compounded rate of return of the stock (which is due only to price appreciation and not dividends) is Y_t, and normally distributed with $\mu\, t$ and $\sigma\,\sqrt{t}$. The total continuously compounded rate of return on the stock will thus be $Y_t + Di$ and hence one plus the rate of return on the stock over a period of a length t is $e^{Y_t + Di\, t}$.

Based on equation (5.24) and our former discussion, under the risk-neutral probability the expected rate of return on the stock should equal the risk-free rate. Hence, we obtain equation (6.10),

$$E_{risk-neutral}\left(e^{Y_t + Di\, t}\right) = e^{r_f\, t}, \tag{6.10}$$

[6] See [5] and [4].

which the risk-neutral probability must now (when a dividend is paid) satisfy. Equation (6.10) is the counterpart of the equation

$$E_{risk-neutral}\left(e^{Y_t}\right) = e^{r_f\, t} \tag{6.11}$$

that we obtained for a stock that pays no dividend.

Since the Di is deterministic we can take e^{Dit} outside of the expectation operator (the "E") and obtain (using subindex f for E to note expectation under risk-neutral probability f)

$$E_f\left(e^{Y_t} e^{Dit}\right) = e^{r_f\, t} \tag{6.12}$$

or equivalently

$$E_f\left(e^{Y_t}\right) = e^{(r_f - Di)\, t}. \tag{6.13}$$

The intuitive explanation is as discussed earlier. Under the risk-neutral probability the expected rate of return should be equal across all securities in the market. Hence, it must equal the risk-free rate. Since the stock also yields a return which is based on dividends, its expected total return is the sum of return due to price appreciation and dividends. The total expected return, under the risk-neutral probability, must equal the risk-free rate. Thus, the expected return, under the risk-neutral probability, that is due to price appreciation must be modified to be the risk-free rate of interest minus the dividend rate.

Following the same steps as for a no-dividend paying stock we obtain a relation, the counterpart of equation (5.41), between the parameters of the risk-neutral probability.

$$\mu_f + \frac{\sigma^2}{2} = r_f - Di \tag{6.14}$$

The risk-neutral probability in this case is in the same family as the risk-neutral probability of a stock that pays no dividend except for the alteration of the value of μ_f. We should use

$$\mu_f = (r_f - Di) - \frac{\sigma^2}{2} \tag{6.15}$$

instead of $\mu_f = r_f - \frac{\sigma^2}{2}$. Thus, the risk-neutral probability will be obtained when, instead of r_f, we substitute $r_f - Di$ in the function **Lnrn**.

```
> Lnrn(s,t,P,r-Di,sigma);
```

$$\frac{1}{2}\,\frac{\sqrt{2}\,e^{\left(-1/2\,\frac{(\ln(s)-\ln(P)-(r-Di-1/2\,\sigma^2)\,t)^2}{\sigma^2\,t}\right)}}{s\,\sqrt{\pi\,\sigma^2\,t}}$$

The price of a call option will thus be, as before, the expected value of the payoff, but the $r - Di$ is substituted for r in **Lnrn** as above. But first we let MAPLE know about some of our assumptions again.

```
> assume(t>0,sigma>0,Di>0,P>0);
> CallPriceDividend:=simplify(exp(-r*t)*int((s- K)*\
> Lnrn(s,t,P,r-Di,sigma),s=K..infinity));
```

$$CallPriceDividend := e^{(-r\,t)} \int_K^{\infty} \frac{1}{2} \frac{(s-K)\sqrt{2}\, e^{(-1/8\, \frac{(2\ln(s)-2\ln(P)-2\,r\,t+2\,t\,Di+t\,\sigma^2)^2}{t\,\sigma^2})}}{s\sqrt{\pi}\sqrt{t\,\sigma^2}}\, ds$$

In order to show the relation between the pricing formula for the nondividend case and the dividend case we replace P in the above equation with $P\,e^{Di\,t}$ and simplify it.

```
> simplify(subs(P=P*exp(Di*t),CallPriceDividend));
```

$$e^{(-r\,t\tilde{})} \int_K^{\infty} \frac{1}{2} \frac{(s-K)\sqrt{2}\, e^{(-1/8\, \frac{(-2\ln(s)+2\ln(P)+2\,r\,t-t\,\sigma^2)^2}{t\,\sigma^2})}}{s\sqrt{\pi}\,\sigma\sqrt{t}}\, ds$$

Note that the above expression is no longer a function of Di. Moreover, compare this expression to the price of a call option on a stock that pays no dividend:

```
> simplify(exp(-r*t)*int((s-K)*Lnrn(s,t,P,r,sigma),\
> s=K..infinity));
```

$$e^{(-r\,t\tilde{})} \int_K^{\infty} \frac{1}{2} \frac{(s-K)\sqrt{2}\, e^{(-1/8\, \frac{(-2\ln(s)+2\ln(P)+2\,r\,t-t\,\sigma^2)^2}{t\,\sigma^2})}}{s\sqrt{\pi}\,\sigma\sqrt{t}}\, ds$$

The reader can verify that these expressions are identical to each other. This can also be confirmed by MAPLE.

```
> %-%%;
```

$$0$$

We therefore conclude that the price of a call option, maturing in t units of time, on a stock that pays no dividend with an initial price of P, will be the same as the price of a call option on a stock with an initial price $P\,e^{-Di\,t}$ and a dividend payment of Di.

We derive the option pricing formula for a dividend-paying stock using the methodology followed in the derivation of the non-dividend-paying security. The only difference is that for a dividend-paying security we derive the option pricing formula using an initial price of $P\,e^{-Di\,t}$ when the initial price is P. The resulting formula is

$$P\,e^{-Dit}\mathrm{N}(d_1) - K\,e^{-r\,t}\,\mathrm{N}(d_2), \tag{6.16}$$

where N (as before) stands for the cumulative normal distribution with $\mu = 0$ and $\sigma = 1$:

$$\mathrm{N}(s) = \int_{-\infty}^{s} \frac{e^{-\frac{x^2}{2}}}{\sqrt{2\pi}}\, dx, \tag{6.17}$$

$$d_1 = \frac{\ln\left(\frac{P}{K}\right) + \left(r - Di + \frac{\sigma^2}{2}\right) t}{\sigma\sqrt{t}}, \tag{6.18}$$

and

$$d_2 = d_1 - \sigma\sqrt{t}. \tag{6.19}$$

It is useful, at this time, to provide an intuitive explanation for why we value dividend-paying options using an initial price of $P\,e^{-Di\,t}$ instead of the "true" initial price of P. The market price of the stock P represents the two components of the return over the interval $[0, t]$ that are implicit in the stock: the price appreciation and the dividend payment. Hence, if we use the true price and assume that at time t the price of the stock will be $P\,e^{Y_t}$, that price includes the return that a stockholder will gain from both the dividend payment and the price appreciation. But when the holders of the option exercise at time t, they will not be entitled to the dividends that are paid throughout the life of the option. Hence, if the option is exercised at time t the relevant value of the asset underlying the option will have a value of $P\,e^{Y_t}$ minus the dividend payment.

To obtain the value net of the dividend payment, we discount the value of the underlying security at the maturity time of the option, using the dividend rate payment. Hence the value of the asset the option holders receive is $P\,e^{Y_t}\,e^{-Di\,t}$, and thus when calculating the expected value of the payoff from the option, we should value the underlying security as $P\,e^{Y_t}\,e^{-Di\,t}$. The holder of the call option has a contingent claim on this value, not on $P\,e^{Y_t}$. Alternatively, we obtain the same result by discounting the initial security price, P, by the expected dividends. Perceiving the initial value from this perspective we obtain a relevant initial price of $P\,e^{-Di\,t}$.

We see that either way of subtracting the dividend component results in an expected value of the security at time t of $P\,e^{Y_t}\,e^{-Di\,t}$. Since the value of the option depends on only the value of the underlying security at maturity, both methodologies work. The above discussion is the intuitive explanation of the relationship between the dividend option pricing formula and the no-dividend option pricing formula demonstrated by MAPLE above.

6.5 Volatility and Implied Volatility

The Black–Scholes option pricing formula requires σ as one of its inputs. It is the standard deviation of the continuously compounded rate of return over a unit time interval. The pricing formula was developed under the assumption that the continuously compounded rate of return over a period of time of length t follows a normal distribution with mean $\mu\, t$ and standard deviation of $\sigma\sqrt{t}$. This section is dedicated to two issues which relate to the determination of σ:

1. Estimating σ based on historical sample data of the continuously compounded rate of return.

2. Estimating σ using the σ which is implicit in the market option prices, taking for granted that the assumptions made in the Black–Scholes option pricing formula are satisfied.

6.5.1 Estimating Volatility from Historical Data

The derivation of the Black–Scholes option pricing formula assumed that the continuously compounded rate of return over a time period of length t is distributed normally with mean $\mu\, t$ and variance $\sigma^2\, t$. Consider a stock whose price at t_1 is S_1. The price of the stock at time t_2 (where $t_1 + \Delta\, t = t_2$) will be $S_2 = S_1\, e^{Y_{\Delta\, t}}$, where $Y_{\Delta\, t} \sim \mathrm{N}(\mu\, \Delta\, t,\ \sigma\sqrt{\Delta\, t})$. It follows that $\ln\left(\frac{S_2}{S_1}\right)$ is a realization of $Y_{\Delta\, t}$. Hence, if we have a sample of the stock prices at equally spaced time intervals $\Delta\, t$ we can convert this sample to realizations of the variable $Y_{\Delta\, t}$. Since we also assume that $Y_{\Delta\, t}$ are independent over disjoint intervals of time, our sample represents *i*ndependent and *i*dentically *d*istributed[7] (*i.i.d.*) observations of $Y_{\Delta\, t}$. Let us use the following notation:

- n, number of observations
- S_i, the security price at time i
- $Y_i = \ln \frac{S_i}{S_{i-1}}$,the continuously compounded rate of return between $i-1$ and i

[7] The issue of estimation of parameters and the properties of the estimators is a statistics topic. Here we only summarize a way of estimating the variance of random variables given a sample of *i.i.d.* observations.

- $Y_{avn} = \frac{\sum_{i=1}^{n} Y_i}{n}$, the arithmetic mean of the sample set of $Y_{\Delta t}$ (an unbiased estimator of the expected value[8])
- $\sigma_{estimate}$, the estimation of σ
- Δt, the time interval between the observations of S_i

The estimation of the standard deviation of the random variable $Y_{\Delta t}$, i.e., the estimate of $\sigma\sqrt{\Delta t}$, will be given[9] by

$$\sigma_{estimate}\sqrt{\Delta t} = \sqrt{\frac{\sum(Y_i - Y_{avn})^2}{n-1}}; \tag{6.20}$$

hence

$$\sigma_{estimate} = \frac{\sqrt{\frac{\sum(Y_i - Y_{avn})^2}{n-1}}}{\sqrt{\Delta t}}. \tag{6.21}$$

This can be restated[10] as

$$\sigma_{estimate} = \sqrt{\frac{\sum Y_i^2 - nY_{avn}^2}{(n-1)\Delta t}}. \tag{6.22}$$

MAPLE can be used to perform many of the calculations discussed above. To demonstrate the use of MAPLE, let us consider the following example.

Consider a sample of observations of Y_i with the following values (we use the conventional notation that Y stands for the random variable and y for its realization): Let the sample of observations be $y_1 = 0.002$, $y_2 = 0.005$, $y_3 = -0.001$, $y_4 = -0.002$, $y_5 = 0.006$, $y_6 = -0.005$, $y_7 = 0.003$, and $y_8 = -0.004$, where $\Delta t = \frac{1}{250}$ year. We then define the above values as an array in MAPLE

```
> ysample:=[.002,.005,-.001,-.002,.006,-.005,.003,-.004];
```

$$ysample := [.002, .005, -.001, -.002, .006, -.005, .003, -.004]$$

[8] An unbiased estimator is an estimator whose expected value is the estimated parameter.

[9] Note that the estimator is the square root of the deviation of the sample mean from each observation divided by $n-1$. Dividing by n would be the arithmetic mean of the deviations. It is divided by $n-1$ to allow the unbiased property. Readers not familiar with this concept should consult a statistics textbook.

[10] One should simplify $\sum_{i=1}^{m}(y_i - y_{avm})^2 = \sum_{i=1}^{m}\left(y_i^2 - 2\,y_{avm}y_i + y_{avm}^2\right) = \sum_{i=1}^{m} y_i^2 - 2y_{avm}\left(\sum_{i=1}^{m} y_i\right) + my_{avm}^2$ but $\sum_{i=1}^{m} y_i = my_{avm}$, and substituting it in the last expression yields the expression in the text.

and ask MAPLE to evaluate the estimator of the standard deviation of[11] $Y_{\Delta t}$:

```
> sigma[estimate]:=stats[describe,standarddeviation[1]]\
> ysample/sqrt(1/250);
```

$$\sigma_{estimate} := .02052872552\,\sqrt{10}$$

```
> evalf(sigma[estimate]);
```

$$.06491753010$$

The reader is encouraged to change the observation values in the above template and witness how the estimation changes accordingly.

6.5.2 Implied Volatility

We can also estimate the volatility by determining the σ implicit in the market option prices. If we know the value of an option, and we know the value of t, P, K, and r, we can solve for the value of σ. Since there is no analytical solution, we must use numerical procedures to determine the estimate of σ. We can do so by attempting to input different values of σ until we find a value for σ that results in the option price observed in the market.

Alternatively, we can use the "solve" command of MAPLE to perform this numerical procedure. We used this command in Section 1.4 to find the solution to a system of equations. Here we have only one equation: the market price of the option equals its value based on the Black–Scholes formula, and one unknown $\sigma_{implied}$. To demonstrate this point we use our procedure **BstCall** which was introduced in the first section of this Chapter.

For example, let us consider a situation where the exercise price is 95, the time to maturity is half a year, the stock price is 90, the annual interest rate is 6%, and the market price of the call is 4.87219. We ask MAPLE to solve for σ, which equates the market price of the option to its theoretical value based on the Black–Scholes equation. First we ensure, in the next command, that $\sigma_{implied}$ is qualified to be the variable in the equation to be solved and is not assigned a numerical value.

```
> sigma[implied]:='sigma[implied]';
```

$$\sigma_{implied} := \sigma_{implied}$$

[11] We inform MAPLE that we are evaluating sample data (and not population data) using the notation **standarddeviation**[1] . Alternatively, placing 0 in the brackets informs MAPLE that we are using population data.

We then ask MAPLE to solve for $\sigma_{implied}$ by executing the following command:

```
> solve({BstCall(95,.5,90,sigma[implied],.06)=4.87219},\
> {sigma[implied]});
```

$$\{\sigma_{implied} = .2299997969\}$$

The derivation of the Black–Scholes formula assumes, as we have seen, that the volatility of a stock is a deterministic constant. It does not depend on time and is definitely not stochastic. Thus, solving for the implied volatility using options on the same stock but with a different maturity time or different exercise price should result in the same solution. Let us look at some real data to see if indeed this is the situation.

Consider a stock with a current price of 37.20. We collected a sample of 10 options written on this stock: 5 options with a maturity of 2 months, 3 options with a maturity of 3 months, and 2 options with a maturity of 9 months. To facilitate solving for the implied volatility in one command we define the following in MAPLE:

- An array of the times to maturity for a series of option contracts written on the stock,

```
> times:=[seq(.167,i=1..5),seq(.417,i=1..3),\
> seq(.75,i=1..3)];
```

$$times := [.167, .167, .167, .167, .167, .417, .417, .417, .75, .75, .75]$$

- An array of the exercise prices of those same options,

```
> K:=[38,46,48,50,52,48,50,52,46,50,50];
```

$$K := [38, 46, 48, 50, 52, 48, 50, 52, 46, 50, 50]$$

- An array of the option prices,

```
> CallPrice:=[12.25,4.60,2.95,2.30,1.25,1.45,\
> 2.50,2.00,5.90,3.50,3.05];
```

$$CallPrice := [12.25, 4.60, 2.95, 2.30, 1.25, 1.45, 2.50, 2.00, 5.90, 3.50, 3.05]$$

The data for the ith option is specified by the ith component of each array. Thus for the options with a maturity of two months we have a variety of exercise prices ranging form \$38 to \$52. The interest rate in the market at that time was 9% for the maturity specified; i.e., the term structure was flat for the time segment of two to nine months.

We determine the implied volatility in each of the above options by solving for sigma via the command below, but first we ensure that sigma has no numerical value assigned to it.

```
> sigma:='sigma';
```

$$\sigma := \sigma$$

```
> seq(solve({BstCall(ExPrice[i],times[i],37.20,\
> sigma[i],9/100)=CallPrice[i]},{sigma[i]}),i=1..10);
```

$$\{\sigma_1 = 2.097092493\}, \{\sigma_2 = 1.208390590\}, \{\sigma_3 = 1.006384663\}, \{\sigma_4 = .9587982897\}, \{\sigma_5 = .8058808416\}, \{\sigma_6 = .4342159751\}, \{\sigma_7 = .6056562879\}, \{\sigma_8 = .5873106678\}, \{\sigma_9 = .6209690787\}, \{\sigma_{10} = .5096387007\}$$

Examining the result above suggests that the assumption made by the Black–Scholes formula regarding the volatility being constant is inconsistent with market prices. The Black–Scholes formula assumes that the standard deviation of the rate of return of the stock per unit of time is a constant. Hence, the same estimation value should result if this standard deviation is imputed from another option on the same stock, but with a different exercise price or different time to maturity. This situation where options with different strike prices yield different implied volatilities is commonly known as a "volatility smile". The name suggests the common resultant "U" shape of the graph of the implied volatility against the strike price.

You can proceed in the above manner to solve for the implied volatility from price quotations in other markets. Just fill in the arrays with the appropriate numerical values and reexecute the command above. You can verify if the same conclusion holds for the market that you examine: Do you get a flat graph or do you also find the "volatility smile"?

Various strategies have been suggested in response to this problem of divergent implied volatilities. Generally, the literature (see, for example, [3] and [31]) argues that more weight should be put on options which are at or close to at-the-money, i.e., to options for which the exercise price is close to the current price of the stock.

The above solutions via the MAPLE **solve** procedure may not work for certain configurations of the exercise price and the stock price. There may, therefore, be a need to use a different numerical procedure. Such a procedure named **ImpliedVol** is offered below. It is programmed based on work of Corrado and Miller [12].

Corrado and Miller derive the formula through a quadratic approximation of the normal distribution function. The formula is

$$\sigma\sqrt{t} = \left(\frac{\sqrt{2\pi}}{P+K\,e^{-r\,t}}\right)\left(C-\frac{P-K\,e^{-r\,t}}{2}\right)+ \\ +\sqrt{\left(C-\frac{P-K\,e^{-r\,t}}{2}\right)^2-\frac{(P-K\,e^{-r\,t})^2}{\pi}} \tag{6.23}$$

and therefore

$$\sigma = \frac{\left(\frac{\sqrt{2\pi}}{P+K\,e^{-r\,t}}\right)\left(C-\frac{P-K\,e^{-r\,t}}{2}\right)}{\sqrt{t}}+ \\ \frac{\sqrt{\left(C-\frac{P-K\,e^{-r\,t}}{2}\right)^2-\frac{(P-K\,e^{-r\,t})^2}{\pi}}}{\sqrt{t}}, \tag{6.24}$$

where t is the time to maturity, r is the risk-free interest rate, P is the initial price of the security, K is the exercise price, and C is the price of the call option.

We can use MAPLE to perform the above calculation using the procedure **ImpliedVol**. The inputs to this procedure are, in order, C, P, K, t,and r.

```
> ImpliedVol(3.176,100,110,.5,.06);
```

$$.2037048755$$

We can compare this result to the σ estimated using the methodology described earlier by issuing the command

```
> solve({BstCall(110,.5,100,sigma[implied],.06)=3.1764},\
> {sigma[implied]});
```

$$\{\sigma_{implied} = .2043591628\}$$

and verify that indeed the results are quite similar. We can utilize the procedure **ImpliedVol** to visualize, in a three-dimensional graph (Figure 6.19), the effect of changes in the exercise price and the time to maturity on the estimated volatility.

```
> plot3d('ImpliedVol'(3.176,100,Exercise,timetomaturity,\
> .02),timetomaturity=.1..1,Exercise=90..110,\
> labels=[`Time to Maturity`,`Exercise Price`,Volatlity],\
> axes=normal,orientation=[130,44],title=`Figure 6.19:\
> The Change in Volatility Relative to Changes in t and K`);
```

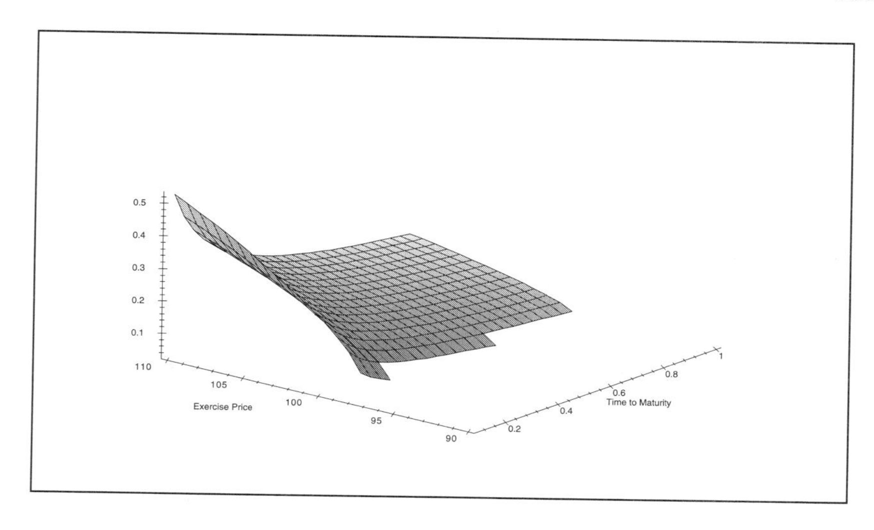

Figure 6.19: Estimated Volatility as a Function of K and t

We see that if the price of the option and the exercise price stay fixed, the volatility estimation increases the closer we are to the expiration of the option. If the price of the option and the time to maturity stay fixed, the volatility estimation increases as the exercise prices increases. Can you explain why this is the case? It, of course, has to do with the sensitivity of the price of a call option to the above parameters. The sensitivity measures of the price of a call option to these parameters and the induced hedging strategies are the topic of the next Chapter.

At times, both of the procedures discussed above may not work. In such a situation we can use a numerical algorithm to search for the value of sigma. The appendix to this Chapter discusses this methodology and introduces a numerical procedure solving for the implied sigma.

6.6 Concluding Remarks

Chapter 5 was dedicated to the risk-neutral price process and pricing mechanism in a continuous time setting. Consequently, this Chapter began with the understanding that a European contingent claim with a payoff $h(s)$ would be priced by equation (5.7). The Chapter investigated the Black–Scholes formula for put and call options which implies that $h(S)$ is of the form $\max(s - K, 0)$ and $\max(K - s, 0)$, where K is the exercise price of the option. Hence, pricing these options amounts to calculating the value of the integral in equation (5.7), where $h(s)$ is replaced with $\max(s - K, 0)$ for a call option and $\max(K - s, 0)$ for a put option. These options are referred to as plain vanilla options, in contrast to exotic options, examples of which are presented in Section 16.3.

The payoff at maturity of different combinations of these options was explored in Chapter 4, which was independent of the price process of the stock. Having established the value of these options facilitates the investigation of the value of different combinations of these options prior to maturity. This was done in Section 6.3, which of course depends on our assumption about the price process of the stock.

The Chapter concluded with the effect of dividends on the Black–Scholes formula and the issue of implied volatility. Assuming that indeed the price of puts and calls in the market follows the Black–Scholes formula, and given the risk-free rate, one can solve for the value of the volatility that equates the Black–Scholes price to the observed market price. Alternatively, as outlined in one of the exercises, one can estimate the volatility from historical data and estimate the risk-free rate from the prices of the options.

Acquiring an insight into the prices of puts and calls allows us to continue to study their sensitivity to various parameters and the related issues of hedging these options. These issues are discussed in Chapters 7 and 8, respectively.

6.7 Questions and Problems

Problem 1. Consider a market with a stock and three puts and calls written on it. The calls and the puts mature at the same time and have exercise prices of K_1, K_2, and K_3 such that $K_1 < K_2 < K_3$. How many portfolios with a payoff function $h(s)$ satisfying the condition $h(s) > 0$ if $K_1 < s < K_3$, $h(s) = 0$ if $s \notin (K_1, K_3)$, and $\max h(s) = c > 0$ can you compose?

Problem 2. Assume that the price of the stock is \$100, its volatility is 0.23 a year, and the continuously compounded risk-free rate is 10%. Utilize equation (5.7) to find the price of the contingent claim $h(s)$ (defined in Problem 1) in 3 months if $K_1 = 20$, $K_2 = 60$, $K_3 = 90$, and $c = 50$.

Problem 3. Consider the payoff described in Problem 2 and the prices of the options that composed one of the portfolios you solved for in Problem 1.

1. Value a payoff of the type $h(s)$ as defined in Problem 2 by the replication method.
2. Assume that the risk-free rate is 9%. Find the implied volatility in the prices of each of the options that composed your portfolio.
3. Find the implied volatility in the price of the contingent claim $h(s)$.
4. Compare and explain your results.

Problem 4. Given the prices of two options maturing in t units of time that are written on the same stock, solve for the risk-free rate spanning the time interval $[0, t]$ and for the volatility of the stock. Prepare a worksheet in MAPLE (or use the MATLAB GUI that solves for the above).

Problem 5. The former question suggests a methodology of estimating a term structure of interest rates.

1. Outline this methodology.
2. What do you think the relation will be between a term structure estimated from the prices of government bonds and a term structure estimated by the above methodology?

Problem 6. A warrant is a security issued by a firm which gives the holder the option to exchange it, on a certain date, for π number of stocks at a price of $\$K$ per stock. It is thus similar to a call option on a stock. However, since it is issued by the firm on its own securities it also affects (dilutes) the value of each stock. It is not only a "side bet" on the price of the stock. The firm's equity, $V(t)$ at time t, is equal to the total value of its outstanding stocks and its outstanding warrants at time t. Assume that $V(t)$ follows a geometric Brownian motion and that all the assumptions under which the Black–Scholes formula is correct are satisfied. Assume that the current time is $t = 0$, the warrant can be exercised at time T, and there are N outstanding stocks and M outstanding warrants. Develop the price of a warrant. Show each step in your calculations.

Problem 7. Explain why the calendar spread in Figure 6.14, has negative value (for some values of the underlying stock price) during the time period from zero to one.

6.8 Appendix

6.8.1 Estimating Implied Volatility Using Trial and Error

In the final section of this Chapter we estimated the implied volatility using two methods: the MAPLE **solve** command and the **ImpliedVol** procedure. However, at times neither of these procedures is able to solve for an estimate of σ. In such situations we use a trial and error numerical algorithm to determine a value for σ. In this appendix the reader is introduced to the procedure **IVtry**. The numerical algorithm used by **IVtry** to determine σ is described below.

The basis behind the search for an implied volatility is the fact that the price of an option, as we shall soon see in the next Chapter, is an increasing function of sigma. Given the value of the option C, the search begins with a guess of an interval $[L_0, U_0]$ that hopefully includes the true value of the implied volatility. The value of the given option is calculated for a volatility of U_0, denoted as $C(U_0)$, and for a volatility of L_0, denoted as $C(L_0)$. It proceeds by following the steps below.

If the condition $C(L_0) \leq C \leq C(U_0)$ is not satisfied then

- If $C < C(L_0)$ we assign to $C(U_0)$ the value of $C(L_0)$ and change L_0 to $\frac{L_0}{2}$.
- If $C > C(U_0)$ we assign to L_0 the value of U_0 and change U_0 to $2U_0$.

We now calculate the value of the option for the new end points of the interval and return to the first "if" statement. Eventually this loop will identify two end points that satisfy the condition in the "if" statement above. Once two such end points are identified, say L_n and U_n in the nth iteration, the value of an option with a volatility of $\frac{L_n+U_n}{2}$ is calculated.

- If $C(L_n) \leq C \leq C\left(\frac{L_n+U_n}{2}\right)$ then U_{n+1} is set equal to $\frac{L_n+U_n}{2}$ and L_{n+1} is set equal to L_n.
- If $C\left(\frac{L_n+U_n}{2}\right) \leq C \leq C(U_n)$ then L_{n+1} is set equal to $\frac{L_n+U_n}{2}$ and U_{n+1} is set equal to U_n.

The process continues in this manner until the first iteration, k, such that $U_k - L_k$ is smaller than a prespecified tolerance error. The process is guaranteed to converge because the value of the option is an increasing function of volatility.

The procedure **IVtry** performs the above process using the parameters (CallPrice, StockPrice, StrikePrice, InterestRate, TimeToMaturity). For example, consider an option with a sigma of 0.45 where the stock price is \$30, the strike price is \$28, the interest rate is 0.05, and the time to maturity is 0.5. We easily value such an option using the **BstCall** procedure:

```
> BstCall(28,.5,30,.45,.05);
```

$$5.123258217$$

Consider a situation where we know (CallPrice, StockPrice, StrikePrice, InterestRate, TimeToMaturity) = (5.123258218, 30, 28, 0.05, 0.5) but we do not know σ. In such a situation we use the procedure **IVtry**. We input the parameters and execute the procedure **IVtry**. **IVtry** then asks us to specify

1. the upper bound on the value of the volatility U_0,
2. the lower bound on the value of the volatility L_0,
3. the number of iterations **IVtry** should attempt before giving up, and
4. the tolerance error.

Finally **IVtry** asks the user to specify whether or not the trials should be observable. (If the user wishes the trials to be observable, the details of the iterations carried out by **IVtry** will be printed on the screen.) We conclude this appendix with a demonstration of **IVtry** applied to the example, where $U_0 = .3$ and $L_0 = .2$.

```
> IVtry(5.123258218,30,28,.05,.5);
```

Welcome to the IVtry procedure. Please input the following information.
Type ; after each entry.

Please enter the lower bound–>.2;
Please enter the upper bound–>.3;
Please enter maximum number of iterations–>1000;
Please enter the tolerance–>.001;
Input 1 **to view trials or** 0 **to run quietly–>**1;

$$High =, .3, \ Low =, .2, \ Iteration\ number =, 1$$

$$High =, .6, \quad Low =, .2, \; Iteration\ number =, 2$$
$$High =, .6, \quad Low =, .40, \; Iteration\ number =, 3$$
$$High =, .6, \quad Low =, .500, \; Iteration\ number =, 4$$
$$High =, .500, \quad Low =, .2500, \; Iteration\ number =, 5$$
$$High =, .500, \quad Low =, .37500, \; Iteration\ number =, 6$$
$$High =, .500, \quad Low =, .437500, \; Iteration\ number =, 7$$
$$High =, .500, \quad Low =, .4687500, \; Iteration\ number =, 8$$
$$High =, .4687500, \quad Low =, .23437500, \; Iteration\ number =, 9$$
$$High =, .4687500, \quad Low =, .351562500, \; Iteration\ number =, 10$$
$$High =, .4687500, \quad Low =, .4101562500, \; Iteration\ number =, 11$$
$$High =, .4687500, \quad Low =, .4394531250, \; Iteration\ number =, 12$$
$$High =, .4687500, \quad Low =, .4541015625, \; Iteration\ number =, 13$$
$$High =, .4541015625, \quad Low =, .2270507813, \; Iteration\ number =, 14$$
$$High =, .4541015625, \quad Low =, .3405761720, \; Iteration\ number =, 15$$
$$High =, .4541015625, \quad Low =, .3973388673, \; Iteration\ number =, 16$$
$$High =, .4541015625, \quad Low =, .4257202150, \; Iteration\ number =, 17$$
$$High =, .4541015625, \quad Low =, .4399108888, \; Iteration\ number =, 18$$
$$High =, .4541015625, \quad Low =, .4470062257, \; Iteration\ number =, 19$$
$$High =, .4541015625, \quad Low =, .4505538942, \; Iteration\ number =, 20$$
$$High =, .4505538942, \quad Low =, .2252769471, \; Iteration\ number =, 21$$
$$High =, .4505538942, \quad Low =, .3379154207, \; Iteration\ number =, 22$$
$$High =, .4505538942, \quad Low =, .3942346575, \; Iteration\ number =, 23$$
$$High =, .4505538942, \quad Low =, .4223942759, \; Iteration\ number =, 24$$
$$High =, .4505538942, \quad Low =, .4364740851, \; Iteration\ number =, 25$$
$$High =, .4505538942, \quad Low =, .4435139897, \; Iteration\ number =, 26$$
$$High =, .4505538942, \quad Low =, .4470339420, \; Iteration\ number =, 27$$
$$High =, .4505538942, \quad Low =, .4487939181, \; Iteration\ number =, 28$$
$$High =, .4505538942, \quad Low =, .4496739062, \; Iteration\ number =, 29$$
$$High =, .4505538942, \quad Low =, .4501139002, \; Iteration\ number =, 30$$
$$implied\ sigma =$$
$$.4501139002$$

IVtry has found the implied σ in 30 iterations. Running **IVtry** quietly increases its speed.

The procedure **Ivtry** (see the HELP file for details) searches for the volatility (σ) implicit in the Black-Scholes European call option pricing formula using a binary search without prompting the user for inputs.

Chapter 7

Sensitivity Measures

Now that we have developed an understanding of the option pricing formula it is time to take a deeper look at it. This Chapter investigates the effects[1] of "small changes" in the value of the parameters on the price of the option. By investigating these effects we will gain a better insight into the underlying concepts of the Black–Scholes formula. Our investigation will concentrate mainly on the price of call[2] options.

At every instant of time, the value of a call option depends on a few parameters. Some of these parameters are fixed and inherent in the specified option definition, e.g., the exercise price. Some are assumed to be fixed but may indeed change, e.g., the interest rate. Other parameters depend on time or are changing in a random manner. For example, the time to maturity depends on time in a very obvious and deterministic way, while the price of the stock follows a random process over time. According to the assumptions employed in developing the option pricing formula, the risk-free rate of interest and the standard deviation of the stock price are assumed to be fixed constants, independent of time and the price of the stock. As we will see below, while we are able to measure the sensitivity of the value of the call option to these parameters, it is more natural to start from the other two parameters.

The effect of a "small change", in a certain parameter on the price of the call option, is captured by its partial derivative (calculus derivative) with respect to the parameter in question. In the text we will motivate the effect of the changes in the different parameters on the price of the call option

[1] See [10].

[2] The reader will be asked later to verify the counterpart results for the case of a put option. Some of the results mentioned in the Chapter will be derived in the Appendix and some will be assigned as exercises.

and will display the algebraic expression for the changes investigated. The appendix develops these notions more rigorously.

7.1 The Theta Measure

The value of the call option, as we have seen in equation (6.2), is a function of the stock price or the state of nature. Given the random nature of the stock price, at each instant of time we perceive the call price as a function of the state of nature. This exposition is consistent with the way in which we described the payoff function for call options and for put options at the maturity time of the options. If the time under examination is not the maturity time, the payoff concept is replaced with the value of the particular derivative security. This is indeed[3] the payoff one can get if the security is sold in the market at that time.

Let us first look at the value of a call option on a stock whose price has a standard deviation of 0.25, when the risk-free rate of interest is 0.10%, the exercise price of the option is $50, and the time to maturity is one year. We plot in the same plane the value of the option at maturity and also the value of the same option with time to maturity of 0.5 years. The longer the time to maturity, the higher the value of the call and thus the higher the graph. The piecewise-linear graph, the lower graph, in Figure 7.1 is (obviously) the value of the call[4] at maturity.

```
> plot({'BstCall(50,1,P,.25,.10)',\
> 'BstCall(50,0,P,.25,.10)'},\
> P=20..70,labels=[`Stock Price`, `Call Value`],\
> thickness=2, title=`Figure 7.1:  Changes in Option\
> Value for Different Times to Maturity`);
```

We use animation to describe how the value of a call option, the graph in Figure 7.1, evolves as the time to maturity decreases. We already know that when the time remaining until the maturity of the option is zero, the value of the option (its payoff) is as the piecewise-linear graph in Figure

[3] There is of course one crucial distinction. While the payoff at maturity is a function of the stock price, it is independent of the model and assumptions that are utilized to calculate the value of options prior to maturity. The payoff at maturity depends on the price of the stock at that time and the exercise price, both of which are known then.

[4] The procedure **Bs** is a variation of the procedure **BstCall,** except **Bs** does not accept a value of zero for the time to maturity parameter. It is easier to work with **Bs** for the purpose of calculation and plotting when there is no need to use a value of zero for the time to maturity.

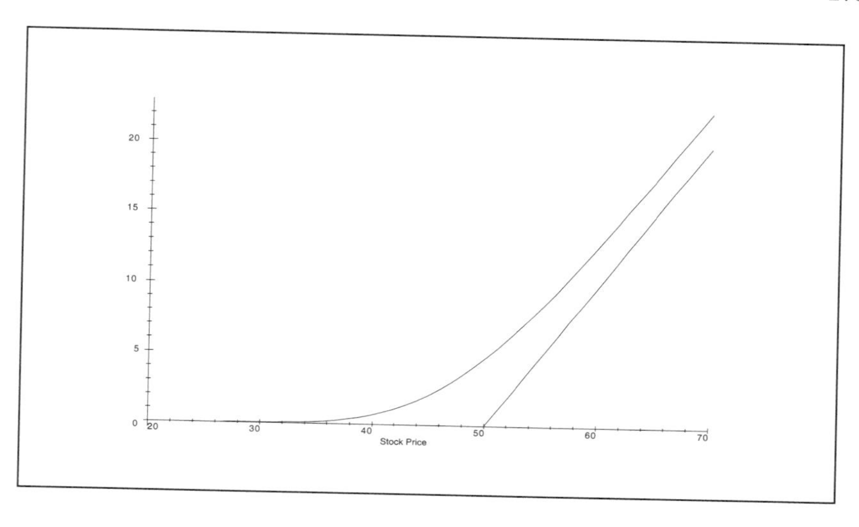

Figure 7.1: Option Values for Different Times to Maturity

7.1. How does this graph evolve through time? How does it approach the piecewise-linear graph as time to maturity runs out? Let us run the following animation for an option on a stock for which the standard deviation of the price is 0.25, and where the risk-free rate is 10% and the exercise price of the option is \$50. We will look at the value of the option at different times to maturity for a stock price range from \$40 to \$70 (Figure 7.2). Such an animation was viewed before in Section 6.3.

```
> plots[animate]('BstCall(50,1-t,P,.25,.10)',P=20..70,\
> t=0..1,frames=30,thickness=3,title=`Figure 7.2:\
> Evolution of Call Option Value as Time to\
> Maturity Decreases`);
```

The reader is encouraged to try other values for the parameters suggested here and run the animation for various combinations. Watching the animation, it is apparent that as time runs out, the graph approaches the piecewise-linear graph in a smooth manner. As the time approaches maturity, for the call to have a "significant" value, the current price of the stock should be larger and larger. That is, the call should be deeper and deeper in the money. To further enhance the graphs we use MAPLE to calculate the sequence of values for this call option as time approaches maturity and when the call price is fixed at \$40, \$50, and \$60.

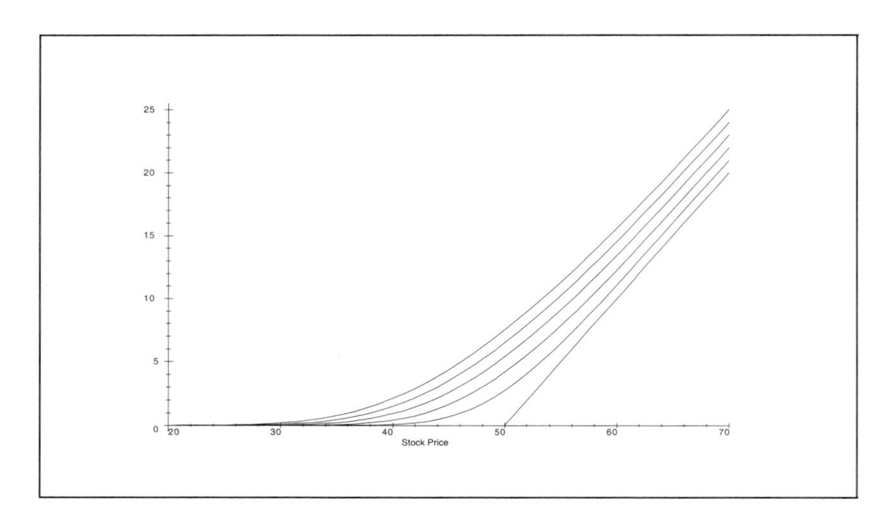

Figure 7.2: Evolution of Call Option Value as Time to Maturity Decreases

We start with an out-of-the-money call option (stock price is smaller than the exercise price) and look at its value for different maturity dates. The maturity time increases in increments of a month, starting with a maturity of a month and continuing until a maturity of one year.

```
> seq(BstCall(50,t/12,40,.25,.10),t=1..12);
```

.00133351, .03608598, .13268303, .27939666,
.46091703, .66629991, .88826045, 1.12191870,
1.36392937, 1.61193996, 1.864255702, 2.119628922

We repeat the same calculations for an at-the-money option (stock price equals the exercise price) and then for an in-the-money option (stock price is greater than the exercise price).

```
> seq(BstCall(50,t/12,50,.25,.10),t=1...12);
```

1.650282700, 2.458181575, 3.127247806, 3.724044585,
4.274283830, 4.79111753, 5.282338799, 5.753012919,
6.206650107, 6.645800572 7.072385664, 7.487895384

```
> seq(BstCall(50,t/12,60,.25,.10),t=1..12);
```

10.41997326, 10.88018177, 11.37100007, 11.86970379,
12.36568032, 12.85451211, 13.33448014, 13.80506546,

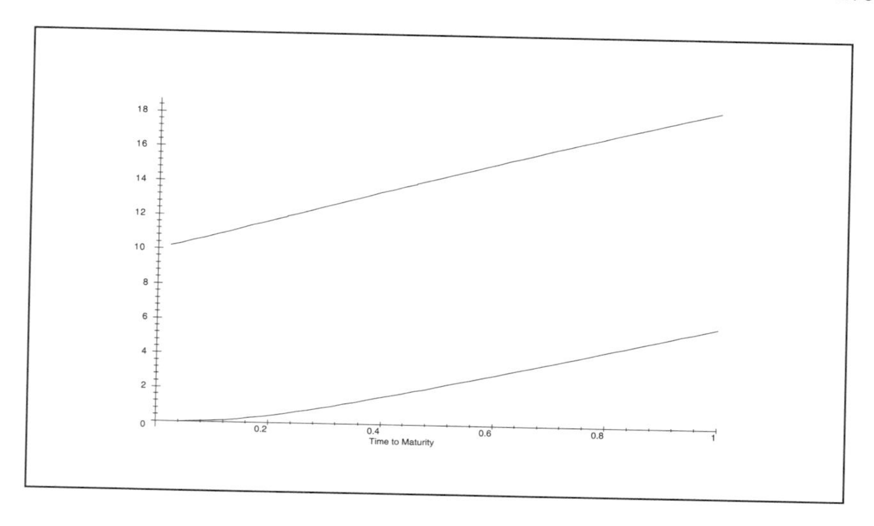

Figure 7.3: The Values of Two Options as Time to Maturity Increases

14.26629503, 14.71844202, 15.16188306, 15.59702831

The trend is (should be) clear: as the time to maturity increases, the value of the option increases regardless of the current value of the stock. This occurs due to two factors. As the time to maturity increases, the current cost of exercising the call decreases. The cost of exercising the call is the present value of the exercise price, which obviously decreases as time increases.

In addition, as the time to maturity increases, the volatility of the stock price at maturity increases: remember that the volatility is $\sigma\sqrt{t}$, where t is the time until maturity. We shall soon demonstrate that indeed the option value increases as a function of the volatility of the stock price. The animation above illustrates this intuition in reverse: the time to maturity is decreasing in that animation. The reader can also visualize this phenomenon by examining Figure 7.3, in which the values of two options are plotted as time approaches maturity, and in which the price of the stock is assumed to be fixed at selected levels.

```
> plot({Bs(90,1-t,80,0.18,0.12),Bs(70,1-t,80,0.18,0.12)},t\
> =0..1,labels=[‘Time to Maturity‘,‘ ‘],title=‘Figure 7.3:\
> The Values of Two Options as Time to Maturity Increases);
```

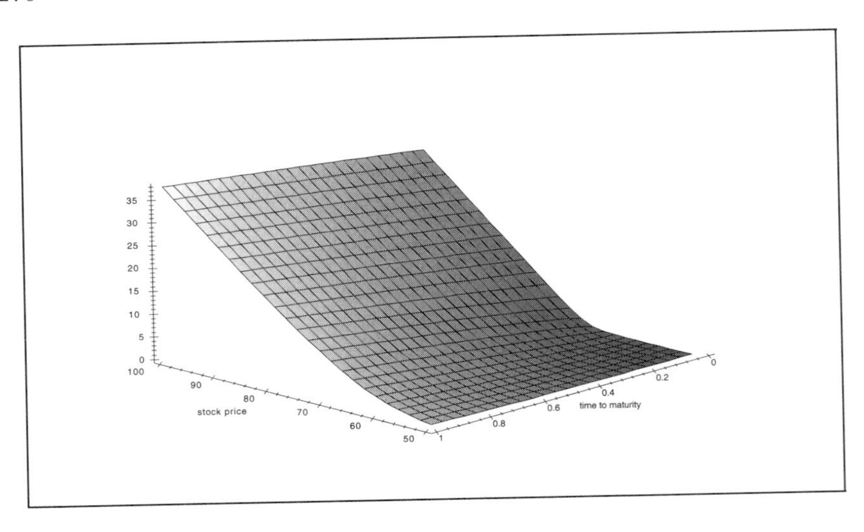

Figure 7.4: Call Option's Value as a Function of the Stock Price and Time to Maturity

Finally, the reader may wish to examine a three-dimensional graph in which the value of the call option is graphed against the value of the stock and the time to maturity t. Note how at $t = 0$ the graph is the familiar piecewise-linear graph of the payoff. At any other fixed time to maturity, the graph is smooth but approaching the piecewise-linear graph at maturity. One can thus think about the three-dimensional graph as a collection of graphs displaying the value of the option for a range of stock prices at a given t. Lining these graphs up, one against the other, generates the three-dimensional graph in Figure 7.4. A similar graph but from a different perspective was displayed in Section 6.3.

```
> plot3d(Bs(70,t,S,0.18,0.12),S=50..100,t=0..1,\
> axes=normal,orientation=[82,53],labels=[‘Stock\
> Price‘,Time,‘Call Value‘], title=‘Figure 7.4:\
> Call Option’s Value as a Function of the \
> Stock Price and Time to Maturity‘);
```

Figure 7.4 also displays the fact that for a fixed time to maturity, the higher the value of the stock, the higher the option value. We shall come back to this point shortly. The reader may again wish to try other combinations by changing the numerical values of the parameters in Figure 7.4.

The measure of the sensitivity of the call price to time is the derivative of the call price with respect to time, i.e., the calculus derivative $\frac{\partial}{\partial t}\, BstCall(K,\, t,\, P,\, \sigma,\, r)$. This derivative is the precise measure of the instantaneous rate of change of the call option's value that is due to the passage of time. It is customary to assign Greek names to the sensitivity measures of options. The sensitivity with respect to time is called Theta and represented by the Greek letter Θ. It is defined to be the rate of change of the call price as the time to maturity **increases**.

Hence, it is the partial derivative[5] of the call price with respect to the time to maturity. In terms of the notation in equation (6.3) for the call price it is calculated as

$$\frac{\partial\left(P\mathrm{N}\left(\frac{\ln\left(\frac{P}{K}\right)+\left(r+\frac{\sigma^2}{2}\right)t}{\sigma\sqrt{t}}\right)-Ke^{-rt}\mathrm{N}\left(\frac{\ln\left(\frac{P}{K}\right)+\left(r-\frac{\sigma^2}{2}\right)t}{\sigma\sqrt{t}}\right)\right)}{\partial t}. \tag{7.1}$$

The Appendix shows that the expression for Theta is given by

$$\begin{aligned}\Theta(K,t,P,\sigma,r) \quad = \quad & \frac{P\sigma}{2\sqrt{t}}\mathrm{N}'\left(\frac{\ln\left(\frac{P}{K}\right)+\left(r+\frac{\sigma^2}{2}\right)t}{\sigma\sqrt{t}}\right)+ \\ & rKe^{-rt}\mathrm{N}\left(\frac{\ln\left(\frac{P}{K}\right)+\left(r-\frac{\sigma^2}{2}\right)t}{\sigma\sqrt{t}}\right),\end{aligned} \tag{7.2}$$

where N' is the (calculus) derivative of N. That is, N' is the density function of the normal distribution function and is given by equation (7.3).

$$\mathrm{N}'(x) = \frac{e^{-\frac{x^2}{2}}}{\sqrt{2\pi}} \tag{7.3}$$

Clearly this expression is always positive since it is the sum of positive components, as both N and N' are positive functions.

The sensitivity measure Θ depends also on the values of the other parameters in the Black–Scholes valuation formula. Let us first define the function Θ in MAPLE.

---

[5] Recall that the definition of the partial derivative assumes that all other parameters but the one in question (t in this case) remain constant at their current values. Thus

$$\lim_{\Delta t\to 0}\frac{\mathrm{BstCall}(K,\, t+\Delta t,\, Pr,\, \sigma,\, r)-\mathrm{BstCall}(K,\, t,\, Pr,\, \sigma,\, r)}{\Delta t}=\Theta(K,\, t,\, Pr,\, \sigma,\, r).$$

```
> Theta:=unapply((diff(Bs(K,t,Pr,sigma,rf),t)),\
> K,t,Pr,sigma,rf):
```

Armed with this function we can calculate Θ for given parameter values. For example, for an exercise price of \$90, a time to maturity of one year, a stock price of \$50, $\sigma = 0.18$, and a risk-free rate of 12 percent, we obtain the following:

```
> evalf(Theta(90,1,50,0.18,0.12));
                          .111497897
```

Since Theta is the first derivative of the call value with respect to time, it can be used to generate a linear approximation of the value of the call option as time progresses. Indeed, this is nothing but using the first term in a Taylor series expansion[6] with respect to the time variable. Assume an option with the following parameters: $Bs(90, 0.5, 80, 0.18, 0.12)$. Then the linear approximation to the value of the option as the time to maturity parameter, t, decreases from 0.5 (and all other parameters remain fixed) is given by

$$\begin{aligned} Bs(90, t, 80, 0.18, 0.12) \cong\ & Bs(90, 0.5, 80, 0.18, 0.12) + \\ & (t-0.5)\,\Theta(90, 0.5, 80, 0.18, 0.12). \end{aligned} \tag{7.4}$$

Equation (7.4) should be interpreted as follows. Recall that we assume the current time to be zero, and thus the maturity time of the option, 0.5, is also the time until the maturity of the option. Hence at some future time, where the time to maturity is t, the time to maturity of the option will be shorter by $(0.5 - t)$. The approximation to the change in the option value is therefore $-(0.5 - t)\Theta(70, 0.5, 80, 0.18, 0.12)$. Here, we are interested

[6]Taylor approximation is a polynomial approximation of a function based on information on the function at one point. It is the value of the function at that point and the value of the derivatives at that point. The formula of the Taylor approximation of order $n = 4$ is given by $f(x) = f(x_0) + (x - x_0)D(f)(x_0) + \frac{D^2(f)(x_0)}{2}(x - x_0)^2 + \frac{D^3(f)(x_0)}{3!}(x - x_0)^3 + o((x - x_0)^4)$, where $D^i(f)(x_0)$ is the MAPLE notation for the ith order derivative of the function f at the point x_0, and $o(h)$ is the notation for an expression that satisfies $\lim_{h \to 0} \frac{o(h)}{h} = 0$. That is, $o(h)$ goes to zero faster than h; fast enough for $\frac{o(h)}{h}$ to go to zero as h goes to zero. Section 15.11.1 elaborates on the meaning of $o(h)$.

The linear approximation to the function is obtained with a first-order Taylor approximation, i.e., as $f(x) = f(x_0) + D(f)(x_0)(x - x_0)$. The intuitive explanation for this approximation is based on the interpretation of the the first-order derivative $D(f)$ as the increase in the function value, per an infinitesimal increase in the variable. Hence an increase in the variable from x_0 to x, $x - x_0$, induces an increase of $D(f)(x_0)(x - x_0)$ in the function value, resulting in a function value of $f(x_0) + D(f)(x_0)(x - x_0)$.

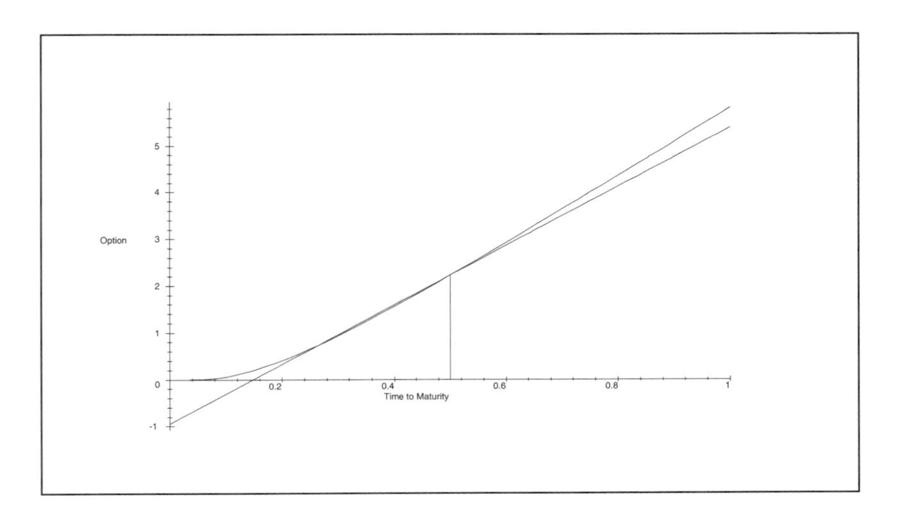

Figure 7.5: Approximation of the Call's Value with Theta around Time to Maturity of 0.5

in approximating a change in a call value that is due to a decrease in the parameter (time to maturity), and thus the minus sign.

This linear approximation is demonstrated in Figure 7.5: the graph of the option value as a function of time to maturity is shown on the same set of axes along with the linear approximation. The bar crosses the graphs at the point around which the approximation is done, i.e., at 0.5. This is the point where the linear graph is tangent to the actual graph of the option value. At this point, the value of the approximation and the true value coincide. As one moves away from that point, the approximation becomes less and less accurate.

```
> plot({Bs(90,t,80,0.18,0.12),Bs(90,.5,80,0.18,0.12)\
> +(t-0.5)*Theta(90,.5,80,0.18,0.12),\
> [[0.5,0],[0.5,Bs(90,0.5,80,0.18,0.12)]]},t=0..1,\
> labels=['Time to Maturity','Option'],thickness=2,\
> title='Figure 7.5:Approximation of the Call's Value\
> with Theta around Time to Maturity of 0.5');
```

We can also plot, Figure 7.6, the graph of Theta for an option which is in-the-money (the red middle graph), at-the-money (the blue upper graph), and out-of-the-money (the green lower graph).

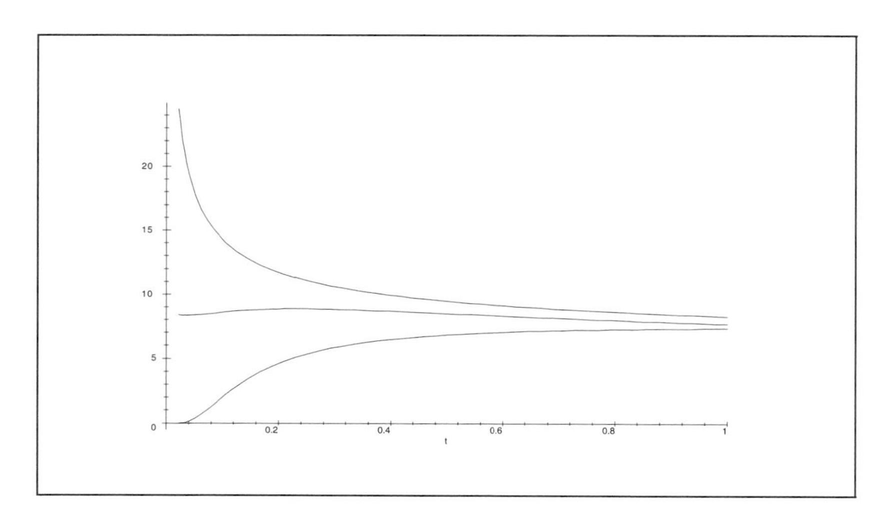

Figure 7.6: Theta for In-the-Money, At-the-Money, and Out-of-the-Money Options

```
> plot([Theta(70,t,80,.18,.12),Theta(80,t,80,.18,.12),\
> Theta(90,t,80,0.18,0.12)],t=0..1,thickness=2,\
> labels=['Time to Maturity', Theta],color=[red,blue,\
> green],title='Figure 7.6:  Theta for In-the-Money,\
> At-the-Money, and Out-of-the-Money Options');
```

A three-dimensional graph of Theta vs. the stock price and time to maturity can be produced by executing the following command (this graph is not produced in the hard copy of the book):

```
> plot3d(Theta(70,t,s,0.18,0.12),t=0..1,s=0..120,\
> orientation=[-8,46],axes=frame, style=patch,\
> labels=['Time to Maturity','Stock Price',Theta],\
> title='Theta vs.Time to Maturity and Stock Price');
```

The reader is invited to experiment with different values and to replicate these results for a put option. The formula for the Θ measure of a put option is given by

$$\Theta \;=\; \frac{P\,\sigma}{2\,\sqrt{t}} \mathrm{N}'\left(\frac{\ln\left(\frac{P}{K}\right)+\left(r-\frac{\sigma^2}{2}\right)\,t}{\sigma\,\sqrt{t}}\right) - \qquad (7.5)$$

$$r\,K\,e^{-r\,t}\,\mathrm{N}\left(\frac{\ln\left(\frac{P}{K}\right)+\left(r+\frac{\sigma^2}{2}\right)\,t}{\sigma\sqrt{t}}\right).$$

7.2 The Delta Measure

It makes sense that, other things being equal, the value of the call is an increasing function of the stock price. The higher the stock price, the larger the difference between the exercise price and the stock price, and thus the larger the value of the option at maturity. This is indeed also the case prior to the maturity time. The higher the current value of the stock the larger is the likelihood for a higher stock price at maturity.

Recall that the price of the option is the discounted expected value under the risk-neutral probability. The higher the current stock price is, the more skewed to the right is the risk-neutral distribution. This has been demonstrated already in Section 5.8, Figure 5.11, where we investigated some properties of the lognormal distribution. Hence, when calculating the expected value, which is really just a weighted average (weighted by the probabilities), higher weights are assigned to higher stock values.

It is therefore not surprising that the expected value of the call's payoff at maturity and consequently the discounted expected value are increasing functions of the stock price. The Delta measure is simply the first derivative of the call value with respect to the stock price. The Delta measure therefore represents the instantaneous rate of change in the call price due to an infinitesimal change in the stock price. It is represented by the Greek letter Δ and is defined below.

$$\Delta=\frac{\partial\left(P\,\mathrm{N}\left(\frac{\ln(\frac{P}{K})+\left(r+\frac{\sigma^2}{2}\right)t}{\sigma\sqrt{t}}\right)-Ke^{-rt}\mathrm{N}\left(\frac{\ln(\frac{P}{K})+\left(r-\frac{\sigma^2}{2}\right)t}{\sigma\sqrt{t}}\right)\right)}{\partial P} \tag{7.6}$$

The appendix to this Chapter shows that the formula for Delta of a call option on a non-dividend-paying stock is given by

$$\Delta(K,t,P,\sigma,r)=\mathrm{N}\left(\frac{\ln\left(\frac{P}{K}\right)+\left(r+\frac{\sigma^2}{2}\right)\,t}{\sigma\sqrt{t}}\right). \tag{7.7}$$

In order to visualize properties of Delta we define the Delta function in MAPLE.

```
> Delta:=unapply((diff(Bs(K,t,Pr,sigma,rf),Pr)),\
```

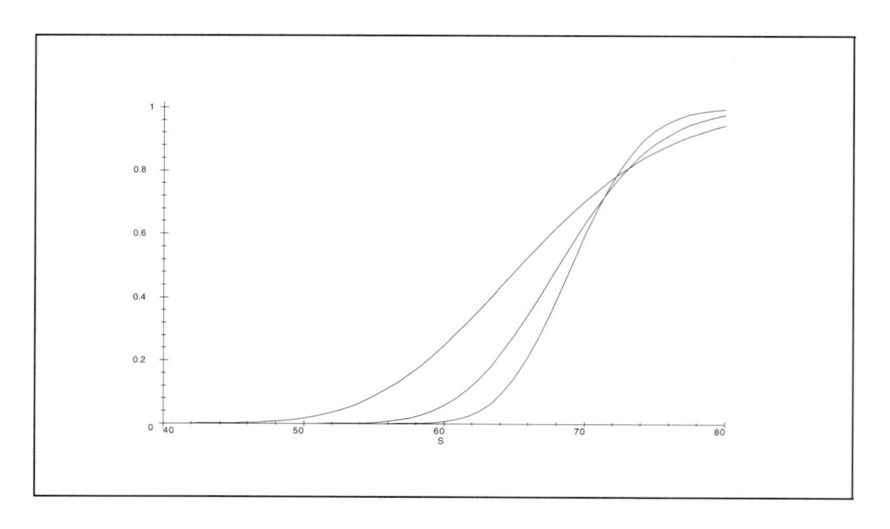

Figure 7.7: Delta of a Call Option for Maturities of $t = 0.5$, $t = 0.2$, and $t = 0.1$

```
> K,t,Pr,sigma,rf):
```

The graph in Figure 7.7 illustrates the function Delta for time to maturity of 0.5 years (the green graph with the smallest x-intersection), for time to maturity of 0.2 years (the black graph with the middle x-intersection), and for time to maturity of 0.1 years (the red graph with the largest x-intersection) in the \$40 to \$60 stock price region.

```
> plot([Delta(70,.5,S,0.18,0.12),Delta(70,.2,S,.18,.12),\
> Delta(70,.1,S,0.18,0.12)],S=40..80,thickness=2,\
> color=[green,black,red],labels=[Stock,Delta],\
> title='Figure 7.7:  Delta of a Call Option\
> for Maturities of t=0.5, t=0.1, t=0.2');
```

Delta is the increase, in the price of the call, per infinitesimal increase in the price of the stock. As indicated by Figure 7.7 and our discussion above, the Delta of a call option is always positive. This is confirmed by the algebraic expression for Delta, equation (7.7). Since N is the commutative distribution function of a standard normal random variable, $\mathrm{N}(x)$ is positive for every x. Hence, Delta is also smaller than one, which means that an infinitesimal increase in the stock value will cause a smaller increase in the call value.

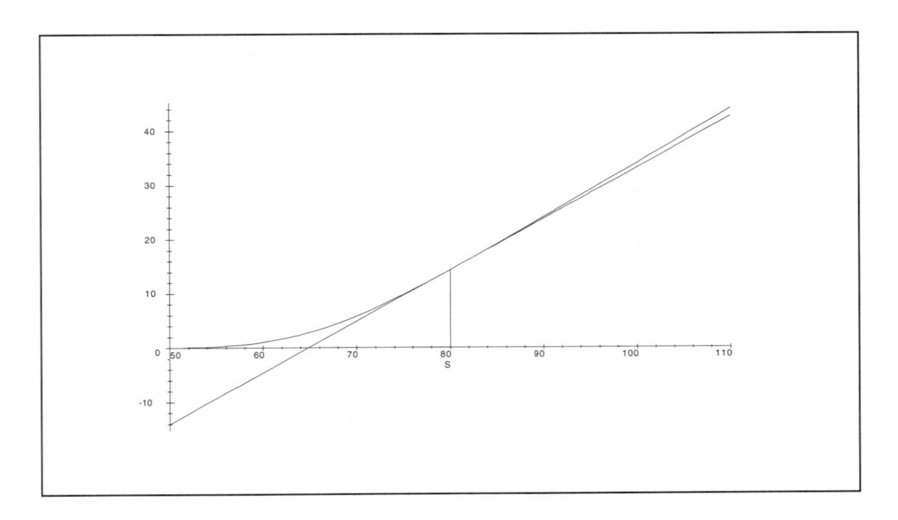

Figure 7.8: Approximation of the Price of the Call with Delta

Delta can also be visualized as Theta before, since it appears in the Taylor series approximation with respect to changes in the price of the stock. Thus, we can approximate the price of the call option via a linear function which is tangent to the call price function at, for example, a stock price of \$80, i.e., as

$$\begin{aligned} Bs(90, 0.5, S, 0.18, 0.12) &\cong Bs(90, 0.5, 80, 0.18, 0.12) + \\ &\quad (S-80)\,\Delta(90, 0.5, 80, 0.18, 0.12). \end{aligned} \tag{7.8}$$

This is illustrated in Figure 7.8.

```
> plot([Bs(70,.5,S,0.18,0.12),(Bs(70,.5,80,0.18 ,0.12)+\
> (S-80)*Delta(70,.5,80,0.18,0.12)),[[80,0],\
> [80,Bs(70,.5,80,0.18,0.12)]]],S=50..110,\
> labels=['StockPrice','Call Value'],thickness=2,\
> color=[green,blue,red],title='Figure 7.8:\
> Approximation of the Price of the Call');
```

In order to approximate (for hedging or replicating purposes) the price of the call option using Delta, it is assumed that the graph of the value of the call option is linear in a small neighborhood of the point. Observe how the value of the call option is approximated around a certain point, using the slope of the graph (Delta) at that point. Figure 7.8 illustrates how close the linear approximation is to the true graph in a certain neighborhood of the current value of the stock.

There is a major distinction between the sensitivity of an option to the passage of time and to a change in the stock price. While the latter is random, the former is not. Given the current stock price one knows exactly the change in the option value due to the passage of time. Hence, no uncertainty is involved in judging the risk in one's position that is due only to the passage of time.

Certainly though, measuring sensitivity using differentiation techniques is done with some reservation. It is correct only for an instant of time and assumes that all other parameters stay constant. However, the changes in the stock price are random in nature and their direction is not known. Thus, if one is interested in neutralizing the (instantaneous) effect of changes in the stock price on the call value, one needs to identify a financial asset which is affected by the change of prices of the stock in an opposite way to the call option. We emphasize again the distinction between Theta and Delta. Theta measures the sensitivity of the option price with respect to time. The direction of the time movement is known with certainty. The time to maturity will definitely decrease as time moves on.

Conversely, changes in the stock price are random. Hence, to hedge against these random changes, the uncertain changes in the direction of price movement, we must hold an asset whose value moves in the opposite direction to the call in response to price movement of the stock. Since Delta is positive a short stock is such an asset.

Hedging against changes in the call value that are results of changes in the stock price is a similar concept to the arbitrage portfolios we composed in the first Chapter. There the idea was to hold assets in a portfolio such that in each state of nature the long portfolio produces more than (or the same as) the short portfolio.

However, in the case of Delta hedges, the offsetting position is correct only for small changes around the current price of the stock and not for every state of nature. Therefore this hedging mechanism is effective only for small changes.

Consider the example in Figure 7.8 where the current stock price is \$80. Recall that Delta is the increase in the call value per small increase in the stock value. Hence, if the stock value will increase to S, the increase in the stock price will be $S - 80$, and the approximation of the increase in the call value will be $(S - 80)\Delta$. Thus, if we subtract $(S - 80)\Delta$ from the call and

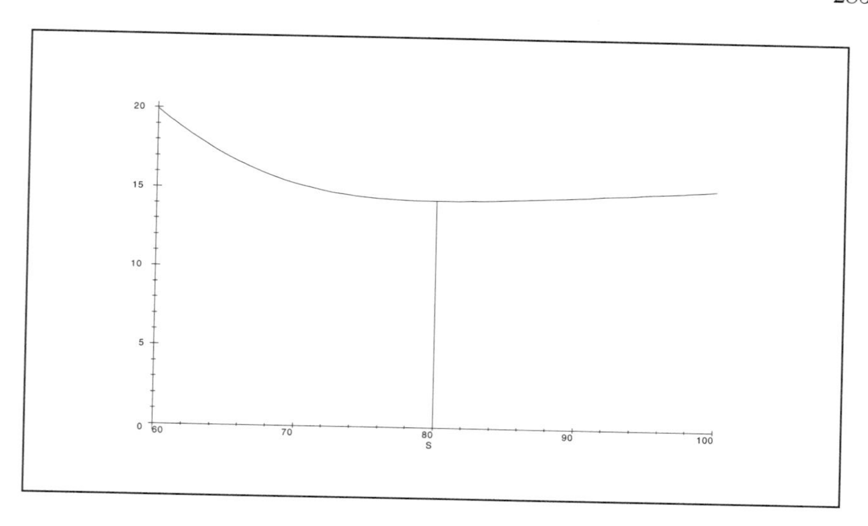

Figure 7.9: Hedging with Delta

then plot the value of the portfolio

$$Bs(70, 0.5, S, 0.18, 0.12) - (S - 80)\Delta(70, 0.5, 80, 0.18, 0.12) \tag{7.9}$$

against the price of the stock, we should observe a U-type function. This is because in the vicinity of $S = \$80$ the expression in equation (7.9) should be insensitive to changes in the stock price. Indeed, it is easily verified by noting that the partial derivative of this expression with respect to S is zero. In order to visualize the meaning of Delta we graph, Figure 7.9, the function in equation (7.9).

```
>   plot([Bs(70,.5,S,0.18,0.12)-\
>   (S-80)*Delta(70,.5,80,0.18,0.12),[[80,0],\
>   [80,Bs(70,.5,80,0.18,0.12)]]],S=60..100,thickness=2,\
>   color=[green,blue,red],labels=['Stock Price',\
>   'Call Value'],title='Figure 7.9:  Hedging with Delta');
```

Around the current stock price, \$80, the function $Bs(70, 0.5, S, 0.18, 0.12) - (S - 80)\Delta(70, 0.5, 80, 0.18, 0.12)$, in Figure 7.9, is relatively flat. This is the vicinity in which the hedging is effective. The portfolio $Call - (S - 80)\Delta = Call + 80\Delta - S\Delta$ can be interpreted as being composed of one long call, 80Δ in cash, and Δ units of short position in the stock. (We shall investigate the issue of how to compose hedge portfolios in the next Chapter.)

Recall that Delta is smaller than one. Therefore it can also be interpreted as the fraction of the short stock position we should hold in order to neutralize the effect of price movement on the call value. Delta helps in hedging by offsetting the change in the call value due to changes in the price of the stock. The fraction of a stock being held short in order to offset the effect[7] of a price change on one long position in the call is referred to as the *hedge ratio*.

We can also numerically illustrate the changes in the value of this portfolio. These changes show how around the current value of the stock, \$80, the curve is nearly flat. The value of the call for a stock price of \$80 is given by

```
> Bs(70,.5,80,0.18,0.12);
```

$$14.33456607$$

The call value variability in the range of $[80-8, 80+8]$ is displayed below:

```
> seq(Bs(70,.5,80+S,0.18,0.12),S=-8..8);
```

7.34341357, 8.13452644, 8.95518517, 9.80227135, 10.67276779, 11.56381009, 12.47272433, 13.39705249, 14.33456607, 15.28327047, 16.24140143, 17.20741547, 18.17997620, 19.15793750, 20.14032547, 21.12631948, 22.11523365

The value of a portfolio, composed of one long call option, 80Δ in cash, and a short position in Δ units of the stock, in the same range of $[80-8, 80+8]$ is given below.

```
> seq(evalf(Bs(70,.5,S,0.18,0.12)+\
> 80*Delta(70,.5,80,0.18,0.12)\
> -S*Delta(70,.5,80,0.18,0.12)),S=72..88);
```

[7] The reservation mentioned before holds true here also. Sensitivity measures devised through partial differentiation techniques are valid only for a small range around the current value of the variable in question, and such measures assume that all other variables are fixed. These types of results are referred to as comparative statics, or comparative static analysis, in the economics literature. As the name suggests, it is assumed that all variables, but the one under examination, remain constant or static while the analysis is conducted. There is another issue concerning the sensitivity measures with respect to a random variable. As we will see in Chapter 15, these measures should involve more than just the first partial derivative as is the case in deterministic calculus. The Delta and other measures developed here can also be used in the context of other derivatives such as puts, futures, and forwards. The definition is simply the partial derivative of the value of the asset in question with regards to the parameter explored.

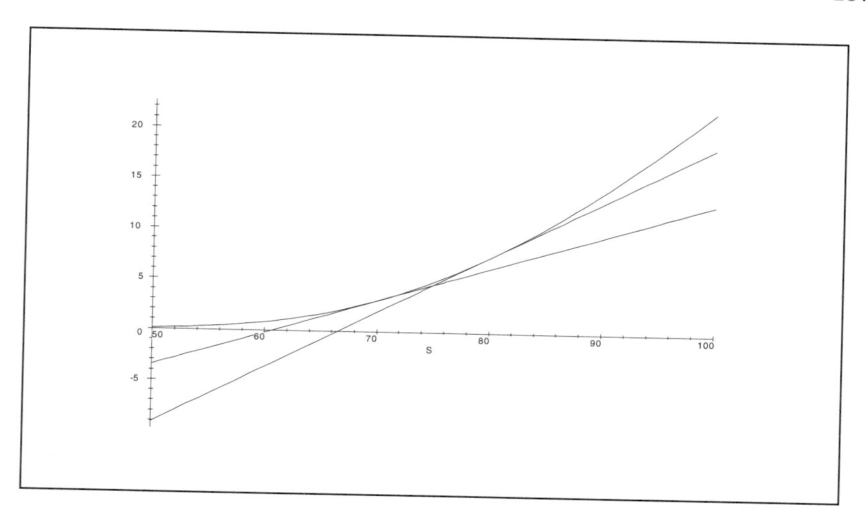

Figure 7.10: Delta Approximation at Different Points of Tangency

14.89078962, 14.73848049, 14.61571722, 14.51938139, 14.44645583, 14.39407610, 14.35956836, 14.34047450, 14.33456607, 14.33984847, 14.35455742, 14.37714946, 14.40628816, 14.44082747, 14.47979343, 14.52236543, 14.56785759

We now see that the hedging strategy is, indeed, effective around that current price of the stock (\$80 in our case). The reader is encouraged to experiment with the effects of different parameters on the hedging activity.

We can demonstrate how the approximation by Delta depends on the current value of the stock in Figure 7.10. We observe the approximation at two points. At each point the approximation is based on the slope of the graph of the value of the call option. Thus, since the slope of the graph changes, so does the Delta approximation. One can see how the approximations differ by observing the two lines in Figure 7.10.

```
> plot([Bs(90,1,S,0.23,0.12),Bs(90,1,70,0.23,0.12)+\
> (S-70)*Delta(90,1,70,0.23,0.12),Bs(90,1,80,0.23,0.12)\
> +(S-80)*Delta(90,1,80,0.23,0.12)],S=50..100,\
> colour=[red,green,navey],thickness=2,labels=\
> ['Stock Price', 'Call Value'],title='Figure 7.10:\
> Delta Approximation at Different Points of Tangency');
```

The change in Delta can also be observed by graphing a three-dimensional curve of the value of Delta as a function of the stock price and of time to maturity. This allows us to observe how Delta changes as the option moves from out-of-the-money, to at-the-money, to in-the-money. This is illustrated in Figure 7.11.

```
> plot3d(Delta(70,t,S,0.18,0.12),t=0..1,S=40..80,labels=\
> [time,Stock,Delta],axes=normal,orientation=[-54,66],\
> title='Figure 7.1:  The Value of Delta as a Function\
> of S and t');
```

Note how Delta changes for options which are deep in-the-money and deep out-of-the-money, as the maturity time approaches. For a deep in-the-money option, in a region where the stock price is above \$70, the curve is very steep. The closer the time to maturity, the steeper the curve is. Thus, changes in Delta are more pronounced the closer Delta is to the maturity time ($t = 0$) and the further the stock price is from \$70. One can observe a triangle-like shape where for a large t and for prices below \$70 Delta is nearly zero. That means that in this region, changes in stock price do not affect the price of the call much. Indeed, this notion can be verified in a more precise manner. The reader may wish to calculate the derivative of Delta with respect to time and stock price for different values of the stock price and for different times to maturity. This calculation can be done via MAPLE. Can you, the reader, do it?

7.3 The Gamma Measure

We already mentioned that there is a major distinction between the Delta and the Theta measure. Delta measures sensitivity of the call option value with respect to the random movements of the stock price. We shall later see that we may need to take account of a more precise approximation to the call value. The Delta approximation, or hedging, amounts to utilizing the slope of the graph of the call option value at a certain point to produce a linear approximation to the true value of the call option. It is, in fact, the Taylor series approximation to the function. However, if the function is very "nonlinear" around the point in question, the approximation will not be very accurate and may not be adequate for the purpose intended.

To increase the quality of the approximation one can take account of the second derivative of the call option value with respect to the stock price. The second derivative of the call price with respect to the stock allows us to include some information about the curvature of the graph around that

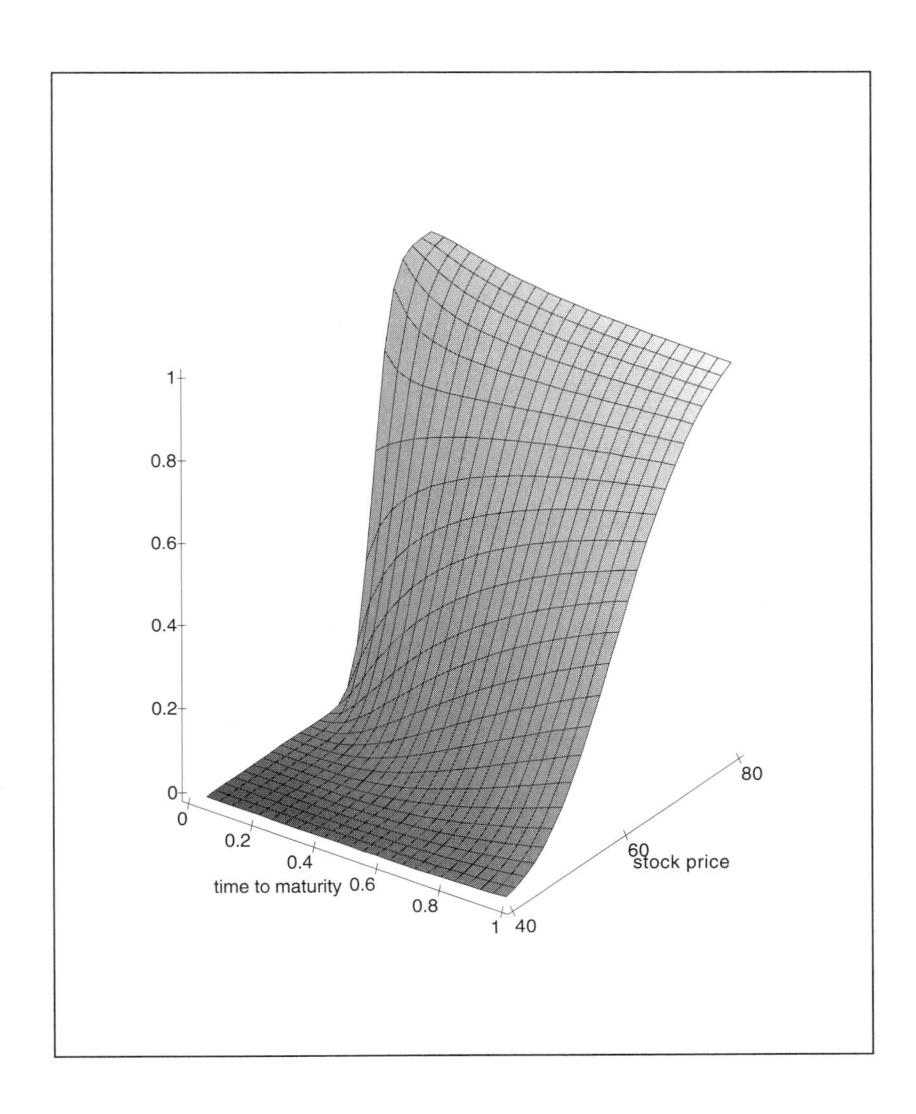

Figure 7.11: The Value of Delta as a Function of S and t

point. The Gamma measure is the second derivative of the call price with respect to the stock price. Correspondingly, Gamma is the first derivative of Delta with respect to the stock price. Gamma is represented by the Greek letter Γ and can be stated as

$$\Gamma(K, t, P, \sigma, r) = \frac{\partial}{\partial P}\Delta(K, t, P, \sigma, r) = \frac{\partial \mathrm{N}\left(\frac{\ln\left(\frac{P}{K}\right)+\left(r+\frac{\sigma^2}{2}\right)t}{\sigma\sqrt{t}}\right)}{\partial P}. \quad (7.10)$$

The appendix to this Chapter shows that the equation for Gamma is

$$\Gamma(K, t, P, \sigma, r) = \frac{1}{P\sigma\sqrt{t}}\mathrm{N}'\left(\frac{\ln\left(\frac{P}{K}\right)+\left(r+\frac{\sigma^2}{2}\right)t}{\sigma\sqrt{t}}\right). \quad (7.11)$$

We use MAPLE to define Gamma as the second derivative of the call option value with respect to the stock price as below:

```
> Gamma:=unapply((diff(Bs(K,t,Pr,sigma,rf),Pr,Pr)),\
> K,t,Pr,sigma,rf):
```

By taking Gamma into account, we can produce a better approximation to the call option value than was possible using Delta alone. This time our approximation will be quadratic rather than linear. Our approximation is, as before, based on Taylor series approximation. It has been expanded to include the second term of the Taylor series approximation. The approximating function of the call option value is given by

$$\begin{aligned}&Bs(70, 0.5, 80, 0.18, 0.12) + (S-80)\Delta(70, 0.5, 80, 0.18, 0.12)\\&+\frac{(S-80)^2}{2}\Gamma(70, 0.5, 80, 0.18, 0.12).\end{aligned} \quad (7.12)$$

Equation (7.12) and Figure 7.12 illustrate this second-order approximation. In this figure the true value of the call option (the green curve with the positive y-intercept) is graphed in the same plane along with the second-order approximation to the value of the call option which is calculated (the red curve with the negative y-intercept)) using Gamma. Note, that this time, unlike the approximation which used Delta, the approximating function is quadratic.

```
> plot([Bs(70,.5,S,0.18,0.12),\
> [[80,0],[80,Bs(70,.5,80,0.18,0.12)]],\
> Bs(70,.5,80,.18,.12)+(S-80)*Delta(70,.5,80,.18,.12)+\
```

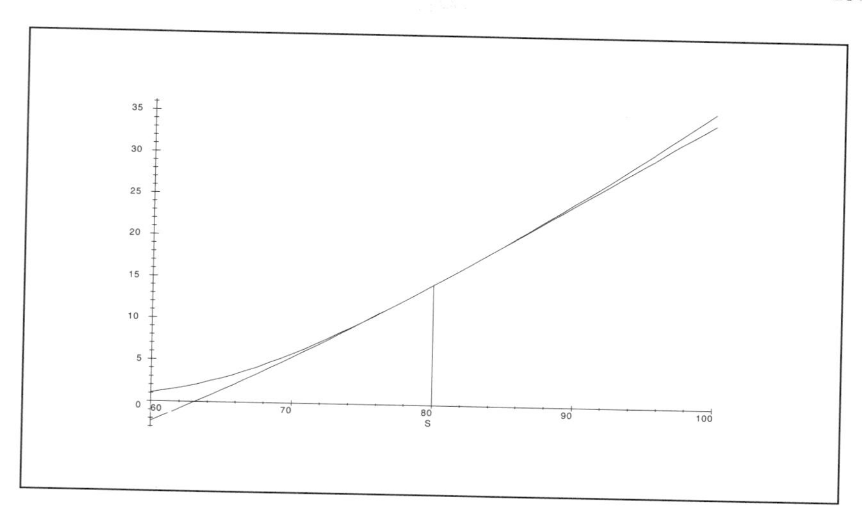

Figure 7.12: Quadratic Approximation Using Gamma and Delta

```
> ((S-80)^2)*Gamma(70,.5,80,0.18,0.12)/2],S=60..100,\
> color=[green,blue,red],thickness=2,labels=\
> ['Stock Price','Call Value'],title='Figure 7.12:\
> Quadratic Approximation Using Gamma and Delta');
```

The reader can also execute the command below to plot the linear approximation with Delta only (the lower brown curve with the largest x-intercept) and the quadratic approximation with Delta and Gamma (the middle red curve with the smaller positive x-intercept) in the same plane as the value of the call option (the upper green curve with positive y-intercept). This is demonstrated in Figure 7.13. Note that the approximation with both Gamma and Delta is better than the one with Delta only.

```
> plot([Bs(70,.5,S,0.25,0.08),\
> [[80,0],[80,Bs(70,.5,80,0.25,0.08)]],\
> Bs(70,.5,80,0.25,0.08)+(S-80)*Delta(70,.5,80,0.25,0.08)\
> +((S-80)^2)*Gamma(70,.5,80,0.25,0.08)/2,\
> Bs(70,.5,80,.25,.08)+(S-80)*Delta(70,.5,80,.25,.08)],\
> S=60..100,color=[green,blue,red,brown],thickness=2,\
> labels=['Stock Price','CallValue'],view=[70..90,0..27],\
> title='Figure 7.13:  Quadratic Approximation vs.\
> Linear Approximation');
```

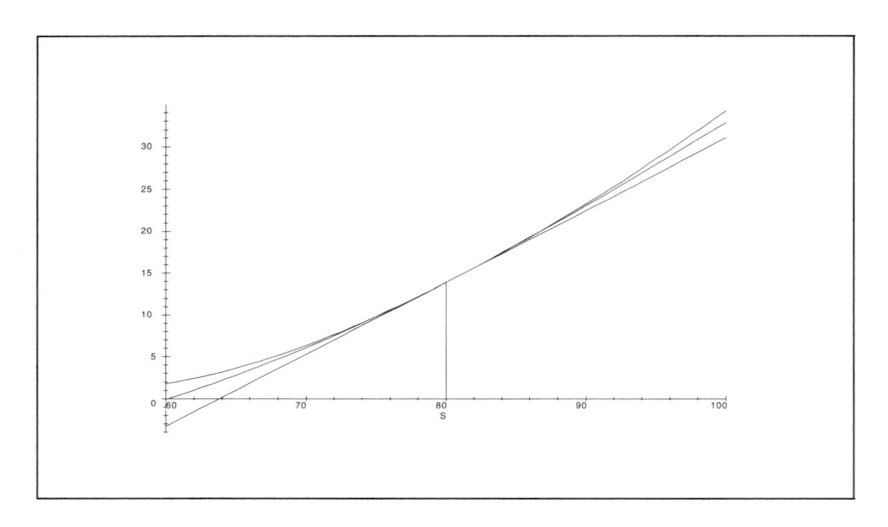

Figure 7.13: Quadratic Approximation vs. Linear Approximation

We can also plot, Figure 7.14, the graph of Gamma for an option which is in-the-money (the lowest red graph), at-the-money (the upper blue graph that does not intersect the axes), and out-of-the-money (the middle green graph that intersects the at-the-money graph).

```
> plot([Gamma(70,t,80,.18,.12),Gamma(80,t,80,.18,.12),\
> Gamma(90,t,80,0.18,0.12)],t=0..1,labels=['Time to\
> Maturity',Theta],color=[red, blue,green],thickness=2,\
> title='Figure 7.14:  Gamma of an In-the-Money,\
> At-the-Money, and Out-of-the-Money Option');
```

We now move on to the investigation of other measures of sensitivity. Given the detailed explanation provided for these first three sensitivity measures, the exposition and description for the others can be shortened. The measures to be defined henceforth are all based on the first-order Taylor series expansion. In other words, these measures use only the slope (not any measure of curvature, as was the case with Gamma) of the graph of the option value with respect to the parameter in question. Hence, these measures induce a linear approximation to the true option value.

The sensitivity measures to be discussed in the rest of this Chapter are with respect to parameters, which are assumed to remain constant in the

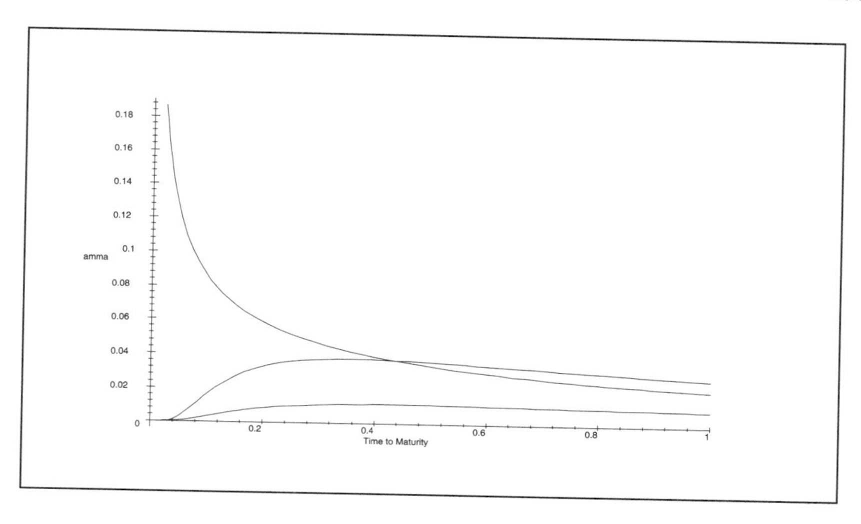

Figure 7.14: Gamma of an In-the-Money, At-the-Money, and Out-of-the-Money Option

Black–Scholes derivation. They even differ from the measure Theta. Theta measures the sensitivity with respect to time. The time indeed proceeds to maturity, albeit in a very deterministic way. The Delta and Gamma measures were with respect to random movements of the stock price.

The remaining sensitivity measures are defined with respect to interest rate and volatility. They differ from Delta, Gamma, and Theta since the parameters are assumed to be fixed. These measures will be informative regarding how to approximate errors in the calculation of the option price if, for example, there were an error in estimation of either the volatility or the risk-free rate. These measures can also be used as an *ad hoc* technique for calculating the value of the option if the interest rate is not fixed or if the volatility is assumed to be nonstationary, but rather varies with time.

7.4 The Vega Measure

Vega, represented by the Greek letter V, measures the sensitivity of the call price with respect to the value of the volatility. Perhaps the best way to appreciate this effect is to look at a graph or an animation of the value of the call option as a function of the time to maturity and as a function

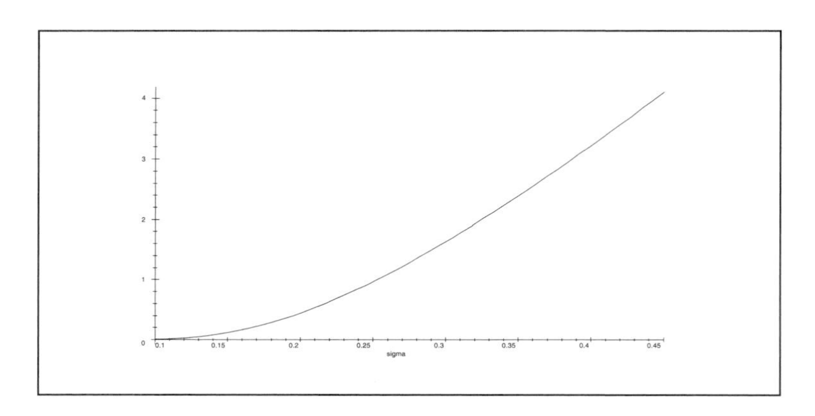

Figure 7.15: The Value of a Call as a Function of the Volatility

of the stock price for different values of volatility. Let us first take a look at a two-dimensional graph (Figure 7.15) of the value of a call option as a function of the volatility.

```
> plot(Bs(90,.5,70,sigma,0.12),sigma=0.1..0.45,\
> axes=normal,labels=[sigma,‘Call Value‘],\
> title=‘Figure 7.15:  The Value of a Call as a Function\
> of the Volatility‘);
```

It is evident that the value of the call option increases as σ (volatility) increases. This is true regardless of the time to maturity. However, the larger the time to maturity, the larger is the effect of increase in σ. This is demonstrated in the three-dimensional animation below. The three-dimensional graph illustrates the value of the call option as a function of the stock price and the volatility, and is animated with respect to t, the time to expiration of the option.

Note that at $t = 0$ the graph is not affected by σ at all. At $t = 0$ the payoff is not stochastic anymore — it is simply $\max(0, S - K)$ — and hence, it is not dependent on σ. The larger the time to maturity, the larger the effect of σ. The deeper in-the-money the call option is, the larger the effect of σ. The reader should run the animation frame by frame and pay attention to the point mentioned above. Figure 7.16 in the hard copy displays one frame of the animation.

```
> plots[animate3d]('BstCall'(90,t,S,sigma,0.12),\
```

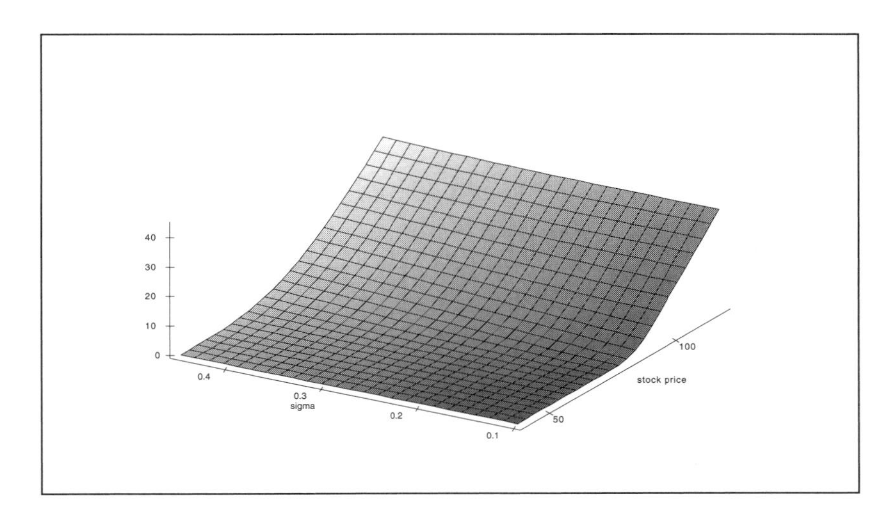

Figure 7.16: The Value of the Call as a Function of S and σ

```
> S=40..120,sigma=0.1..0.45,t=0..1,axes=normal,\
> orientation=[-149,44],labels=[`stock price`,sigma,\
> ` `],frames=20, title=`Figure 7.16:\
> The Value of the Call as a Function of S and Sigma`);
```

This graph conveys the following message: The higher the volatility or the higher the time to maturity, the higher the value of the call option.

It is apparent from Figures 7.15 and 7.16 that Vega is positive. This can also be understood from an intuitive point of view. As sigma increases, the distribution becomes wider and wider (see Figure 7.17). Since a lognormally distributed random variable has only positive realizations, the increase in volatility will make the right tail end of the distribution fatter than the left tail. As the volatility increases, we therefore observe the effect of increased probability of the realization of higher and lower stock price values.

The price of the call is the discounted weighted average of the possible realizations of the value of the call. The weights are the probabilities of the different realizations. However, for stock values of less than the exercise price, the value of the call is zero. Hence, increasing the probability for a realization of stock price less than the value of the exercise price has no direct effect on the value of the call. Moreover, while we observe increased probability of the left tail of the realization, this phenomenon also affects the

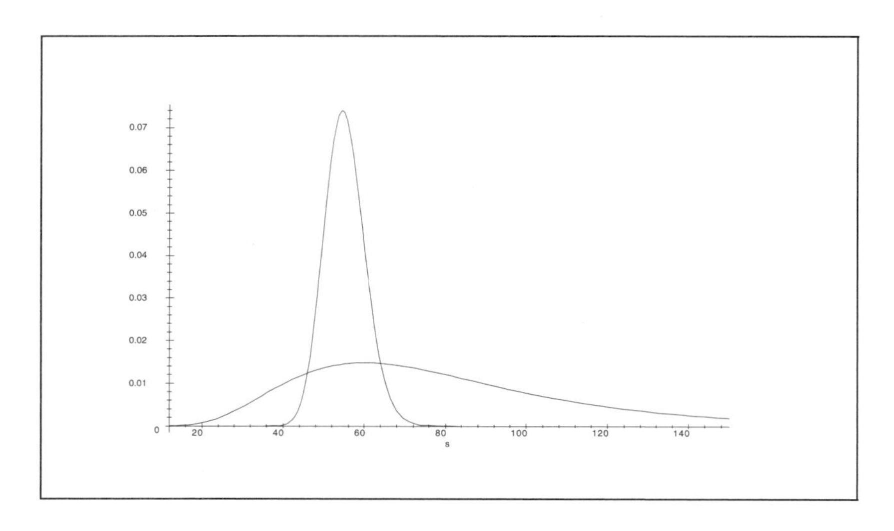

Figure 7.17: The (Risk-Neutral) Lognormal Density for $\sigma = 0.9$ and $\sigma = 0.45$

price but to a lesser extent. Thus, since the right tail grows fatter to a much greater degree than the left tail, the dominant effect is that of increasing the value of the call. This is described in Figure 7.17, where the upper graph corresponds to $\sigma = 0.09$ and the lower graph to $\sigma = 0.45$.

```
> Plot([Lnrn(60,1,s,.10,.09),Lnrn(60,1,s,.10,.45)],\
> s=12..150, title=‘ Figure 7.17:  The Lognormal\
> Distribution as a Function of Sigma‘,thickness=3,\
> titlefont=[TIMES, BOLD, 10],colour=[red,green]);
```

The measure Vega is defined in a similar manner to the other measures. It is the derivative of the value of the call option with respect to a particular parameter, in this case sigma, i.e.,

$$V = \frac{\partial\left(P\mathrm{N}\left(\frac{\ln\left(\frac{P}{K}\right)+\left(r+\frac{\sigma^2}{2}\right)t}{\sigma\sqrt{t}}\right) - Ke^{-rt}\mathrm{N}\left(\frac{\ln\left(\frac{P}{K}\right)+\left(r-\frac{\sigma^2}{2}\right)t}{\sigma\sqrt{t}}\right)\right)}{\partial\sigma}. \tag{7.13}$$

The appendix to this Chapter shows that the equation for Vega of a call option is

$$V(K,t,P,\sigma,r) = P\sqrt{t}\mathrm{N}'\left(\frac{\ln\left(\frac{P}{K}\right)+\left(r+\frac{\sigma^2}{2}\right)t}{\sigma\sqrt{t}}\right). \tag{7.14}$$

The expression in equation (7.14) indeed confirms that Vega is positive and is an increasing function of the time to maturity t and the current stock price P. The effect of time to maturity and stock prices on Vega can be visualized in a three-dimensional graph. To this end we define the function Vega in MAPLE.

```
> Vega:=unapply((diff(Bs(K,t,Pr,sigma,rf),sigma)),\
> K,t,Pr,sigma,rf):
```

In the figure below, only in the on-line version of the book, one can clearly see that the larger the time to maturity, the larger the effect of changes in the volatility. This has already been explained. We also see in this graph that the effect is more pronounced in a larger interval of prices the longer the time to maturity.

```
> Plot3d(Vega(90,t,S,0.1,0.12),t=0..1,S=30..120,\
> axes=normal, labels=[Time,'StockPrice',Vega])
```

We can observe the value of Vega for different values of σ. Note that the expression for Vega is the same for a put and a call option. One of the exercises at the end of this Chapter asks the reader to prove it. In Figure 7.18 we can view Vega for an in-the-money call or put option (the lower red graph), for an at-the-money call or put option (the middle green graph), and for an out–of-the-money call or put option (the upper blue graph).

```
> plot([Vega(70,t,80,0.18,0.12),Vega(80,t,80,0.18,0.12),\
> Vega(90,t,80,0.18,0.12)],t=0..1,thickness=2,\
> labels=['Time to Maturity',Vega],color=[red,blue,green]\
> ,title='Figure 7.18:  Vega as a Function of Sigma\
> for an In-the-Money, At-the-Money, and\
> Out-of-the-Money Option');
```

We can define a Gamma-like measure for Vega — say Vega2:

```
> Vega2:=unapply((diff(Bs(K,t,Pr,sigma,rf),\
> sigma,sigma)),K,t,Pr,sigma,rf):
```

This Gamma-like measure for Vega is the approximation to the change in the value of a call (or put) option as a result of changes in Vega. It is illustrated in Figure 7.19. This Gamma-like measure is simply the second derivative of the option value with respect to the volatility. Hence, we can use the first two terms of the Taylor series expansion (approximation), with respect to sigma, to the option value. The blue parabola is the second-order Taylor approximation and the (by now familiar shape of the call value) red graph is the true value of the call option.

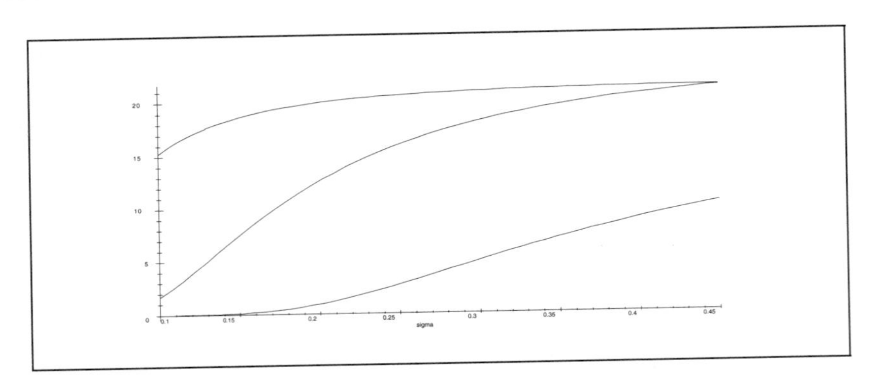

Figure 7.18: V as a Function of σ for an In-the-Money, At-the-Money, and Out-of-the-Money Option

```
> plot([Bs(90,1,60,vol,0.12),Bs(90,1,60,0.16,0.12)+\
> (vol-0.16)*Vega(90,1,60,0.16,0.12)+\
> ((1/2)*(vol-0.16))^2*Vega2(90,1,60,0.16,0.12)],\
> vol=0.01..0.25,colour=[red,blue],thickness=2,\
> labels=[sigma,'Call Value'],title='Figure 7.19:\
> Vega and Gamma-like Approximation of the\
> Call Value as a Function of Sigma');
```

In a similar manner to the definition of other measures of sensitivity of option values, we define the measure of sensitivity of the call option value to changes in the interest rate. This measure is called rho.

7.5 The Rho Measure

The sensitivity measure investigated in this section belongs to the same "group" as Vega. This is in contrast to Theta, Delta, and Gamma that can be considered as sensitivity measures with respect to "state variables". The price of the stock and the time to expiration are state variables: they denote the state of nature and keep changing.

This section defines the measure of sensitivity of the call option value to changes in the interest rate. This measure is called rho and is denoted by the Greek letter ρ. At the end of this Chapter we will also define the sensitivity to changes in the exercise price. The interest rate and the exercise price, like the volatility, are also parameters that are assumed to be fixed and constant

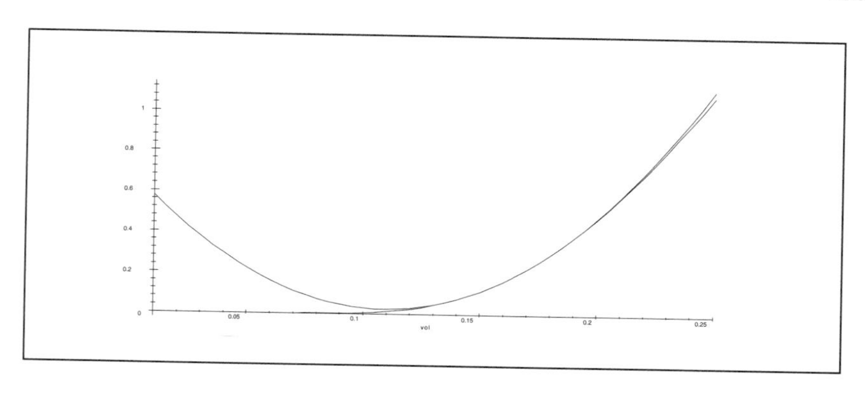

Figure 7.19: Vega and Gamma-like Approximation of the Call Value as a Function of σ

throughout the life of the option. As time moves on, the stock price usually changes and of course the time to expiration becomes shorter. Hence only Theta, Delta and Gamma are used to hedge the changes in the option value due to the passage of time.

The measure rho is defined again as the first partial (calculus) derivative of the call value with respect to the interest rate r. In other words,

$$\rho = \frac{\partial\left(P\,\mathrm{N}\left(\frac{\ln\left(\frac{P}{K}\right)+\left(r+\frac{\sigma^2}{2}\right)t}{\sigma\sqrt{t}}\right) - Ke^{-rt}\mathrm{N}\left(\frac{\ln\left(\frac{P}{K}\right)+\left(r-\frac{\sigma^2}{2}\right)t}{\sigma\sqrt{t}}\right)\right)}{\partial r}. \tag{7.15}$$

The appendix to this Chapter shows that the formula for rho is

$$\rho = Kt\,e^{-r\,t}\,\mathrm{N}\left(\frac{\ln\left(\frac{P}{K}\right)+\left(r-\frac{\sigma^2}{2}\right)t}{\sigma\sqrt{t}}\right). \tag{7.16}$$

Hence its definition in MAPLE is given below:

```
> rho:=unapply((diff(Bs(K,t,Pr,sigma,rf),rf)),\
> K,t,Pr,sigma,rf):
```

We can observe the value of rho for different values of r. In Figure 7.20 we consider rho for the case of an in-the-money option (the red graph with the largest y-intercept), for an at-the-money option (the blue graph with the middle y-intercept), and for an out–of-the-money option (the green graph with the lowest y-intercept).

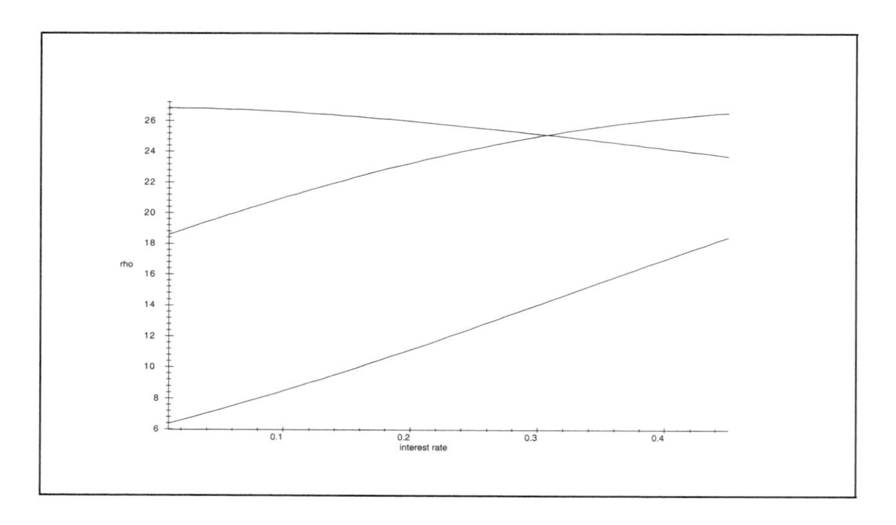

Figure 7.20: ρ as a Function of r for an In-the-Money, At-the-Money, and Out-of-the-Money Option

```
> plot([rho(60,.5,80,.3,r),rho(80,.5,80,.3,r),\
> rho(100,.5,80,.3,r)],r=0.01..0.45,thickness=2,labels=\
> ['interest rate',rho],color=[red,blue,green],title=\
> 'Figure 7.20:  Rho as a Function of r (for S = 80) for an\
> , In-the-Money,At-the-Money, and Out-of-the-Money Option');
```

ρ can be interpreted as the slope of the linear approximation to the graph of the value of the call. This is demonstrated in the following command. The graph appears only in the on-line version of the book.

```
> plot([Bs(90,1,60,.16,r),[[0.12,0],[0.12,Bs(90,1,60,\
> 0.16,.12)]],Bs(90,1,60,0.16,.12)+(r-0.12)*rho(90,1,\
> 60,0.16,0.12)],r=0.01..0.25,colour=[red,green,blue],\
> thickness=2,labels=['interest rate','Call Value'],\
> title='Linear Approximation to the Value of the\
> Call as a Function of r');
```

One can view how rho changes for different stock values, for a fixed interest rate, by plotting rho against the stock price. This can be done by issuing the following MAPLE command (Figure 7.21):

```
> plot(rho(90,1,S,0.16,0.12),S=60..140,\
```

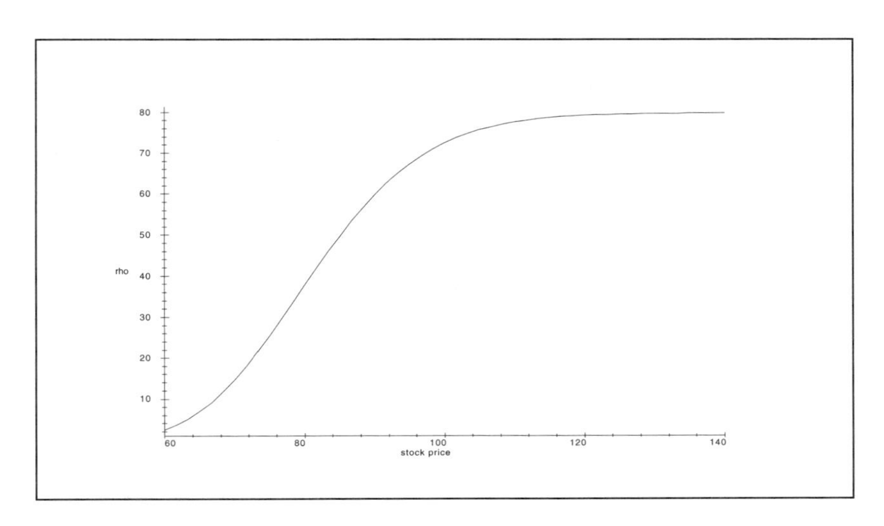

Figure 7.21: The Measure ρ as a Function of the Stock Price

```
> title='Figure 7.21:  The Measure Rho as\
> a Function of the Stock Price');
```

One can also investigate how the measure rho changes as the volatility changes for a different range of stock price. This is demonstrated in a three-dimensional graph in Figure 7.22.

```
> plot3d(rho(90,1,S,sigma,0.12),S=60..140,\
> sigma=0.01..0.45,axes=normal, orientation=[143,53],\
> labels=['Stock Price',sigma,rho],title='Figure 7.22:\
> rho as a Function of the Stock Price and sigma.');
```

Finally, in the on-line version of the book, one can play the animation of Figure 7.22 to see the effect of time to maturity.

```
> plots[animate3d](rho(90,t,S,sigma,0.12),S=60..140,\
> sigma=0.01..0.45,t=0...9,axes=normal,\
> labels=['stock price',sigma,rho],orientation=[143,53],\
> title='Rho as a Function of Sigma and the Stock Price\
> for Various Times to Maturity');
```

The reader is invited to produce a similar set of figures to enhance the understanding of the sensitivity of the option price, with respect to the exercise price. One can easily verify (see also the Appendix) that this measure

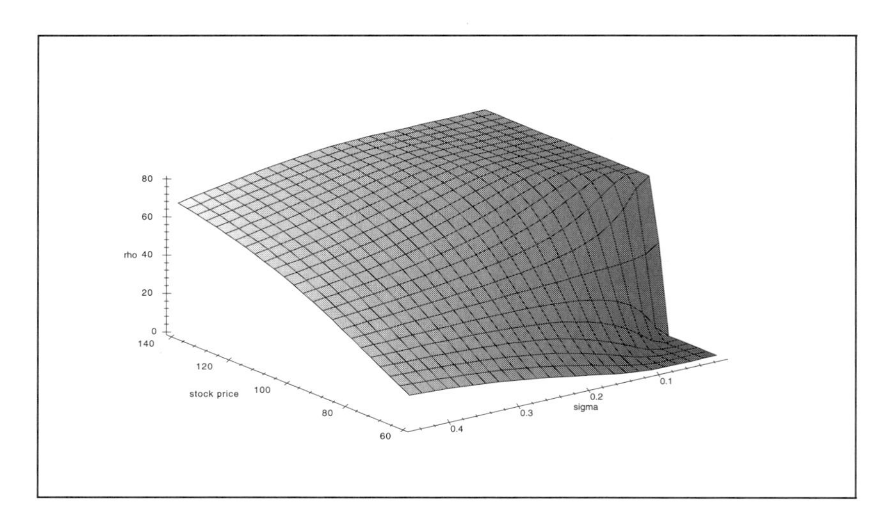

Figure 7.22: ρ as a Function of the Stock Price and σ

is given by

$$-e^{-rt}\mathrm{N}\left(\frac{\ln\left(\frac{P}{K}\right)+\left(r-\frac{\sigma^2}{2}\right)t}{\sigma\sqrt{t}}\right). \tag{7.17}$$

As seen by the expression and supported by our intuition, this expression is always negative. The higher the exercise price, the lower the value of the option.

7.6 Concluding Remarks

In this Chapter, sensitivity measures of a call option with respect to various parameters were investigated. A derivative security, as its name implies, is contingent on an underlying security. Hence, the source of the risk of these two securities is the same. It is therefore conceivable that if these two securities are being held in a portfolio in a correct proportion over an infinitesimal time interval, the portfolio might be risk-free.

As time passes, the value of a derivative security changes due to two factors: the passage of time that is connected to the time value of money and the changes in the price of the underlying asset. Hence, utilizing the sensitivity measures Theta and Delta, one might suspect that a portfolio that is risk-free over an instant could be composed. In the Chapter we concentrated mostly on the approximation of changes to the option price that are due to changes in only one of its parameters. It is also possible to approximate the changes in the option price that are due to changes in several parameters. As might be anticipated, such an approximation will be induced by a Taylor approximation for a function of several variables.

A first-order Taylor approximation to a function of several variables can be roughly[8] described as the aggregate of the approximations for each of the parameters. Hence, if we are interested in approximating the changes in a call price that are due to the passage of time and to the changes in the stock price we combine equations (7.4) and (7.8) to arrive at equation (7.18).

$$-(t-.5)\Theta(70,.5,80,0.18,0.1,2)+(S-80)\Delta(70,.5,80,0.18,0.1,2) \quad (7.18)$$

Therefore the approximation to the change in the option price which is due to both the progress of time and the change in the stock price is given by equation (7.18). To demonstrate the effectiveness of the approximation, we plot in a three-dimensional plane the value of the call option, $Bs(70,t,S,0.18,0.12)$, as a function of the stock price and time, together with the value of the approximation of the call value. The approximation to the call value will be the current call value plus the approximation to the changes in the call price as in equation (7.18). This is illustrated in Figure 7.23.

```
> plot3d(Bs(70,t,S,0.18,0.12)-((.5-t)*Theta(70,.5,80,\
> 0.18,0.1,2)+(S-80)*Delta(70,.5,80,0.18,0.1,2)),\
> S=50..100,t=0...1,axes=normal,orientation=[-51,45],\
> labels=[‘stock price‘,time to maturity,‘‘],title=‘Figure\
> 7.23:  The Value of the Call and Its Approximation\
> with Theta and Delta‘);
```

In the MAPLE version, the reader can change the perspective of these figures by "dragging" the graph to see exactly how the approximation is related to the true value. The effectiveness (or lack thereof, in certain regions) of the approximation is demonstrated by the gaps between these two graphs. Note that the two graphs coincide at the point $t = 0.5$, $S = 80$, and that they are very close in that vicinity. These types of arguments help in composing hedged portfolios as explored in the next Chapter. Moreover, a modified version of these approximations, that is required since S is not a deterministic variable, is used to derive the option pricing formula in Chapter 15.

[8] We will revisit this issue more precisely in Chapter 15.

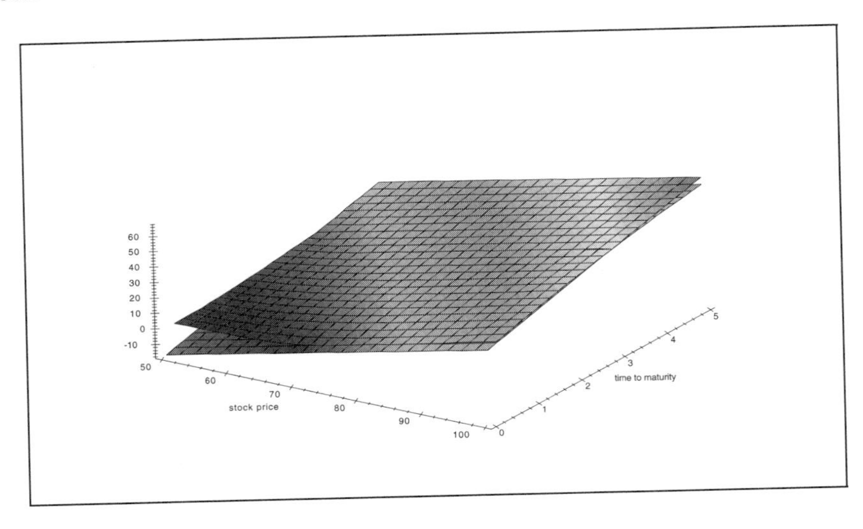

Figure 7.23: The Value of the Call and Its Approximation with Theta and Delta

7.7 Questions and Problems

Problem 1. Consider the data in the table below.

Stock Price	Exercise Price	Maturity	σ	r
100	97	30 days	0.12	12%

1. Set up two portfolios composed of a bond and a stock such that one portfolio has the same Delta as the call option and the other has the same Delta and Gamma as the call option. Use the MATLAB GUI or MAPLE to simulate a realization of the continuously compounded rate of return over 10 days. Recalculate the value of the option, its Delta and Gamma, and the value of the portfolios. Record the difference between the value of the portfolios and the value of the option after 10 days. Revise your initial portfolios to satisfy again the constraints on Gamma and Delta. Repeat the same procedure every 10 days until the maturity of the option. What are your conclusions?

2. Repeat the above for a binary option with a payoff function

$$f(x) = \begin{cases} 10 \text{ if } x \geq 97 \\ 0 \text{ otherwise.} \end{cases}$$

Problem 2. Prove that the expression for Vega is the same for a put and a call option (on the same underlying asset, with the same maturity and the same strike price).

Problem 3. The sensitivity measures of a put option can be derived directly, as in Appendix 7.8.2, or by utilizing the put-call parity relation. Use the latter approach to verify the results reported in Appendix 7.8.2 for a put option on a stock that pays no dividends.

Problem 4. Consider the data in the table below.

Stock Price	Exercise Price	Maturity	σ	r
100	110	30 days	0.18	10%

1. Estimate, utilizing Vega, the effect on the price of a call option that is the result of a slightly inaccurate estimate of the volatility of the stock return.
2. Reestimate part 1 using both Vega and Vega2.
3. How would your answers to parts 1 and 2 above change if you were asked to estimate the changes to the price of a portfolio composed of:
   (a) a put and a call,
   (b) a short put and a long call.

Problem 5. Assume the data are as in Problem 4. Provide the answers to all parts of Problem 4 with respect to the interest rate, where Vega and Vega2 are replaced with ρ and the second derivative (calculus) of the asset with respect to the interest rate, respectively.

Problem 6. The fact that the Γ of a call option is positive means that the option value is a convex function of the stock price. For a convex function $f(\cdot)$ and a fixed point x_0 the following is true: $f(x) \geq f(x_0) + \frac{\partial f}{\partial x}(x_0)(x - x_0)$, for every x. Does it mean, therefore, that the approximation to the changes in the price of a call option as a function of the stock price, approximated by a first-order Taylor approximation, underestimates the changes in the price of the call for every realization of the stock price?

Problem 7. The sensitivity measures are simply partial derivatives with respect to certain parameters. Hence, a sensitivity measure of a portfolio is a linear combination of the sensitivity measures of the assets composing the portfolio. The coefficients of the combination are the same as those of the assets comprising the portfolio. Assume the data as in the table below.

	Call 1	*Put 2*	*Stock*
Interest Rate	0.12	0.12	0.12
σ	0.18	0.18	0.18
Maturity	0.16	0.16	
Price	21.52	4.65	100.00
Exercise Price	80.00	105.00	

1. Given the above put and call, written on the same stock and having the same maturity date, is it possible to combine these assets into a portfolio such that the Δ of the portfolio is zero? If yes, does the portfolio include a short position?

2. Is it possible to combine the assets in part 1 to generate a portfolio such that the Γ of the portfolio is zero?

3. Would it be possible to combine the assets in part 1 such that both Δ and Γ of the portfolio are zero?

4. Repeat the analysis of parts 1, 2, and 3 for a general case where the parameters are given symbolically. You may use the MATLAB GUI or MAPLE to solve it symbolically.

5. Repeat parts 1 to 4 when Γ is replaced with Θ.

6. Assume now that the portfolio also includes the stock. Given the parameters above, solve (if it is feasible) for a portfolio such that its Δ, Γ, and Θ are zero. What do you conclude?

7. Repeat the above question in a symbolic manner (you may use the MATLAB GUI or MAPLE).

Problem 8. The sensitivity measures Δ, Γ, and Θ of a call option are related to each other via the price of a call option, equation (6.16). The price of a call option can be written as $\Delta P r_f - \Theta + \frac{1}{2}\Gamma\sigma^2 P^2$, where Δ, Γ, and Θ are as defined above and r_f is the risk-free rate. Prove the above relation, the significance of which will be reveled in Chapter 15.

7.8 Appendix

7.8.1 Derivation of Sensitivity Measures

In this section we derive the formulas for call option sensitivity measures. To begin our derivation we recall that MAPLE uses the **erf** function to express the standard normal distribution. For our derivation below we will need to make use of the relation between the **erf** function and the normal distribution. Thus we redefine the function **newerf** which expresses the **erf** function in terms of the standard normal distribution.

```
> newerf:= t -> 2*N(t*sqrt(2))-1;
```

$$newerf := t \to 2\,\mathrm{N}(t\,\sqrt{2}) - 1$$

The sensitivity measures are derived through differentiation and simplification of the Black–Scholes formula. We use the function **Eqbs**, defined in Chapter 6, and show again that it is an equivalent form of the Black–Scholes formula.

```
> Eqbs(K,t,P,sigma,r);
```

$$\frac{1}{2}P - \frac{1}{2}K\,e^{(-r\,t)} + \frac{1}{2}\operatorname{erf}\left(\frac{1}{4}\,\frac{\sqrt{2}\,(-2\ln(K) + 2\ln(P) + 2\,r\,t + \sigma^2\,t)}{\sigma\,\sqrt{t}}\right)P$$
$$- \frac{1}{2}\operatorname{erf}\left(\frac{1}{4}\,\frac{\sqrt{2}\,(-2\ln(K) + 2\ln(P) + 2\,r\,t - \sigma^2\,t)}{\sigma\,\sqrt{t}}\right)K\,e^{(-r\,t)}$$

```
> subs(erf=newerf,Eqbs(K,t,P,sigma,r));
```

$$\frac{1}{2}P - \frac{1}{2}K\,e^{(-r\,t)} + \frac{1}{2}\operatorname{newerf}\left(\frac{1}{4}\,\frac{\sqrt{2}\,(-2\ln(K) + 2\ln(P) + 2\,r\,t + \sigma^2\,t)}{\sigma\,\sqrt{t}}\right)P$$
$$- \frac{1}{2}\operatorname{newerf}\left(\frac{1}{4}\,\frac{\sqrt{2}\,(-2\ln(K) + 2\ln(P) + 2\,r\,t - \sigma^2\,t)}{\sigma\,\sqrt{t}}\right)K\,e^{(-r\,t)}$$

```
> expand(%);
```

$$P\,\mathrm{N}\left(\frac{1}{2}\,\frac{-2\ln(K) + 2\ln(P) + 2\,r\,t + \sigma^2\,t}{\sigma\,\sqrt{t}}\right) - \frac{K\,\mathrm{N}\left(\frac{1}{2}\,\frac{-2\ln(K) + 2\ln(P) + 2\,r\,t - \sigma^2\,t}{\sigma\,\sqrt{t}}\right)}{e^{(r\,t)}}$$

Derivation of Delta

We begin the derivation by making the following assumptions:

```
> assume(P>0,K>0,r>0,t>0,sigma>0);
```

We then differentiate the **Eqbs** function with respect to the security price, P, as required based on the definition of Delta:

```
> Delta:=diff(Eqbs(K,t,P,sigma,r),P);
```

$$\Delta := \frac{1}{2} + \frac{1}{2} \frac{e^{\left(-1/8 \frac{(-2\ln(K)+2\ln(P)+2rt+\sigma^2 t)^2}{\sigma^2 t}\right)} \sqrt{2}}{\sqrt{\pi}\,\sigma\,\sqrt{t}} + \frac{1}{2}\operatorname{erf}\left(\frac{1}{4}\frac{\sqrt{2}\,(-2\ln(K)+2\ln(P)+2rt+\sigma^2 t)}{\sigma\sqrt{t}}\right) - \frac{1}{2}\frac{e^{\left(-1/8 \frac{(-2\ln(K)+2\ln(P)+2rt-\sigma^2 t)^2}{\sigma^2 t}\right)} \sqrt{2}\,K\,e^{(-r\,t)}}{\sqrt{\pi}\,P\,\sigma\,\sqrt{t}}$$

We next replace **erf** with **newerf** in the above expression.

```
> Delta:= subs(erf=newerf,Delta);
```

$$\Delta := \frac{1}{2} + \frac{1}{2} \frac{e^{\left(-1/8 \frac{(-2\ln(K)+2\ln(P)+2rt+\sigma^2 t)^2}{\sigma^2 t}\right)} \sqrt{2}}{\sqrt{\pi}\,\sigma\,\sqrt{t}} + \frac{1}{2}\operatorname{newerf}\left(\frac{1}{4}\frac{\sqrt{2}\,(-2\ln(K)+2\ln(P)+2rt+\sigma^2 t)}{\sigma\sqrt{t}}\right) - \frac{1}{2}\frac{e^{\left(-1/8 \frac{(-2\ln(K)+2\ln(P)+2rt-\sigma^2 t)^2}{\sigma^2 t}\right)} \sqrt{2}\,K\,e^{(-r\,t)}}{\sqrt{\pi}\,P\,\sigma\,\sqrt{t}}$$

We finally expand the above equation to obtain the required result.

```
> Delta:=expand(Delta);
```

$$\Delta := \mathrm{N}\left(\frac{1}{2}\frac{-2\ln(K)+2\ln(P)+2rt+\sigma^2 t}{\sigma\sqrt{t}}\right)$$

Derivation of Gamma

In order to derive Gamma we differentiate Delta with respect to P. This is because Gamma is the second derivative of the value of the call with respect to P, while Delta is the first derivative of the value of the call with respect to P.

```
> Gamma:=diff(Delta,P);
```

$$\Gamma := \frac{\mathrm{D}(N)(\frac{1}{2}\frac{-2\ln(K)+2\ln(P)+2\,r\,t+\sigma^2\,t}{\sigma\sqrt{t}})}{P\,\sigma\sqrt{t}}$$

where $D(\mathrm{N}) = \mathrm{N}'$ is the derivative of N.

Derivation of Rho

Recall that rho is the first derivative of the option value with respect to r. It follows:

```
> Rho:=diff(Eqbs(K,t,P,sigma,r),r);
```

$$R := \frac{1}{2}K\,t\,e^{(-r\,t)} + \frac{1}{2}\frac{e^{(-1/8\,\frac{(-2\ln(K)+2\ln(P)+2\,r\,t+\sigma^2\,t)^2}{\sigma^2\,t})}\sqrt{2}\sqrt{t}\,P}{\sqrt{\pi}\,\sigma} - \frac{1}{2}\frac{e^{(-1/8\,\frac{(-2\ln(K)+2\ln(P)+2\,r\,t-\sigma^2\,t)^2}{\sigma^2\,t})}\sqrt{2}\sqrt{t}\,K\,e^{(-r\,t)}}{\sqrt{\pi}\,\sigma} + \frac{1}{2}\,\mathrm{erf}(\frac{1}{4}\frac{\sqrt{2}\,(-2\ln(K)+2\ln(P)+2\,r\,t-\sigma^2\,t)}{\sigma\sqrt{t}})\,K\,t\,e^{(-r\,t)}$$

We replace **erf** with **newerf:**

```
> Rho:= subs(erf=newerf,Rho);
```

$$R := \frac{1}{2}K\,t\,e^{(-r\,t)} + \frac{1}{2}\frac{e^{(-1/8\,\frac{(-2\ln(K)+2\ln(P)+2\,r\,t+\sigma^2\,t)^2}{\sigma^2\,t})}\sqrt{2}\sqrt{t}\,P}{\sqrt{\pi}\,\sigma} - \frac{1}{2}\frac{e^{(-1/8\,\frac{(-2\ln(K)+2\ln(P)+2\,r\,t-\sigma^2\,t)^2}{\sigma^2\,t})}\sqrt{2}\sqrt{t}\,K\,e^{(-r\,t)}}{\sqrt{\pi}\,\sigma} + \frac{1}{2}\,\mathrm{newerf}(\frac{1}{4}\frac{\sqrt{2}\,(-2\ln(K)+2\ln(P)+2\,r\,t-\sigma^2\,t)}{\sigma\sqrt{t}})\,K\,t\,e^{(-r\,t)}$$

We expand the above expression and obtain rho

```
> expand(%);
```

$$\frac{K\,t\,\mathrm{N}(\frac{1}{2}\frac{-2\ln(K)+2\ln(P)+2\,r\,t-\sigma^2\,t}{\sigma\sqrt{t}})}{e^{(r\,t)}}$$

Derivation of Theta

Theta is the first derivative of the option value with respect to t. It follows:

```
> Theta:=diff(Eqbs(K,t,P,sigma,r),t);
```

$$Q := \frac{1}{2}\,K\,r\,e^{(-r\,t)} + e^{(-1/8\,\frac{(-2\ln(K)+2\ln(P)+2\,r\,t+\sigma^2\,t)^2}{\sigma^2\,t})}$$
$$(\frac{1}{4}\,\frac{\sqrt{2}\,(2\,r+\sigma^2)}{\sigma\,\sqrt{t}} - \frac{1}{8}\,\frac{\sqrt{2}\,(-2\ln(K)+2\ln(P)+2\,r\,t+\sigma^2\,t)}{\sigma\,t^{(3/2)}})$$
$$P\,/\sqrt{\pi} - \frac{e^{(-1/8\,\frac{\%1^2}{\sigma^2\,t})}\,(\frac{1}{4}\,\frac{\sqrt{2}\,(2\,r-\sigma^2)}{\sigma\,\sqrt{t}} - \frac{1}{8}\,\frac{\sqrt{2}\,\%1}{\sigma\,t^{(3/2)}})\,K\,e^{(-r\,t)}}{\sqrt{\pi}}$$
$$+\frac{1}{2}\,\mathrm{erf}(\frac{1}{4}\,\frac{\sqrt{2}\,\%1}{\sigma\,\sqrt{t}})\,K\,r\,e^{(-r\,t)}$$
$$\%1 := -2\ln(K)+2\ln(P)+2\,r\,t-\sigma^2\,t$$

We expand the expression and replace **erf** with **newerf**:

```
> Theta:=expand(Theta);
```

$$\Theta := \frac{1}{2}\,\frac{K\,r}{e^{(r\,t)}} + \frac{1}{4}\,\frac{K^{(\frac{\ln(P)}{\sigma^2\,t})}\,K^{(\frac{r}{\sigma^2})}\,\sqrt{K}\,\sqrt{P}\,\sqrt{2}\,\sigma}{\sqrt{\pi}\,\sqrt{e^{(\frac{\ln(K)^2}{\sigma^2\,t})}}\,\sqrt{e^{(\frac{\ln(P)^2}{\sigma^2\,t})}}\,P^{(\frac{r}{\sigma^2})}\,\sqrt{e^{(\frac{t\,r^2}{\sigma^2})}}\,\sqrt{e^{(r\,t)}}\,(e^{(\sigma^2\,t)})^{1/8}\,\sqrt{t}}$$
$$+\frac{1}{2}\,\frac{\mathrm{erf}(-\frac{1}{2}\,\frac{\sqrt{2}\ln(K)}{\sigma\,\sqrt{t}} + \frac{1}{2}\,\frac{\sqrt{2}\ln(P)}{\sigma\,\sqrt{t}} + \frac{1}{2}\,\frac{\sqrt{2}\,\sqrt{t}\,r}{\sigma} - \frac{1}{4}\,\sqrt{2}\,\sigma\,\sqrt{t})\,K\,r}{e^{(r\,t)}}$$

```
> Theta:= subs(erf=newerf,Theta);
```

$$\Theta := \frac{1}{2}\,\frac{K\,r}{e^{(r\,t)}} + \frac{1}{4}\,\frac{K^{(\frac{\ln(P)}{\sigma^2\,t})}\,K^{(\frac{r}{\sigma^2})}\,\sqrt{K}\,\sqrt{P}\,\sqrt{2}\,\sigma}{\sqrt{\pi}\,\sqrt{e^{(\frac{\ln(K)^2}{\sigma^2\,t})}}\,\sqrt{e^{(\frac{\ln(P)^2}{\sigma^2\,t})}}\,P^{(\frac{r}{\sigma^2})}\,\sqrt{e^{(\frac{t\,r^2}{\sigma^2})}}\,\sqrt{e^{(r\,t)}}\,(e^{(\sigma^2\,t)})^{1/8}\,\sqrt{t}}$$
$$+\frac{1}{2}\,\frac{\mathrm{newerf}(-\frac{1}{2}\,\frac{\sqrt{2}\ln(K)}{\sigma\,\sqrt{t}} + \frac{1}{2}\,\frac{\sqrt{2}\ln(P)}{\sigma\,\sqrt{t}} + \frac{1}{2}\,\frac{\sqrt{2}\,\sqrt{t}\,r}{\sigma} - \frac{1}{4}\,\sqrt{2}\,\sigma\,\sqrt{t})\,K\,r}{e^{(r\,t)}}$$

```
> expand(%);
```

$$\frac{1}{4}\,\frac{K^{(\frac{\ln(P)}{\sigma^2\,t})}\,K^{(\frac{r}{\sigma^2})}\,\sqrt{K}\,\sqrt{P}\,\sqrt{2}\,\sigma}{\sqrt{\pi}\,\sqrt{e^{(\frac{\ln(K)^2}{\sigma^2\,t})}}\,\sqrt{e^{(\frac{\ln(P)^2}{\sigma^2\,t})}}\,P^{(\frac{r}{\sigma^2})}\,\sqrt{e^{(\frac{t\,r^2}{\sigma^2})}}\,\sqrt{e^{(r\,t)}}\,(e^{(\sigma^2\,t)})^{1/8}\,\sqrt{t}}$$
$$+\frac{K\,r\,\mathrm{N}((-\frac{1}{2}\,\frac{\sqrt{2}\ln(K)}{\sigma\,\sqrt{t}} + \frac{1}{2}\,\frac{\sqrt{2}\ln(P)}{\sigma\,\sqrt{t}} + \frac{1}{2}\,\frac{\sqrt{2}\,\sqrt{t}\,r}{\sigma} - \frac{1}{4}\,\sqrt{2}\,\sigma\,\sqrt{t})\,\sqrt{2})}{e^{(r\,t)}}$$

The above equation is equivalent to the formula for Theta,

$$\frac{P\sigma}{2\sqrt{t}}\mathrm{N}'\left(\frac{\ln\left(\frac{P}{K}\right)+\left(r+\frac{\sigma^2}{2}\right)t}{\sigma\sqrt{t}}\right)+ \qquad (7.19)$$
$$rKe^{-rt}\mathrm{N}\left(\frac{\ln\left(\frac{P}{K}\right)+\left(r-\frac{\sigma^2}{2}\right)t}{\sigma\sqrt{t}}\right).$$

Derivation of Sensitivity to Exercise Price

Using the above methodologies we can also derive a formula for the sensitivity measure with respect to the exercise price:

```
> diff(Eqbs(K,t,P,sigma,r),K);
```

$$-\frac{1}{2}e^{(-r\,t)}-\frac{1}{2}\frac{e^{\left(-1/8\,\frac{(-2\ln(K)+2\ln(P)+2\,r\,t+\sigma^2\,t)^2}{\sigma^2\,t}\right)}\sqrt{2}\,P}{\sqrt{\pi}\,K\,\sigma\,\sqrt{t}}$$
$$+\frac{1}{2}\frac{e^{\left(-1/8\,\frac{(-2\ln(K)+2\ln(P)+2\,r\,t-\sigma^2\,t)^2}{\sigma^2\,t}\right)}\sqrt{2}\,e^{(-r\,t)}}{\sqrt{\pi}\,\sigma\,\sqrt{t}}$$
$$-\frac{1}{2}\operatorname{erf}\left(\frac{1}{4}\frac{\sqrt{2}\,(-2\ln(K)+2\ln(P)+2\,r\,t-\sigma^2\,t)}{\sigma\,\sqrt{t}}\right)e^{(-r\,t)}$$

We replace **erf** with **newerf** and then expand:

```
> subs(erf=newerf,%);
```

$$-\frac{1}{2}e^{(-r\,t)}-\frac{1}{2}\frac{e^{\left(-1/8\,\frac{(-2\ln(K)+2\ln(P)+2\,r\,t+\sigma^2\,t)^2}{\sigma^2\,t}\right)}\sqrt{2}\,P}{\sqrt{\pi}\,K\,\sigma\,\sqrt{t}}$$
$$+\frac{1}{2}\frac{e^{\left(-1/8\,\frac{(-2\ln(K)+2\ln(P)+2\,r\,t-\sigma^2\,t)^2}{\sigma^2\,t}\right)}\sqrt{2}\,e^{(-r\,t)}}{\sqrt{\pi}\,\sigma\,\sqrt{t}}$$
$$-\frac{1}{2}\operatorname{newerf}\left(\frac{1}{4}\frac{\sqrt{2}\,(-2\ln(K)+2\ln(P)+2\,r\,t-\sigma^2\,t)}{\sigma\,\sqrt{t}}\right)e^{(-r\,t)}$$

```
> expand(%);
```

$$-\frac{\mathrm{N}\left(\frac{1}{2}\,\frac{-2\ln(K)+2\ln(P)+2\,r\,t-\sigma^2\,t}{\sigma\,\sqrt{t}}\right)}{e^{(r\,t)}}$$

7.8.2 Sensitivities of Other Options

This Chapter discussed the sensitivity measures of call options on non-dividend-paying stocks, which will now be denoted by Θ_c, Δ_c, Γ_c, $Vega_c$, and ρ_c. In this section we extend the analysis to call options on stocks which pay dividends, and to puts on both non-dividend-paying and dividend-paying stocks. The sensitivities of other derivative securities are developed in exactly analogous ways: they are the partial derivatives of the value of the derivative in question with respect to the parameter of interest.

As was the case for call options on non-dividend-paying stocks, the sensitivity measures represent the first (calculus) derivative of the option pricing function (or the Δ function by Γ) with respect to one of its parameters. The following sensitivities differ from the call option sensitivities in that the option pricing functions for put options and for options written on dividend-paying assets are different than those sensitivity measures with which we are already familiar.

Call Options on a Dividend-Paying Stock

We evaluate the sensitivities of call options on a dividend-paying stock, Θ_{Dc}, Δ_{Dc}, Γ_{Dc}, $Vega_{Dc}$, and ρ_{Dc}, in the same way as those of call options written on a stock that pays no dividends. There is however one distinction: in the first case we discount the initial price, P, by the expected dividends over the life of the option, resulting in a current price parameter of $P\,e^{-Dit}$ instead of P. The MAPLE procedure **BstCall** can therefore be used to price call options on a dividend-paying stock as follows:

$$\mathrm{BstCall}(K,\, t,\, P\,e^{-Dit},\, \sigma,\, r). \tag{7.20}$$

In other words, we make use of the same MAPLE procedure, but input a different value for the price of the underlying asset. We can use the unapply command and define the following functions in MAPLE.

```
> ThetaDc:=unapply((evalf(-diff(Bs(K,t,P*exp(-D*t),\
> sigma,r),t))),K,t,P,D,sigma,r):
> DeltaDc:=unapply((evalf(diff(Bs(K,t,P*exp(-D*t),\
> sigma,r),P))),K,t,P,D,sigma,r):
> GammaDc:=unapply((evalf(diff(Bs(K,t,P*exp(-D*t),\
> sigma,r),P,P))),K,t,P,D,sigma,r):
> VegaDc:=unapply((evalf(diff(Bs(K,t,P*exp(-D*t),\
> sigma,r),sigma))),K,t,P,D,sigma,r):
> rhoDc:=unapply((evalf(diff(Bs(K,t,P*exp(-D*t),\
```

```
> sigma,r),r))),K,t,P,D,sigma,r):
```

These are just the sensitivity measures for a call option on a dividend-paying stock. Using the above sensitivities we can determine Θ_{Dc}, Δ_{Dc}, Γ_{Dc}, $Vega_{Dc}$, and ρ_{Dc} for a call option on a dividend-paying stock.

For example, consider a call option with the following characteristics:

$$K = 100,\ t = 1,\ P = 110,\ D = 0.05,\ \sigma = 0.3,\ r = 0.08.$$

Having defined the functions above, we can use them to evaluate each of the sensitivities numerically. We calculate the sensitivities of the above dividend call option:

```
> Theta[Dc]:=(ThetaDc(100,1,110,.05,.3,.08));
```

$$\Theta_{Dc} := -6.061368031$$

```
> Delta[Dc]:=(DeltaDc(100,1,110,.05,.3,.08));
```

$$\Delta_{Dc} := .680015707$$

```
> Gamma[Dc]:=(GammaDc(100,1,110,.05,.3,.08));
```

$$\Gamma_{Dc} := .009788098414$$

```
> Lambda[Dc]:=(VegaDc(100,1,110,.05,.3,.08));
```

$$\Lambda_{Dc} := 35.53079720$$

```
> rho[Dc]:=(rhoDc(100,1,110,.05,.3,.08));
```

$$\rho_{Dc} := 55.8979354$$

The equations for the sensitivities of a call option on a dividend-paying stock are as follows:

$$\Theta_{Dc} = -\frac{P\,\sigma\,e^{-Dit}}{2\sqrt{t}}\mathrm{N}'(d_1) + Di\,P\,\mathrm{N}(d_1)\,e^{-Dit} - r\,K\,e^{-r\,t}\,\mathrm{N}(d_2) \tag{7.21}$$

$$\Delta_{Dc} = e^{-Dit}\,\mathrm{N}(d_1) \tag{7.22}$$

$$\Gamma_{Dc} = \frac{e^{-Dit}}{P\,\sigma\,\sqrt{t}}\mathrm{N}'(d_1) \tag{7.23}$$

$$Vega_{Dc} = P\,\sqrt{t}\,e^{-Dit}\mathrm{N}'(d_1) \tag{7.24}$$

$$\rho_{Dc} = t\,K\,e^{-r\,t}\,\mathrm{N}(d_2). \tag{7.25}$$

For all of the above, d_1 and d_2 are as they were for the case of dividend options, equations (6.18) and (6.19).

Put Options on a Stock That Pays No Dividend

We evaluate the sensitivities of a put option on a stock that pays no dividend, Θ_p, Δ_p, Γ_p, $Vega_p$, and ρ_p, utilizing the procedure **BstPut**.

$$\text{BstPut}(K,\ t,\ P,\ \sigma,\ r) \tag{7.26}$$

Alternatively, we can use put–call parity to state

$$Put\,Option\,Price = \text{Bs}(K,\ t,\ P,\ \sigma,\ r) + K\,e^{-r\,t} - P \tag{7.27}$$

and the **unapply** procedure to define the following functions:

```
> Thetap:=unapply((evalf(diff(BsPut(K,t,P,sigma,\
> r),t))),K,t,P,sigma,r):
> Deltap:=unapply((evalf(diff(BsPut(K,t,P,sigma,\
> r),P))),K,t,P,sigma,r):
> Gammap:=unapply((evalf(diff(BsPut(K,t,P,sigma,\
> r),P$2))),K,t,P,sigma,r):
> Vegap:=unapply((evalf(diff(BsPut(K,t,P,sigma,\
> r),sigma))),K,t,P,sigma,r):
> rhop:=unapply((evalf(diff(Bs(K,t,P,sigma,r),r\
> ))),K,t,P,sigma,r):
```

Using the above sensitivities we can determine Θ_p, Δ_p, Γ_p, $Vega_p$, and ρ_p for a put option on a stock that pays no dividend. For example, consider a put option with the following characteristics:

$$K = 70,\ t = \frac{1}{12},\ P = 60,\ \sigma = 0.15,\ r_f = 0.10.$$

We calculate the sensitivities of the above put option:

```
> Theta[p]:=evalf(Thetap(70,1/12,60,.15,.10));
```

$$\Theta_p := -6.916418659$$

```
> Delta[p]:=evalf(Deltap(70,1/12,60,.15,.10));
```

$$\Delta_p := -.9995898563$$

```
> Gamma[p]:=evalf(Gammap(70,1/12,60,.15,.10));
```

$$\Gamma_p := .0005693083044$$

```
> Lambda[p]:=evalf(Vegap(70,1/12,60,.15,.10));
```

$$\Lambda_p := .025618875$$

```
> rho[p]:=evalf(rhop(70,1/12,60,.15,.10));
```

$$\rho_p := .0020278358$$

The equations for the sensitivities of a put option on a stock that pays no dividend are as follows:

$$\Theta_p = -\frac{P\,\sigma}{2\,\sqrt{t}}\mathrm{N}'(d_1) + r\,K\,e^{-r\,t}\,\mathrm{N}(-d_2) \tag{7.28}$$

$$\Delta_p = \mathrm{N}(d_1) - 1 \tag{7.29}$$

$$\Gamma_p = \Gamma_c = \frac{1}{P\,\sigma\,\sqrt{t}}\mathrm{N}'(d_1) \tag{7.30}$$

$$Vega_p = Vega_c = P\,\sqrt{t}\mathrm{N}'(d_1) \tag{7.31}$$

$$\rho_p = -t\,K\,e^{-r\,t}\,\mathrm{N}(-d_2) \tag{7.32}$$

Put Options on a Dividend-Paying Stock

We evaluate the sensitivities of a put options on a dividend-paying stock, Θ_{Dp}, Δ_{Dp}, Γ_{Dp}, $Vega_{Dp}$, and ρ_{Dp}, using the methodology we used before. Since the put option is written on a dividend-paying stock we discount the initial stock price, P, by the dividend yield. Hence, the current price parameter is $P\,e^{-Di\,t}$ instead of P as it is in the case for a put option on a stock that pays no dividend. Pricing such a put option with **BstPut** requires executing it as follows:

$$\mathrm{BstPut}(K,\,t,\,P\,e^{-Di\,t},\,\sigma,\,r). \tag{7.33}$$

Alternatively, the value of such a put option can be calculated based on the put–call parity as

$$\mathrm{Bs}(K,\,t,\,P\,e^{-Di\,t},\,\sigma,\,r) + K\,e^{-r\,t} - P\,e^{-Di\,t}. \tag{7.34}$$

We use the unapply procedure to define the functions for each of the sensitivity measures below:

```
> ThetaDp:=unapply(-diff(BsPut(K,t,P*exp(-D*t),\
> sigma,r),t),K,t,P,D,sigma,r):
> DeltaDp:=unapply(diff(BsPut(K,t,P*exp(-D*t),\
> sigma,r),P),K,t,P,D,sigma,r):
> GammaDp:=unapply(diff(BsPut(K,t,P*exp(-D*t),\
> sigma,r),P$2),K,t,P,D,sigma,r):
> VegaDp:=unapply(diff(BsPut(K,t,P*exp(-D*t),\
> sigma,r),sigma),K,t,P,D,sigma,r):
> rhoDp:=unapply(diff(BsPut(K,t,P*exp(-D*t),\
```

```
> sigma,r),r),K,t,P,D,sigma,r):
```

Using the above sensitivities we can determine Θ_{Dp}, Δ_{Dp}, Γ_{Dp}, $Vega_{Dp}$, and ρ_{Dp} for a put option on a dividend-paying stock. For example, consider a put option with the following characteristics:

$$K = 30,\ t = \frac{3}{12},\ P = 33,\ Di = 0.02,\ \sigma = 0.3,\ r_f = 0.08.$$

The sensitivities of the above put option are thus given by

```
> Theta[Dp]:=evalf(ThetaDp(30,3/12,33,.02,.3,.0 8));
```

$$\Theta_{Dp} := -2.368298396$$

```
> Delta[Dp]:=evalf(DeltaDp(30,3/12,33,.02,.3,.0 8));
```

$$\Delta_{Dp} := -.2078131460$$

```
> Gamma[Dp]:=evalf(GammaDp(30,3/12,33,.02,.3,.0 8));
```

$$\Gamma_{Dp} := .05774600221$$

```
> Lambda[Dp]:=evalf(VegaDp(30,3/12,33,.02,.3,.0 8));
```

$$\Lambda_{Dp} := 4.716404715$$

```
> rho[Dp]:=evalf(rhoDp(30,3/12,33,.02,.3,.08));
```

$$\rho_{Dp} := -1.870940999$$

The formulas for the sensitivities of a put option on the stock that pays known dividend are given below:

$$\Theta_{Dp} = -\frac{P\sigma e^{-Dit}}{2\sqrt{t}} \mathrm{N}'(d_1) - DiP\mathrm{N}(d_1)\, e^{-\,Dit} + r\,K\,e^{-r\,t}\,\mathrm{N}(-d_2) \tag{7.35}$$

$$\Delta_{Dp} = e^{-Di\,t}\,(\mathrm{N}(d_1) - 1) \tag{7.36}$$

$$\Gamma_{Dp} = \Gamma_{Dc} = \frac{e^{-Di\,t}}{P\sigma\sqrt{t}} \mathrm{N}'(d_1) \tag{7.37}$$

$$Vega_{Dp} = Vega_{Dp} = P\sqrt{t}\,e^{-Dit}\mathrm{N}'(d_1) \tag{7.38}$$

$$\rho_{Dp} = -tK\ e^{-r\,t}\,\mathrm{N}(-d_2). \tag{7.39}$$

For all of the above, d_1 and d_2 are defined as they were for an option on a dividend-paying stock, i.e., as in equations (6.18) and (6.19).

7.8.3 Signs of the Sensitivities

To further develop an understanding of option sensitivities we consider the sign of each of the sensitivities. An intuitive explanation for the signs of some of the sensitivities in the case of a call option has been provided in the text. In this appendix we provide a mathematical argument for the signs of each of the sensitivities. To facilitate this explanation we note the signs of the following parameters:

- $P \geq 0$, obviously, since security prices cannot be negative due to the limited liability.
- $K \geq 0$, since the value of the underlying asset cannot be negative, the exercise price also must be nonnegative.
- $r > 0$, interest rates are assumed to be positive
- $D \geq 0$, all dividend yields are assumed to be nonnegative.
- $t \geq 0$, as a future time.
- $\sigma \geq 0$, by definition of variance and standard deviation.
- $e^x \geq 0$, by definition of the exponent function.
- $\mathrm{N}(x) \geq 0$, as $\mathrm{N}(\cdot)$ is the normal cumulative distribution function.
- $\mathrm{N}'(x) \geq 0$, since $\mathrm{N}(\cdot)$ is an increasing function.

The signs of the sensitivities are easily determined given the signs of the parameters above:

$$\Theta_c = -\left(\frac{P\,\sigma}{2\,\sqrt{t}}\mathrm{N}'(d_1) + r\,K\,e^{-r\,t}\,\mathrm{N}(d_2)\right) < 0 \tag{7.40}$$

$$\Delta_c = N(d_1) > 0 \tag{7.41}$$

$$\Gamma_c = \frac{1}{P\,\sigma\,\sqrt{t}}\mathrm{N}'(d_1) > 0 \tag{7.42}$$

$$Vega_c = P\,\sqrt{t}\mathrm{N}'(d_1) > 0 \tag{7.43}$$

$$\rho_c = t\,K\;e^{-r\,t}\,\mathrm{N}(d_2) > 0. \tag{7.44}$$

The sign of Θ_{Dc} is ambiguous.

$$\Theta_{Dc} = -\frac{P\,\sigma\,e^{-Dit}}{2\,\sqrt{t}}\,\mathrm{N}'(d_1) + Di\,P\,\mathrm{N}(d_1)\,e^{-Dit} - r\,K\,e^{-r\,t}\,\mathrm{N}(d_2) \tag{7.45}$$

$$\Delta_{Dc} = e^{-Di\,t} \mathrm{N}\,(d_1) > 0 \tag{7.46}$$

$$\Gamma_{Dc} = \frac{e^{-Di\,t}}{P\,\sigma\,\sqrt{t}} \mathrm{N}^{'}\,(d_1) > 0 \tag{7.47}$$

$$Vega_{Dc} = P\,\sqrt{t}\,e^{-Di\,t} \mathrm{N}^{'}\,(d_1) > 0 \tag{7.48}$$

$$\rho_{Dc} = t\,K\,e^{-r\,t}\,\mathrm{N}(d_2) > 0 \tag{7.49}$$

The sign of Θ_p, as that of Θ_{Dc} is also ambiguous.

$$\Theta_p = -\frac{P\,\sigma}{2\,\sqrt{t}} \mathrm{N}^{'}\,(d_1) + r\,K\,e^{-r\,t}\,\mathrm{N}(-d_2) \tag{7.50}$$

$$\Delta_p = \mathrm{N}(d_1) - 1 < 0 \tag{7.51}$$

$$\Gamma_p = \frac{1}{P\,\sigma\,\sqrt{t}} \mathrm{N}^{'}\,(d_1) > 0 \tag{7.52}$$

$$Vega_p = P\,\sqrt{t} \mathrm{N}^{'}\,(d_1) > 0 \tag{7.53}$$

$$\rho_p = -t\,K\,e^{-r\,t}\,\mathrm{N}(-d_2) < 0. \tag{7.54}$$

We observe again that the sign of Θ_{Dp} is ambiguous.

$$\Theta_{Dp} = -\frac{P\,\sigma\,e^{-Di\,t}}{2\,\sqrt{t}} \mathrm{N}^{'}\,(d_1) - Di\,P\,\mathrm{N}(d_1)\,e^{-Di\,t} + r\,K\,e^{-r\,t}\,\mathrm{N}(-d_2) \tag{7.55}$$

$$\Delta_{Dp} = e^{-Di\,t}\,(\mathrm{N}(d_1) - 1) < 0 \tag{7.56}$$

$$\Gamma_{Dp} = \frac{e^{-Di\,t}}{P\,\sigma\,\sqrt{t}} \mathrm{N}^{'}\,(d_1) > 0 \tag{7.57}$$

$$Vega_{Dp} = P\,\sqrt{t}\,e^{-Di\,t} \mathrm{N}^{'}\,(d_1) > 0 \tag{7.58}$$

$$\rho_{Dp} = -t\,K\,e^{-r\,t}\,\mathrm{N}(-d_2) < 0. \tag{7.59}$$

We see that most of the signs of the sensitivities are not ambiguous. Let us investigate the following three situations where ambiguity exists:

$$\Theta_{Dc} = -\frac{P\,\sigma\,e^{-Di\,t}}{2\,\sqrt{t}} \mathrm{N}^{'}\,(d_1) + Di\,P\,\mathrm{N}(d_1)\,e^{-Di\,t} - r\,K\,e^{-r\,t}\,\mathrm{N}(d_2)$$

$$\Theta_p = -\frac{P\,\sigma}{2\,\sqrt{t}} \mathrm{N}^{'}\,(d_1) + r\,K\,e^{-r\,t}\,\mathrm{N}(-d_2)$$

$$\Theta_{Dp} = -\frac{P\,\sigma\,e^{-Di\,t}}{2\,\sqrt{t}} \mathrm{N}^{'}\,(d_1) - Di\,P\,\mathrm{N}(d_1)\,e^{-Di\,t} + r\,K\,e^{-r\,t}\,\mathrm{N}(-d_2).$$

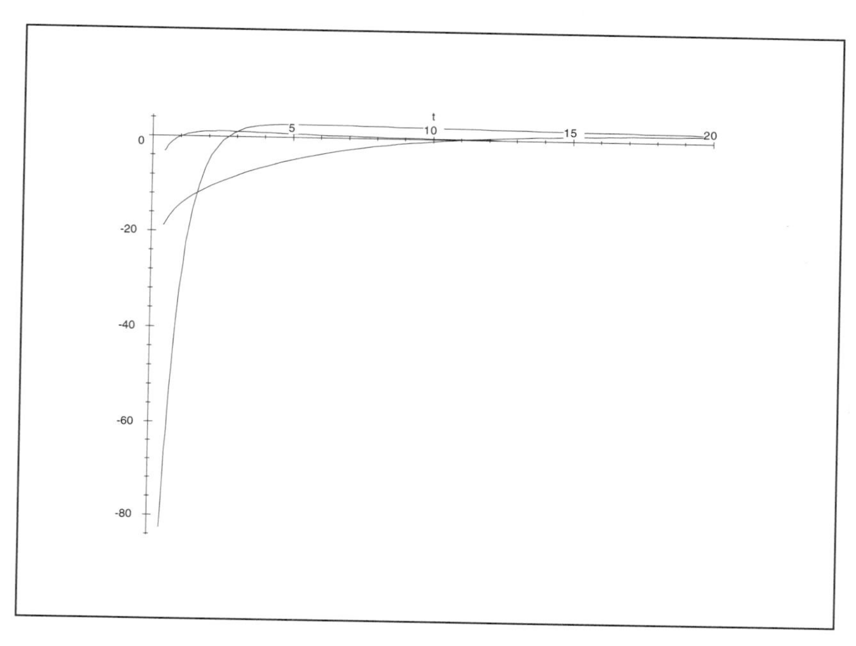

Figure 7.24: Different Thetas for an At-the-Money Call Option

Note that all three of the above sensitivities are Θ sensitivities. In addition, all of the above combine positive and negative expressions. We demonstrate graphically that no absolute sign exists in the case of each of the above three expressions. We first graph in Figure 7.24 Θ_{Dc} for an at-the-money call with a very high interest rate (red), a "typical" interest rate (blue), and a very high dividend (green).

```
> plot([ThetaDc(110,t,110,.03,.3,1.5),
> ThetaDc(110,t,110,.03,.3,.2),\
> ThetaDc(110,t,110,.30,.3,.2)],t=0..20,\
> labels=[t, ThetaDc],color=[red,blue,green],\
> thickness=3,title=`Figure 7.24:  Different\
> Theta's for an At-the-Money Call Option`);
```

We see that Θ_{Dc} may be positive at combinations of very high values of r, D, or t. In order to further demonstrate this point, the reader may wish to observe the effect of high values of r , D, and t on the equation for Θ_{Dc}.

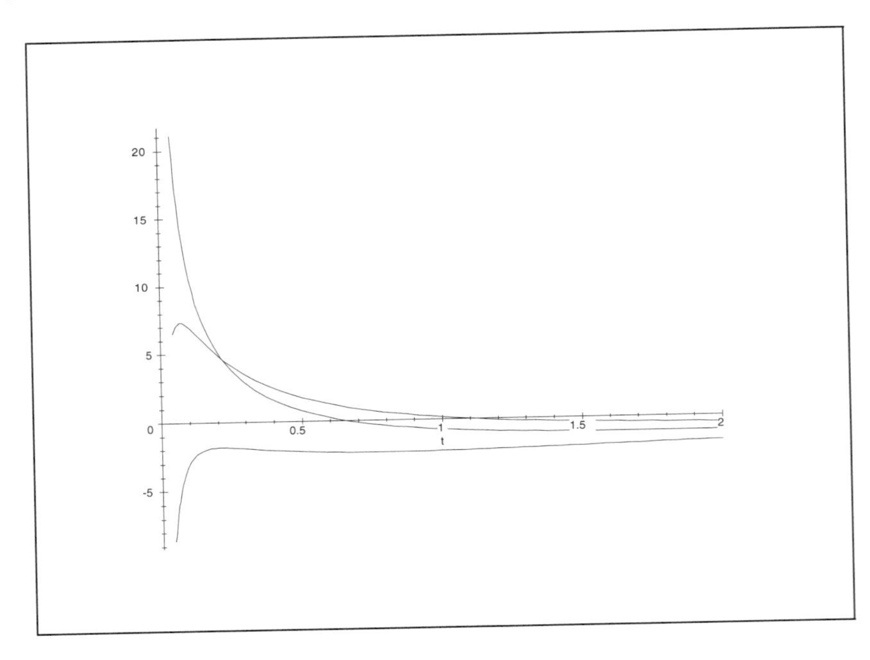

Figure 7.25: Theta for an In-, Out-, and At-the-Money Put Option

We next graph in Figure 7.25 Θ_p for in-the-money (red), at-the-money (blue), and out-of-the-money (green) option.

```
> plot([Thetap(110,t,100,.3,.2),\
> Thetap(110,t,110,.3,.2),\
> Thetap(100,t,110,.3,.2)],t=0..2,\
> color=[red,blue,green],thickness=3,\
> labels=[t, Thetap],title=`Figure 7.25:  Theta \
> for an In, Out and At-the-Money Put Option`);
```

We see that no absolute sign exists for Θ_p. We finally graph in Figure 7.26 Θ_{Dp} for in-the-money (the red graph that coincide with the x-axis and is hardly noticeable), at-the-money (the blue graph with the hump closer to the y-axis), and out-of-the-money (the third green graph with the hump further away from the y-axis) put option.

```
> plot([ThetaDp(110,t,110,.03,.3,1.5),\
> ThetaDp(110,t,110,.03,.3,.2),\
> ThetaDp(110,t,110,.30,.3,.2)],t=0..20,\
```

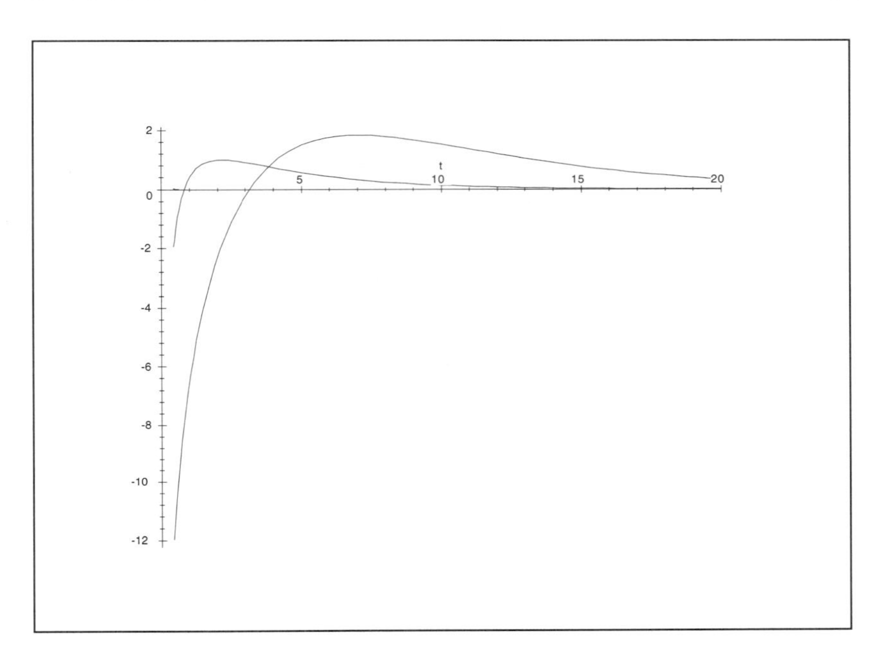

Figure 7.26: Theta for an In-, At-, and Out-of-the-Money Put Option on a Stock That Pays Dividends

```
>   color=[red,blue,green],thickness=3,\
>   labels=[t, ThetaDp],title='Figure7.26:\
>   Theta for an In-, At-, and Out-of-the-Money Put\
>   Option on a Stock That Pays Dividends');
```

Once again, we see that we cannot make a statement regarding the absolute sign of Θ_{Dp}.

Chapter 8

Hedging with the Greeks

8.1 Hedging: The General Philosophy

Having developed the sensitivity measures, we can now use them for hedging purposes. As we have mentioned before, these sensitivity measures are accurate only for infinitesimal changes that occur over an "instant" of time. They are induced by a first-order Taylor approximation (see footnote 6 in the former Chapter) and can be utilized to compose portfolios that over the "next instant" will be hedged against particular changes.

Employing these hedging strategies over a time interval longer than an instant should be executed with some care. For a larger change, or for hedging over a long time period, a "better" approximation can be obtained by utilizing a higher order Taylor approximation. This, in turn, will require the use of higher order (calculus) derivatives (with respect to the parameters in question) as exemplified in equation (7.12) and displayed in Figures 7.12 and 7.13. Alternatively, a hedge over a "long period" can be carried by reshuffling the portfolio every so often. This will maintain its insensitivity to changes in certain parameters, as the sensitivity measures are effective only over an instant. The main theme of this Chapter is the development of some hedging strategies that are based on the measures developed in the former Chapter.

The preceding Chapter has overlooked the fact that we are actually dealing with some parameters that are stochastic in nature, such as the price of a stock. Hence, the approximation will not be exact even over the next instant. An appropriate modification taking account of the randomness does make the approximation exact over the next instant. Chapter 15 takes another look at the derivation of the Black–Scholes formula utilizing exact hedging over an instant. Appendix 15.11.1, "A Change over an Instant," also offers a review of the precise meaning of the "next instant", limits, and their relation to the Taylor approximation. Readers that are not familiar with this terminology may find it advantageous to read this appendix before proceeding with the current Chapter.

When approaching hedging with respect to changes in certain parameters, one should recall the portfolios of Chapter 1 which were composed to have the same payoffs in every state of nature. There, the uncertainty was reflected by the different states of nature that might be realized in the next time period. Thus in trying to determine the risk-free rate implicit in the market prices we composed a (hedged) portfolio with a payoff of $1 in every state. The price of such a portfolio would have to be the discounted value of the payoff. Such a portfolio was composed in Chapter 1 in the discussion preceding equation (1.8). Portfolios such that their payoffs in every state of nature were the same, e.g., zero, were instrumental in defining the no-arbitrage condition. In this former case, it was accomplished by holding two different portfolios, one short and the other long. The future obligations due to the short part were equal to the payoffs due to the long part. The combined portfolio (the long part and the short part) had a payoff of zero in every state of nature: it was a riskless portfolio. Viewed alternatively, it was just a portfolio the value of which was not sensitive to the state of nature. Such a portfolio was a hedge against the realization of the state of nature.

Here we are interested in portfolios the values of which are insensitive to changes in certain parameters. The risk here is implicit in the changes of the parameters in question. Each parameter plays a role in a way similar to a state of nature in the model of Chapter 1 — it induces a certain constraint that the portfolio should satisfy. Each such constraint ensures that the sensitivity of the hedged portfolio with respect to that parameter is zero. This constraint acts exactly as in the model of Chapter 1: the payoff in a state of nature was constrained to be zero or set up to be a constant. Thus, as in Chapter 1, the hedging requires solving a system of equations. The variables of this (linear) system are the weights, which determine the portfolio composition.

The change over the next instant has a precise meaning in terms of a limit, as explored in Appendix 15.11.1. The "instant" here is the equivalent of the length of the time period, from time zero to time one, in the model of Chapter 1. The uncertainty here is reflected in the possible values of the parameters. The value of the portfolio in different states of nature, in Chapter 1, corresponds to the value of the portfolio here induced by different values of the parameters. The system of equations (1.1) and (1.2) solved in Chapter 1 are equivalent here to a system of equations which ensure that the sensitivity of the portfolio to a change in the parameters in question is zero.

The number of equations will depend on the number of parameters against which we would like to hedge. This is much in the same manner as in the model of Chapter 1: the number of equations was then the number of states. If we are interested in hedging against more than one parameter, we will have to allow the use of more than one security. The intuition here is much the same as in the discussion of complete markets from Chapter 1. Since our sensitivity measures are partial derivatives, they obey the rule that the sensitivity measure of a portfolio is the linear combination of the sensitivity measures of the components of the portfolio. Consequently, the induced system of equations is linear (in the portfolio composition). These sensitivity measures, however, by virtue of being partial derivatives, measure sensitivity with respect to one parameter, assuming that all other parameters remain constant.

We now consider a few examples which will help to clarify these ideas. Since the concept is applied in much the same way regardless of the parameter in question, we investigate in detail hedging utilizing Delta. We then move on to discuss general hedging and optimized hedged portfolios. However, before doing so, we point out one more important similarity between the ideas expressed here and those of Chapter 1. If, having made the appropriate modifications for the instantaneous change in the random variable, one manages to produce a portfolio which is risk-free over that instant, the rate of return on the portfolio must be equal to the risk-free rate. Were this not so, arbitrage opportunities would exist in the financial market. We will make use of this observation to generate a differential equation for the price of the contingent claim. The reader may therefore anticipate that there exists a functional relation between the different sensitivity measures. This approach to the Black–Scholes formula is presented in Chapter 15, where we come back to expose the functional relation, equation (15.61), mentioned there. The reader is encouraged to start thinking of an "instant" as a time period from the model of Chapter 1. Recollection of and use of the intuition from Chapter 1 will be of tremendous use in understanding the more advanced material.

8.2 Delta Hedging

Consider first hedging with respect to the changes in the value of the security, e.g., a stock, underlying the derivative security. For example, one could write a call option and simultaneously hold a portfolio whose value offsets any changes in the value of the call option. We shall refer to such a portfolio as the *hedge portfolio*. Alternatively, one could hold a portfolio and then buy a call option to offset changes in the value of the portfolio which result from uncertainty about the price of the underlying asset. The states of nature, or the uncertainty, are the possible values of the stock price. These prices fall in the interval $[0, \infty)$. The value of the call option will increase or decrease as the stock price increases or decreases.

Consequently, in order to hedge, we seek a hedge portfolio constructed in such a way that when the stock price increases, the portfolio value will decrease by the same amount that the call option value increased, and vice versa. If such a hedge portfolio can be found, the combination of that portfolio and the call option position will not be sensitive to changes in the price of the stock. The Delta of the total portfolio will be zero and we shall refer to such a portfolio as a *hedged portfolio*. For this reason, such combinations, or portfolios, are often referred to as Delta neutral portfolios. Recall that the Delta of a portfolio will be the combination of the Deltas of the assets composing it. In what follows, we first find a Delta neutral portfolio and then connect the solution to the definition of Delta itself already presented.

8.2.1 Solving for a Delta Neutral Portfolio

From the point of view of the writer of a call option, an increase (decrease) in the value of the stock increases (decreases) the potential liability of the writer. Placed in such a position, the writer of a call option might like to hedge. Such a writer is therefore looking for a portfolio the value of which will increase when the value of the stock increases and vice versa.

It is natural to include the stock itself in such a portfolio. It should be clear that the source of uncertainty in the value of a call option, or of any contingent claim, is the uncertainty about the price of the underlying asset. Furthermore, from the former Chapter we know that the Delta of a call option is positive. Hence the price of the option is an increasing function of the price of the underlying asset.

This suggests that holding the underlying asset long will offset change in the value of a call option to the writer. The ratio of calls to stocks one needs to hold in a hedged portfolio remains to be determined. This provides the intuition for the inclusion of the underlying[1] asset in the hedge (or hedged) portfolio.

If the option writer is only interested in offsetting the change in the value of the option, holding the stock alone in the correct proportion will be sufficient. As has already been discussed in Chapter 7, the correct proportion is given by the Delta measure. If the option writer would like to ensure that the short position in the call option and the hedge portfolio are equal in value at today's prices, another asset should be included in the total position. In such a case, the option writer would like the total portfolio to be a self-financing portfolio. (The concept of a self-financing portfolio was explored in Chapter 1.)

The missing asset to create a self-financing hedge portfolio might be the risk-free asset. As long as this asset is not sensitive to the price of the underlying asset it will not impair the Delta neutrality of the portfolio. The straightforward way is to include a short position in a bond[2] or cash to finance the purchasing of the underlying asset.

We now attempt to form a self-financing hedged portfolio composed of the stock (underlying asset), a bond, and the call option. We will construct the portfolio in such a way that both its Delta and its current value will be zero. The corresponding system of equations which we would like to satisfy can thus be seen to be induced by these two conditions. Equation (8.1) stipulates that the value of the portfolio should be zero,[3]

$$SXS + XB - P_c = 0, \tag{8.1}$$

[1] We have mentioned that, in dealing with changes which are uncertain, and in attempting to hedge such changes, one should look for another asset whose changes will originate from the same source of uncertainty. More precisely, the two assets should have perfect correlation. Since the changes are random, one must find an asset which changes in the opposite way. This can be achieved by taking a short position. We will revisit this issue from different points of view. We have already discussed this in a discrete-time framework, such as the binomial model. We will also consider the continuous-time framework in Chapter 15.

[2] Theoretically, another asset that is not correlated with the underlying asset might be chosen. In that case, however, the portfolio will be exposed to risks that are avoided when a short position is being held in a bond. One might also argue that in a one-period model, as shown in Chapter 2, a bond and a stock can replicate the option.

[3] In solving for the hedge portfolio, we assume for simplicity that a call option, on one unit of the underlying asset (one share), can be purchased. This is not a standard option contract. A standard call option contract is for the purchase of 100 shares. This is not a conceptual difference since 100 calls on one unit of a stock are the same as one call on 100 units of stock. That is, a call option price (and put, too) is homogeneous of degree one in the underlying asset. The exercises pertaining to this Chapter elaborate on this issue.

where XS represents the number of units of the stock held, P_c is the price of the call, and XB stands for the dollar amount of the holdings in the bond (or cash). Multiplying equation (8.1) by negative one we obtain equation (8.2).

$$P_c - SXS - XB = 0 \tag{8.2}$$

This equation can be interpreted in the following way. The cost of the position in the stock is XS times the stock price, and the current cost of the position in the bond is XB. (This will be recalled from Chapter 1.) In the same way, shorting the call option generates an income of P_c at the time the transaction is undertaken.

Recall that Delta obeys the rules of partial derivatives: the Delta of a portfolio is merely the combination of the Deltas of the assets in the portfolio. Hence, since $\frac{\partial}{\partial S} XB = 0$, and $\frac{\partial}{\partial S} S\ XS = XS$ (because $\frac{\partial}{\partial S} S = 1$ and XS is not a function of S), we have $\frac{\partial}{\partial S}(S\ XS + XB - P_c) = XS - \frac{\partial}{\partial S} P_c$. Since $\frac{\partial}{\partial S} P_c = \Delta(K,\ t,\ S,\ \sigma,\ r)$, we obtain equation (8.3), which ensures that the Delta of the portfolio is zero.

$$XS - \Delta(K,\ t,\ S,\ \sigma,\ r) = 0 \tag{8.3}$$

Consider the following example. Assume a given call option with an exercise price of \$70, a time until expiration of 60 days (i.e., $\frac{60}{365}$ years), and a volatility of 18 percent, written on a stock with a current price of \$80. Assume that the risk-free rate in the market is 12 percent. The first equation which must be satisfied by the self-financing hedged portfolio (for which we are trying to solve) is that the current cost of the portfolio is zero. This is given by equation (8.4), which is simply equation (8.1).

$$\text{Bs}\left(70,\ \frac{60}{365},\ 80,\ 0.18,\ 0.12\right) - XB - XS\,80 = 0 \tag{8.4}$$

The current value of the hedged portfolio as a function of S (the price of the underlying asset) is thus given by equation (8.5).

$$-\text{Bs}\left(70,\ \frac{60}{365},\ S,\ 0.18,\ 0.12\right) + XB + SXS \tag{8.5}$$

Now, to ensure that the Delta of the portfolio is zero, we require that the partial derivative of equation (8.5) with respect to S be zero.

$$\frac{\partial}{\partial S}\left(-\text{Bs}\left(70,\ \frac{60}{365},\ S,\ 0.18,\ 0.12\right) + XB + SXS\right) = 0 \tag{8.6}$$

Equation (8.6) ensures that the self-financing hedged portfolio is insensitive to changes in the price of the stock. This second equation is of course equation (8.3) written in terms of our example.

We solve the system of equations (8.4) and (8.5) in order to find the required hedge portfolio. First, though, we make sure that XB and XS do not have assigned numerical values and so can serve as variables in a system of equations. (If you want to run this part again for different values, you must execute the next line, which unassigns whatever numerical values have been assigned to the variables XB and XS.)

```
> XB:='XB';XS:='XS';
```

$$XB := XB$$

$$XS := XS$$

The equations to be submitted to MAPLE are therefore

```
> solve({Bs(70,60/365,80,0.18,0.12)-XB-XS*80=0,\
> subs(S=80,diff(-Bs(70,60/365,S,0.18,0.12)+XB+XS*S,S))\
> =0},{XB,XS});
```

$$\{XB = -67.29217590,\ XS = .9836803467\}$$

Alternatively, solving the system of equations below should provide the same result as was already obtained. (You might need to redefine Delta, as in the next MAPLE command, for it to work.)

```
> Delta:=unapply((diff(Bs(K,t,S,sigma,rf),S)),\
> K,t,S,sigma,rf):
> solve({Bs(70,60/365,80,0.18,0.12)-XB-XS*80=0,\
> -Delta(70,60/365,80,0.18,0.12)+XS=0},{XB,XS});
```

$$\{XB = -67.29217590,\ XS = .9836803468\}$$

We now assign the numerical solution values to the variables. (We will unassign them at the end of this section. Assigning them allows the user to alter the parameters of the problem and rerun this part without interfering with the MAPLE commands.)

```
> assign(%);
> XB;
```

$$-67.29217590$$

```
> XS;
```

$$.9836803468$$

The current value of the combined portfolio (the hedged portfolio: short call option, cash, and the stock) as a function of S is given by

$$-\mathrm{Bs}\left(70, \frac{60}{365}, S, 0.18, 0.12\right) + XB + SXS.$$

The short call option represents a liability to the writer, the current value of which is

```
> -Bs(70,60/365,80,0.18,0.12);
```

$$-11.40225184$$

The current value of the bond and stock combination is

```
> XB+XS*80;
```

$$11.40225184$$

For a current stock price of $80, the two parts of the hedged portfolio we have constructed indeed offset each other. This is a self-financing portfolio. Let us examine Figure 8.1 in which the values of the two parts the hedged portfolio are graphed separately on the same set of axes.

```
> plot([-Bs(70,60/365,S,0.18,0.12),(XS*S+XB),\
> [[80,-Bs(70,60/365,80,0.18,0.12)],[80,XS*80+XB]]],\
> S=60..90,thickness=2,labels=['StockPrice','Call Value'],\
> colour=[red,green,blue],title='Figure 8.1:  The Values \
> of the Two Parts of the Hedged Portfolio'):
```

The reader may wish to compare this figure to Figure 4.22 where the value of a hedged portfolio at the expiration date is displayed. In that case, the hedge was perfect. Figure 4.22 showed the sum of two graphs, one of which was the negative of the other, such that their combined values were zero regardless of the stock price. Here, we see that this is nearly true again, but only in the vicinity of the current stock price (as marked by the vertical blue line). This phenomenon is due to the nature of the approximation and it being exact only in the limit.

Let us hold the time constant at $t = \frac{60}{365}$ (i.e., 60 days) and examine the graph, Figure 8.2, of the value of the hedged portfolio.

```
> plot(-Bs(70,60/365,S,0.18,0.12)+(XS*S+XB),S=60...120,\
> thickness=2,labels=['Stock Price','Portfolio Value'],\
> title='Figure 8.2:  The Graph of the Hedged Portfolio \
> as a Function of S for a Fixed t'):
```

We see that close to the current stock price of $80 (i.e., within a small interval of the current stock price), the value of the combined portfolio is

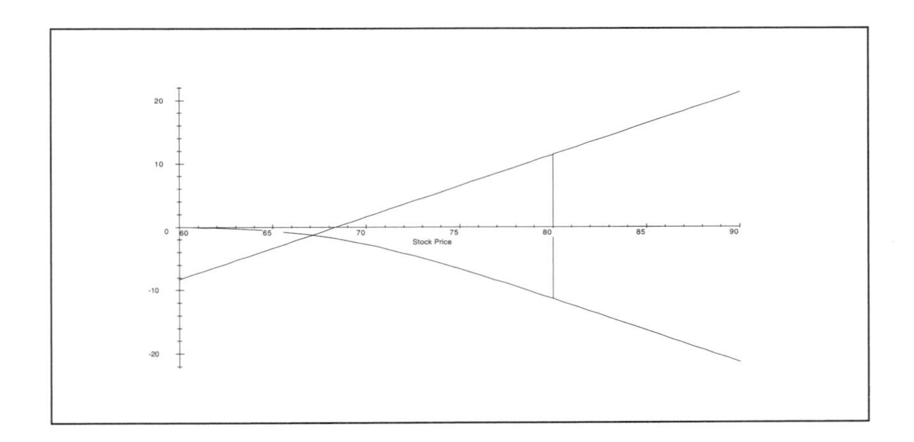

Figure 8.1: The Values of the Two Parts of the Hedged Portfolio

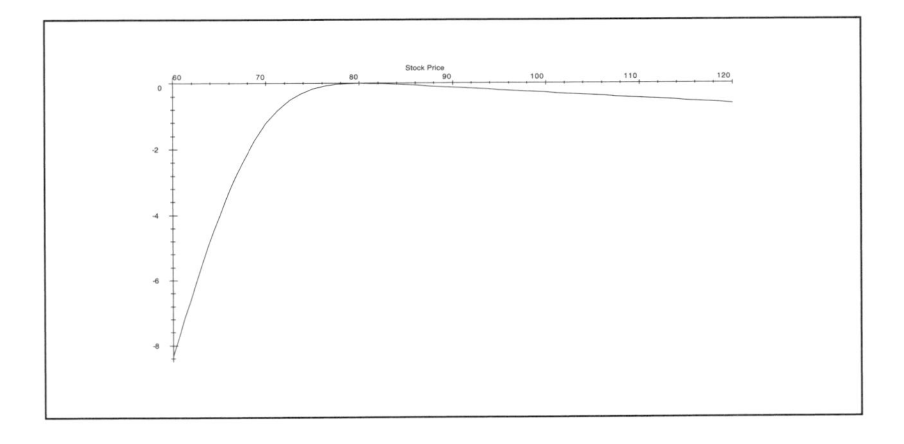

Figure 8.2: The Graph of the Hedged Portfolio as a Function of S for a Fixed t

indeed zero. As the price of the underlying asset moves away from its current value of \$80, the value of the call option does not perfectly offset the value of the stock and the bond. Thus in contrast to the perfect hedging discussed in Section 4.3.1, and displayed in Figure 4.22, here the value of the combined portfolio as a function of S is not the constant function[4] zero.

We preview some aspects of the approximation in the following discussion. If the stock price changes, it must be the case that some time has passed. However, approximations that are built on partial derivatives assume that all other parameters, but the one in question, do not change. Thus, the hedging above assumes that the passage of time had no effect on the hedged portfolio.

As time progresses the value of the call changes due to two factors: the time to expiration is getting shorter and the price of the underlying security changes. In addition, the amount invested in the bond (or money market account) also changes as it accumulates interest. Recall that we assume that the term structure is flat and deterministic, and thus this effect is easily incorporated. Assuming that the 12 percent is an annual continuously compounded risk-free rate, in t days the value of XB will change to $XB\,e^{\frac{0.12\,t}{365}}$. In contrast, the price of the underlying asset moves in a stochastic fashion. Thus, we can only state the price of the portfolio, equation (8.7), in t days, where the time to maturity is now $60 - t$ days, as a function of S at that time.

$$-\mathrm{Bs}\left(70,\, \frac{60-t}{365},\, S, 0.18,\, 0.12\right) + XB\,e^{\frac{0.12\,t}{365}} + S\,XS \tag{8.7}$$

As time progresses the stock price changes and the hedged portfolio and its Delta deviate from their ideal value of zero. We present the spreadsheet[5] below (in the on-line version of the book, and Table 8.1 in the hard copy) to exemplify this phenomenon. Each cell (S, t) has two numbers: the upper number is the value of the portfolio for a stock price S in t days, i.e., equation (8.7). The lower number, equation (8.8) below, is the Delta of the portfolio for a stock price S in t days.

$$-\frac{\partial}{\partial S}\,\mathrm{Bs}\left(70,\, \frac{60-t}{365},\, S,\, 0.18,\, 0.12\right) + XS \tag{8.8}$$

[4] The reader may be curious as to why the graph in Figure 8.2 shows a maximum at the point $S = 80$. We will address the question of the graph's maximum point at the current stock price in the next section.

[5] Alternatively, one can display these changes by issuing a command like

```
> seq(seq(-Bs(70,(60-t)/365,S,0.18,0.12)+XB+XS*(S),S=60..90),t=0..4):
```

$S\backslash t$	0	1	2	3	4
76	$\begin{bmatrix} -.0941 \\ .0595 \end{bmatrix}$	$\begin{bmatrix} -.0890 \\ .0585 \end{bmatrix}$	$\begin{bmatrix} -.0839 \\ .0575 \end{bmatrix}$	$\begin{bmatrix} -.0788 \\ .0564 \end{bmatrix}$	$\begin{bmatrix} -.0737 \\ .0554 \end{bmatrix}$
78	$\begin{bmatrix} -.0180 \\ .0204 \end{bmatrix}$	$\begin{bmatrix} -.0147 \\ .0196 \end{bmatrix}$	$\begin{bmatrix} -.0115 \\ .0188 \end{bmatrix}$	$\begin{bmatrix} -.0082 \\ .0180 \end{bmatrix}$	$\begin{bmatrix} -.0050 \\ .0172 \end{bmatrix}$
79	$\begin{bmatrix} -.0039 \\ .0084 \end{bmatrix}$	$\begin{bmatrix} -.0260 \\ .0077 \end{bmatrix}$	$\begin{bmatrix} -.0482 \\ .0071 \end{bmatrix}$	$\begin{bmatrix} -.0703 \\ .0064 \end{bmatrix}$	$\begin{bmatrix} -.0925 \\ .0057 \end{bmatrix}$
80	$\begin{bmatrix} 0 \\ -.8\ 10^{-9} \end{bmatrix}$	$\begin{bmatrix} -.0221 \\ -.0005 \end{bmatrix}$	$\begin{bmatrix} -.0442 \\ -.0010 \end{bmatrix}$	$\begin{bmatrix} -.0664 \\ -.0015 \end{bmatrix}$	$\begin{bmatrix} -.0885 \\ -.0020 \end{bmatrix}$
81	$\begin{bmatrix} -.0030 \\ -.0057 \end{bmatrix}$	$\begin{bmatrix} -.0252 \\ -.0061 \end{bmatrix}$	$\begin{bmatrix} -.0473 \\ -.0065 \end{bmatrix}$	$\begin{bmatrix} -.0694 \\ -.0069 \end{bmatrix}$	$\begin{bmatrix} -.0916 \\ -.0073 \end{bmatrix}$
82	$\begin{bmatrix} -.0109 \\ -.0096 \end{bmatrix}$	$\begin{bmatrix} -.0330 \\ -.0099 \end{bmatrix}$	$\begin{bmatrix} -.0551 \\ -.0102 \end{bmatrix}$	$\begin{bmatrix} -.0773 \\ -.0105 \end{bmatrix}$	$\begin{bmatrix} -.0994 \\ -.0107 \end{bmatrix}$
83	$\begin{bmatrix} -.0219 \\ -.0121 \end{bmatrix}$	$\begin{bmatrix} -.0440 \\ -.0123 \end{bmatrix}$	$\begin{bmatrix} -.0662 \\ -.0125 \end{bmatrix}$	$\begin{bmatrix} -.0883 \\ -.0127 \end{bmatrix}$	$\begin{bmatrix} -.1104 \\ -.0129 \end{bmatrix}$
84	$\begin{bmatrix} -.0350 \\ -.0138 \end{bmatrix}$	$\begin{bmatrix} -.0571 \\ -.0139 \end{bmatrix}$	$\begin{bmatrix} -.0792 \\ -.0140 \end{bmatrix}$	$\begin{bmatrix} -.1014 \\ -.0142 \end{bmatrix}$	$\begin{bmatrix} -.1235 \\ -.0143 \end{bmatrix}$

Table 8.1: A Spreadsheet of XB and XS as a Function of S and t

Note that XB and XS must be assigned for the spreadsheet to show numerical results. Calculating Delta and the portfolio value for additional values of S and t can be done via the “fill” command. See the help page for instructions, but you must be careful to fill it along the rows.

The spreadsheet, Table 8.1, contains the range of zero to six days and the range of \$76 to \$84 dollars for the stock price. To expose the unseen part of the array the reader should place the cursor in the array and use the page down or left arrow to move within the array. The reader is also invited to calculate values of other combinations by following the instructions in the MAPLE help page.

We can visualize the changes in the hedge portfolio as a function of S and t utilizing a three-dimensional plot. This is demonstrated in Figure 8.3, where we plot the portfolio values for a range of stock prices from \$60 to \$90. Figure 8.2 is a two-dimensional version of Figure 8.3; i.e., it is the slice corresponding to $t = 0$.

```
> plot3d(-Bs(70,(60-t)/365,S,0.18,0.12)+XS*S+\
> XB*exp(.12*(t/365)),S=60..90,t=0..14,thickness=1,\
> labels=['StockPrice','Days After','PortfolioValue'],\
```

```
> orientation=[-116,65],axes=frame,title=‘Figure 8.3:\
> The Value of the Hedge Portfolio t Days from the \
> Current Time‘):
```

In order to display the parts composing the hedge portfolio, i.e., the call option versus the stock and the cash, the reader may execute the command below. (This graph appears only in the on-line version of the book.)

```
> plot3d({-Bs(70,(60-t)/365,S,0.18,0.12),XS*S+\
> XB*exp(.12*(t/365))},S=60..90,t=0..14,\
> labels=[‘Stock Price‘,‘DaysAfter‘,‘Portfolio Value‘],\
> orientation=[65,52],thickness=1,axes=frame,\
> title=‘The Value of the Two Parts of the Hedge \
> Portfolio t Days from the Current Time‘):
```

We choose to display, in Table 8.1, a range of stock prices around the exercise price of the call. It can be seen from the spreadsheet that the portfolio sensitivity to the changes in the stock price is more pronounced at that range. That this is the case can also be confirmed by the three-dimensional plot of Delta, Figure 8.4. Note how for stock prices in the upper part of the range, Delta of the portfolio is close to zero, while at the lower part of the range the Delta of the portfolio is close to one. Can you explain why this is the case?

```
> plot3d(diff(-Bs(70,(60-t)/365,S,0.18,0.12)+XS*S+\
> XB*exp(.12*(t/365)),S),S=40..100,t=0..14,thickness=1,\
> labels=[‘stock price‘,‘days after‘,‘ ‘],axes=frame,\
> orientation=[-47,71], title=‘Figure 8.4:  Delta of the\
> Hedge Portfolio‘);
```

The sensitivity of Delta to changes in the stock price is more pronounced in a range close to the exercise price. This can be verified by either plotting or calculating the value of Gamma at this range. Gamma measures the sensitivity of Delta, and it is indeed nearly zero at this range. This is confirmed by the three-dimensional graph of Gamma in Figure 8.5.

```
> plot3d(diff(-Bs(70,(60-t)/365,S,0.18,0.12)+XS*S+\
> XB*exp(.12*(t/365)),S$2),S=50..90,t=0..14,\
> style=patchcontour,shading=ZHUE,axes=frame,\
> labels=[‘StockPrice‘,‘DaysAfter‘,‘Gamma‘],\
> orientation=[-94,58],title=‘Figure 8.5:  Gamma of the\
> Hedge Portfolio‘):
```

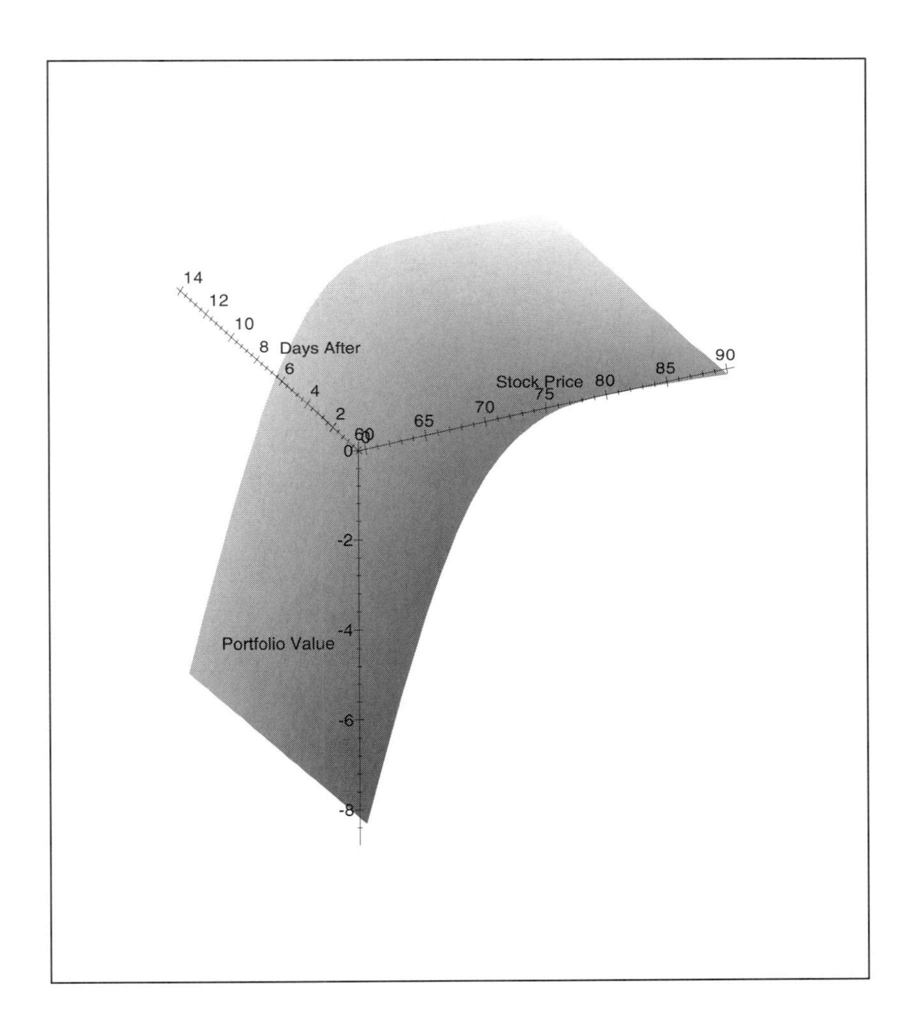

Figure 8.3: The Value of the Hedged Portfolio t Days from the Present Time

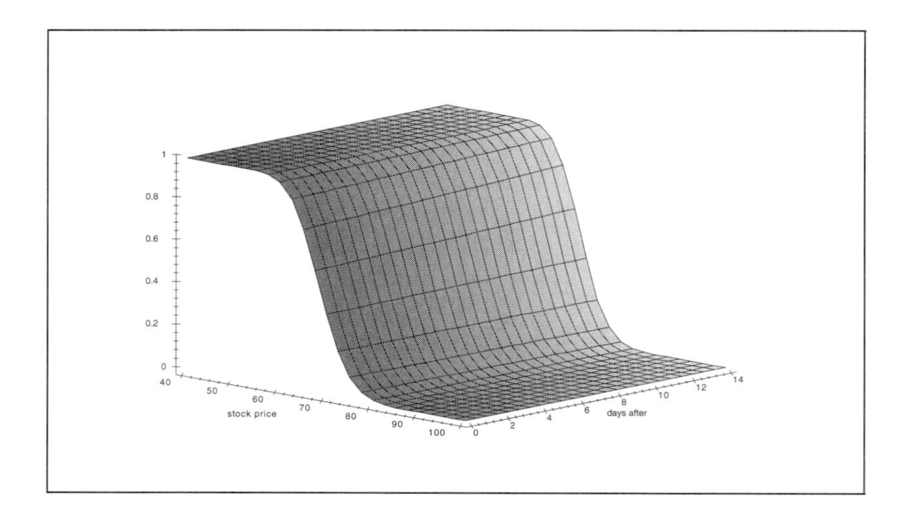

Figure 8.4: Delta of the Hedged Portfolio

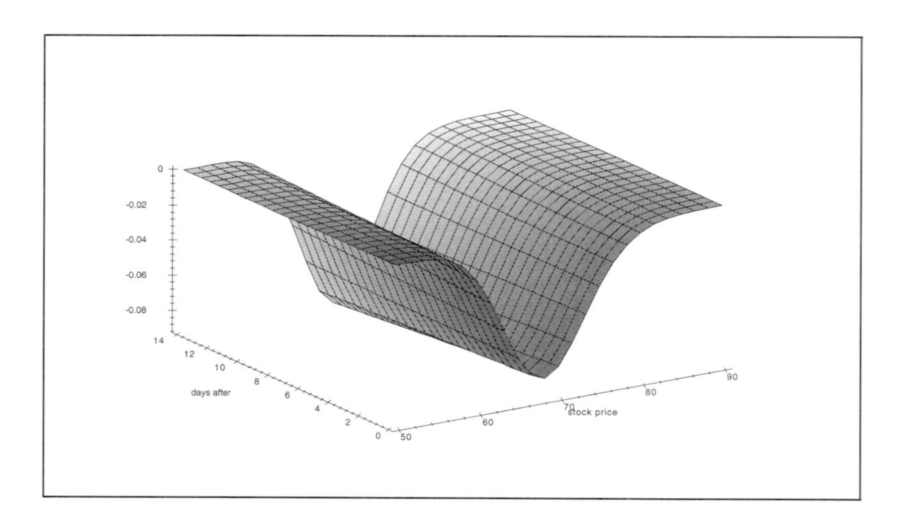

Figure 8.5: Gamma of the Hedged Portfolio

The exercises corresponding to this section will walk the reader through a simulated environment like the one presented in Table 8.2. This environment will convey the different hedging strategies and the frequencies of portfolio reshuffling needed to keep up with the changes of the prices. Obviously, Figures 8.4 and 8.5 convey the notion that in the vicinity of the exercise price, portfolio revisions should be carried out more frequently.

We demonstrate this phenomenon for a call option with an exercise price of \$40 and a time to expiration of $60 - 12$ days, i.e., 12 days after the initial position was established. Consequently, we solve the following system of equations to determine the stock and bond position in the hedge portfolio, XS_{12} and XB_{12}, respectively, as a function of S. Note that S is an unassigned parameter and thus the solution set (**sol12** defined by the command below) is parameterized by S. The reader might like to experiment with the solution, trying it out for different times to expiration by changing the $t = 12$ of the **subs** command to another value.

```
> sol12:=solve(subs(t=12,{Bs(40,(60-t)/365,50,0.18,0.12)\
> -XB12-XS12*S=0,diff(-Bs(40,(60-t)/365,S,0.18,0..12)+XB12\
> +XS12*S,S)=0}),{XB12,XS12}):
```

The solution value for XS_{12} and XB_{12} is presented below. The on-line version of the book also displays "live" graphs of XS_{12} and XB_{12} versus S immediately following the algebraic expression for the solution.

```
> op(1,select(has,sol12,XS12));
```

$$XS12 = .1128379167\,10^{-7}(.4431134627\,10^{8}\; \mathrm{erf}(10.83273978\ln(.02500000000\,S) + .1940277248)\,S + .4431134627\,10^{8}\,S + .541636989\,10^{9}\; e^{(-.7510288064\,10^{-18}\,(.1250000000\,10^{11}\ln(.02500000000\,S)+.223890411\,10^{9})^2)}\,S - .2132626406\,10^{11}\; e^{(-.7510288064\,10^{-18}\,(.1250000000\,10^{11}\ln(.02500000000\,S)+.170630137\,10^{9})^2)})/ S$$

```
> op(1,select(has,sol12,XB12));
```

$$XB12 = 10.62636674 - .5000000000\; \mathrm{erf}(10.83273978\ln(.02500000000\,S) + .1940277248)\,S - .5000000000\,S - 6.111718945\; e^{(-.7510288064\,10^{-18}\,(.1250000000\,10^{11}\ln(.02500000000\,S)+.223890411\,10^{9})^2)}\,S + 240.6411208\; e^{(-.7510288064\,10^{-18}\,(.1250000000\,10^{11}\ln(.02500000000\,S)+.170630137\,10^{9})^2)}$$

S	32	33	34	35	36
$XS12$	.0008	.0037	.0133	..0382	.0901
$XB12$	10.5997	10.5027	10.1718	9.2873	7.3802
S	37	38	39	40	41
$XS12$	.1787	.3045	.4548	..6081	.7430
$XB12$	4.0108	−.9458	−7.1120	−13.6980	−19.8374
S	42	43	44	45	46
$XS12$	.8465	.9165	.0008	..9811	.9921
$XB12$	−24.9297	−28.7859	−31.5515	−33.5268	−35.0122

Table 8.2: The Value of XS_{12} and XB_{12} as a Function of S

The portfolio composition for a range of stock prices is demonstrated in Table 8.2 of the hard copy and in the spreadsheets below in the on-line version of the book.

The numerical results in the spreadsheet and Table 8.2 make it apparent that in the neighborhood of the exercise price, the portfolio composition is most sensitive to the change in the price of the underlying asset. In order to visualize the change in the composition of the hedge portfolio we plot (a parametric plot) the triplets $XB_{12}(S), XS_{12}(S)$, and S for various values of S. This is presented in Figure 8.6.

```
> plot3d([rhs(op(select(has,sol12,XS12))),\
> rhs(op(select(has,sol12,XB12))),S],\
> dummy=-10..10,S=0..120,axes=box,color=blue,\
> thickness=3,labels=[XS12,XB12,S],style=wireframe,\
> orientation=[99,67], title='Figure 8.6:  A Parametric \
> Visualization of the Triplet XS12(S),XB12(S), and S):
```

Note again how for stock prices away from the exercise price of \$40 the graph in Figure 8.6 is close to being vertical. That means that $XS12$ and $XB12$ do not change much in this range of the stock price. In contrast, the parametric plot makes it apparent that in the neighborhood of \$40, $XS12$ ranges from nearly zero to one and the graph is close to being horizontal. The exercise pertaining to this section will elaborate further on this point. The interested reader may change the parameters in the above sequence of commands and investigate the effect on the portfolio composition in different circumstances.

The discussion above implies that Delta hedging is not a static strategy. To keep the portfolio Delta neutral over time, we have to, in theory, continuously change its composition, i.e., resolve for XB and XS. This indeed

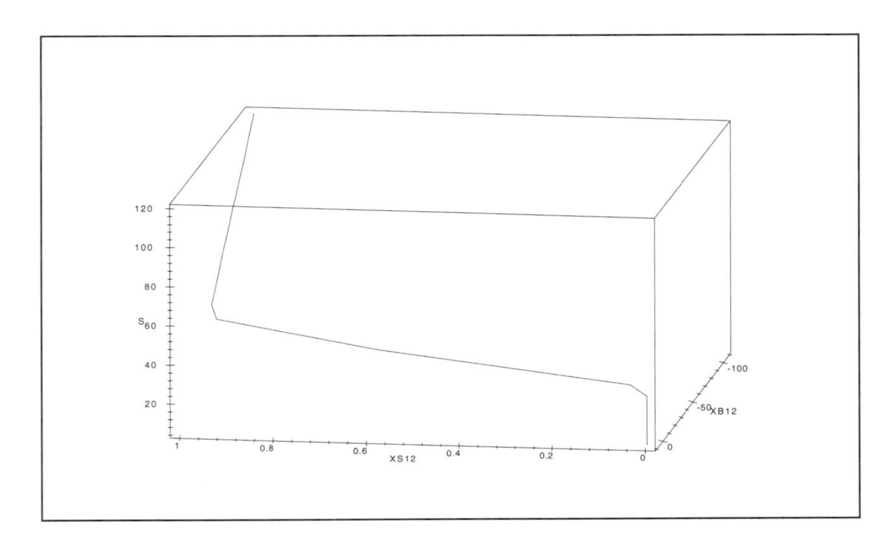

Figure 8.6: A Parametric Visualization of the Triplet $XS_{12}(S)$, $XB_{12}(S)$, and S

is the approach we investigate in Chapter 15 to arrive at the Black–Scholes formula. In practice the portfolio composition is adjusted every so often based on the time to expiration and the stock price at that time. Delta hedging is therefore not a static strategy but rather a dynamic one. The hedge portfolio must be readjusted as time progresses, e.g., every Δt units of time. To demonstrate the practical applications of the strategy we display, in the on-line version of the book, the spreadsheet below. (A static version of it is displayed in Table 8.3 of the hard copy.) This spreadsheet presents a dynamic simulated environment for Delta hedging of a call option. The parameters of the call appear in the first row, and are defined below. The parameters of this simulation appear in the second row, and are defined below.

- S, current price of the stock
- μ, continuously compounded (annual) rate of return of the stock
- σ, the volatility of the stock price
- t, the revision period

- T, the expiration time
- K, exercise price
- r, risk-free rate

The cells that correspond to the following are being reevaluated each time you click them.

- t, the current time measured in days from time zero when the first hedge portfolio was constructed
- T, the expiration time
- Δt, the revision period
- XB_t and XS_t, the composition of the hedge portfolio at time t
- S_t, A simulated price for the stock at t, based on the lognormal distribution as in equation (5.32)
- $C(S_t, t)$, the price of the call at time t where the stock price is S_t. As shorthand notation we also use C_t

The spreadsheet (Table 8.3) presents the hedging of a call option that matures in 180 days where the hedge portfolio is readjusted every month. Our time unit is a year. We display T and t as days but in the calculation use it as $\frac{T}{365}$ and $\frac{t}{365}$. The interest rate, r, and the standard deviation, σ, are annual figures. The user simply needs to initiate the value of the parameters by clicking the cells in the rectangle A2 to G4. Clicking on cell B5 and proceeding down the column will generate the results for the first period. Continuing in the same manner from cell C5 down the column will produce the results for the second time period. The last cell in the column is the cost of adjusting the portfolio; hence a negative number indicates cash inflow.

To find out the net result of hedging this call, one needs to calculate the future value, as of time T, of the cash flows listed in the "*Cost at t*" row. The row "*Cost up to t*" reports the cumulative cost up to time t. Consequently, the cell in that row that corresponds to the last time period reports the total hedging cost in terms of dollars of time T.

The user may consult the explanation in the MAPLE help page in order to alter this spreadsheet, prolong the time period or change it. The formulas are already entered in the cells and one needs to use the fill command to

S_0	μ	σ	Δt	T	K	r
45	.15	.23	30	180	40	.07
Period	0	1	2	3	4	5
t	0	30	60	90	120	150
Time to Expiration	180	150	120	90	60	30
S_t	45	46.32	49.52	53.99	49.34	48.27
$C(S_t, T-t)$	6.98	7.81	10.52	14.69	9.81	8.50
XB	−31.13	−33.73	−37.45	−39.19	−39.15	−39.70
$XS = \Delta$	.847	.897	.968	.998	.992	..998
$XB_t + XS_t\ S_t$	6.98	6.82	10.52	14.68	9.81	8.50
$XB_t + XS_t\ S_t - C_t$	10^{-8}	0	10^{-8}	0	0	0
Cost at t	10^{-8}	−.11	.03	.0513	−.0035	−.02
Cost up to t	10^{-8}	−.11	−.08	−.06	−.11	−.13

Table 8.3: A Simulation of Delta Hedging

extend them to other cells. A similar spreadsheet can be constructed, by the user, to hedge other derivative securities and/or with other sensitivity measures.

8.3 Delta Neutral Portfolios and the Definition of Delta

The definition of Delta and its graphical exposition (Figure 7.8) has a strong tie to Delta neutral portfolios that were investigated in the last section. This is being explored in this section together with a property of the hedge portfolio. Let us look again at Figure 7.8 (it is reproduced here only in the on-line version of the book), where the value of the call, as given in equation (8.9),

$$\mathrm{Bs}\left(70,\ \frac{60}{365},\ S,\ 0.18,\ 0.12\right), \tag{8.9}$$

and its linear approximation, equation (8.10),

$$\mathrm{Bs}\left(70,\ \frac{60}{365},\ 80,\ 0.18,\ 0.12\right) + (S-80)\,\Delta\left(70,\ \frac{60}{365},\ 80,\ 0.18,\ 0.12\right) \tag{8.10}$$

are displayed as functions of S.

```
> plot([Bs(70,60/365,S,.18,.12),(Bs(70,60/365,80,.18,.12)\
```

```
> +(S-80)*subs(S=80,diff(Bs(70,60/365,S,0.18,0.12),S))),\
> [[80,0],[80,Bs(70,60/365,80,0.18,0.12)]]],S=50...110,\
> color=[green,blue,red],thickness=2,\
> labels=['StockPrice','Call Value']):
```

The current value of the call is $\mathrm{Bs}\left(70, \frac{60}{365}, 80, 0.18, 0.12\right)$ and the approximation of the changes in its value due to changes in the stock price is $(S-80)\,\Delta(70, \frac{60}{365}, 80, .18, .12)$. This is why in Figure 7.9 when we plot

$$\mathrm{Bs}\left(70, \frac{60}{365}, 80, 0.18, 0.12\right) - (S-80)\,\Delta\left(70, \frac{60}{365}, 50, 0.18, 0.12\right)$$

we obtained a U-type graph being flat around $S = 80$. (If you are reading this on-line you may want to reexecute the plot command below.)

```
> plot(Bs(70,60/365,S,0.18,0.12)-(S-80)*subs(S=80,\
> diff(Bs(70,60/365,S,0.18,0.12),S)),S=60..100,\
> thickness=2,labels=['StockPrice','Call Value']):
```

Assume that we have written the call option, and so have a short position in the call option denoted by $-\mathrm{Bs}\left(70, \frac{60}{365}, 80, 0.18, 0.12\right)$. Let us rewrite equation (8.10) as equation (8.11).

$$\begin{aligned} &\mathrm{Bs}\left(70, \frac{60}{365}, 80, 0.18, 0.12\right) - 80\,\Delta\left(70, \frac{60}{365}, 80, 0.18, 0.12\right) \\ &+S\,\Delta\left(70, \frac{60}{365}, 80, 0.18, 0.12\right). \end{aligned} \tag{8.11}$$

The expression in equation (8.11) can be interpreted as investing

$$\mathrm{Bs}\left(70, \frac{60}{365}, 80, 0.18, 0.12\right) - 80\,\Delta\left(70, \frac{60}{365}, 80, 0.18, 0.12\right) \tag{8.12}$$

in a bond or a money market account, and holding $\Delta\left(70, \frac{60}{365}, 80, 0.18, 0.12\right)$ units of the stock. We shall soon see that the two parts of this portfolio correspond to XB and XS of the Delta neutral portfolio (solved for in equations (8.4) and (8.5) of the last section). As a result of writing the call we collected its price, which is

```
> Bs(70,60/365,80,.18,.12);
                    11.40225184
```

Given the interpretation of the linear approximation, our position in the bond is the value of equation (8.12). This is being calculated in the MAPLE command below. (Note that Delta is being recalculated here rather than assuming that it has been defined in this worksheet.)

```
> evalf(Bs(70,60/365,80,.18,.12)-80*\
> subs(S=80,diff(Bs(70,60/365,S,.18,.12),S)));
```

$$-67.29217598$$

Therefore, we have taken a short position in the bond (or borrowed the money). The other component of the linear approximation is interpreted as holding $\Delta(70, .5, 80, 0.18, 0.12)$ units of the stock. The cost of this part of the portfolio is given by

```
> evalf(subs(S=80,diff(Bs(70,60/365,S,.18,.12),S)))*80;
```

$$78.69442780$$

Altogether, we collected approximately \$11.40 for writing the call option, we borrowed \$67.29, and we invested the total proceeds in 0.983 units of the underlying asset. The current stock value is \$80, so the portfolio actually costs nothing. The proceeds received were enough to finance the stock position. This can be confirmed by the following:

```
> simplify(subs(S=80,-Bs(70,60/365,S,0.18,0.12)+\
> Bs(70,60/365,80,0.18,0.12)+(S-80)*\
> Delta(70,60/365,80,0.18,0.12)));
```

$$0$$

The reader may compare this portfolio to the one which resulted from solving the system of equations (8.4) and (8.5), solved below again.

```
> XB:='XB';XS:='XS';
```

$$XB := XB$$
$$XS := XS$$

```
> solve({Bs(70,60/365,80,.18,.12)-XB-XS*80=0,\
> subs(S=80,diff(-Bs(70,60/365,S,.18,.12)+XB+XS*S,S))=0},\
> {XB,XS});
```

$$\{XS = .9836803467,\ XB = -67.29217590\}$$

Indeed the same portfolio is obtained, demonstrating that the Delta approximation and the solution for a Delta neutral portfolio are actually two sides of the same coin. Specifically, the value of XS is

```
> evalf(subs(S=80,(diff(Bs(70,60/365,S,0.18,0.12),S))));
```

$$.9836803475$$

and the value of XB is

```
> evalf(Bs(70,60/365,80,0.18,0.12)-80*\
> subs(S=80,(diff(Bs(70,60/365,S,0.18,0.12),S))));
```

$$-67.29217598$$

which confirms our claim.

We continue now to investigate a property of the hedge portfolio. Take a second look at Figure 7.8. The approximation to the change in the value of the call option, (the line which is tangent to the value of the call option at $S = \$80$) is always below the curve of the value of the call option. The approximation always understates the change in the value of the call option. This is because the value of the call option, as a function of the price of the underlying asset, is a convex[6] function.

If we combine the linear approximation to the price of the call together with the short call position we have, as we just saw, a self-financing portfolio. Consider now how this portfolio reacts to changes in the price of the stock. This is demonstrated in Figure 8.2 (repeated below only in the on-line version of the book).

At the current time, the price of the stock is \$80 and the value of the portfolio is zero. The expression plotted in Figure 8.2 has a maximum at the current stock price of \$80. It is actually the same phenomenon of the linear approximation being less than the value of the call for every S. It can be easily verified that indeed the Delta of this self-financing portfolio is zero at $S = \$80$. This is done by asking MAPLE to calculate the partial derivative of the value of the portfolio with respect to S, and then substituting $S = 80$ into the expression.

```
> subs(S=80,diff(-Bs(70,60/365,S,0.18,0.12)+\
> (Bs(70,60/365,80,0.18,0.12)+(S-80)*\
> subs(S=80,(diff(Bs(70,60/365,S,0.18,0.12),S)))),S));
```

$$0$$

Viewing Figure 8.2 and the graphs referred to in this section, it is tempting to believe that Delta hedging is "guaranteed" to underestimate the value of the call option. These graphs convey the notion that at nearly every stock price the value of the hedged portfolio is negative. Furthermore, the graphs demonstrate that the value of the hedged portfolio is close to its theoretical

[6]Note that a convex function is of a U shape, or, more precisely, the line connecting any two points on the graph either coincides with the graph or is higher than the graph. The second derivative of a twice differentiable function is positive if, and only if, the function is convex. Indeed for this reason the Gamma of a call, which is nothing but the second derivative of the call with respect to the stock price, is positive.

value of zero only in a neighborhood of the current stock price. Figure 7.8 makes it clear to the eye that this approximation is only good for a range of stock values close to the current stock price. However, will the hedge portfolio indeed be nonpositive regardless of the value of the stock?

The analysis that led to Figure 8.2 used linear approximation and assumed that a change in the stock price occurred, but that there was no movement in time. Clearly, it is not possible for the stock price to change without time moving forward. Figure 8.2, showing the value of the hedge portfolio, ignored the impact of the passage of time. It also did not consider[7] the time value of money, or the money earned on that invested in the risk-free asset.

In what follows we demonstrate that when this effect is being taken into account, the hedge position is no longer a "sure loser". Indeed, if we would have identified a self-financing portfolio with a payoff as described in Figure 7.8 it would have been an arbitrage portfolio. The exercises pertaining to this section elaborate on this point.

We demonstrate that a Delta hedge portfolio may have a positive value by turning to yet another example. (The reader might also like to rerun a few simulations of Delta hedging utilizing the spreadsheet in Section 8.2.1 to realize that this is not the case.) Consider a call with an exercise price of \$40 where the current stock price is \$50 and the rest of the parameters are as above. The value of a self-financing hedge portfolio t days after its construction, when the stock price is S, is given by equation (8.13).

$$-\mathrm{Bs}\left(40,\,\frac{60-t}{365},\,S,\,0.18,0.12\right)+\mathrm{Bs}\left(40,\,\frac{60}{365},\,50,\,0.18,\,0.12\right)\,e^{\frac{0.12\,t}{365}}$$
$$+\left(S-50\,e^{\frac{0.12\,t}{365}}\right)\,\Delta\left(40,\,\frac{60}{365},\,50,\,0.18,0.12\right) \tag{8.13}$$

In Figure 8.7 we plot the value of the hedged portfolio as a function of S and t, together with the plane $f(S,\,t)=0$. Figure 8.7 clearly demonstrates that when the time value of money is taken into account the hedged portfolio does not always have a negative value. The "slice" of Figure 8.7 that corresponds to $t=0$ is exactly Figure 8.2.

```
> plot3d({-Bs(40,(60-t)/365,S,0.18,0.12)+\
```

[7]This is true not only of time. Given the definition of Delta, as a partial derivative, it implicitly assumed that the other parameters influencing the value of a call option remained at their current value. This is the usual way these sorts of approximations are done. The reader should bear in mind that solving for a hedge portfolio is done separately for each parameter in turn, assuming each time that the others remain constant.

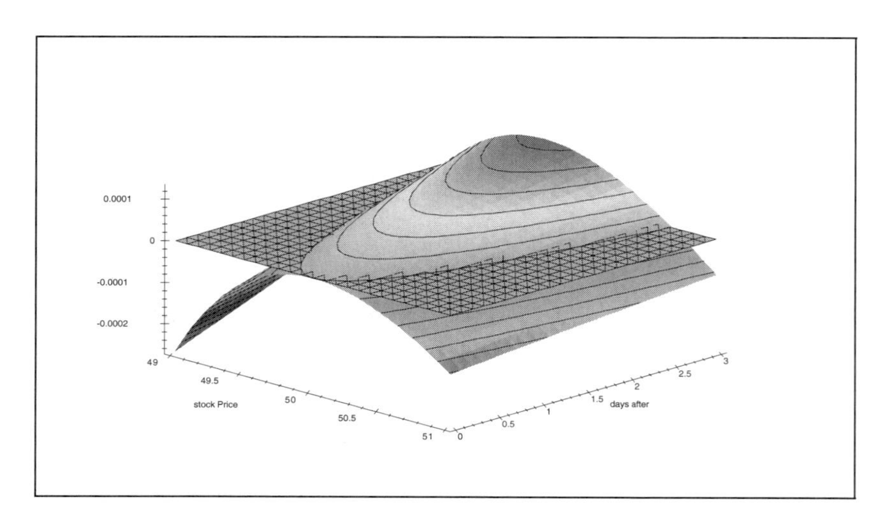

Figure 8.7: Value of the Hedged Portfolio as Stock Price and Time Change

```
> Bs(40,60/365,50,0.18,0.12)*exp(.12*t/365)+\
> (S-50*exp(.12*t/365))*subs(S=50,\
> diff(Bs(40,60/365,S,0.18,0.12),S)),0},S=49..51,\
> t=0..3,thickness=2 ,labels=[‘Stock Price‘,‘Days After‘,\
> ‘Port.Value‘],style=patchcontour,shading=ZHUE,\
> axes=frame,orientation=[-46,57],title=‘Figure 8.7:\
> Value of the Hedged Portfolio as Stock Price\
> and Time Change‘):
```

Putting it differently, when the time value of money is **not** taken into account we have that

$$\begin{aligned} &\text{Bs}\left(40, \frac{60}{365}, 50, 0.18, 0.12\right) + (S-50)\Delta\left(40, \frac{60}{365}, 50, 0.18, 0.12\right) \\ \leq\ &\text{Bs}\left(40, \frac{60}{365}, 50, 0.18, 0.12\right) \end{aligned} \tag{8.14}$$

for every[8] value of S. This represents the fact that the tangent line in Figure 7.8 is always under the graph of the value of the call. When the time value

[8]This inequality holds because the price of the call is a convex function of S. Given a convex function f, $f(x_0) + \frac{\partial f}{\partial x}(x_0)(x - x_0) \leq f(x)$ for every x.

of money is taken into account, the left-hand side of the inequality in (8.14) should be replaced with

$$\mathrm{Bs}\left(40, \frac{60}{365}, 50, 0.18, 0.12\right) e^{\frac{0.12\,t}{365}} + \left(S - 50\, e^{\frac{0.12\,t}{365}}\right) \Delta \left(40, \frac{60}{365}, 50, 0.18, 0.12\right)$$

and there might exist some S for which

$$\begin{aligned} & \mathrm{Bs}\left(40, \frac{60-t}{365}, S, 0.18, 0.12\right) \\ < \; & \mathrm{Bs}\left(40, \frac{60}{365}, 50, 0.18, 0.12\right) e^{\frac{0.12\,t}{365}} + \\ & \left(S - 50\, e^{\frac{0.12\,t}{365}}\right) \Delta \left(40, \frac{60}{365}, 50, 0.18, 0.12\right). \end{aligned}$$

This can also be confirmed by referring to the spreadsheet just before equation (8.8). For example, the cell that corresponds to $S = 79$ and $t = 5$ clearly demonstrates that the value of the call may exceed the value of the stock and the bond.

8.4 General Hedging

Hedging in general with respect to the sensitivity measures is done in much the same way as we have already discussed in the context of the Delta measure. Perhaps the best way to formalize the framework is to relate it to what was presented in Chapter 1. We have already mentioned the similarities between forming an arbitrage portfolio and forming a hedged portfolio. We have also mentioned that the states of nature of Chapter 1 are the counterparts of the sensitivity measures (parameters) against which we would like to hedge. The universe of securities from Chapter 1 is the counterpart of the stock (the underlying security), the bond, and all the contingent claims which are written on that stock.

The array which is composed of the security payoffs for each state of nature of Chapter 1 (see Table 1.1) also has a counterpart. We can generate an array where each row of the array corresponds to a security and each column corresponds to a parameter against which we would like to hedge. The entry in row i and column j of the array is the sensitivity measure of security i with respect to parameter j. If, for example, column 1 corresponds to the change in the price of the derivative security with respect to a change in the price of the underlying security (the stock) then that column will contain the Delta of each security[9] in the universe. These sensitivity measures are partial derivatives and hence obey the linear structure. A portfolio is a linear combination of assets, and a sensitivity measure of a portfolio is the same linear combination of the sensitivity measures of the assets in the portfolio.

We have explored the intuition behind Delta hedging from a few viewpoints. The generalization to hedging with respect to any sensitivity measure or combination thereof is, we believe, best presented as the solution to a system of linear equations. The reasoning for this remains as it was for the explanation of the case for Delta hedging. The system of equations in this generalized framework is much the same as the one we solved to find an arbitrage portfolio or a risk-free portfolio in the one-period model of Chapter 1. Hedging, in this framework, is simply a search for a portfolio whose sensitivity measures are equal to the sensitivity measures of the hedged instrument. A sensitivity measure of a portfolio is a linear combination of the sensitivity measure of the assets making up the portfolio. We therefore solve for a combination of assets which generates (replicates) the sensitivity of the hedged instrument.

We saw an example of such a system of equations, equations (8.5) and (8.6), when we solved for a hedge portfolio with respect to Delta. Moreover, the reader should recall the discussion concerning replication from Chapter 1 and approach the discussion here in the same manner. The idea is to replicate the sensitivity measure(s) of a given asset. The sensitivity measures here play the part of the payoffs of a replicated portfolio in Chapter 1. In the search for a replicating portfolio, we equate those payoffs to the one specified by the portfolio which we wish to replicate. It is much the same here: we have the specification of the hedged instrument and thus have the value of its sensitivity measure with respect to each[10] parameter.

We proceed to explain the above ideas by a way of an example. Let us assume that we would like to hedge a call option which we have written

---

[9] We will only use the bond, the underlying security, and derivative securities written on that underlying security, although in principle this need not be the case. If we have another derivative security which is sensitive to those parameters in which we are interested, we can include it as a row in the array with its sensitivity measures.

[10] There is some connection between these parameters in a way that will be explored in Chapter 15, equation (15.61). Also, given a certain universe (or financial market), a certain hedge might not be possible in much the same way as a certain cash flow cannot be generated in an incomplete market.

	Call Written	*Call* 1	*Call* 2	*Put* 1	*Put* 2	*Stock*
Interest Rate	0.12	0.12	0.12	0.12	0.12	0.12
σ	0.18	0.18	0.18	0.18	0.18	0.18
Maturity	0.24	0.16	0.08	0.24	0.16	
Price	12.86	21.56	2.57	16.67	4.65	100.00
Exercise Price	90.00	80.00	100.00	120.00	105.00	

Table 8.4: General Hedging

on a particular stock. Let us also assume that there are two other call options and two other put options on the same stock with different maturity dates[11] available in the market. These call and put options are available in the market for purchase or for sale. Since all these derivative securities are contingent on the same stock, they all have the same volatility of the underlying asset. They are each subject to the same value of the risk-free rate of interest[12] since they are available in the same market. Table 8.4 summarizes the relevant information regarding these securities.

Let us verify that the prices noted in Table 8.4 are consistent with the Black–Scholes option pricing formula. (We have assumed here that the underlying asset does not pay dividends). The price of the written call is given by

```
> Bs(90,90/365,100,0.18,0.12);
                              12.86325746
```

for call 1, for call 2 it is given by

```
> Bs(80,60/365,100,0.18,0.12);
                              21.56335855
> Bs(100,30/365,100,0.18,0.12);
                              2.576405603
```

and for the two puts it is given by

```
> BsPut(120,90/365,100,0.18,0.12);
                              16.67389558
```

[11] If we would like to keep these call options in the hedge portfolio until the maturity date of a written call option, the maturities of the former ones should be longer than those of the latter.

[12] This really assumes that the interest rates are the same for every maturity, which may not be an acceptable assumption. However, this assumption is easily relaxed and one can vary the level of the risk-free rate of interest according to the maturity date of the option.

```
> BsPut(105,60/365,100,0.18,0.12);
```

$$4.65933808$$

We denote the number of call options of type 1 and 2 to be held in the hedge portfolio by $C1$ and $C2$, respectively, and by $P1$ and $P2$ the number of put options to be held in the hedge portfolio. We denote the holdings of the underlying stock by XS and the amount invested in the risk-free asset by XB. Assume now that we would like to compose a hedge portfolio which will last for the next three days. In order to solve for the required hedge portfolio, we must specify the sensitivity measures (the parameters) against which we wish to be hedged. Each parameter specified will induce an equation which needs to be satisfied by our variables.

Each constraint stipulates that a sensitivity measure of the hedge portfolio, plus the same sensitivity measure of the written call, vanishes. A sensitivity measure of a written call is the negative of the same measure of a long position in the call. Hence, the constraints below are written in a form that imposes equality between the sensitivity measures of the hedge portfolio and those of a long position in the call. The hedge portfolio together with the written call is what we termed a hedged portfolio, since its sensitivity measures in question are all zeros. An additional equation will be needed if we would like the hedged portfolio to be self-financing.

We begin with the self-financing equation (constraint), pausing first to ensure that the variables are unassigned.

```
> XS:='XS';XB:='XB';C1:='C1';C2:='C2';P1:='P1';P2:='P2';
```

$$XS := XS$$
$$XB := XB$$
$$C1 := C1$$
$$C2 := C2$$
$$P1 := P1$$
$$P2 := P2$$

This self-financing equation will ensure that the price of the hedge portfolio under current conditions is equal to the price of the written call option. The constraint will be

```
> Selfin:=C1*Bs(80,60/365,100,0.18,0.12)+\
> C2*Bs(100,30/365,100,0.18,0.12)+\
> (P1*BsPut(120,90/365,100,0.18,0.12)+\
> P2*BsPut(105,60/365,100,0.18,0.12)+\
> XS*100+XB=Bs(90,90/365,100,0.18,0.12);
```

$$\begin{aligned} Selfin := {} & 21.56335855\, C1 + 2.576405603\, C2 + 16.67389558\, P1 \\ & + 4.65933808\, P2 + 100\, XS + XB = 12.86325746 \end{aligned}$$

Each of the following equations matches the sensitivity measure of the written call option in question to the same sensitivity measure of the hedge portfolio.

Hedging against the sensitivity of the call option to changes in the price of the underlying stock S requires equating the partial derivative of the call price with respect to S to the partial derivative of the value of the hedge portfolio with respect to S.

```
> DeltaHedg:=evalf(subs(S=100,\
> C1*diff(Bs(80,60/365,S,0.18,0.12),S)+\
> C2*diff(Bs(100,30/365,S,0.18,0.12),S)+\
> P1*diff(BsPut(120,90/365,S,0.18,0.12),S)+\
> P2*diff(BsPut(105,60/365,S,0.18,0.12),S)=\
> diff(Bs(90,90/365,S,0.18,0.12),S)));
```

$$\begin{aligned} \mathit{DeltaHedg} := {} & .9996164469\, \mathit{C1} + .585868391\, \mathit{C2} \\ & - .9519519295\, \mathit{P1} - .641234358\, \mathit{P2} = .939968249 \end{aligned}$$

The equation which matches the Gamma of the written call option to the Gamma of the hedge portfolio equates the second derivative of the call option price with respect to S to the second derivative of the value of the hedge portfolio with respect to S.

```
> GammaHedg:=evalf(subs(S=100,
> C1*diff(Bs(80,60/365,S,0.18,0.12),S$2)+\
> C2*diff(Bs(100,30/365,S,0.18,0.12),S$2)+\
> P1*diff(BsPut(120,90/365,S,0.18,0.12),S$2)+\
> P2*diff(BsPut(105,60/365,S,0.18,0.12),S$2)=\
> diff(Bs(90,90/365,S,0.18,0.12),S$2)));
```

$$\begin{aligned} \mathit{GammaHedg} := {} & .0001904495775\, \mathit{C1} + .07551004918\, \mathit{C2} \\ & + .01117752648\, \mathit{P1} + .05120238860\, \mathit{P2} = .01333300649 \end{aligned}$$

Further, if we would also like the hedged portfolio to be protected against changes in the volatility of the asset underlying the written call option a constraint is needed. It equates the partial derivative of the value of the written call option with respect to σ to the partial derivative of the value of the hedge portfolio with respect to[13] σ.

[13]These last two constraints are different than the rest. The other constraints were induced by parameters (state variables) of the model. Our assumption in developing the model was that the interest rate and the volatility are fixed constants. Thus the last two constraints may be interpreted as hedging against the risk induced by misspecification of the model.

```
> VegaHedg:=evalf(subs(sigma=0.18,\
> C1*diff(Bs(80,60/365,100,sigma,0.12),sigma)+\
> C2*diff(Bs(100,30/365,100,sigma,0.12),sigma)+\
> P1*diff(BsPut(120,90/365,100,sigma,0.12),sigma)+
> P2*diff(BsPut(105,60/365,100,sigma,0.12),sigma)=\
> diff(Bs(90,90/365,100,sigma,0.12),sigma)));
```

$$\begin{aligned} VegaHedg := {} & .056352209\, C1 + 11.17134976\, C2 + \\ & 4.96098410\, P1 + 15.15029579\, P2 = 5.91766314 \end{aligned}$$

Finally, in order to hedge against fluctuations in the level of interest rates, we equate the partial derivative of the value of the written call option with respect to interest rates to the partial derivative of the value of the hedge portfolio with respect to interest rates.

```
> RhoHedg:=evalf(subs(r=.12,\
> C1*diff(Bs(80,60/365,100,.18,r),r)+\
> C2*diff(Bs(100,30/365,100,.18,r),r)+\
> P1*diff(BsPut(120,90/365,100,.18,r),r)+\
> P2*diff(BsPut(105,60/365,100,.18,r),r)=\
> diff(Bs(90,90/365,100,.18,r),r)));
```

$$\begin{aligned} RhoHedg := {} & 12.88738951\, C1 + 4.60359728\, C2 - 11.30675734\, P2 \\ & = 20.00553718 \end{aligned}$$

To actually find the hedge portfolio which meets our specifications, we solve a system of equations. The equations are determined by the parameters against which we would like the portfolio to be hedged. In order to solve for the hedged portfolio which is protected in terms of Delta and, Gamma and is self-financing, we solve the three equations noted above: **Selfin**, **Delta-Hegd,** and **GammaHedg**. Below we submit them to MAPLE and request that MAPLE solve these three equations.

```
> solve({Selfin,DeltaHedg,GammaHedg}, {C2,C1,XB,XS,P1,P2});
```

$$\begin{aligned} \{ & C2 = C2,\ P1 = P1,\ P2 = P2, XB = \\ & -31027.58933\, C2 - 4713.081288\, P1 - 21146.17011\, P2 + 5407.378781, \\ & C1 = \\ & -396.4831541\, C2 - 58.69021411\, P1 - 268.8501034\, P2 + 70.00806547, \\ & XS = \\ & 395.7452133\, C2 + 59.61965522\, P1 + 269.3882194\, P2 - 69.04124541\} \end{aligned}$$

The general solution is presented by MAPLE. The meaning of $C2 = C2$, $P1 = P1$, and $P2 = P2$ is that $C2$, $P1$, and $P2$ are free variables. They can be set at any value and the resultant solution will still satisfy the set of equations above.

One can also choose to specify a hedged portfolio which is protected in terms of all the sensitivity measures. This can be done by submitting the set of equations below to MAPLE and requesting the solution.

```
> solve({Selfin,DeltaHedg,GammaHedg,VegaHedg,RhoHedg},\
> {C2,C1,XB,XS,P1,P2});
```

$$\begin{aligned}\{P1 &= -2.257739198\, P2 + 1.158475821,\\ C1 &= -3.835728562\, P2 + 4.033752249,\\ C2 &= -.3342063251\, P2 - .005086926710,\\ XS &= 2.522033450\, P2 - 1.986443279,\\ XB &= -135.6451342\, P2 + 105.2231405,\ P2 = P2\}\end{aligned}$$

In this case the only free variable is $P2$. The reason there are free variables is because we began with a universe containing six different securities (two put options, two call options, a stock, and a bond) and we only had five constraints which those securities were required to satisfy. Let us assign the variables and check what the value of XB is:

```
> assign(%);
> XB;
```

$$-135.6451342\, P2 + 105.2231405$$

We can examine the value of the hedge portfolio in a few days, say, t days. As before, we assume that there are 365 days in a year, and therefore investing a dollar at the risk-free rate for t days will result in $e^{\frac{rt}{365}}$ dollars in t days.[14] Hence, the value of the portfolio after t days, as a function of the stock price S, will be given by the function **PortValue1**

$$\begin{aligned} PortValue1(S, t, \sigma, r) &= C1\, Bs\left(80, \frac{60-t}{365}, S, \sigma, r\right) + \\ & C2\, Bs\left(100, \frac{30-t}{365}, S, \sigma, r\right) + \\ & P1\, BsPut\left(120, \frac{90-t}{365}, S, \sigma, r\right) + \\ & P2\, BsPut\left(105, \frac{60-t}{365}, S, \sigma, r\right) + SXS \end{aligned} \tag{8.15}$$

[14]The general practice in the financial industry is to use 365 days in the year. The calculation performed here may also be performed using discrete compounding.

$$+XB\,e^{\frac{r\,t}{365}} - Bs\left(90,\, \frac{90-t}{365},\, S,\, \sigma,\, r\right),$$

which is defined in MAPLE below, where the variable $P2$ is replaced with $P21$.

```
> P21:='P21':
> PortValue1:=unapply(evalf(subs(P2=P21,\
> C1*Bs(80,(60-t)/365,S,sigma,r)+\
> C2*Bs(100,(30-t)/365,S,sigma,r)+\
> P1*BsPut(120,(90-t)/365,100,sigma,r)+\
> P2*BsPut(105,(60-t)/365,S,sigma,r)+\
> XS*S+XB*exp(r*t/365)-Bs(90,(90-t)/365,S,sigma,r))),\
> S,t,sigma,r):
```

This function allows us to check how the value of the portfolio varies as the parameters change. Note that $P21$ was not yet assigned. Hence we can present the value of the portfolio after, say, one day, for a range of stock values as a function of $P21$. This is presented below for a range of stock price between \$95 and \$105, where σ and r remain unchanged.

```
> seq([`Stock Price`=S,`PortfolioPrice`=\
> PortValue1(S,1,0.18,0.12)],S=95..105);
```

$$[\mathit{Stock\ Price} = 95,\ \mathit{Portfolio\ Price} = 10.99455930\,\mathit{P21} - 5.72022469],$$
$$[\mathit{Stock\ Price} = 96,\ \mathit{Portfolio\ Price} = 8.768094475\,\mathit{P21} - 4.53712510],$$
$$[\mathit{Stock\ Price} = 97,\ \mathit{Portfolio\ Price} = 6.551663608\,\mathit{P21} - 3.37556983],$$
$$[\mathit{Stock\ Price} = 98,\ \mathit{Portfolio\ Price} = 4.348993595\,\mathit{P21} - 2.23364659],$$
$$[\mathit{Stock\ Price} = 99,\ \mathit{Portfolio\ Price} = 2.163814725\,\mathit{P21} - 1.10917817],$$
$$[\mathit{Stock\ Price} = 100,\ \mathit{Portfolio\ Price} = -.0000727511\,\mathit{P21} + .00008953],$$
$$[\mathit{Stock\ Price} = 101,\ \mathit{Portfolio\ Price} = -2.138861224\,\mathit{P21} + 1.09635613],$$
$$[\mathit{Stock\ Price} = 102,\ \mathit{Portfolio\ Price} = -4.248902866\,\mathit{P21} + 2.18168209],$$
$$[\mathit{Stock\ Price} = 103,\ \mathit{Portfolio\ Price} = -6.326927570\,\mathit{P21} + 3.25793350],$$
$$[\mathit{Stock\ Price} = 104,\ \mathit{Portfolio\ Price} = -8.370279658\,\mathit{P21} + 4.32676134],$$
$$[\mathit{Stock\ Price} = 105,\ \mathit{Portfolio\ Price} = -10.37710789\,\mathit{P21} + 5.38959128]$$

The reader might like to reexecute the above command specifying different values for t. The symbolic result above demonstrates the value of three portfolios as a function of S for three different choices of $P21$. It is displayed in Figure 8.8, where the (blue) graph with the highest y-intercept corresponds to $P21 = -3$. The (red) graph with the middle y-intercept corresponds to

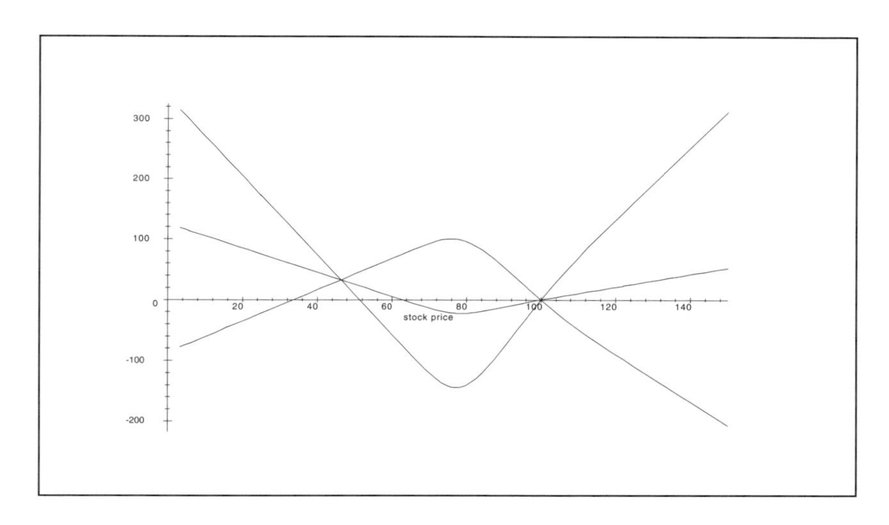

Figure 8.8: **PortValue1** as a Function of the Stock Price for $P21 = -3, 0, 3$ after One Day

$P21 = 0$, and the (green) graph with the lowest y-intercept corresponds to $P21 = 3$.

```
> plot([subs(P21=3,PortValue1(S,1,0.18,0.12)),\
> subs(P21=0,PortValue1(S,1,0.18,0.12)),\
> subs(P21=-3,PortValue1(S,1,0.18,0.12))],\
> S=0..150,colour=[green,red,blue],thickness=2,\
> thickness=2,labels=[`StockPrice`,`Port.  Value`],\
> title=`Figure 8.8:  PortValue1 as a Function of \
> the Stock Price for P21=-3,0,3 after One Day`,\
> titlefont=[TIMES,BOLD,8]);
```

The three-dimensional graph in Figure 8.9 presents the value of the portfolio as a function of S and $P21$.

```
> plot3d({PortValue1(S,1,0.18,0.12)},\
> S=70..125,P21=-10..10,axes=frame,style=PATCHCONTOUR,\
> orientation=[50,43],shading=ZHUE,title=`Figure 8.9:\
> PortValue1 as a Function of the Stock Price for a \
> Range of P21 after One Day`,titlefont=[TIMES,BOLD,8]);
```

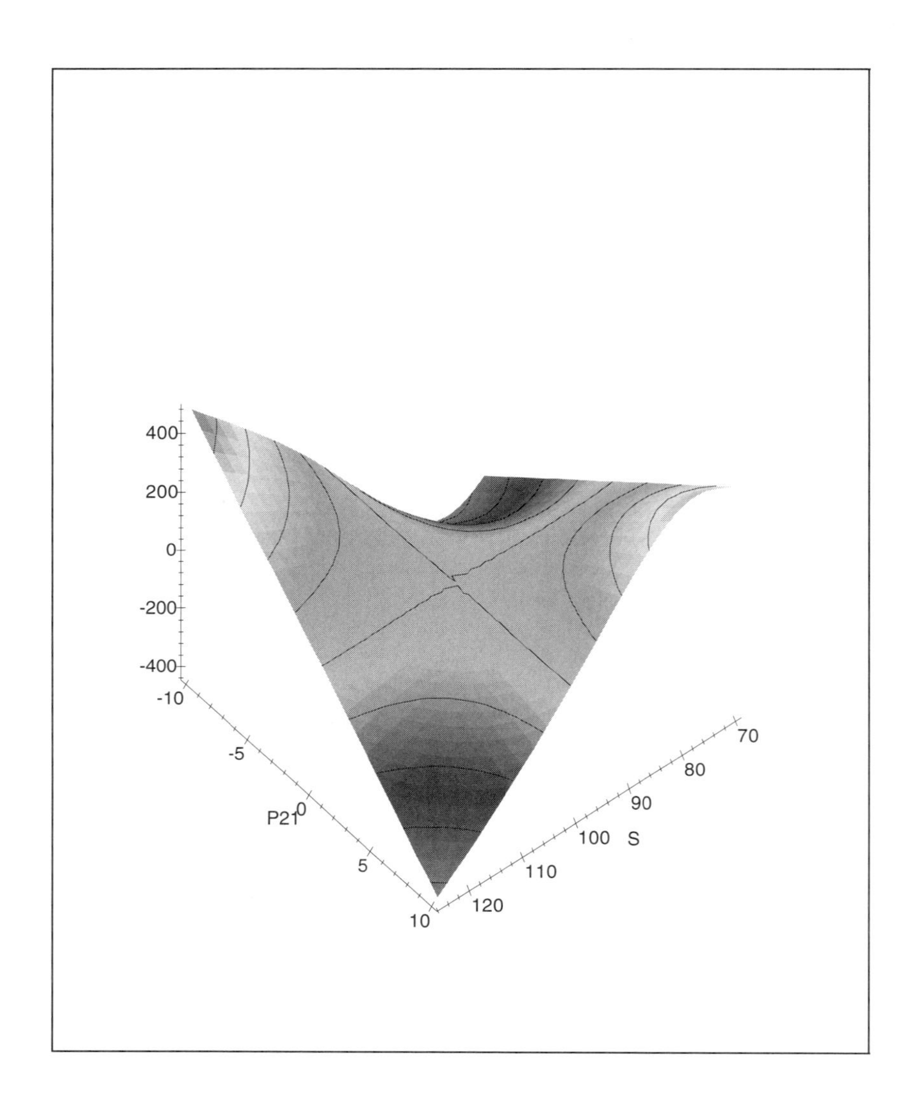

Figure 8.9: **PortValue1** as a Function of the Stock Price for a Range of $P21$ after One Day

The contours in Figure 8.9 demonstrate a saddle-like function exhibiting the phenomenon that, at every stock price, for $P21$ close to 1, the portfolio value is close to zero. For S close to 100 the function in $P21$ is nearly flat and close to zero, around $P21 = 1$.

The reader might also like to try the numerical calculations presented (only in the on-line version of the book) in the spreadsheet below. It displays the value of the portfolio after one day for a range[15] of stock values between \$95 and \$105, and a range of $P21$ values between -4 and 4, where σ and r remain unchanged.

Since our universe is composed of six securities, and the variable $P21$ was actually free, we can attempt to impose another constraint. We could, for example, devise a hedged portfolio which is more accurately hedged against changes in the volatility of the underlying asset. Does the reader recall our definition of Vega2? It is the second partial derivative of the value of a derivative security (or a portfolio of such securities) with respect to the volatility of the underlying asset.

We now experiment with imposing a sixth constraint, one which concerns Vega2. This constraint equates the measure Vega2 of the written call and of the hedge portfolio value. In order to accomplish this and to examine the solution, we unassign the variables.

```
>  XS:='XS'; XB:='XB'; C1:='C1';C2:='C2';P1:='P1'; P2:='P2';
```

$$XS := XS$$
$$XB := XB$$
$$C1 := C1$$
$$C2 := C2$$
$$P1 := P1$$
$$P2 := P2$$

Next we define the new constraint concerning Vega2 and call it **Veg2Hedg.**

```
>  Vega2Hedg:=evalf(subs(sigma=0.18,\
>  C1*diff(Bs(80,60/365,100,sigma,0.12),sigma$2)+\
>  C2*diff(Bs(100,30/365,100,sigma,0.12),sigma$2)+\
>  P1*diff(BsPut(120,90/365,100,sigma,0.12),sigma$2)+\
>  P2*diff(BsPut(105,60/365,100,sigma,0.12),sigma$2)=\
```

[15] Remember that we have arbitrarily specified a few values for the free variable. Each choice of the free variable will result in a different hedged portfolio, even though all such portfolios are hedged against the same set of parameters. These portfolios will perform in different ways. The reader might like to assign a different value to $P21$ and t and reevaluate the spreadsheet as well as plot again the above figures to visualize the different performances.

```
> diff(Bs(90,90/365,100,sigma,0.12),sigma$2)));
```

$$Vega2Hedg := 3.4668023\,C1 + 2.2258246\,C2 \\ +80.4204588\,P1 + 13.2372778\,P2 = 74.8764660$$

Now we are ready to solve the new system of six equations. This system resembles the one already solved, but it has one additional constraint.

```
> solve({Selfin,DeltaHedg,GammaHedg,VegaHedg,\
> RhoHedg,Vega2Hedg},{C2,C1,XB,XS,P1,P2});
```

$$\{P2 = .1768991885,\ C1 = 3.355214979, \\ XB = 81.22762631,\ XS = -1.540297609, \\ P1 = .7590835890,\ C2 = -.06420775442\}$$

The additional constraint causes the solution to be unique. Let us see what will be the result if we define another constraint and add it to the list of constraints. We define below the constraint related to Theta and try to solve the new system of equations.

```
> ThetaHedg:=evalf(subs(t=0,\
> C1*diff(Bs(80,60/365-t/365,100,.18,.12),t)+\
> C2*diff(Bs(100,30/365-t/365,100,.18,.12),t)+\
> P1*diff(BsPut(120,90/365-t/365,100,.18,.12),t)+\
> P2*diff(BsPut(105,60/365-t/365,100,.18,.12),t)=\
> diff(Bs(90,90/365-t/365,100,.18,.12),t)));\
```

$$ThetaHedg := -.02585930731\,C1 - .0519284384\,C2 \\ + .03181789420\,P1 - .00011192912\,P2 = -.03259171272$$

```
> solve({Selfin,DeltaHedg,GammaHedg,ThetaHedg,\
> VegaHedg,RhoHedg,Vega2Hedg},{C2,C1,XB,XS,P1,P2});
```

MAPLE's solution or reply to the request to solve this system of seven equations was **NULL**. This means that there does not exist a feasible solution to this system of seven equations. As mentioned before, the meaning of this is analogous to the situation of incomplete markets discussed earlier in the book. This infeasibility depends on the structure of the market and on the number of securities available in the market. In practice, one could include in the universe of available securities all securities which have as their underlying asset the same stock, and then attempt to find a solution to the resultant system of equations.

Alternatively, we could attempt to assess which of the constraints is most important for us to be satisfied. For example, let us see what happens if we omit the **Vega2Hedg** constraint.

```
> solve({Selfin,DeltaHedg,GammaHedg,VegaHedg,\
> RhoHedg,ThetaHedg},{C2,C1,XB,XS,P1,P2});
```

$$\{P2 = .7757236673,\ C1 = 1.058286823,\ P1 = -.5929059095,\ C2 = -.2643386829,\ XS = -.03004224224,\ XB = -.4545952605\,10^{-6}\}$$

A feasible solution exists in this case. We therefore assign the variables and define the function **PortValue**. **PortValue** is the counterpart of the function **PortValue1** but it is based on the current solution.

```
> assign(%);
> PortValue:=unapply(evalf(\
> C1*Bs(80,(60-t)/365,S,sigma,r)+\
> C2*Bs(100,(30-t)/365,S,sigma,r)+\
> P1*BsPut(120,(90-t)/365,S,sigma,r)+\
> P2*BsPut(105,(60-t)/365,S,sigma,r)+XS*S+\
> XB*exp(r*t/365)-Bs(90,(90-t)/365,S,sigma,r)),\
> S,t,sigma,r):
```

We can compare, after one day, the performance of **PortValue1** when $P21 = 0$ to that of **PortValue**. This is demonstrated in Figure 8.10.

```
> plot([subs(P21=0,PortValue1(S,1,0.18,0.12)),\
> PortValue(S,1,0.18,0.12)],S=0..150,colour=[red,black],\
> thickness=2,labels=[`Stock Price`,`Port.Value`],title=\
> `Figure 8.10:  PortValue and PortValue1 as a Function \
> of the Stock Price for P21=-3,0,3 after One Day`,\
> titlefont=[TIMES,BOLD,8]);
```

In Figure 8.10 the graph that is closer to the x-axis corresponds to the function **PortValue**. It therefore provides a better hedge against changes in the price of the stock.

Let us see what happens when the constraint regarding Theta is eliminated but the constraint induced by Vega2 is present. We unassign the variables and try to solve for yet another hedged portfolio.

```
> XS:='XS':XB:='XB':C1:='C1':C2:='C2':P1:='P1':P2:='P2':
> solve({Selfin,DeltaHedg,GammaHedg,VegaHedg,\
> Vega2Hedg,RhoHedg},{C2,C1,XB,XS,P1,P2});
```

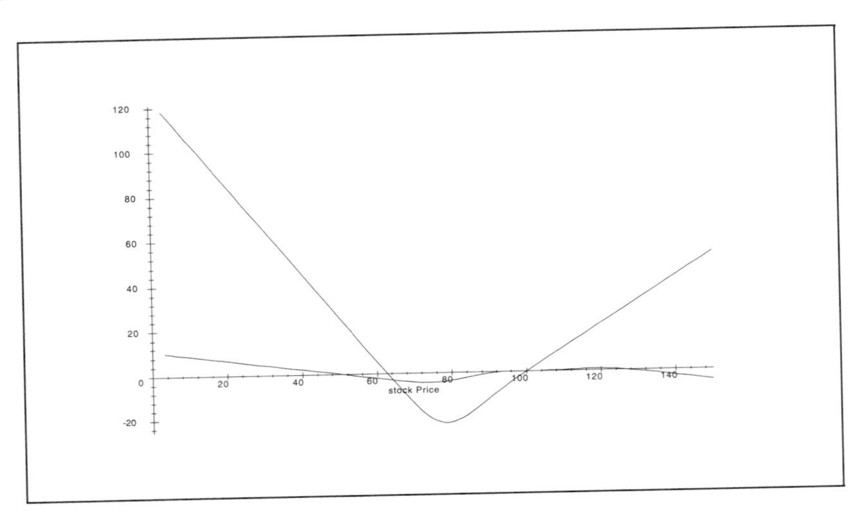

Figure 8.10: **PortValue** and **PortValue1** as a Function of the Stock Price for $P21 = -3, 0, 3$ after One Day

$$\{P2 = .1768991885,\ XS = -1.540297609,\\ P1 = .7590835890,\ XB = 81.22762631,\\ C2 = -.06420775442,\ C1 = 3.355214979\}$$

We can, as before, assign the variables and define the function **PortValue2**.

```
> assign(%);
> PortValue2:=unapply(evalf(\
> C1*Bs(80,(60-t)/365,S,sigma,r)+\
> C2*Bs(100,(30-t)/365,S,sigma,r)+\
> P1*BsPut(120,(90-t)/365,S,sigma,r)+\
> P2*BsPut(105,(60-t)/365,S,sigma,r)+XS*S+\
> XB*exp(r*t/365)-Bs(90,(90-t)/365,S,sigma,r)),S,t,sigma,r):
```

Having defined the **PortValue2** function we can compare, after one day, the three portfolios we composed. This is done in Figure 8.11.

```
> plot([subs(P21=0,PortValue1(S,1,0.18,0.12)),\
> PortValue(S,1,0.18,0.12),PortValue2(S,1,0.18,0.12)],\
> S=0..150,colour=[red,black,green],thickness=2,
> labels=['StockPrice','Port.Value'],title='Figure 8.11:\
> The Three Portfolios after One Day as a Function of\
```

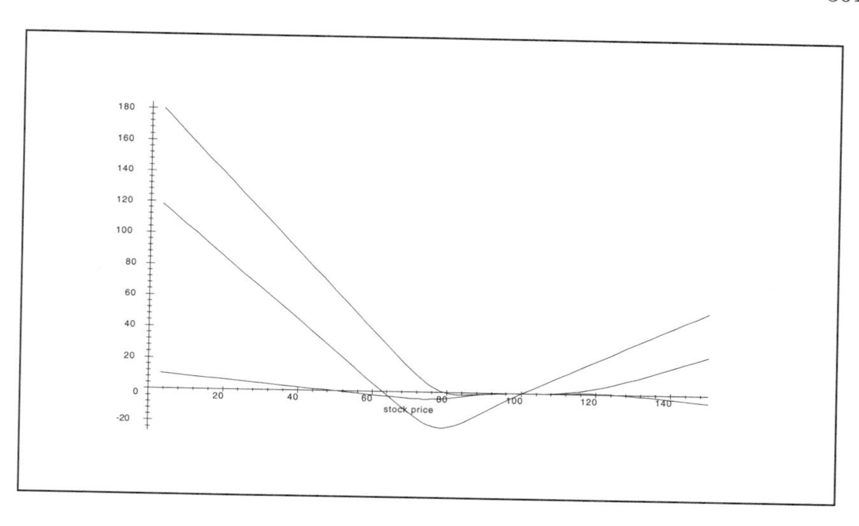

Figure 8.11: The Three Portfolios after One Day as a Function of the Price of the Stock

```
>  the Price of the Stock',titlefont=[TIMES,BOLD,8]);
```

In Figure 8.11, the (green) graph with the highest y-intercept corresponds to **PortValue2**. The (red) graph with the middle y-intercept corresponds to **PortValue1** (which is defined where $P21 = 0$), and the (black) graph with the lowest y-intercept corresponds to **PortValue**. We see again that the portfolios corresponding to **PortValue** are a better hedge.

The three portfolios are protected against changes in the volatility of the underlying asset, presumably **PortValue2,** in a better way. The hedged portfolio **PortValue2** includes a constraint involving the second partial derivatives with respect to volatility: the **Vega2Hedge** constraint. This is visualized in Figure 8.12, where the three portfolios are graphed for a σ ranging from 0 to 0.32 a day after the construction of the hedge for a stock price of 100.

```
>  plot([subs(P21=0,PortValue1(100,1,sigma,0.12)),\
>  PortValue(100,1,sigma,0.12),\
>  PortValue2(100,1,sigma,0.12)],\
>  sigma=0.1..0.32,colour=[red,black,green],thickness=2,\
>  labels=['Sigma','Port.Value'],title='Figure 8.12:  The\
>  Three Portfolios after One Day as a Function of Sigma',\
```

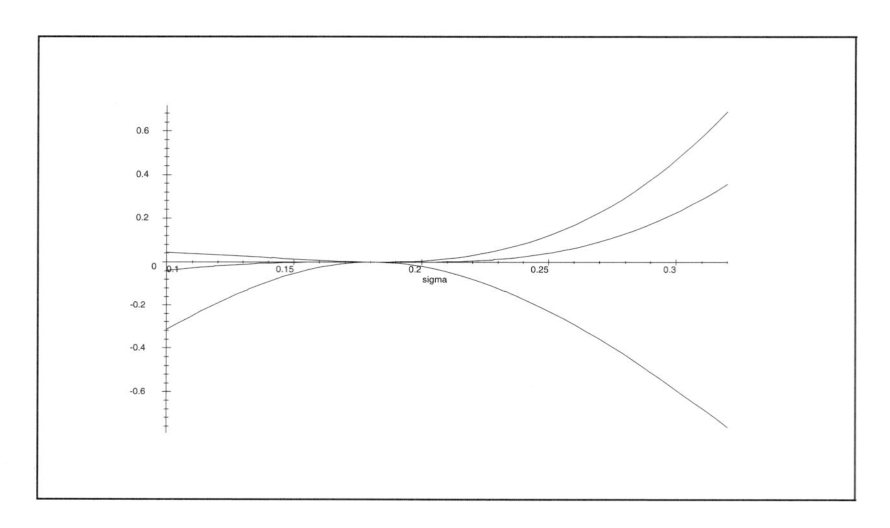

Figure 8.12: The Three Portfolios after One Day as a Function of Sigma

```
> titlefont=[TIMES,BOLD,8]);
```

In Figure 8.12 the graph of **PortValue1** is the highest, and the graph corresponding to the **PortValue** is the lowest. As expected, the graph corresponding to **PortValue2** is closer to zero when viewed as a function of sigma. We can also visualize in a three-dimensional setting the performance of the portfolios as a function of sigma and the price of the stock. In Figure 8.13, the functions **PortValue** and **PortValue2** are graphed for a sigma ranging from 0 to 0.32 and for a price ranging from 80 to 120, a day after the construction of the hedge. We see that around a volatility of 0.18 and a stock price of 100 the graphs are very close to each other. The upper manifold in Figure 8.13 corresponds to **PortValue2** and, as expected, it varies less than the lower manifold with respect to sigma.

```
> plot3d({PortValue2(S,1,sigma,0.12),\
> PortValue(S,1,sigma,0.12)},S=80..120,sigma=0.17..0.19,\
> labels=['StockPrice','Volatility','Port.Value'],\
> axes=frame,style=PATCH,orientation=[-37,76],title=\
> 'Figure 8.13:  Value of the Hedged Portfolios for\
> Different Values of the Stock Price and Volatility',\
> titlefont=[TIMES,BOLD,8]);
```

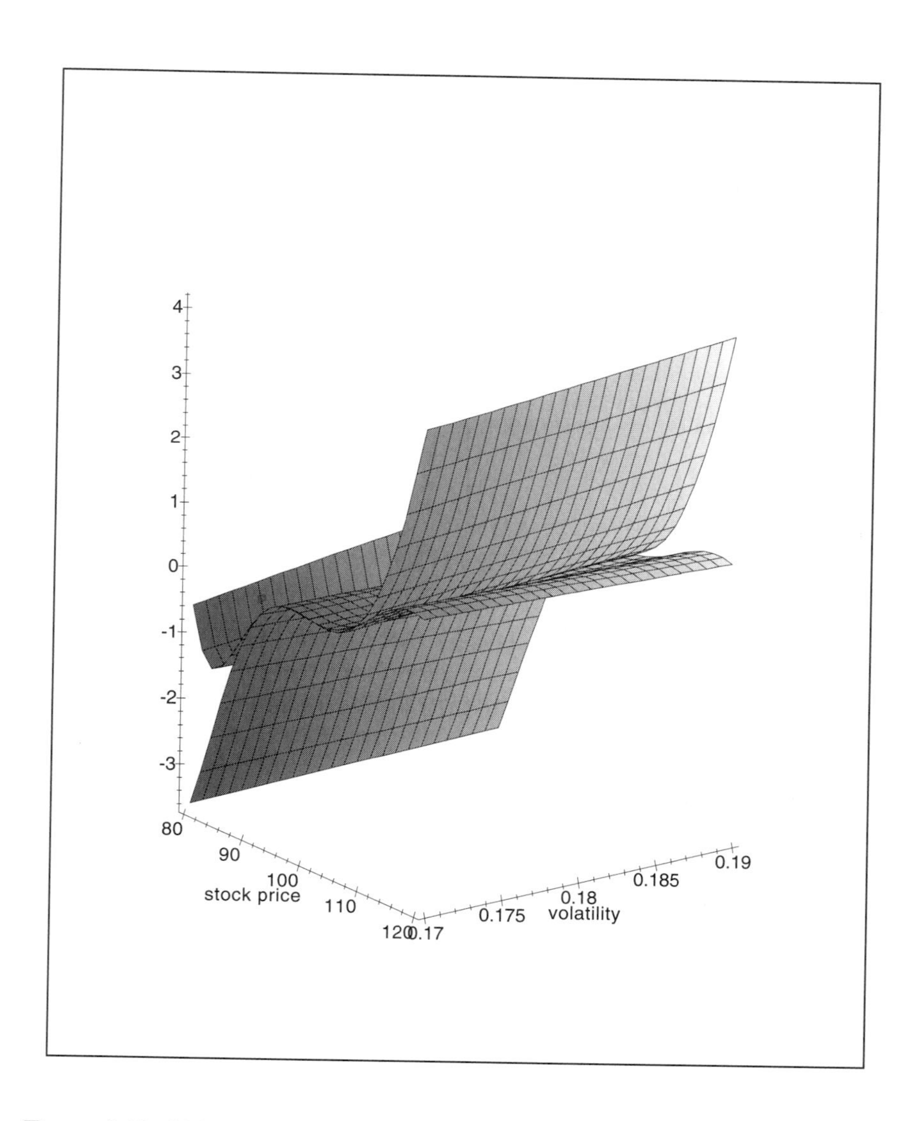

Figure 8.13: Value of the Hedged Portfolios for Different Values of the Stock Price and Volatility

The reader may wish to examine the values of these hedged portfolios for different values of the stock price or different values of the other underlying variables. This can be done visually by changing the values of the parameters and executing the commands above. Alternatively, the reader may examine it in a spreadsheet framework as was done before.

We have already seen that with the current universe of securities it is not possible to satisfy all of the constraints **Selfin**, **DeltaHedg**, **GammaHedg**, **ThetaHedge, VegaHedg**, **RhoHedg**, and **Vega2Hedg** simultaneously. There are a few possibilities for constructing a hedged portfolio in this case. We can eliminate one of the constraints and see if a solution can be found. Alternatively, we can proceed by using optimization to identify a second-best, in a certain sense, solution to the problem. This is done in the next section. Readers who wish to skip the next part can do so without loss of continuity.

The reader is encouraged to use the framework supplied here[16] to try different hedging strategies. This can be done by simply reexecuting the commands above for different values of the parameters or different hedging strategies.

8.5 Optimizing Hedged Portfolios

In the absence of a feasible hedging solution we may wish to attempt to find a second-best solution. In the last section we attempted to satisfy six of the seven constraints from the last example. That attempt resulted in a feasible solution, while satisfaction of all seven constraints was not possible. We can, however, try to select a solution via optimization rather than omitting, as was done before, one of the constraints arbitrarily. For example, we could choose to find a hedged portfolio which satisfies the first six constraints above, and for which the value of Vega2 is as close as possible to its target value. This requires solving an optimization problem.

The hedging constraints are linear. Hence, if the criteria used for a second-best solution can be written as a linear function, the simplex method, as in Chapter 1, can be utilized. We set up a problem for which we satisfy the six constraints **Selfin, DeltaHedg, GammaHedg, VegaHedg**, **Rho-Hedg,** and **ThetaHedg,** and minimize the absolute deviations of Vega2 from its target value. As we shall soon see, this can be formulated as a

[16] The reader is alerted to the order of execution. If the commands are not executed in order and numerical values assigned to variables are not unassigned, the desired correct result will not be obtained.

linear programming problem.

```
> XS:='XS'; XB:='XB'; C1:='C1';C2:='C2';P1:='P1'; P2:='P2';
```

$$XS := XS$$
$$XB := XB$$
$$C1 := C1$$
$$C2 := C2$$
$$P1 := P1$$
$$P2 := P2$$

In order to formulate this problem, we must[17] alter the **Vega2Hedge** constraint and define the deviation of Vega2 from its target value. This requires two additional variables.

```
> Neg:='Neg';Pos:='Pos';
```

$$Neg := Neg$$
$$Pos := Pos$$

The new definition of the **Vega2Hedge** constraint is given below.

```
> Vega2Hedgapp:=3.4668023*C1+2.2258249*C2+\
> 80.4204579*P1+13.2372779*P2-74.8764639=Neg-Pos;
```

$$\begin{aligned} Vega2Hedgapp :=\ & 3.4668023\ C1 + 2.2258249\ C2 + \\ & 80.4204579\ P1 + 13.2372779\ P2 - 74.8764639 \\ & = Neg - Pos \end{aligned}$$

Neg and Pos are the two sides of the absolute deviation of Vega2 from its target value.[18] The optimization problem which we solve is defined below.

$$\begin{aligned} & \min_{Neg \geq 0, Pos \geq 0} Neg + Pos \\ \text{Subject to}\ :\ & Selfin, DeltaHedg, GammaHedg, \\ & VegaHedg, RhoHedg, ThetaHedg, \\ & Vega2Hedgapp \end{aligned}$$

[17] First we unassign the variables from the last section and make sure that the additional variables needed are unassigned. Readers who start reading the on-line version from this section will need to operate the commands in the former section. This will define the constraints as well as the functions **PortValue** and **PortValue2** that correspond to the portfolios constructed in the former section.

[18] Every number can be written as the difference between two positive numbers, denoted here by Pos and Neg. Thus, the deviation of Vega2 from its target value can be written as $Pos - Neg$. In the optimal solution, one of Neg and Pos will be positive and the other will have a zero value. This follows from the properties of linear programming. Hence, at the optimal solution, $Pos + Neg$ is the absolute value of $Pos - Neg$, which we try to minimize.

$Port\backslash S$	95	96	97	98	99	100	101
$PortValue$	$-.0778$	$-.0373$	$-.0133$	$-.0020$	.0010	.0000	$-.0010$
$PortValue2$	$-.0851$	1.4690	$-.0150$	$-.0023$	.0012	.0000	$-.0012$
$PortValue3$	2.8085	2.2644	1.7067	1.1399	.5693	.0000	$-.5628$

Table 8.5: $PortValue$, $PortValue2$, and $PortValue3$ as a function of S

Hence the problem which is submitted to MAPLE is

```
> simplex[minimize](Neg+Pos,{Selfin,DeltaHedg,GammaHedg,\
> VegaHedg,RhoHedg,ThetaHedg,Vega2Hedgapp,\ ;
> Neg>=0,Pos>=0});
```

$$\{C1 = 1.058286823,\ XS = -.03004224224,\ XB = -.4545950000\,10^{-6},\ P2 = .7757236673,\ P1 = -.5929059095,\ C2 = -.2643386829,\ Pos = 109.2092593,\ Neg = 0\}$$

We can assign the variables, as before, and proceed to examine the performance of the hedged portfolio, say, after one day. (The reader may wish to compare the performance of this hedged portfolio to that of those which we have already examined.)

```
> assign(%);
```

We define the function **PortValue3**, as below, in order to examine the performance of the newest hedged portfolio.

```
> PortValue3:=unapply(evalf(\
> C1*Bs(80,(60-t)/365,S,sigma,rf)+\
> C2*Bs(100,(30-t)/365,S,sigma,rf)+\
> P1*BsPut(120,(90-t)/365,100,sigma,rf)+\
> P2*BsPut(105,(60-t)/365,S,sigma,rf)+XS*S+\
> XB*exp(rf*t/365)-Bs(90,(90-t)/365,S,sigma,rf)),\
> S,t,sigma,rf):
```

The value of this hedged portfolio, after one day, is compared in the spreadsheet below (Table 8.5) to the value of the first and second portfolio.

The difference in the performance of the three hedged portfolios as a function of S is clearest from a graph. But, first, the values of the three portfolios, for a stock price of \$90, after one day, are plotted against different values of σ in Figure 8.14. It should be clear that the second portfolio is the most protected against changes in the volatility. After all, this portfolio satisfied both the Vega and Vega2 constraints.

```
> plot([PortValue(90,1,sigma,.12),\
```

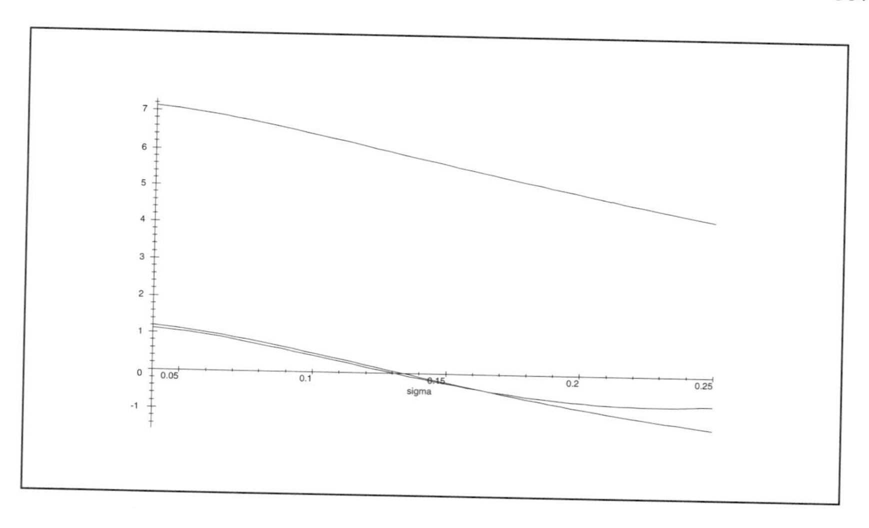

Figure 8.14: The Values of the Three Portfolios at $S = 90, t = 1$, for Different Values of Sigma

```
> PortValue2(90,1,sigma,.12),PortValue3(90,1,sigma,.12)],\
> sigma=0..0.32,color=[red,blue,green],thickness=2,\
> labels=['sigma',' '],title='Figure 8.14:\
> The Values of the Three Portfolios at\
> S=90, t=1, for Different Values of Sigma');
```

The highest (green) graph is of **PortValue3**, the lowest (red) graph is of **PortValue,** and the middle (blue) graph is of **PortValue2**. Again we see that **PortValue2** is the most protected against changes in the volatility. This is not surprising as this portfolio satisfies the two volatility constraints (**VegaHedg** and **Vega2Hedg**).

In order to compare the values of the three portfolios after one day as a function of the stock price, we plot them in Figure 8.15 on the same set of axes. The graph of **PortValue** is represented by the red curve (lowest y-intercept), the graph of **PortValue2** is the blue curve (middle y-intercept), and the graph of **PortValue3** is the green curve (highest y-intercept).

```
> plot([PortValue(S,1,.18,.12),PortValue2(S,1,.18,.12),\
> PortValue3(S,1,.18,.12)],S=85..115,thickness=2,\
> labels=['stock price',' '],\
> color=[red,blue,green],title='Figure 8.15:\
```

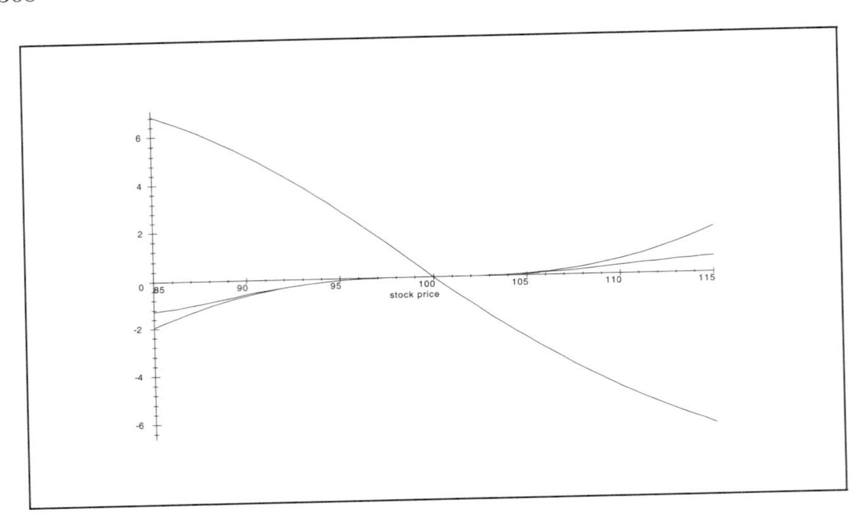

Figure 8.15: The Values of the Three Portfolios for Different Values of the Stock Price

```
> The Values of the Three Portfolios for Different\
> Values of the Stock Price'):
```

The graphs in Figure 8.15 are plotted using the original values of σ and of the risk-free rate. The plotting command may, however, be reexecuted with some other parameter values changed, if the reader wishes to examine the performance of the three hedged portfolios in these altered circumstances. The reader should be able to deduce from these graphs that the market does not allow arbitrage opportunities. The portfolios cost nothing (they are self-financing). Had one portfolio dominated the others, arbitrage opportunities would have existed. Note though that there is a trade-off between the three hedged portfolios: the red hedged portfolio will be the most valuable if the stock price increases, and also lead to the greatest losses if the stock price decreases. The performance of these three hedged portfolios can also be examined after two days using a volatility measure of 0.15 (instead of 0.18) and a risk-free rate of interest of 10 percent (rather than 12 percent). This can be viewed by executing the command below. The resultant graph appears only in the on-line version of the book.

```
> plot([PortValue(S,3,.15,.10),PortValue2(S,3,.15,.10),\
> PortValue3(S,3,.15,.10)],S=85..115,thickness=2,\
```

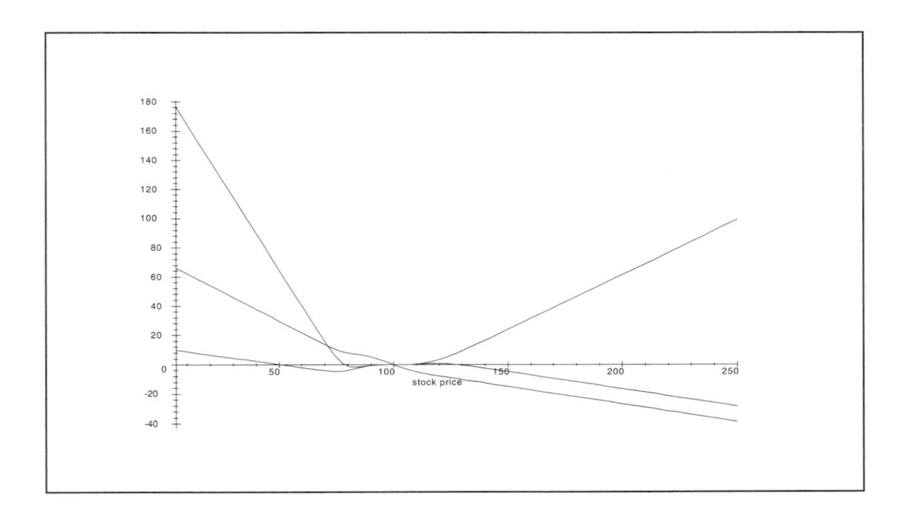

Figure 8.16: The Hedged Portfolios after the Passage of 15 Days

```
> labels=['StockPrice','PortfolioValue'],\
> color=[red,blue,green]):
```

There is no great difference in the ranking of the two hedged portfolios **PortValue** and **PortValue2**. If hedging in the purest sense of the word is what is of interest, i.e., staying as close to zero as possible, then it should be clear that the red hedged portfolio (the first) is the best of the three. It is the portfolio **PortValue**, which resulted from the solution to the system of equation having Theta in the constraints. This conclusion will be more obvious if we examine the hedged portfolios after the passage of 15 days as illustrated in Figure 8.16.

```
> plot([PortValue(S,15,.18,.12),PortValue2(S,15,.18,.12),\
> PortValue3(S,15,.18,.12)],S=5..250,thickness=2,\
> labels=['stock price',' '],\
> color=[red,blue,green],title='Figure 8.16:  The\
> Hedge Portfolios after the Passage of 15 Days'):
```

The on-line version of the book includes a three-dimensional graph that demonstrates how the hedged portfolios, **PortValue**, **PortValue2**, and **PortValue3**, behave when both the passage of time and changes in the stock price are taken into account.

```
> plot3d({PortValue(S,t,.18,.12),PortValue2(S,t,.18,.12),\
```

```
> PortValue3(S,t,.18,.12)},S=5..250,t=0..20,axes=normal,\
> shading=ZHUE,axes=frame,orientation=[127,59],\
> labels=[`StockPrice`,`Time`,`Portfolio Value`]):
```

One final point before leaving this Chapter: There is a connection between Delta, Gamma, and Theta. In the limiting case, i.e., hedging over an instant in time, there is no need to be concerned about the Theta measure. A self-financing portfolio which is Delta and Gamma neutral is also Theta neutral. This point will be revisited later in the book. The point is also connected closely to the derivation of the Black–Scholes option pricing formula in Chapter 15. Note though that it is only a limiting argument. Thus, the inclusion of a constraint which ensures Theta neutrality might improve the hedging ability of the portfolio over an interval of time which is longer than an instant. Indeed such was the case as can be seen in Figure 8.14.

We unassign the values of the variables before leaving the section.

```
> Neg:='Neg';Pos:='Pos';XS:='XS'; XB:='XB'; C1:='C1';
> C2:='C2';P1:='P1'; P2:='P2';
```

$$Neg := Neg$$
$$Pos := Pos$$
$$XS := XS$$
$$XB := XB$$
$$C1 := C1$$
$$C2 := C2$$
$$P1 := P1$$
$$P2 := P2$$

The reader is encouraged to make use of the framework presented here and experiment with different hedging strategies.

8.6 Concluding Remarks

This Chapter investigated the details of Delta hedging and examined a variety of other hedging strategies. The exercises of this Chapter provide an environment that will aid the user in formulating different hedging strategies and simulating the price process of the underlying asset. The reader may wish to explore this environment and so gain experience with different hedging possibilities.

For example, one possibility would be to choose a particular hedging strategy for a few periods, and then switch to another strategy over the subsequent few periods. The simulation provides different combinations of hedging strategies with respect to a variety of given securities. The topic of static hedging has received some attention in the recent literature. You might like to explore these issues by consulting some readings, e.g., [9] and [18].

The reader, however, should be aware of transaction costs. They have been overlooked in this book. The more assets there are in the hedge portfolio, the more expensive it will be in terms of transaction costs to rebalance the hedge. This might make a very good hedge portfolio (one which is most precise) relatively more expensive. However, it should be remembered that financial institutions trade at favorable levels of transaction costs, and the benefit of the extra hedging precision might be worth the extra cost.

As we have mentioned numerous times in the text, a linkage exists between the hedging strategies explored here and the differential equation satisfied by a derivative security. This topic requires a continuous-time model allowing trading every instant and is investigated in Chapter 15. The order of the topics in this book follow a natural progression of the models explored herein. European options were investigated in a framework of a one-period model with a continuum of possible states of nature. We proceed now to an investigation of the term structure and its estimation, which are explored in a multiperiod discrete-time model. This will facilitate the pricing of swaps, futures, and forwards in a multiperiod setting.

8.7 Questions and Problems

Problem 1. For a stock price in the upper part of the range, Delta of the portfolio in Figure 8.4 is close to zero, while at the lower part of the range the Delta of the portfolio is close to one. Explain why this is the case.

Problem 2. Prepare a worksheet (in MAPLE or the MATLAB GUI) that simulates a hedging environment. Use the spreadsheet (Table 8.3) as a guideline. Allow the reader to choose interactively the period Δt, the security to hedge (based on its payoff at maturity), and the hedging methods (one of the methods explained in the text). Utilize your worksheet to examine hedging a put and a call option maturing in 30 days. Assume the same parameters as in Table 8.3 and experiment with the different methods for $\Delta t = 10$ days and $\Delta t = 5$ days.

Problem 3.

1. Repeat the calculations of $XS12$ and $XB12$, illustrated in Figure

8.6, for a binary option that pays \$10 if the price of the stock is above \$40. Use the same parameters as in Figure 8.6.

2. Repeat the question above where the solution is parameterized by sigma. What is your conclusion?

Problem 4. Explain why, if we would have identified a self-financing portfolio with a payoff as described in Figure 7.8, it would have been an arbitrage portfolio. Identify an arbitrage position that would have existed in the market if a portfolio with the above payoff could be constructed.

Problem 5.

1. Formulate and solve a linear programming hedging problem in which the optimization criteria is the satisfaction of all the constraints as defined in Sections 8.4 and 8.5, i.e., *Selfin*, *DeltaHedg*, *GammaHedg*, *VegaHedg*, *RhoHedg*, and *Vega2Hedgapp*.
2. Formulate and solve a linear programming hedging problem that uses a second-best solution for the satisfaction of all the constraints. (Hint: use the a second-best criteria like the one utilized for *Vega2*.) Compare the solutions of these two problems. Add the constraint regarding Theta hedge and resolve the optimization problem. What is your conclusion?

Problem 6. Formulate a *Vega2*-like constraint for rho. Repeat Problem 5 with this added constraint and compare the hedging strategies based on the problems you solved.

Chapter 9

The Term Structure, Its Estimation, and Smoothing

In the last few chapters we investigated the use of the stochastic discount factors (SDFs) in valuing equity options. Some financial instruments that also fall into the class of derivative securities are valued based on risk-free discount factors. The valuation concept of these instruments is induced, as before, by the no-arbitrage condition. These securities are valued either by replication or by applying to their cash flow the discount factors. The latter is done essentially in the same manner as the SDFs were applied to equity derivatives. The risk-free discount factors are implicit in bond prices, as we saw in Chapter 3. They are consequences of the no-arbitrage condition in the same manner that the SDFs are. However, before we dwell on the valuation issue we introduce the reader to some estimation matters, terminology such as yield, yield curve, spot curve, and zero-coupon curve, and their interrelation.

The reader who is interested only in valuation concepts may skip the following subsections and continue reading on from Section 9.4. We do recommend reading on since the similarities of the valuation concepts in the debt and equity markets are highlighted. The valuation concepts can be followed if the reader

1. takes as given the discount factor $d(t)$ for time t, which is sometimes expressed as $e^{-r(t)t}$, where for each t, the rate $r(t)$ is termed the *continuously compounded interest rate*, and

2. realizes that the function $r(t)$ from the time domain $[0, T]$ to the positive real line is termed the *term structure of interest rates*.

9.1 The Term Structure of Interest Rates

The discount factors d's reflect the discounting due to the time value of money. Thus, they have implicit in them the interest rate that prevails in the market. For example, if $d_1 = 0.9$ it means that in this market a dollar at time one is worth 0.9 of a dollar today. The difference in value between a dollar today and a dollar received in the future results from the opportunity cost of investing that dollar at the risk-free rate of interest. A dollar today can be invested in a risk-free asset from time zero to time one and will earn a rate of interest which we denote by $R1$. Hence a dollar today will be equal to $1+R1$ dollars at time one. Consequently, $\frac{1}{1+R1}$, if invested at the risk-free rate from time zero to time one, will grow to be one dollar. It thus follows that

$$d_1 = \frac{1}{1+R1} \tag{9.1}$$

and we can solve for $R1$ in terms of d_1.

```
> solve(d[1]=1/(1+R1),R1);
```

$$-\frac{d_1 - 1}{d_1}$$

The rate $R1$ is called the *spot rate* spanning the time interval zero to one. It is the rate applied to loans which commence now (at time zero) and which mature at some time in the future — time one here. In the same manner we can calculate the rate, $R2$, spanning the time interval zero to two, i.e.,

```
> solve(d[2]=1/(1+R2),R2);
```

$$-\frac{d_2 - 1}{d_2}$$

With complete precision of notation, the rates $R1$ and $R2$ should have a second index. After all, these rates are payable from time zero to times one and two, respectively. For simplicity, however, we omit the index reference to time zero, and thus instead of using $R01$, we use $R1$, with the understanding that if only one index is mentioned, it is the second one, and the first is (implicitly) zero. These rates, as mentioned, are applicable to a loan which commences now (on the spot) and lasts until some future time. Hence they are termed *spot rates*.

Equivalently, we can solve for $R1$ by utilizing the function **Vdis** defined by **NarbitB**. Consider, for simplicity, a bond market with three bonds in which coupons are paid annually, the payments dates of the bonds coincide, and the longest maturity is three years. The structure of the market is described in **NarbitB** below.

```
> NarbitB([[105,0,0],[8,108,0],[3,3,103]],[96,97,89]);
```

The no − arbitrage condition is satisfied.

The discount factor for time, 1, *is given by*, $\frac{32}{35}$

The interest rate spanning the time interval, [0, 1], *is given by*, .09375

The discount factor for time, 2, *is given by*, $\frac{3139}{3780}$

The interest rate spanning the time interval, [0, 2], *is given by*, .2042

The discount factor for time, 3, *is given by*, $\frac{21109}{25956}$

The interest rate spanning the time interval, [0, 3], *is given by*, .2296

The function $Vdis([c1, c2, ..])$, *values the cashflow* $[c1, c2, ..]$

The rate $R1$, for example, applicable to money invested for one year, can be found by solving **Vdis**$([1 + R1, 0, 0]) = 1$. The reader may wish to try and solve for some of the R_i's and compare them to the values reported by **NarbitB**.

It is customary to report interest rates for different terms in such a way that they can be compared. Obviously, $R2$ is a larger number than $R1$ since the former represents the rate earned over the longer period of time [0, 2] for which the first period [0, 1] is a subperiod. If this were not the case, arbitrage opportunities would exist. See the exercises at the end of this Chapter.

In order to report the rates in a comparable manner, the rates are reported (usually) on a per annum basis. It is therefore natural to measure the time in units of a year. The spot rate for time n (n years from now) is reported as the annual rate which, if compounded semiannually, will produce the rate applicable to the period $[0, n]$. Hence, the spot rate applicable to time n is the solution r_n to equation (9.2),

$$d_n = \frac{1}{\left(1 + \frac{r_n}{2}\right)^{2\,n}}. \tag{9.2}$$

The use of the annual rate (and semiannual compounding) gives rise to the term $\frac{r_n}{2}$, and since time is measured in units of a year, the relevant exponent is $2n$. This is where special care should be given to the assumption regarding the length of the time intervals [0, 1], [1, 2], and [2, 3] that are used in the input to **NarbitB**.[1] Thus, for example, the spot rate for time one (a year in our example), r_1, is given by

[1] The procedure **NarbitB** assumes that time 1 is the time of the first payment and that the payments are equally spaced. Thus time 2 is the time of the second payment,

```
> solve(d[1]=1/(1+r[1]/2)^2,r[1]);
```

$$\frac{-2\,d_1 + 2\sqrt{d_1}}{d_1},\ \frac{-2\,d_1 - 2\sqrt{d_1}}{d_1}$$

As we see, MAPLE reports two solutions and we choose the positive one, $2\frac{\sqrt{d_1}}{d_1} - 2$. In our example the value of the spot rate is obtained by substituting $d_1 = 0.9$ in the chosen solution.

```
> subs(d[1]=0.9, -2+2/sqrt(d[1]));
```

$$.1081851067$$

In the same manner, r_2 will be given by

```
> solve(d[2]=1/(1+r[2]/2)^4,r[2]);
```

$$2\frac{1}{d_2{}^{(1/4)}} - 2,\ 2\frac{I}{d_2{}^{(1/4)}} - 2,\ -2\frac{1}{d_2{}^{(1/4)}} - 2,\ -2\frac{I}{d_2{}^{(1/4)}} - 2$$

We choose again the positive real solution $2d_2^{-\frac{1}{4}} - 2$ and substitute $d_2 = 0.8$ in it to get the value of $r2$ in our example.

```
> subs(d[2]=0.8, -2+2*d[2]^(-1/4));
```

$$.1147425275$$

The discount factor for time t is the present value (price) of one dollar obtainable at time t, in as much as in the case of the equity market the price of one dollar contingent on state one was d_1. Hence, the discount factor for time three, for example, will be given by

```
> evalf(Vdis([0,0,1]));
```

$$.8132609031$$

We can therefore solve numerically for r_3 by combining the **solve** command with the value of d_3 given in terms of **Vdis** as below.

```
> solve({evalf(Vdis([0,0,1]))=1/(1+r[3]/2)^(6),\
> r[3]>=0,r[3]<=1});
```

$$\{r_3 = .07010169007\}$$

---

and time k is the time of the kth payment. Consequently, the output of **NarbitB,** when reporting a rate spanning the time interval $[0, k]$, pertains to the time interval from the current time to that time of the kth payment. It is therefore important to notice the time period entered into **NarbitB** and the way in which the spot rates are reported. Similarly, when **NarbitB** defines a discount factor function for time t it uses the same units as entered by the user. If the user wishes to change the time unit of the function, this should be done manually by scaling the function. For further help on **NarbitB** see the help page on **NarbitB**.

If the length of the time intervals $[0, 1], [1, 2]$, etc., used in **NarbitB** is half a year, the spot rate applicable to the time interval $[0, n]$ will be the solution r_n to the equation $d_n = \left(\frac{1}{1+r_n}\right)^n$. In that case, the spot rate for one year is r_2.

Continuing in this manner, we obtain a schedule, a structure of spot rates for each payment date in the future. This structure is referred to as the *term structure of interest rates* or, in short, the *term structure.* In general, the spot rate r_n spanning the time interval $[0, n]$ based on compounding m times per unit of time will be the solution to

$$d_n = \left(\frac{1}{1+\frac{r_n}{m}}\right)^{nm}.$$

Thus, if the unit of time is half a year and the rates are reported per annum based on semiannual compounding, a dollar invested from time zero to time three is invested for 1.5 years and will grow to $\left(1+\frac{r_3}{2}\right)^{2(1.5)} = \left(1+\frac{r_3}{2}\right)^3$.

9.1.1 Zero-Coupon, Spot, and Yield Curves

The Term Structure, Zero-Coupon, and Spot Curves

The term structure is actually a function, which we will denote by $r(t)$, that relates time to the interest rate prevailing in the market. For each t, $r(t)$ is the interest rate per period paid for a dollar invested from time zero to time t. **We keep our units of time measured in years and report the interest rate based on semiannual compounding.** Hence, $r(t)$ is the rate such that a dollar invested from time zero to time t will grow to be

$$\left(1+\frac{r(t)}{2}\right)^{2t}$$

at time t. This function $r(t)$ is referred to as the *term structure of interest rates,* and for each t the rate $r(t)$ is called a *spot rate.*

These are the rates paid on an investment over a period that starts now (on the spot) and ends at some future time. The graph of this function is sometimes referred to as the *spot curve* or the *zero-coupon curve*. In what follows, we justify this name followed by a graphical representation of the term structure which will be explored shortly afterward.

Certain bonds pay only one payment at the maturity time of the bond. These bonds are referred to as *zero-coupon bonds*, since they make no coupon payments. Furthermore, zero-coupon bonds are sold at a discount. The interest rate offered by such bonds is implicit in the difference between the face value of the bond and its market price. Namely, if the price of a zero-coupon bond maturing at time t, t years from now, is P and its face value is \$1, the interest it pays is the solution r to the equation

$$P = \left(1+\frac{r}{2}\right)^{-2t}. \qquad (9.3)$$

Hence r will be given by

```
> solve(P=(1+r/2)^(-2t),r);
```

$$2\,e^{\left(-\frac{\ln(P)}{t}\right)} - 2$$

where we assume again that t is measured in years and compounding is done semiannually.

The zero-coupon bonds are bonds which repay the principal at some time t and pay nothing at any other time. The zero-coupon bonds, therefore, are the counterparts of the elementary securities of the equity market. In exactly the same manner that in the equity market risky securities are interpreted as a portfolio of elementary securities, coupon bonds can be interpreted as a portfolio of zero-coupon bonds.

Consider a bond that pays a coupon of $\frac{c}{2}$ at times $t_1, t_2, ..., t_{n-1}$, and matures at time t_n where it pays $100 + \frac{c}{2}$. This bond may be replicated by a portfolio of zero-coupon bonds each with a face value of \$1. A portfolio composed of $\frac{c}{2}$ units of zero-coupon bonds maturing at times $t_1, t_2, ..., t_{n-1}$, and $100 + \frac{c}{2}$ of a zero-coupon bond which matures at time t_n has the same cash flow as the coupon-paying bond.

If the principal of a zero-coupon bond maturing at time t is \$1, the price of the bond is the discount factor for time t. Hence, the spot rates extracted from coupon-paying bonds or those extracted from zero-coupon bonds, which are the building blocks of the coupon bonds, will be the same. Thus, the term structure of interest rates is sometimes referred to as the *zero-coupon curve*.

Let us consider an example in order to demonstrate this idea. We will start with a market of coupon bonds and estimate the discount factors. Then we will replace the bonds in this market by zero-coupon bonds with face values of \$1. The prices we will assign to these zero-coupon bonds will be the discount factors obtained in the first market. We then reestimate the discount factors. Of course, the discount factors in these two markets should be the same.

Let us modify slightly the first example so the coupon offered by the bonds now is exactly half the coupon offered, and the price of bond one is \$94 instead of \$94.5. Hence, we get a market as specified in **NarbitB** below, where executing **NarbitB** confirms that the no-arbitrage condition is satisfied, and the discount factors and the function **Vdis** are solved for.

```
> NarbitB([[105,0,0],[5,105,0],[4,4,104]],[94,9 7,89]);
```

The no − arbitrage condition is satisfied.

$$\textit{The discount factor for time, } 1, \textit{ is given by, } \frac{94}{105}$$

The interest rate spanning the time intreval, [0, 1], *is given by*, .1170

$$\textit{The discount factor for time}, 2, \textit{is given by}, \frac{1943}{2205}$$

The interest rate spanning the time intreval, [0, 2], *is given by*, .1348

$$\textit{The discount factor for time}, 3, \textit{is given by}, \frac{180577}{229320}$$

The interest rate spanning the time intreval, [0, 3], *is given by*, .2699

The function Vdis([*c1*, *c2*, ..]), *values the cashflow* [*c1*, *c2*, ..]

The discount factors are specified above, but of course the discount factor for time one will be the value of a zero-coupon bond maturing at time one and paying $1. Hence the discount factor for time one is also given by **Vdis**([1, 0, 0]).

```
> Vdis([1,0,0]);
```

$$\frac{94}{105}$$

In the same manner, the discount factors for times two and three will be specified by **Vdis**([0, 1, 0]) and **Vdis**([0, 0, 1]), respectively. Let us now re-run **NarbitB** on a market composed of only these three zero-coupon bonds. The prices are specified by **Vdis** and applied to the corresponding zero-coupon bond.

```
> NarbitB([[1,0,0],[0,1,0],[0,0,1]],\
> [Vdis([1,0,0]), Vdis([0,1,0]),Vdis([0,0,1])]);
```

The no − arbitrage condition is satisfied.

$$\textit{The discount factor for time}, 1, \textit{is given by}, \frac{94}{105}$$

The interest rate spanning the time intreval, [0, 1], *is given by*, .1170

$$\textit{The discount factor for time}, 2, \textit{is given by}, \frac{1943}{2205}$$

The interest rate spanning the time intreval, [0, 2], *is given by*, .1348

$$\textit{The discount factor for time}, 3, \textit{is given by}, \frac{180577}{229320}$$

The interest rate spanning the time intreval, [0, 3], *is given by*, .2699

The function Vdis([*c1*, *c2*, ..]), *values the cashflow* [*c1*, *c2*, ..]

We indeed confirm that the resultant market and discount factors are the same as those in the first market. This should come as no surprise to the reader. We have already seen a similar example at work in the equity market case.

If we replace the primary securities in the market by securities which are replicable by those primary securities and attach to the new securities the prices of the replicating portfolios, then the same market is obtained. The reader will recall the example from Section 2.1 where the risk-free security was removed from an equity market and another security (replicated by the primary securities) was added in its place. When the risk-free rate in the new market was calculated, it turned out to be identical to the one removed.

Yield and Spot Curves

The financial press reports a measure of a bond's return called a *yield*. The yield of a bond, maturing in n years, is defined as the rate y that solves the equation

$$\sum_{i=1}^{2n} \frac{\frac{c}{2}}{\left(1+\frac{y}{2}\right)^{i}} + \frac{FC}{\left(1+\frac{y}{2}\right)^{2n}} = P, \tag{9.4}$$

where c is the coupon of the bond, FC is its face value, and P is its price. Therefore the yield of a bond is a rate such that if payments from the bond obtained in i years are discounted to its present value by $\frac{1}{\left(1+\frac{y}{2}\right)^{2i}}$, the price of the bond equals its present value. Readers who are already familiar with the concept of *internal rate of return* will immediately recognize that the yield is nothing but the same. From equation (9.3) it follows that the yield of a zero-coupon bond, maturing at time t and having a face value of \$1, is the spot rate for time t. Hence, if the market is populated only with such zero-coupon bonds, the concept of a yield would be equivalent to that of a spot rate. Let us calculate the yield for some zero-coupon bonds of the examples used above.

Consider the bond with the semiannual cash flow of $(0, 0, 1)$. Since, for this bond, $c = 0$ and $FC = 1$, equation (9.4) reduces to equation (9.3). Hence, we can solve for the yield of this bond by submitting to MAPLE the command below.

```
> solve(Vdis([0,0,1])=(1+y/2)^(-3));
```

$$\frac{4}{180577}\,934709837209785^{1/3} - 2,$$
$$-\frac{2}{180577}\,934709837209785^{1/3} - 2 + \frac{2}{180577}\,I\,\sqrt{3}\,934709837209785^{1/3},$$

$$-\frac{2}{180577}\,934709837209785^{1/3} - 2 - \frac{2}{180577}\,I\,\sqrt{3}\,934709837209785^{1/3}$$

As we can see, there are multiple possible solutions to this equation, and not all solutions are positive numbers or even real numbers. (The I is for the imaginary number.) The next line chooses the solutions which are both real and positive, and the subsequent line calculates the solutions in decimal form.

```
> map(proc (x) if type(x,'realcons') and evalf(0 <= x) \
> then x else NULL fi end,[%]);
```

$$[\frac{4}{180577}\,934709837209785^{1/3} - 2]$$

If no positive solution exists, MAPLE will return a NULL and in such a case there is no point in continuing to the next line and executing its command.

```
> evalf(op(%));
```

$$.165824001$$

To compare, and to confirm that this is the same result as that obtained from **NarbitB**, we can calculate the interest rate which spans the time interval $[0, 1.5]$. This is done below.

```
> (1+%/2)^3-1;
```

$$.269929173$$

We confirm that the results are those reported by **NarbitB**, albeit 1.5 years is three units of half a year and thus reported as time three in **NarbitB**. The yield of a zero-coupon bond coincides with the spot rate for the length of time corresponding to the maturity of the bond.

A different result is obtained for the case of yields of coupon-paying bonds. We examine now the yield of the bond maturing in a year with a cash flow of $(5, 105, 0)$ and a current price of \$97. To solve for the yield of this bond, equation (9.4) is solved, using values of $\frac{c}{2} = 5$ and $FC = 100$. Again, we submit this to MAPLE and choose the real and positive solutions, if they exist.

```
> solve(5/(1+y/2)+105/(1+y/2)^2=97);
```

$$-\frac{189}{97} + \frac{1}{97}\sqrt{40765},\ -\frac{189}{97} - \frac{1}{97}\sqrt{40765}$$

```
> map(proc (x) if type(x,'realcons') and evalf(0<= x) \
> then x else NULL fi end,[%]);
```

$$[-\frac{189}{97} + \frac{1}{97}\sqrt{40765}]$$

```
> evalf(op(%));
```

$$.133025178$$

This time if we solve for the one-year spot interest rate, based on this yield, we obtain

```
> (1+%/2)^2-1;
```

$$.137449103$$

while, on the other hand, the result from **NarbitB** is 0.1348. This is not due to round-off error. The reader can perform the conversion of the two numbers using MAPLE as below. The exact solution for the two-period spot rate is given by

```
> solve(Vdis([0,1,0])=(1+r/2)^(-2));
```

$$-2 + \frac{42}{1943}\sqrt{9715},\ -2 - \frac{42}{1943}\sqrt{9715}$$

The positive solution can now be compared to the positive solution obtained for the yield of the bond. One can simply copy and paste it, and then ask MAPLE to compare the two solutions.

```
> is(-189/97+(1/97)*40765^(1/2)=-2+(42/1943)*97 15^(1/2));
```

$$\mathit{false}$$

This example serves to demonstrate a few key points. The yields of two bonds which mature at the same time may be different even though their maturity dates coincide. It also highlights that the yield measure suffers from some deficiencies. The price of a bond is the discounted value of its future cash flows. This is a consequence of the no-arbitrage condition. Hence, the contribution of each future payment to the price of the bond is the present value of that payment.

The price of a sure dollar to be obtained in a future time period is independent of the bond from which it is to be obtained. The bonds, as explained above, can all be constructed from the same basic building blocks: one dollar obtained at time i. These building blocks must therefore also have the same value in the current time period, regardless of the bonds they are used in. The yield concept, however, is a result of using different discount factors for payments obtained from different bonds at the same time.

As was demonstrated by the example, a dollar obtained from the coupon bond with a maturity of one year was discounted using the yield of the bond y, which was different from the one-year spot rate. There is no economic rationale for using different discount factors for a dollar which has the same risk characteristics just because it is obtained from a different financial instrument, i.e., a zero-coupon vs. a coupon-paying bond. Hence, the calculation of present values using yields (unless it is a zero-coupon bond) is an incorrect procedure.

One should think about the yield to maturity of a bond as some sort of "average" or mixture of returns on the bond. We know that instead of $\left(\frac{1}{1+\frac{y}{2}}\right)^{2i}$ in equation (9.4) we should have used $\left(\frac{1}{1+\frac{r_i}{2}}\right)^{2i}$, where r_i is the spot rate for i years. Solving for the yield as is done in equation (9.4), we are actually constraining the spot rate to be the same for all time (referred to as a *flat term structure* of interest rates). The y is really some sort of a "mixture" of the different spot rates r_i.

We demonstrate this below by asking MAPLE to solve for y in terms of r_1 and r_2. Given a bond which matures in one year, pays a coupon of $\frac{c}{2}$, and has a face value of FC, we display the first of the solutions produced (this is the meaning of the "[1]" below).

```
>   assume (r1>0,r2>0,c>0,FC>0);

>   simplify(solve((c/2)/(1+r0.5)+(FC+(c/2))/(1+r1/2)^2=\
>   (c/2)/(1+y/2)+(FC+(c/2))/(1+y/2)^2,y)[1]);\
```

$$\begin{aligned}
&-(16\,FC\,r05 + 4\,c\,r05 + 16\,FC + c\,r1^2 + 4\,c\,r1 + 12\,c - 4\,c\,r05\,r1 \\
&\quad - c\,r05\,r1^2 - 2\sqrt{1+r05}\mathrm{sqrt}(c^2\,r05\,r1^2 + 8\,FC\,c\,r1^2 \\
&\quad + 5\,c^2\,r1^2 + 4\,c^2\,r05\,r1 + 32\,FC\,c\,r1 + 20\,c^2\,r1 + 20\,c^2\,r05 \\
&\quad + 64\,FC\,c\,r05 + 64\,FC^2\,r05 + 64\,FC^2 + 96\,FC\,c + 36\,c^2) \\
&\quad - \sqrt{1+r05}\mathrm{sqrt}(c^2\,r05\,r1^2 + 8\,FC\,c\,r1^2 + 5\,c^2\,r1^2 \\
&\quad + 4\,c^2\,r05\,r1 + 32\,FC\,c\,r1 + 20\,c^2\,r1 + 20\,c^2\,r05 \\
&\quad + 64\,FC\,c\,r05 + 64\,FC^2\,r05 + 64\,FC^2 + 96\,FC\,c + 36\,c^2) \\
&\quad r1)/(8\,FC\,r05 + c\,r1^2 + 8\,c + 8\,FC + 4\,c\,r1 + 4\,c\,r05)
\end{aligned}$$

The next section elaborates on the term structure of interest rates and on its estimation.

9.2 Smoothing of the Term Structure

Indeed one can think of the term structure as a function from the time domain to the interest rate domain. A realistic market requires the knowledge of the discount factors and of the spot rates not only for times one, two, and three, as in our examples, but for every possible t in some interval $[0, T]$. Sometimes the value of $r(t)$ is required even for a time period which extends beyond the maximum maturity, time three in our example. The common practice in the marketplace is to use the knowledge of the spot rate of interest for the discrete points (1, 2, and 3 in our examples) and generate from these an estimation of a discount factor or a spot rate for each t in a certain time interval. This process is referred to as *smoothing*.

In the scope of this book, we will give only the gist (overview) of these techniques. The Appendix will explore these ideas further. Since spot rates are recovered from discount factors, and vice versa, the smoothing can be done in terms of either of these rates. Our discussion will be done with respect to discount factors. Equation (9.2) shows how one determines the other.

The general form of the smoothing techniques is based on the assumption that the discount factor for time t, $d(t)$, is a function which can be approximated by some structure. Usually, the structure, $d_{es}(t)$, which is adopted is of the form

$$d_{es}(t) = \sum_{i=0}^{K} \alpha_i f_i(t), \tag{9.5}$$

where the f_i are certain functions and the α_i are numerical coefficients. Hence an assumption is made that the discount factor function can be written in such a form. The different smoothing techniques are based on the choices of the approximating functions.

The first step is to choose the set of $K+1$ functions $f_0, ..., f_K$. Next, an optimization problem is solved in order to uncover the values of the α_i such that the present value of the cash flow from bond i, $\sum_{t=1}^{T_i} d_{es}(t)a_{it}$, where a_{it} is the payment from bond i at time t, is as close as possible to P_i, the market price of bond i. To this end, one must choose how to measure the distance between the present value and the market price of the bond. Here again, methods differ from one another. Some selection criteria are based on the sum of squared differences and some on the sum of the absolute value of the differences.

The procedure **NarbitB** is also capable of producing the smoothing. It uses the criteria of absolute values for distance. It is sometimes possible to think of and display the smoothing process as being done in two phases. In the first phase the discount factors for the discrete payment dates of the bonds are identified. In the second phase an optimization problem is solved. It uncovers the values of the α_i such that the value of the continuous curve $d_{es}(t)$ is as close as possible to the value of the discount factors at the payment dates. Whenever this is possible, **NarbitB** will demonstrate it with a graph. The reader will find more information about these techniques and the specification of the functions $f_i(t)$ in the Appendix.

Let us examine now an example of smoothing the discount factors. *For simplicity we assume a market in which coupons are paid annually. Hence in* **NarbitB** *and its output the time unit is a year.* In order to accomplish this, we need to run the procedure **NarbitB** as before, but with the addition of two parameters. We need to specify the number of approximating functions and the name which will be assigned to our new discount factors function. This function is termed the *continuous approximation of the discount factors.*

Suppose we would like to have six approximating functions, and we would like to call the continuous approximation **ConApp**. We thus need to run **NarbitB** as is demonstrated below. The output of the procedure also generates a graph. This is illustrated in Figure 9.1, which, in the hard copy version of this text, does not appear directly following the output. (You can add a zero as a parameter at the end to suppress the graph of the discount factor.)

```
> NarbitB([[105,0,0],[10,110,0],[8,8,108]],[94,97,85], \
> 6,ConApp);
```

$$\mathit{The\ no-arbitrage\ condition\ is\ satisfied.}$$

$$\mathit{The\ discount\ factor\ for\ time},\ 1,\ \mathit{is\ given\ by},\ \frac{94}{105}$$

$$\mathit{The\ interest\ rate\ spanning\ the\ time\ interval},\ [0,\ 1],\ \mathit{is\ given\ by},\ .1170$$

$$\mathit{The\ discount\ factor\ for\ time},\ 2,\ \mathit{is\ given\ by},\ \frac{1849}{2310}$$

$$\mathit{The\ interest\ rate\ spanning\ the\ time\ interval},\ [0,\ 2],\ \mathit{is\ given\ by},\ .2493$$

$$\mathit{The\ discount\ factor\ for\ time},\ 3,\ \mathit{is\ given\ by},\ \frac{82507}{124740}$$

$$\mathit{The\ interest\ rate\ spanning\ the\ time\ interval},\ [0,\ 3],\ \mathit{is\ given\ by},\ .5119$$

$$\mathit{The\ function\ Vdis}([c1,c2,..]),\ \mathit{values\ the\ cashflow}\ [c1,c2,..]$$

$$\mathit{The\ continuous\ discount\ factor\ is\ given\ by\ the\ function},\ '\mathit{ConApp}',\ (.)$$

Figure 9.1 demonstrates the smoothing process. The (blue) boxes are the values of the discount factors for the times at which the bonds are making coupon payments or are repaying the face value. The continuous graph is the result of our approximation. The function **ConApp** estimates the discount factor associated with every t in the interval $[0, 3]$. It is being used[2] to value cash flows payable at times other than 1, 2, or 3. Hence, to value $4 obtainable at time 1.75, we perform the following calculation:

```
> ConApp(1.75)*4;
```

$$3.328855996$$

Similarly, to value the cash flow $4 at time 1.75, $4 at time 2.25, and $100 at time 2.75, we calculate

```
> ConApp(1.75)*4+ConApp(2.25)*4+ConApp(2.75)*100;
```

$$77.21438618$$

In these last examples, we see that the graph passes through the value of the discount factors obtained directly from the market. Hence, in this case we manage to fit the term structure estimation defined by equation (9.5) to coincide with the discrete factors obtained from the market. This, of course, is not always possible and will depend on the type and number of approximating functions chosen for the approximation. Suppose we attempt to do the same fitting with only three approximation functions. Examine the resultant discount factor function, graphed in Figure 9.2.

```
> NarbitB([[105,0,0],[10,110,0],[8,8,108]],[94,97,85], \
> 3,ConApp1);
```

$$\textit{The no}-\textit{arbitrage condition is satisfied.}$$

$$\textit{The discount factor for time}, 1, \textit{is given by}, \frac{94}{105}$$

$$\textit{The interest rate spanning the time interval}, [0,\ 1], \textit{is given by}, .1170$$

$$\textit{The discount factor for time}, 2, \textit{is given by}, \frac{1849}{2310}$$

$$\textit{The interest rate spanning the time interval}, [0,\ 2], \textit{is given by}, .2493$$

$$\textit{The discount factor for time}, 3, \textit{is given by}, \frac{82507}{124740}$$

$$\textit{The interest rate spanning the time interval}, [0,\ 3], \textit{is given by}, .5119$$

---

[2] By interpolation, it is possible to obtain an estimation for discount factors beyond $t = 3$ as well. To this end, one needs to specify the range to which the function should be extrapolated. The procedure **NarbitB** is also capable of producing the output without the graph. See also the Help for **NarbitB**. This is done when a fifth parameter is added to the input. This added parameter can be either zero for surpassing the graph or one to produce the graph. The default value of this parameter is one. To produce the extrapolation, a sixth parameter must be added specifying the right end point of the range. However, in this case the fifth parameter must be specified. Hence to extrapolate the example in the text for 6.5 years with the graph one should run

```
NarbitB([[105,0,0],[10,110,0],[8,8,108]],[94.5,97,89],6,ConApp,1,6.5);
```

To produce the extrapolation without a graph one should run

```
NarbitB([[105,0,0],[10,110,0],[8,8,108]],[94.5,97,89],6,ConApp,0,6.5);
```

$$The\ function\ Vdis([c1, c2, ..]),\ values\ the\ cashflow\ [c1, c2, ..]$$
$$The\ continuous\ discount\ factor\ is\ given\ by\ the\ function,\ 'ConApp1',\ (.)$$

It is now apparent from Figure 9.2 that we do not have a perfect fit. The procedure **NarbitB** also defines a local variable **SumAbsDiv.** It measures the "goodness-of-fit" in terms of the sum of the absolute deviations of the approximation from the actual observations. Thus, if the value of this variable is zero the fit is perfect. The Appendix and the Help for **NarbitB** elaborate on this point. Each time **NarbitB** is run and a continuous approximation of the term structure is solved for, the variable **SumAbsDiv** is redefined. To check its value issue the following command:

```
> SumAbsDiv;
```

$$\frac{464839}{111363}$$

Sometimes a perfect fit is not possible and we have to make do with the best approximation possible. In fact, the situation in real-world markets is somewhat more complicated. We highlight some of these features in the Appendix.

As mentioned before, once we have the continuous approximation of the discount factors we can generate the continuous approximation of the term structure of interest rates. Let us try now to run an example, using only five approximating functions.

```
> NarbitB([[105,0,0],[10,110,0],[8,8,108]],[94,97,85], \
> 5,ConApp2);
```

$$The\ no-arbitrage\ condition\ is\ satisfied.$$
$$The\ discount\ factor\ for\ time,\ 1,\ is\ given\ by,\ \frac{94}{105}$$
$$The\ interest\ rate\ spanning\ the\ time\ interval,\ [0,\ 1],\ is\ given\ by,\ .1170$$
$$The\ discount\ factor\ for\ time,\ 2,\ is\ given\ by,\ \frac{1849}{2310}$$
$$The\ interest\ rate\ spanning\ the\ time\ interval,\ [0,\ 2],\ is\ given\ by,\ .2493$$
$$The\ discount\ factor\ for\ time,\ 3,\ is\ given\ by,\ \frac{82507}{124740}$$
$$The\ interest\ rate\ spanning\ the\ time\ interval,\ [0,\ 3],\ is\ given\ by,\ .5119$$
$$The\ function\ Vdis([c1, c2, ..]),\ values\ the\ cashflow\ [c1, c2, ..]$$
$$The\ continuous\ discount\ factor\ is\ given\ by\ the\ function,\ 'ConApp2',\ (.)$$

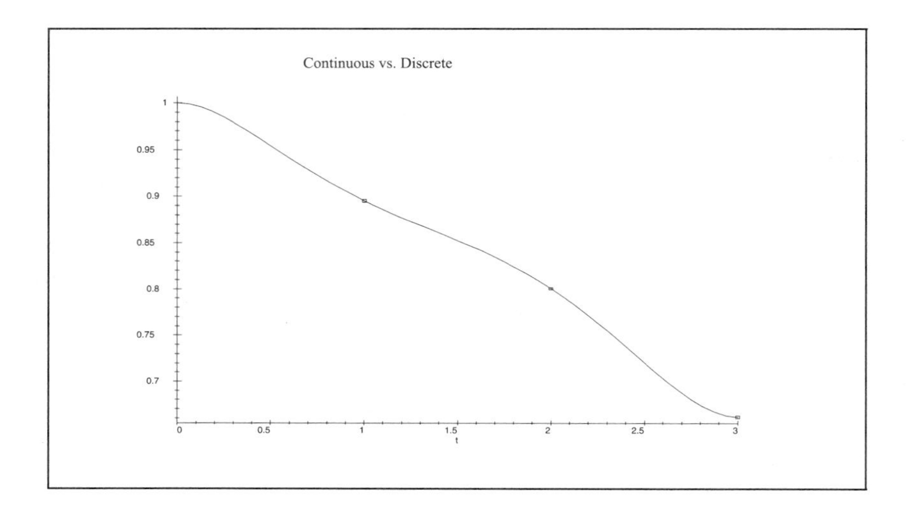

Figure 9.1: The Graph of the Function **ConApp**

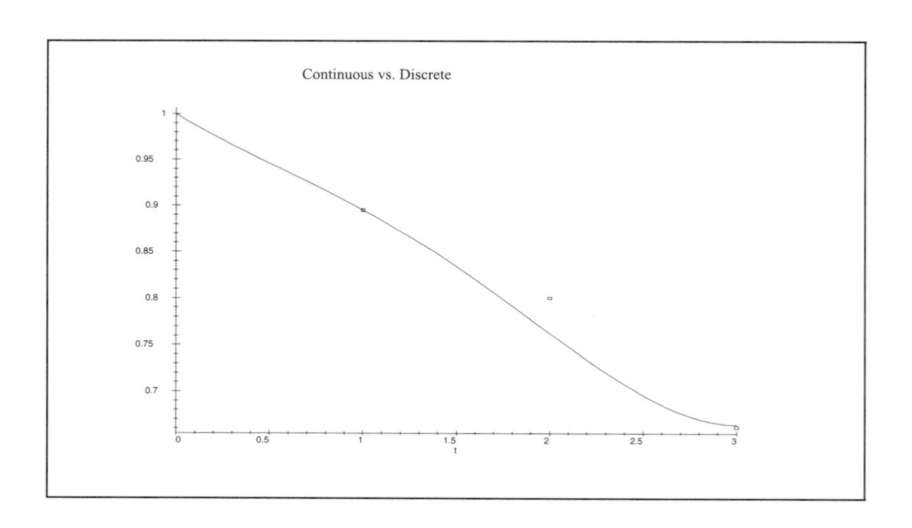

Figure 9.2: The Graph of the Function **ConApp1**

```
> SumAbsDiv;
```

$$0$$

As we see in this example (Figure 9.3), a perfect fit was possible.[3] The relation between the discount factors and the interest rates is given by equation (9.2). We use the MAPLE **solve** and **unapply** commands to find and define the term structure function $r(t)$ by substitution of **ConApp2**(t) for $d(t)$ in (9.2). Specifically, the spot rate for time t, $r(t)$ will be given by the function $r(t)$ defined below:

```
> r:=unapply(solve(op(2,ConApp2(t))= \
> (1/(1+(r/2)))^(2*t),r),t);
```

$$r := t \to -2\,\frac{e^{\left(1/2\,\frac{\ln(1-\frac{193}{1320}\,t+\frac{105733}{3129840}\,t^2+\frac{1868729}{77463540}\,t^3-\frac{1633507}{103284720}\,t^4-\frac{6467}{3521070}\,t^5+\frac{26}{21735}\,t^6)}{t}\right)}-1}{e^{\left(1/2\,\frac{\ln(1-\frac{193}{1320}\,t+\frac{105733}{3129840}\,t^2+\frac{1868729}{77463540}\,t^3-\frac{1633507}{103284720}\,t^4-\frac{6467}{3521070}\,t^5+\frac{26}{21735}\,t^6)}{t}\right)}}$$

We can now plot this function, Figure 9.4, and see how the spot rate in this market behaves.

```
> plot(r(t),t=0..3,labels=[Time,`Int.Rate`],thickness=2,\
> title=`A Continuous Approximation of a Term Structure`);
```

The fact that the rates are reported on an annual basis allows us to compare the rates for different maturities. We see that the rates are decreasing as time increases and then, at about 1.5 years, the rates increase. Different shapes of the term structure are possible. There are a few theories which explain the shape of the term structure of interest rates. These theories are summarized in the Appendix.

9.2.1 Smoothing and Continuous Compounding

Modern finance uses continuous-time models of the term structure as opposed to the discrete-time models which have been discussed thus far. It is therefore customary and convenient to treat the compounding period as continuous, i.e., every instant of time as in Section 3.4.1, rather than semiannually. As a consequence, the term structure of interest rates and its estimation are also often reported based on the continuous compounding assumption. Thus, another step toward generalization of the discrete-period simple model of Chapter 1 to a more realistic continuous-time model has been taken.

---

[3] In certain cases multiple solutions exist for the approximation. Hence running the same procedure with the same parameters may result in a different answer.

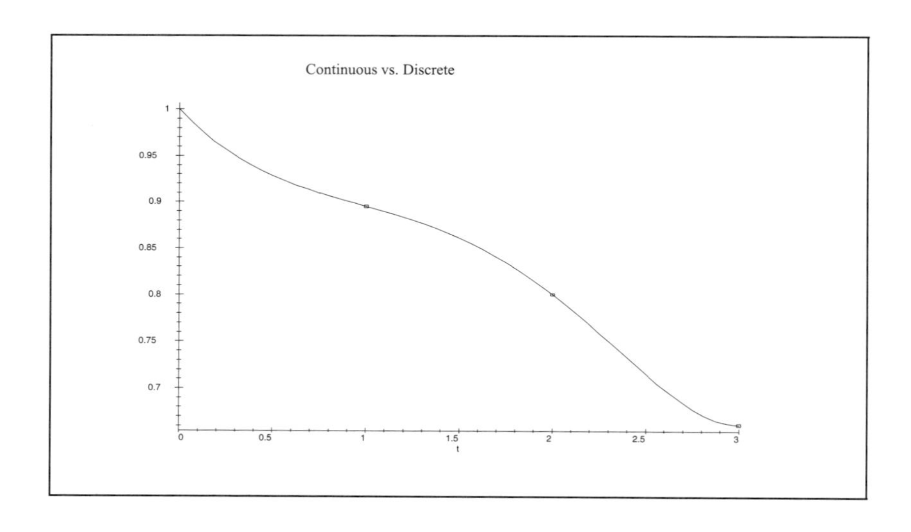

Figure 9.3: The Graph of the Function **ConApp2**

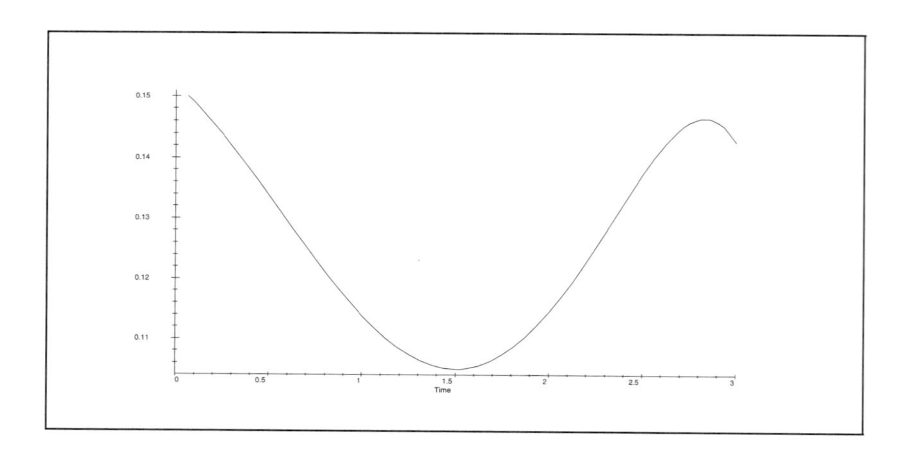

Figure 9.4: A Continuous Approximation of a Term Structure

If one is given the function $r(\cdot)$, based on equation (9.2), the value of \$1 obtained in t years is

$$\left(\frac{1}{1+\frac{r(t)}{2}}\right)^{2t}. \tag{9.6}$$

Using equation (3.23) we can solve for the continuously compounded spot rate of interest. Given the discount factor for time t, $d(t)$ (we will measure time in years henceforth), the continuously compounded rate $r(t)$ is the solution to equation (9.7),

$$d(t) = e^{-r(t)t}. \tag{9.7}$$

Hence $r(t)$ is given by

$$r(t) = -\frac{\ln(d(t))}{t}. \tag{9.8}$$

Previously we calculated the discount factor function, **ConApp2**, when the units of time were measured in years. Therefore, the continuous approximation of the term structure based on continuous compounding will be given by $-\frac{ln(\mathbf{ConApp2}(t))}{t}$. Let us define the function **Ts** to be the continuously compounded term structure in our example.

```
> Ts:=unapply(-ln(op(2,ConApp2(t)))/t,t);
```

$$Ts := t \to -\ln(1 - \frac{193}{1320}\,t + \frac{105733}{3129840}\,t^2 + \frac{1868729}{77463540}\,t^3 - \frac{1633507}{103284720}\,t^4 - \frac{6467}{3521070}\,t^5 + \frac{26}{21735}\,t^6)/t$$

The graphs of the continuous approximation of the term structure based on semiannual compounding (the blue upper graph) and based on continuous compounding (the green lower graph) are plotted together in Figure 9.5.

```
> plot([r(t),Ts(t)],t=0..3,color=[blue,green], \
> thickness=2, labels=[ 'Time in Years','Int.  Rate']);
```

The discount for time t in terms of **Ts** is given by the expression $e^{-Ts(t)\,t}$ and is plotted in Figure 9.6. Note that indeed the discount factor is a decreasing function for which the range is between zero and one — as it should be.

```
> plot(exp(-Ts(t)*t),t=0..3,color=[blue],thickness=2, \
> labels=[ 'Time in Years','Int.  Rate']);
```

Note that we have not discussed or specified the evolution of the term structure across time. Rather, this Chapter has been devoted to the term structure or set of discount factors that prevail in the market at a certain

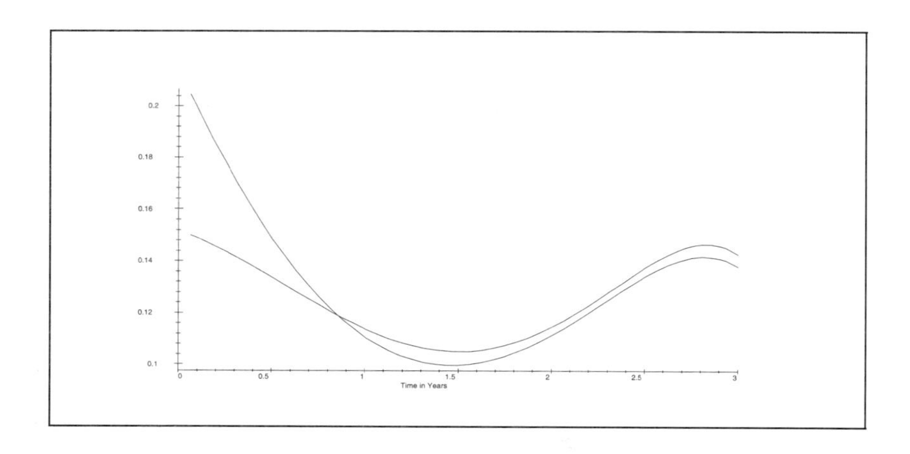

Figure 9.5: Continuous Approximation of the Term Structure: Semiannual Compounding vs. Continuous Compounding

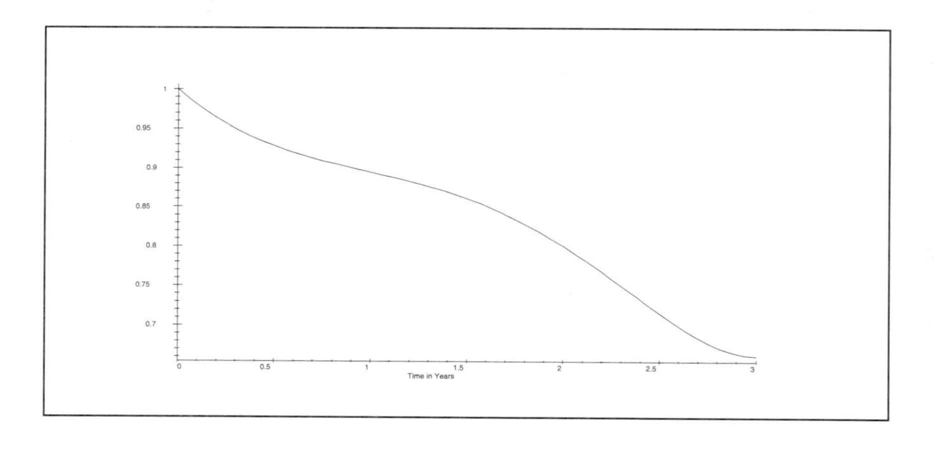

Figure 9.6: The Discount Factor Function Implied by the Term Structure **Ts**

point. It is as if we take a "snapshot" of bond prices in the market at a certain moment, and extract the term structure prevailing in the market at that time. As time passes, the term structure may, and usually does, change. Many valuation methods, as we will see later, require some specification of a model governing the evolution of the term structure.

Yet, there are many instruments and types of cash flows that could be valued given only the current discount factor $d(t)$. The rest of this Chapter is devoted to two such examples. Some other instruments of these types are valued in the next Chapter. Not surprisingly, most of the cash flows we are able to value in this manner (and their replicating strategies) are static. They do not require (optimal) actions in the future and the valued cash flows are fixed and known as of the current time. These cash flows do not include any random or stochastic components that depend on a realization of some future interest rates that are not known at the present time. There is, however, an exception to this rule and this is the cash flow valued in our second example in this Chapter.

9.3 Forward Rate

We offer two explanations for the concept of forward rate: a classical explanation that is more theoretical, and a practical one. The former is related to valuation by arbitrage. We therefore utilize the opportunity to review valuation both by replication and via discount factors. We begin our examination of the classical approach to the discussion of forward rates with an example.

9.3.1 Forward Rate: A Classical Approach

Let $r(t)$ be the interest rate which prevails in the market from time zero to time t. For simplicity, we assume that the rates are quoted based on continuous compounding and that time is measured in years. Hence, a dollar invested at time zero for one year will grow into $e^{r(1)1}$ and a dollar invested at time zero for two years will grow into $e^{r(2)2}$.

There is another way to invest money from time zero to time two. The dollar could be invested from time zero to time one, and then invested from time one to time two at the rate that will prevail in the market at that time. At the current time this rate is unknown, as it is a random variable. Under this investment strategy, a dollar invested from time zero to time one will be worth $e^{r(1)}$ at time one. This amount is then invested for one further time period at the rate prevailing in the market at that time. Had the dollar been invested for two time periods at the prevailing rate of interest covering the interval from time zero to time two it would have grown to a value of $e^{2r(2)}$.

Let us denote by $x_1(1,2)$ the (as yet unknown) interest rate from time one to time two. The subindex 1 of x means that this is the interest rate that will prevail at time one and the $(1,2)$ means that this rate is spanning the time interval $[1,2]$. Hence, the dollar at time zero amounts to $e^{r(1)}$ at time one, which grows to $e^{r(1)}e^{x_1(1,2)}$ at the end of two time periods. The value of $r(1,\ 2)$ which solves

$$e^{r(1)}e^{r(1,2)} = e^{2r(2)} \tag{9.9}$$

is the *forward rate* from time one to time two, as of time zero. As has been our custom before, we omit the index for zero and use two indexes in order to denote the span of time for which this forward rate is applicable. That is, we use $r(1,\ 2)$ instead of $r_0(1,\ 2)$ in the same manner we use $r(t)$ instead of $r(0,\ t)$ when the span is $[0,\ t]$.

The intuitive meaning of the forward rate is that it is a rate implicit in the spot rate which will prevail in the market from time one to time two. If an investor perceives that $r(2)$ is not high enough to induce the investor to pursue a two-period investment horizon, then that investor will prefer to use a roll-over strategy. The investor invests for one period from time zero to time one, and then "rolls over" the investment for another period, from time one to time two. The rates prevailing in the market are those which induce an equilibrium: $r(1,2)$ is the value which makes these two strategies equivalent. Hence, these rates convey information about the market's anticipation of future rates which will prevail from time one to time two.

This discussion was based on an example. One can extract the forward rate as of time zero for any future time period t_1 to t_2 $(t_2 > t_1)$. A dollar invested from time zero to time t_1 will grow to be $e^{r(t_1)t_1}$ dollars. If the dollar is invested from time zero to time t_2, it will grow to $e^{r(t_2)\,t_2}$ and $r(t_1,\ t_2)$ will be the solution to

$$e^{r(t_1)t_1}e^{r(t_1,t_2)(t_2-t_1)}{=}e^{r(t_2)\,t_2}. \tag{9.10}$$

We can ask MAPLE to solve for $r(t_1,\ t_2)$ as below:

```
> solve(exp(r(t1)*t1)*exp(r(t1,t2)*(t2-t1))=\
> exp(r(t2)*t2),r(t1,t2));
```

$$\frac{-\mathrm{r}(t1)\ t1 + \mathrm{r}(t2)\ t2}{t2 - t1}.$$

In our example we defined the continuously compounded term structure of interest rates as the function **Ts**(t). We can also now define the function **Fr**(t_1, t_2), based on equation (9.10), as the forward rate from time t_1 to time t_2.

```
> Fr:=(t1,t2)->(Ts(t2)*t2-Ts(t1)*t1)/(t2-t1);
```

$$Fr := (t1,\, t2) \rightarrow \frac{\mathrm{Ts}(t2)\, t2 - \mathrm{Ts}(t1)\, t1}{t2 - t1}$$

The function **Fr** should always have the second argument t_2 greater than the first argument t_1.

Armed with the function so defined, we can plot it together with the term structure and see how the term structure is related to the forward rate. In Figure 9.7 the forward rate (the red graph with the lower y-intercept) is graphed together with the term structure (the green graph with the higher y-intercept), where the time domain is 0.5 to 3 and the forward rate is graphed from time 0.5 to any time t in the interval $[0.5, 3]$.

```
> plot([Ts(t),Fr(.5,t)],t=0.5...1.5,color=[green,red],\
> thickness=2,labels=[Time,Rate],title='Figure 9.7:\
> The Forward Rate and the Term Structure');
```

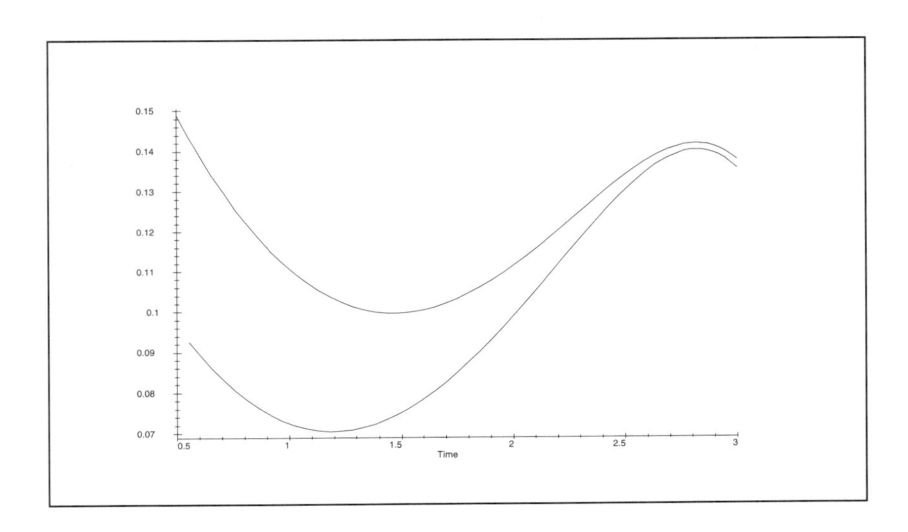

Figure 9.7: The Forward Rate and the Term Structure

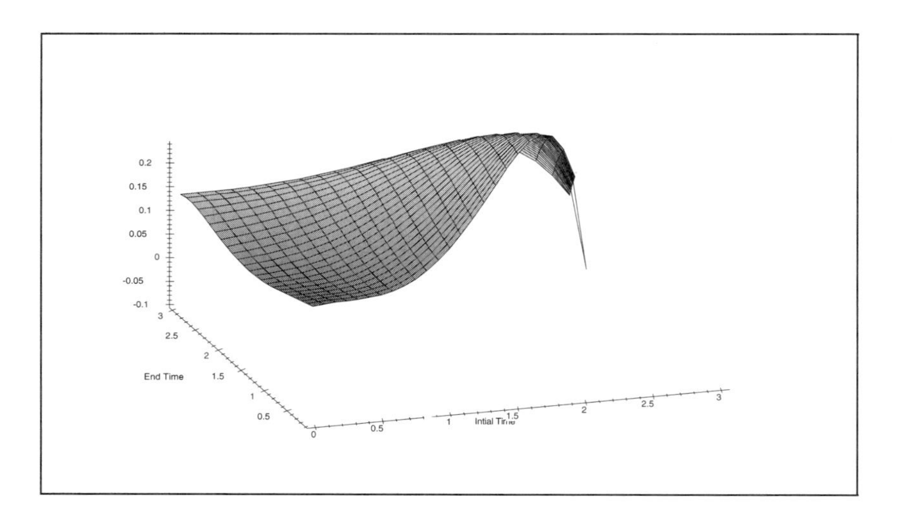

Figure 9.8: The Forward Rate from Time t_1 to t_2: $0 < t_1 < 3,\ t_1 < t_2 < 3$

We can also display the forward rate from time t_1 to t_2 where we let t_1 range from 0 to 3 and t_2 from t_1 to 3. This is displayed in Figure 9.8. Note that in Figure 9.7 the graph of the forward rate is a slice of the graph in Figure 9.8 for $t_1 = 0.5$.

```
> plot3d(Fr(t1,t2),t1=0..1.5,t2=t1..1.5,axes=normal,\
> labels=['Intial Time','End Time',Rate],\
> orientation=[-108,65],title='Figure 9.8:  The Forward\
> Rate from Time t1 to t2:  0<t1<3, t1<t2<3');
```

9.3.2 Forward Rate: A Practical Approach

A forward rate (not to be confused with a forward agreement or with a forward price) is not only a theoretical concept, but an implied rate to prevail in the future. It is actually possible to borrow at this rate. It is the rate at which an investor can secure or commit to a loan now (at time zero) which will be taken at some future time t_1 and be repaid at a later time t_2. Hence, in a sense, the forward rate resembles a forward contract and perhaps its name originates from the similarity.

A forward contract is an agreement made now for delivery of a certain amount of money x at time t_1 and repayment of x plus some interest at time t_2. Thus the amount paid back at time t_2 will be $x(1 + r(t_1, t_2))$, where we denote by $r(t_1, t_2)$ the interest rate paid over the time interval $[t_1, t_2]$. (Indeed we somewhat abuse the notation as in the former subsection where $r(t_1, t_2)$ denoted the continuously compounded interest rate.) Let us demonstrate such a position with an example. Consider a bond market with three periods and three bonds as specified in **NarbitB** below:

```
> NarbitB([[105,0,0],[10,110,0],[8,8,108]],[94,97,85]);
```

The no − arbitrage condition is satisfied.

The discount factor for time, 1, *is given by*, $\frac{94}{105}$

The interest rate spanning the time intreval, [0, 1], *is given by*, .1170

The discount factor for time, 2, *is given by*, $\frac{1849}{2310}$

The interest rate spanning the time intreval, [0, 2], *is given by*, .2493

The discount factor for time, 3, *is given by*, $\frac{82507}{124740}$

The interest rate spanning the time intreval, [0, 3], *is given by*, .5119

The function Vdis([c1, c2, ..]), values the cashflow [c1, c2, ..]

First we would like to check if it is possible to have a certain self-financing portfolio in this market. The reader will recall that a self-financing portfolio is one with a zero cash flow at time zero. We are interested in a self-financing portfolio with a cash inflow of $1 at time two followed by a cash outflow at time three. If such a portfolio exists in this market, we would like to know what the cash outflow will be at time three.

Since there are no arbitrage opportunities in this market, as was just confirmed, every self-financing portfolio which has a zero cash flow at time one and a cash inflow of $1 at time two should have the same cash outflow at time three. To this end, we solve the system of equations (9.11) where B_1, B_2, and B_3 stand for the holdings of bonds one, two, and three in the portfolio, respectively. The cash outflow at time three is denoted by RF and is also one of the variables for which we seek a solution.

$$\begin{aligned} 94\,B_1 + 97\,B_2 + 85\,B_3 &= 0 \\ 105\,B_1 + 10\,B_2 + 8\,B_3 &= 0 \\ 110B_2 + 8B_3 &= 1 \\ 108B_3 &= RF. \end{aligned} \tag{9.11}$$

The request to solve system (9.11) is submitted to MAPLE below.

```
> solve({B1*94+B2*97+B3*85=0,B1*105+B2*10+B3*8=0,\
```

```
> B1*0+B2*110+B3*8=1,B1*0+B2*0+B3*108=RF},{B1,B2,B3,RF});
```

$$\{B2 = \frac{8173}{825070}, B3 = \frac{-1849}{165014}, B1 = \frac{-37}{412535}, RF = \frac{-99846}{82507}\}$$

Hence, we see that if we hold a long position in bond two of $\frac{8173}{825070}$ units, a short position in bond one of $-\frac{37}{412535}$ units, and a short position in bond three of $-\frac{1849}{165014}$ units, the cost of this portfolio is zero. It produces a cash inflow of \$1 at time two and a cash outflow of $-\frac{99846}{82507}$ dollars at time three. The portfolio is purchased at time zero, at which time no cash changes hands. At time two, the buyer receives \$1 and then repays the loan with $\frac{99846}{82507}$ dollars at time three. The buyer is essentially "buying" this portfolio at time zero to secure a loan which will be in effect from time two to time three. Let us see what interest rate the borrower who purchases this portfolio is paying.

We refer to the loan as synthetic since it is the same sort of cash flows as a loan, but it is not a loan in the conventional sense. The cash flows are those of a loan which is agreed upon today but which will be transacted at time two. It is a forward contract. The interest rate implicit in this loan is the value of r which solves the equation

$$1 + r = \frac{99846}{82507}. \tag{9.12}$$

Hence, the implicit rate is

```
> solve((1+r)=99846/82507);
```

$$\frac{17339}{82507}$$

Thus, given the current market conditions, the investor can secure, at time zero, a loan which will be transacted at time two and repaid at time three with an interest rate of $\frac{17339}{82507}$. This rate is referred to as the *forward rate* from time two to time three. In the same manner, the forward rate from time one to time two could be calculated. This is left as an exercise for the reader.

The above argument demonstrates the concept of a forward rate by the replication argument. Let us see how the same concept will be explained utilizing the discount factor valuation approach. Consider the cash flow generated above. It costs zero to generate, produces \$1 at time two, and $\$(-(1+r))$ at time three. We would, thus, like to find the numerical value of r such that the value of the cash flow $(0, 1, -(1+r))$ will be zero at time zero. That is, valuing from the point of view of time zero, what should a person return at time three for a dollar he or she receives at time two. This can be calculated easily by the **Vdis** function. We want to solve for the value of r such that **Vdis**$([0,\ 1,\ -(1+r)]) = 0$. This is submitted to MAPLE below.

```
> solve(Vdis([0,1,-(1+r)])=0);
```

$$\frac{17339}{82507}$$

Indeed, the same value for the forward rate is obtained. The concept of a forward rate is thus the rate, $r(t_1,\ t_2)$, at which one can secure at time zero a loan from time $0 < t_1$ to some time t_2, $t_1 < t_2$. If the market is not complete, we cannot be sure that these types of loans can actually be executed by purchasing certain portfolio combinations. Nevertheless, the concept of a forward rate has been extended for every t_1 and t_2 making use of the continuous approximation of the discount factors.

We can then offer yet another explanation for forward rates to justify equation (9.9). Given an approximation of the continuously compounded interest rate $r(t)$ spanning the time interval $[0,\ t]$, proceed in the following manner. Consider an investor who borrows a dollar amount at time zero such that a dollar will be returned at time t_2. In other words, the investor receives $e^{-r(t_2)\,t_2}$ dollars at time zero. This amount is immediately invested until time t_1, where $t_1 < t_2$. Hence at time t_1 the investor will have $e^{r(t_1)\,t_1}\,e^{-r(t_2)\,t_2}$ dollars and will have to give back one dollar at time three. The interest rate implicit in such a loan, reported as a continuously compounded rate, is the $r(t_1,\ t_2)$ that solves

$$e^{r(t_1)t_1}e^{-r(t_2)t_2}e^{r(t_1,t_2)(t_2-t_1)} = 1. \tag{9.13}$$

Multiplying (9.13) by $e^{r(t_2)\,t_2}$ results in equation (9.9), and hence the same solution for the forward rate $r(t_1,\ t_2)$ is obtained.

9.4 A Variable Rate Bond

A *variable rate bond*, occasionally referred to as a *floater*, is a bond that has no fixed rate of interest. Instead, it pays the interest rate prevailing in the market at the relevant time. Hence, as opposed to the bonds we dealt with until now, we cannot specify at the current time the future cash flow of this bond. Instead, the future cash flow will vary with the interest rate that prevails in the market.

Consider a variable rate bond with a face value of FC that pays a coupon at every specified period. At the beginning of each time period the interest rate spanning the period is known. The bond will thus pay, at the end of each period, the interest rate spanning that period: a certain percentage of FC. If, for example, the bond was issued now, at time t_0, and pays at times $t_1, t_2, ..., t_n$, then at time t_0 the interest rate spanning the time interval $[t_0, t_1]$ would be known. Let us denote it by $r_{t_0}(t_0, t_1)$. The interest rate that will prevail in the market at time t_1 spanning the time interval $[t_1, t_2]$ is not known at time t_0. Let us denote it by $r_{t_1}(t_1, t_2)$, not to be confused with the forward rate $r(t_1, t_2)$, which in this notation will be $r_{t_0}(t_1, t_2)$. The bond will therefore pay $FCr_{t_0}(t_0, t_1)$ at time t_1, at time t_2 it will pay $FC\, r_{t_1}(t_1, t_2)$, and so on. Hence, we do not know with certainty the complete payment schedule from the bond in the future. We only know with certainty the next payment of the bond. The last coupon payment at time t_n will be $FCr_{t_{n-1}}(t_{n-1}, t_n)$.

While we only know with certainty the payoff at the end of period one, it is still possible to value such a bond. We demonstrate below two arguments for valuing such a bond. The first argument is built upon the replication and discount factor valuation. It exemplifies the type of nondeterministic (random) cash flows we can value without specifying a model for the behavior of the term structure. The second argument is based on logic.

Recall the discussion of an equity swap in Section 2.5.2, equation (2.25). There, we discussed methods of replicating the return on a certain index. The problem we face here is quite similar. This issue is also related to our explanation of the forward rate. Let us see how we can replicate now at time zero, the return on FC dollars invested at the risk-free rate that will prevail in the market from time t_1 to time t_2, where $t_1 < t_2$. That is, we want to generate now at time t_0 a cash flow of $FCr_{t_1}(t_1, t_2)$ to be received at time t_2. Note that we do not know what the actual cash flow will be since the value of $r_{t_1}(t_1, t_2)$ is a random variable that is currently unknown. We nevertheless can find a replicating strategy that requires actions only at the current time and whose cost is known with certainty at the current time. Such a strategy is referred to as a "static" or "buy and hold" strategy.

Not only are we able to replicate at t_0 the return on FC dollars invested at the risk-free rate from time t_1 to time t_2 and payable at time t_2, but we already derived the cost of such a strategy. This is explained in the discussion prior to equation (2.25) on page 75. We repeat this explanation here in terms of the risk-free rate.

Consider the following investment strategy. At time t_0 borrow the amount $FCd(t_2)$ to be repaid at time t_2 and immediately invest an amount of $\$FCd(t_1)$ until time t_1. Since $d(t_2) < d(t_1)$ the amount invested, $\$FCd(t_1)$, is larger then the amount borrowed, $\$FCd(t_2)$. The out-of-pocket cost of this strategy is thus $FCd(t_1) - FCd(t_2)$. At time t_1 the amount invested at t_0, $FCd(t_1)$, will be worth FC.

At time t_1 the $\$FC$ will be invested until time t_2 at the rate prevailing in the market at that time, $r_{t_1}(t_1, t_2)$. Thus at time t_2 it will be worth $FC(1 + r_{t_1}(t_1, t_2))$. At t_2 the loan that was initiated at time t_0 should be paid back. Since the loan principal was $\$FCd(t_2)$ the amount to be paid back is FC.

Note that the borrowing and lending discussed here do not necessarily require borrowing from a bank. Instead, one may interpret "lending" or "borrowing" as the purchase of a portfolio of bonds that produces the required cash flow. Thus "borrowing" is a short position in a certain bond portfolio and "lending" is a long position. Of course it may be the case that in order to generate a borrowing-like cash flow we may require a portfolio that is composed of both short and long positions. One of the exercises to this Chapter elaborates on this point.

Let us consider the cost and payoff from the above described strategy. The cost of this strategy is the amount in excess of the loan that we had to invest at t_0, namely, $FCd(t_1) - FCd(t_2)$. The payoff from the strategy is the payoff at the end of time t_2, $FC(1+r_{t_1}(t_1, t_2))$, minus the cost of paying back the loan, FC. The net payoff is therefore $FC - FC(1 + r_{t_1}(t_1, t_2))$, which is equal to $FCr_{t_1}(t_1, t_2)$. The payoff at t_2 is the return on FC invested from time t_1 to time t_2, i.e., $FCr_{t_1}(t_1, t_2)$. In terms of $d(t)$ the cost at t_0 of replicating this return is

$$FCd(t_1) - FCd(t_2). \tag{9.14}$$

By the same argument, the cost of replicating the return on $\$FC$ invested at the risk-free rate from time t_2 to time t_3, i.e., replicating $FCr_{t_2}(t_2, t_3)$ payable at time t_3, will be

$$FCd(t_2) - FCd(t_3). \tag{9.15}$$

Assume that the bond matures at time t_3. Hence at time t_3 the buyer of such a bond (playing the same role as the lender in the strategy described above) will also get back $\$FC$ whose cost at t_0 is

$$FCd(t_3). \tag{9.16}$$

In addition, at time t_1 the buyer will get the return on FC invested from time t_0 to time t_1 whose cost, based on the same argument as above, will be

$$FC - FCd(t_1). \tag{9.17}$$

Hence the cost of producing this cash flow at times t_1, t_2, and t_3 is the sum of equations (9.14), (9.15), and (9.16), and equation (9.17). The total is thus

```
> (FC*d(t1)-FC*d(t2))+(FC*d(t2)-FC*d(t3))+(FC*d(t3))+\
> (F-F*d(t1));
```

$$FC$$

The initial cost of the variable rate bond is thus its face value. The assumption of the bond maturing at time t_3 does not affect the end result. One can easily see that this will be the case for a bond maturing after n payment periods as well. Note, however, that we assumed that the bond was issued at t_0. A slight modification is needed to calculate the value of the bond as of some time between payment periods after the issuing time. For example, if the value of the bond is calculated at some midpoint v, between t_0 and t_1, its value will be calculated based on the discount factor function estimated at time v. If we denote this discount function by $d_v(t)$ the bond value will be

$$FCd_v(t_1) + FCr_{t_0}(t_0, t_1)d_v(t_1). \tag{9.18}$$

We leave the derivation of this result as an exercise.

There is another approach for calculating the value of a variable rate bond. This approach applies a simple logical argument. Recall that the bond is paying at times $(t_1, ..., t_n)$. At time t_{n-1} (immediately after the coupon payment) the value of that bond must be its face value of FC. An investor holding that bond at time t_{n-1} will receive, at time t_n, the face value back and the interest rate prevailing at that time in the market.

Thus, when we discount the future value of the bond at time t_{n-1} we are discounting $FC(1+r)$ (where r is the interest rate from time t_{n-1} to time t_n) by the discount factor, $\frac{1}{1+r}$. Hence, the value of the bond at time t_{n-1} is FC. Now, at time t_{n-2} the value of the bond must again be FC. This is because the value of the bond at time t_{n-1} is FC and the investor holding the bond at time t_{n-2} will receive an interest payment at time t_{n-1}.

At time t_{n-1} the interest payment is according to the rate prevailing in the market from time t_{n-2} to t_{n-1} and the investor will also have the bond whose value at that time is FC. Thus, by the same argument as above, the value of the bond at time t_{n-2} is FC. Proceeding in the same manner we can show that the value of the bond at time zero is again FC.

9.5 Concluding Remarks

This Chapter applied the framework of the simple model of our very first Chapter to the debt market. Essentially we still employed the same model but for the modifications needed to describe the distinct features of the debt market. The discount factors defined here were used in the same way as the stochastic discount factors in the equity market. In fact no new concept was introduced in this Chapter. The no-arbitrage condition was shown to be the source of the existence of the discount factors and thus of the term structure of interest rates. This Chapter also explained how the term structure is estimated from the bond prices and how a continuous approximation is generated. The concept of a forward rate was introduced and was linked to a valuation by arbitrage.

The discount factor function $d(t)$ or equivalently the term structure of interest rates $r(t)$ is used to value cash flows across time. This is performed using the same methodology as with the stochastic discount factor to value cash flows across the states of nature. The discount factor implied in the prices of government bonds can be used to discount only the time value of money.

The convention is to perceive government bonds as free from default risk. In contrast, the stochastic discount factors (introduced in the simple model of the first Chapter) incorporate both the risk and the time value of money. Certain bonds, like corporate bonds, are subject to default risk. Thus their market prices reflect the possibility of default. There are, of course, different categories of risk and most markets in the world have rating agencies that classify the default risk of bonds.

The existence of a market populated with enough bonds in a certain grade allows us to estimate the discount factor for the grade of bonds. This is performed in exactly the same manner as we have demonstrated in the "risk-free" government bond market. The discount factor of a grade of bonds is the price of a dollar at time t but subject to the risk of default. Therefore, if $d(t)$ is the discount factor obtained from the government bonds and $(d^f)(t)$ is the discount factor one obtains from a certain group of risky bonds, the latter should be smaller than the former; otherwise arbitrage opportunities will arise. Can you delineate an arbitrage portfolio if the relation would have been reversed and $d(t) < (d^f)(t)$? Note, that in this case the arbitrage portfolio produces a sure profit (with no initial investment) at the initiation time, and a positive probability of making further profit in the future. We leave this as an exercise for the reader. Hence, if we need to value certain cash flows across time that are not of the risk-free category, we can do so, provided we have the term structure for this type of risk. We shall see an example of such an application in a case of a forward rate agreement discussed in Section 10.4.1. The next two Chapters are devoted to valuation of derivative securities requiring knowledge only of the discount factors, implied by the prices of bonds, and not the evolution of the term structure (discount factor function) across time.

9.6 Questions and Problems

Problem 1: Consider the following bond market:

	Time 1	*Time 2*	*Time 3*	*Prices*
Bond 1	105	0	0	96
Bond 2	8	108	0	97
Bond 3	3	3	103	89

1. Check whether this is an arbitrage-free market by using the **NarbitB** procedure.
2. Explain why the rate $R1$ spanning the time interval $[0, 1]$ is determined by solving $\mathbf{Vdis}([1 + R1, 0, 0]) = 1$.
3. What are the interest rates spanning the time intervals $[0, 1]$, $[0, 2]$, and $[0, 3]$?
4. What are the interest rates, as of the current time, spanning the time intervals $[1, 2]$, $[2, 3]$, and $[1, 3]$? (You may solve it utilizing the function **Vdis** and setting an equation that each rate must satisfy.)
5. Identify a portfolio that costs zero, pays a certain amount at time 2, and requires a payment of \$1 at time 3. What must the payment of the portfolio be at time 2? Is it necessary for such a portfolio to include a short position?
6. What is the relation between your answer to the above and the rate spanning the time interval $[2, 3]$?
7. Generate a continuous approximation of the discount factor function that perfectly fits the discrete discount factors. You may use the procedure **NarbitB**.
8. What is your estimation for the forward rate from time 0.5 to time 2.5?

Problem 2. Assume that $R1$ and $R2$ are the spot rates corresponding to times t_1 and t_2 ($t_1 < t_2$), respectively, where the units of time are measured in years.

1. Show that $R1 < R2$.
2. Identify an arbitrage strategy in a market where $R1 > R2$.

3. Determine the corresponding annual interest rates assuming semi-annual compounding.
4. Let $r1$ and $r2$ be the rates solved for in part 3 of this question. Does the no-arbitrage condition imply that $r1 < r2$ (prove or provide a counterexample)?

Problem 3. Show that if r_n is the annual rate corresponding to time n (measured in years) and is compounded m times per year, the spot rate can be determined as $R_n = \left(1 + \frac{r_n}{m}\right)^{nm} - 1$.

Problem 4. Assume the following structure for the bond market.

	Time 1	*Time 2*	*Time 3*	*Prices*
Bond 1	107	0	0	94
Bond 2	9	109	0	100
Bond 3	5	5	105	93

1. Verify in two different ways that the market is arbitrage-free.
2. Consider the three zero-coupon bonds stipulated in the table below:

	Maturity Time	*Face Value*	*Price*
Bond 1	1	1000	940
Bond 2	2	1090	1000
Bond 3	3	1000	854

   Are these prices the "no-arbitrage" prices of the bonds? Justify your answer by pricing these cash flows utilizing the replication method. Identify an arbitrage position if the price(s) admits arbitrage.
3. Construct a new market consisting of only zero-coupon bonds with face values of \$100, maturing at times 1, 2, and 3 such that the cash flows of the primary bonds have the same prices as in the original market.
4. Prove that the new market, constructed in part 3 of this question, actually implies that every cash flow has the same price in both markets.

Problem 5. Consider the following bond market where the time is measured in years and bonds are paying semiannually:

	Time 1	*Time 2*	*Prices*
Bond 1	108	0	102.6
Bond 2	2	102	93.7

1. Give an example of two coupon bonds maturing at time 2 that have different yields.
2. What would be a yield on a zero-coupon bond maturing at time 2?
3. Can you give an example of a coupon-paying bond that has a yield the same as the one from a zero-coupon bond?
4. In general, is it possible that a zero-coupon bond and a coupon bond having the same face value, both maturing at time 2, will have the same yield? Justify your answer.

Problem 6. Let r_n and r_m be two annual interest rates, in the same market, corresponding to the same maturity but being compounded n and m times per year, respectively.

1. Show that $r_n > r_m$ if and only if $n < m$.
2. Show that, for all $n < m$, the term structure of interest rates plotted in terms of compounding n times per year is always above the one plotted in terms of compounding m times per year.
3. Show that a term structure of interest rates plotted in terms of continuous compounding is always below the one plotted for any finite compounding.

Problem 7. Use the data of Problem 1 to find the value at $t = 0.5$ of a \$100 face value variable rate bond that pays coupons at times 1, 2, and 3.

Problem 8. Provide an example of a bond market with three bonds and three discrete payoff periods such that

- the no-arbitrage condition is satisfied
- the market is complete; and
- the term structure is flat, i.e., the spot rate (based on annual compounding) is the same for every time t.
- Confirm your results by plotting the term structure in this market.

Problem 9. Consider the market in Problem 1 but where the prices of the bonds are 93, 96, and 90 for bond 1, 2, and 3, respectively. How would you answer parts 3, 4, 5, 6, and 8 of Problem 1 given the new prices?

Problem 10. Delineate an arbitrage strategy if the relation between the discount function estimated from the government bonds, $d(\cdot)$, and the discount factor obtained from a certain group of risky bonds, $d^f(\cdot)$, was $d(t) < (d^f)(t)$. Note that in this case the arbitrage portfolio produces a sure profit at the initiation time, with no initial investment, and a positive probability of making further profit in the future.

Problem 11. Consider the bond market introduced in Problem 4. Devise a portfolio that would replicate a loan of $1 commencing at time 1 to be repaid at time 2. What is the price of this portfolio now? What is the interpretation of the price of such a portfolio?

Problem 12. Consider the bond market introduced in Problem 1.

1. Devise a portfolio that would have a cash flow of $(-c, -c, -100-c)$ for some positive number c.
2. Find such c that would make the value of this cash flow $100 today.
3. What is an interpretation of c found in 2?
4. Construct a portfolio that would constitute a loan of $200 commencing at time 1 to be repaid at time 3 with an intermediate interest payment at time 2.
5. Identify the short and the long position of the portfolio.
6. What should be the interest on this loan in order for this portfolio to cost nothing now?
7. What is the value of this portfolio now if the interest on the loan is chosen to be 10 percent?

Problem 13. Suppose you have a bond market and at every point in time in the future a continuous approximation of the discount function is available. Consider a variable rate bond with the face value FC that pays a coupon at some prespecified times $t_1 < t_2 < \cdots < t_n$. Assume it is now time v such that $t_i < v < t_{i+1}$, for some $i = 1, 2, ..., n-1$. Prove that the value of this bond now is

$$FCd_v(t_{i+1}) + FCr_{t_i}(t_i, t_{i+1})d_v(t_{i+1}),$$

where d_v is the discount function as of time v and $r_{t_i}(t_i, t_{i+1})$ is the interest rate spanning the time $[t_i, t_{i+1}]$.

9.7 Appendix

9.7.1 Theories of the Shape of the Term Structure

There are several theories which help describe the shape of the term structure of interest rates. Some compete with one another. Others can be viewed more reasonably as complementary. Taken all together, they provide a sensible description of the underlying determinants of the shape of the yield curve.

The Unbiased Expectations Theory

The unbiased expectations theory posits that investors are risk-neutral. In choosing, investors, regardless of their investment time horizon, will choose the instruments with the highest return. They require no additional compensation for any perceived risk associated with the time frame involved.

There is a direct relationship between the spot rates in the market place and the forward rates of interest. Let r_2 be the one-period spot interest rate (that is, the rate of interest which prevails from time period zero to time period one), and let r_2 be the two-period spot interest rate (prevailing from time zero to time two). The corresponding forward rate, f_{12}, is the one-period rate of interest which prevails from time one to time two which is implicit in these two spot rates. It can be calculated from the two spot rates as follows: $\frac{(1+r_2)^2}{1+r_1} - 1$. More generally, any forward rate can be calculated according to the following formula:

$$\frac{(1+r_n)^n}{(1+r_{n-1})^{n-1}} - 1.$$

The expectations theory is based on the idea that these risk-neutral investors set interest rates in a manner so that the forward rate is equal to the spot rate expected in the market one year from now. The theory is expressed in terms of expected one-period spot rates. In terms of bonds, the yield on a two-year bond is set in such a way that the return on that two-year bond is equal to the return on a one-year bond plus the expected return on another one-year bond purchased in one year. This notion is not unique to one- and two-year bonds. If all investors in the marketplace operate this way, then prices will adjust until the expected return from holding a two-year bond is the same as the expected return from holding two one-year bonds.

According to the expectations theory of the term structure, the yield curve can be derived from a series of expected one-year spot rates. Consider, for example, that the one-year spot rate is 5 percent from year zero to year one, the expected spot rate from year one to year two is 6 percent, and the expected spot rate from year two to year three is 7 percent. A two-year bond would earn the spot rate over the period from time zero to time two, which, according to this theory, is the same as investing in a one-year bond from time zero to time one, and then investing in another bond for one year from time one to time two.

Thus, the two-year spot rate from time zero to time two is calculated as $\left(1+\frac{r_2}{2}\right)^2 = \left(1+\frac{0.05}{2}\right)\left(1+\frac{0.06}{2}\right)$. The two-year spot rate is 5.5 percent. The three-year spot rate from time zero to time three can be calculated as $\left(1+\frac{r_3}{2}\right)^3 = \left(1+\frac{0.05}{2}\right)\left(1+\frac{0.06}{2}\right)\left(1+\frac{0.07}{2}\right)$. The three-year spot rate is 6 percent. Furthermore, the market's belief about the future of one-year spot rates can be read easily from an observed yield curve. Note that this theory assumes that the expected future spot rate is equal to the corresponding forward rate. This assumption does not hold for some of the other theories of the term structure.

The Liquidity Preference Theory

Under the liquidity preference theory of the term structure, investors again examine the returns from holding bonds of differing maturities. This theory does not assume that investors are risk-neutral. Investors are assumed to demand extra compensation to be induced to hold a bond of a relatively long maturity over a bond of relatively short maturity. Furthermore, the market is populated with relatively more short-term investors, which requires that investors receive additional inducement to hold long-term bonds.

In the example of the section which described the expectations theory of the term structure, we calculated that if the one-period rate from time zero to time one were 5 percent and the one-period rate from time one to time two were 6 percent, then the two-period rate prevailing from time zero to time two would be expected to be 5.5 percent. Under the liquidity preference theory, the two-period rate would have to be higher than 5.5 percent to induce investors to hold relatively longer-term instruments. For an investor with a one-year investment horizon, there is risk associated with the two-year investment.

The liquidity preference theory leads us to different conclusions about the shape of the term structure. Even if expectations are such that there will be no change in one-period rates, it would still be the case that the yield curve would be upward sloping in the presence of a liquidity premium. Even if one-period spot rates were expected to decline, if the liquidity premium were sufficiently large, there could still be an upward sloping term structure. A flat or downward sloping yield curve, under the liquidity preference theory, can only be possible in an environment of decreasing one-period spot rates.

The Market Segmentation Theory

The origin of the market segmentation theory stems from the observation that some investors apparently prefer debt of a particular maturity. This preference is so pronounced that these investors are insensitive to the yield differential of their preferred debt maturity over the debt of another maturity. The theory posits that investors are so risk-averse that they remain in their desired maturity spectrum and cannot be induced by yield differentials to change maturities. In this way, long-term rates are determined by the supply of and demand for long-term debt instruments, and similarly for short-term interest rates. Proponents of this theory closely watch the flow of funds into different segments of the bond market to determine changes in the yield curve.

Consider examples of investors who can reasonably be expected to have strong preference for debt of a particular maturity. An insurance company facing liabilities which are in the distant future will choose to invest for a long time horizon. There may be considerable risk to the insurance company in investing in a series of short-term instruments, compared to the known return and the predictability of available long-term debt. Similarly, corporations may have strong preference for issuing debt of a particular maturity depending on the use to which the funds will be put. Corporations will generally prefer to pay for long-term investment projects over a long period of time and so the corresponding debt issued for those projects is likely to be of long maturity.

The market segmentation theory of the term structure is popular with practitioners. Academics maintain that the market is more likely composed of both investors with definite maturity preferences and those who invest on the basis of relative yields. The predictions of the shape of the term structure based on the market segmentation theory will be offset if there are in fact enough investors who fall into the second category.

9.7.2 Approximating Functions

The text states that the discount factor for time t, $d(t)$, is approximated by the structure

$$d(t) = \sum_{j=0}^{K} \alpha_j \, f_j(t),$$

where the $f_j(t)$ are certain functions and the α_j are numerical coefficients. In this Appendix we provide more detailed insight into the nature of $f_j(t)$ and the corresponding function $d(t)$.

The approximation technique described in the text, taken from [44], uses a set of polynomials resembling Bernstein polynomials. The use of these polynomials is advantageous as they result in an approximation that is always a decreasing function of the interval [0, 1]. For this reason we always scale the time interval to [0, 1]. This is done by dividing the time by the largest maturity. Hence, the longest maturity is always one. We undo this scaling when reporting the discount factors. We know that the monotonicity property should be satisfied by the discount factors or else arbitrage opportunities would exist.

The functions $f_j(t)$, for $j = 1, ..., K$, are defined as

$$f_j(t) = \sum_{l=0}^{K-j} (-1)^{l+1} \binom{K-j}{l} \frac{t^{j+l}}{j+l},$$

where $\binom{n}{k} = \frac{n!}{k!\,(n-k)!}$, while for $j = 0$, $f_0(t) = 1$ and $\alpha_0 = 1$.

The discount function $\mathbf{dis}(t)$, in terms of the approximating function, is given by

$$dis(t) = \sum_{j=0}^{K} \alpha_j f_j \left(\frac{t}{nst} \right),$$

where nst is the longest maturity and $\frac{t}{nst}$ is the scaling mentioned earlier.

To solve for the value of coefficients α_j, we solve an optimization problem. We minimize, with respect to the α_j, the sum of absolute deviations of the bond price p_i from its discounted cash flow, where $a_i(t)$ is the cash flow from bond i at time t. Thus, we solve the optimization problem below with respect to $\alpha_1, ..., \alpha_K$ and $\epsilon_1, ..., \epsilon_n$:

$$\begin{array}{ll} & \min_{\epsilon_1,...,\epsilon_n,\alpha_1,...,\alpha_K} \sum_{i=1}^{n} |\epsilon_i| \\ \text{subject to} & \sum_{t=1}^{nst} a_i(t) \sum_{j=0}^{K} \alpha_j \, f_j(\frac{t}{nst}) = p_i + \epsilon_i, \quad \text{for } i = 1, ..., n. \end{array}$$

Since the above problem can be transformed into an equivalent linear programing problem, the procedure **NarbitB** solves it using the simplex method.

Indeed, the situation in a real market is more complicated. Due to both the nonsynchronization of markets and the fact that not every bond is traded on every day, the prices of the bonds include some statistical noise. Hence, in almost all cases the data will allow for arbitrage opportunities. Moreover, most bond markets are incomplete. In a typical bond market the number of bonds is about a third of the number of payment dates. Consequently, even when the discount factors are not approximated by $dis(t)$, there does not exist a set of discount factors according to which the present value of the bonds equals the prices. Therefore, the constraints of the optimization problem above are not feasible. Hence, the term structure is estimated as a second-best solution by the optimization problem above. The ϵ are thus a combination of statistical noises and misspecification of the approximating function.

Chapter 10

Forwards, Eurodollars, and Futures

The simple model of Chapter 1 has now been extended in two slightly different directions. In Chapter 2 the one-period model was extended to a multiperiod model with a finite number of states in order to examine the bond market. Chapter 4 extended the one-period model with a finite number of states to the continuum of states of nature. This last extension facilitates the investigation of equity options. The current Chapter combines the two extensions of the simple model into a framework of a multiperiod model for which a continuum of states of nature can be realized in each time period. In this framework, this Chapter takes a second and more realistic look at forwards and futures.

We now analyze these contracts in a multiperiod model with a continuum of states of nature. We also examine the difference between forward contracts and futures contracts, a difference that was moot in the setting of a one-period model. We still suppress the effect of possible default on the valuation of these contracts since we focus on the fundamental concepts. As we will see, given the structure of the two contracts, the default issue is of greater concern in the case of a forward contract than in the case of a futures contract.

This multiperiod setting also allows the value of a forward contract to be investigated at some point in time after the initiation of the contract and before its maturity. The investigation of a multiperiod scenario facilitates the analysis of the pricing of forward contracts on assets that pay known cash flows during the life of the contract. A very similar argument allows us to find the forward price of a stock that pays a known and fixed dividend yield during the life of the contract. This Chapter also analyzes forward rate agreements as they are very similar to forward contracts.

10.1 Forward Contracts: A Second Look

Section 2.4.1 introduced *forward contracts* in the setting of a one-period model with finite states of nature. The payoff, at maturity, of such a contract in a one-period setting but with a continuum of states of nature was investigated in Section 4.3.1 and displayed in Figure 4.20. That figure displayed the payoff resultant from a forward contract on a stock. The same graphical display will show the payoff from a forward contract on a different financial asset or a commodity. The only difference is that the x-axis will display the possible states of nature with respect to the underlying asset in question. Thus, for a forward contract on an exchange rate, the x-axis will display the possible realizations of the exchange rates. The actual shape of the graph will stay intact as well as the point at which the payoff structure intersects the x-axis. This point remains the forward price of the contract under investigation.

We begin by first analyzing a forward contract in the setting of a multiperiod model. From a conceptual perspective, a forward contract in a multiperiod setting is not much different from a forward contract in a one-period setting. We can think about the multiperiod as one long time period. We are only concerned with the spot price of the asset on which the contract is written and the price of that same asset on the maturity date of the contract. Thus, if we think about the time until the maturity of the contract as one (longer) time period we can still apply the results obtained in Section 2.4.1. Namely, the forward price is equal to the future value of the asset. The forward price can be written as $e^{r(t)\,t}\,S(0)$ or as $\frac{S(0)}{d(t)}$, where $S(0)$ is the spot price of the asset, t is the time to maturity, and $r(t)$ and $d(t)$ are, respectively, the risk-free rate and the discount factor spanning the time interval $[0,\, t]$. We see that we perceive the forward price as the future value of the asset, based on the term structure and the spot price of the asset. Following the argument in Section 2.4.1, it is easy to see that the same result will hold even with a continuum of states of nature.

We can visualize the payoff from a forward contract in a multiperiod setting in the following way. Let us consider a forward contract in which one party is committed to deliver a certain good (or a financial asset) at some future time T for a price of FF agreed upon now and to be paid at time T. Assume that the current time is 0, the forward contract matures at time one, and the forward price is \$7. We consider the payoff from the point of view of the party with a long position in the contract. The cash flow resultant from this contract occurs only at time $T = 1$ and will be $S(T) - FF$, where $S(T)$ is the spot price at time T.

At any other time, of course, the cash flow from this contract is zero. Hence, if we visualize the cash flows from this forward contract in the time span zero to six, the result will be as in Figure 10.1 where the black linear graph represents the payoff from the contract. The procedure **PlotForPyf** generates the payoff. The parameters for **PlotForPyf** are, in order, the maturity time of the contract, the forward price, and the right end point of the time span to be displayed in the graph.

```
> PlotForPyf(1,7,6);
```

The reader can experiment with different values of the parameters and drag the graph to look at it from different perspectives in the on-line version. This ought to convey to the reader that considering the forward contract in a multiperiod setting really does not change any of its characteristics relative to a forward contract in a one-period model.

At any point in time prior to the maturity of the contract, the cash flow from the forward contract, as seen in Figure 10.1, is zero. Of course, the contract still has some value. This value can be positive or negative, depending on the spot price of the asset and on the term structure (or equivalently on the discount factor function) at that point in time. Examining forward contracts in a multiperiod setting facilitates the investigation of their value prior to maturity, which is our next topic.

10.2 Valuation of Forward Contracts Prior to Maturity

Consider a forward contract requiring the delivery of a certain asset at time T. As usual, we assume that the current time is zero; hence the contract matures in T units of time. At the time the contract is written its value is zero. Indeed, this is how the forward price is calculated. We will refer to this contract as the "old" forward contract. Suppose that some time has passed, say t units of time, and the time to maturity of the old contract is now $T - t$. If a forward contract on the same asset maturing in $T - t$ units of time were written now, its value would be zero. The new forward price may differ (and most likely would) from the price the one who holds the old contract long pays in order to take delivery of the underlying asset. The

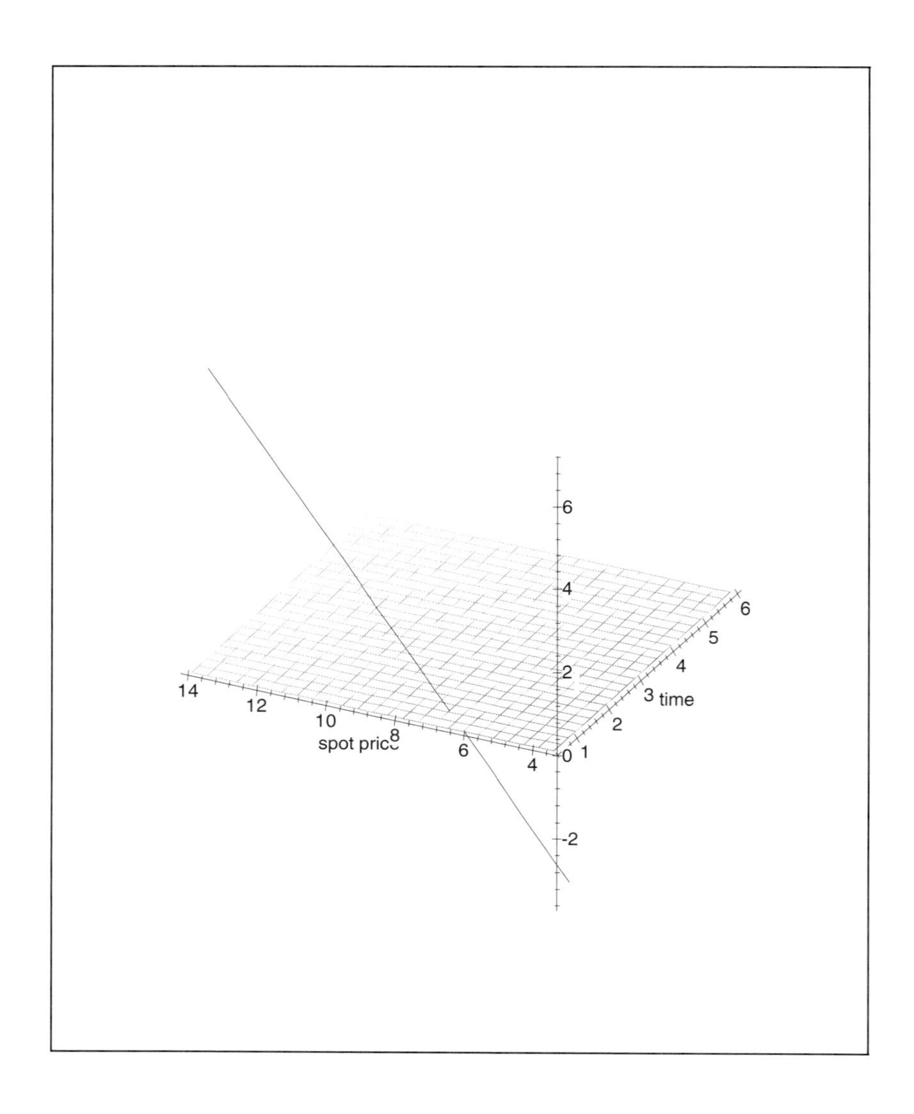

Figure 10.1: Forward's Payoff

current value of the "old" contract, however, is not necessarily zero. The "old" forward price specified on the "old" contract, the *delivery price*, does not reflect the circumstances of the market today. The distinction between a forward price and a delivery price was moot in the setting of a one-period model.

Each forward contract specifies the price to be paid on the delivery date for the asset in question, the underlying asset. The delivery price is equal to the forward price on the date the forward contract is initiated. The forward price is set up in such a way that the initial value of the contract is zero. As market conditions evolve, the spot price changes and so does the forward price. The forward price, which is the future value of the spot price of the asset, changes also, while the delivery price remains the same. Hence, the forward price and the delivery price, though identical on the initiation date of the contract, need not be identical at any other point in time over the life of the forward contract.

Let us now get back to the question of the value of the old contract. This contract obligates the one who holds it short to deliver the underlying assets in $T-t$ units of time. Thus it is equivalent to paying out the price of the asset $S(T)$, in $T-t$ units of time, and receiving the delivery price, which we denote by $F0$. Hence, holding a short position in the forward contract is equivalent to the cash flow $F0 - S(T)$ in $T-t$ units of time. Therefore, following the same guidelines as in the one-period model, we have to discount this cash flow to the present to obtain its price.

The value of $S(T)$ is stochastic, and to get its current value we must therefore apply an appropriate stochastic discount factor. However, we know the current price, $S(t)$, of the underlying asset and therefore $S(t)$ must be the value of $S(T)$ at the current time, time t. That the present value of $S(T)$ is the spot price of the asset can also be explained in a slightly different way. Replicating $S(T)$ at time T costs $S(t)$ today (assuming the asset provides no income) since buying it now guarantees having exactly $S(T)$ at time T. Suppressing default risk, the value of $F0$ is simply its present value based on the risk-free discount factor. Denoting the current, at time t, discount factor function by d we arrive at equation (10.1) for the value of the forward contract.

$$S(t) - d(T-t)F0 \tag{10.1}$$

Given a discount factor function and the delivery price we can visualize the value of the forward contract as a function of time and of the spot price in a three-dimensional graph. Assume a forward contract that was initiated in the past with a delivery price of \$70 and which matures in three units of time. Let us generate a discount factor function **disf** by running **NarbitB** for a certain market. We will suppress the graphing of the discount factor function by adding a fifth input parameter and setting it equal to zero.

```
> NarbitB([[110,0,0],[8,108,0],[6,6,106]],[90,80,75],\
> 3,disf,0);
```

$$The\ no-arbitrage\ condition\ is\ satisfied.$$

$$The\ discount\ factor\ for\ time,\ 1,\ is\ given\ by,\ \frac{9}{11}$$

$$The\ interest\ rate\ spanning\ the\ time\ interval,\ [0,\ 1],\ is\ \ given\ by,\ .2222$$

$$The\ discount\ factor\ for\ time,\ 2,\ is\ given\ by,\ \frac{202}{297}$$

$$The\ interest\ rate\ spanning\ the\ time\ interval,\ [0,\ 2],\ is\ \ given\ by,\ .4703$$

$$The\ discount\ factor\ for\ time,\ 3,\ is\ given\ by,\ \frac{6535}{10494}$$

$$The\ interest\ rate\ spanning\ the\ time\ interval,\ [0,\ 3],\ is\ \ given\ by,\ .6058$$

$$The\ function\ Vdis([c1, c2, ..]),\ values\ the\ cashflow\ [c1, c2, ..]$$

$$The\ continuous\ discount\ factor\ is\ given\ by\ the\ function,\ 'disf',\ (.)$$

If we were interested in checking the goodness-of-fit of our continuous approximation, we would just need to inquire about the value of the variable **SumAbsDiv** that is defined by the procedure.

```
> SumAbsDiv;
```

$$0$$

A value of zero means that the continuous approximation coincides with the value of the discount factors on the payment dates of the bonds.

Figure 10.2, produced by the procedure **ForVal**, demonstrates the value of the forward contract as a function of the time to maturity and of the spot price. The parameters for this procedure are, in order, the discount factor function, the range of the time to maturity in the plot, the range of the spot price in the plot, and the delivery price. The horizontal color-shaded plane defines the $Value = 0$ plane. The other manifold is the value of the forward contract for different combinations of the spot price and the time to maturity. The emphasized black line is the intersection of the plane and the manifold. It emphasizes the locus of points $(Spot\ Price, Time\ To\ Maturity)$ at which the value of the forward contract is zero. Note of course, that when the time to maturity is zero, the forward contract has a zero value if the spot price and the delivery price coincide (at \$70). At some time t prior to maturity the forward contract will have a value of zero if $70 disf(t) = SpotPrice$. Thus the black line, the intersection of the plane and the manifold, is the graph of the function $SpotPrice = 70disf(t)$ in the *Spot Price - Time To Maturity* plane.

```
> ForValue(disf,0..3,30..80,70);
```

Each point on the emphasized black line represents a combination of its coordinates — t, time to maturity, and S, spot price — at which the value of the forward contract is zero. Hence, it is the combination at which a forward contract, on the same asset, with a forward price of $70 would be issued if the time to maturity were t and the spot price were S. If the delivery price of the "old" forward contract happens to coincide with the forward price in the market, then the value of the "old" forward contract is zero. The exercises at the end of the Chapter ask the reader to demonstrate this algebraically. Note that when the time to maturity is zero, the value of the forward contract is zero if the delivery price equals the spot price. In our case, that spot price is $70.

In Figure 10.2 we study the value of the forward contract for a given discount factor function as the time to maturity approaches the current time. Once the discount factor function has been assumed the only source of uncertainty regarding the value of the contract is the spot price. It would be more realistic to visualize the evolution of the value of the forward contract as time moves forward and the current time approaches the maturity time. In doing so, however, there is another source of uncertainty besides the spot price that has to be considered. This is the interest rate that will prevail in the market from the current time to the maturity of the contract. Therefore, the state of nature is summarized by two numbers: the spot price and the interest rate.

We assume, as before, a forward contract with a delivery price of $70 and maturity in three units of time. A three-dimensional animation can be utilized to visualize the evolution of the value of the forward contract as a function of the states of nature (the interest rate and the spot price) as time approaches the maturity time. A static version of one frame of the animation is depicted in Figure 10.3.

```
> plots[animate3d]({70*exp(-r*t)-Spt,0},\
> r=0.01..0.18,Spt=30..100,t=0..3,axes=normal,title=\
> 'Figure 10.3:  Value of Forwards That Mature in t Units\
> of Time as a Function of Spot Price and Interest Rate',\
> labels=['rate','spot price',Value],orientation=[32,69],\
> titlefont=[TIMES,BOLD,10], style=PATCHCONTOUR);
```

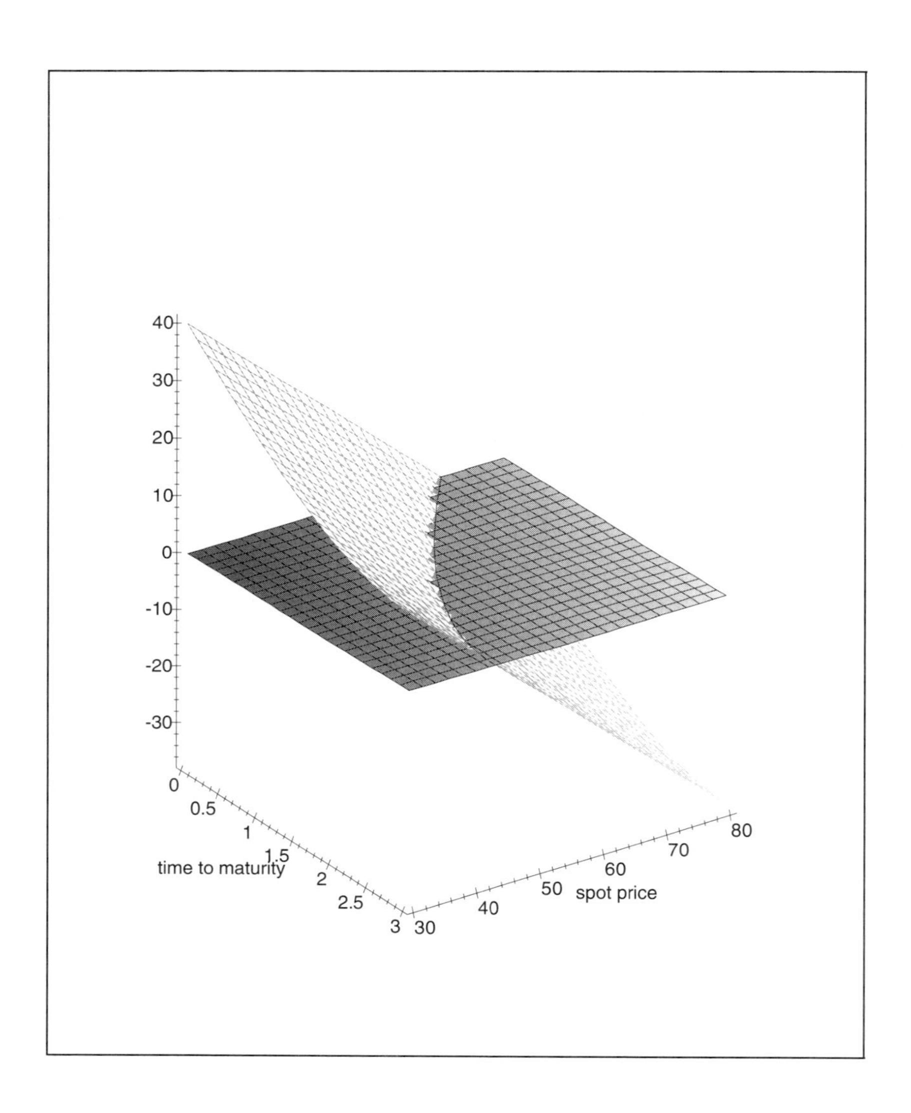

Figure 10.2: Forward's Value as a Function of Spot Price and Time to Maturity

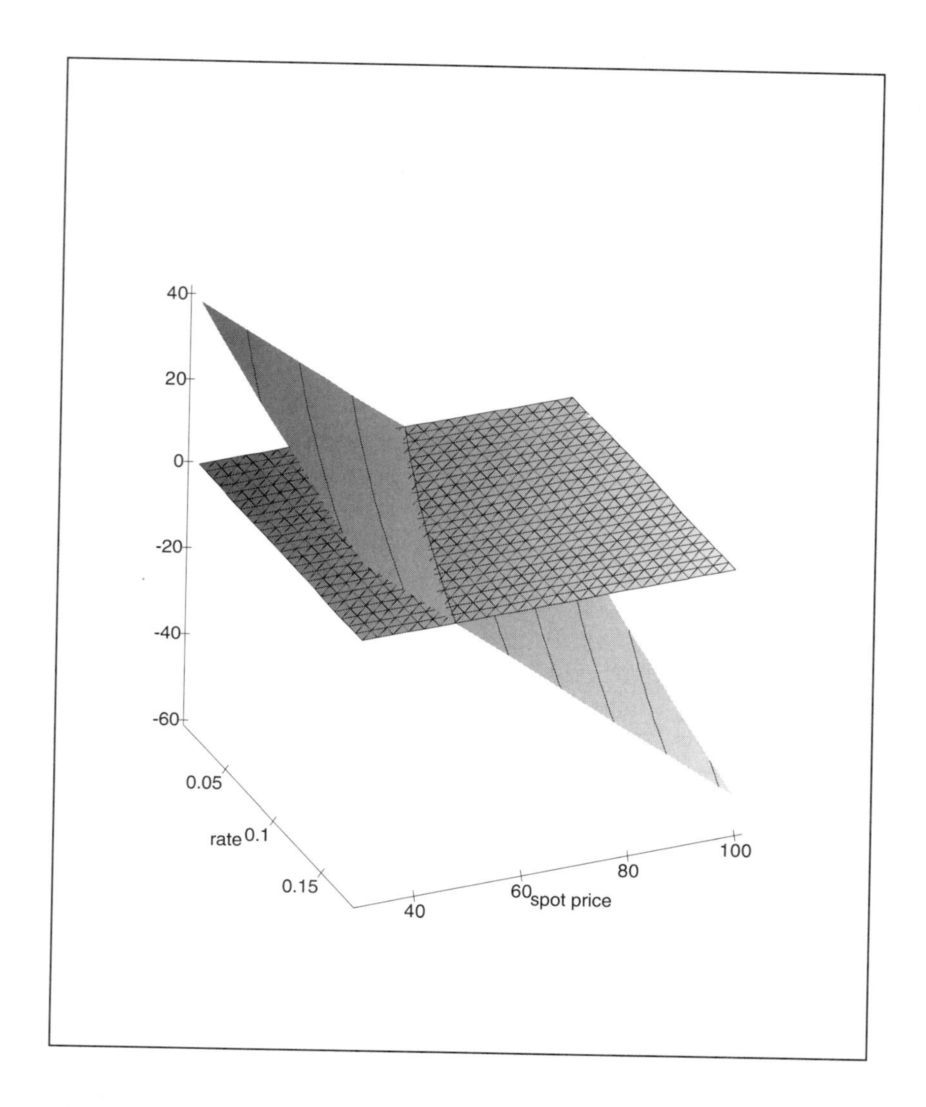

Figure 10.3: Value of Forwards That Mature in t Units of Time as a Function of Spot Price and Interest Rate

Note that when the time to maturity is zero (the last frame in the animation), the value of the forward contract is independent of the rate of interest.

In equation (10.1) the value of an "old" forward contract at time t, maturing in $T-t$ units of time, was stipulated in terms of the spot price of the asset $S(t)$, the delivery price, and the discount factor as $S(t)-d(T-t)F0$. Let us denote the forward price of a forward contract written on the same asset today, at time t, and maturing in $T-t$ units of time by $F(t)$. We can substitute for the spot price $S(t)$ in terms of the discount factor and the forward price of a contract maturing in t units of time. According to the conclusion following the system of equations (2.9), the forward price $F(t)$ is $\frac{S(t)}{d(T-t)}$. Hence, $S(t) = F(t)d(T-t)$ and we arrive at another expression, equation (10.2), for the value of an "old" forward contract at time t.

$$d(T-t)(F(t) - F0) \tag{10.2}$$

Thus the value of the "old" contract today is the present value of the difference between the delivery price of the "new" and "old" contracts. It could be positive or negative, depending on whether the forward price today is smaller or larger than the forward price in the past. We therefore see that the value of the contract is positive if the delivery price is smaller than the forward price, i.e., $F0 < F(t)$ and vice versa.

As in most cases, there is more to the result in equation (10.2) than merely a mechanical substitution. The exercise actually points out some insights and certain financial courses of action open to an investor who wants out of a forward contract. Suppose you have a long position in an "old" forward contract on a particular asset. If you enter an opposite forward contract now on the same asset with the same maturity date, then you no longer need to deliver the asset. To understand this, note that you had a long position in the contract which meant you would be receiving the asset at the maturity date. If you hold a short position, then you would be obligated to deliver what you would have received. At maturity, you will be paying the old forward price and receiving the new forward price .

To enter a forward contract costs nothing. Hence, the cash flow which is a consequence of the transaction of taking an opposite position in another forward contract is the difference between the two forward prices. This cash flow, however, will be transacted at the maturity time of the opposing contracts. The value of this cash flow now is, thus, the discounted value of the difference between the two forward prices. Hence, the value of the "old" forward contract is given by equation (10.2).

The investigation of a multiperiod scenario facilitates an analysis of pricing forward contracts on assets that pay known cash flows during the life of

Price/Time	1	2	3	Security
\$94	\$105	\$0	\$0	B1
\$97	\$10	\$110	\$0	B2
\$85	\$8	\$8	\$108	B3

Table 10.1: A Simple Bond Market Specification

the contract. A very similar argument allows us to find the forward price of a stock that pays a known and fixed dividend yield during the life of the contract. These topics are examined next, starting with a forward contract on a coupon-paying bond, where we also use the opportunity to exemplify the value of a forward contract prior to maturity.

10.3 Forward Price of Assets That Pay Known Cash Flows

Consider the bond market outlined in Table 10.1. We use this market as an example through which we will explain the effect of the cash flows obtained from the asset during the life of the forward contract. The setting of the bond market is also as specified below by **NarbitB**. We use five approximating functions to estimate the term structure and define the discount function as **dis**.

```
> NarbitB([[105,0,0],[10,110,0],[8,8,108]],[94,97,85],\
> 5,dis);
```

The no − arbitrage condition is satisfied.

$$\textit{The discount factor for time, 1, is given by, } \frac{94}{105}$$

$$\textit{The interest rate spanning the time interval, [0, 1], is given by, } .1170$$

$$\textit{The discount factor for time, 2, is given by, } \frac{1849}{2310}$$

$$\textit{The interest rate spanning the time interval, [0, 2], is given by, } .2493$$

$$\textit{The discount factor for time, 3, is given by, } \frac{82507}{124740}$$

$$\textit{The interest rate spanning the time interval, [0, 3], is given by, } .5119$$

The function Vdis([c1, c2, ..]), values the cashflow [c1, c2, ..]

The continuous discount factor is given by the function, 'dis', (.)

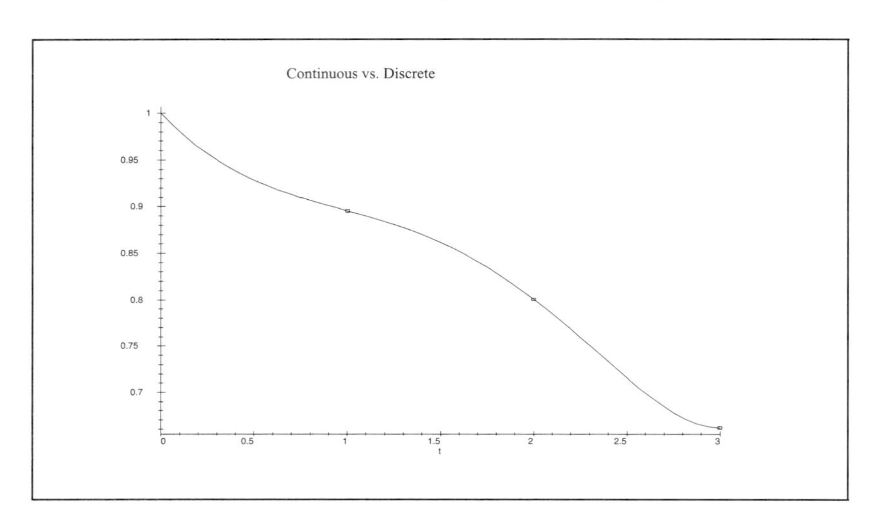

Figure 10.4: The Graph of the Function *dis*

As we can see from **NarbitB**, there are no arbitrage opportunities. As well, the continuous discount factor function, as seen in Figure 10.4, fits perfectly with the discrete discount factors. In this market there are three time periods and three different bonds. For simplicity one may think about the time period segments as years and assume that this is a market in which the bonds make annual coupon payments.

Consider a forward contract written at time zero specifying delivery of bond three at time two, immediately after the second coupon is paid. Bond three's cash flows are (8, 8, 108). According to the contract, the party with a long position will pay the forward price, F, at time two and will receive an instrument similar to a zero-coupon bond that will pay \$108 at time three. Hence, we can calculate the forward price based on the present value of the cash flow $(0, -F, 108)$ being equal to zero. Once we have solved for F we can also graph the payoff, as of time two, from such a contract. We leave this derivation as an exercise for the reader.

Recall our discussion using the cost-of-carry model to calculate the forward price. The idea was to buy the asset now, hold it until maturity, and deliver it at that time. The purchase of the asset was financed by a loan and hence, at the delivery time, the loan had to be repaid. The amount of money borrowed was equal to the spot price of the asset, and hence the amount to be repaid was the future value of the spot price. Consequently, a forward price, to avoid arbitrage opportunities, must be equal to the future value of the spot price. The only difference is that the purchase of an asset that pays known cash flows does not require taking a loan for the full amount of the spot price. The spot price of the bond is \$85, but the bond pays a coupon of \$8 at time one and at time two.

When the coupons are received, they can be reinvested at the risk-free rate of interest until the maturity time of the forward contract, time two. Hence, the cost of delivering the bond at time two will be less than the future value of the spot price. At time zero we can secure the rate that will be paid on the \$8 invested from time one to time two. This will be exactly the forward rate $r_0(1, 2)$, as we explained in Section 9.3. Similarly, the \$8 obtained at time two can be used at that time. Consequently, the net cost of delivering the bond at time three will be reduced by the future value of the \$8 obtained at time one and at time two. Hence, the cost will be

$$\frac{85}{dis(2)} - 8\,(1 + r_0(1,\, 2)) - 8. \tag{10.3}$$

We have already mentioned in equation (9.9) that, in general,

$$1 + r_0(t_1,\, t_2) = \frac{dis(t_1)}{dis(t_2)}.$$

Thus, (10.3) can be rewritten in terms of the function **dis** by applying this substitution. However, we wish to reinterpret this relationship here. This will also be useful in understanding how to derive these types of relationships.

One should always remember that a particular cash flow has two characteristics: its magnitude and its timing. The discount factors allow us to move cash flows through time, converting[1] cash of one time period to that of another. A useful methodology when converting cash in this manner is to "move" all the cash flows involved to a mutual point in time, and then to "move" this amount to the required point in time.

Proceeding in this manner, to find the cost of delivering the bond at time three, we first calculate the cost as of time zero. The bond costs \$85 at time zero but pays \$8 at times one and two. The \$8 paid at time one is worth

[1] One can conceptualize two different points in time as two different markets, similar to a foreign market. For example, there is a market at time t_1 and a market at time t_2 with different currency much the same as the U.S. and Canadian dollar. In the foreign market situation the exchange rate is the factor that allows us to convert from one currency to another. Similarly, the discount factor allows us to convert dollars of time t_1 to dollars of time t_2.

\$8 $d(1)$ at time zero. Similarly, the \$8 paid at time two is worth \$8 $d(2)$ at time zero. Hence, the total cost of delivering the bond at time three is

$$85 - 8\,dis(1) - 8\,dis(2) \tag{10.4}$$

in terms of dollars at time zero. In our case this will be

```
> 85-8*dis(1)-8*dis(2);
```

$$\frac{82507}{1155}$$

Therefore, the cost of delivering the bond at time two, in terms of dollars of time two, will be the future value of this amount, i.e., $\frac{85-8\,dis(1)-8\,dis(2)}{d(2)}$. Making use of equation (9.9), we see that this is exactly equation (10.3). In our example we calculate it as follows:

```
> (85-8*dis(1)-8*dis(2))/dis(2);
```

$$\frac{165014}{1849}$$

Or in a decimal form,

```
> evalf(%);
```

$$89.24499730$$

We can therefore make a general statement about the forward price of a forward contract on an asset that pays a known cash flow throughout the life of the contract. The forward price is the future value of the spot price, minus the future value, as of time zero, of the cash flow obtained during the life of the contract.

Let $S(0)$ be the spot price of the asset, and T the maturity of the contract, and assume that the asset pays a cash flow of $c(t)$ at time t. Let the discount factor for time t be denoted as $d(t)$. The general result for the future price F is thus

$$F = \frac{S(0)}{d(T)} - \frac{\sum_{t=1}^{T} c(t)\,d(t)}{d(T)}, \tag{10.5}$$

or in terms of the forward rates in equation (10.6),

$$F = S(0)(1 + r(T)) - \left(\sum_{t=1}^{T} c(t)\,(1 + r_0(t, T)) \right). \tag{10.6}$$

We can see that a positive cash flow obtained during the life of the contract reduces the forward price. We have alluded to a similar effect earlier in our discussions of forward contracts on commodities, Section 2.4.2, and on foreign currencies, Section 2.4.1. It is best understood utilizing the cost-of-carry model.

If the deliverable good produces positive cash flows, these cash flows can help with financing the purchase of the good on the spot market. Hence the amount of money borrowed to buy the good on the spot is reduced by the present value of the cash flow produced by the good to be delivered. In the foreign currency case, the positive cash flow was due to the interest earned on the currency in the foreign market.

In the case of a commodity, the cash flow produced (or its equivalent) can be positive or negative depending on the storage cost and the convenience yield. A similar argument, see Section 10.3.2, applies to a forward contract on a stock that pays a fixed dividend yield. To cement the ideas explored in the last two sections we investigate in the next section the forward price, prior to maturity, of assets that pay a known cash flow — a coupon-paying bond.

10.3.1 Forward Contracts, Prior to Maturity, of Assets That Pay Known Cash Flows

Consider the same forward contract as in Section 10.3. Let us see what the forward price will be if the delivery time is $t = 1.5$ instead of time $t = 2$. The maturity date of this contract does not coincide with any cash flow payment date of the asset. Indeed, there is no conceptual change. One needs only to be careful with the timing and the specification of the cash flows. In our example, we can apply equation (10.5) to obtain the forward price as

```
> (85-8*dis(1))/dis(15/10);
```

$$\frac{55305448704}{611850385}$$

or in decimal form

```
> (85-8*dis(1))/dis(1.5);
```

$$90.39047790$$

Suppose that a year has passed and we take another snapshot of the same market. The first bond has matured and we will assume that no new bonds have been issued. The second bond will mature in a year and will pay \$110. The third bond will mature in two years and will pay \$8 in one year, and \$108 in two years. The prices of these two outstanding bonds are now assumed to be different than they were one year ago. They are \$90 and \$80, respectively. We run **NarbitB** based on this specification to get an estimate of the term structure. The current time zero is of course time one of the last model. (We suppress the graph by adding a fifth parameter and assigning it a zero value.) If you would like to see the graph of the discount factor function, rerun **NarbitB** as below but without the zero as the last input parameter.

```
> NarbitB([[110,0],[8,108]],[90,80],3,dis1,0);
```

The no − arbitrage condition is satisfied.

$$\textit{The discount factor for time, 1, is given by, } \frac{9}{11}$$

The interest rate spanning the time interval, [0, 1], *is given by,* .2222

$$\textit{The discount factor for time, 2, is given by, } \frac{202}{297}$$

The interest rate spanning the time interval, [0, 2], *is given by,* .4703

The function Vdis([c1, c2, ..]), values the cashflow [c1, c2, ..]

The continuous discount factor is given by the function, 'dis1', (.)

We name the current discount factor as **dis1** and we use three estimating functions for the continuous approximation of the discount function. As before, we see that there are no arbitrage opportunities and we obtain a perfect fit for the estimation.

Consider the forward contract discussed above. We will refer to this contract as the "old" forward contract. The holder of the short position of this contract needs to deliver the 8 percent bond in one year, immediately after the coupon payment. The value of the short position in the "old" forward contract is no longer zero. The value of the long position in the same contract is not zero either: it is the negative of the value to the party with the short position. The short position in the forward contract obligates the investor to deliver the specified bond for $\$\frac{165014}{1849}$. We can use the function **dis1** to calculate the cash flows which are consequent from this contract, and so derive the value of this contract today.

In other words, the party with the long position in the bond will receive the bond. This is equivalent to obtaining \$108 in two years and paying out $\$\frac{165014}{1849}$ in one year. Hence the value of the cash flow $\left(-\frac{165014}{1849}, 108\right)$ can be calculated as

```
> dis1(1)*(-165014/1849)+dis1(2)*108;
```

$$\frac{806}{1849}$$

or in decimal form

```
> evalf(%);
```

$$.4359113034$$

The value is not zero because the "old" forward price of the contract, the delivery price, does not reflect the circumstances of the market today.

Let us see what the value of the forward contract on that bond would be if the contract were issued today. Following the same arguments as above, equation (10.5), we can find the forward price as

```
> (80/dis1(1))-8;
```

$$\frac{808}{9}$$

or alternatively as

```
> (108*dis1(2))/dis1(1);
```

$$\frac{808}{9}$$

Can you explain this alternative way? We leave this explanation as an exercise. This idea is also explored in the end-of-Chapter exercises.

We know already from equation (10.2) the relation which holds between the forward price of the "old" and "new" contracts. We repeat the explanation here with reference to the contracts we are examining. We calculate the difference in the forward prices of the two contracts.

```
> (808/9)-(165014/1849);
```

$$\frac{8866}{16641}$$

If an investor holds a short position now in the "new" contract, and a long position in the "old" contract, it is equivalent to the cash flow of $\frac{8866}{16641}$ one year from now. Consider why this is the case. The long position in the "old" contract obligates the investor to pay $\frac{165014}{1849}$ in a year and receive the bond. The short position in the "new" contract obligates the investor to deliver the bond (which the investor received as a consequence of the "old" contract) and to receive $\frac{808}{9}$. Therefore, the portfolio composed of a long position in the "old" contract and a short position in the "new" contract amounts to the above cash flow received in one year. Thus, the value of this portfolio today will be given by

```
> dis1(1)*((808/9)-(165014/1849));
```

$$\frac{806}{1849}$$

This value happens to be the value of the "old" forward contract now. This result is not a coincidence, but rather a general result as we have seen before. It is a consequence of the fact that the price of the portfolio will be the value of the old forward contract since the value of the new contract is now zero.

10.3.2 Forward Price of a Stock That Pays a Known Dividend Yield

Consider a stock that pays a fixed and known dividend yield. Usually we assume that the dividend is paid in a continuous manner, at every instant of time. The dividend is a certain percentage of the stock's value. Thus, if the stock price at time t is denoted by $S(t)$, the dividend payment, at every instant t, is $y\,S(t)$. We use y to denote the continuous dividend yield — the percentage of the stock price that is paid as dividends. As the argument depicted below shows, the case of the forward contract on a stock which pays a fixed dividend yield follows an already known derivation. It is in the same vein and uses the same logic as utilized in the calculation of the future value based on the continuously compounded rate of interest.

As usual, let time zero denote the current time and assume a forward contract obligating the holder of the short position to deliver one unit of the stock at time T. We would like to apply the cost-of-carry model to calculate the forward price of the stock as of time zero. The asset, the stock in this case, produces income between now and time T. Therefore, we do not have to borrow $S(0)$, the spot price of the stock, in order to purchase one unit of the stock. We can take advantage of the income stream.

There is, however, a slight difference between this case and the other cases of assets that produce income streams. In this case the income stream at time t is not known. The stock price at time t, $S(t)$, is a random variable. This is in contrast to the example of a coupon-paying bond for which the income is known with certainty.

Nevertheless, we can still analyze this case without assuming the price process or making any other assumptions regarding a random nature of the price process. The forward contract commits the party with the short position to deliver one unit of the stock. If there is a certain (nonrandom) strategy that will produce a unit of the stock at time T and for which the cost of the strategy is known now, then the future value of that cost must be the forward price. If this were not true, arbitrage opportunities would exist.

The key idea is to have an investment strategy which is certain to produce one unit of the stock at time T. We do not actually care about the value of the stock at time T, or at any time t prior to time T. As long as our strategy has a known cost and is certain to produce one unit of the stock at time T, that is what we require.

Consequently, we are looking for an investment vehicle for which the value of the portfolio is perfectly correlated with (or mimics exactly) the movement of the stock price. If such an investment strategy can be found (and followed by the party with the short position in the forward contract), then there is no further need to be concerned about the value of the stock price. The commitment of the party with the short position is to deliver one unit of the stock, not a specified dollar value of stock.

Let us now examine if we can find such a strategy, given the structure of the problem. Assume that at time zero we buy X units of the stock. The writer of the contract is obligated to deliver a unit of the stock at time T and is guaranteed an income in the form of dividends from the stock. Hence, applying a variant of the cost-of-carry model, the writer should buy less than a unit of the stock, i.e., $X < 1$, and use the dividend income to purchase more units of the stock along the way.

The main point is the fact that the dividend is a certain fixed percentage of the stock value. Hence, regardless of the value of the stock, if at time zero X units of the stock are purchased and the dividends are immediately reinvested in the stock, the holding of stocks, (number of units, not value) at time t becomes deterministic.

Assuming that the above strategy is followed, let $X(t)$ be the holding of the stock at time t, so $X(0) = X$. At each time t the growth rate of $X(t)$ is y. This very much resembles the notion of continuous compounding at a known risk-free rate r. In fact it is exactly the same situation. We know, therefore, from Section 3.4.1, that at time T the holding of the stock will be $X(T) = X(0)e^{yT}$, and thus for $X(T)$ to be 1, $X(0)$ should be e^{-yT}. Thus, if the investor purchases e^{-yT} units of the stock at time $t = 0$ at a cost of $e^{-yT}S(0)$ and follows the above strategy, the investor is guaranteed to have a unit of the stock at time T.

Consequently, applying the usual arguments of the cost-of-carry model, which indeed is the pricing by replication argument, the writer of the contract can borrow $e^{-yT}S(0)$, to be paid back at time T, and can buy the stock. Following this strategy, the writer will have a unit of the stock at time T and will have to repay the loan. The loan repayment will be $e^{-yT}S(0)e^{r(T)T}$, where $r(T)$ is the continuously compounded interest rate from time zero to time T. Hence, to avoid arbitrage opportunities, the forward price must be the future value of $e^{-yT}S(0)$, which is

$$e^{-yT}S(0)e^{rT} = S(0)e^{(r-y)T}. \tag{10.7}$$

10.4 Eurodollar Contracts

A *Eurodollar* is a U.S.-dollar-denominated deposit in banks outside the USA. Eurodollars are similar to *Treasury bills or zero-coupon bonds,* as they are discount instruments. The interest rate earned on a Eurodollar deposit is higher than the Treasury bill rate. The Eurodollar is not a risk-free instrument since it is invested with a bank and not backed by a government. Thus, it should offer a higher rate than government bonds in order to compensate for the risk. The interest rate associated with a Eurodollar is known as *LIBOR*, or ***London Interbank Offer Rate,*** as it is the rate earned by one bank on Eurodollars deposited with another bank.

As the term LIBOR implies, most Eurodollar deposits are issued in London. A LIBOR term structure of interest rate can be estimated and used to value Eurodollars belonging to a different risk category than government bonds. This is much the same as the explanation regarding a term structure for municipal or corporate bonds offered in Section 9.4.

10.4.1 Forward Rate Agreements (FRAs)

A *forward rate agreement* is a contract very similar to the equity swap explained in Section 2.5.2. It can be perceived as a swap of a variable return for a fixed return on some future period. The FRA is a contract obligating the one who holds it short to pay the LIBOR rate at time t_2, on a notional principal invested from some future time t_1 to time t_2, where zero is the current time and $0 < t_1 < t_2$.

The one who holds the contract long pays at time t_2 a fixed return on the notional principal, but the return is agreed upon at time zero. This return is referred to as the FRA rate. As with other forward contracts, no cash changes hands at time zero. The FRA rate is fixed so that the value of the contract at time zero, the initiation date, is zero.

The underlying asset in this contract is usually the LIBOR rate. The contract will be realized to be profitable to the one who holds it short, if the LIBOR rate spanning the time interval (t_1, t_2) will be smaller than the FRA rate. The payoff to the one who holds it long at time t_2 is thus

$$N\,(L_{t_1}(t_1,\, t_2) - F), \tag{10.8}$$

where N is the notional principal, F is the FRA rate, and $L_{t_1}(t_1, t_2)$ is the LIBOR rate,[2] as of time t_1, spanning the time interval t_1 to t_2. Note that the

[2] The LIBOR term structure is actually estimated based on FRA agreements. The FRA

rate $L_{t_1}(t_1, t_2)$ is known at time t_1, but not at time zero (when the parties enter the contract). Hence, in this situation F plays the role of a forward price, which is a rate in this case.[3] In order to value F we proceed to seeking a portfolio replicating the LIBOR return from time t_1 to time t_2. We have, in fact, seen how this is done before in Section 2.5.2. Here, however, the return is on some future period. We have to be careful about the two types of risk categories involved. If we would like to have no risk in replicating the LIBOR rate from time t_1 to time t_2, we should use risk-free instruments to insure having $\$N$ to invest at time t_1 at the LIBOR rate.

Let us apply a cost-of-carry-like model which, as we mentioned before, is really the replication argument. We will approach the replication from the point of view of the short seller and set up a portfolio with cash flow only at time t_2. As explained above, in order to have $\$N$ with no risk of default at time t_1, one needs to have $\$\frac{N}{1+r(t_1)}$ at time zero, where $r(t_1)$ is the risk-free interest rate from time zero to t_1.

Investing $\$\frac{N}{1+r(t_1)}$ at the risk-free rate from time zero to time t_1 will give rise to $\$N$ at time t_1, which will be invested at the LIBOR rate to time t_2. Hence, at time t_2, the holder of the short position will have $N\left(1 + L_{t_1}\left(t_1, t_2\right)\right)$, will deliver $NL_{t_1}(t_1, t_2)$, and will receive NF. The remaining issue (which is a bit sticky) is how to finance having $\$\frac{N}{1+r(t_1)}$ at time zero, but let us look first at the current pattern of cash flow. We define the following array in MAPLE where the first component is the cash flows at time zero, the second is that at time t_1, and the third is that at time t_2 (for simplicity we omit the subindex t_1 in the MAPLE calculations).

```
> [time0=-N/(1+r(t1)),timet1=0,\
> timet2=N*(1+L(t1,t2))-N*L(t1,t2)+N*F];
```

---

data convey to us the forward rate in the LIBOR market. Hence in order to estimate the term structure of LIBOR we have to utilize equation (10.8) to derive the term structure from the forward rate. In our discussion regarding the forward rate in the government bond market we derive the forward rate from the term structure. In the LIBOR market we have to do the reverse, as demonstrated by an exercise at the end of this Chapter

[3] These agreements are usually settled in cash on the initiation of the forward period. For example, if the agreement is signed on date t_0 for a forward rate from date t_1 to t_2, it is settled on date t_1 in cash. That is, the payoff to the holder of the short position is the present value of (10.8), namely, $\frac{N\left(L_{t_1}(t_1, t_2) - F\right)}{L_{t_1}(t_1, t_2)}$.

If the FRA is done with respect to the term structure of government bonds and both parties are federal agencies with no risk of default, the rate of the FRA must be the forward rate implied in the term structure. The reader is asked, at the end-of-Chapter exercises, to prove this in two ways — through replication and through valuation by discount factors.

$$[time0 = -\frac{N}{1 + \mathrm{r}(t1)},\ timet1 = 0,$$
$$timet2 = N\,(1 + \mathrm{L}(t1,\ t2)) - N\,\mathrm{L}(t1,\ t2) + N\,F]$$

We simplify it and then collect terms based on N.

```
> simplify(%);
```

$$[time0 = -\frac{N}{1 + \mathrm{r}(t1)},\ timet1 = 0,\ timet2 = N + N\,F]$$

```
> collect(%,N);
```

$$[time0 = -\frac{N}{1 + \mathrm{r}(t1)},\ timet1 = 0,\ timet2 = (1 + F)\,N]$$

It is apparent now that without loss of generality N can be set to one. Furthermore, for the value of the contract to be zero at the initiation time, F should be chosen such that the present value, as of time zero, of $1 + F$ obtained at time t_2 will be $\frac{1}{1+r(t_1)}$. Let us substitute $N = 1$ in the above and investigate the issue further.

```
> subs(N=1,%);
```

$$[time0 = -\frac{1}{1 + \mathrm{r}(t1)},\ timet1 = 0,\ timet2 = 1 + F]$$

To solve for F we need to know how to value either $1 + F$ at time zero or $\frac{1}{1+r(t_1)}$ at time t_2. Consider the $1 + F$ obtained at time t_2. The issue of how to value it at time zero amounts to a question of determining its risk category. Should we discount it with the risk-free rate or with the LIBOR rate?

In our investigation above we suppress the default risk by one of the parties, and assume that they are in the same risk category. Since the holder of the short position of the contract pays interest on Eurodollars at time t_2, we should treat $1 + F$ as being in the same risk category. Hence, moving the $1+F$ to time zero amounts to discounting it at the LIBOR rate. Consequently, F should satisfy equation (10.9) since entering the contract costs nothing.

$$-\frac{1}{1 + r(t_1)} + \frac{1 + F}{1 + L_0(0,\ t_2)} = 0 \tag{10.9}$$

We can call upon MAPLE to solve for F as below.

```
> solve (-1/(1+r(t1))+(1+F)/(1+L[0](0,t2))=0,F);
```

$$-\frac{-L_0(0,\ t2) + \mathrm{r}(t1)}{1 + \mathrm{r}(t1)}$$

Alternatively, we can determine how to finance the $\frac{1}{1+r(t_1)}$ needed at time zero. This amounts to valuing it at time t_2. Again, we consider the issue of what its risk category is. Since we want to have N at time t_1 with no default risk we must invest it at the risk-free rate from time zero to time t_1. However, if we assume that both parties to the contract are at the LIBOR risk category, the commitment of the one who holds the contract short to pay $NL_{t_1}(t_1, t_2)$ should also be replicated based on the Eurodollar risk category. Therefore the alternative way is to finance the funds needed at time zero by borrowing (or taking a deposit of) Eurodollars at time zero to be held until time t_2. By arbitrage arguments the cash flow at time t_2 should also be zero,which gives rise to equation (10.10).

$$1 + F = \frac{1 + L_0(0, t_2)}{1 + r(t_1)} \tag{10.10}$$

Namely, what is being received from holding the contract long should equal the loan repayment taken to create this position. Clearly, equation (10.10) yields the same value for F as does equation (10.9).

10.5 Futures Contracts: A Second Look

Futures contracts are very similar instruments to forward contracts. There are nevertheless some important distinctions between the two instruments. We have outlined some of these features below, particularly those essential for the valuation of futures contracts. A description of the institutional details surrounding futures contracts can be found in the web cite of the CBOT (www.cbot.com/ourproducts/financial/index.html). If you are reading the book on-line, click on the above URL to be transferred to the above site.

A futures contract is exchange traded and is consequently a highly standardized instrument. It is easier to get out of a certain position in a futures contract than the same position in a forward contract, making use of the strategy outlined in Section 10.2 after equation (10.2). Taking the opposite position to a currently held position is relatively straightforward to accomplish in the futures market.

Conversely, getting out of a forward contract position is potentially much more difficult since it is generally the case that forward contracts are tailor-made for a particular client. As a result of this tailoring, it may be difficult to find a party with whom to establish the opposite position to that originally taken.

In order to reduce the risk of default implicit in these types of contracts (a point which we have ignored thus far in our discussion) and to protect the exchange, a certain procedure is followed. Profit or loss from a particular position in a futures contract is calculated and settled for each investor's account at the end of each trading day. This process is called *marking to market*, and will be explained shortly.

By virtue of marking to market, the consequent cash flows from a futures contract are a series of cash flows. These occur every trading day from the day the position is initiated until its maturity (or until the contract position is closed out). In contrast, the consequent cash flows of a forward contract occur only at one point in time, the maturity date.

A futures contract commits two parties to an exchange of a certain good or financial asset (the underlying asset) at a certain agreed-upon date at a certain agreed-upon price. This price is the *futures price*. In this respect, a futures contract is parallel to a forward contract. Also, as in the case of forward contracts, the futures price is decided on the initiation date in such a way that the value of the contract at that time vanishes.

The mechanism of marking to market may be thought of as writing a new futures contract at the end of each trading day. The new contract has a new futures price which makes the value of the contract zero as of that day. The difference between the futures price from the previous day and the new one is paid to the party whose position has positive value that day. It is subtracted from the account of the party whose position had negative value that day. The two values perfectly offset each other since the net value of the two positions must be zero.

If the futures price today is higher than the futures price yesterday, the party with the long position profits as the contract requires a higher price in order to take delivery of the underlying asset. The value of this futures contract today is positive. The party with the short position pays the difference between the futures price of yesterday and that of today to the holder of the long position. The direction of cash flows is reversed if the futures price today is lower than that of yesterday.

In any event, unless there is no change in the futures price, money changes hands and a new futures contract with the same details as the original, but for a different price, is written today. This process repeats itself every trading day until the maturity date of the contract. At maturity, the party with the long position pays the writer of the contract the spot price of the underlying asset and takes[4] possession of the security.

---

[4]Very few futures contracts end in delivery of the underlying asset at maturity. Most futures contracts are settled in cash by the investors close to or prior to maturity. Updated information regarding characteristics of exchange-traded futures contracts in terms of standardization, margin requirements, daily settlement, the clearinghouse mechanism, and

Let us consider a symbolic example to clarify the description. We will examine the cash flows from the point of view of the party with a long position in the futures contract. A negative cash flow therefore means that the party with the long position is making a payment to the party with a short position. Let us denote the futures price of the underlying security on date t by $\mathrm{fr}(t)$. The payment on day t to the party with the long position in the futures contract is therefore $\mathrm{fr}(t) - \mathrm{fr}(t-1)$. Let us generate a sequence of futures prices from day 1 to day 9, assuming that the futures contract matures on day 10. The futures prices for each day are noted below. The **seq** command generates what is referred to in MAPLE as an expression sequence.

```
> seq(fr(i),i=1..10);
```

$$\mathrm{fr}(1),\ \mathrm{fr}(2),\ \mathrm{fr}(3),\ \mathrm{fr}(4),\ \mathrm{fr}(5),\ \mathrm{fr}(6),\ \mathrm{fr}(7),\ \mathrm{fr}(8),\ \mathrm{fr}(9),\ \mathrm{fr}(10)$$

On the maturity day of the contract, $t = 10$, the futures price is the spot price of the security $s(10)$, that is,

```
> fr(10):=s(10);
```

$$\mathrm{fr}(10) := \mathrm{s}(10)$$

On day one, the party with the long position enters the contract and no cash changes hands since the value of the contract to both parties on that date is zero. On the second trading day, the party with the long position receives payment of $\mathrm{fr}(2) - \mathrm{fr}(1)$ and the futures price of the contract is changed to $\mathrm{fr}(2)$. We denote the payment to the party with the long position of the futures contract on day i, for $i = 2, ..., 10$, by **PayDay**(i) and define this in MAPLE. The value of **PayDay**(1) is zero so we just omit it from the discussion. This is done via the MAPLE **do** command below.

```
> for i from 2 to 10 do;
> PayDay(i):=fr(i)-fr(i-1);
> od;
```

$$\mathrm{PayDay}(2) := \mathrm{fr}(2) - \mathrm{fr}(1)$$
$$\mathrm{PayDay}(3) := \mathrm{fr}(3) - \mathrm{fr}(2)$$
$$\mathrm{PayDay}(4) := \mathrm{fr}(4) - \mathrm{fr}(3)$$
$$\mathrm{PayDay}(5) := \mathrm{fr}(5) - \mathrm{fr}(4)$$
$$\mathrm{PayDay}(6) := \mathrm{fr}(6) - \mathrm{fr}(5)$$
$$\mathrm{PayDay}(7) := \mathrm{fr}(7) - \mathrm{fr}(6)$$
$$\mathrm{PayDay}(8) := \mathrm{fr}(8) - \mathrm{fr}(7)$$

---

regulations can be found at www.cbot.com/ourproducts/financial/index.html.

$$\text{PayDay}(9) := \text{fr}(9) - \text{fr}(8)$$
$$\text{PayDay}(10) := \text{s}(10) - \text{fr}(9)$$

On the maturity date of the futures contract, the holder of the contract (the party with the long position) pays the party with the short position the futures price of yesterday, fr(9), and receives the underlying security. The cash flow on the date on which the futures contract matures is thus $s(10) - \text{fr}(9)$. We ignore for a moment the concern of time value of money. Hence, we sum the payments received by the party with the long position in the futures contract on each trading day between the initiation of the contract and its maturity. This is done with the next MAPLE command.

```
>  sum('PayDay(i)','i'=2..10);
```

$$-\text{fr}(1) + \text{s}(10)$$

So we see that, ignoring the time value of money, the party with the long position in the futures contract pays fr(1), the futures price agreed upon on the day the contract was written, and in exchange receives the underlying security. This exchange is identical to that of a forward contract written on the same underlying asset and maturing on the same date. The cash flows from a forward contract and from a futures contract are, of course, not the same since we have ignored the time value of money.

Note also that the $\text{fr}(i)$ of the futures contract are not known with certainty on the date the futures contract is initiated. Nevertheless, ignoring the time value of money, the result is always that the cash flows of a futures contract are like those of a forward contract. As we have mentioned already, if there is one time period before the expiration of the futures contract (i.e., one trading day), then, ignoring the institutional details of these two contracts, they are identical.

Due to the different cash flow profiles which result from a forward contract and a futures contract, the forward price is not generally equal to the futures price. In order to calculate the value of a futures contract and the futures price, some assumptions regarding the properties of the evolution of the term structure must be delineated. The valuation of a futures contract utilizes the concept of risk-neutral valuation, studied in the first Chapter, but this time applied to the market for fixed income securities. This will be reinvestigated in a later Chapter.

The calculation of the futures price and the valuation of a futures contract is simpler, however, if we assume that the term structure is deterministic. In fact, in such an environment, forward and futures prices are equal if the default risk is assumed away. We thus first explain what is meant by a deterministic term structure, and then proceed to prove this equality in such an environment.

10.6 Deterministic Term Structure (DTS)

In a market where the term structure is deterministic one knows with certainty the spot interest rate which will prevail in the market from any future time t_1 to time t_2, $0 < t_1$. As usual, we assume that time zero is the current time. Of course, one always knows the prevailing term structure at the current time. It turns out that this certainty assumption together with the absence of arbitrage opportunities means that the spot interest rate for future time periods must be equal to the forward interest rate $r_0(t_1, t_2)$. The reader is asked in an exercise at the end of the Chapter to show that this must be the case in a market with a deterministic term structure.

Graphically, we can visualize the evolution of the term structure or the discount factor function as shown below. We first recalculate the discount factor function for a new market and define it to be **dis**$(\cdot)$. This is done by rerunning **NarbitB**.

```
> NarbitB([[105,0,0],[5,105,0],[4,4,104]],[94,97,85],\
> 10,dis,0);
```

The no − arbitrage condition is satisfied.

$$\textit{The discount factor for time, } 1\textit{, is given by, } \frac{94}{105}$$

$$\textit{The interest rate spanning the time interval, } [0,\ 1]\textit{, is given by, } .1170$$

$$\textit{The discount factor for time, } 2\textit{, is given by, } \frac{1943}{2205}$$

$$\textit{The interest rate spanning the time interval, } [0,\ 2]\textit{, is given by, } .1348$$

$$\textit{The discount factor for time, } 3\textit{, is given by, } \frac{171757}{229320}$$

$$\textit{The interest rate spanning the time interval, } [0,\ 3]\textit{, is given by, } .3351$$

The function Vdis([c1, c2, ..]), values the cashflow [c1, c2, ..]

The continuous discount factor is given by the function, 'dis', (.)

```
> SumAbsDiv;
```

0

The reader may note that we have chosen 10 approximating functions and that the discount function **dis** perfectly fits the discount factors of times one, two, and three. If that term structure is deterministic, then the discount factor function which will prevail from time t_1 to time t, $t_1 < t$, must be given by

$$d(t_1, t) = \frac{dis(t)}{dis(t_1)}. \tag{10.11}$$

The reader should recognize that (10.11) is a consequence of perceiving the discount factors as factors which convert dollars from one time period to dollars of another. The value at time zero of one dollar to be obtained at time t is $d(t)$. The conversion in equation (10.11) can be perceived as being composed of two separate steps. First, convert a dollar from time t to its value in terms of dollars of time t_1, where $0 < t_1 < t$, using the discount factor (conversion factor) $d(t_1, t)$. Then convert that dollar value at time t_1 to a dollar value at time zero. If there are no arbitrage opportunities, the equality

$$d(t) = d(t_1)d(t_1, t) \tag{10.12}$$

must hold. Consequently, the discount factor as of time t_1, $d(t_1, t)$, which converts dollars received at time t to dollars of time t_1, is given by (10.11).

We can now graph the evolution of the discount factor function. The original discount function is displayed (i.e., the discounted dollar of any time period t at time zero) in Figure 10.5. The bottom (red) graph represents the discount factor function as of time $t = 0$, the middle (green) graph represents the discount factor function as of time $t = 0.3$, and the top (blue) graph represents the discount factor function as of time $t = 0.5$.

```
> plot([dis(t),dis(t)/dis(.3),dis(t)/dis(.5)],t=0.5..2,\
> titel=`Figure 10.5:  Discount Factor Function\
> for t=0, t=0.3 and t=0.5`,\
> colour=[red,green,blue],thickness=2,\
> labels=[time, `disc.fac.`]);
```

The same idea can be viewed in three dimensions, as shown in Figure 10.6 and plotted below.

```
> plot3d(dis(t)/dis(t0),t0=0..3,t=t0..3,title=`Figure 10.6:\
> The Discount Factor as of Time t0, for t0=0,...,3`\
> axes=normal,orientation=[61,48],style=patch,\
> labels=[`Time t0`,`Time t`,`discfac.`]);
```

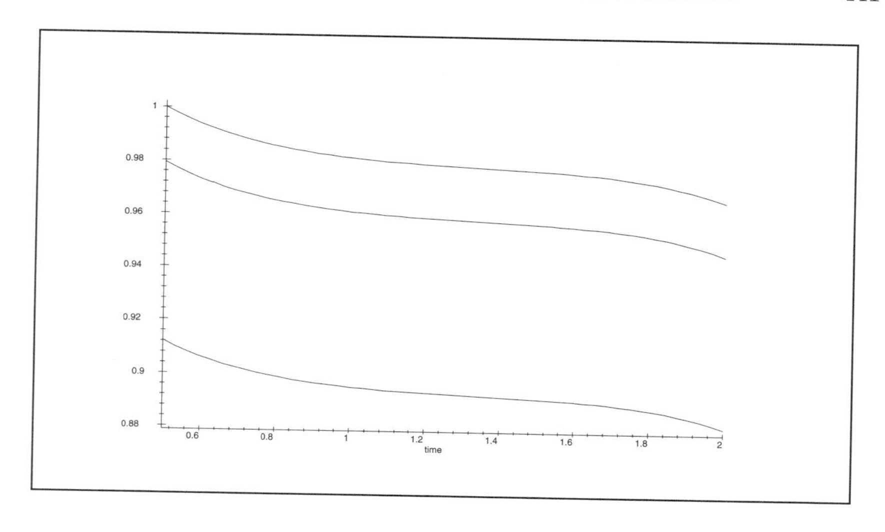

Figure 10.5: Discount Factor Functions for $t = 0$, $t = 0.3$, and $t = 0.5$

10.7 Futures Contracts in a DTS Environment

Assume that the discount factor function is $d(t)$. When the term structure is deterministic, what can be said about the relationship between a futures contract and a forward contract? To answer this question, as in [14] and [26], we try to find a portfolio composed of futures and forward contracts from which the required result can be deduced via arbitrage arguments.

A futures contract generates cash flows during its life while a forward contract does not. The cash flow from a forward contract is obtained only at maturity of the contract. Using this fact, we can attempt to mimic a forward contract by investing the cash flow obtained during the life of the futures contract until its maturity date. We maintain the notation of $S(t)$ for the price of the underlying asset at time t, T as the maturity time, $t = 1$ as the day on which the contract is initiated, and F as the forward price. As we demonstrated above (ignoring the time value of money), the cash flow from the futures contract to the party with the short position in the contract is $\mathrm{fr}(1) - S(T)$. The forward contract generates a cash flow of $S(T) - F$ to the party with the long position in the contract, only at T.

Since we have assumed that the term structure is deterministic, perhaps we will be able to devise a strategy whereby at each marking to market time an investment in a futures contract is also possible together with investing

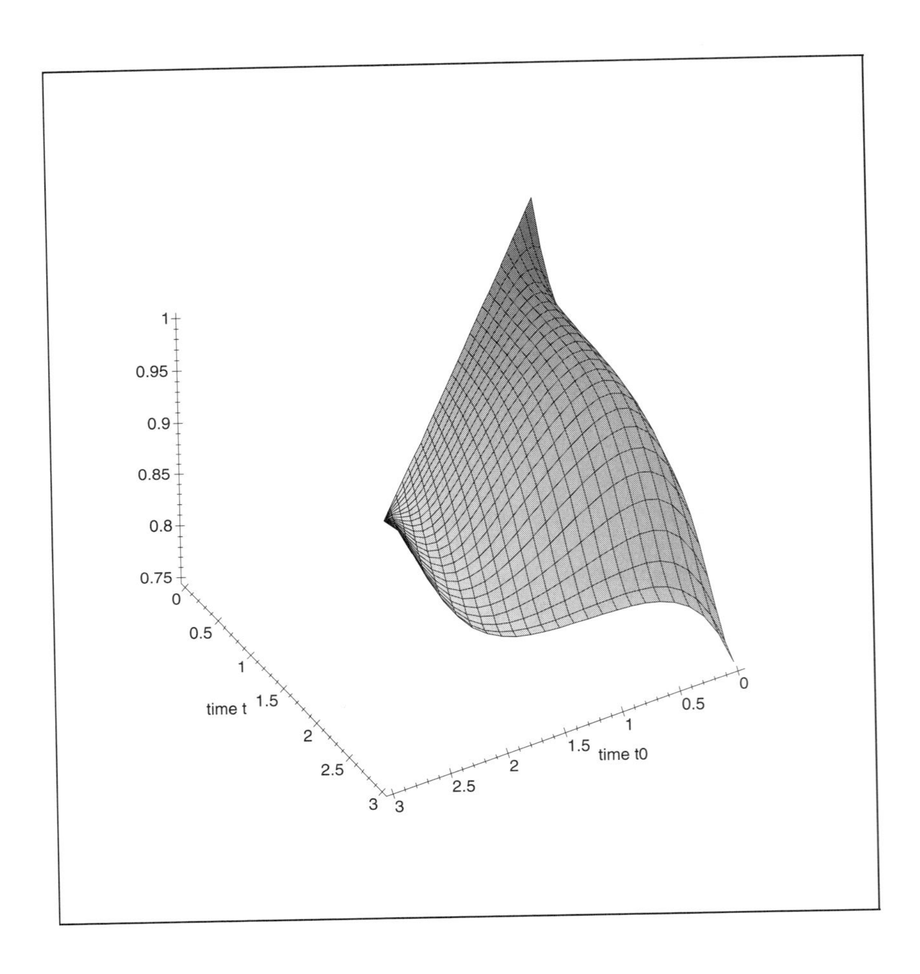

Figure 10.6: The Discount Factor as of Time $t0$, for $t0 = 0, ..., 3$

and borrowing at the risk-free rate of interest. Our aim will be to find an investment strategy that offsets the cash flow from the marking to market by investing those cash flows at the risk-free rate until the maturity time of the contract. Because the term structure is deterministic, we know the interest rate that will prevail in the market in the future. This knowledge may allow us to reenter the futures contract, knowing how many units of the original contract to buy so as to offset the payoff at maturity from the investment made at the risk-free rate. Since entering a futures contract requires no cash up front, we are aiming to generate an investment strategy that results in a cash flow only on the maturity date of the futures contract. We would like to have a portfolio and an investment strategy involving the futures contract and the risk-free rate so that the cash flow during the time interval $[1, T]$ is obtained only at time T and is equal to $-\mathrm{fr}(1) + S(T)$. This is the same cash flow that is obtained from the futures contract if the time value of money is ignored.

Since entering into a futures contract costs nothing (as is also the case for a forward contract), does this mean that if we find a portfolio of futures contracts such as described above we can conclude that equation (10.13) is true?

$$F - S(T) - \mathrm{fr}(1) + S(T) = 0 \tag{10.13}$$

This must be the case, since equation (10.13) will be the cash flow from a long position in a forward contract and from the portfolio we found above. Neither entering a forward contract nor entering the portfolio found above costs anything. Hence, the resultant cash flows from this combined position must also be zero on the date on which they are initiated. Otherwise arbitrage opportunities exist. Equation (10.13), however, leads to the conclusion that the futures price and the forward price are equal in this deterministic interest rate environment.

Assume that we take a long position in a forward contract that matures two days from today. As before, let F denote the forward price of the contract, $S(i)$ the spot price of the asset at date i, and $\mathrm{fr}(i)$ the futures price at date i. Let time one be the current time, the first day of the contract. The discount factor function is known and deterministic, where $d(i)$ is the discount factor from time one to time i. Our method is simply to try and work backward and use the same approach of the simpler model of Chapter 1. If we treat each one-day time period as if it were the simple one-period model, two consecutive models will generate a two-time-period model.

Consider an example where a contract is initiated on day one and thus there is no marking to market for the futures contract on this day. On day two marking to market occurs. The contract matures on day three.

We will work through this example carefully to demonstrate the marking to market process. A long position in the forward contract gives rise to a payoff of $S(3) - F$ on day three. Our aim is to try and find a portfolio consisting of futures contracts and risk-free investments that will cost nothing to establish and will produce a cash flow of $\mathrm{fr}(1) - S(3)$ only on day three. Suppose on day one we take a position in x_1 futures contracts. As usual, our notation convention is that a positive value of x_1 stands for a long position while a negative value of x_1 refers to a short position. The first marking to market occurs a day later, at the end of day two, and we have the cash flow of $x_1(\mathrm{fr}(2) - \mathrm{fr}(1))$. We define this object in MAPLE as **PayDay(2)**.

```
> PayDay(2):= x1*(fr(2)-fr(1));
```

$$\mathrm{PayDay}(2) := x1\ (\mathrm{fr}(2) - \mathrm{fr}(1))$$

If $\mathrm{fr}(2)$ is greater than $\mathrm{fr}(1)$ there is a positive cash flow to the party with the long position and vice versa. In any event, we invest this amount at the risk-free rate of interest. That is, we borrow if the cash flow of our position is negative and we lend if the cash flow of our position is positive. We do this immediately to accomplish our goal of having actual cash flows happen only at the end of day three.

We can also enter, at the end of this day (day two), into a futures contract that matures one day later (day three). We may need it to offset some cash flow. Suppose we took a position in x_2 units of a futures contract on the same underlying asset. This costs zero and hence does not alter the cash flow at the end of day one.[5]

Now consider the cash flow at the end of day three. The original contract, entered into on day one, requires delivering $x_1 S(3)$ of the underlying asset and receiving $x_1 \mathrm{fr}(2)$. We also took x_2 positions in a futures contract on day two which results in a cash flow of $x_2\,(S(3) - \mathrm{fr}(2))$. Finally, the amount invested at the risk-free rate on day two will have grown to $x_1(\mathrm{fr}(2) - \mathrm{fr}(1))(1 + r_2(2,\,3))$, where $r_2\,(2,3)$ is the spot rate prevailing at time two spanning the time interval $[2,3]$. Indeed, in a deterministic interest rate environment

$$1 + r_2(2,3) = \frac{d(3)}{d(2)} \tag{10.14}$$

and thus it is known as of day one.

[5] "Day" does not necessarily refer to a calendar day. Instead, it should be interpreted as the time period that elapses between consecutive marking to markets.

However, for the time being we will continue using $1 + r_2(2,3)$. In the upcoming MAPLE calculation we will further simplify this expression as $1+r$. We revert to the substitution as in (10.14) at the end of our argument. This will emphasize the need for the assumption of a deterministic term structure.

In total, therefore, the cash flow at day three is as defined below by **PayDay(3)**.

```
> PayDay(3):=x1*(S(3)-fr(2))+x2*(S(3)-fr(2))+\
> x1*(fr(2)-fr(1))*(1+r);
```

$$\mathrm{PayDay}(3) := \mathit{x1}\,(\mathrm{S}(3) - \mathrm{fr}(2)) + \mathit{x2}\,(\mathrm{S}(3) - \mathrm{fr}(2)) \\ + \mathit{x1}\,(\mathrm{fr}(2) - \mathrm{fr}(1))\,(r + 1)$$

Since we try to have an end cash flow of $\mathrm{fr}(1) - S(3)$, we simplify it and collect the terms in the above expression with respect to fr(2). Later we will equate to zero the coefficient of fr(2). Hence, collecting the coefficients yields

```
> PayDay(3):=collect(simplify(PayDay(3)),fr(2));
```

$$\mathrm{PayDay}(3) := \\ (-\mathit{x2} + \mathit{x1}\,r)\,\mathrm{fr}(2) + \mathit{x1}\,\mathrm{S}(3) + \mathit{x2}\,\mathrm{S}(3) - \mathit{x1}\,\mathrm{fr}(1)\,r - \mathit{x1}\,\mathrm{fr}(1)$$

Since our aim is to **not have** fr(2) appearing as part of the cash flow at the end of day two, we will try to purchase futures contracts in quantities x_1 and x_2 such that the coefficient of fr(2) is zero. We ask MAPLE to extract the coefficient of fr(2) in the expression **PayDay(3)**.

```
> coeff(PayDay(3),fr(2));
```

$$-\mathit{x2} + \mathit{x1}\,r$$

Hence x_1 and x_2 should satisfy equation (10.15).

$$x_1 = \frac{x_2}{r} \tag{10.15}$$

We thus substitute it in the expression **PayDay(3)** and obtain the following new expression for **PayDay(3)**:

```
> PayDay(3):=subs(-x2+x1*r=0,PayDay(3));
```

$$\mathrm{PayDay}(3) := -\mathit{x1}\,\mathrm{fr}(1)\,r - \mathit{x1}\,\mathrm{fr}(1) + \mathit{x1}\,\mathrm{S}(3) + \mathit{x2}\,\mathrm{S}(3)$$

The following MAPLE command solves for x_1 in terms of x_2 and r, and substitutes the solution based on equation (10.15) to reach a new expression for **PayDay**(3).

```
> PayDay(3):=subs(x1=solve(-x2+x1*r=0,x1,PayDay(3));)
```

$$\text{PayDay}(3) := -x2\,\text{fr}(1) - \frac{x2\,\text{fr}(1)}{r} + \frac{x2\,\text{S}(3)}{r} + x2\,\text{S}(3)$$

Finally we see if the expression **PayDay**(3) can be factored.

```
> PayDay(3):=factor(PayDay(3));
```

$$\text{PayDay}(3) := -\frac{x2\,(\text{fr}(1) - \text{S}(3))\,(r+1)}{r}$$

Thus, we see that in order to have a cash flow of $\text{fr}(1) - S(3)$ at the end of day three, we need to choose x_2 such that the coefficient of $\text{fr}(1) - S(3)$ is one. Hence we extract the coefficient of $\text{fr}(1) - S(3)$ in the MAPLE command

```
> coeff(PayDay(3),fr(1)-S(3));
```

$$-\frac{x2\,(r+1)}{r}$$

and solve for x_2:

```
> solve(%=1,x2);
```

$$-\frac{r}{r+1}$$

Consequently, we should choose x_2 to satisfy equation (10.16),

$$x_2 = -\frac{r}{1+r}, \tag{10.16}$$

and since by equation (10.15) $x_1 = \frac{x_2}{r}$, x_1 should satisfy

$$x_1 = -\frac{1}{1+r}. \tag{10.17}$$

In order to confirm the reception of the required cash flow at the end of day three we substitute the value of x_2, based on equation (10.16), in **PayDay**(3). This is performed in the next MAPLE command.

```
> subs(x2=-r/(1+r),PayDay(3));
```

$$\text{fr}(1) - \text{S}(3)$$

We therefore see that by shorting (since x_1 and x_2 are negative) $\frac{1}{1+r}$ futures contracts on day one and $\frac{r}{1+r}$ futures contracts on day two, the required cash flow as far as magnitude and timing is obtained. As explained at the outset of these arguments, we would like to examine this portfolio and the investment strategy described more closely. At time one, we take a long position in a forward contract and we short $\frac{1}{1+r}$ futures contracts. On the next trading day, the proceeds from the marking to market process are invested at the risk-free rate and we take a short position in $\frac{r}{1+r}$ units of the futures contract. We have no intermediate cash flows and the cash flow at time three (at the end of the third trading day) from the forward contract and the futures contracts is

$$\mathrm{fr}(1) - S(2) + S(2) - F. \tag{10.18}$$

However, the combined cost of this portfolio and of the investment strategy is zero. Hence, to avoid arbitrage opportunities the payoff must be zero as well; thus,

$$\mathrm{fr}(1) - F = 0. \tag{10.19}$$

Consequently, the futures price fr(1) is equal to the forward price F.

There is only one small problem: on day one, how can we short $\frac{1}{1+r}$ futures contracts when r, the spot rate from day two to day three, is not known[6] on day one? Indeed this is why we need the assumption of the term structure being deterministic. As explained above, in such an environment, the spot rate $r_2(2,3)$ (we used r in the MAPLE calculation) is known on day one and is given by

$$1 + r_2(2, 3) = \frac{d_2}{d_3}. \tag{10.20}$$

Hence x_1 is given by $-\frac{d_3}{d_2}$ and x_2 is given by

```
> simplify((-(d2/d3)-1)/(d3/d2));
```

$$-\frac{(d2 + d3)\, d2}{d3^2}$$

[6] We cannot instruct an investor to buy x futures contracts, where x is the interest rate that will prevail in the market from tomorrow to the day after tomorrow. The information the investor has now does not include the value of this rate unless the term structure is deterministic. In a more technical notation this issue is related to measurability.

Our example considered a forward and a futures contract with two days to maturity. It will work for any arbitrary number of days until maturity. The reader is asked in an exercise to develop a similar template for finding an investment strategy – a strategy which will demonstrate the same relationship between the forward price and the futures price for a pair of contracts with n days until maturity.

The argument presented here (and its proof or a slight modification of it) will reappear in several other situations related to valuation. The idea of solving for the required position by working backward (in a recursive manner) one period at a time is reminiscent of a concept known as *dynamic programming*.

Given a sequential decision problem, one starts with penultimate decisions to be made, assuming that an optimal decision was taken at every previous step. An optimal strategy is solved for as a function of the decision taken a step before. In the example above we assumed that at the initial time an optimal number of futures contracts x were purchased.

We then solved for the optimal decision, x_2, a day prior to maturity assuming that x_1 was optimal, thereby receiving the equation for the payoff at maturity as a function of x_1 and x_2. This payoff was then equated to the required payoff and the values of x_1 and x_2 were solved for. A very similar approach is taken in Chapter 13, Section 13.4, for numerical valuation of assets where an analytical solution does not exist.

10.8 Concluding Remarks

This Chapter takes a second look at forward contracts. These contracts were analyzed in Chapter 2 in the framework of a one-period model. Here they are investigated in a multiperiod setting with a continuum of states of nature. The multiperiod model presents yet another extension to the one-period model of Chapter 1. It also accommodates the differences between forward and futures contracts that remain indistinguishable in a one-period model. The Chapter investigates Eurodollar contracts and forward rate agreements, as well as the effect of known cash flows paid by an asset on its forward price.

There is a common feature of financial assets considered in this Chapter. They can be priced knowing only the realization of the current term structure and do not require a specification of the evolution, across time, of the term structure. The Chapter concludes by proving the equivalency between forward and futures contracts in a DTS environment. The technique used to demonstrate this equivalency paves the ground for the numerical valuation of American options that is attended to in Chapter 1. The next Chapter takes a second look at swaps, the pricing of which again requires knowledge only of the current term structure, and not of its evolution through time. Chapter 16 will investigate some assets, the valuation of which necessitates the specifications of the evolution of the term structure of interest rates.

10.9 Questions and Problems

Problem 1. Consider a forward contract that was issued some time ago for a delivery at time T with the forward price of $\$F$. Suppose it is now time t ($t < T$) and the conditions in the market are such that the forward price of the same underlying asset for delivery at time T is also $\$F$.

1. Assuming that the continuously compounded risk-free interest rate prevailing in the market is r, recover the price of the underlying asset today.
2. What is the current value of the forward contract that was issued some time ago for a delivery at time T with the forward price of $\$F$?

Problem 2. Assume the bond market outlined in Table 10.1 and consider a forward contract written at time zero specifying delivery of bond three at time two (immediately after the second coupon is paid). According to the contract, the party holding it long will pay the forward price F, at time two, and will receive an instrument similar to a zero-coupon bond that will pay \$108 at time three.

1. Calculate the forward price F.
2. Graph the payoff, as of time two, from such a contract.

Problem 3. Suppose you have a bond market where bonds pay an annual coupon. Assume further that the discount factor function is known and is denoted by $d(\cdot)$. Consider a bond with a face value of \$100 that just paid its coupon, $\$C$, and matures in two years. The remaining cash flows from this bond are $\$C$ in a year from now and $\$(100+C)$ in two years from now. As it has been discussed in the text, the forward price , F, of this bond, for delivery in one year from now, can be found utilizing equation (10.5) and is given by

$$F = \frac{P}{d(1)} - C,$$

where P is the current price of the bond. Show that, alternatively, the forward price of this bond can be found as

$$F = \frac{(100 + C)\, d(2)}{d(1)}.$$

Interpret these results.

Problem 4. Assume that the current term structure of interest rates is known. Show that a forward price, F, of a coupon - paying bond that matures in n units of time from now and has n coupon payments remaining for a delivery in m ($m < n$) units of time (immediately after a coupon payment at time m) can be determined as

$$F = \frac{1}{d(m)} \left(\sum_{i=m+1}^{n} C d(i) + FV d(n) \right),$$

where $d(i)$ is the discount factor for time i, FV is the face value of the bond, and C is a coupon payment. How would this result have to be modified if the delivery would occur just before the coupon payment at time m?

Problem 5. The LIBOR term structure is actually estimated based on FRA agreements. The FRA data conveys to us the forward rate in the LIBOR market. Hence in order to estimate the term structure of LIBOR we have to utilize equation (10.8) to derive the term structure from the forward rate. In our discussion regarding the forward rate, in the government bond market, we derive the forward rate from the term structure. In the LIBOR market we have to do the reverse. Delineate the steps of extracting the term structure from the forward rate structure.

Problem 6. Consider a FRA in which the underlying rate is risk-free rate, the rate implied by government bonds. Assume that both parties are federal agencies with no risk of default. Prove that the rate of the FRA must be the forward rate implied in the term structure. Prove this statement in two ways — through replication and through valuation by discount factors.

Problem 7. Assume a market with a deterministic term structure and that time zero is the current time; that is, the spot interest rate which will prevail in the market from any future time t_1 to time t_2 $(0 < t_1 < t_2)$ is known with certainty. Show that in such a market the spot interest rate for future time periods must be equal to the forward interest rate $r_0(t_1, t_2)$ – otherwise arbitrage opportunities would exist.

Problem 8. Assume a market with a deterministic term structure. Find an investment strategy which will demonstrate that the forward price and the futures price are the same for a pair of contracts with $n > 3$ days until maturity.

Chapter 11

Swaps: *A Second Look*

We have already been introduced to swaps in Chapter 2 where they were investigated in the framework of a one-period model. Here we take a second look at swaps in a more realistic multiperiod setting for which we have a continuum of states of nature in each period. This setting allows the investigation of one of the most common forms of swaps that could have not been analyzed in Chapter 2. It will be followed by a revisit of currency and equity swaps as in Sections 2.5 and 2.6, but in a multiperiod setting. As before, the risk of default by the parties is ignored.

11.1 A Fixed-for-Float Swap

One of the most common swaps is a "fixed-for-float" swap. In this type of swap, one party pays a fixed rate of interest on a certain principal every period, e.g., semiannually. The other party pays a floating rate of interest, commonly LIBOR, according to the rate that prevails in the market at that time. The parties do not swap the principal and hence it is termed a notional principal.

Assume a *notional* principal of N, and assume that the parties enter into the swap agreement at time t_0, and that the payment dates are at times $t_1, ..., t_n$. Thus, for example, at time t_1, the party who pays the fixed rate, r, pays the amount Nr. The party who pays the floating rate pays the amount $Nr_{t_0}(t_0, t_1)$ at time t_1. The rate $r_{t_0}(t_0, t_1)$ is the prevailing rate in the market from time t_0 to t_1 and is known at time t_0. At time t_i, this party will pay the amount $N\, r_{t_{i-1}}(t_{i-1}, t_i)$, $i = 1, ..., n-1$. At time t_0 the value of $r_{t_0}(t_0, t_1)$ is known but the other rates are not. These rates will be known only at the beginning of the period to which they apply; e.g., the rate $r_{t_2}(t_2, t_3)$ will be known at time t_2. In practice, the payments are netted and only the difference is paid to the appropriate party. If we look at the swap from the point of view of the party who pays the fixed rate, for example, at time t_1, this party pays (or, if the expression in (11.1) is negative, receives)

$$Nr - Nr_{t_0}(t_0, t_1). \tag{11.1}$$

Consider the swap at its initiation time. We would like to find the fixed rate that will make this swap fair. We are ignoring the credit risk of the parties in order to simplify our discussion. Therefore, on the initiation date of the contract, the fixed cash flows are nearly those of a regular bond. The only difference is that at maturity the principal is not paid. (In an exercise, the reader will be asked to find the fixed rate that will make the swap fair, assuming that the principal value of the swap is also exchanged at maturity.) The floating cash flows are like those of a floating (variable) rate bond, but again the principal is not paid at maturity. As we have already seen in Section 9.4, the value of a variable rate bond immediately after a coupon payment is its face value — the principal. Hence, in our case, the value of the floating payments will be N minus the present value of N, the notional principal.

Let $d(t)$ be the discount factor for time t, where time is measured in years, and assume that the payment dates are $t = \frac{1}{2}, \frac{2}{2}, \frac{3}{2}, \ldots, \frac{2T}{2}$; i.e., payments are made semiannually and the current time is $t = 0$. The value of the floating payments today is

$$N - Nd(T). \tag{11.2}$$

The value of the fixed payments is the present value of the future cash flows. We assume that the rate[1] is reported as an annual rate based on semi-annual compounding. The value of the fixed payments is thus given by equation (11.3).

$$\sum_{t=1}^{2T} N\frac{r}{2} d\left(\frac{t}{2}\right) \tag{11.3}$$

Therefore, the value of the swap to the party who pays the fixed rate is

$$N - Nd(T) - \sum_{t=1}^{2T} N\frac{r}{2} d\left(\frac{t}{2}\right). \tag{11.4}$$

We say that the swap is a "fair" deal if the present value of the fixed payments equals that of the variable payments. The value of r which makes

[1] In practice the rates are reported annually but based on the number of days in each period. Hence instead of having $\frac{r}{2}$ we will actually have $r\left[\frac{\textit{number-of-days-in-the-period}}{365}\right]$.

the swap a fair deal, i.e., makes the expression in equation (11.4) vanish, is given by

$$r = 2\,\frac{1 - d(T)}{\sum_{t=1}^{2T} d\left(\frac{t}{2}\right)}. \tag{11.5}$$

This is confirmed by MAPLE below. First let us make sure that the variables, N, T , r, t and d are not assigned any value.

```
> N:='N':d:='d':T:='T':r:='r':
> solve(N-N*d(T)-sum('N*(r/2)*d(t/2)','t'=1..2* T)=0,r);
```

$$-2\,\frac{-1 + \mathrm{d}(T)}{\sum_{t=1}^{2\,T} \mathrm{d}(\frac{1}{2}\,t)}$$

The procedure **FxFlswap** values a swap at its initiation from the point of view of the party who pays the fixed rate. The input parameters, in order, are the notional principal, the discount factor function, the number of payments until the swap matures, and the fixed rate of interest at which the swap was initiated. (We maintain the assumption that the payments are made semiannually.) Given the values of N, the notional principal, and of $d(t)$, $t = \frac{1}{2}, \frac{2}{2}, \frac{3}{3}, ..., \frac{2T}{2}$, where $2T$ is the number of payments and T is the maturity date (we assume that the swap is initiated at time zero), the procedure **FxFlswap** can be utilized to solve, symbolically, for the fixed rate r. This is shown in the command below.

```
> solve(FxFlswap(N,d,T,r)=0,r);
```

$$-2\,\frac{-1 + \mathrm{d}(T)}{\sum_{t=1}^{2\,T} \mathrm{d}(\frac{1}{2}\,t)}$$

Now let us look at a structure of a bond market as defined by **NarbitB** below and name the discount factor function $d(\cdot)$. This allows the execution of the procedure **FxFlswap** in order to find the value of r. Consider the market outlined in **NarbitB** below in which, for simplicity, we assume that[2] **coupons are paid annually**. (We add the zero parameter at the end to

[2] The reader may try to run **NarbitB** for a smaller number of approximating functions to see that a perfect fit is not possible with a smaller number of approximating functions. The first time that the absolute deviations are reported to be zero is for seven approximating functions. Each time you run **NarbitB** it defines a global variable **SumAbsDiv**. If its value is zero it means that a perfect fit has been achieved. To verify the value of **SumAbsDiv** just issue the command '**SumAbsDiv;**' on the command line. If a coupon would have been paid semiannually we would have had to take another step prior to the

suppress the graph of the discount factor function. If you would like to see it, rerun **NarbitB** and omit the last parameter.)

```
> NarbitB([[5,5,5,105],[3,103,0,0],[6,6,106,0],\
> [110,0,0,0]],[60,90,85,101],7,d,0);
```

The no − arbitrage condition is satisfied.

The discount factor for time, 1, *is given by,* $\frac{101}{110}$

The interest rate spanning the time interval, [0, 1], *is given by,* .08911

The discount factor for time, 2, *is given by,* $\frac{9597}{11330}$

The interest rate spanning the time interval, [0, 2], *is given by,* .1806

The discount factor for time, 3, *is given by,* $\frac{84305}{120098}$

The interest rate spanning the time interval, [0, 3], *is given by,* .4246

The discount factor for time, 4, *is given by,* $\frac{163553}{360294}$

The interest rate spanning the time interval, [0, 4], *is given by,* 1.203

The function Vdis([*c1*, *c2*, ..]), *values the cashflow* [*c1*, *c2*, ..]

The continuous discount factor is given by the function, '*d*', (.)

Assume a fixed-for-float swap initiated at time zero that matures in four years. Since the payments are semiannual the swap will have eight payments. To solve for the value of the fixed rate we should execute the command below:

```
> evalf(solve(FxFlswap(N,d,4,r)=0,r));
```

.1783857932

Note that the value of r is independent of N. Indeed, this is easily verified by equating the value of the swap, equation (11.4), to zero. The fixed rate that will make this swap a fair deal is about 17.8%.

The annual interest rate that applies to a loan given from time zero to time i, based on semiannual compounding, in a decimal form, is the ith element of the array below.

```
> [seq(evalf(((1/d(i/2)^(1/i))-1)*2),i=1..4)];
```

[.06805395576, .087207618, .086963241, .084748161]

---

valuation of the swap. Since our unit of time is a year and **NarbitB**, in such a market, uses half a year as the unit of time, we first need to scale the function $d(t)$ to conform to the time unit of a year. This is done by issuing the command $d := unapply(d(\frac{t}{2}), t)$:.

The fixed rate determined by **FxFlswap**, given the discount factor function $d(t)$, is only a function of T, namely, the maturity time of the swap. We can calculate, as in the array below, the fixed rate that will apply to a swap with maturity dates of T, $T = 1, ..., 4$. This is the Tth element of the following array.

```
> [seq(evalf(solve(FxFlswap(N,d,T,r)=0,r)),T=1..4)];
```

$$[.08679709887, .08468322068, .1166149015, .1783857932]$$

We therefore see the relation between the fixed rate that applies to the swap and the term structure of interest rates in the market. In fact, we can further analyze this relation utilizing the connection among forward rates, the term structure of interest rates, forward rate agreements and swaps.

A fixed-for-float swap can be decomposed into some basic building blocks. The first block is the exchange of cash at time t_1, which is exactly identical to the case of the swap investigated in a one-period setting. The remaining blocks are much the same as the forward rate agreement (FRA) dealt with in Section 10.4.1. At time t_0 one party agrees to pay the other the return on N dollars invested from, for example, time t_1 to t_2 in terms of LIBOR. In exchange, the other party pays a fixed rate at that time. Similarly the exchange of cash at time t_3 and so on can be thought of in the same way. Indeed, the fact that a swap is built from basic building blocks is a generic property of most types of swaps. The idea can be visualized using the three-dimensional capabilities of MAPLE. We defer this visualization to Section 11.4,by which time we will have had the opportunity to familiarize ourselves with other types of swaps.

Based on what we already know from the section dealing with FRAs, Section 10.4.1, we are in a position to calculate the fixed rate that should apply to each of these building blocks. The catch, though, is that in a swap agreement such as this, the same fixed rate should apply to all the building blocks. Hence, the value of each block may not be zero but the value of the blocks in aggregate is zero if the swap is fair. The exercise at the end of the Chapter will return to this point and will ask the reader to analyze this relation further.

Of course the **FxFlswap** procedure can also calculate a swap that is initiated at some fixed value r. For example, consider a swap that matures in two years where the notional principal is $\$10,000$ and the fixed rate is initially set at 15 percent. Since we already know that the fair fixed rate should be 8.468322068%, the value of such a swap should be negative to the party who pays the fixed rate. This is confirmed below.

```
> FxFlswap(10000,d,2,.15);
```

$$-1179.766468$$

Occasionally, a customer will ask for a swap in which the fixed rate is below the fair fixed rate. This is referred to as a *buy down swap.* Such a customer (who pays fixed and receives LIBOR) will have to pay a certain amount at the initiation of the swap. The reverse situation is called a *buy up swap.* Both of these situations are referred to in the exercises at the end of the Chapter. A swap in which the value of the fixed cash flow equals that of the floating cash flow is called a *par swap.* After the initiation time, the situation in the market may (and usually does) change and a new term structure of interest rates should be used to value the swap.

You may like to try and change the structure of the market and re-run the commands above to investigate the sensitivity of the fixed rate to the structure of the market and the implied term structure. You, however, must be careful and first unassign the values of the variable by operating the command *T*:='*T*', etc. If we value the swap after its initiation at a some point in time between two consecutive payment dates, we must pay attention to the timing. This is done in the same manner as valuing a variable rate bond between two consecutive payments, equation (9.18). The next subsection deals with valuing an existing swap.

11.1.1 Valuing an Existing Swap

Consider an existing swap. Assume that the current time is zero and that the length of time until the next payment is v. The value of an outstanding variable bond immediately after a payment, say, at time v, is N. Thus, the value of such a bond at time zero is $(N + \tau N)d(v)$, where τN is the payment at time v. In other words, τ is the interest rate applicable to the payment at time v. However, the floating payment in a swap does not include the payment of the principal on the last payment date. Hence, we should subtract the present value of the principal in order to arrive at the value of the floating payment in a swap. If we assume that the maturity date is T and, as before, that the units of time are years, then the value of the floating payments is given by equation (11.6) below.

$$(N + \tau N)d(v) - Nd(T) \tag{11.6}$$

One must remember that the discount factor we use is the one in effect at the current time, time zero. The value of the fixed payments will be given by

$$\sum_{t=0}^{2(T-v)} N \frac{r}{2} d\left(v + \frac{t}{2}\right). \tag{11.7}$$

In equation (11.7) r is the fixed rate agreed upon at the initiation of the swap. The value of the swap to the party who pays the fixed interest is given by equation (11.8) below.

$$(N + \tau\, N)d(v) - N d(T) - \sum_{t=0}^{2\,(T-v)} N \frac{r}{2} d\left(v + \frac{t}{2}\right) \tag{11.8}$$

The procedure **FxFlswap** is also capable of valuing a swap already in existence[3]. As before we value it from the point of view of the party that pays the fixed rate. The input parameters, in order, are the notional principal, the discount factor function, the maturity date of the swap, the fixed rate of interest at which the swap was initiated, the time until the next payment (measured in years), and the variable rate applicable to the first payment. Consider the swap we valued above. Now assume that we keep the same parameters as before, but that the swap was initiated prior to the current time. The time until the next payment is $\frac{3}{12}$ (i.e., three months) and the floating rate applicable to the next payment is 7 percent. The value of such a swap will be given by executing the command below.

```
> FxFlswap(10000,d,21/12,.15,3/12,0.07);
```

$$-822.9342530$$

Let us look at another example in a market with a different structure. In order to define a term structure we run **NarbitB** for another market[4] structure. Suppose we look at the market specified by **NarbitB** below, where **coupons are paid annually,** and we define the discount factors to be the function **Dis**.

```
> NarbitB([[5,5,105],[8,108,0],[6,6,106]],[90,102,95],\
> 2,Dis,0);
```

The no − arbitrage condition is not satisfied

An arbitrage portfolio is :

[3] It is easy to see that, as for the par swap, the value of a swap after its intiaiton with a notional principal of N is also N times the value of a swap with a notional principal of \$1. That is, the value of the swap is a homogeneous (of degree one) function of N. Hence we could have programmed the procedure **FxFlswap** to value a swap of a \$1 notional principal.

[4] Note that the no-arbitrage condition is not satisfied here, but we are using an estimation of the term structure which is "best" in a certain sense. This is usually the case in real markets since the prices of the bonds include some "noise". Moreover, note that the arbitrage portfolio generates a negative cash flow, $-\$\frac{50}{67}$, at time one and the same amount of positive cash flow is produced at time zero. The arbitrage profit is thus due only to the interest earned investing $\$\frac{50}{67}$ from time zero to time one.

$$Buy,\ \frac{348}{335},\ of\ Bond,\ 1$$

$$Buy,\ \frac{1}{134},\ of\ Bond,\ 2$$

$$Short,\ 1,\ of\ Bond,\ 3$$

$$Buying\ this\ portfolio\ produces\ income\ of,\ \frac{50}{67},\ at\ time\ 0$$

$$This\ portfolio\ produces\ income\ of,\ \frac{-50}{67},\ \ at\ time,\ 1$$

$$This\ portfolio\ produces\ income\ of,\ 0,\ \ at\ time,\ 2$$

$$This\ portfolio\ produces\ income\ of,\ \frac{206}{67},\ \ at\ time,\ 3$$

$$The\ continuous\ discount\ factor\ is\ given\ by\ the\ function,\ 'Dis',\ (.)$$

Consider a swap in existence where the notional principal is $10,000, the discount factor is the function **Dis**, the maturity time is $\frac{13}{12}$ years, the fixed rate of the swap is 12 percent, the time until the next payment is one month, and the rate of the next floating payment is 10 percent. The value of the swap is calculated below:

```
> FxFlswap(10000,Dis,13/12,.12,1/12,.10);
```

$$-146.1283520$$

Such a swap, if it had been initiated today (with six months until the next payment), would have had a different fixed rate. We can actually solve for the rate that would apply if such a swap were initiated today by executing the next command.

```
> solve(FxFlswap(10000,Dis,1.5,r)=0);
```

$$.07259337189$$

We can also utilize the **FxFlswap** procedure in the case where a swap is renegotiated. Consider a swap in existence and assume that the parties would like to change the fixed rate of the swap so that the value of the swap will be zero. This of course would trigger some cash transfer. Such a transaction might be motivated by tax reasons. Consider the swap we just valued at −$146.1283520. Suppose the parties would like to change the fixed rate so that the swap value will be zero. This is solved for by the command below.

```
> solve(FxFlswap(10000,Dis,13/12,r,1/12,.10)=0, r);
```

$$.1099852543$$

Thus, the party who pays the fixed rate should now pay \$146.1283520 and in the future pay a fixed rate of 10.99852543% instead of the 12 percent at which the swap was initiated. We can verify this result via the command below.

```
> FxFlswap(100,Dis,13/12,.1099852543,1/12,.10);
```

$$-.100\,10^{-7}$$

The value of the fixed rate of interest that will make the value of the existing swap, equation (11.8), zero is given in equation (11.9) below.

$$\frac{2\left(\tau(1+d(v))-d(T)\right)}{\sum_{t=0}^{2(T-v)} d(v+\frac{t}{2})} \tag{11.9}$$

This is confirmed by MAPLE by solving for r such that the expression in equation (11.9) is zero. We first make sure the variables have no assigned values.

```
> N:='N':d:='d':T:='T':r:='r':v:='v':tau:='tau' :
> solve(((N+tau*N)*d(v)-N*d(T)-sum('N*(r/2)*d(v+t/2)',\
> 't'=0..2*(T-v))=0,r));
```

$$2\,\frac{\mathrm{d}(v)+\mathrm{d}(v)\,\tau-\mathrm{d}(T)}{\displaystyle\sum_{t=0}^{2\,T-2\,v}\mathrm{d}(v+\frac{1}{2}\,t)}$$

11.2 Currency Swaps

In Chapter 2, Section 2.5.1, we investigated a currency swap, albeit in the setting of a one-period model. The situation in a multiperiod environment is very similar. The vast majority of these type of swaps involve swapping not only interest payments, but also the principals. It is therefore essentially a situation where a company, in a domestic market, agrees to swap two cash flows. It exchanges a cash flow identical to the one resulting from issuing a foreign bond in a foreign market for a cash flow identical to one resulting from issuing a domestic bond in the domestic market.

For the sake of consistency, we keep the same notation as in Section 2.5.1. Hence, we consider a domestic firm which receives an amount N_1 of foreign currency now (at time zero) and is required to pay an amount of foreign currency $N_1\,R_1$ at time $t=1,...,T-1$, and $N_1(1+R_1)$ at time T, where T is the maturity time. This cash flow profile could be a loan repayment for a loan that the firm obtained in the foreign market at a rate R_1. N_1 would then be the principal amount of the loan. This cash flow is being swapped with a foreign firm that has to pay an amount N_0 of domestic currency at the current time, $N_0\, r_0$ at time $t = 1, ..., T-1$, and $N_0\,(1 + r_0)$ at time T. Again, this could be a loan payment on a loan obtained in the domestic market at a rate r_0, where N_0 is the principal amount.

In practice, the swap might be arranged for the domestic firm by a financial institution. The institution either finds a foreign firm like above or "warehouses" this cash flow until it finds an optimal match for it. In certain cases, the institution must assume some risk as a result of the absence of an exact matching. Hence, it requires a fee, a "finder's fee", as well as compensation for the risk taken. This fee is usually obtained in terms of an increase in the rate, r_0, that the domestic firm is asked to pay. (The end-of-Chapter exercises ask the reader to analyze such a situation.) The motivation for entering into such a swap is as it was in the case of the one-period model. As before, we still ignore the risk of default by the parties involved.

Given the exogenous circumstances, i.e., the domestic term structure the foreign term structure, and the spot exchange rate, valuing such a swap is a straightforward matter. Assume that the domestic discount factor is given by the function $disL(t)$, and that the foreign discount factor function is given by $disF(t)$. The present value, in terms of the local currency, of the domestic cash flow is like the present value of the cash flow from a bond and is thus given by equation (11.10) below.

$$N_0 - \left(\sum_{t=1}^{T} N_0 r_0 disL(t) \right) - N_0 disL(T) \tag{11.10}$$

The present value of the foreign cash flow in terms of the foreign currency is, by the same token, given by equation (11.11) below.

$$N_1 - \left(\sum_{t=1}^{T} N_1\, R_1 disF(t) \right) - N_1\, disF(T) \tag{11.11}$$

As in Section 2.5.1, we denote the cost of one unit of the foreign currency in terms of units of the domestic currency by $F0$. Hence, the spot exchange rate of one unit of the domestic currency is $\frac{1}{F0}$ units of the foreign currency. At the initiation of the swap, the domestic firm receives the domestic currency N_0 and pays out the foreign currency, N_1. At each coupon payment time the domestic firm pays the domestic coupon, $N_0 r_0$, and receives the foreign coupon, $N_1 r_1$. At the maturity date of the swap, the principals are swapped back: the domestic firm pays out N_0, the domestic principal, and receives back N_1, the foreign principal. Therefore the value of the swap, from the point of view of the domestic firm in the domestic currency, is given by

$$N_0 - \left(\sum_{t=1}^{T} N_0 r_0 disL(t) \right) - N_0\, disL(T)$$
$$- \left[N_1 - \left(\sum_{t=1}^{T} N_1 R_1 disF(t) \right) - N_1 disF(T) \right] F0. \qquad (11.12)$$

There is another way of valuing this type of a swap. Rather than discounting the foreign cash flow to the current time period and then converting it to the local currency via the spot exchange rate, we can convert each component of the future foreign cash flow at the time it occurs. This requires the use of a series of forward contracts. We essentially repeat the same argument as in Section 2.6.3 that was employed for the one-period model. This time, however, we think of each time period $[0, t_1]$, $[0, t_2]$, etc., as a separate one-period model. Thus, if we denote the forward exchange rate for time t, as of time zero, by $FF(t)$, applying equation (2.15) separately to each time period yields equation (11.13) below.

$$FF(t) = \frac{F0\, disF(t)}{disL(t)} \qquad (11.13)$$

$FF(t)$ is the forward price of one unit of the foreign currency in terms of the domestic currency.[5] The foreign discount factors function can be written in terms of the domestic discount factor function and the spot exchange rate, as in equation (11.14).

$$disF(t) = \frac{FF(t) disL(t)}{F0} \qquad (11.14)$$

Substituting equation (11.14) into equation (11.12) results in the relation (11.15).

$$N_0\,(1 - disL(T)) - \sum_{t=1}^{T} N_0 r_0\, disL(t) \qquad (11.15)$$

[5] The reader may recall that equation (11.13) is a consequence of the replication argument, that in this context is referred to as the *cost-of-carry model*. To guarantee a unit of a foreign currency at time t the investor can buy the present value, as of time zero, of a unit of a foreign currency obtainable at time t. In terms of the local currency this will cost disF(t) $F0$, which must equal the present value of the forward exchange rate.

$$- \left[N_1 \left(F0 - FF(T)\, disL(T) \right) - \sum_{t=1}^{T} N_1 R_1\, FF(t) disL(t) \right]$$

Clearly, this substitution is more than just the pure mathematical operation. It actually points out to us the building blocks of foreign currency swaps. One can actually "engineer" a homemade foreign currency swap by entering into a sequence of forward contracts, maturing at times $t = 1, ..., T$. If the investors have a liability of $N_1 R_1$ units of foreign currency at time t, they can enter into a forward agreement to receive that amount at time t. In exchange, the investor would then pay the amount $N_1 R_1 FF(t)$ in local currency at time t. Thus, the present value, as of time zero, of the $N_1 R_1$ units of the foreign currency obtainable at time t is $N_1 R_1 FF(t) disL(t)$ in terms of the domestic currency. Equation (11.15) follows directly from this argument and can also be presented in a manner that relates more closely to this interpretation, i.e., as equation (11.16):

$$\begin{aligned} &\sum_{t=1}^{T} (N_1 R_1 FF(t) - N_0\, r_0) disL(t) + \\ &+ (N_1 FF(T) - N_0)\, disL(T) - N_1 F0 + N_0. \end{aligned} \tag{11.16}$$

Let us look at an example to clarify this idea and to see the interrelation between the parameters involved in this valuation. To this end we make use of the procedure **FXswap** that values a currency swap from the point of view of the domestic firm in the domestic currency. The input parameters to the procedure **FXswap** are, in order, N_0, r_0, $disL$, N_1, R_1, $disF$, and $F0$ as defined in equation (11.12). These are followed by the parameters, T, and per, where T is the time to the last payment and per is the payment period (usually semiannual). Solving symbolically for the value of the swap produces the expression below, which is a generalized version of equation (11.16).

```
> FXswap(N0,r0,disL,N1,R1,disF,F0,T,per);
```

$$N0 - N1\ F0 - \left(\sum_{t=1}^{\frac{T}{per}} N0\ r0\ \mathrm{disL}(t\, per) \right) - N0\ \mathrm{disL}(T)$$

$$+\left(\left(\sum_{t=1}^{\frac{T}{per}} N1\ R1\ \mathrm{disF}(t\,per)\right) + N1\ \mathrm{disF}(T)\right)\ F0$$

It is also possible to generate the symbolic relation between r_0 and the other parameters by running **FXswap** as below.

```
> solve(FXswap(N0,r0,disL,N1,R1,disF,F0,T,per)= 0,r0);
```

$$-\frac{-N0 + N1\ F0 + N0\ \mathrm{disL}(T) - F0\ N1\ R1\left(\sum_{t=1}^{\frac{T}{per}} \mathrm{disF}(t\,per)\right) - F0\ N1\ \mathrm{disF}(T)}{N0\left(\sum_{t=1}^{\frac{T}{per}} \mathrm{disL}(t\,per)\right)}$$

Let us consider now a numerical example. Assume a foreign exchange swap in a market where the domestic term structure is solved for by the **NarbitB** procedure below and is defined to be **disL**. For simplicity, we assume that in this market coupons are paid annually, and hence the units of time are years. Figure 11.1 depicts the continuous approximation of the discount factor in this market.

```
> NarbitB([[107,0,0,0],[5,5,5,105],[8,108,0,0],\
> [6,6,106,0]],[97,90,102,95],8,disL);
```

The no − arbitrage condition is satisfied.

The discount factor for time, 1, *is given by*, $\frac{97}{107}$

The interest rate spanning the time interval, [0, 1], *is given by*, .1031

The discount factor for time, 2, *is given by*, $\frac{5069}{5778}$

The interest rate spanning the time interval, [0, 2], *is given by*, .1399

The discount factor for time, 3, *is given by*, $\frac{40589}{51039}$

The interest rate spanning the time interval, [0, 3], *is given by*, .2575

The discount factor for time, 4, *is given by*, $\frac{4722407}{6430914}$

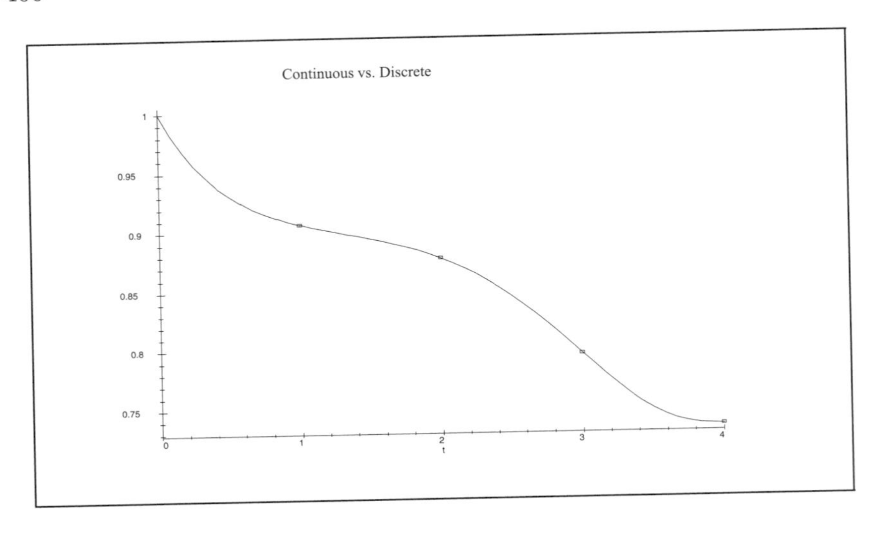

Figure 11.1: The Graph of the Function $DisL$

The interest rate spanning the time interval, [0, 4], *is given by*, .3618
The function Vdis([*c1*, *c2*, ..]), *values the cashflow* [*c1*, *c2*, ..]
The continuous discount factor is given by the function, '*disL*', (.)

Let us assume that the spot exchange rate is $F0$ and is assigned the numerical value of $\frac{11}{10}$, and that the term structure of forward exchange rates is given as below:

```
> F0:=11/10;
```

$$F0 := \frac{11}{10}$$

We use the **do** command to generate the term structure of forward exchange rates.

```
> for i from 1 to 4 do;
> FF(i):=F0-i/7;
> od;
```

$$\mathrm{FF}(1) := \frac{67}{70}$$
$$\mathrm{FF}(2) := \frac{57}{70}$$

$$\mathrm{FF}(3) := \frac{47}{70}$$
$$\mathrm{FF}(4) := \frac{37}{70}$$

The purpose of the next command it to unassign the value of i, which is still equal to 5 from the end of the loop above.

```
> i:=evaln(i);
```

$$i := i$$

Hence, $F0$ is the spot exchange rate and $FF(i)$ is the forward exchange rate for time i, as of the current time. That is, an investor can enter into a forward contract now to buy a unit of the foreign currency at time i for the price of $FF(i)$. According to equation (11.14) the discount factor in the foreign market is given by

$$disF(t) = \frac{FF(t)disL(t)}{F0}. \tag{11.17}$$

The numerical value of the foreign discount factor **disF** at times one, two, and three is evaluated below[6] based on equation (11.14).

```
> FF(1)*disL(1)/F0;
```

$$\frac{6499}{8239}$$

```
> FF(2)*disL(2)/F0;
```

$$\frac{96311}{148302}$$

```
> FF(3)*disL(3)/F0;
```

$$\frac{1907683}{3930003}$$

We can utilize **NarbitB** to generate a continuous estimation of the discount factors in the foreign market in the following way. Consider an artificial foreign market populated only by four zero-coupon bonds with a face value of one unit that mature at the same time as the bonds in the domestic market. The prices of such bonds must be the value of the discount factor for their maturity time. (We have already used a similar argument in Section 9.1.1.) Thus running **NarbitB** as below will produce **disF**.

[6] Alternatively we could value it by the command

```
> seq(FF(i)*disL(i)/F0,i=1..3);
```

$$\frac{6499}{8239}, \frac{96311}{148302}, \frac{1907683}{3930003}$$

```
> NarbitB([[1,0,0,0],[0,1,0,0],[0,0,1,0],[0,0,0,1]],\
> [FF(1)*disL(1)/F0,FF(2)*disL(2)/F0,FF(3)*\
> disL(3)/F0,FF(4)*disL(4)/F0],5,disF,0);
```

$$The\ no-arbitrage\ condition\ is\ satisfied.$$

$$The\ discount\ factor\ for\ time,\ 1,\ is\ given\ by,\ \frac{6499}{8239}$$

$$The\ interest\ rate\ spanning\ the\ time\ interval,\ [0,\ 1],\ is\ \ given\ by,\ .2677$$

$$The\ discount\ factor\ for\ time,\ 2,\ is\ given\ by,\ \frac{96311}{148302}$$

$$The\ interest\ rate\ spanning\ the\ time\ interval,\ [0,\ 2],\ is\ \ given\ by,\ .5398$$

$$The\ discount\ factor\ for\ time,\ 3,\ is\ given\ by,\ \frac{1907683}{3930003}$$

$$The\ interest\ rate\ spanning\ the\ time\ interval,\ [0,\ 3],\ is\ \ given\ by,\ 1.060$$

$$The\ discount\ factor\ for\ time,\ 4,\ is\ given\ by,\ \frac{174729059}{495180378}$$

$$The\ interest\ rate\ spanning\ the\ time\ interval,\ [0,\ 4],\ is\ \ given\ by,\ 1.834$$

$$The\ function\ Vdis([c1,c2,..]),\ values\ the\ cashflow\ [c1,c2,..]$$

$$The\ continuous\ discount\ factor\ is\ given\ by\ the\ function,\ 'disF',\ (.)$$

```
> SumAbsDiv;
```

$$0$$

We can now value a foreign exchange swap. The two term structures of interest rates and the term structure of forward exchange rates are shown in Figure 11.2. The two graphs starting from one are the foreign and domestic term structures, where the domestic graph (green) is above the foreign graph (red). The third graph (black) is that of the forward exchange rate.

```
> plot([disL(t),disF(t),disF(t)*(11/10)/disL(t)],\
> t=0..4,colour=[green,red,black],thickness=2,\
> title=`TS of Foreign and Domestic Interest Rates and\
> TS of Forward Exchange Rate`,labels=[time,rates]);
```

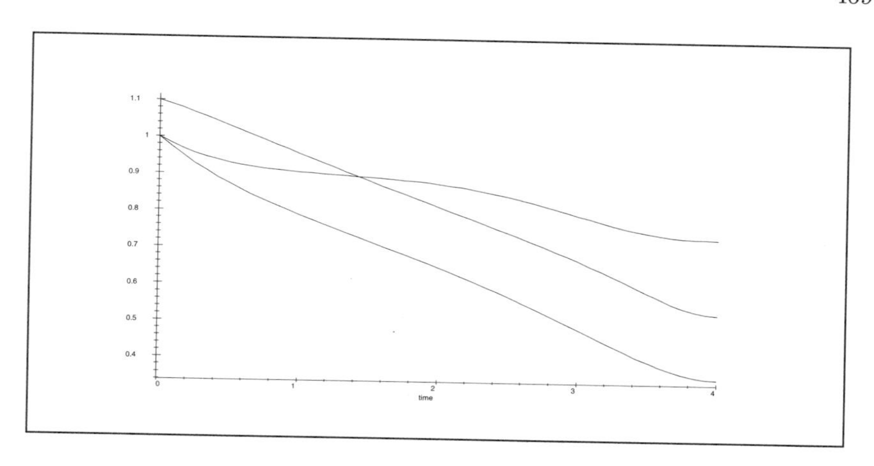

Figure 11.2: TS of Foreign and Domestic Interest Rates and TS of Forward Exchange Rate

Consider a foreign currency swap in which both principals are \$10,000, the foreign rate of interest is 15 percent, and the domestic rate is 6 percent. Assume that the swap is initiated at the current time with maturity in four years, that the spot exchange rate is $\frac{11}{10}$, and that the payments are made semiannually. The procedure **FXswap** values such a swap as demonstrated below. The **time parameters**, i.e., the maturity of the swap, the length of the period between payments, and the time to the first payment, **must be entered as fractions**. MAPLE checks that the number of periods is an integer and a decimal parameter will trigger an error message.

```
> FXswap(10000,.06,disL,10000,.15,disF,11/10,4, 1/2);
```

$$-505.82398$$

It is possible to solve numerically for some of the parameters, given values for others. Consider a two-year swap at initiation with 10,000 units of domestic principal, 11,000 units of foreign principal, payments that are made semiannually, a foreign interest rate of 13 percent, and a spot exchange rate of 0.6. Given the above, the domestic rate can be solved for as below. Remember that if the unit of time is a year then the time to maturity will be 4 and the period will be $\frac{1}{2}$ for semiannual payments.

```
> FXswap(10000,.06,disL,10000,.15,disF,11/10,4,1/2);
```

$$\{r = .04206519491\}$$

Alternatively, one can solve for the possible R and r given the exogenous circumstances as below.

```
> solve({FXswap(10000,r,disL,11000,R,disF,0.6,2,1/2)=0},\
> {r});
```

$$\{r = -.03010704306 + .5551710612\, R\}$$

The reader may try different combinations of parameters and different settings of foreign and domestic markets.

The **FXswap** procedure can also be used if instead of the two term structures, the domestic term structure and the term structure of forward exchange rates are supplied. A direct substitution in the **FXswap** procedure will generate the value of a foreign exchange swap. Given the term structure of forward exchange rates and the foreign term structure, one should proceed in a similar manner as in the text above. Utilizing equation (11.13), the domestic term structure can be inputted and substituted in the procedure **FXswap** to value a foreign exchange swap. Such an exercise appears at the end of the Chapter.

In the other swaps investigated in this Chapter, no principal was exchanged at the time the swap was initiated. Therefore, when the additional parameter was added to the procedures, in order to value an existing swap, it actually allowed one to value a *deferred swap*. In other words, this is a swap that is agreed upon now and that will start at some specified time in the future. This future time is not necessarily a starting point, but coincides with the beginning of a payment period. (See the exercises at the end of the Chapter which elaborate on deferred swaps.) This is not the case for a foreign currency swap.

The value of an existing foreign currency swap does not (**and should not**) take into account cash flows that were exchanged in the past. When valuing an existing foreign currency swap, only the principals that will be exchanged at maturity and all of the future regular cash flows ("coupons") should be considered. Hence, the procedure that values an existing foreign currency swap cannot be applied to value a deferred foreign currency swap. Fortunately, there is a way of valuing such a swap, as is explained in the end-of-Chapter exercises.

Consider a foreign exchange swap where the principals are 11,000, the domestic interest rate is 0.03, the foreign rate is 0.13, and the spot exchange rate is 0.6. The swap matures in $\frac{18}{12}$ years and payments occur semiannually. The value of such a swap is calculated as below:

```
> FXswap(11000,0.03,disL,11000,.13,disF,.6, 3/2,1/2);
```

$$427.700104$$

Assume that two months passed ($\frac{2}{12}$ units of time) from the initiation of the swap and that the exchange rate and the discount factor functions stayed the same. Thus, the exchange of interest payments will occur in $\frac{4}{12}$, $\frac{10}{12}$, and $\frac{16}{12}$ years, and in $\frac{16}{12}$ years the principals will be exchanged. Let us value this existing swap step by step and then verify the result by running the procedure **FXswap**.

```
> FXswap(11000,.03,disL,11000,0.13,disF,.6,16/12,1/2,4/12);
```

$$-3796.591614$$

One should remember that at the initiation of the swap the principals are exchanged, and then they are exchanged again at the maturity of the swap. The present value of the domestic interest and principal payments is calculated below and is assigned to the variable **Dome**

```
> Dome:=add(11000*.03*disL(4/12+i/2),i=0..2)+\
> 11000*disL(16/12);
```

$$Dome := 10796.68020$$

The present value of the foreign interest and principal currency (in terms of the foreign currency) is calculated below and is assigned to the variable **Fore**.

```
> Fore:=add(11000*.13*disF(4/12+i/2),i=0..2)+\
> 11000*disF(16/12);
```

$$Fore := 11666.81431$$

Thus, the value the foreign payment in terms of the domestic currency is

```
> Fore*0.6;
```

$$7000.088586$$

and hence the value of the swap (to the domestic firm) is given below.

```
> Fore*0.6-Dome;
```

$$-3796.591614$$

If we use our procedure to calculate the swap's value, we have to execute the procedure with an additional parameter at the end. This parameter is the time until the first payment: $\frac{4}{12}$ years in our case. The time to maturity of the swap, now that 2 months have passed since the initiation of it, is $\frac{16}{12}$ years instead of $\frac{18}{12}$ years.

```
> FXswap(11000,.03,disL,11000,.13,disF,.6,16/12,1/2,4/12);
```

$$-3796.591614$$

Indeed we have confirmed that the same result is produced.

11.3 Commodity and Equity Swaps

Commodity and equity swaps follow the same basic structure as the other swaps we have investigated. One party agrees to pay another the spot price of a certain underlying asset and receives, in return, a fixed price agreed upon at the initiation of the swap.

In case of a commodity swap the parties agree on the quantity, referred to as the *notional quantity*, and on the time schedule $t = t_1, t_2, ..., t_k$, where t_k is the maturity time of the swap and k is the number of exchanges. For example, the notional quantity could be a number N of barrels of oil. The payments exchanged at time t are $S(t)N$ and FpN, where the spot price of a barrel of oil at time t is $S(t)$, and the fixed price agreed upon is Fp. The net payment at time t, to the party who pays the fixed rate, is therefore

$$S(t)\,N - FpN. \tag{11.18}$$

The present value of $S(t)$, which is the value at time zero of having $S(t)$ at time t, is of course the spot price at time zero, $S(0)$: buying the good at time zero and holding it until time t will produce $S(t)$ at time t. This is just another way of looking at the cost-of-carry model through the present value concept. (See also footnote 2 in Chapter 6.) Hence, the present value of equation (11.18) is

$$S(0)\,N - d(t)\,Fp\,N, \tag{11.19}$$

where $d(t)$ is the discount factor for time t.

There is another way of explaining the discounting of this cash flow. The investor can remove the uncertainty from $S(t)$ and can fix today the price of a barrel of oil a time t by entering into a forward contract. Hence, the investor knows that paying the forward price as of today, $Fr(t)$, at time t ensures the delivery of the oil. We can therefore replace $S(t)$, which is a random variable, with the certain (or deterministic) amount $Fr(t)$. This in turn allows us to calculate the value of the contract by simply discounting the net cash flow based on the current term structure of interest rates. Therefore, the value of the commodity swap is given by equation (11.20).

$$\sum_{i=1}^{k} d(t_i)(Fr(t_i)\,N - FpN) \tag{11.20}$$

It is easy to see that each element in the summation of equation (11.20) is equal to equation (11.19). To this end, one simply needs to substitute for $Fr(t)$ its value in terms of $S(0)$ and $d(t)$. Since the forward price of an asset is the future value of its spot price, $Fr(t)$ can be replaced with $\frac{S(0)}{d(t)}$. Hence, the value of the swap can also be written as in equation (11.21) below.

$$\sum_{i=1}^{k}(S(0)N - FpNd(t_i)) = kS(0)N - \sum_{i=1}^{k} FpNd(t_i) \qquad (11.21)$$

Given N, $d(t)$, and $S(0)$ we can solve for the numerical value of Fpthat makes the value of the swap zero. This is done in the command below where we also make sure that d is unassigned. However, we assume, as is generally the case, that for a swap at its initiation the payment schedule is equally spaced from the current time (time zero) to time k.

```
> d:='d':solve(sum(S(0)*N-d(t[i])*Fp*N,i=1..k)= 0,Fp);
```

$$\frac{k\,\mathrm{S}(0)}{\sum_{i=1}^{k}\mathrm{d}(t_i)}$$

Hence, Fp, as in equation (11.22), will indeed be the value agreed upon by the two parties.

$$Fp = \frac{kS(0)}{\sum_{i=1}^{k} d(t_i)} \qquad (11.22)$$

Valuing a swap at some time other than its initiation follows the same argument as discussed in Section 11.1.1. One should remember to make sure that $d(t)$ is the current term structure and that the time until the first payment may be different than the length of the time period between two consecutive payments.

The procedure **COmswap** values a commodity swap from the point of view of the party who pays the fixed price. The input parameters to this procedure are, in order, N, Fp, *dis*, $S(0)$, T, and *per*. T is the maturity time of the swap and *per* is the time between each consecutive payment. The rest of the parameters are as defined for equation (11.21). We can use the procedure in a symbolic way to solve for the value of the fixed rate given the rest of the parameters, as below.

```
> solve(COmswap(N,Fp,d,S0,T,per)=0,Fp);
```

$$\frac{T\,S0}{\left(\sum_{t=1}^{\frac{T}{per}}\mathrm{d}(t\,per)\right) per}$$

We can also use the procedure to value an equity swap at its initiation or to solve for the fixed price. Suppose we would like to find the fixed price that makes the value of a swap zero. Consider a swap of 1000 barrels of oil to be delivered (received) every 6 months for a period of 2 years when the spot price per barrel is $15. We first solve for the discount factor d utilizing **NarbitB** in a market where *coupons are paid semiannually.* Using **NarbitB** in this manner means that we keep a time unit of 6months.

```
> NarbitB([[107,0,0,0],[5,5,5,105],[3,103,0,0],\
> [6,6,106,0]],[97,90,95,94],16,d,0);
```

The no − arbitrage condition is satisfied.

$$\textit{The discount factor for time, 1, is given by, } \frac{97}{107}$$

The interest rate spanning the time interval, [0, 1], *is given by,* .1031

$$\textit{The discount factor for time, 2, is given by, } \frac{9874}{11021}$$

The interest rate spanning the time interval, [0, 2], *is given by,* .1162

$$\textit{The discount factor for time, 3, is given by, } \frac{458392}{584113}$$

The interest rate spanning the time interval, [0, 3], *is given by,* .2743

$$\textit{The discount factor for time, 4, is given by, } \frac{9002797}{12266373}$$

The interest rate spanning the time interval, [0, 4], *is given by,* .3625

The function Vdis([c1, c2, ..]), values the cashflow [c1, c2, ..]

The continuous discount factor is given by the function, 'd', (.)

```
> SumAbsDiv;
```

0

Since we maintain a time period of six months, the maturity of the swap will be at time 4 and payments will be exchanged every one unit of time. The price of a barrel of oil for this swap is solved for below:

```
> evalf(solve(COmswap(1000,Fp,d,15,4,1)=0,Fp));
```

18.06589418

Consider the same swap but with a fixed price of $18; its value in this case is calculated by the next command.

```
> evalf(COmswap(1000,18,d,15,4,1));
```

218.8461088

The procedure **COmswap** is also capable of valuing a swap after its initiation. Suppose we would like to value the above swap but that the time until the first payment is three months from initiation, rather than six. The only parameter we add to the procedure, as the last parameter, is the length of time until the next payment. Hence, the swap above, three months after its initiation, assuming the same d, will be valued as below.

```
> evalf(COmswap(1000,25,d,18,3.5,1,1/2));
```

$$-13389.69735$$

The fixed price that makes this swap a fair deal is given below.

```
> evalf(solve(COmswap(1000,Fp,d,15,3.5,1,1/2)=0,Fp));
```

$$17.56652203$$

In the above valuation we assumed that the discount factor function **d** has the same structure as at the initiation time. If one would like to value the swap for a different structure one needs to run **NarbitB** again with the new data of the bond market.

In the commodity swap cases, and in the equity swap case as we will see below, the valuation makes use of the replication argument. In the commodity swap case, this replication argument is nothing more than the cost-of-carry model.

11.3.1 Equity Swaps

In case of an equity swap, the party agrees on a notional principal and on an index to which the floating rate is linked. At times $t = t_1, ..., t_k$ one party pays the other a fixed rate of interest on the notional principal N. In exchange, the party receives, at time t, the return on N invested in the index (including dividends if such are paid out) over the period $[t-1,\ t]$. An equity swap is thus essentially a fixed-for-float swap where the floating rate is linked to an index, e.g., the S&P 500 index, rather than to a variable interest rate. The net payment at time t to the party who pays the fixed rate is thus

$$FRN - I(t)N, \tag{11.23}$$

where $I(t)$ is the return on the index over the time period from $t-1$ to t, and FR is the fixed rate agreed upon at the initiation time $t = 0$.

The valuation of the equity swap follows the same guidelines as that of the commodity swap. We first need to find the current value (present value) of $I(t)\,N$. In Section 2.5.2 we explained how we were able to arrive at the required value. Using the replication argument, we repeat it here for convenience. We borrow, at the current time, an amount $N\,d(t-1)$ to be repaid at time t. This amount is invested at the risk-free rate of interest until time $t-1$, at which point it will be worth N. At time $t-1$, the amount N is invested in the index for one period until time t. Let us denote the value of this investment at time t (including the dividend paid at that time) by $V(t)$. At time t the loan needs to be paid back. The amount to be paid back is $N\,\frac{d(t-1)}{d(t)}$. However, since we only need to replicate the return on the index, which is $V(t)-N$, we can use N to subsidize the loan repayment. Hence, replicating the return on the index results in a certain (deterministic) payment equal to $N\,\frac{d(t-1)}{d(t)} - N$ at time t. Therefore, we can replace the random quantity $N\,I(t)$ in equation (11.24) with the deterministic quantity $N\,\frac{d(t-1)}{d(t)} - N$, arriving at equation (11.26) for the net payment at time t.

$$FRN - N\frac{d(t-1)}{d(t)} + N \tag{11.24}$$

Equation (11.24) does not involve random quantities and hence its present value can be calculated by discounting it using the risk-free rate of interest. The value of this swap is then the summation, as in equation (11.25), of the present value of the net payments across $t = 1,...,k$.

$$\sum_{t=1}^{k}(FRNd(t) - Nd(t-1) + Nd(t)) \tag{11.25}$$

The value of $d(0)$ is one and since the terms $N\,d(t-1) + Nd(t)$ cancel each other for t such that $1 < t < k$, we arrive at equation (11.26) for the value of the swap

$$\left[\sum_{t=1}^{k} FRN\,d(t)\right] - N\,(1-d(k)). \tag{11.26}$$

We can therefore solve for the numerical value of FR which makes the swap value zero. The next command uses dis for the discount factor instead of our usual notation d. This is because d is assigned a value and we would soon like to use it for numerical calculations.

```
> solve(sum(FR*N*dis(t),t=1..k)-N*(1-dis(k))=0, FR);
```

$$-\frac{-1+\mathrm{dis}(k)}{\sum_{t=1}^{k}\mathrm{dis}(t)}$$

At the initiation time, the value of the swap is zero and the parties will agree on FR such that

$$FR = \frac{1 - d(k)}{\sum_{t=1}^{k} d(t)}. \tag{11.27}$$

The procedure **EQswap** values an equity swap. The input parameters to this procedure are, in order, N, FR, and d, as defined in equation (11.26). We can verify the solution for FR by executing the command below:

```
> solve(EQswap(N,Fp,dis,T,per)=0,Fp);
```

$$-\frac{-1 + \mathrm{dis}(T)}{\displaystyle\sum_{t=1}^{\frac{T}{per}} \mathrm{dis}(t\, per)}$$

We can make use of the procedure **EQswap** to value an equity swap at its initiation and to value an equity swap which is already in existence. The numerical value of FR, the fixed rate of interest, that will make the value of the swap vanish is independent of the particular index to which the variable return is pegged. What matters is only the term structure of interest rates. Equivalently, what matters is the discount factor function. We already have explained that phenomenon at the end of Chapter 2 in Section 2.5.2. The reader may wish to review the strategy which replicates the return on the index, making sure to verify that the return is independent of the actual index being replicated. This is the driving force behind this phenomenon.

Consider a swap at its initiation with a maturity of two years for which payments are exchanged once each year. The fixed rate of interest is 8 percent and the notional principal is \$1000. The discount factor function is d, as defined above.

Since the units of time in the term structure estimation are half-years, to value such a swap we execute the command with the maturity time $T = 4$ and a period of two, i.e., $per = 2$.

```
> EQswap(1000,.08,d,4,2);
```

$$-135.6694598$$

The fixed rate that will make the value of this swap zero is given by

```
> evalf(solve(EQswap(1000,Fp,d,4,2)=0,Fp));
```

$$.1632395333$$

If this swap had been in existence for six months, the next payment would occur in six months and its value would be calculated by simply adding another input parameter. This parameter would be the time until the next payment. The value would then be calculated by executing the command below.

```
> EQswap(1000,.070741961591,d,4,2,1);
```

$$-201.9281975$$

This calculation of course assumes that in six months from now the discount factor is still $d(t)$. In reality one would have to reestimate the discount factor.

Consider a swap with the same parameters as above but for which the first exchange of payments will be in one month, rather than in one year. The fixed rate of interest that would make this swap a fair deal for both parties is found by executing the MAPLE command below.

```
> evalf(solve(EQswap(1000,Fp,d,4,2,1/6)=0,Fp));
```

$$.2796156552$$

Note that when we changed the time until the first payment we effectively altered the present value calculation in equation (11.25). We can investigate the sensitivity of the fixed rate of interest to the time until the first payment via the MAPLE command below. In this command, the time until the next payment is allowed to range from $\frac{1}{10}$ to 1, in increments of $\frac{1}{10}$. The sequence is a variety of fixed rates of interest that would make the value of the swap zero.

```
> seq(evalf(solve(EQswap(1000,Fp,d,4,2,i/10)=0,Fp)),\
> i=1..10);
```

.2751183891, .2814796606, .2858187784, .2887079248, .2905920785,
.2917982222, .2925571447, .2930269608, .2933136697, .2934874990

The last section of this Chapter visualizes the relation between swaps and forwards.

11.4 Forwards and Swaps: A Visualization

A swap can actually be thought of as a portfolio of forward contracts. We have already alluded to this relation at the end of Section 11.1, which discussed fixed-for-float swaps. We can visualize this interpretation with the aid of the three-dimensional graphing capability of MAPLE. We examine the payoff structure which is a consequence of a forward contract, and then demonstrate that the aggregate of these payoffs across time is equivalent to the payoff structure of a swap contract. We examine the commodity swap in our example, but any of the other swaps would serve the purpose equally well.

Let us consider a swap contract in which one party is committed to swap a barrel of oil at some future times $t = t_1, t_2, ..., t_k$, for a price of Fp agreed upon now and payable at those times. Assume that the current time is zero, that $t_i = i$ for $i = 1, ..., 4$, and that Fp is \$7. We consider the payoff from the point of view of the party who pays the fixed price for the barrel of oil. The cash flows resultant from this contract occur at times $t = 1, 2, 3$, and 4, and are equal to $S(t) - FF$, where $S(t)$ is the spot price at time t of a barrel of oil. At any other time, of course, the cash flow from this contract is zero. We visualize the cash flow from this swap contract by executing the procedure **PlotSwPyf.** The parameters for this procedure are, in order, the sequence of time at which a swap payment occurs and the price agreed upon for the swap. The points in time are entered into the procedure in the form $[t_1, t_2, ..., t_k]$, or, in our example, as $[1, 2, 3, 4]$.

Executing the command **PlotSwPyf** $([1, 2, 3, 4], 7)$ will generate the graph in Figure 11.3. The reader can, and is encouraged to, experiment with different values of the parameters and change the viewing perspective of the graph by dragging it in the on-line version.

```
> PlotSwPyf([1,2,3,4],7);
```

It is now visually apparent that the cash flows from a swap are like the cash flows from four forward contracts with a forward price of \$7 and maturity dates of 1, 2, 3, and 4. Economically, the swap is a portfolio composed of the forward contracts. To visualize this phenomenon we can instruct MAPLE to plot, on the same set of axes, the cash flows of these four forward contracts over the time span $[0, 6]$ and then examine the resultant graphs. This is done with the next MAPLE command. This last graph, however, is displayed only in the on-line version of the book.

```
> plots[display](seq(PlotForPyf(k,7,6),k=1..4));
```

11.5 Concluding Remarks

This Chapter takes a second look at swaps. Swaps can be categorized as assets the valuation of which does not require knowledge of the evolution of the term structure over time. The setting used in this second look is characterized by a multiperiod time with a continuum of states of nature. It is shown here that swaps can be viewed as a portfolio of forward contracts

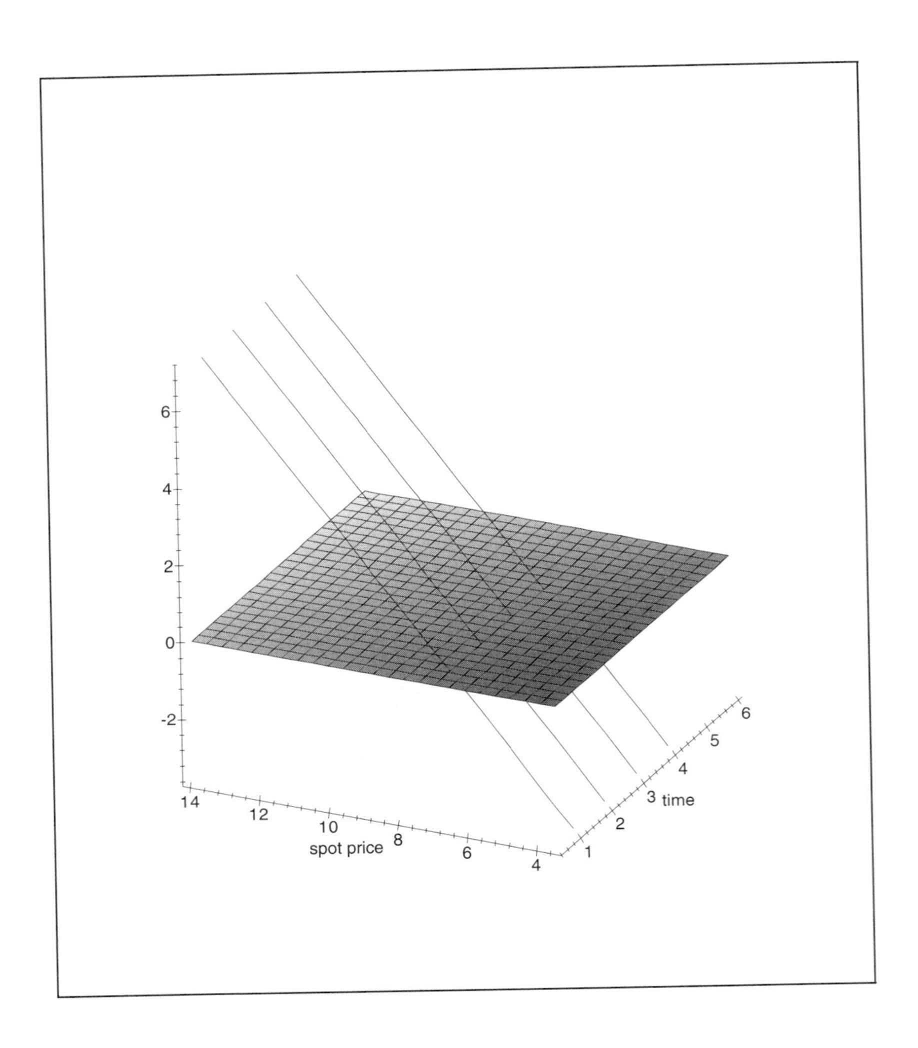

Figure 11.3: Swaps' Building Blocks

and thereby a link is established to the analysis of swaps in the setting of a one-period model in Chapter 2.

The key idea in the valuation of swaps is essentially the calculation of the net present values of the swapped cash flows. The present values are calculated utilizing the discount factor function estimated from the prices of the bonds in the market. At the initiation of the swap, its parameters are set so the values of the swapped cash flows offset each other. For example, the fixed-for-float swap is set in such a way that the value of the swap is zero at initiation. After initiation of the swap, market conditions usually change and the value of the swap is no longer zero. The value of a swap, either at initiation or later, is the difference between the current value of the swapped cash flows.

The investigation of equity swaps highlights the effect of the absence of arbitrage opportunities in a market. Such a market "prices" the risk embedded in each asset by adjusting the returns on the different assets available in the market. Hence, the return on one dollar invested in asset A should be worth the same as the return on one dollar invested in asset B. Thus swapping the returns on any two assets, as long as the amounts invested in them are the same, should be a fair swap.

The next Chapter explores the properties of American options in the same setting used in this Chapter. It is followed by Chapters 13 and 14 for which the setting is a multiperiod model with only two states of nature in each period: a binomial model. Such a model facilitates, among other things, the numerical valuation of American options and prepares us for the continuous-time, continuous-states model of Chapter 15.

11.6 Questions and Problems

Problem 1. Consider a fixed-for-float swap at its initiation time. Assume that the swap has semiannual payments for the next m years and has a principal of $\$N$. Assume further that the term structure of interest rates prevailing in the market is known.

1. Ignoring the credit risk, find the fixed rate, r, that will make this swap fair.

2. Ignoring the credit risk, find the fixed rate, r_F, that will make the party that pays the fixed rate pay $\$M$ at the initiation of the swap.

3. Ignoring the credit risk, find the fixed rate, r_L, that will make the party that pays LIBOR pay \$$M$ at the initiation of the swap.

Problem 2. Assume that the term structure of interest rates is known. Based on the discussion of FRAs, in Section 10.4.1, calculate the fixed rate that would make a fixed-for-float swap a fair deal. (*Hint:* As it has been discussed in the text, the swap can be decomposed into building blocks that are similar to the FRAs. The catch, though, is that in a swap the same fixed rate should apply to all the building blocks. Hence, the value of each block may not be zero, but the value of the blocks in aggregate is zero if the swap is fair.)

Problem 3. Suppose a customer would like to enter into a *buy down swap* and would like to pay a fixed rate of $r - \delta$, where r is the fixed rate that would make an otherwise equivalent regular swap a fair deal. Determine the amount that this customer would have to pay at the initiation of the swap. Assume that the discount function is known and the discount factor for time t is denoted by $d(t)$.

Problem 4. Suppose a customer would like to enter into a *buy up swap* and would like to pay a fixed rate that of $r + \delta$, where r is the fixed rate that would make an otherwise equivalent regular swap a fair deal. Determine the amount that this customer would have to pay at the initiation of the swap. Assume that the term structure of interest rates is known and the continuously compounded risk-free rate for time t is denoted by $r(t)$.

Problem 5. In practice, a currency swap might be arranged for a domestic firm by a financial institution. The institution either finds a foreign firm or "warehouses" this cash flow until it finds an optimal match for it. In certain cases, the institution must assume some risk as a result of the absence of an exact match. Hence, it requires a fee, a "finder's fee", as well as compensation for the risk taken. This fee is usually obtained in terms of an increase in the rate, r_0, that the domestic firm is asked to pay. Consider the swap in the example on page 465. Assume that a financial institution agrees to arrange this swap for a domestic firm but requires the firm to pay $r_0 = 0.047$ instead of $r_0 = 0.042$. What is the "finder's fee" that the institution charged the firm?

Problem 6. The **FXswap** procedure can also be used if instead of the two term structures, the domestic term structure and the term structure of forward exchange rates are supplied. A direct substitution in the **FXswap** procedure will generate the value of a foreign exchange swap. Use the term structure of forward exchange rates and the foreign term structure given in the example in the Chapter (on page 465), to value the foreign exchange swap in the Chapter.

Problem 7. Suppose an investor would like to enter now in a deferred swap that will commence in n years from now, will last for m years after and will make semiannual payments on the principal of $\$K$. Assuming that at the current time the term structure of interest rates is known, find a fixed rate r that would make this deferred swap a fair deal.

Problem 8. The value of an existing foreign currency swap does not (**and should not**) take into account cash flows that were exchanged in the past. When valuing an existing foreign currency swap, only the principals that will be exchanged at maturity and all of the future regular cash flows ("coupons") should be considered. Hence, the procedure **FXswap,** that values an existing foreign currency swap, cannot be applied to value a deferred foreign currency swap. However, valuing such a swap should not present a problem. One needs simply to value also the principals that will be exchanges in the future date, when the swap is initiated. This should be done in exactly the same manner the principals (of a regular swap) that are swapped the termination time are valued. Consider the example of a foreign currency swap in Section 11.2 and value it, if it would have been a deferred swap starting in six months.

Chapter 12

American Options

Until this point we focused our attention solely on European options. We are now ready to embark on another and actually more common type of option: an American option. While a European option grants the holder rights that can be exercised **on** a specific day, an American option allows for the exercising of those rights **until** a specific day. It is obvious then that given two options that differ only in that one is American and the other is European, the former must be worth at least as much as the latter. A simple arbitrage argument may be used to show that the additional freedom offered by an American option cannot have a negative value. Putting it differently, an American option must be worth at least as much as its counterpart European option.

There are, however, certain circumstances whereby it is never profitable to exercise an option prior to its maturity. Consequently, the additional feature offered by an American option is worthless. Thus, in these circumstances the value of the American and the European options must be the same. These situations are characterized by the saying that the option is worth more when it is "alive" than "dead". This conveys the notion that, whatever the state of nature, it is not optimal to adopt an early exercise strategy.

Such is the case for a call option on a stock that does not pay dividends, which is the topic of the next subsection. The upcoming section will also derive arbitrage bounds on call options that are independent of the underlying price process. It will be followed by the investigation of put options on non-dividend-paying stocks. We shall then attend to the characteristics of options on dividend-paying stocks, followed by the put–call parity. This Chapter will not be dealing with the valuation of American options but rather with their characterizations and arbitrage pricing bounds. The valuation issue is deferred to Chapters 13 and 14, and will also be touched on in Chapter 15.

12.1 American Call Option

12.1.1 Arbitrage Bounds

The fact that an American option can be exercised at any time prior to maturity induces a lower bound on its value. At any time prior to maturity, the value of an American option cannot be lower than the amount it grants the holder, if exercised. This follows from a basic arbitrage argument. Can you, the reader, describe it?

An American call option can be exercised at any time, t, prior to the maturity time, T, and generate a cash flow of $\max(0, S_t - K)$. We therefore arrive at a lower bound for the value of an American call as specified in equation (12.1),

$$P_{ca(S_t,K)} \geq \max(0, S_t - K), \tag{12.1}$$

where $P_{ca(S_t,K)}$ denotes the price of an American call with exercise price of K when the underlying price is S_t. The reader may however recall that a European call option has a lower bound that is more refined than equation (12.1). Equation (4.4) in Chapter 4, reproduced here as equation (12.2), stipulated that a European call option satisfy

$$P_{c(S_t,K)} \geq \max(0, S_t - Ke^{-r(T-t)}). \tag{12.2}$$

Can you reconstruct the arbitrage argument that led to this bound?

The bound we just produced in equation (12.1) did not in fact refine what was known to us about options from our investigation of European options. As we argued above,

$$P_{ca} \geq P_c, \tag{12.3}$$

where P_c denotes the price of a European call. Combining equation (12.2) and equation (12.3) yields that

$$P_{ca(S_t,K)} \geq \max(0, S_t - Ke^{-r(T-t)}). \tag{12.4}$$

Assuming that $r > 0$, and since $t < T$, we have that $e^{-r(T-t)} < 1$ and thus that $Ke^{-r(T-t)} < K$ and that $\max(0, S_t - Ke^{-r(T-t)}) \geq \max(0, S_t - K)$. Consequently, equation (12.4) displays a tighter bound on the lower value of the American option than that of (12.1). Of course the stock price serves as an upper bound for both American and European call options as neither can exceed the price of the underlying asset.

12.1.2 Early Exercise Decision

Equation (12.1) is indeed only a lower bound on the value of an American option and so is equation (12.2). Although the lower bound on the European option is higher than the lower bound on the American option, the latter could still have a higher value. Nonetheless, there might be some meaning to the fact that utilizing the possibility to exercise the American call at each moment does not yield a refinement of the bound in equation (12.4) (obtained for a European call). It suggests that if this additional feature, of early exercising, did not help in producing a tighter bound on American options, perhaps such does not exist. American and European call options on non-dividend-paying stocks may have the same value. Or, in other words, it is indicated that it might never be optimal to exercise an American call option early.

This is, indeed, the case, as was virtually proven by the above discussion. We shall soon provide a more formal argument, but let us first take a second look at the following animation. The animation below, in the on-line version of the book, (Figure 7.1 of Section 7.1) demonstrates the evolution of the value of a European call option as a function of S_t, as time approaches maturity, together with the value of the American call if exercised at that time. The latter graph is, of course, just the value of a call (American and or European) at maturity and clearly is a function of S_t only. Hence, the piecewise-linear graph stays static as time progresses to maturity.

Observe that, regardless of the stock price, S_t, at each moment of time the European call value is larger than its value at maturity: the piecewise linear graph. The piecewise-linear graph is also the cash flow obtained from exercising an American option at time t. It thus follows that it would be more advantageous to sell the American option than to exercise it. If you were holding a European call option and at a certain time t prior to maturity you were allowed to exercise it, it would be better not to do so. The additional flexibility offered by the American call is therefore valueless and it must have the same value as a European call.

For a more formal argument, but virtually the same as above, consider the following. Exercising the option now (even if it produces a negative cash flow) costs K and produces S_t for a total cash flow of $S_t - K$. However, the value of the call by equation (12.4) is greater then or equal to $S_t -$

$Ke^{-r(T-t)}$,which for $r > 0$ is greater then $S_t - K$. It is therefore clearly not optimal to exercise the call and one would be better off selling the option then exercising it.

Consider a yet more direct argument. Compare the value of the call now to its value if exercised an instant prior to maturity (even if it produces a negative cash flow). An instant prior to maturity, exercising the call produces $S_T - K$, the present value of which is $S_t - Ke^{-r(T-t)}$. Since following such a strategy may not be optimal, the value of the call must be at least $S_t - Ke^{-r(T-t)}$. As we just noted (when $r > 0$), it is greater than the cash flow from exercising the option now. It thus follows that it is not optimal to exercise early, and that the price of an American call is the same as that of a European call. This conclusion is a consequence of the time value of money one loses on K when exercising early, as is essentially induced by the relation between $Ke^{-r(T-t)}$ and K.

12.2 American Put Options

12.2.1 Arbitrage Bounds

As its counterpart, the American call option, an American put option cannot be worth less than a European put. The flexibility to exercise early cannot have a negative value. It can only be worthless as we saw above. That flexibility induces a lower bound on the value of the American put. Its value must be greater than or equal to the cash flow generated if it is being exercised. Equation (12.5),

$$P_{pa(S_t,K)} \geq \max(0, K - S_t), \tag{12.5}$$

where $P_{pa(S_t,K)}$ stands for the value of an American put, stipulates the obtained lower bound. This follows again from a basic arbitrage argument. Can you, the reader, delineate an arbitrage portfolio that would exist if relation (12.5) would have been violated?

Recall the lower bound on the value of a European put option. Equation (4.9) in Chapter 4 provides the inequality

$$P_{p(S_t,K)} \geq \max(0, Ke^{-r(T-t)} - S_t),$$

where P_p stands for the value of a European put. The lower bound, provided by equation (12.5), indeed refines the lower bound that could be obtained on P_{pa} from the fact that $P_{pa} \geq P_p$. Figure 12.1 demonstrates that there

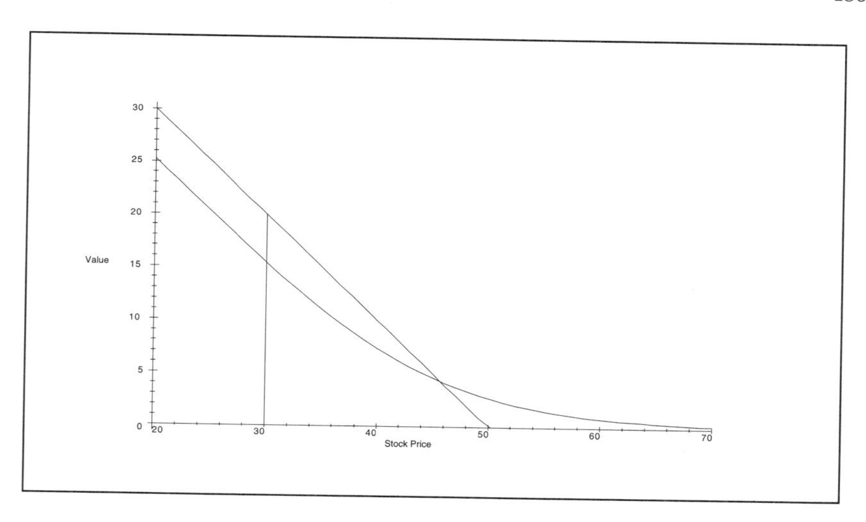

Figure 12.1: t and S_t where the value of a European put is below the lower bound of an American put

exists a time t and a stock price S_t, e.g., \$30, where the value of a European put is below the lower bound, equation (12.5), of an American put.

```
> plot({Put(P,50),BsPut(50,0.5,P,.25,.10),[[30,0],[30,20]]}\
> ,P=20..70,thickness=3,title=‘Figure 12.1:\
> t and St where the value of a European put\
> is below the lower bound of an American put‘ ,labels=\
> [‘Stock Price‘,‘Put Value‘],colour=green);
```

This is in contrast to the relation between American and European calls. Here we compare $(K - S_t)$ to $(Ke^{-r(T-t)} - S_t)$ as opposed to $(S_t - K)$ to $(S_t - Ke^{-r(T-t)})$ in the case of a call option. Since for $r > 0$, $Ke^{-r(T-t)} - S_t < K - S_t$, the lower bound on the value of the put options is refined. Following the same logic as above, this may suggest that an American put is valued more than its European counterpart. A lower bound on an American put option is therefore $\max(P_{p(S_t,K)}, K - S_t)$. It is displayed in Figure 12.2, which appears as an animation in the on-line version of the book.

```
> plots[animate](max(Put(P,50),BsPut(50,1-t,P,0.25,0.10)),\
> P=20..70,t=0..1,frames=30,thickness=3,title=‘Figure\
> 12.2:  Evolution of a Lower Bound on the Value of an\
> American Put Option‘,labels=[‘Stock Price‘,‘Put Value‘]);
```

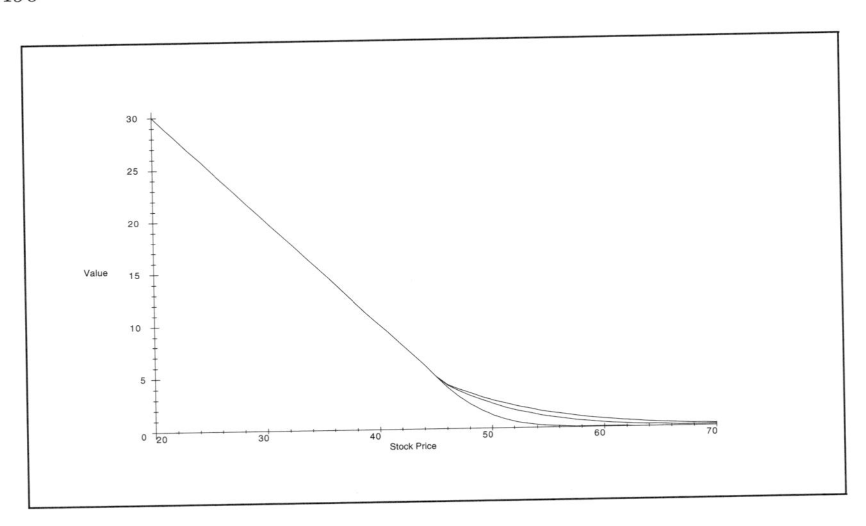

Figure 12.2: Evolution of a Lower Bound on the Value of an American Put Option

Like its counterpart, the European put, the cash flow generated by an American put cannot exceed the exercise price. Therefore, we have the upper bound specified in equation (12.6).

$$P_{pa(S_t,K)} \leq K \tag{12.6}$$

If this bound would be violated, then writing such a put option and investing the received premium at the risk-free rate would guarantee (strictly) positive cash flow at maturity (or at the exercising time).

12.2.2 Early Exercise Decision

The situation in the case of an American put is somewhat different than that of an American call. It might be optimal to exercise an American put early. In fact, the arbitrage bound just developed in equation (12.6) points out an extreme situation where early exercise is optimal.

If the stock price is zero, i.e., the company is in bankruptcy, clearly the value of an American option is the exercise price K. If that occurs prior to the maturity of the option, one might as well exercise immediately. Waiting will only cause losing the time value of money.

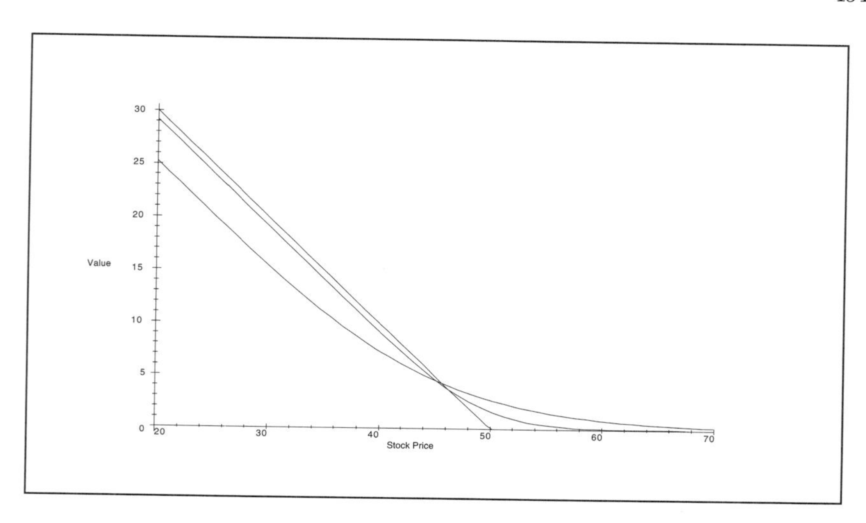

Figure 12.3: Evolution of Put Option Value as Time to Maturity Decreases

Let us look at the early exercise decision more carefully. As was alluded to before, the lower bound on an American put option refines that of the European put option. Observe the animation demonstrated in the on-line version of the book below. It describes the evolution of the value of a European put as a function of the stock price when time approaches maturity together with its value at maturity. Figure 12.3 displays three frames of the animation.

```
> plot([BsPut(50,1,P,.25,.10),BsPut(50,2/12,P,.25,.10)\
> ,BstPut(50,0,P,.25,.10)],P=20..70,labels=['Stock\
> Price', 'Put Value'],color=[green,red,blue], thickness\
> =2,title='Figure 12.3:Evolution of Put Option Value \
> as Time to Maturity Decreases');
```

Both the animation and Figure 12.3 clearly demonstrate that there are situations at some time t where the graph of the value of a European option, as a function of S_t, falls below the same graph at T. Thus, if at such a time holders of a European put option would be allowed to exercise the option, they would have done so. It follows that having the flexibility to exercise early must have some value.

Note that this does not mean that an American put option would be exercised in such circumstances. It might be optimal to wait a little longer for the price of the stock to be even lower as it increases the cash flow from exercising the put. We shall attend to the question of the optimal exercising time in Chapters 13, and 14, and somewhat in 15. Here we only point out that the price at which exercising is optimal is a function of the time to maturity. Furthermore, let S_v^* be a price at which exercising should take place if the time to maturity is $T-v$. Then if the time to maturity is $T-v$ and the price of the stock is $S < S_v^*$, it is obviously optimal to exercise the put. Can you, the reader, explain why this is true?

Indeed, as mentioned before the lower bound on the value of an American put, given by equation (12.5), can be refined to equation (12.7).

$$\begin{aligned} P_{pa(S_t,K)} &\geq \max(P_{p(S_t,K)}, \max(0, K - S_t)) \\ &= \max(P_{p(S_t,K)}, K - S_t) \end{aligned} \tag{12.7}$$

The upper and lower bounds on an American put, as a function of S and t, are displayed in Figure 12.4.

```
> plot3d({max('BstPut'(50,t,P,.25,.10),50-P),50},P=0\
> ..70,t=0..1,labels=[`Stock Price`,`Time to Maturity`,\
> `Put Value`],orientation=[-78,71],axes=frame,\
> title=`Figure 12.4:  Upper and Lower Bounds\
> for an American Put`);
```

Intuitively, the main factor that makes it optimal to exercise a put early and not a call is the existence of an upper bound on the payoff from a put. The payoff from a put will never exceed the exercise price and will equal it when the stock price is zero. Thus, when the value of the stock is "close to zero", relative to the exercise price, it might be optimal to exercise a put.

The payoff from a call, on the other hand, has no upper bound, as there is no upper bound on the value of the stock. Hence, there exists no benchmark, like the zero in the put case, to state a counterpart rule to the one stated in the put case. What should the stock price be close to, relative to the exercise price, in order for exercising an American call to be optimal?

12.3 Put–Call Parity

The flexibility of an early exercise induces some change to the put–call parity relation. We still assume a stock, or an underlying asset, that pays no dividend, but it is no longer sufficient to only compare the payoffs at maturity. In the case of European options, the put-call parity with exercise price K was directly deduced from the cash flow at maturity. It is a consequence

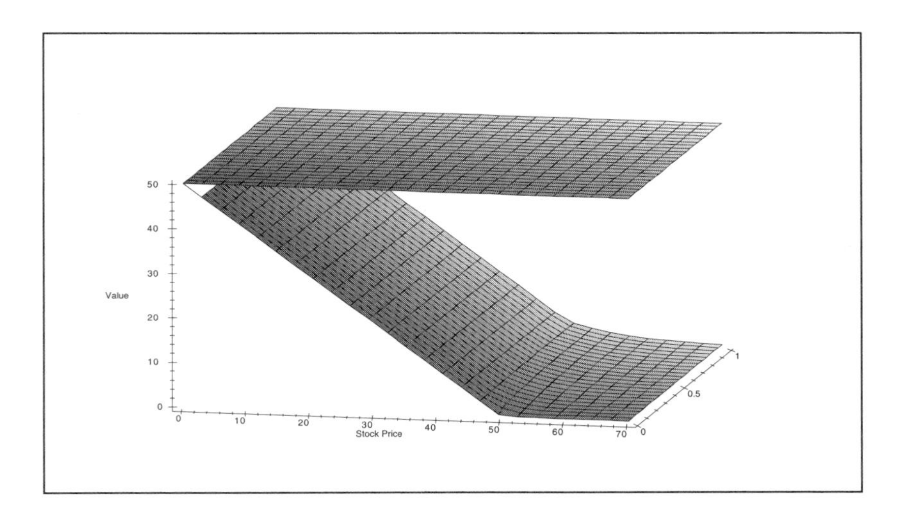

Figure 12.4: Upper and Lower Bounds for an American Put

of the payoff from a long call position and a short put position, at maturity, being exactly the value of a stock less the exercise price.

We already established that the value of an American call option is equal to that of a European call option, on a stock that pays no dividend, and that an American put is more valuable than its European counterpart. Rearranging equation (4.12), we obtain equation (12.8) for European put options.

$$P_{p(S_t,K)} = P_{c(S_t,K)} - S_t + Ke^{-r(T-t)} \tag{12.8}$$

Substituting P_{ca} for P_c and utilizing the fact that $P_{pa} > P_p$ yields the inequality in (12.9).

$$P_{pa(S_t,K)} > P_{ca(S_t,K)} - S_t + Ke^{-r(T-t)} \tag{12.9}$$

Rearranging the inequality in (12.9) so that it will be in the form of equation (4.12), we arrive at the relation in (12.10).

$$P_{pa(S_t,K)} - P_{ca(S_t,K)} + S_t \geq Ke^{-r(T-t)} \tag{12.10}$$

Of course, relation (12.9) can also be proven directly by arbitrage consideration rather than by a mathematical substitution. Can you, the reader, try to formulate this argument?

Note that the left-hand side of the expression in (12.9) alludes to a portfolio composed of long positions in the stock and the put, and a short position in the call. It claims that at any time prior to maturity, the holder of this portfolio can liquidate it for at least the present value of the exercise price. While the call can be exercised against the portfolio owner, it would not be optimal to do so. Hence if it would be exercised it would only benefit the call writer's position and "strengthen" the inequality rather than violate it.

Let us see what happens if we reverse the positions in the above portfolio. Suppose we own a portfolio composed of short positions in a put and a stock and a long position in a call. The put can be exercised against us, forcing us to buy the stock for $\$K$. This can happen an instant after we established our position. In that case we pay $\$K$, buy the stock, and close our short position in the stock. We are left with a long position in the call, which is not optimal to exercise, but its value is positive. Hence, the value of such a portfolio, any instant prior to maturity, is at least -$\$K$.

It is easy to see that, at maturity, if the call is in-the-money ($S_t > K$), the put is worthless and the portfolio value is $S_t - K - S_t = -K$. If the call is out-of-the-money ($S_t < K$) then the put will be exercised against us; we buy the stock for $\$K$ and close the short position. Hence, the portfolio value is again -$\$K$. We therefore obtain the inequality

$$-P_{pa(S_t,K)} + P_{ca(S_t,K)} - S_t > -K \qquad \text{for any time } t \leq T. \qquad (12.11)$$

Thus, in contrast to the European case, here we have a wedge and cannot derive an equality relation, but rather two inequalities. Multiplying equation (12.11) by negative one and utilizing relation (12.9) we obtain relation (12.12).

$$K > P_{pa(S_t,K)} - P_{ca(S_t,K)} + S_t \geq Ke^{-r(T-t)} \qquad (12.12)$$

Hence, we cannot equate the value of the portfolio to the present value of K but only bound its value between K and the present value of K.

Intuitively, this is the result of the fact that American options can be exercised at any instant prior to maturity, causing us a loss of $\$K$. The worst-case scenario is when this loss occurs an instant after the portfolio position was established. In that case, the loss will be $\$K$, causing the wedge of $K(1 - e^{-r(T-t)})$ in the portfolio value.

12.4 The Effect of Dividends

The dividends affect our conclusion above via their effect on the price of the stock. We assume that the time and amount of the dividends during the life of the option are known[1] deterministic constants. The payment of a dividend causes a jump in the continuous path of the price process of the stock and makes it discontinuous.

The reader might like to recall our discussion of the path of stock prices in Sections 5.5 and 5.6; in particular footnote 15 on page 202 and the explanation prior to it. Perhaps it would be a good idea to return and review the explanation here after having a second look at these processes in Section 15.2. If the price would not have been affected by the dividends, then buying the stock immediately before the payment and selling it immediately[2] after would generate arbitrage profit. It is therefore possible to show that arbitrage arguments imply that when a dividend is declared, the price of the stock falls by the amount of the dividend.

When dividends are declared they are going to be paid to the stockholders on record by a certain date: the ex-dividend date. Thus, on the ex-dividend date, the stock falls by the amount of the dividend. In fact we are ignoring some frictions, e.g., the effect of taxes and the fact that the payment date might be some time away from the ex-dividend date. However, these will not significantly influence our analysis. Since the stock price is affected, the price of a claim contingent on such a stock must also be affected by the dividends. It may therefore change the optimal policy of exercising American calls and/or puts.

12.4.1 A Call Option

In order to clarify the discussion to follow, we start with an example that highlights the effects of dividends on the early exercise decision of an American call. Consider an American call an instant prior to the stock going ex-dividend the last time before the option matures. The instant the stock price falls by the dividend amount, the option is traded as if it were a European call option. In the absence of dividends, as we saw above, it is not optimal to exercise the option early and thus American and European

[1] The opinion in the literature is that this is a reasonable assumption given the duration of the life of options.

[2] The assumed price process is such that over an instant, with probability one, there is no change in the price process and hence profit is indeed riskless. This is also the reason why the path is a continuous one.

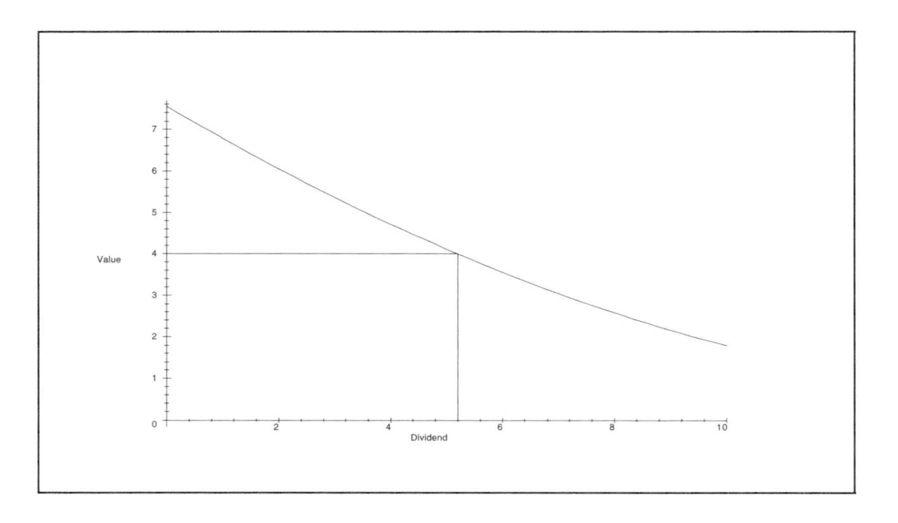

Figure 12.5: The Value of a Call as a Function of the Dividend

options have the same value.

Assume an American option, which is \$4 in-the-money, an instant prior to the ex-dividend date. Hence if exercised at that time it will give rise to a cash flow of \$4. In particular we assume the current price of the stock is 100 and the following parameters: $\sigma = 0.18$, $r = 0.10$, $K = \$90$, and the time to maturity is $\frac{3}{12}$ years. Consider first the value of the option an instant after the stock went ex-dividend. At that time we can calculate the value of the option as if it were a European option. We can therefore use the Black–Scholes formula, but the price of the stock should be adjusted to be $100 - Div$. The graph of the value of the call as a function of the dividend level, Div, is present in Figure 12.5.

```
> plot([Bs(96,3/12,100-Div,.18,.10),[[0,4],[5.198048647,4]\
> ],[[5.198048647,0],[5.198048647,4]]],Div=0...10,labels=\
> [Dividend,Value],thickness=2,colour=[red,green,blue],\
> thickness=3, title='Figure:12.5:  The Value of a Call \
> as a Function of the Dividend');
```

The value of a call option is a monotonically increasing function of the price of the stock. Therefore the value of the option, an instant after the dividend payment, is a decreasing function of the dividend level. This is apparent from Figure 12.5. We can therefore solve for the unique value of

$r\backslash Div$	0	1	2	3	4	5	6	7
1	6.0320	5.3510	4.7111	4.1144	3.5625	3.0567	2.5974	2.1846
2	6.1931	5.5022	4.8520	4.2446	3.6818	3.1648	2.6945	2.2708
3	6.3561	5.6556	4.9951	4.3771	3.8033	3.2753	2.7940	2.3594
4	6.5211	5.8110	5.1404	4.5118	3.9272	3.3882	2.8958	2.4503
5	6.6880	5.9685	5.2879	4.6489	4.0535	3.5034	2.9999	2.5434
6	6.8567	6.1279	5.4375	4.7881	4.1820	3.6210	3.1064	2.6389
7	7.0272	6.2893	5.5892	4.9295	4.3128	3.7409	3.2152	2.7366
8	7.1995	6.4526	5.7429	5.0731	4.4458	3.8630	3.3263	2.8367

Table 12.1: A Spreadsheet of the Call's Value as a Function of Div and r

the dividend such that an instant after the stock goes ex-dividend the value of the option is \$4. (See the vertical and horizontal bars in Figure 12.5.)

Remember that at that instant we can calculate the value of the call based on the Black–Scholes formula, taking advantage of the procedure **Bs**. Below we request MAPLE to solve numerically for the value of Div such that

$$Bs(96, \frac{3}{12}, 100 - Div, .18, .10) = 4.$$

```
> fsolve(Bs(96,3/12,100-Div,.18,.10)=4);
```

$$5.198048647$$

Utilizing the procedure **Eqbs**, we can produce the symbolic solution for Div.

```
> solve({S-K=Eqbs(K,t,S-Div,sigma,r)},{Div});
```

$$\{Div = -e^{\mathrm{RootOf}(e^{-Z}+e^{-Z}\,\mathrm{erf}(1/4\,\frac{\sqrt{2}\,(-2\ln(K)+2\_Z+2\,r\,t+\sigma^2\,t)}{\sigma\,\sqrt{t}})-2\,S+2\,K-K\,e^{(-r\,t)}}$$
$$-\mathrm{erf}(1/4\,\frac{\sqrt{2}\,(-2\ln(K)+2\_Z+2\,r\,t-\sigma^2\,t)}{\sigma\,\sqrt{t}})\,K\,e^{(-r\,t)})+S\}$$

It is therefore apparent that if $Div > 5.198048647$ it is optimal to exercise the option an instant prior to the time the stock goes ex-dividend. The spreadsheet presented in Table 12.1 demonstrates the value of the call for combinations of stock price and risk-free rates.

The reader might like to consult the Help file and see how to change the parameters written in the cells. This will facilitate the investigation of the changes in the call value for other parameters. These changes are also visualized in Figure 12.6 as a three-dimensional curve, where the horizontal hyperplane corresponds to the horizontal bar in Figure 12.5.

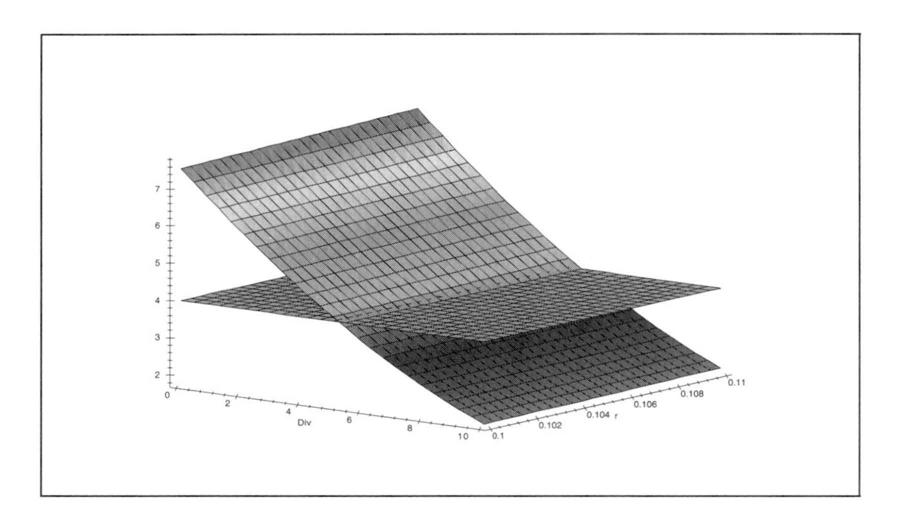

Figure 12.6: The Value of a Call Option as a Function of *Div* and r

```
> plot3d({Bs(96,3/12,100-Div,.18,r),4},Div=0..10,\
> r=.10..0.11,axes=frame,title=`Figure 12.6:  The\
> Value of a Call Option as\
> a Function of Div and r`,orientation=\
> [-97,67],shading=zhue,orientation=[-52,73]);
```

We choose to display the value of a call option as a function of the dividends and the rate of interest. This is because given K, as the argument below shows, the decision of whether to exercise or not depends on the risk-free rate r and the magnitude of Div. We already know that after the last dividend payment it is not optimal to exercise the option prior to maturity. The example above makes it clear that under certain condition it will be optimal to exercise the option an instant prior to the ex-dividend date.

The intuition behind this phenomenon is as follows. If the option is exercised, one avoids the loss of the option value that is due to the dividend payment. However, one forgoes the time value of money since the exercise price is paid at present instead of (perhaps) at maturity. The decrease of Div in the price of the stock causes a decrease in the price of the option. The former, however, is larger than the latter. It is a consequence (see Section 7.2), of the sensitivity measure Delta (Δ) being less than unity.

Consider the inequality

$$K(1 - d(T)) > Div, \tag{12.13}$$

where $d(T)$ is the discount factor, as of the current time, per dollar paid at T, the maturity time of the option. Its left-hand-side is simply the interest lost if one pays K now and not in t units of time. If the inequality in (12.13) holds, the loss due to time value of money is larger than Div, which is larger than the loss in the value of the option. Consequently, if the inequality in (12.13) holds, it will not be optimal to exercise the option. We shall come back to investigate the general case of n dividend dates after we refine some of the bounds in the presence of dividends.

Assume that the option matures at time T, that the current time is t, and that there are n dividend dates $0 < t < t_1 < ... < t_n < T$. Then, in addition to the bounds in equation (12.4), we have that

$$P_{ca(S_t,K)} \geq S_t - \left[\sum_{i=1}^{j} Div_{t_i} d(t_i)\right] - Kd(t_j), \quad \text{for } j = 1, ..., n-1, \tag{12.14}$$

and

$$P_{ca(S_t,K)} \geq S_t - \left[\sum_{i=1}^{n} Div_{t_i} d(t_i)\right] - Kd(T), \tag{12.15}$$

where $d(t_i)$ is as defined above. Each of the bounds above corresponds to an exercising decision, and hence the optimal expressing decision must produce a value greater than or equal to the one of the above bounds.

Consider, for example, the bound obtained for $j = 3$ in inequality (12.14). Exercising the option an instant prior to the ex-dividend date t_3 will produce $S_{t_3} - K$ at time t_3. The present value of K, as of the current time t, is $Kd(t_3)$. The present value of S_{t_3} as of time t_2 (after the dividend was paid) is $S_{t_2} - Div_{t_2}$, where S_{t_2} is the price of the stock an instant before the ex-dividend date. The present value of $S_{t_2} - Div_{t_2}$, an instant after t_1, is $S_{t_1} - Div_{t_2}\frac{d(t_2)}{d(t_1)} - Div_{t_1}$. Following in the same way, the present value of $S_{t_1} - Div_{t_2}\frac{d(t_2)}{d(t_1)} - Div_{t_1}$ at time t is

$$S_t - d(t_1)\left[Div_{t_2}\frac{d(t_2)}{d(t_1)} - Div_{t_1}\right] = S_t - Div_{t_2}d(t_2) - Div_{t_1}d(t_1),$$

which implies equation (12.14) for $j = 3$. The rest of the inequalities are followed applying essentially the same argument.

Alternatively, applying the replication approach can prove these bounds. For example, consider the following portfolio: take a long position in the call; invest $\$d(t_1)$ in a bond maturing at time t_1, and $\$K$ in a bond maturing at t_2; and take a short position in the stock. At time t_2, just before the stock goes ex-dividend, liquidate the portfolio. Do not exercise the call prior to that time regardless of the price of the stock. The cost of such a portfolio is

$$-P_{ca(S_t,K)} - Kd(t_2) - Div_{t_1}d(t_1) + S_t. \tag{12.16}$$

At time t_1 the holder of the portfolio needs to pay Div_{t_1}, which is covered by the bond maturing at that time. Thus the cash flow at time t_2, an instant before the stock goes ex-dividend, is

$$Call(S_{t_2}, K) + K - S_{t_2}. \tag{12.17}$$

It is easy to verify that regardless of the state of nature ($S_{t_2} > K$ or $S_{t_2} < K$) the cash flow at time t_2 is positive. This is also demonstrated by the MAPLE commands below.

```
> assume(S>K);
> Call(S,K)+K-S;
```

$$0$$

```
> assume(S<K);
> Call(S,K)+K-S;
```

$$-S\tilde{} + K\tilde{}$$

```
> is(Call(S,K)+K-S>0);
```

$$true$$

```
> S:='S':K:='K':
```

It thus follows that the price of the portfolio at initiation must be nonnegative, or else arbitrage opportunities would exist. Hence, the expression in (12.16) must be nonnegative, which yields the bounds in (12.14) for $j = t_2$. The rest of the bounds are proved in a similar way.

The above discussion leads to the conclusion that only an instant before the stock goes ex-dividend it might be optimal to exercise early an American call on a dividend-paying stock. Consider a time v such that $t_l < v < t_{l+1}$ and let $d_v(\tau)$ be the discount factor applicable from time v to some future time τ. Exercising the option at v gives rise to $S_v - K$ and exercising it an instant before t_{l+1} produces $S_{t_{l+1}} - K$, the present value of which is $S_v - Kd_v(t_{l+1})$. As long as

$$d_v(t_{l+1}) < 1 \tag{12.18}$$

we have that

$$S_v - K > S_v - Kd_v(t_{l+1}). \tag{12.19}$$

Hence, it is never optimal to exercise between ex-dividend dates.

Equation (12.19) is a consequence of the same type of argument that led (in Section 12.1.2) to the conclusion that it is never optimal to exercise early an American option on a stock that pays no dividends. This conclusion could have been reached by a slightly different approach; noticing that at time v the value of the call is greater than or equal to $S_v - Kd_v(t_{l+1})$. This can be proven again by the replication argument and we leave it as an exercise for the reader. Since equation (12.18) holds true, we have that (12.19) is satisfied, which in turn means that the option is worth more when it is "alive" than "dead". Consequently we also demonstrated that if there are multiple dividend payments, that option might be exercised just prior to the stock going ex-dividend.

12.4.2 A Put Option

As we have already seen above, things are different in the case of an American put option. It could be exercised early, even when it is contingent on a stock that pays no dividend. We can prove the following bounds on the value of a put option, P_{pa}, for a stock that pays dividends.

$$P_{pa(S_t,K)} \geq -S_t + \left[\sum_{i=1}^{j} Div_{t_i} d(t_i)\right] + Kd(t_j), \quad \text{for } j = 1, ..., n-1, \tag{12.20}$$

and

$$P_{pa(S_t,K)} \geq -S_t + \left[\sum_{i=1}^{n} Div_{t_i} d(t_i)\right] + Kd(T). \tag{12.21}$$

We leave the proof of these bounds as an exercise for the reader. The required argument is very similar to the one proving relations (12.14) and (12.15) for an American call option.

The reader should be able to justify the result that dividends will have an effect of delaying early exercising. Furthermore, the strategy of exercising a put option just after an ex-dividend date dominates that of exercising just prior to the ex-dividend date. Hence, the latter strategy should never be optimal. We leave the proof of these two propositions to the reader.

12.5 Concluding Remarks

This Chapter has investigated bounds and optimal strategies of exercising American options. Almost all of the results presented here were independent of the stochastic process assumed for the price of the stock. These groups of results are referred to as "rational option pricing" and a classic reference to such results is the notable paper by Merton [33].

One such result that has not been mentioned in this Chapter and which we asked the reader to prove is the put–call parity for American options on a dividend-paying stock. Adopting the same notation as in Section 12.4.1, we ask the reader to prove the validity of the relations in (12.22).

$$K + \sum_{i=1}^{n} Div_{t_i} d(t_i) > P_{pa(S_t,K)} - P_{ca(S_t,K)} + S_t \geq Ke^{-r(T-t)} \tag{12.22}$$

We have not yet touched upon the issue of valuation of these types of American options. Even the example in Section 12.4.1 showing how dividends can cause early exercise of an American call did not value an American option. Such a valuation must take into account the possibility of an early exercise. Hence, in order to apply to it a stochastic discount factor, one first needs to know if it is optimal to exercise the option. But this can only be decided if one knows the value of the option if it is not being exercised. We therefore see that there is a vicious circle element in calculating the value of such options. If we would know when it is optimal to exercise we would know the value of it, but in order to know if it is optimal to exercise we must know its value.

For this reason, there is no analytical solution to the pricing of American options on dividend-paying stocks and for American puts. We, therefore, have to resort to numerical valuation of these options. One way of doing it is by utilizing[3] the binomial model. These topics and some other issues that will be investigated in the framework of the binomial model are dealt with in the next Chapter.

12.6 Questions and Problems

Problem 1. At any time prior to maturity, the value of an American call option cannot be lower than the amount it grants the holder if exercised. Prove the above statement utilizing a basic arbitrage argument.

[3] We will be touching on this issue also in Chapter 15.

Problem 2. Reconstruct the arbitrage argument that led to the lower bound, equation (12.2), on the value of a European option.

Problem 3. Delineate an arbitrage portfolio that would exist if relation (12.5) would have been violated.

Problem 4. Let S_v^* be a price at which exercising an American put should take place if the time to maturity is $T - v$. Prove that if the time to maturity is $T - v$ and the price of the stock S_v is such that $S_v < S_v^*$ it is optimal to exercise the put.

Problem 5. Prove relation (12.9) directly by arbitrage consideration rather than by a mathematical substitution, as done in the text.

Problem 6. Prove the bounds in equations (12.20) and (12.21) on the value of a put option, P_{pa}, for a stock that pays dividends.

Problem 7. Utilize the replication argument to prove that it is never optimal to exercise an American call between ex-dividend dates. (See the end of Section 12.4.1.)

Problem 8. Prove the validity of the relations in (12.22) concerning put–call parity on a dividend-paying stock.

Problem 9. Prove that the strategy of exercising a put option just after an ex-dividend date dominates that of exercising just prior to the ex-dividend date.

Chapter 13

Binomial Models I

13.1 Setting the Premises

This Chapter utilizes the ingredients of the simple model of Chapter 1 to build a discrete multiperiod model. We actually have encountered such models before in our investigations, e.g., when we investigated forwards, futures, swaps, and a variable rate bond. However, only in one case have we actually looked more carefully into an environment where we applied a dynamic strategy as opposed to a buy-and-hold strategy. The latter strategy is a static one. A portfolio is composed at the beginning of the period, and although the investor has the possibility of trading in the interim, the portfolio is not revised until liquidation at the end of the period.

The exception mentioned was in Chapter 10 where we investigated the equality of futures and forward contracts in a deterministic term structure environment. In proving that in such an environment these two contracts are the same, we made use of a dynamic strategy: the portfolio revision at the end of each period was dependent on the realization of the price of the underlying asset. There we alluded to the use of dynamic programming and recursive techniques which will be used in the context of the multiperiod binomial model investigated here.

The binomial model is not new to us. It was introduced at the very beginning of this book in Chapter 2, Section 2.2. In fact, the one-period version of the binomial model can be regarded as the simplest model of the one-period type, first discussed in Chapter 1. The multiperiod binomial model is simply a sequence of one-period binomial models "joined together". Recall that the one-period binomial model is composed of two states of nature. The states correspond to the (possible) values of the stock referred

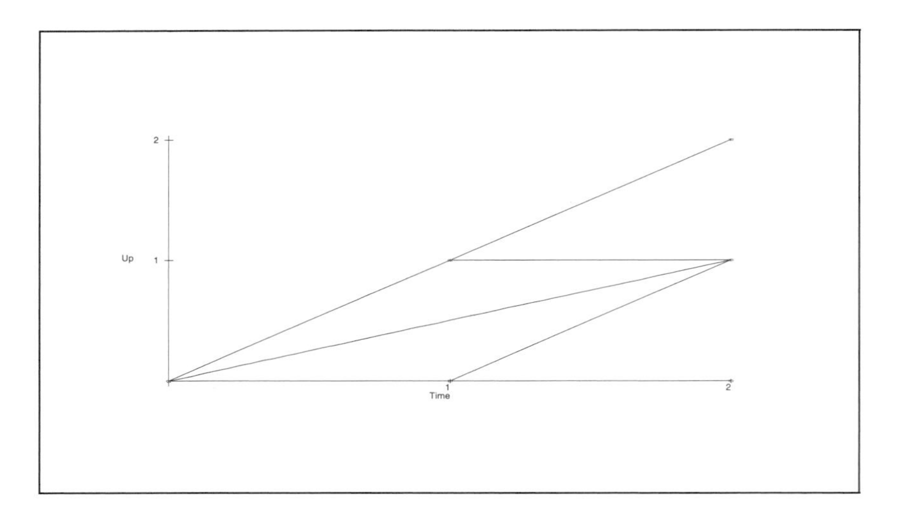

Figure 13.1: A Two-Period Binomial Model

to as **Up** and **Down** and two securities: a bond and a stock. Hence, it is a model of a complete market. The binomial model can also be used to describe the evolution of other underlying assets or processes such as the evolution of interest rates in a model of stochastic interest rates. We will address some of these situations later. In what follows we are concerned with the evolution of the price of a stock.

Imagine that at the end of the first period, another period begins which is again modeled as a one-period binomial model. A two-period binomial model is depicted in Figure 13.1.

```
> BinTree(2);
```

This figure is generated by the procedure **BinTree** that uses only one input parameter — the number of one-period models joined together. More precisely, this parameter is the number of time periods. In fact, it denotes the time span of the model. Time in these types of models is discrete and can take on only the values $0, 1, 2, ...$. Thus, it is customary to denote it by N, playing the role of the symbol T in continuous-time models. If the length of a time period is Δt, time $t = 2$ is $2\Delta t$ and time T is $N\Delta t$.

The graphical representation of the model is displayed in an (X, Y) plane, where X denotes the discrete time $t = 0, 1, ..., N$, and Y represents the number of up movements. In the on-line version of the book, a pink line

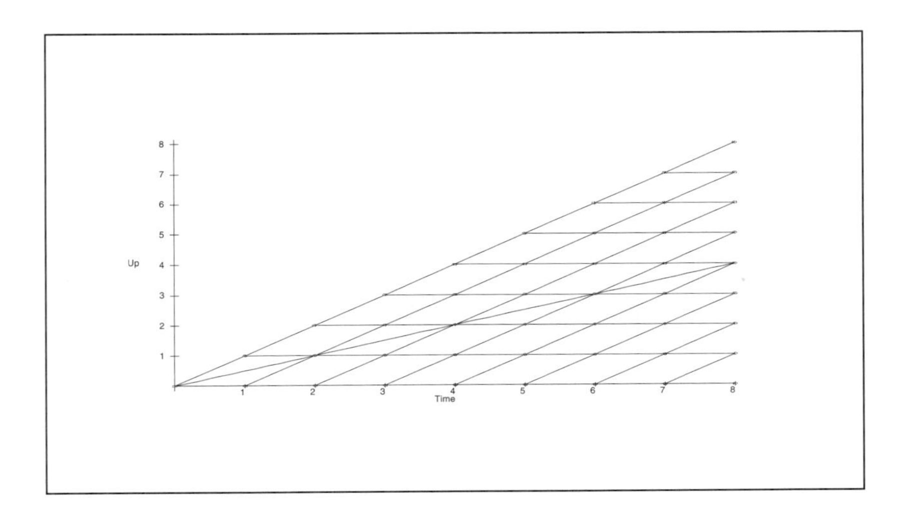

Figure 13.2: An Eight-Period Binomial Model

designates an up movement and a green line designates a down movement. The yellow line passes through the coordinates of times and states where the number of up movements is equal to the number of down movements.

Note that in Figure 13.1 at the end of the second period there are three possible states of nature corresponding to the evolution of states: **Up** then **Up**, **Up** then **Down** or **Down** then **Up**, and **Down** then **Down**. Figure 13.1 displays one way of modeling a binomial model or a binomial tree, as it is often called. The tree in Figure 13.1 is called a *recombining tree* since the evolutions **Up** then **Down,** and **Down** then **Up,** result in the same state of nature. Furthermore, this property holds true for all recombining binomial trees. Namely, all paths (evolutions) with the same number of steps and up movements, regardless of the order in which they took place, result in the same stock price. In a non-recombining tree, the realization of **Up** then **Down** does not necessarily coincide with that of **Down** then **Up**. Let us take a look at a binomial tree with a larger number of time periods to emphasize the features of this model. Consider the tree in Figure 13.2 with $N = 8$.

```
> BinTree(8);
```

Intuitively, we may interpret the coordinates $(x = i, y = j)$ as being associated with time i and state j. We will omit the "$x =$" and "$y =$" in the above notation and will use only (i, j). One may think about the state of nature in this model as being identified by two indexes. Indeed, knowing the number of up movements before time i is equivalent to knowing the price of the stock at time i, which is always associated with the state of nature.

In a recombining tree, we can therefore refer to a state of nature at a certain time by a set of coordinates. For example, $(2, 1)$ stands for time 2 where one up movement has occurred. Let us denote, as in Section 2.2, one plus the return if state **Up** occurs by Up, and one plus the return[1] if state **Down** occurs by Do. The spreadsheet, table 13.1 in the hard copy, demonstrates in a symbolic fashion the price evolution for a binomial tree with six time periods.

The spreadsheet displays the binomial tree in a slightly different manner than the procedure **BinTree**. Here the numbers in the headings of the columns designate the time period, while the numbers in the first column are the net up or down movements. The 0 in the first column indicates that there have occurred an equal number of up and down movements. The number above the zero indicates the net number of up movements and the number below the zero indicates the net number of down movements. In this approach, the price of the stock in the forthcoming period is displayed above the prevailing price if an up movement has occurred, and below otherwise.

The reader may insert numerical values for the initial stock price and the Up and Do factors in the cells (see the on-line version of the book) $B3$, $C3$, and $D3$, respectively. Reevaluating[2] the spreadsheet will generate a numerical evolution of the price process. See the spreadsheet in Table 13.2 for an example.

The nature of a recombining tree, as can be seen in Tables 13.1 and 13.2, is such that an up movement from a state (i, j) coincides with a down movement from the state $(i, j+1)$. In a model with N time periods there are $N+1$ different states of nature at time N. If these up and down movements do not coincide, then in a model of N time periods there may be as many as 2^N different states of nature at time N. These types of trees are sometimes referred to as "bushy" trees and, of course, are less efficient numerically. However, in certain modelling situations, one has no alternative but to use them. We will almost exclusively deal with recombining trees.

---

[1] We use the notation **Do** for the state of nature as well as for one plus the return if a down movement occurs. While this may constitute some abuse of notation it is consistent with our convention of relating the state of nature to the price of the underlying asset.

[2] **Intructions for initiating the reevaluation:** Change the required cell. Place the cursor somewhere in the spreadsheet. Click the right button of the mouse and choose "Evaluate Spreadsheet" from the menu.

U&D\t	0	1	2	3	4	5	6
6	**S**	**Up**	**Do**				$S\,Up^6$
5	S	Up	Do			$S\,Up^5$	
4					$S\,Up^4$		$S\,Up^5\,Do$
3				$S\,Up^3$		$S\,Up^4\,Do$	
2			$S\,Up^2$		$S\,Up^3\,Do$		$S\,Up^4\,Do^2$
1		$S\,Up$		$S\,Up^2\,Do$		$S\,Up^2\,Do^3$	
0	S		$S\,Up\,Do$		$S\,Up^2\,Do^2$		$S\,Up^3\,Do^3$
1		$S\,Do$		$S\,Up\,Do^2$		$S\,Up^3\,Do^2$	
2			$S\,Do^2$		$S\,Up^3\,Do$		$S\,Up^2\,Do^4$
3				$S\,Do^3$		$S\,Up^4\,Do$	
4					$S\,Do^4$		$S\,Up\,Do^5$
5						$S\,Do^5$	
6							$S\,Do^6$

Table 13.1: A Symbolic Price Evolution in a Six-Period Binomial Tree

U&D/t	0	1	2	3	4	5	6
6	**S**	**Up**	**Do**				177.1561
5	100	1.1000	0.9000			161.0510	
4					146.4100		144.9459
3				133.1000		131.7690	
2			121.0000		119.7900		118.5921
1		110.0000		108.9000		107.8110	
0	100		99.0000		98.0100		97.0299
1		90.000		89.1000		88.2090	
2			81.0000		80.1900		79.3881
3				72.9000		72.1710	
4					65.6100		64.9539
5	**Δt**	**qu**				59.0490	
6	$\frac{1}{6}$	0.5824					53.1441

Table 13.2: A Numeric Price Evolution in a Six-Period Binomial Tree

The multiperiod binomial model possesses many of the components of a model in which both the time space and the state space are continuous, but yet it does not require very sophisticated mathematical knowledge. Nevertheless, it is very intuitive and can be utilized to generate numerical pricing when an analytical solution does not exist, e.g., an American put. It also serves as a methodological tool to present a derivation of the Black–Scholes formula using a limiting argument.

Conceptually, a multiperiod model[3] may be thought of in the following way. At time zero we know that at the end of the model there are $N+1$ possible outcomes. As time progresses the uncertainty is gradually resolved and our information set is refined. At time zero, referring to Figure 13.2, we know that the state of nature that will be realized at time 8 is one out of nine possible states: $(8,0),(8,1),...,(8,8)$. This realized state of nature, which is unknown at time zero, is sometimes referred to as the true state of the economy. As time progresses, we eliminate the possibility of some of these states. For example, if at time 7 we are in coordinate $(7,2)$, only $(8,2)$ or $(8,3)$ is the possible realization at time 8. Similarly, if at time 1 the **Up** state was realized, it will be impossible to get to the node $(8,0)$. Finally, at time 8, all uncertainty is resolved and the true state of nature is revealed. Hence, we can think of the uncertainty in a multiperiod time as being resolved gradually. In certain cases the payoffs from derivative securities that we will be valuing will depend on the path taken by the evolution of the uncertainty. These types of derivatives are *path dependent*, (see the exercises at the end of this Chapter).

In what follows we will be using both replication and stochastic discount factors to value securities in this environment. In certain cases, the replication argument will require us to specify a dynamic strategy. These strategies, however, cannot depend on information which is not known at the particular time given the state of nature. For example, in specifying such a strategy we cannot stipulate that in coordinate $(3,3)$ one should follow the set of instructions A if at time four there would[4] be four up steps.

[3]This way of thinking extends to the continuous-time, continuous-state type models. However, one needs to be careful in such an infinite setting. The components of a model with infinite states of nature must be defined in a rigorous way. Readers familiar with probability theory, beyond a fundamental level, would recognize the need to define a probability space, as a triplet of the probability measure, the sample space and the sigma field. The evolution of the information set, the parameterization of the sigma fields by time, should be modeled by a filtration. Still, conceptually, finite and infinite models possess the same ingredients.

[4]The technical term related to this restriction is "measurability".

13.2 No-Arbitrage and SDFs

13.2.1 No-Arbitrage

Until now we have not yet specified the properties of the price process of the stock nor the probability of the different states. As in Section 2.2 we will continue to investigate some of the properties of this model and maintain our reference to the prices process and to the state probabilities in a symbolic manner. We will also adhere to the notation used in that section, repeated here for convenience.

Consider a discrete grid of the time interval $[0, T]$, where T is measured in years. Assume that this time interval is divided into N equal subintervals by $N-1$ knot points,

$$0 = t_0 < t_1 < t_2 < \cdots < t_{N-1} < t_N = T, \tag{13.1}$$

such that

$$\Delta t = t_i - t_{i-1}, \quad i = 1, ..., N. \tag{13.2}$$

This is much like the discretization we used in Section 5.6 when we took the first look at the path of the price process. Given the similarity, we can use a binomial tree with N time periods, the length of each being Δt years, to investigate the time span $[0, T]$. Within the context of the binomial tree we can think about the time as discrete where time 0 is the beginning of the period, time i is time $i\Delta t$,and time N is time T which equals $N\Delta t$.

For now we assume that the risk-free rate is a deterministic constant, r_f, quoted as a continuously compounded annual figure. Hence, in a binomial tree with N periods modeling a time span of T years, the interest rate per period r, will be the solution (assuming continuous compounding) to

$$e^{rN} = e^{r_f T}, \tag{13.3}$$

that is,

$$r = \frac{r_f T}{N} \quad \text{or} \quad r = r_f \Delta t.$$

In the binomial model, given a time i, only one of two possible states of nature can occur in the next time: **Up** or **Down**. Assume further that at each time i there exists in the market a risk-free zero-coupon bond maturing in the next time period. An investment of \$1 in such a bond is worth \$ e^r in the next time period, regardless of which state of nature occurs. We will denote e^r by R so that the bond will have value \$$R$ in either state of nature in the next time period.

Recall that we denoted one plus the return if state **Up** occurs by Up and one plus the return if state **Down** occurs by Do. If the stock value is $\$\, S(i)$ at time i it will be worth $\$\, S(i)Up$ in the next time period if state **Up** occurs, and $\$\, S(i)Do$ if state **Down** occurs. We can therefore express $S(i+1)$ as

$$S(i+1) = S(i)e^{Z_{\Delta t}}, \tag{13.4}$$

where the continuously compounded rate of return over the period Δt is the random variable $Z_{\Delta t}$. The random variable $Z_{\Delta t}$ takes a positive value, $\ln(Up)$, if state **Up** occurs and a negative value, $\ln(Do)$, if state **Down** occurs. This is a simplified version of the price process modeled in Section 5.6, where $S(i+1)$ was given by

$$S(i+1) = S(i)e^{Z_{\Delta t}} \qquad \text{and} \qquad Z_{\Delta t} \sim N(\mu\Delta t, \sigma\sqrt{\Delta t}). \tag{13.5}$$

Equation (13.5) describes a model in which the states of nature are continuous. Each of the infinitely many possible realizations of $Z_{\Delta t}$ corresponds to a state of nature. In contrast, the environment modeled in equation (13.4) has only two states of nature.

The factors Up, Do, and R are modeled here as independent of the time period, and as mentioned in Section 2.2, footnote 2, must satisfy the inequalities $Do < R < Up$ to avoid arbitrage opportunities. While we do not yet specify numerical values for these factors, we do assume that the no-arbitrage condition is satisfied. If the above inequality is satisfied, then there are no arbitrage opportunities during the time period $[i, i+1]$, $i = 0, ..., N-1$. This leads to a stronger statement summarized in the theorem below. We leave the proof as an exercise to the reader.

Theorem 11 *In an N-period binomial model the no-arbitrage condition is satisfied if and only if there are no arbitrage opportunities between time i and $i+1$ for all $i = 0, ..., N-1$.*

13.2.2 SDF

Assuming that the no-arbitrage condition is satisfied we can, as we did in Section 2.2, solve for the stochastic discount factors from time i to time $i+1$. The stochastic discount factors will have to satisfy the same set of equations as in Section 2.2, i.e., equation (2.1). A slightly modified version of this system is repeated here as the system of equations in (13.6),

$$\begin{aligned} 1 &= du\, R + dd\, R \\ S(i) &= du\, S(i)\, Up + dd\, S(i)\, Do, \end{aligned} \tag{13.6}$$

where du and dd are the SDFs, e.g., du is the price of a dollar contingent on state **Up** occurring. Below we submit the request to solve the system of equations in (13.6) to MAPLE.

```
> SolBin:=solve({1=du*R+dd*R,S(i)=du*S(i)*Up+dd*S(i)*Do},\
> {du,dd});
```

$$SolBin := \{du = \frac{R - Do}{R\,(Up - Do)},\ dd = \frac{Up - R}{R\,(Up - Do)}\}$$

As might be expected, after a casual inspection of the equations in (13.6), the one-period stochastic discount factors are independent of time and of the state.

The multiperiod binomial model can also be analyzed as a one-period model — a model where the length of the period is the time interval $[0, N\Delta t]$ with $N+1$ states of nature, (N, j) for $j = 0, ..., N$. As mentioned before the states of nature could be considered the number of up movements since these are in one-to-one correspondence with the price of the stock.

The no-arbitrage condition, Theorem 3 in Section 1.4 is equivalent to the existence of a set of discount factors. Those factors tell us the value of \$1 contingent on (N, j) as of time zero. A set of discount factors $d(N, j)$, $j = 0, ..., N$, is then a consequence of approaching the binomial model in this way. A discount factor $d(N, j)$ is the value of a dollar contingent on having j up movements from time zero to time N. According to this view, based on Theorem 3, there are no arbitrage opportunities if and only if there exists a set of SDF, $d(N, j)$, $j = 0, ..., N$, satisfying the system of equations in (13.7).

$$\begin{aligned}\sum_{j=0}^{N} d\,(N, j)\,Up^j\,Do^{N-j}S &= S \\ \sum_{j=0}^{N} d\,(N, j)\,e^{rN} &= 1\end{aligned} \tag{13.7}$$

The system of equations in (13.7) is the counterpart[5] of the first system of this kind introduced in the very first Chapter of this book, namely, (1.6). It does not, however, unambiguously determine the stochastic discount factors in an N-period binomial model where $N > 2$.

If $N > 2$, viewing a multiperiod binomial model as a one period model might give the impression that it is an incomplete market. There are $N+1$

[5] Equivalently, the second equation in (13.7) can be written as $\sum_{j=0}^{N} d(N, j)e^{r_f T} = 1$, which is the counterpart of equation (1.8) in Chapter 1.

states of nature and only two assets. There is of course some structure lost when such a market is viewed as a one-period model. After all, the facts that this one-period is actually composed of N subperiods and that trading is allowed in the interim impose a structure that may reduce the ambiguity[6] in determining the stochastic discount factors.

Let us consider an example of a five-period binomial tree and the value, as of time zero, of a dollar contingent on the state $(5, 3)$. That is, the payoff is one dollar if by the end of the fifth period there were three up movements, and the payoff is zero otherwise. The value of this contingent claim is the value of the discount factor $d(5, 3)$. Recall that the realization of $(5, 3)$ must have followed either the realization of $(4, 3)$ and a down movement, or the realization of $(4, 2)$ and an up movement. There is no other way to reach $(5, 3)$. Thus this contingent claim will have a value of zero at time 4 if there were $0, 1$, or 4 up movements until that time. In such circumstances it is already known that the true state of nature could not be $(5, 3)$. Furthermore, the present value, as of time 0, of a dollar contingent on $(5, 3)$ must be equivalent to the sum of two contingent claims:

- $\$du$ contingent on state $(4, 2)$ and
- $\$dd$ contingent on $(4, 3)$.

This can be verified by analyzing $(5, 3)$ as a possible outcome in the one-period model consisting of $(4, 2)$ as an initial state, with possible realization of $(5, 2)$ or $(5, 3)$, and also as an outcome in the one-period model consisting of $(4, 3)$ as an initial state, with possible realization of $(5, 4)$ or $(5, 3)$. The portfolio of $\$du$ contingent on state $(4, 2)$ and $\$dd$ contingent on $(4, 3)$ guarantees that whatever path along which the uncertainty evolves, at time 5 the value of this portfolio will be worth \$1 in state $(5, 3)$ and zero otherwise.

Proceeding in the same manner, it follows that

1. $\$du$ contingent on $(4, 2)$ must be equivalent to $\$du\ dd$ contingent on $(3, 2)$ plus $\$du\ du$ contingent on $(3, 1)$, and
2. $\$dd$ contingent on $(4, 3)$ must be equivalent to $\$dd\ dd$ contingent on $(3, 3)$ plus $\$dd\ du$ contingent on $(3, 2)$.

Hence, a dollar contingent on $(5, 3)$ must be equivalent to the sum of the claims in (1) and (2) above. The reader probably realizes by now that

[6] The issue at hand is related to the subject of a dynamically complete market. See also our discussion of a complete and incomplete market in the appendix to the first Chapter.

the phrase "the value must be equivalent to" implies "otherwise arbitrage opportunities would exist". One of the exercises that corresponds to this Chapter asks the reader to outline the arbitrage arguments proving the statements above.

Extending the logic above in a recursive manner from time 4 to time 3, etc., will result in a value for $d(5,3)$. The idea can be displayed graphically making use of the procedure **AllWays**. This procedure visualizes all the possible evolutions of uncertainty that will start at time zero and end at time N with j up movements. Indeed, this procedure has two input parameters, N and j. In our case $N = 5$ and $j = 3$. The procedure is executed by the command **AllWays**$(5,3)$ and its output is shown in Figure 13.3.

```
> AllWays(5,3);
```

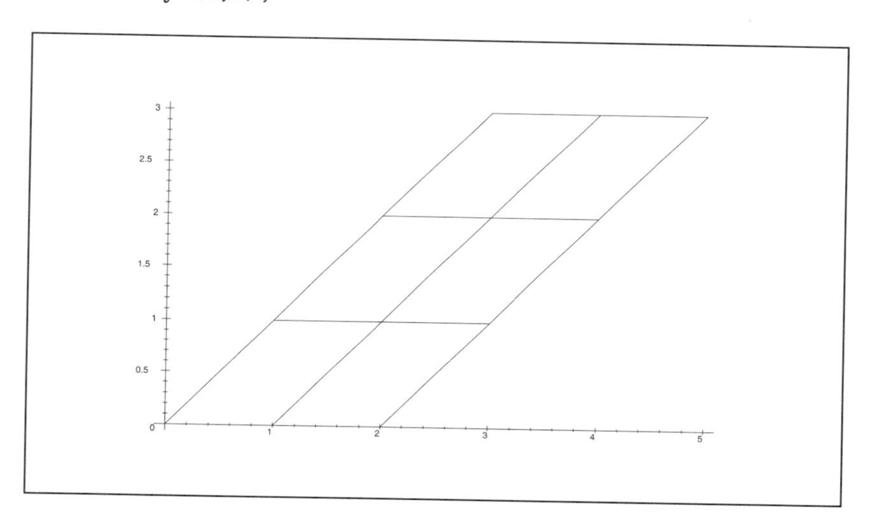

Figure 13.3: The Possible Evolutions of Uncertainty That Will End at Time 5 with Three Up Movements

In order to superimpose the produced graph (Figure 13.3) on the binomial tree, the following command is issued (Figure 13.4 displays the output of the MAPLE command below):

```
> plots[display](BinTree(5),AllWays(5,3));
```

The emphasized vertices in Figure 13.4 display all the possible evolutions from time zero to three up movements at time 5. As the uncertainty is resolved in a gradual manner, it may be the case that prior to time 5 the

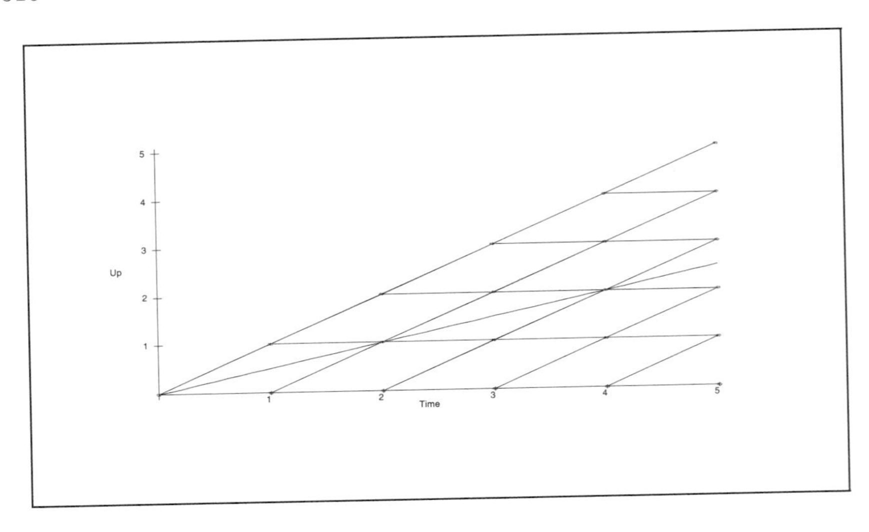

Figure 13.4: The Possible Evolutions of Uncertainty That Will End at Time 5 with Three Up Movements Superimposed on the Tree

contingent claim becomes worthless. Such is the case if, for example, by time 4 there was only one up movement. That is why $(4, 1)$ is not an emphasized vertex.

In other words, once the evolution of the uncertainty steps outside of the emphasized vertices, the contingent claim becomes worthless. Let us now see how this valuation idea can be implemented in MAPLE. We define below the function $F53$ that equals 1 if $x = 3$ and zero otherwise. It represents the payoff at time 5 of \$1 if there were three up movements until time 5 and a payoff of zero otherwise.

```
> F53:=x->if x=3 then 1 else 0 fi:
```

We use a two-dimensional array $\mathbf{Claim}(i, j)$ in which we store the value of the contingent claim at time i for $i = 0, ..., 5$. We initialize N to 5 and $\mathbf{Claim}(5, j)$ to have the value of 1 if $j = 3$ and zero otherwise, i.e., to be the payoff at time 5:

```
> j:='j':N:=5:
> for j from N to 0 by -1 do;
> Claim[N,j]:=F53(j);
> od;
```

$$Claim_{5,5} := 0$$

$$Claim_{5,\,4} := 0$$
$$Claim_{5,\,3} := 1$$
$$Claim_{5,\,2} := 0$$
$$Claim_{5,\,1} := 0$$
$$Claim_{5,\,0} := 0$$

We then follow the argument presented by (1) and (2) on page 514 to get the contingent claim as of time 4 that is equivalent to the contingent claim represented by the function $F53$. The group of MAPLE commands immediately following this paragraph does this.

```
> j:='j':
> for j from N-1 to 0 by -1 do;
> Claim[N-1,j]:=dd*Claim[N,j]+du*Claim[N,j+1];
> od;
```

$$Claim_{4,\,4} := 0$$
$$Claim_{4,\,3} := dd$$
$$Claim_{4,\,2} := du$$
$$Claim_{4,\,1} := 0$$
$$Claim_{4,\,0} := 0$$

Proceeding in the same manner we find the contingent claim as of time 3 that is equivalent to the contingent claim of time 4 found above. This is done by the group of MAPLE commands below.

```
> j:='j':
> for j from N-2 to 0 by -1 do;
> Claim[N-2,j]:=dd*Claim[N-1,j]+du*Claim[N-1,j+1];
> od;
```

$$Claim_{3,\,3} := dd^2$$
$$Claim_{3,\,2} := 2\,dd\,du$$
$$Claim_{3,\,1} := du^2$$
$$Claim_{3,\,0} := 0$$

We then continue in the same fashion to find the value of the claim as of time zero.

```
> j:='j':
> for j from N-3 to 0 by -1 do;
> Claim[N-3,j]:=dd*Claim[N-2,j]+du*Claim[N-2,j+1];
> od;
```

$$Claim_{2,\,2} := 3\,dd^2\,du$$

$$Claim_{2,\,1} := 3\; dd\; du^2$$
$$Claim_{2,\,0} := du^3$$

```
> j:='j':
> for j from N-4 to 0 by -1 do;
> Claim[N-4,j]:=dd*Claim[N-3,j]+du*Claim[N-3,j+1];
> od;
```

$$Claim_{1,\,1} := 6\; dd^2\; du^2$$
$$Claim_{1,\,0} := 4\; dd\; du^3$$

```
> j:='j':
> for i from N-5 to 0 by -1 do;
> Claim[N-5,j]:=dd*Claim[N-4,j]+du*Claim[N-4,j+1];
> od; j:='j':N:='N':
```

$$j := j$$
$$Claim_{0,\,0} := 10\; dd^2\; du^3$$

Note that the elements of **Claim** that have a value of zero are exactly the coordinates of the vertices that are not emphasized in Figure 13.4. The calculations above verify that the value of such a claim is $10dd^2du^3$. Thus the stochastic discount factor $d(5,3)$ in this model must satisfy equation (13.8),

$$d(5,3) = 10dd^2du^3. \tag{13.8}$$

Despite the market seeming to be incomplete, we have managed to obtain an unambiguous[7] value for the discount factor $d(5,3)$.

The reader is encouraged to use the template that is provided here to value other stochastic discount factors. One may change the definition of the function $F53$ and/or the number of time periods in the model by assigning a different value to N. Reexecuting the MAPLE commands above will generate the value of the newly defined stochastic discount factors. In fact, the procedure above can be followed to value different types of contingent claims[8] not only of the form of an elementary security that pays one dollar if a certain state occurs and zero otherwise.

---

[7] This is due to the fact that we are allowed to utilize dynamic strategies and are not constrained to a buy-and-hold strategy. These phenomena are known as dynamically completing the market and will be visited again shortly. (Readers that are interested in an elaborate discussion of this issue may consult [21].)

[8] This approach can be applied with a double "do loop" or by a recursive procedure. The procedure followed above will also come in handy in valuing contingent claims in the binomial framework.

The approach taken above to calculate the discount factor $d(5,3)$ was in the spirit of valuation via the discount factor methodology. An exercise at the end of this Chapter asks the reader to value $d(5,3)$ via the replication approach. The next subsection discusses both of these approaches within context of the binomial model.

Let us review this result from another point of view that will lead us to a generalization of the answer obtained above. There are a few paths of the evolution of the states of nature that, starting at time zero, will end at state 3 in time 5, i.e., at $(5,3)$. Observing the structure of a binomial tree, one can verify that to end at $(5,3)$ there must be three up movements somewhere between time 0 and time 5. Moreover, as long as there are exactly three up movements, regardless of the timing of these movements, the end result will be $(5,3)$. In general, if the true state of nature is (N,j) for $j = 0, ..., N$, there should have been j up movements, and $N-j$ down movements, occurring in any order.

Reviewing the example above demonstrates that we calculate the value of $1 contingent on $(5,3)$, by valuing an equivalent claim(s) that is contingent on the state of nature at time 4. This latter claim is valued by considering equivalent claims contingent on the state of nature at time 3. Continuing in this manner we obtained the desired value of the original claim at time zero.

Traversing back in this way we actually map all the paths that lead from time zero to $(5,3)$. This is not a coincidence, as we will soon see. Since the ordering of the up movements is irrelevant, the number of paths that lead from time 0 to $(5,3)$ is like the number of possible ways of choosing three objects out of five, which is $\frac{5!}{3!2!}$. In general, the number of paths leading from time 0 to (N,j) is

$$\frac{N!}{j!(N-j)!}. \tag{13.9}$$

Let us follow one of the $\frac{5!}{3!2!} = 10$ paths that leads from $(5,3)$ back to time zero. Along each such path there will be three up movements and two down movements. Hence, traversing back from $(5,3)$, at which time the dollar is obtained, to time zero, the contribution of each such path to the present value is $\$du^3\ dd^2$. Since there are $\frac{5!}{3!2!} = 10$ such paths, the present value of $1 contingent on $(5,3)$ is $\$10\ du^3\ dd^2$. Applying the same argument in general yields that the present value of $1 contingent on (N,j) is $\frac{N!}{j!(N-j)!} du^j dd^{N-j}$. Hence, $d(N,j)$, the stochastic discount factor for state

(N, j), is given by equation (13.10).

$$d(N,j) = \frac{N!}{j!(N-j)!} du^j dd^{N-j} \tag{13.10}$$

The stochastic discount factors and the risk-neutral probability, as already seen in Chapter 1, are very much related. As always, normalizing the SDF will result in a probability distribution. From equations (1.8) and (1.12) it follows that $\sum_{j=0}^{N} d(N,j) e^{r_f T} = 1$ and thus $\sum_{j=0}^{N} d(N,j) e^{rN} = 1$, which implies, in view of equation (13.3), that

$$\sum_{j=0}^{N} d(N,j) = \frac{1}{e^{rN}} = \left(\frac{1}{R}\right)^N. \tag{13.11}$$

The normalization of $d(N,j)$ is obtained by dividing it by $\sum_{j=0}^{N} d(N,j)$. Thus, based on equations (13.11) and (13.10), $q(N,j)$, the normalized $d(N,j)$, is given by

$$q(N,j) = \frac{N!}{j!(N-j)!} (du\ R)^j (dd\ R)^{N-j}, \tag{13.12}$$

and of course

$$\sum_{j=0}^{N} q(N,j) = 1. \tag{13.13}$$

The one-period counterpart of equation (13.11) is equation (1.8) of Chapter 1, which stipulates that $dd + du = \frac{1}{R}$. Thus denoting $qu = duR$ and $qd = ddR$ yields

$$q(N,j) = \frac{N!}{j!(N-j)!} qu^j qd^{N-j}.$$

Consequently, the risk-neutral probability is the binomial distribution where qu and qd are the probabilities for an up and a down movement, respectively. This result can also be obtained utilizing a probabilistic approach. Such an approach will be taken in Chapter 14. Chapter 14 also offers a discussion of a certain property satisfied by a price process that does not admit arbitrage opportunities. The next section is dedicated to valuation within the context of the binomial model.

13.3 Valuation

Calculating the value of the stochastic discount factor as in equation (13.10) alluded to the mechanics of recursive valuation in the framework of a binomial tree. The value of $d(5,3)$ above was calculated recursively in the spirit of valuation via the discount factor methodology. However, once the stochastic discount factors are known, valuing any other European-type options does not require a recursive procedure. Viewing the binomial model as a one-period model induces the valuation formula, as we shall soon see. Valuation via the replication approach requires the knowledge of the replicating portfolio at each time and state of nature. Once it has been solved for symbolically, as shown below, it also does not require the use of a recursive procedure. In contrast, the valuation of an American-type option requires a recursive procedure because of the possibility of exercising early.

13.3.1 Valuation with SDFs

We first attend to valuation using the stochastic discount factors approach. As always, of course, there is another approach to valuation that is essentially identical. It is based on the risk-neutral distribution. In this approach the value of a security is the discounted expected value of the payoff. In the context of the binomial model the risk-neutral distribution is specified by equation (13.12).

Let S be the current price of the stock and F the payoff function of a conventional[9] European-type derivative. The value of such a derivative security maturing at the end of the Nth period and paying $F(SUp^j Do^{N-j})$ is calculated straightforwardly. Applying the risk-neutral valuation concept, we may calculate it as the expected value of $F(SUp^j Do^{N-j})$ discounted by the risk-free rate. Hence, the value of such a security will be

$$e^{-r_f T} \sum_{j=0}^{N} \frac{N!}{j!\,(N-j)!} qu^j (1-qu)^{N-j} F(SUp^j Do^{N-j}). \tag{13.14}$$

Equation (13.14) follows naturally if we think, as we already did, about the N-period binomial tree in terms of a one-period model. It is exactly as equation (1.11) in the one-period model of Chapter 1, albeit here there are $N+1$ states of nature and the risk-neutral probabilities are specified

[9] By conventional we mean a contingent claim that depends only on the value of the underlying asset at maturity and not on the historical path of its price process.

in equation (13.12). Alternatively, such a derivative security might be valued utilizing the stochastic discount factors found in equation (13.10). The reader is asked in one of the exercises pertaining to this Chapter to show that

$$d(N,j) = e^{-r_f T} \frac{N!}{j!\,(N-j)!} qu^j (1-qu)^{N-j},$$

and hence that equation (13.14) can be written as equation (13.15):

$$\sum_{j=0}^{N} d(N,j)\, F\left(SUp^j Do^{N-j}\right). \tag{13.15}$$

13.3.2 Valuation by Replication

We can also value a derivative security by applying the replication argument. This approach must be used when valuing an American-type option, as will be demonstrated in Section 13.4.4. Valuing by replication requires a recursive argument. At the end of the last period, time N, the option matures and its value is simply the payoff at maturity.

Consider the situation at time $N-1$, where the stock price is $S(N-1)$, and assume that there were j ($j = 0, ..., N-1$) up movements. In the next time period, time N, there will have been either $j+1$ up movements, in which case the price of the stock will be $S(N-1)\,Up$, or j up movements, in which case the price of the stock will be $S(N-1)\,Do$.

In each of these two possible states of nature the value of the derivative security is known. We can solve a system of two equations with two unknowns wherein Xs represents the holding of the stock and Xb the amount invested at the risk-free rate in a portfolio that replicates the value of the derivative security in each of the two states of nature. The value of the derivative security at time $N-1$, if there were j up movements, will be the value of this replicating portfolio.

This is just the same old argument we encountered in the very first Chapter of this book. Following in the same way, we can find the value of the derivative security at time $N-1$ for the N possible values of j. Having accomplished this, we can now move back one time period. We can repeat the same argument for the value of the derivative security at time $N-2$ for each possible value of j. After all, whether it is time N or time $N-1$, or any other time, the analysis is not altered.

At any time when we look ahead we are simply applying the analysis of a one-period binomial model. We know the current value of the stock, its value in the two possible states of nature in the next time period, and the value of the derivative security in each of these two states. Proceeding in this manner and traversing backward to time zero, we will recover the value of the derivative security at time 0.

Let us look at a symbolic example. Assume that the current price of the stock is S and that the derivative security will have a value of Cu if state **Up** occurs and Cd if state **Down** occurs. As before, let Xs denote the holding of the stock and Xb the amount invested in the risk-free rate in the replicating portfolio. Thus Xs and Xb should satisfy the system of equations

$$\begin{aligned} Xs\ S\ Up + Xb\ R &= Cu \\ Xs\ S\ Do + Xb\ R &= Cd, \end{aligned} \tag{13.16}$$

which is submitted to MAPLE to be solved below.

```
> solve({Xs*S*Do+Xb*R=Cd,Xs*S*Up+Xb*R=Cu}, {Xs,Xb});
```

$$\{Xs = \frac{-Cd + Cu}{S\,(Up - Do)},\ Xb = -\frac{Do\ Cu - Cd\ Up}{(Up - Do)\,R}\}$$

The value of the replicating portfolio will be

$$XsS + Xb,$$

where

$$Xs = \frac{Cu - Cd}{S\,(Up - Do)} \qquad \text{and} \qquad Xb = \frac{UpCd - DoCu}{(Up - Do)\,R}. \tag{13.17}$$

After some algebraic simplification, as shown in the MAPLE command below, the value of the replicating portfolio can be written as

$$\frac{R - Up}{(Do - Up)\,R}Cd + \frac{Do - R}{(Do - Up)\,R}Cu \tag{13.18}$$

```
> simplify(-(-Cu+Cd)/(S*(-Do+Up))*S+Up*Cd-Cu*Do)/\
> (((-Do+Up)*R));
```

$$\frac{-R\ Cu + R\ Cd - Up\ Cd + Cu\ Do}{(Do - Up)\ R}$$

```
> collect(%,{Cu,Cd},recursive);
```

$$\frac{(R - Up)\ Cd}{(Do - Up)\ R} + \frac{(-R + Do)\ Cu}{(Do - Up)\ R}$$

Recall our discussion about Δ — the sensitivity measure of a call option to changes in the prices of the underlying asset. In the continuous time context of Section 7.2, Δ measured the increase in the price of the call, per infinitesimal increase in the price of the stock. Section 8.2 reinforces the other interpretation of Δ as the holding of the stock in a portfolio that replicates the call movements over an instant. The value of Xs, in the discrete time setting of the binomial model, is therefore the counterpart of Δ. An exercise at the end of this Chapter will elaborate on this issue.

The portfolio value displayed in equation (13.18) is a linear combination of the two possible values of the derivative securities, Cu and Cd. The coefficients are independent of the current value of the underlying asset. Hence, they would apply equally to each of the one-period binomial models that comprise the N-period binomial model. Let us denote the coefficients of Cd and Cu by η_d and η_u, respectively. Thus

$$\eta_d = \frac{R - Up}{(Do - Up)\,R} \tag{13.19}$$

and

$$\eta_u = \frac{Do - R}{(Do - Up)\,R}. \tag{13.20}$$

Let us consider an example to cement the ideas represented above. Assume a five-period binomial model is utilized to value European options, where the price of the underlying asset at time zero is S. Denote the value of the option at time i, if there were j up movements to that time, by $C_{i,j}$. That is, $C_{i,j}$ is the value of the option at node (i, j).

Below we make sure that j is unassigned, and then assign it the value of N and initialize the value of $C_{5,j}$, $j = 0, ..., 5$, to the payoff of the derivative at maturity in state j.

```
> j:='j':  N:=5:  K:='K': S:='S': Do:='Do':  Up:='Up':
> for j from N to 0 by -1 do;
> C[N,j]:=max(0,S*Up^j*Do^(N-j)-K);
> od;
```

$$C_{5,5} := \max(0,\ S\ Up^5 - K)$$
$$C_{5,4} := \max(0,\ S\ Up^4\ Do - K)$$
$$C_{5,3} := \max(0,\ S\ Up^3\ Do^2 - K)$$
$$C_{5,2} := \max(0,\ S\ Up^2\ Do^3 - K)$$
$$C_{5,1} := \max(0,\ S\ Up\ Do^4 - K)$$
$$C_{5,0} := \max(0,\ S\ Do^5 - K)$$

```
> j:='j';
```

$$j := j$$

Now that we have recovered the value of the derivative security at time 5, we can step backward and apply equation (13.18) at each point $(4, j)$, $j = 0, ..., 4$. This is demonstrated below.

```
> eta:='eta':
> for j from 0 to 4 by 1 do;
> C[4,j]:=eta[u]*C[5,j+1]+eta[d]*C[5,j];
> od;
```

$$C_{4,0} := \eta_u \max(0,\ S\ Up\ Do^4 - K) + \eta_d \max(0,\ S\ Do^5 - K)$$
$$C_{4,1} := \eta_u \max(0,\ S\ Up^2\ Do^3 - K) + \eta_d \max(0,\ S\ Up\ Do^4 - K)$$
$$C_{4,2} := \eta_u \max(0,\ S\ Up^3\ Do^2 - K) + \eta_d \max(0,\ S\ Up^2\ Do^3 - K)$$
$$C_{4,3} := \eta_u \max(0,\ S\ Up^4\ Do - K) + \eta_d \max(0,\ S\ Up^3\ Do^2 - K)$$
$$C_{4,4} := \eta_u \max(0,\ S\ Up^5 - K) + \eta_d \max(0,\ S\ Up^4\ Do - K)$$

```
> j:='j';
```

$$j := j$$

We can now proceed backward, using the same technique, and calculate the value of $C_{3,j}$, $j = 0, ..., 3$.

```
> for j from 0 to 3 by 1 do;
> C[3,j]:=expand(eta[u]*C[4,j+1]+eta[d]*C[4,j]);
> od;
```

$$C_{3,0} := {\eta_u}^2 \max(0,\ S\ Up^2\ Do^3 - K) + 2\,\eta_u\,\eta_d \max(0,\ S\ Up\ Do^4 - K) + {\eta_d}^2 \max(0,\ S\ Do^5 - K)$$

$$C_{3,1} := {\eta_u}^2 \max(0,\ S\ Up^3\ Do^2 - K) + 2\,\eta_u\,\eta_d \max(0,\ S\ Up^2\ Do^3 - K) + {\eta_d}^2 \max(0,\ S\ Up\ Do^4 - K)$$

$$C_{3,2} := {\eta_u}^2 \max(0,\ S\ Up^4\ Do - K) + 2\,\eta_u\,\eta_d \max(0,\ S\ Up^3\ Do^2 - K) + {\eta_d}^2 \max(0,\ S\ Up^2\ Do^3 - K)$$

$$C_{3,3} := {\eta_u}^2 \max(0,\ S\ Up^5 - K) + 2\,\eta_u\,\eta_d \max(0,\ S\ Up^4\ Do - K) + {\eta_d}^2 \max(0,\ S\ Up^3\ Do^2 - K)$$

The value of the derivative security can be determined using the same mechanism at each node at time 2 followed by the same calculation at time 1.

```
> j:='j';
```

$$j := j$$

```
> for j from 0 to 2 by 1 do;
> C[2,j]:=expand(eta[u]*C[3,j+1]+eta[d]*C[3,j]);
> od;
```

$$C_{2,0} := {\eta_u}^3 \max(0,\ S\ Up^3\ Do^2 - K) + 3\,{\eta_u}^2\,\eta_d \max(0,\ S\ Up^2\ Do^3 - K) + 3\,\eta_u\,{\eta_d}^2 \max(0,\ S\ Up\ Do^4 - K) + {\eta_d}^3 \max(0,\ S\ Do^5 - K)$$

$$C_{2,1} := {\eta_u}^3 \max(0,\ S\ Up^4\ Do - K) + 3\,{\eta_u}^2\,\eta_d \max(0,\ S\ Up^3\ Do^2 - K) + 3\,\eta_u\,{\eta_d}^2 \max(0,\ S\ Up^2\ Do^3 - K) + {\eta_d}^3 \max(0,\ S\ Up\ Do^4 - K)$$

$$C_{2,2} := {\eta_u}^3 \max(0,\ S\ Up^5 - K) + 3\,{\eta_u}^2\,\eta_d \max(0,\ S\ Up^4\ Do - K) + 3\,\eta_u\,{\eta_d}^2 \max(0,\ S\ Up^3\ Do^2 - K) + {\eta_d}^3 \max(0,\ S\ Up^2\ Do^3 - K)$$

```
> j:='j';
```

$$j := j$$

```
> for j from 0 to 1 by 1 do;
> C[1,j]:=expand(eta[u]*C[2,j+1]+eta[d]*C[2,j]);
> od;
```

$$C_{1,0} := {\eta_u}^4 \max(0,\ S\ Up^4\ Do - K) + 4\,{\eta_u}^3\,\eta_d \max(0,\ S\ Up^3\ Do^2 - K) + 6\,{\eta_u}^2\,{\eta_d}^2 \max(0,\ S\ Up^2\ Do^3 - K) + 4\,\eta_u\,{\eta_d}^3 \max(0,\ S\ Up\ Do^4 - K) + {\eta_d}^4 \max(0,\ S\ Do^5 - K)$$

$$C_{1,1} := {\eta_u}^4 \max(0,\ S\ Up^5 - K) + 4\,{\eta_u}^3\,\eta_d \max(0,\ S\ Up^4\ Do - K) + 6\,{\eta_u}^2\,{\eta_d}^2 \max(0,\ S\ Up^3\ Do^2 - K) + 4\,\eta_u\,{\eta_d}^3 \max(0,\ S\ Up^2\ Do^3 - K) + {\eta_d}^4 \max(0,\ S\ Up\ Do^4 - K)$$

Finally, we arrive at the calculation of the derivative security at time zero that produces the required result. Of course a "do loop" command is not needed for that, but to maintain the uniformity we have retained it here.

```
> for j from 0 to 0 by 1 do;
> C[0,j]:=expand(eta[u]*C[1,j+1]+eta[d]*C[1,j]);
> od;
```

$$C_{0,0} := {\eta_u}^5 \max(0,\ S\ Up^5 - K) + 5\,{\eta_u}^4\,\eta_d \max(0,\ S\ Up^4\ Do - K) + 10\,{\eta_u}^3\,{\eta_d}^2 \max(0,\ S\ Up^3\ Do^2 - K) + 10\,{\eta_u}^2\,{\eta_d}^3 \max(0,\ S\ Up^2\ Do^3 - K) + 5\,\eta_u\,{\eta_d}^4 \max(0,\ S\ Up\ Do^4 - K) + {\eta_d}^5 \max(0,\ S\ Do^5 - K)$$

In a more general setting, the payoff from a derivative security can be stipulated to be a function of the underlying asset, e.g., as $F\left(SUp^j Do^{N-j}\right)$. Thus, initializing the value of $C_{5,j},\ j = 0, ..., 5$, will be done as follows:

```
> i:='i':  j:='j' :N:=5:
> for j from N to 0 by -1 do;
> C[N,j]:=F(S*Up^j*Do^(N-j));
> od;
```

$$C_{5,5} := \mathrm{F}(S\ Up^5)$$
$$C_{5,4} := \mathrm{F}(S\ Up^4\ Do)$$
$$C_{5,3} := \mathrm{F}(S\ Up^3\ Do^2)$$
$$C_{5,2} := \mathrm{F}(S\ Up^2\ Do^3)$$
$$C_{5,1} := \mathrm{F}(S\ Up\ Do^4)$$
$$C_{5,0} := \mathrm{F}(S\ Do^5)$$

The value of $C_{0,0}$ can be calculated, utilizing a "do loop" within a "do loop", as below:

```
> i:='i';j:='j';
```

$$i := i$$
$$j := j$$

```
> for j from N-1 to 0 by -1 do;
> for i from 0 to j by 1 do;
> C[j,j]:=eta[u]*C[j+1,j+1]+eta[d]*C[j+1,j];
> od;
> od;
> i:='i';j:='j';
```

$$i := i$$
$$j := j$$

The value of $C_{0,0}$ is given below:

```
> expand(C[0,0]);
```

$$\eta_u{}^5\,\mathrm{F}(S\ Up^5) + 5\,\eta_u{}^4\,\eta_d\,\mathrm{F}(S\ Up^4\ Do) + 10\,\eta_u{}^3\,\eta_d{}^2\,\mathrm{F}(S\ Up^3\ Do^2)$$
$$+ 10\,\eta_u{}^2\,\eta_d{}^3\,\mathrm{F}(S\ Up^2\ Do^3) + 5\,\eta_u\,\eta_d{}^4\,\mathrm{F}(S\ Up\ Do^4) + \eta_d{}^5\,\mathrm{F}(S\ Do^5)$$

A careful inspection of the above expression reveals that it satisfies equation (13.21).

$$\sum_{j=0}^{N} \frac{N!}{j!\,(N-j)!} \eta_u^j \eta_d^{N-j} F\left(SUp^j Do^{N-j}\right). \tag{13.21}$$

The sum $\eta_u + \eta_d$, when η_u and η_d are as specified in equations (13.20) and (13.19), satisfies $\eta_u + \eta_d = \frac{1}{R}$.

```
> simplify(subs(eta[d]=(R-Up)/((Do-Up)*R),
> eta[u]=(-R+Do)/((Do-Up)*R),eta[u]+eta[d]));
```

$$\frac{1}{R}$$

Hence, based on equation (1.12) $qu = \eta_u R$, and $qd = \eta_d R$, and thus

$$\frac{qu}{R} = \eta_u$$

and

$$\frac{qd}{R} = \eta_d.$$

Substituting the values of η_u and η_u in terms of qu and qd in equation (13.21) yields

$$\left(\frac{1}{R}\right)^N \sum_{j=0}^{N} \left(\frac{N!}{j!\,(N-j)}\right) qu^j qd^{N-j} F\left(SUp^j Do^{N-j}\right)$$

for a derivative security with a payoff $F\left(SUp^j Do^{N-j}\right)$. Since $\left(\frac{1}{R}\right)^N = e^{-r_f T}$, equations (13.14) and (13.21) are, as expected, consistent. Consequently, this proves that, as in the one-period binomial model, valuation by replication generates the same result as with the stochastic discount factors.

If the derivative security in question is of an American type, at each node the value obtained by replication should be compared with the value of the derivative if it were exercised immediately. Consequently, its value cannot be presented by an equation like (13.21) or (13.14) but should be solved for numerically. We shall investigate the value of an American-type derivative in the next subsection, Numerical Valuation, after we have presented the examples for European-type options. The European examples will point out the instances where early exercise is optimal and thereby prepare the reader for the discussion concerning American-type derivatives.

13.4 Numerical Valuation

13.4.1 Price Evolution

Let us look at a numerical example to clarify the price process. Assume a European call option that matures in a year ($T = 1$) in a market where the annual continuously compounded risk-free rate is $r_f = 0.10$. We divide the year into time periods of 2 months in length and so generate a binomial tree with $N = 6$, i.e., $\Delta\ t = \frac{1}{6}$. As per equation (13.3), $r = \frac{r_f}{6}$. Let us fix the factors Do and Up according to the specification of equation (13.4) so that $Z_{\Delta\ t}$ takes the value $u = \frac{\sqrt{\frac{1}{6}}}{5}$ if state **Up** occurs or the value $v = -\frac{\sqrt{\frac{1}{6}}}{5}$ if state **Down** occurs. For the time being we do not explain how these factors were determined[10], but we will dedicate Section 14.1 to this issue. Hence, $Up = e^{\frac{1}{5}\sqrt{\frac{1}{6}}}$

```
>   evalf(exp((1/5)*sqrt(1/6)));
```

$$1.085075596$$

and $Do = e^{-\frac{1}{5}\sqrt{\frac{1}{6}}}$

```
>   evalf(exp(-(1/5)*sqrt(1/6)));
```

$$.9215947754$$

We verify below that these choices for Up and Do satisfy the no-arbitrage condition. The no-arbitrage condition in a one-period binomial model is satisfied if and only if the system of equations in (2.1) in Chapter 2 is consistent. In our context it can be translated to the system of equations and inequalities represented in (13.22) being consistent.

$$\begin{array}{l} e^{-r\Delta t}\left(S Up\, qu + S Do\,(1 - qu)\right) = S \\ 0 < qu < 1 \end{array} \tag{13.22}$$

The consistency of (13.22) guarantees the existence of a risk-neutral probability under which the expected value of the stock future price, discounted by the risk-free rate, is its current price. The first equation in (13.22) is independent of S, the price of the stock at the beginning of the period, and can be rewritten as

$$e^{-r\Delta t}\left(Upqu + Do\,(1 - qu)\right) = 1.$$

---

[10]As might be suspected, the $\frac{1}{6}$ under the square root sign is indeed $\Delta\ t$ and the meaning of the $\frac{1}{5}$ will be uncovered soon.

U&D\t	0	1	2	3	4	5	6
6	**S**	**Up**	**Do**	**r**			163.2149
5	100	1.0850	0.9215	0.100		150.4180	
4					138.6244		138.6244
3				127.7556		127.7556	
2			117.7389		117.7389		117.7389
1		108.5075		108.5075		108.5075	
0	100		100.00		99.9999		99.9999
1		92.1594		92.1594		92.1594	
2			84.9336		84.9336		84.9336
3				78.2744		78.27444	
4					72.1373		72.1373
5		**Δt**	**qu**			66.4813	
6		$\frac{1}{6}$	0.5824				61.2688

Table 13.3: A Numeric Price Evolution in a Six-Period Binomial Tree

Hence, if the system in (13.22) is consistent, there are no arbitrage opportunities in each of the one-period binomial models composing the N-period binomial tree spanning the full year. Therefore, by Theorem 11, the N-period binomial model is arbitrage-free. We ask the reader to verify that $qu = 0.5824$ solves the equation above.

With these values for Up and Do the evolution of the price of the stock is displayed in the spreadsheet below (Table 13.3 in the hard copy). In the on-line version, the reader may change the values of S, the current price of the stock, of r, the annual continuously compounded risk-free rate, and of Δt, the length of a time period (in years), and reevaluate the spreadsheet. Note that for the prepared spreadsheet the maturity time will be $6\Delta t$ but the reader may increase the number of periods and thereby extend the maturity time. Reevaluating the spreadsheet will also trigger the recalculation of the factors Up, Do, and of the risk-neutral probability qu.

13.4.2 European Call

The value of a European-type contingent claim that pays $f(S)$ at its maturity time (time 6 in our example) can be calculated utilizing the replication approach as in equation (13.21). Alternatively, it can be done via the stochastic discount factors as in equation (13.15). In this section we value the call option via the replication approach. We do it in two ways — once based on equation (13.15), and once showing the price evolution and the replicating portfolio in each time and state.

Let us assign values to the following variables:

```
> i:='i';j:='j';T:=1;N:=6;r[f]:=.1;
```

$$i := i$$

$$j := j$$

$$T := 1$$

$$N := 6$$

$$r_f := .1$$

```
> Up:=evalf(exp(1/5*sqrt(T/N)));Do:=1/Up:S:=100;
```

$$Up := 1.085075596$$

$$S := 100$$

```
> qu:=(exp(r[f]/N)-Do)/(Up-Do);
```

$$qu := .5824019866$$

With these assignments the value of the stochastic discount factor in state j at time N is given by

```
> d(N,j):=(N!/(j!*(N-j)!))*exp(-r[f]*T)*qu^j*(1-qu)^(N-j);
```

$$\mathrm{d}(6, j) := 651.4829410 \frac{.5824019866^j \, .4175980134^{(6-j)}}{j! \, (6-j)!}$$

We also specify the payoff (value) at maturity that is defined by the function f. The function defined below is the payoff from a European call option with an exercise price of \$80. (The reader may choose to define the function f differently in order to value another contingent claim.)

```
> f:=S->max(S-80,0);
```

$$f := S \to \max(S - 80, 0)$$

The value of such a contingent claim, utilizing the stochastic discount factor approach, is based on equation (13.15). It is given by the command below:

```
> 'sum(d(N,j)*f(S*Up^j*Do^(N-j)),j=0..N)'\
> =sum(d(N,j)*f(S*Up^j*Do^(N-j)),'j'=0..N);
```

$$\sum_{j=0}^{N} \mathrm{d}(N, j) \, \mathrm{f}(S \, Up^j \, Do^{(N-j)}) = 28.01861454$$

We can also calculate the value of the derivative security and its replicating portfolio at each intermediate node. This is not necessary if one is only interested in the value of the European option. However, for an American-type option the situation is different and the value of the option in each state must be calculated. To pave the ground for our discussion of American options we present the value of the option and its value if exercised at each node.

The value of the option at each node is calculated in a recursive way as was done in the symbolic example of Section 13.2.2. We use the two-dimensional array **DerPrice** to store the value of the option at each node, where the first index of the array will denote time and the second index denotes the number of up movements since time zero. Hence we assign the value of $DerPrice_{6,j}$ for $j = 0, ..., 6$, as below.

```
> j:='j':i:=6;
```

$$j := 6$$

```
> for j from i to 0 by -1 do;(i-j)
> DerPrice[6,j]:=f(S*Up^j*Do^(i-j));
> od;
```

$$DerPrice_{6,6} := 83.2149650$$

$$DerPrice_{6,5} := 58.6244973$$

$$DerPrice_{6,4} := 37.7389049$$

$$DerPrice_{6,3} := 19.99999999$$

$$DerPrice_{6,2} := 4.93369296$$

$$DerPrice_{6,1} := 0$$

$$DerPrice_{6,0} := 0$$

As was have seen in the symbolic example at the end of Subsection 13.3.2, we can run a "do loop" within a "do loop" to calculate the value of the option at time 0. This is done below but we will also run it step-by-step and observe the evolution of the price of the option and the replicating portfolios.

```
> i:='i';j:='j';
```

$$i := i$$

$$j := j$$

```
> for i from 5 to 0 by -1 do;
> for j from i to 0 by -1 do;
> DerPrice[i,j]:=exp(-r[f]*T/N)*\
> (qu*DerPrice[i+1,j+1]+(1-qu)*DerPrice[i+1,j]);
> od;
> od;
> DerPrice[0,0];
```

$$28.01861454$$

We therefore conclude that the value calculated in this manner leads to the same price, 28.01861454.

We proceed now to calculate the option value step-by-step and its evolution. Having calculated the value of the option at maturity we can calculate the value of the option at each node at time 5. Its price at time 5 is the discounted expected value of the option price at time 6. Hence,

```
> j:='j':i:=5;
```

$$i := 5$$

```
> for j from i to 0 by -1 do;
> DerPrice[i,j]:=exp(-r[f]*T/N)*\
> (qu*DerPrice[i+1,j+1]+(1-qu)*DerPrice[i+1,j]);
> od;
```

$$DerPrice_{5,5} := 71.74034264$$
$$DerPrice_{5,4} := 49.07789606$$
$$DerPrice_{5,3} := 29.82984325$$
$$DerPrice_{5,2} := 13.48176118$$
$$DerPrice_{5,1} := 2.825899579$$
$$DerPrice_{5,0} := 0$$

This option is of European-type and cannot be exercised at time 5. Let us see, however, if it would have been optimal to exercise the option in some of the states at time 5, if that possibility had been available to us. We should realize that since the underlying asset pays no dividend, early exercise would not have been optimal. Nevertheless, we present, in the group of Maple commands below, the value of the option at time 5 at each node $DerPrice_{5,j}$ for $j = 0, ..., 5$, compared with its value if exercised at time 5, $f\left(SUp^{j}Do^{5-j}\right)$. Furthermore, we display the composition of the replicating portfolio, where Xs is the holding of the stock, and Xb is the amount invested in the risk-free rate. The on-line version of the book presents the output of these commands and also summarizes the output in a spreadsheet. The hard copy of the book only presents the spreadsheet[11] in the table below.

```
> for j from i to 0 by -1 do;
> ['j'=j,'S'=S*Up^j*Do^(i-j),'DerPrice'=DerPrice[i,j], \
> 'f(S)'=f(S*Up^j*Do^(i-j)),\
> 'Xs'=(-DerPrice[i+1,j]+DerPrice[i+1,j+1])\
> /(S*Up^j*Do^(i-j)*(Up-Do)),\
> 'Xb'=-(Do*DerPrice[i+1,j+1]-DerPrice[i+1,j]*Up) \
> /((Up-Do))*exp(r[f]*T/N)];
> od;j:='j':
```

i	*equqls*	5			
j	S	*DerPrice*	f(S)	*Xs*	*Xb*
5	150.4180590	71.74034264	70.4180590	.9999999982	−81.34450606
4	127.7556124	49.07789606	47.7556124	1.000000001	−81.34450650
3	108.5075596	29.82984325	28.5075596	1.000000000	−81.34450644
2	92.15947751	13.48176118	12.15947751	1.000000001	−81.34450638
1	78.27444767	2.825899579	0	.3855540745	−28.28027016
0	66.48137906	0	0	0	0

The on-line version of the book proceeds with the recursive solution showing the output of the Maple commands finding the value of the option at time 4, and continuing to time 3 until we reach time zero. The hard copy of the book will show only the spreadsheet. At each time period we compare the value of the option to its value if exercised immediately and display the replicating portfolio.

```
> i:=4;
```

$$i := 4$$

```
> for j from i to 0 by -1 do;
> DerPrice[i,j]:=exp(-r[f]*T/N)*\
> (qu*DerPrice[i+1,j+1]+(1-qu)*DerPrice[i+1,j]);
> od;j:='j':
```

$$DerPrice_{4,4} := 61.24720914$$

[11] The spreadsheet uses some global variables that are defined by the Maple commands. However, the reader may change the template presented above and modify the spreadsheet to summarize the new results.

$$DerPrice_{4,3} := 40.36161677$$

$$DerPrice_{4,2} := 22.62271189$$

$$DerPrice_{4,1} := 8.882610458$$

$$DerPrice_{4,0} := 1.618606690$$

```
> for j from i to 0 by -1 do;
> ['j'=j,'S'=S*Up^j*Do^(i-j),'DerPrice'=DerPrice[i,j], \
> 'f(S)'=f(S*Up^j*Do^(i-j)),\
> 'Xs'=(-DerPrice[i+1,j]+DerPrice[i+1,j+1])\
> /(S*Up^j*Do^(i-j)*(Up-Do)),\
> 'Xb'=-(Do*DerPrice[i+1,j+1]-DerPrice[i+1,j]*Up) \
> /((Up-Do))*exp(r[f]*T/N)];
> od;j:='j':
```

i	*equqls*	4			
j	S	*DerPrice*	f(S)	*Xs*	*Xb*
4	138.6244973	61.24720914	58.6244973	.9999999988	−79.99999990
3	117.7389049	40.36161677	37.7389049	1.000000000	−80.00000002
2	99.99999999	22.62271189	19.99999999	.9999999994	−79.99999990
1	84.93369297	8.882610458	4.93369297	.7674352685	−58.20674500
0	72.13732201	1.618606690	0	.2396237716	−16.19825235

```
> i:=3;
```

$$i := 3$$

```
> for j from i to 0 by -1 do;
> DerPrice[i,j]:=exp(-r[f]*T/N)*\
> (qu*DerPrice[i+1,j+1]+(1-qu)*DerPrice[i+1,j]);
> od;j:='j':
```

$$DerPrice_{3,3} := 51.65725831$$

$$DerPrice_{3,2} := 32.40920551$$

$$DerPrice_{3,1} := 16.60579043$$

$$DerPrice_{3,0} := 5.752498524$$

```
> for j from i to 0 by -1 do;
> ['j'=j,'S'=S*Up^j*Do^(i-j),'DerPrice'=DerPrice[i,j], \
> 'f(S)'=f(S*Up^j*Do^(i-j)),\
> 'Xs'=(-DerPrice[i+1,j]+DerPrice[i+1,j+1])\
> /(S*Up^j*Do^(i-j)*(Up-Do)),\
> 'Xb'=-(Do*DerPrice[i+1,j+1]-DerPrice[i+1,j]*Up) \
> /((Up-Do))*exp(r[f]*T/N)];
> od;j:='j':
```

i	*equqls*	3			
j	S	*DerPrice*	f(S)	*Xs*	*Xb*
3	127.7556124	51.65725831	47.7556124	.9999999994	−78.67771634
2	108.5075596	32.40920551	28.5075596	.9999999988	−78.67771616
1	92.15947752	16.60579043	12.15947752	.9119754034	−69.72731964
0	78.27444768	5.752498524	0	.5676612372	−39.99196378

```
> i:=2;
```

$$i := 2$$

```
> for j from i to 0 by -1 do;
> DerPrice[i,j]:=exp(-r[f]*T/N)*\
> (qu*DerPrice[i+1,j+1]+(1-qu)*DerPrice[i+1,j]);
> od;j:='j':
```

$$DerPrice_{2,\,2} := 42.89834592$$

$$DerPrice_{2,\,1} := 25.38313344$$

$$DerPrice_{2,\,0} := 11.87392026$$

```
> for j from i to 0 by -1 do;
> ['j'=j,'S'=S*Up^j*Do^(i-j),'DerPrice'=DerPrice[i,j], \
> 'f(S)'=f(S*Up^j*Do^(i-j)),\
> 'Xs'=(-DerPrice[i+1,j]+DerPrice[i+1,j+1])\
> /(S*Up^j*Do^(i-j)*(Up-Do)),\
> 'Xb'=-(Do*DerPrice[i+1,j+1]-DerPrice[i+1,j]*Up) \
> /((Up-Do))*exp(r[f]*T/N)];
> od;j:='j':
```

i	*equqls*	2			
j	S	*DerPrice*	f(S)	Xs	Xb
2	117.7389049	42.89834592	37.7389049	.9999999994	−77.37728809
1	100.0000000	25.38313344	20.0000000	.9666831254	−73.70139829
0	84.93369297	11.87392026	4.93369297	.6638877794	−56.36264491

Before we continue with the next steps of the calculation we would like to alert the reader again to the evolution and meaning of Xs. We have mentioned before that Xs plays the role of the Δ sensitivity measure. Indeed, Xs, the holding of the stock in the replicating portfolio, takes care of the changes in the value of the derivative security that are the result of changes in the price of the stock. In our discussion of the Δ sensitivity measure, Chapter 7, the measure was defined as the (calculus) derivative of the derivative security's price with respect to the stock price. In this model, though, the change in the stock price is discrete and Δ cannot be defined as the (calculus) derivative. In the current setting, Xs is the solution to the system of equations in (13.16) stipulating that holding of Xs units of the stock ensures that the portfolio value is equal to the value of the derivative security in the next time period. The investigation of hedging with Δ in Chapter 8 enforces the similarity between Xs, in the binomial model, and Δ, in a model where the stock price can take any value in $[0, \infty)$.

Looking at the evolution of Xs and Xb exemplifies this interpretation. When the call option is in the money it is "closer" in nature to a stock than to a bond and Xs should be closer to 1. Indeed the closer $f(s)$ is to *DerPrice*, the value of the derivative, the closer Xs is to 1 and the smaller is Xb. The reader is encouraged to pay attention to this phenomenon, which essentially holds for a put option as well. However, in the case of a put, the larger the value of the stock, the smaller the value of the put option and hence we expect Xs to be negative and close to -1. We proceed now with the final steps of the calculation. Observe the evolution of Xs, keeping in mind the explanation above.

```
> i:=1;
```

$$i := 1$$

```
> for j from i to 0 by -1 do;
> DerPrice[i,j]:=exp(-r[f]*T/N)*\
> (qu*DerPrice[i+1,j+1]+(1-qu)*DerPrice[i+1,j]);
> od;j:='j':
```

$$DerPrice_{1,1} := 34.99587574$$
$$DerPrice_{1,0} := 19.41541104$$

```
> for j from i to 0 by -1 do;
> ['j'=j,'S'=S*Up^j*Do^(i-j),'DerPrice'=DerPrice[i,j], \
> 'f(S)'=f(S*Up^j*Do^(i-j)),\
> 'Xs'=(-DerPrice[i+1,j]+DerPrice[i+1,j+1])\
> /(S*Up^j*Do^(i-j)*(Up-Do)),\
> 'Xb'=-(Do*DerPrice[i+1,j+1]-DerPrice[i+1,j]*Up) \
> /((Up-Do))*exp(r[f]*T/N)];
> od;j:='j':
```

i	*equqls*	1			
j	S	*DerPrice*	f(S)	*Xs*	*Xb*
1	108.5075596	34.99587574	28.5075596	.9873897275	−74.58868168
0	92.15947752	19.41541104	12.15947752	.8966505960	−65.36226947

```
> i:=0;
```

$$i := 0$$

```
> for j from i to 0 by -1 do;
> DerPrice[i,j]:=exp(-r[f]*T/N)*\
> (qu*DerPrice[i+1,j+1]+(1-qu)*DerPrice[i+1,j]);
> od;j:='j':
```

$$DerPrice_{0,0} := 28.01861454$$

```
> for j from i to 0 by -1 do;
> ['j'=j,'S'=S*Up^j*Do^(i-j),'DerPrice'=DerPrice[i,j], \
> 'f(S)'=f(S*Up^j*Do^(i-j)),\
> 'Xs'=(-DerPrice[i+1,j]+DerPrice[i+1,j+1])\
> /(S*Up^j*Do^(i-j)*(Up-Do)),\
> 'Xb'=-(Do*DerPrice[i+1,j+1]-DerPrice[i+1,j]*Up) \
> /((Up-Do))*exp(r[f]*T/N)];
> od;j:='j':i:='i':
```

i	*equqls*	0			
j	S	*DerPrice*	f(S)	*Xs*	*Xb*
0	100.	28.01861454	20.	.9530454168	−69.56659122

We confirm that the call price is indeed 28.01861454 as calculated before and that it was not optimal to exercise the option prior to maturity. The latter follows since there was no node at which $f(S)$, the payoff from the option if exercised, was larger than the option's value. This, as we know from Chapter 12, is not necessarily the situation for a put option, even a put written on a stock that pays no dividends. The next subsection exemplifies the optimality of early exercising of a put option within the framework of the binomial model.

13.4.3 European Put

We repeat the same sort of calculations, only this time for a put option. The only change required for this calculation is the definition of the payoff function at maturity. Nevertheless, for convenience we redefine all the parameters below.

```
> i:='i';j:='j';T:=1;N:=6;r[f]:=.1;
```

$$i := i$$
$$j := j$$
$$T := 1$$
$$N := 6$$
$$r_f := .1$$

```
> Up:=evalf(exp(1/5*sqrt(T/N)));Do:=1/Up:S:=100;
```

$$Up := 1.085075596$$
$$S := 100$$

```
> qu:=(exp(r[f]/N)-Do)/(Up-Do);
```

$$qu := .5824019866$$

```
> d(N,j):=(N!/(j!*(N-j)!))*exp(-r[f]*T)*qu^j*(1-qu)^(N-j);
```

$$\mathrm{d}(6,\,j) := 651.4829410\,\frac{.5824019866^{i}\,.4175980134^{(6-j)}}{j!\,(6-j)!}$$

Here is where the value of f is defined to reflect a put payoff at maturity.

```
> f:=S->max(80-S,0);
```

$$f := S \to \max(80 - S,\, 0)$$

The value of the put based on equation (13.15) is given by the command below.

```
> 'sum(d(N,j)*f(S*Up^j*Do^(N-j)),j=0..N)'\
> =sum(d(N,j)*f(S*Up^j*Do^(N-j)),'j'=0..N);
```

$$\sum_{i=0}^{N} \mathrm{d}(N,\,j)\,\mathrm{f}(S\,Up^{j}\,Do^{(N-j)}) = .4056082068$$

We proceed to calculate step-by-step the value of the put option based on the replication approach. We start, as we did for the call option, with the value of the put at maturity.

```
> j:='j':i:=6;
```

$$i := 6$$

```
> for j from i to 0 by -1 do;
> DerPrice[i,j]:=f(S*Up^j*Do^(i-j));
> od;
```

$$DerPrice_{6,6} := 0$$
$$DerPrice_{6,5} := 0$$
$$DerPrice_{6,4} := 0$$
$$DerPrice_{6,3} := 0$$
$$DerPrice_{6,2} := 0$$
$$DerPrice_{6,1} := 7.86267799$$
$$DerPrice_{6,0} := 18.73110841$$

The value of the put can be calculated as before using a "do-loop" within a "do-loop". This is done below:

```
> i:='i';j:='j';
```

$$i := i$$
$$j := j$$

```
> for i from 5 to 0 by -1 do;
> for j from i to 0 by -1 do;
> DerPrice[i,j]:=exp(-r[f]*T/N)*\
> (qu*DerPrice[i+1,j+1]+(1-qu)*DerPrice[i+1,j]);
> od;
> od;
```

```
> DerPrice[0,0];
```

$$.4056082067$$

We confirm that the value of the put coincides with its value calculated using the stochastic discount factor approach. We continue with the step-by-step calculation, displaying the replicating portfolio and the optimal early exercise situations. As expected, in the case of a put, it might be optimal to exercise prior to maturity. At certain states and times $DerPrice$ might be less than f, which means exercising would be optimal.

```
> i:=5;
```

$$i := 5$$

```
> for j from i to 0 by -1 do;
> DerPrice[i,j]:=exp(-r[f]*T/N)*\
> (qu*DerPrice[i+1,j+1]+(1-qu)*DerPrice[i+1,j]);
> od;j:='j':
```

$$DerPrice_{5,5} := 0$$

$$DerPrice_{5,4} := 0$$

$$DerPrice_{5,3} := 0$$

$$DerPrice_{5,2} := 0$$

$$DerPrice_{5,1} := 3.229168241$$

$$DerPrice_{5,0} := 12.19633727$$

```
> for j from i to 0 by -1 do;
> ['j'=j,'S'=S*Up^j*Do^(i-j),'DerPrice'=DerPrice[i,j], \
> 'f(S)'=f(S*Up^j*Do^(i-j)),\
> 'Xs'=(-DerPrice[i+1,j]+DerPrice[i+1,j+1])\
> /(S*Up^j*Do^(i-j)*(Up-Do)),\
> 'Xb'=-(Do*DerPrice[i+1,j+1]-DerPrice[i+1,j]*Up) \
> /((Up-Do))*exp(r[f]*T/N)];
> od;j:='j':
```

i	*equqls*	5			
j	*S*	*DerPrice*	f(*S*)	*Xs*	*Xb*
5	150.4180590	0	0	0	−81.34450606
4	127.7556124	0	0	0	−81.34450650
3	108.5075596	0	0	0	−81.34450644
2	92.15947751	0	0	0	−81.34450638
1	78.27444767	3.229168241	1.72555233	−.6144459253	−28.28027016
0	66.48137906	12.19633727	[**13.51862094**]	−1.000000001	0

The reader's attention is drawn to the values displayed in the row that orresponds to $j = 0$. The value of $f(S)$ is larger than the value of the option, meaning it would have been optimal to exercise the option, were it f the American-type. The spreadsheet displays the values of $f(S)$ in those ases where early exercise is optimal, i.e., if $f(S) > DerPrice$, in [ ]. The able in the hard copy also displays these values of $f(S)$ in bold fonts.

The value of Xs is negative since the put option behaves like a short position in the stock. Recall the payoff graph of a put option (see Figure 4.7 in Chapter 4) in comparison to the payoff from a short position in a stock. In the region when the put option is in the money, its payoff graph has a slope like that of the payoff from a short position in the stock, i.e., of −1. Thus the deeper in the money the put is, the closer is Xs to −1. Observe the evolution of Xs in comparison to the value of the derivative and to the instances where early exercise, if possible, would have been optimal.

```
> i:=4;
```

$$i := 4$$

```
> for j from i to 0 by -1 do;
> DerPrice[i,j]:=exp(-r[f]*T/N)*\
> (qu*DerPrice[i+1,j+1]+(1-qu)*DerPrice[i+1,j]);
> od;j:='j':
```

$$DerPrice_{4,4} := 0$$
$$DerPrice_{4,3} := 0$$
$$DerPrice_{4,2} := 0$$
$$DerPrice_{4,1} := 1.326205593$$
$$DerPrice_{4,0} := 6.858572774$$

```
> for j from i to 0 by -1 do;
> ['j'=j,'S'=S*Up^j*Do^(i-j),'DerPrice'=DerPrice[i,j], \
> 'f(S)'=f(S*Up^j*Do^(i-j)),\
> 'Xs'=(-DerPrice[i+1,j]+DerPrice[i+1,j+1])\
> /(S*Up^j*Do^(i-j)*(Up-Do)),\
> 'Xb'=-(Do*DerPrice[i+1,j+1]-DerPrice[i+1,j]*Up) \
> /((Up-Do))*exp(r[f]*T/N)];
> od;j:='j':
```

i	*equqls*	4			
j	S	*DerPrice*	f(S)	Xs	Xb
4	138.6244973	0	0	0	0
3	117.7389049	0	0	0	0
2	99.99999999	0	0	0	0
1	84.93369297	1.326205593	0	−.2325647318	21.7932549
0	72.13732201	6.858572774	[**7.86267799**]	−.7603762282	63.8017476

Here we see again that for a value of $j = 0$, early exercise would have been optimal. The value of the option is \$6.85 and by exercising it one collects \$7.86. If the option had been American it would have been exercised. This would trigger a domino effect causing a change in the value of the option at time $i = 3$, which in turn would cause a change in the value of the option at time $i = 2$. We shall examine this in the next subsection where an American option is valued. The rest of this subsection presents the evolution of the value of the European put and its replicating portfolios.

```
> i:=3;
```

$$i := 3$$

```
> for j from i to 0 by -1 do;
> DerPrice[i,j]:=exp(-r[f]*T/N)*\
> (qu*DerPrice[i+1,j+1]+(1-qu)*DerPrice[i+1,j]);
> od;j:='j':
```

$$DerPrice_{3,3} := 0$$
$$DerPrice_{3,2} := 0$$
$$DerPrice_{3,1} := .5446669680$$
$$DerPrice_{3,0} := 3.576404895$$

```
> for j from i to 0 by -1 do;
> ['j'=j,'S'=S*Up^j*Do^(i-j),'DerPrice'=DerPrice[i,j], \
> 'f(S)'=f(S*Up^j*Do^(i-j)),\
> 'Xs'=(-DerPrice[i+1,j]+DerPrice[i+1,j+1])\
> /(S*Up^j*Do^(i-j)*(Up-Do)),\
> 'Xb'=-(Do*DerPrice[i+1,j+1]-DerPrice[i+1,j]*Up) \
> /((Up-Do))*exp(r[f]*T/N)];
> od;j:='j':
```

i	*equqls*	3			
j	S	*DerPrice*	f(S)	*Xs*	*Xb*
3	127.7556124	0	0	0	0
2	108.5075596	0	0	0	0
1	92.15947752	.5446669680	0	−.08802459622	8.950396663
0	78.27444768	3.576404895	1.72555232	−.4323387623	38.68575252

```
> i:=2;
```

$$i := 2$$

```
> for j from i to 0 by -1 do;
> DerPrice[j,i]:=exp(-r[f]*T/N)*\
> (qu*DerPrice[i+1,i+1]+(1-qu)*DerPrice[i+1,i]);
> od;j:'j':
```

$$DerPrice_{2,\,2} := 0$$

$$DerPrice_{1,\,2} := 0$$

$$DerPrice_{0,\,2} := 0$$

```
> for j from i to 0 by -1 do;
> ['j'=j,'S'=S*Up^j*Do^(i-j),'DerPrice'=DerPrice[i,j], \
> 'f(S)'=f(S*Up^j*Do^(i-j)),\
> 'Xs'=(-DerPrice[i+1,j]+DerPrice[i+1,j+1])\
> /(S*Up^j*Do^(i-j)*(Up-Do)),\
> 'Xb'=-(Do*DerPrice[i+1,j+1]-DerPrice[i+1,j]*Up) \
> /((Up-Do))*exp(r[f]*T/N)];
> od;j:='j':
```

i	*equqls*	2			
j	*S*	*DerPrice*	f(*S*)	*Xs*	*Xb*
3	138.6244973	0	0	0	0
2	117.7389049	0	0	0	0
1	100.0000000	.2236923955	0	−.03331687261	3.675889650
0	84.93369297	1.780786221	0	−.2183457985	21.01464316

```
> i:=1;
```

$$i := 1$$

```
> for j from i to 0 by -1 do;
> DerPrice[i,j]:=exp(-r[f]*T/N)*\
> (qu*DerPrice[i+1,j+1]+(1-qu)*DerPrice[i+1,j]);
> od;j:='j':
```

$$DerPrice_{1,\,1} := .09186951062$$

$$DerPrice_{1,\,0} := .8594868635$$

```
> for j from i to 0 by -1 do;
> ['j'=j,'S'=S*Up^j*Do^(i-j),'DerPrice'=DerPrice[i,j], \
> 'f(S)'=f(S*Up^j*Do^(i-j)),\
> 'Xs'=(-DerPrice[i+1,j]+DerPrice[i+1,j+1])\
> /(S*Up^j*Do^(i-j)*(Up-Do)),\
> 'Xb'=-(Do*DerPrice[i+1,j+1]-DerPrice[i+1,j]*Up) \
> /((Up-Do))*exp(r[f]*T/N)];
> od;j:='j':
```

i	*equqls*	1			
j	S	*DerPrice*	f(S)	*Xs*	*Xb*
1	108.5075596	.09186951062	0	−.01261027086	1.509672167
0	92.15947752	.8594868635	0	−.1033494023	10.73608448

```
> i:=0;
```

$$i := 0$$

```
> for j from i to 0 by -1 do;
> DerPrice[i,j]:=exp(-r[f]*T/N)*\
> (qu*DerPrice[i+1,j+1]+(1-qu)*DerPrice[i+1,j]);
> od;j:='j':
```

$$DerPrice_{0,0} := .4056082067$$

i	*equqls*	0			
j	S	*DerPrice*	f(S)	*Xs*	*Xb*
0	100.	.4056082067	0	−.04695458153	5.273967580

We shall now proceed with the calculation of the value of an American option. The reader probably expects, given the examples above, the methods by which this will be done.

13.4.4 American Options

Because of the possibility of early exercise, there is no analytical solution for the value of an American put option. The nonexistence of an analytical solution follows from the fact that in order to value this option one needs to know when it is optimal to exercise it. On the other hand, in order to know if it is optimal to exercise the option one needs to know its value. This creates a vicious circle that cannot be overcome analytically. The value of an American option for which early exercise is not optimal is equal to the value of an otherwise identical European option. Such is the case, as we have seen in Chapter 12, for an American call option on a stock that pays no dividends. Hence, if it is not possible to prove that early exercise is not optimal we have to resort to numerical solutions.

The previous section and Chapter 12, within the context of the binomial model, demonstrated that early exercise of an American put might be optimal. Those descriptions pointed to the method of identifying, within the context of the binomial model, a numerical solution for the value of American options lacking an analytical solution. The vicious circle, mentioned above, can be overcome within the binomial model framework, again by recursive valuation. Starting from the maturity time and going backward, the option value, if the option is not exercised, is calculated at each possible state of nature and compared to its value if exercised. The "true" value of the option at each node is the maximum of these two values. Hence, when the recursion continues backward, it takes into account the additional value due to the possibility of early exercise.

Calculating the price of an American option, we transverse back from the maturity time N to time $N-1$, and so on. At each time we evaluate whether or not it is optimal to exercise the option, given the time and the realized state. Since the value of the option at maturity is known, we essentially can apply the pricing-by-arbitrage arguments of the one-period model to value the option at each of the coordinates $(N-1, j)$ for $j = 1, ..., N-1$. After all, looking ahead from a coordinate $(N-1, j)$, for some j, one sees exactly the same situation as in a one-period binomial model. Therefore, at that time, we can compare the value of the option if not exercised to its value if exercised. We are not sure which of the $(N-1, 0), ..., (N-1, N-1)$ states will be realized at time $N-1$. Hence we calculate the value of the option at each possible state. Continuing in the same manner we can find the value of the option at time zero.

Let us see how this works. We reassign the following parameters and redefine f to be the payoff from a put option with an exercise price of \$80.

```
> i:='i';j:='j';T:=1;N:=6;r[f]:=.1;
```

$$i := i$$

$$j := j$$

$$T := 1$$

$$N := 6$$

$$r_f := .1$$

```
> Up:=evalf(exp(1/5*sqrt(T/N)));Do:=1/Up:S:=100;
```

$$Up := 1.085075596$$

$$S := 100$$

```
> qu:=(exp(r[f]/N)-Do)/(Up-Do);
```

$$qu := .5824019866$$

```
> f:=S->max(80-S,0);
```

$$f := S \to \max(80 - S, 0)$$

Since we have to take into account the possibility of early exercise it is not possible to value the option via the stochastic discount factor approach as we did for a European option. We will have to do it recursively step-by-step as explained in Section 13.4.3. Let us assign the value of the put option at maturity to $DerPrice_{6,j}$, $j = 0, ..., 6$. This value, needless to say, coincides with the value of a European option at maturity.

```
> j:='j':i:=6;
```

$$i := 6$$

```
> for j from i to 0 by -1 do;
> DerPrice[i,j]:=f(S*Up^j*Do^(i-j));
> od;
```

$$DerPrice_{6,6} := 0$$

$$DerPrice_{6,5} := 0$$

$$DerPrice_{6,4} := 0$$

$$DerPrice_{6,3} := 0$$

$$DerPrice_{6,2} := 0$$

$$DerPrice_{6,1} := 7.86267799$$

$$DerPrice_{6,0} := 18.73110841$$

We then go a step backward and again consider the value of the option if it is held without being exercised. As we have seen before, at a few states in time (nodes), it was profitable to exercise the option. At these nodes the payoff from exercising the option is larger than the value if exercised. The possibility of early exercise does not influence the value of the option at maturity. Thus for both American and European options the value at maturity is the payoff from the option.

Having calculated, as above, the value of the option at time 6 we can calculate its value at each node at time 5 if not exercised. This amounts to calculating the discounted expected value of the option payoff at each node. The next step is to calculate the value of the option if exercised at this point, i.e., $\max\left(80 - SUp^j Do^{5-j}, 0\right)$, for $j = 0, ..., 5$. The value of the option at each node (i, j) is the larger of the two values calculated. Hence, at each node, we should check if it is optimal to exercise the option and value it by

$$\max\left[e^{-\frac{r_f T}{N}} E\left(DerPrice_{j+1}\right), f\left(SUp^j Do^{i-j}\right)\right], \tag{13.23}$$

where

$$E\left(DerPrice_{j+1}\right) = qu\, DerPrice_{i+1,j+1} + qd\, DerPrice_{i+1,j}.$$

Calculating the value at each node based on equation (13.23) affects the value of the option at all the nodes prior in time to this node and so alters the value of the option at time zero. It also follows that a very small change in the code shown above for a European option will suffice to value an American option. This is exemplified below where the value of the American option is calculated by a "do loop" within a "do loop".

```
> i:='i';j:='j';
```

$$i := i$$
$$j := j$$

```
> for i from 5 to 0 by -1 do;
> for j from i to 0 by -1 do;
> DerPrice[i,j]:=max(exp(-r[f]*T/N)*\
> (qu*DerPrice[i+1,j+1]+(1-qu)*DerPrice[i+1,j]),\
> f(S*Up^j*Do^(i-j)));
> od;
> od;

> DerPrice[0,0];
```

.4341748988

The comparison of this value to the counterpart European option below reveals that the European option is cheaper.

```
> 'sum(d(N,j)*f(S*Up^j*Do^(N-j)),j=0..N)'\
> =sum(d(N,j)*f(S*Up^j*Do^(N-j)),'j'=0..N);
```

$$\sum_{j=0}^{N} \mathrm{d}(N,\, j)\, \mathrm{f}(S\ Up^j\ Do^{(N-j)}) = .4056082068$$

This confirms that the possibility of early exercise increases the value of an American option relative to its European equivalent.

To gain a better understanding and to visualize the impact of early exercise, we repeat the above calculation step-by-step. In addition, we would like to keep a record of the coordinates at which it was optimal to exercise the option prior to maturity. We define *OptExer* as an object where we keep these coordinates. We will add another line to the code above that will append the coordinate (i, j) to this object if it was optimal to exercise the option there, that is, if (i, j) satisfies the inequality in (13.24).

$$\begin{aligned} & e^{-\frac{r_f T}{N}}\left(quDerPrice_{i+1,j+1} + qd\,DerPrice_{i+1,j}\right) \\ < \; & f\left(SUp^j Do^{i-j}\right) \end{aligned} \tag{13.24}$$

We initialize *OptExer* to be the empty set, unassign the values of i and j, and then run the modified code.

```
> OptExer:={};
```

$$OptExer := \{\}$$

```
> i:='i':j:='j';
```

$$j := j$$

```
> for i from 5 to 0 by -1 do;
> for j from i to 0 by -1 do;
> DerPrice[i,j]:=max(exp(-r[f]*T/N)*\
> (qu*DerPrice[i+1,j+1]+(1-qu)*DerPrice[i+1,j]),\
> f(S*Up^i*Do^(i-j)));
> if exp(-r[f]*T/N)*(qu*DerPrice[i+1,j+1]+\
> (1-qu)*DerPrice[i+1,j])<f(S*Up^i*Do^(i-j))\
> then OptExer:=[[i,j],op(OptExer)] fi;
> od;
> od;
```

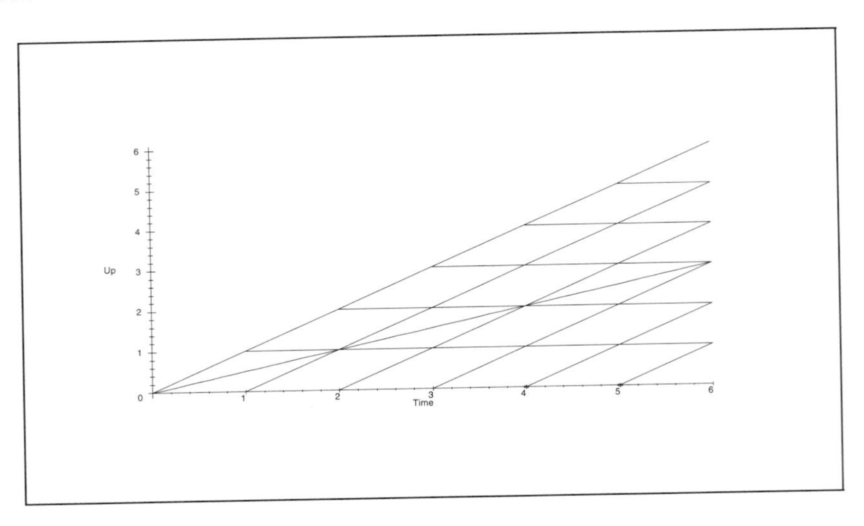

Figure 13.5: A Binomial Tree with a Box Superimposed on Nodes Where Early Exercise Is Optimal

The coordinates at which it was optimal to exercise the option are now contained in *OptExer*.

```
> OptExer;
```

$$[[4, 0], [5, 0]]$$

In order to visualize the optimal coordinates within the binomial tree we define the object *OptPath* below.

```
> OPtPath:=(plottools[curve](OptExer,\
> style=point,color=blue,symbol=box)):
```

The following command superimposes a blue box on each node of the binomial tree at which it is optimal to exercise the option. This is displayed in Figure 13.5.

```
> BinTree(6,OptExer);
```

Let us rerun this same valuation again, but for a higher exercise price. This will make it optimal to exercise prior to maturity more frequently. We thus reassign the variables and change the payoff function f to have an exercise price of \$110.

```
> j:='j':i:=6;
```

$$i := 6$$

```
> f:=S->max(110-S,0);
```

$$f := S \to \max(110 - S, 0)$$

```
> for j from i to 0 by -1 do;
> DerPrice[i,j]:=f(S*Up^j*Do^(i-j));
> od;
```

$$DerPrice_{6,6} := 0$$

$$DerPrice_{6,5} := 0$$

$$DerPrice_{6,4} := 0$$

$$DerPrice_{6,3} := 10.00000001$$

$$DerPrice_{6,2} := 25.06630704$$

$$DerPrice_{6,1} := 37.86267799$$

$$DerPrice_{6,0} := 48.73110841$$

```
> OptExer:={};
```

$$OptExer := \{\}$$

```
> i:='i':j:='j':
```

$$j := j$$

```
> for i from 5 to 0 by -1 do;
> for j from i to 0 by -1 do;
> DerPrice[i,j]:=max(exp(-r[f]*T/N)*\
> (qu*DerPrice[i+1,j+1]+(1-qu)*DerPrice[i+1,j]),\
> f(S*Up^j*Do^(i-j)));
> if exp(-r[f]*T/N)*(qu*DerPrice[i+1,j+1]+
> (1-qu)*DerPrice[j+1,i])<f(S*Up^i*Do^(i-j))\
> then OptExer:=[[i,j], op(OptExer)] fi;
> od;
> od;
```

The price of such an option will be

```
> DerPrice[0,0];
```

$$10.54784840$$

and the nodes at which it is optimal to exercise this option are given below.

```
> OptExer;
```

[[1, 0], [2, 0], [3, 0], [3, 1], [4, 0], [4, 1], [4, 2], [5, 0], [5, 1], [5, 2]]

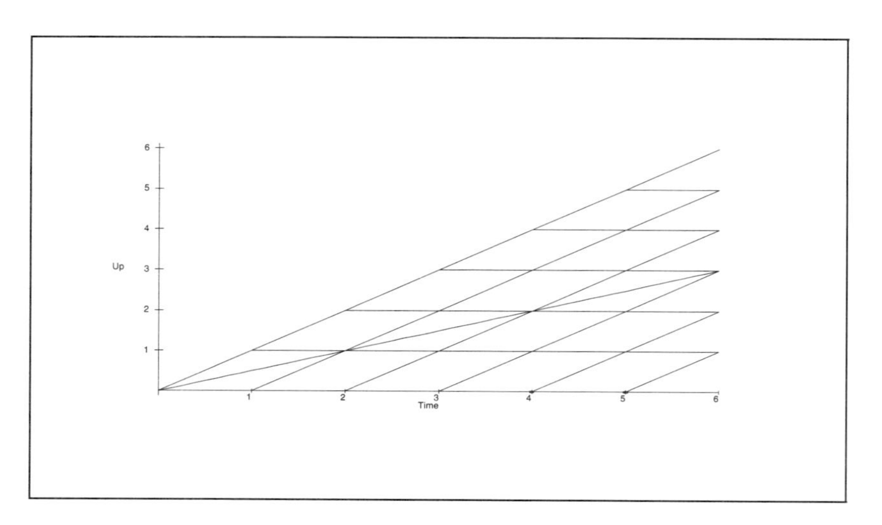

Figure 13.6: A Binomial Tree with a Box Superimposed on Nodes Where Early Exercise Is Optimal for a Binary Option

The next command superimposes the boxes on the nodes at which the option should be exercised. Figure 13.6 in the hard copy displays the output.

```
> BinTree(6,OptExer);
```

We can utilize the template presented above to calculate the prices of different types of American options. Suppose we have an option that pays \$10 if the price of the stock falls below \$90, and zero otherwise. This is an example of a binary option introduced in Section 4.5. In Chapter 4 we only approximated the value of such a European option by replicating its payoff using plain vanilla puts and calls. Since the option is American and the payoff is \$10 or nothing, if the price of the stock falls below \$90 it should be clear that it is optimal to exercise immediately. To value such an option we need only to change the payoff function and define it as below.

```
> f:=S->piecewise(S<90,10);
```

$$f := S \rightarrow \text{piecewise}(S < 90,\ 10)$$

The payoff at maturity is displayed in the on-line version of the book in the figure below, but is omitted from the hard copy.

```
> plot(f(s),s=0..150,thickness=3);
```

We keep the parameters as above and change only the form of f. The valuation code is rerun below.

```
> j:='j':i:=6;
```

$$j := 6$$

```
> for j from i to 0 by -1 do;
> DerPrice[i,j]:=f(S*Up^j*Do^(i-j));
> od;
```

$$DerPrice_{6,6} := 0$$

$$DerPrice_{6,5} := 0$$

$$DerPrice_{6,4} := 0$$

$$DerPrice_{6,3} := 0$$

$$DerPrice_{6,2} := 10$$

$$DerPrice_{6,1} := 10$$

$$DerPrice_{6,0} := 10$$

```
> OptExer:={};
```

$$OptExer := \{\}$$

```
> i:='i':j:='j':
```

$$j := j$$

```
> for i from 5 to 0 by -1 do;
> for j from i to 0 by -1 do;
> DerPrice[i,j]:=max(exp(-r[f]*T/N)*\
> (qu*DerPrice[j+1,i+1]+(1-qu)*DerPrice[i+1,j]),\
> f(S*Up^i*Do^(j-i)));
> if exp(-r[f]*T/N)*(qu*DerPrice[j+1,i+1]+\
> (1-qu)*DerPrice[j+1,i])<f(S*Up^j*Do^(i-j))\
> then OptExer:=[[i,j],op(OptExer)] fi;
> od;
> od;
```

The price of such an option is therefore

```
> DerPrice[0,0];
```

$$2.946942162$$

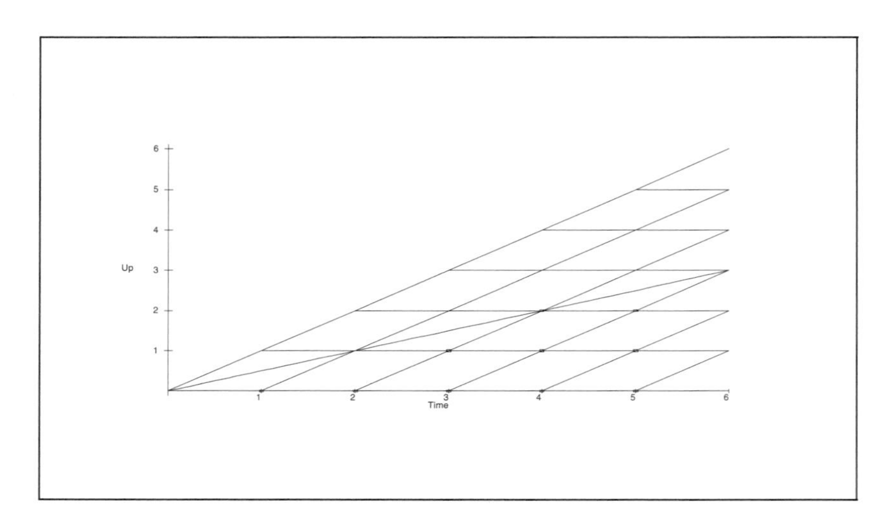

Figure 13.7: A Binomial Tree with a Box Superimposed on Nodes Where Early Exercise Is Optimal

The nodes at which it is optimal to exercise the option are presented below and the binomial tree with these nodes superimposed is displayed in Figure 13.7.

```
> OptExer;
```

$$[[2, 0], [3, 0], [4, 0], [4, 1], [5, 0], [5, 1]]$$

```
> BinTree(6,OptExer);
```

13.5 Concluding Remarks

This Chapter extended the one-period binomial model introduced in Chapter 2 to a multiperiod setting. We have investigated the equivalence of the no-arbitrage condition and the existence of SDFs in this multiperiod framework. The reader is prepared to think in terms of a multiperiod setting. The Chapter also set up the guidelines of valuation in a recursive manner within the binomial model. Valuing in a recursive manner facilitates a numerical solution to the value of American options. We have employed the following guidelines.

- We prove in a symbolic manner that even within the multiperiod model, valuing by replication is equivalent to valuing with the SDF or with the risk-neutral probability,
- We recover the values of the SDF although the model appears to be set in an incomplete market

Numerical valuation of puts and calls within the realm of the binomial model was also introduced. We are still missing the "linkage" between the valuation within the binomial model and as the discounted expected value, where the risk-neutral distribution is assumed to be lognormal. This along with some other issues will be discussed in the next Chapter.

13.6 Questions and Problems

Problem 1. Prove Theorem 11.

Problem 2. Consider the example of a five-period binomial tree on page 514 where we value, as of time zero, a dollar contingent on the state $(5, 3)$. Prove that a dollar contingent on $(5, 3)$ must be equivalent to the sum of the claims in (1) and (2) defined on the same page.

Problem 3. Verify via the replication approach that the value of $d(5, 3)$ is as stated in equation (13.8).

Problem 4. The measure Δ was defined in Section 7.2 as the derivative (calculus) of the price of an option with respect to the value of the underlying asset.

1. Explain why the value of Xs, in the discrete-time setting of the binomial model, is the counterpart of Δ — the sensitivity measure of the derivative security to changes in the price of the underlying asset.
2. What, in the binomial framework, will be your estimates of Γ, the second derivative (calculus) of the price of an option with respect to the value of the underlying asset?

Problem 5. Write a recursive procedure in MAPLE that values American options.

Problem 6. Consider a European call option that matures in a year in a market characterized by a risk-free rate and Up and Do factors as in Section 13.4.1. Assume that model as described in the text but with taxes on capital gains at a rate of $\tau\%$ payable at the maturity of the option. The taxes are paid on the difference between the price at which the option was purchased and the cash flow received at maturity. (Assume a negative tax is a tax rebate, received from the government.)

1. Derive the no-arbitrage pricing of a call option in the above environment.
2. Write a MAPLE procedure that values such options.

Problem 7. Prove that equations (13.14) and (13.15) are equivalent.

Problem 8. Verify that the solution of the system of equations in (13.22) is as reported in the text. Calculate the value of a European call option where the payoff at maturity is given by $\max(S_{\text{max}} - K, 0)$, where S_{max} is the maximum price of the stock during the life of the option. Such an option is called a path-dependent option.

Problem 9. Assume the parameters as in Section 13.4.1 and calculate the value of a chooser option, that is, at maturity the holder can decide if the option was a put or a call. Write a MAPLE procedure that values such options.

Chapter 14

Binomial Models II

This Chapter sets out to accomplish three main goals. The first part of the Chapter completes our treatment of the binomial model. This is where the linkage between the Black–Scholes formula, and the binomial model is explored. The established linkage sheds new light on the Black–Scholes formula, presenting it using limiting arguments when Δt approaches zero. The binomial model does so without requiring mathematics of the stochastic calculus[1] type. Numerical applications of the binomial model along with the effect of dividends are dealt with in this first part as well.

The Chapter then continues and utilizes the realm of the binomial model to investigate other issues which in a more general setting require "heavier mathematics". In the multiperiod binomial model both the states of nature and the discrete-time spaces are finite. Yet, this model facilitates the introduction ("conveys the feel") of many of the properties that are satisfied by sophisticated models, i.e., those in which both the time space and the state space are continuous. We take advantage of the binomial setting to explore the difference between futures and forward contracts and the Martingale property of the price process and its equivalency to the no-arbitrage condition. The final part of the Chapter prepares the reader for the continuous-time model investigated in the next Chapter. It does so by motivating the Brownian motion as a limit of a random process in the binomial model setting.

[1] This type of argument will be introduced and utilized for such a derivation in the next Chapter.

14.1 Binomial Model and the Black–Scholes Formula

Until now we have not explained how to choose the parameters Up and Do of the binomial model. In particular, we did not address the issue of the price process in the binomial setting. If we believe that the distribution of the price process is lognormal, as was alluded to in Section 5.7, this impacts the choice of the parameters in a binomial model.

The guiding principle in choosing these parameters is based on the Central Limit Theorem. The binomial model was presented in Section 13.2 as a consequence of the discretization of the time space and the price process. The Central Limit Theorem directs us in choosing the parameters in such a way that when the grid is refined, the distribution of the price in the binomial model converges to the lognormal distribution.

This theorem does not uniquely determine the parameters. One possibility[2] for the parameters is presented in the next subsection. This subsection also helps us to visualize the reason the Black–Scholes formula can be obtained as the limit of the option pricing formula in the binomial setting.

14.1.1 Binomial vs. Lognormal

Recall our comment in the paragraph preceding equation (13.5) regarding certain models[3] in which the Up and Do factors are specified as $Up = e^u$ and $Do = e^v$, where $u > 0$ and $v < 0$. In such a model, the distribution of the price process in T years (our units of time) is given by

$$Se^{\sum_{i=0}^{N-1} B_i}.$$

In the above equation the length of each period in the binomial model is $\Delta t = \frac{T}{N}$, and thus $N = \frac{T}{\Delta t}$. S is the current stock price, and B_i is a random variable taking the values u and v with (risk-neutral) probabilities qu and $1 - qu$, respectively. The probability qu is defined in equation (13.12).

Let us denote the exponent of the expression above by

$$Y_N = \sum_{i=0}^{N-1} B_i. \tag{14.1}$$

[2] Another possibility is pointed out in Section 14.4 where Brownian motion is revisited in the realm of the binomial model.

[3] This derivation follows the one in [27]. See also [15].

Y_N so defined is the continuously compounded rate of return spanning the time interval $[0, T]$. It is apparent from the above equation that Y_N is the sum of N random variables B_i, $i = 0, ..., N-1$. Furthermore, given the specification of the binomial model as a sequence of one-period binomial models "joined together" and the specification of the risk-neutral probability in each of the one-period models, e.g., equation (13.6), the random variables B_i, $i = 0, ..., N-1$, are independent and identically distributed. The expected value and the variance of each B_i, $0 \leq i \leq N-1$, are given by equations (14.2) and (14.3), respectively.

$$E(B_i) = qu\,u + (1 - qu)v \tag{14.2}$$

$$Var(B_i) = qu\,u^2 + (1 - qu)v^2 - (qu\,u + (1 - qu)v)^2\,. \tag{14.3}$$

The explicit expressions for $E(B_i)$ and $Var(B_i)$ are given by the MAPLE commands below.

```
> E(B[i])=collect(qu*u+(1-qu)*v,u);
```

$$\mathrm{E}(B_i) = qu\,u + (1 - qu)\,v$$

```
> Var(B[k])=factor(qu*u^2+(1-qu)*v^2-(qu*u+(1-qu)*v)^2);
```

$$\mathrm{Var}(B_k) = qu(1 - qu)\,(u - v)^2$$

Since the B_i, $i = 0, ..., N-1$, are independent and identically distributed random variables, the expected value and variance of Y_N are given by equations (14.4) and (14.5), respectively.

$$E(Y_N) = \sum_{i=0}^{N-1} E(B_{i.}) = N\,(qu\,u + (1 - qu)\,v) \tag{14.4}$$

$$Var(Y_N) = \sum_{i=0}^{N-1} Var(B_i) = Nqu\,(1 - qu)\,(u - v)^2\,. \tag{14.5}$$

Hence, as N approaches infinity, i.e., as $\Delta\, t$ approaches zero, the distribution of Y_N, by the Central Limit Theorem, approaches the normal distribution. (Appendix 14.7 elaborates somewhat on this limiting process.) Based on our specifications of the price process given in equation (5.30), i.e.,

$$E(Y_t) = \mu t,$$

and equation (5.31), i.e.,

$$Var(Y_t) = \sigma^2 t,$$

both the expected value and the variance are linear functions of t. The risk-neutral distribution imposes a constraint on μ which we suppress for now.

We would like, therefore, the expected value in equation (14.4) to be μT and the standard deviation in equation (14.5) to be $\sigma\sqrt{T}$. The parameters μ and σ are the expected value and the standard deviation of the rate of return per unit of time (a year).

One way of satisfying these requirements[4] is by constraining the expected value and variance of Y_N to satisfy equations (14.6) and (14.7),

$$\mu T = N\left(qu\, u + (1 - qu)\, v\right) \tag{14.6}$$

$$\sigma^2 T = N qu\,(1 - qu)\,(u - v)^2\,, \tag{14.7}$$

for which

$$u = \mu\,\frac{T}{N} + \sigma\sqrt{\frac{T}{N}}\sqrt{\frac{1-qu}{qu}} \text{ and } v = \mu\,\frac{T}{N} - \sigma\sqrt{\frac{T}{N}}\sqrt{\frac{qu}{1-qu}} \tag{14.8}$$

is a solution, as demonstrated by MAPLE below:

```
> u:='u':  sigma:='sigma':  T:='T': qu:='qu':  N:='N':
> GenSol:=solve({mu*T=N*(qu*u+(1-qu)*v),\
> sigma^2*T=N*qu*(1-qu)*(u-v)^2},{u,v}):
> assign(GenSol);
> simplify(convert(u,radical));
```

$$\frac{\mu\,T\,qu + \sqrt{-\sigma^2\,T\,N\,qu\,(-1+qu)}}{N\,qu}$$

```
> simplify(convert(v,radical));
```

$$\frac{\mu\,T\,qu - \mu\,T + \sqrt{-\sigma^2\,T\,N\,qu\,(-1+qu)}}{N\,(-1+qu)}$$

```
> simplify(N*(qu*u+(1-qu)*v));
```

$$\mu\,T$$

```
> simplify(N*qu*(1-qu)*(u-v)^2);
```

$$\sigma^2\,\mathrm{T}$$

[4]Another way is by ensuring that as N approaches infinity and Δt approaches zero, the right-hand sides of equations (14.6) and (14.7) converge to μT and $\sigma^2 T$, respectively. In the solution demonstrated in the text this is true for every N.

It is therefore apparent that, as N approaches infinity, for each choice of qu such that $0 < qu < 1$, the expected value of the expression in equation (14.1) converges to μT and its standard deviation converges to $\sigma\sqrt{T}$. For simplicity qu is chosen to be $\frac{1}{2}$ and is substituted in equation (14.8). This yields (see the MAPLE commands below) a solution for u and v as specified in equations (14.9) and (14.10) below.

$$u = \frac{\mu T}{N} + \frac{\sigma\sqrt{T}}{\sqrt{N}} = \mu\Delta t + \sigma\sqrt{\Delta t} \tag{14.9}$$

$$v = \frac{\mu T}{N} - \frac{\sigma\sqrt{T}}{\sqrt{N}} = \mu\Delta t - \sigma\sqrt{\Delta t}. \tag{14.10}$$

```
> qu:=1/2;
> simplify(convert(u, radical));
```

$$\frac{\mu T + \sigma\sqrt{TN}}{N}$$

```
> simplify(convert(v, radical));
```

$$\frac{\mu T - \sigma\sqrt{TN}}{N}$$

```
> u:='u':  sigma:='sigma':  T:='T': qu:='qu':  N:='N':
```

If we would like the distribution of the price in the binomial model to approach the risk-neutral distribution, then by Proposition 9 (page 210) and equation (5.42), μ in equations (14.9) and (14.10) should be chosen to be $r - \frac{\sigma^2}{2}$. To visualize how the distribution of the stock price in the binomial model converges to the lognormal (risk-neutral) distribution, we make use of the procedure **Binomvsln**.

The parameters of the procedure are, in order, N, T, S, σ, r, and Op, where N, T, S, and σ are as defined above, and r is the annual continuously compounded risk-free rate. When $Op = 1$, the cumulative lognormal distribution is plotted together with the binomial cumulative distribution. If Op is set to zero then the density of the lognormal distribution is plotted together with the density of the binomial approximation.

To demonstrate the convergence of the price process, Figure 14.1 displays the density function of the binomial approximation beside that of the lognormal distribution for a maturity of one year when $N = 10$. The lower graph is that of the lognormal distribution and the upper graph is the binomial approximation.

```
> Binomvsln(10,1,100,.18,0.10,0);
```

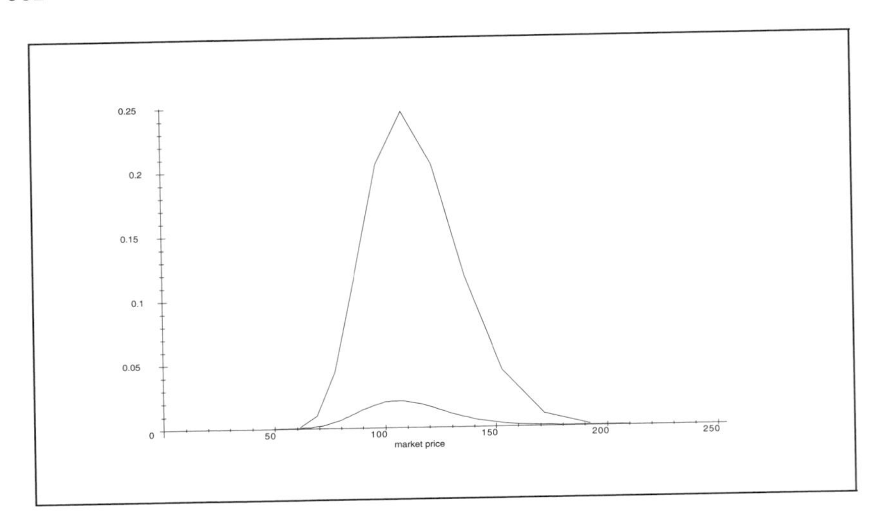

Figure 14.1: A Binomial Approximation to a Lognormal Density Function for $T = 1$ and $N = 10$

Figure 14.2 demonstrates the approximation for the same parameters, but with $N = 50$. Again the upper graph is the binomial approximation and the lower graph is the lognormal one. Note the scale on the y-axis, which implies that the approximation is quite good.

```
> Binomvsln(50,1,100,.18,0.10,0);
```

The quality of the approximation can be judged more clearly when the cumulative distributions are plotted, as in Figure 14.3. The step graph is obviously the cumulative distribution of the binomial approximation, while the smooth graph is the cumulative distribution of the lognormal distribution. The lowest graph around the x-axis is the difference between the two distribution functions.

```
> Binomvsln(50,1,100,.18,0.10,1);
```

14.1.2 Numerical Implementations

Assume that the parameters of the binomial approximation are specified by equations (14.9) and (14.10). Under the risk-neutral probability, if the price of the stock at time t is S, then the price of the stock at time T, where

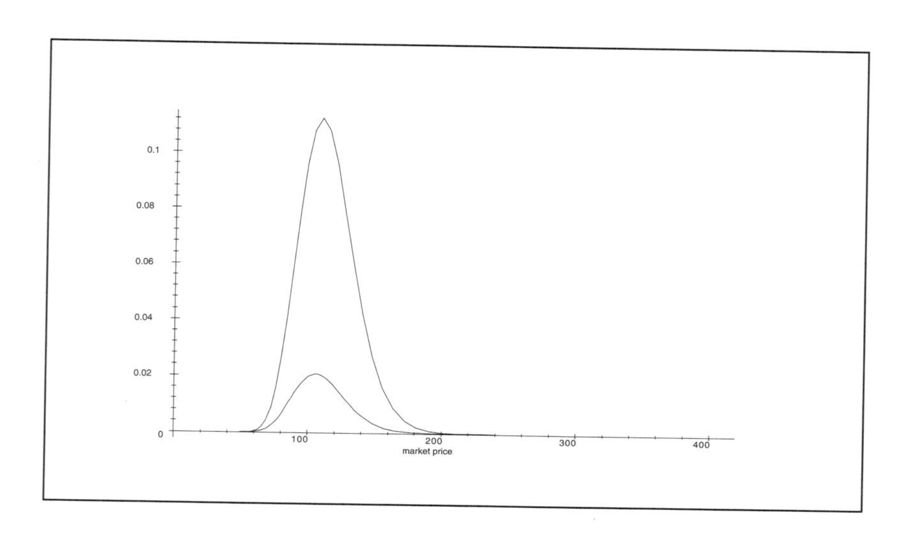

Figure 14.2: A Binomial Approximation to the Lognormal Density Function for $T = 1$ and $N = 50$

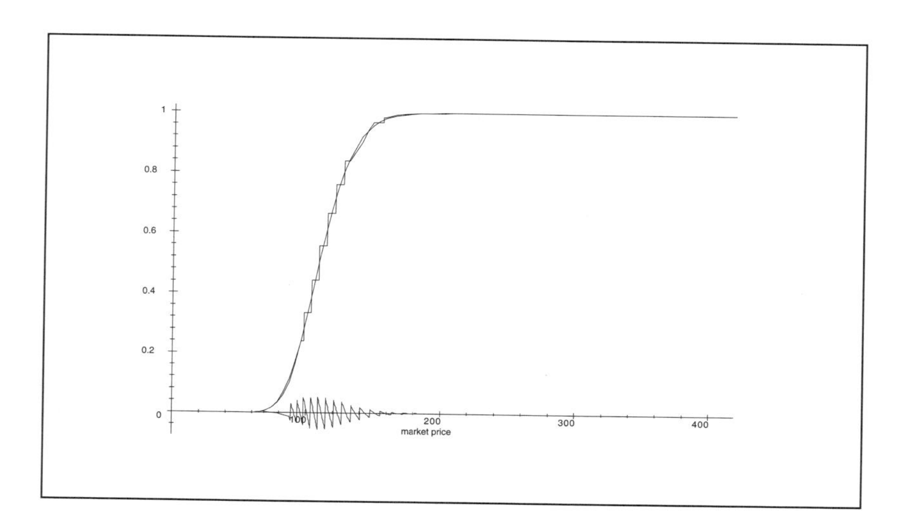

Figure 14.3: A Binomial Approximation to a Lognormal Cumulative Distribution Function for $T = 1$ and $N = 10$

$\Delta t = \frac{T-t}{N}$, can take $N+1$ different values given by

$$S_N(j) = S\, e^{\left(\left(r - \frac{\sigma^2}{2}\right)\Delta t + \sigma\sqrt{\Delta t}\right) j} e^{\left(\left(r - \frac{\sigma^2}{2}\right)\Delta t - \sigma\sqrt{\Delta t}\right)(N-j)}, \tag{14.11}$$

for $j = 0, \ldots, N$. Here $j \sim B(N, \frac{1}{2})$ and thus

$$Pr\left(S = S_N(j)\right) = \frac{N!}{j!(N-j)!}\left(\frac{1}{2}\right)^N.$$

The convergence, as N approaches infinity, of the binomial variable $S_N(j)$ to a random variable $S(T)$ with a lognormal (risk-neutral) density function,

$$f_{risk-neutral}\left(S(T)\right) = \frac{e^{\left(-\frac{\left(ln\left(\frac{S(T)}{S}\right) - \left(r - \frac{\sigma^2}{2}\right)(T-t)\right)^2}{2\sigma^2 (T-t)}\right)}}{S(T)\sqrt{2\pi\sigma^2(T-t)}}, \tag{14.12}$$

is conveyed by Figures 14.1, 14.2, and 14.3.

Having settled the choice of the parameters we can proceed to the valuation of a European option. A European call option with an exercise price of K is contingent on the price of an underlying asset $S(T)$ at time T, and is valued by equation (14.13). We were first introduced to this valuation method by equation (5.7) in Section 5.2.

$$\begin{aligned} & e^{-r_f (T-t)} E_{risk-neutral}\left(\max\left(0, S(T) - K\right)\right) \\ = \; & e^{-r_f (T-t)} \int_0^\infty \max\left(0, s - K\right) f_{risk-neutral}(s) ds. \end{aligned} \tag{14.13}$$

In words, the value of the call option is the discounted expected value of its future payoff, $\max(0, S(T) - K)$, with respect to $f_{risk-neutral}$ — the risk-neutral distribution of $S(T)$. In the setting of the binomial model, equation (13.14) is the counterpart of equation (14.13). The expected value, under the risk-neutral probability in this setting, is given by equation (14.14),

$$\begin{aligned} & E_{Bin}\left(\max\left(0,\, S_N(T) - K\right)\right) \\ = \; & \sum_{j=0}^{N} \frac{N!}{j!(N-j)!}\left(\frac{1}{2}\right)^N \max\left(0,\, S_N(j) - K\right), \end{aligned} \tag{14.14}$$

where $S_N(j)$ is as defined in equation (14.11).

An intuitively appealing property[5] is that as N approaches infinity and the distribution of $S_N(T)$ approaches the lognormal distribution,

$$E_{Bin}\left(\max\left(0, S_N(T) - K\right)\right)$$

approaches

$$E_{risk-neutral}\left(\max\left(0, S(T) - K\right)\right).$$

As demonstrated in Chapter 6, the integral in equation (14.13) is the Black–Scholes formula for a European call option. Thus the Black–Scholes formula for a European call value can be viewed as the limit, as N approaches infinity, of the value of such an option in the binomial model.

As N approaches infinity, the number of possible values for the underlying asset at maturity approaches infinity. At the same time the length of each period in the binomial model approaches zero. Consequently, the binomial model converges to the more realistic continuous-time model.

To further illustrate this property, the reader may like to execute the commands below. These commands calculate the difference between the price of a call based on the lognormal distribution and that based on the binomial approximation for increasing values of N. The procedure **Valbin** values a European contingent claim based on the binomial approximation, equation (13.14). The parameters in order are:

- N, the number of the periods in the binomial model
- G, a function describing the payoff at maturity, where G is explicitly defined in the input field (see the example below)
- T, time to maturity in years
- S, the current price of the stock
- σ, the standard deviation of the stock price (expressed as a percentage per year)
- r, the continuously compounded risk-free rate

---

[5] Under certain regularity conditions, given a function g and a random variable X_N with a density function h_N such that $\lim_{N\to\infty} h_N = h$, we have that $\lim_{N\to\infty} E_{h_N}\left(g(x)\right) = E_h\left(g(x)\right)$. A consequence of this property is the convergence of the binomial valuation to the Black–Scholes formula. This also can be proven directly; see [15] for the case at hand where h_N is the binomial distribution with N nodes and h is the normal distribution.

Thus, for example, the value of a European call with an exercise price of \$80, where $N = 30$, $T = \frac{3}{12}$, $S = 100$, $\sigma = 0.26$, and $r = .09$, is given by

```
>  Valbin(30, x->max(0,x-80), 3/12, 100, 0.25, 0.09);
```

21.87734722

Its value based on the Black–Scholes formula is given by the command below.

```
>  Bs(80,3/12,100,.25,.09);
```

21.88274720

To view the sequence of the price differences between the binomial and the Black–Scholes formulas for $N = 100, ..., 120$, execute the following command:

```
>  seq(Valbin(N,S->max(0,S-80),3/12,100,0.25,0.09)-\
>  Bs(80,3/12,100,.25,.09),N=100..120);
```

−.00023636, −.00275538, −.00030069, −.00185035, −.00059658, −.00114506, −.00111996, −.00063780, −.00186635, −.00032665, −.00283099, −.00020929, −.00241566, −.00028321, −.00164660, −.00054550, −.00104132, −.00099314, −.00059864, −.00162285, −.00031697

We can also visualize such a convergence by plotting the value of the binomial approximation against N, together with the value of the Black–Scholes formula. This is displayed in Figure 14.4 for $N = 20, ..., 50$,where the flat line is the value based on the Black–Scholes formula.

```
>  plotvalbin:=plot([seq([N,Valbin(N,S->max(0,S-80),\
>  3/12,100,0.25,0.09)],N=20..50)],color=green):

>  plotvalBs:=plot(Bs(80,3/12,100,.25,.09),\
>  N=20..50,color=red):

>  plots[display](plotvalbin,plotvalBs, \
>  labels=[N,Value],thickness=2,title= \
>  "Black-Scholes vs.  Binomial");
```

The reader may use the procedure **Valbin** to value other types of contingent claims based on the binomial approximation. This is done by changing the definition of the payoff function at maturity on the input line. For example, to value a claim that pays \$10 if the price of the stock is above \$100 and zero otherwise (with $N = 30$), execute the following command.

```
>  Valbin(30,x->piecewise(x>100,10),3/12,100,0.25,0.09);
```

5.595007650

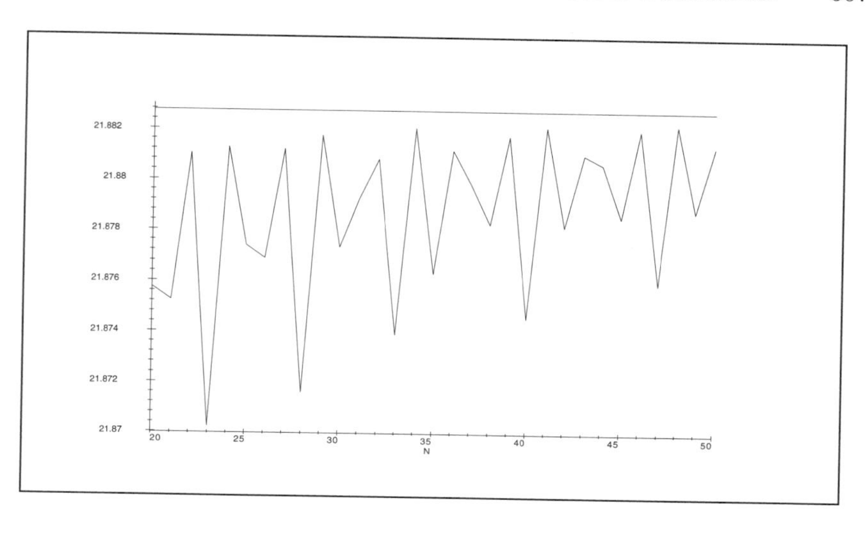

Figure 14.4: The Black–Scholes Value and the Binomial Approximation vs. N

The above discussion also serves as the rationale for using binomial models in order to solve numerically for the value of an American option. The procedure is much the same as described in Section 13.3. One need only change the value of the parameters based on equations (14.9) and (14.10). The procedure **AmerPut** calculates the numerical value of an American put option. The requested parameters in order are:

- N, the number of time periods used in the binomial approximation
- K, the exercise price of the option
- T, years to maturity
- S, the price of the underlying asset at initiation
- σ, the standard deviation of the underlying asset (in percentage per year)
- r, the continuously compounded risk-free rate
- Path, optional parameter used to store the coordinates at which it was optimal to exercise the option

Having specified the "Path" parameter in the **AmerPut** procedure, the reader can generate a graph of the binomial tree with boxes superimposed on the nodes where it was optimal to exercise the option. As a rule of thumb, dividing a year into 40 periods generates an accurate numerical value. Below we compare the price of an American put option valued using $N = 12$, with its value using $N = 40$. Figure 14.5 displays the nodes at which the option should be exercised for the case $N = 40$.

```
> AmerPut(12,40,1,40,.18,.10);
                    1.629171720
> AmerPut(40,40,1,40,.18,.25,opi);
                    .9231089931
> BinTree(40,opi);
```

Note the shape of the set of the superimposed boxes. It suggests that if at time t it is optimal to exercise the put option when the underlying price is S, then it will be optimal to exercise it

1. at t when the underlying price is smaller than S, and
2. at any time after t if the underlying price is smaller than or equal to S.

One of the exercises at the end of the Chapter asks the reader to prove these properties.

The upper envelope of the (blue) superimposed region in Figure 14.5 is referred to as a free boundary. If the price of the stock at time t is above this price, the option should not be exercised and vice versa. It is called a "free" boundary, as it is unknown prior to the option being valued numerically. Had it been know at the outset, then an analytical solution would have been possible. In order to determine the boundary, the value of the option at each time for every possible relation of the underlying asset's price must be known. The lack of knowledge of the free boundary is the cause for the "vicious circle" mentioned at the beginning of Section 13.4.4. We will touch upon the free boundary again in Section 15.8, where the valuation of options as a solution to partial differential equations will be explained briefly.

14.1.3 The Effect of Dividends

Incorporating the effect of dividends in the binomial model may cause some numerical complications. If the dividend is modeled as a constant continuous

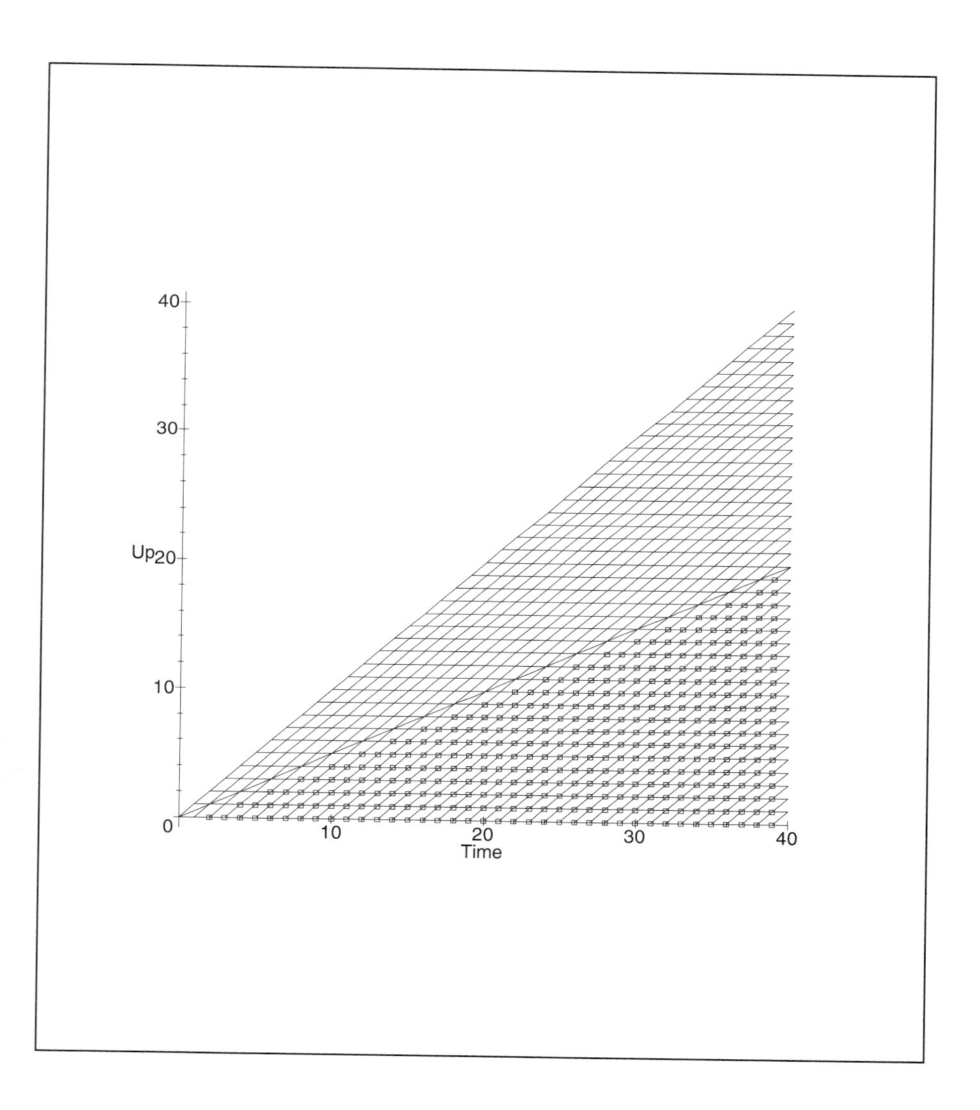

Figure 14.5: The Collections of (t, Up) Where It Is Optimal to Exercise Early an American Put Option

dividend yield δ, such that the dividend payment at time $i+1$ is

$$Di\,S(i) = \left(e^{\delta\,\Delta\,t} - 1\right) S(i),$$

then we require only[6] a slight modification to our existing methods. Utilizing the same arguments as in Section 6.4, the equations determining the one-period stochastic discount factors should now be

$$\begin{aligned} 1 &= du\,R + dd\,R \\ S(i) &= du\,(S(i)Up + DiS(i)) + dd\,(S(i)\,Do + S(i)Di) \end{aligned} \qquad (14.15)$$

instead of the equations in (13.6). Hence, the solutions for du and dd will be given as below.

```
> SolBin:=solve( {1=du*R+dd*R,\
> 1=du*(Up+Di)+dd*(Do+Di)},{du,dd});
```

$$SolBin := \{du = \frac{R - Do - Di}{R\,(Up - Do)},\ dd = -\frac{-Up + R - Di}{R\,(Up - Do)}\}$$

However, no other change needs to be made to the algorithm and equations (13.10) and (13.15) are still valid when dd and du are defined as above. If we wish to approximate the Black–Scholes formula in this setting we have to change the parameters of the risk-neutral probability, i.e., replace r with $r-\delta$. This follows from the same arguments as in Section 6.4, which implies that μ in equations (14.9) and (14.10) should be chosen to be $r - \delta - \frac{\sigma^2}{2}$.

Sometimes it is more reasonable to assume a one-time dividend Di, expressed as an absolute dollar amount rather than as a percentage of the price of the stock, paid at time τ. Let m be the time in the binomial tree such that $m\,\Delta\,t < \tau < (m+1)\,\Delta\,t$ and assume that at time m the price of the stock was S. Time $m+1$ is thus after the stock went ex-dividend and hence the stock price, as explained in Section 12.4, will be $(S - Di)Up$ or $(S - Di)Do$. Similarly, at each time $m+j$, S will be replaced by $S - Di$, which will induce a change in the value of the option and in the structure of the tree.

While conceptually[7] the recursive calculations are done as in the no-dividend case, after time m the tree does not recombine. The reader can verify that, by construction, the price of the stock at nodes (m,k) and

[6] In a continuous-time setting, where δ is the dividend yield, Di should be interpreted as $(e^{\delta\,\Delta\,t} - 1)S(i)$, which, for a small $\Delta\,t$, by Taylor sequence, is approximately $S(i)\delta\Delta t$.

[7] One also should decide how to interpret the case where the realized stock evolution is such that the dividend payment is larger then the stock price. Commonly this is interpreted as bankruptcy.

$(m, k+1)$ satisfies $S_m(k)Up = S_m(k+1)Do$ since the tree recombines. This means that $(S_m(k) - Di)\,Up$ will not equal $(S_m(k+1) - Di)\,Do$ and thus the tree will not recombine. Hence the number of nodes at time $m+1$ will be twice the number of nodes at time m instead of increasing by only one. Yet, the valuation concept as mentioned above will be intact.

There is a way to view the situation such that the tree will be recombining. This is obtained by separately considering the stock price process (not including dividends) along its up movements and along its down movements and the deterministic dividend payments. The price of the stock, at each node (i, j) such that $i\Delta t < \tau$, is then $SUp^j Do^{i-j}$ plus the present value of the dividend. One of the end-of-Chapter exercises directs the reader to calculate the price of a call option utilizing such an approach. This model can be extended to incorporate multiple dividend payments in quite a straightforward manner.

14.2 Risk-Neutral Probabilities and Price Processes

The first Chapter of the book (Theorem 3, Section 1.4.1) in a setting of a one-period model, established the equivalency between the no-arbitrage condition and the existence of a risk-neutral probability. In a multiperiod setting a similar property holds. However, the role of the risk-neutral probability is replaced with a stochastic process of the price of the stock at time t.

Once again we take advantage of the binomial setting to establish an equivalency between the no-arbitrage condition and a property of the price process in a multiperiod setting. To this end we should recall our interpretation of the binomial model as comprising many one-period binomial models "joined" together.

We can apply, to each of these one-period models, the result of Theorem 3. This is justified in light of Theorem 11 in Section 13.2.1 that ensures us of the no-arbitrage condition being satisfied in the multiperiod model if and only if it is satisfied in each of the one-period models. Hence, if the no-arbitrage condition holds, there exists a risk-neutral probability over the two states of nature in each of the one-period models. Let us recall how this probability distribution is identified.

Proceeding as we did in Section 2.2, we can normalize the stochastic discount factors solved for in equation (13.6) to generate this risk-neutral probability. The next MAPLE commands calculate the risk-neutral probability of an up movement $qu = \frac{du}{du+dd}$.

```
> SolBin:=solve( {1=du*R+dd*R,\
> S(j)=du*S(j)*Up+dd*S(j)*Do},{du,dd});
```

$$SolBin := \{du = \frac{R - Do}{R\,(Up - Do)},\ dd = \frac{Up - R}{R\,(Up - Do)}\}$$

```
> simplify(rhs(op(select(has,SolBin,du)))\
> /(rhs(op(select(has,SolBin,dd)))\
> +rhs(op(select(has,SolBin,du)))));
```

$$\frac{R - Do}{Up - Do}$$

The command above stipulates that qu, the risk-neutral probability of an up movement, is given by equation (14.16).

$$qu = \frac{R - Do}{Up - Do} \tag{14.16}$$

The probability of a down movement is thus $qd = 1 - qu$. (We would like to remind the reader that these are the risk-neutral probabilities and not the real-life probabilities.)

At time N there are $N+1$ possible states of nature. We can calculate the risk-neutral probabilities of these states given qu. To this end, recall that for the evolution of the price to terminate at (N, j) for $j = 0, ..., N$, there should have been j up movements and $N - j$ down movements. Furthermore, the end result will be (N, j) regardless of the order of these movements. The movements at time i and $i + 1$ are statistically independent of each other. Hence the probability of having the first j movements up and then $N - j$ movements down is the product $qu^j\,(1 - qu)^{N-j}$.

However, there are many possibilities of having j up movements and $N-j$ down movements. The probability of each will be equal to $qu^j\,(1 - qu)^{N-j}$ since a product is unaltered by the order of the multiplication. Thus, the probability of the true state being (N, j) will be the probability of having j up movements, in a certain order, multiplied by the number of possibilities of having j up movements out of the N movements. Since, as explained prior to equation (13.9), there are $\frac{N!}{j!(N-j)!}$ possibilities for j out of the N movements being up movements, the probability of having j up movements at time N is given by the expression in (14.17).

$$\frac{N!}{j!\,(N-j)!} qu^j\, qd^{N-j} \tag{14.17}$$

This is of course the specification of a binomial random variable and, perhaps, one of the reasons for the name “binomial trees.”

The group of MAPLE commands below defines a matrix wherein[8] the probability of having j up movements, $j = 0, ..., 6$, at time $i = 0, ..., 6$, is given by the element $(7 - i,\ j + 1)$. In MAPLE notation the expression $\frac{N!}{j!(N-j)!}$ in equation (13.9) is denoted as *binomial*(N, j).

```
> i:='i':j:='j':
> Pr:=matrix(7,7,0):
> for i from 0 to 6 by 1 do;
> for j from 0 to i by 1 do;
> Pr[6-j+1,i+1]:=binomial(i,j)*qu^(j)*(qd)^(i-j);
> od:od:
> eval(Pr);
```

$$\begin{bmatrix} 0 & 0 & 0 & 0 & 0 & 0 & qu^6 \\ 0 & 0 & 0 & 0 & 0 & qu^5 & 6\,qu^5\,qd \\ 0 & 0 & 0 & 0 & qu^4 & 5\,qu^4\,qd & 15\,qu^4\,qd^2 \\ 0 & 0 & 0 & qu^3 & 4\,qu^3\,qd & 10\,qu^3\,qd^2 & 20\,qu^3\,qd^3 \\ 0 & 0 & qu^2 & 3\,qu^2\,qd & 6\,qu^2\,qd^2 & 10\,qu^2\,qd^3 & 15\,qu^2\,qd^4 \\ 0 & qu & 2\,qu\,qd & 3\,qu\,qd^2 & 4\,qu\,qd^3 & 5\,qu\,qd^4 & 6\,qu\,qd^5 \\ 1 & qd & qd^2 & qd^3 & qd^4 & qd^5 & qd^6 \end{bmatrix}$$

The probability distribution of the number of up jumps is therefore the binomial distribution. It is denoted by $B(N, qu)$, where N is the number of time periods and qu is the probability of an up movement or a "success" in the usual terminology.

Assume that at time 0, the price of the stock is given as $S(0)$, and that the possible values (the sample space) of the price $S(k)$ at time k $(k > 0)$ are $S(0)Up^j Do^{k-j}$ for $j = 0, ..., k$. At the end of Section 5.4 we defined a stochastic price process. The price process $S(k)$ is a stochastic process that satisfies certain properties. The probability that $S(k) = S(0)Up^j Do^{k-j}$, $j = 0, ..., k$, is the probability that from time 0 to time k there will have occurred j up movements. This probability is given by $\frac{k!}{j!(k-j)!} qu^j qd^{k-j}$.

[8] We maintain the classical convention that (j, i) stands for row j and column i and we use a 7×7 matrix, given the page margin constraint. We choose to represent the probabilities and indexing in this particular way so it will coincide with our graphical representation, e.g., in Figure 13.2. The 1 in position $(7, 1)$ of the matrix might be interpreted as the probability that the state price equals its initial price at time zero. At time zero, of course, the price of the stock is deterministic. Note that in Figure 13.2 we use the notation (i, j), which corresponds to (j, i) here.

We thus conclude that the probability distribution of the price at some future time k depends only on the current price and not on the historical path the price has taken to reach the present time. A stochastic process that satisfies such a property is called a *Markovian process* and therefore $S(k)$ is such a process. Furthermore, there is another property that is satisfied by the price process $S(k)$.

The expected value of $S(k)$ given $S(0)$, denoted by $E\left(S(k) \mid S(0)\right)$, satisfies equation (14.18).

$$E\left(S(k) \mid S(0)\right) = \sum_{j=0}^{k} S\left(0\right) Up^j \, Do^{k-j} \frac{k!}{j!\left(k-j\right)!} qu^j \; qd^{k-j} \tag{14.18}$$

Recall that this expectation is being calculated based on the risk-neutral probability, as indeed qu and qd are risk-neutral probabilities. Under the risk-neutral probability, the expected value of $S(k)$ appreciates at the risk-free rate. Viewing the time interval $[0, k\Delta t]$ as a one-period model with a risk-free rate of rk implies, by equation (5.40), that

$$E\left(\frac{S\left(k\right)}{e^{rk}} \;\middle|\; S\left(0\right)\right) = S\left(0\right). \tag{14.19}$$

Hence, $E\left(S\left(k\right) \mid S\left(0\right)\right) = e^{rk} S\left(0\right)$ since at time 0 the interest rate spanning $[0, k\Delta t]$ is a deterministic constant, and can be moved outside of the expectation operator. The expected value of the price in k periods is thus its future value based on the risk-free rate spanning the k periods. Put differently, and in a more general way, we have that given two times $k_0 < k$,

$$E\left(\frac{S\left(k\right)}{e^{r\left(k-k_0\right)}} \;\middle|\; S\left(k_0\right)\right) = S\left(k_0\right), \quad k > k_0. \tag{14.20}$$

In words, equation (14.20) states that the expected value of the discounted price, at some future time k viewed from time k_0, is the prices $S(k_0)$.

There is another way to arrive at equation (14.19) from equation (14.18). It utilizes the structure of the binomial models as being a consequence of one-period models joined together and a law called "iterated expectations". As the name implies, it involves repeated application of the conditional expectation operator. Let us see how it works.

Approaching equation (14.19) from this angle enhances our understanding of the Markovian property. Namely, the price at some future time depends only on the current realization of the price process. Once the current price is known, the distribution of the future price process in k time periods is known too, and is given by

$$\Pr\left(S(k) = S(0)\,Up^j\,Do^{k-j} \;\middle|\; S(0)\right) = \frac{k!}{j!\,(k-j)!}qu^j qd^{k-j} \tag{14.21}$$

for $j = 0, ..., k$.

The price history prior to the current time is irrelevant. All relevant information is summarized in the current price. The probability distribution of the stock's price in the future will not be changed if the price of a stock at some past time, t $(t < 0)$, is given. That is,

$$\begin{aligned} &\Pr\left(S(k) = S(0)\,Up^j\,Do^{k-j} \;\middle|\; S(0), S(t)\right) \\ = \;&\Pr\left(S(k) = S(0)\,Up^j\,Do^{k-j} \;\middle|\; S(0)\right). \end{aligned} \tag{14.22}$$

The above relation is consistent with the concept of market efficiency. It defines the future price path as a function of new information rather than past price changes. There are a few definitions of market efficiency. The one that states that all relevant information (private and/or public) is summarized in the current price is known as the strong form of the efficient market hypothesis. The link of the property summarized in the above equation to market efficiency was also mentioned in Section 5.2, footnote 7.

Let us now see the other approach to obtaining equations (14.19) and (14.20). Consider first $E(S(k) \mid S(k-1))$. When $S(k-1)$ is given, the random variable $S(k)$ behaves like the price process of a one-period binomial model. If at time $k-1$ the price of the stock is $S(k-1)$, at time k its price will be either $S(k-1)Up$ or $S(k-1)Do$ with the appropriate probabilities. We do assume that the no-arbitrage condition is satisfied. Hence, based on equation (2.3), explored in our discussion of the one-period binomial model, we have that

$$E\left(\frac{S(k)}{e^r} \;\middle|\; S(k-1)\right) = \frac{S(k-1)\;e^r}{e^r} = S(k-1). \tag{14.23}$$

Note further that while in our model r is a deterministic constant, equation (14.23) is satisfied even in a binomial model in which the interest rate[9]

---

[9] At time $k-1$, the interest spanning the time interval $[(k-1)\,\Delta t, k\Delta t]$ will be realized. In the parlance of stochastic processes, such a process is said to be *predictable.* That is, given the set of information at time $k-1$, r is a deterministic constant while the price process $S(k)$ (or $S(i)$, $i = k, k+1, ...$) is a random variable, the distribution of which is known (as we mentioned before, $S(k)$ is measurable with respect to the information at time $k-1$). In view of this footnote it might be more appropriate to interpret the notation $E(S(m) | S(n))$ as the expectation of $S(m)$ given the information set at time n, $n < m$.

is stochastic. (See the exercises at the end of this Chapter.) Using the same argument as above,

$$E\left(\frac{S(k-1)}{e^{r}}\ \middle|\ S(k-2)\right)=S(k-2). \tag{14.24}$$

To explain how the iterated expectation "works" we call to our aid a result from probability theory that makes use of conditional expectations. Given two random variables X and Y, the expected value of some function $h(X)$ with respect to the distribution of X, $E_X(h(X))$, is equal to $E_Y(E_X(h(X)\mid Y))$. That is, the expected value of $h(X)$ with respect to X may be calculated in two steps: first, the expected value of X given Y, denoted by $E_X(h(X)\mid Y)$, is calculated, and then the expected value of $E_X(h(X)\mid Y)$ with respect to Y is calculated. That is,

$$E_Y(E_X(h(x)\mid Y))=E_X(h(X)). \tag{14.25}$$

Applying this result in our context we can write that for $k>m$

$$\begin{aligned}
&E_{S(k)}\left(\frac{S(k)}{e^{r(k-m)}}\ \middle|\ S(m)\right)\\
&= E_{S(k-1)}\left(E_{S(k)}\left(\frac{S(k)}{e^{r(k-m)}}\ \middle|\ S(m)\ \middle|\ S(k-1)\right)\right) \qquad (14.26)\\
&= E_{S(k-1)}\left(E_{S(k)}\left(\frac{S(k)}{e^{r(k-m)}}\ \middle|\ S(m),\ S(k-1)\right)\right).
\end{aligned}$$

However, in view of the Markovian property,

$$\begin{aligned}
&E_{S(k)}\left(\frac{S(k)}{e^{r(k-m)}}\ \middle|\ S(m),S(k-1)\right) \qquad (14.27)\\
&= E_{S(k)}\left(\frac{S(k)}{e^{r(k-m)}}\ \middle|\ S(k-1)\right).
\end{aligned}$$

The right-hand side of equation (14.27) is again embedded in a one-period binomial model. By the no-arbitrage condition we have that

$$E_{S(k)}\left(\frac{S(k)}{e^{r}}\ \middle|\ S(k-1)\right)=S(k-1). \tag{14.28}$$

Consequently, $E_{S(k)}(S(k)\mid S(k-1))=S(k-1)\,e^{r}$, and hence the right-hand side of (14.27) equals

$$E_{S(k)}\left(\frac{S(k)}{e^{r(k-m)}}\ \middle|\ S(k-1)\right)=\frac{S(k-1)}{e^{r(k-m-1)}}.$$

Therefore, for $k > l > m$ we obtain that

$$E_{S(k)}\left(\frac{S(k)}{e^{r(k-m)}} \,\middle|\, S(l)\right) = E_{S(k-1)}\left(\frac{S(k-1)}{e^{r(k-m-1)}} \,\middle|\, S(l)\right). \tag{14.29}$$

Hence, proceeding in the same manner, taking the expected value given $k-1$ and then $k-2$, etc., we obtain the desired result, namely, equation (14.20). The reader might like to take account of this structure since it helps with the understanding of the more complicated model discussed in the next Chapter. In addition, a very similar argument facilitates calculating the numerical value of an American option.

Equation (14.20) can be written in a slightly different way. Noting that $e^{r(k-k_0)} = \frac{e^{rk}}{e^{rk_0}}$ and rearranging yields

$$E\left(\frac{S(k)}{e^{rk}} \,\middle|\, S(k_0)\right) = \frac{S(k_0)}{e^{rk_0}}. \tag{14.30}$$

Note that we could have moved e^{rk_0} outside the expectation operator since we assumed a deterministic interest rate. Equation (14.20) written in this form describes the process $\frac{S(k)}{e^{rk}}$, i.e., the process of the "relative price".

Consider a dollar invested at time zero for one period to yield e^r, which is then reinvested, at the same rate, for another period, and so on. At the end of the ith period the value of the investment will be e^{ri}. The process $\frac{S(i)}{e^{ri}}$ is the value of the stock, relative to the value of a dollar invested in the risk-free rate at time zero for one period, rolled over a period at a time, until time i. (In a deterministic term structure environment, investing a dollar from time zero to time i, or rolling it over each period, generates the same ultimate value. This is not the case if interest rates are stochastic.) We therefore see that at each point of time the expected value of the relative price process, at some future time, is its current value.

A stochastic process $v(t)$ satisfying the property $E(v(t_1) \mid v(t_0)) = v(t_0)$, where $t_1 > t_0$, is called a Martingale. The following theorem is not hard to verify, and thus its proof is left for the reader.

Theorem 12 *In the context of the binomial model the no-arbitrage condition is satisfied if, and only if, the relative price process is a Martingale.*

This claim is correct even in a more general setting, e.g., in a nonbinomial environment when the interest rate process is also stochastic. An exercise at the end of this Chapter guides the reader in proving this result in the context of the binomial model when the interest rate process is also stochastic.

One may view the random source behind the price process as being the continuously compounded rate of return. This is the manner in which we presented the price process in a setting where time was continuous. Making reference to the stochastic process of the continuously compounded rate of return facilitates a comparison between the discussion here and that of the rate of return $Y(t)$ in Section 5.4. It thereby prepares the ground for our investigation of a continuous-time model in Chapter 15.

The up and down movements over two disjoint time intervals are statistically independent. Thus the stochastic process

$$Y_d\left(i\right) = \frac{S\left(i\right)}{S\left(0\right)} \tag{14.31}$$

depends on i, the time, only via the length of the time interval $[0, i\Delta t]$. $Y_d(i)$ so defined is one plus the rate of return over the time interval $[0, i\Delta t]$, and its sample space is $Up^j Do^{i-j}$, $j = 0, ..., i$. Consequently, the continuously compounded rate of return

$$Y\left(k\right) = \ln\left(\frac{S\left(k\right)}{S\left(0\right)}\right), \text{ for } k > 0, \tag{14.32}$$

is also dependent on time only via the length of the time interval $[0, k\Delta t]$.

Let us define by $Y\left(k_1, k_2\right)$ the continuously compounded rate of return over the interval $[k_1\Delta t, k_2\Delta t]$. Given the time intervals $[k_1\Delta t, k_2\Delta t]$ and $[k_2\Delta t, k_3\Delta t]$, where $k_1 < k_2 < k_3$, the random variables $Y\left(k_1, k_2\right) = \ln\left(\frac{S(k_2)}{S(k_1)}\right)$ and $Y\left(k_2, k_3\right) = \ln\left(\frac{S(k_3)}{S(k_4)}\right)$ are independent. If $k_2 - k_1 = k_3 - k_2$ then $Y(k_1, k_2)$ and $Y(k_2, k_3)$ are also identically distributed random variables. This is exactly Assumption III made in Section 5.4 for the rate of return $Y(t)$. The only difference is that there the time space was continuous whereas here it is discrete.

It is easy to verify that by equation (14.31) the process $Y(k)$ is a Markovian process. Furthermore, consider the increments of the stochastic process $Y(k)$. The increment of a process is defined as the change in the process value from some time k_1 to time k_2, $k_1 < k_2$. That is, given an increasing sequence of time $k_1 < k_2 < k_3 < ... < k_n$, the elements of the sequence $Y(k_1) - Y(k_2), Y(k_3) - Y(k_2), ..., Y\left(k_n\right) - Y\left(k_{n-1}\right)$ are increments of the process $Y(k)$. Based on equations (14.31) and (14.32), these increments are independent. Hence, the process $Y(k)$ has independent increments.

Furthermore, these equations imply that the changes (increments) in the price process over two disjoint time intervals of equal length are identically distributed and depend only on the length of the interval. This last property is the definition of a process with stationary increments. (See, for example, [40].) That is, the changes in the process from a time k_1 to a time k_2, where $k_1 < k_2$, and from a time k_4 to a time k_5 $(k_4 < k_5)$, where $k_5 - k_4 = k_2 - k_1$, are independent, identically distributed (i.i.d.) random variables.

The fact that $Y(k)$ is a stochastic process that possesses independent and stationary increments allows us to write $Y(k)$ as the sum of i.i.d. random variables. That is, consider $Y(t_N)$ and the grid of the time interval $[0, T]$ as defined by equations (13.1) and (13.2). We can thus define

$$\Delta Y_k = Y(t_k) - Y(t_{k-1}) \text{ for some } 0 < k \leq N \tag{14.33}$$

and then write $Y(t_N)$ as its value at time zero plus the aggregate of the increments as it is presented in equation (14.34).

$$Y(t_N) = Y(t_0) + \sum_{i=1}^{N} \Delta Y_{t_i} \tag{14.34}$$

Equation (14.34) plays a major role in deriving the Black–Scholes formula via limiting arguments starting from a binomial model with a finite number of periods. This derivation is explored in Appendix 14.7. The next section is devoted to the valuation mechanics in the framework of the binomial model.

14.3 Futures and Forwards: A Symbolic Example

Within the realm of the binomial model it is easier than in a continuous-time setting to take a look at the difference between a forward and a futures contract. We already know, based on our discussion in Section 10.6, that in a deterministic term structure environment, the prices of these two contracts are the same. Let us now consider a symbolic example of a two-period binomial tree, where the interest rate is stochastic. We will keep the assumption about the evolution of the price process as in Section 13.1. That is, the same factors Up and Do are applied to the underlying asset at each node (i, j).

We will ensure, of course, that the no-arbitrage condition is satisfied. Based on Theorem 11 (page 512), we need only worry about the satisfaction of the no-arbitrage condition in each of the one-period models which compose the multiperiod binomial model. However, in order to analyze this example we must resort to a non-recombining tree. This will be apparent as the example unfolds.

In a context of a two-period binomial tree, a non-recombining tree means that what used to be node $(2,1)$ in a recombining tree (see Figure 14.6) is now represented by two nodes. The evolution of uncertainty of an up movement followed by a down movement will result in a different state of nature than the evolution of a down movement followed by an up movement.

Since we keep the assumption that the same factors Up and Do apply to the underlying asset at each node, the evolution of an up movement followed by a down movement will result in the same stock price as a down movement followed by an up movement. However, since the interest rate is assumed to be stochastic, as the example unfolds it will be apparent that the forward (and futures) price at time one will not be the same at nodes $(1,1)$ and $(1,0)$.

The cash flows resulting from such contracts in node $(2,1)$ will depend on the preceding node. The evolution $(1,1)$ followed by $(2,1)$ will generate a different cash flow than the evolution of $(1,0)$ followed by $(2,1)$. Therefore knowing that until time 2 there was only one up movement (i.e., the information implicit in the notation $(2,1)$) does not convey all the necessary information. Knowing only the current state, e.g., $(2,1)$, is not sufficient to determine the payoff from all the securities in the market. Hence, there is a need to split $(2,1)$ into two nodes that correspond to the two possible evolutions.

We split what used to be node $(2,1)$ into two nodes: one resulting from an up movement followed by a down movement, and the other resulting from a down movement followed by an up movement. In non-recombining models, it is convenient to label the nodes with a sequence of zeros and ones, where one corresponds to an up movement and zero to a down movement. Since we are only analyzing a two-period model we will label the nodes that correspond to the up – down movement by $(2,1.5)$ and the down – up movement by $(2,0.5)$. These labels correspond to the coordinates of the boxes in Figure 14.6 and replace the node $(2,1)$. The MAPLE command below generates Figure 14.6.

```
> BinTree(2,[[2,.5],[2,1.5]]);
```

Let us denote the one-period interest rate prevailing at time 0 by $r(0)$ and the one-period interest rate prevailing from time i to time $i+1$ at node (i,j) by $r(i,j)$. Recall that the interest rate process is such that at each node the term structure is known. Hence at time zero the rate prevailing from time zero to time two is known as well. (See also footnote 9 in this Chapter.) The stochastic discount factor at time zero for funds obtained at time one is given by the solution to equation (13.6) when R is replaced with

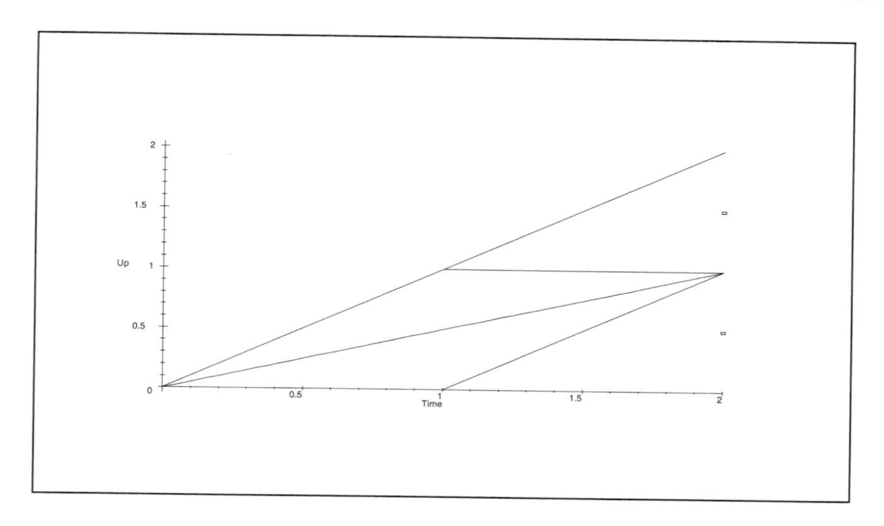

Figure 14.6: A Non-recombining Binomial Tree

$R0 = 1 + r(0)$. Hence,

$$du = \frac{R0 - Do}{R0\,(Up - Do)}, \quad dd = \frac{Up - R0}{R0\,(Up - Do)}, \tag{14.35}$$

and by no-arbitrage arguments, $R2$, one plus the interest rate prevailing from time 0 to 2, must satisfy

$$\frac{1}{R2} = \frac{du}{R\,(1,1)} + \frac{dd}{R\,(1,0)}, \tag{14.36}$$

where $R\,(1,1) = 1 + r\,(1,1)$ and $R\,(1,0) = 1 + r\,(1,0)$. Therefore, $R2$ is given by the output of the next MAPLE command.

```
>  solve({1/R2 = ((R0-Do)/(R0*(Up-Do))/R(1,1))\
>  +((Up-R0)/(R0*(Up-Do))/R(1,0))},{R2});
```

$$\left\{R2 = \frac{R0\,(-Up + Do)\,\mathrm{R}(1,\,1)\,\mathrm{R}(1,\,0)}{-\mathrm{R}(1,\,0)\,R0 + \mathrm{R}(1,\,0)\,Do - \mathrm{R}(1,\,1)\,Up + \mathrm{R}(1,\,1)\,R0}\right\}$$

That is,

$$R2 = \frac{R0\,(Up - Do)\,R\,(1,1)\;\,R\,(1,0)}{R\,(1,0)\,(R0 - D0) + R\,(1,1)\,(Up - R0)}, \tag{14.37}$$

or in terms of $r(1,1)$, $r(1,0)$ and $r(0)$, $R2$ is given as below.

```
> subs(R(1,1)=1+r(1,1), R(1,0)=1+r(1,0),R0=1+r(0),%);
```

$$\{R2 = \frac{(1 + \mathrm{r}(0))\,(Up - Do)\,(1 + \mathrm{r}(1,\, 1))\,(1 + \mathrm{r}(1,\, 0))}{(1 + \mathrm{r}(1,\, 0))\,(1 + \mathrm{r}(0) - Do) + (1 + \mathrm{r}(1,\, 1))\,(Up - 1 - \mathrm{r}(0))}\}$$

We ask the reader to verify that if this relation does not hold, arbitrage opportunities would exist. Furthermore, according to Theorem 11 the no-arbitrage condition is satisfied in this multiperiod model if and only if $Do < 1 + r(i,j)$ and $1 + r(i,j) < Up$ for every i and j.

Consider a forward contract written at time zero with a delivery date of time 2 on an asset with a current price of S. As we have seen in Chapter 10, the forward price is the future value of S; hence the forward price is $SR2$. Now let us see what the futures price of such a contract on the same asset would be. To this end we will again use the recursive technique demonstrated in Sections 13.3.2 and 13.4.4.

Let us denote the futures price as of time 0 for a contract to be delivered at time 2 by $F0$ and the futures price of a contract written at node (i,j) to be delivered at time 2 by $Fu_{i,j}$. Since at time 1 a futures contract with a delivery date of time 2 is like a forward contract, we have that

$$Fu_{1,j} = SUp^j Do^{1-j} R\,(1,j)\,, \quad j = 0, 1. \tag{14.38}$$

The MAPLE commands below initiate the value of some variables and define $Fu_{i,j}$ as in equation (14.38).

```
> j:='j':Up:='Up':Do:='Do':r:='r':S:='S':Cash:='Cash':
> for j from 1 to 0 by -1 do;
> Fu[1,j]:= S*Up^j*Do^(1-j)*R(1,j);
> od;
```

$$Fu_{1,\,1} := S\ Up\,\mathrm{R}(1,\,1)$$
$$Fu_{1,\,0} := S\ Do\,\mathrm{R}(1,\,0)$$

Hence, at time 2, the cash flow to the long holder of the futures contract, denoted by $Cash_{2,j}$, will be as outlined below. Viewed from node $(1,1)$ the possible states of nature could be $(2,2)$ or $(2,1.5)$ with cash flows $Cash_{2,2}$, and $Cash_{2,1.5}$, respectively:

$$Cash_{2,2} = SUp^2 - Fu_{1,1} = SUp^2 - SUpR\,(1,1) \tag{14.39}$$

$$Cash_{2,\ 1.5} = SUpDo - Fu_{1,1} = SUpDo - SUpR\,(1,1)\,. \tag{14.40}$$

Viewed from node $(1,0)$, the possible states of nature could be $(2,0.5)$ or $(2,0)$ with cash flows $Cash_{2,0.5}$ and $Cash_{2,0}$, respectively:

$$Cash_{2,\ .5} = SDoUp - Fu_{1,0} = SDoUp - SDoR\,(1,0) \tag{14.41}$$

$$Cash_{2,0} = SDo^2 - Fu_{1,0} = SDo^2 - SDo(R(1,0)). \qquad (14.42)$$

At each node $(1, j)$ the present value of the above cash flows is, of course, zero. Remember that the futures price is chosen in such a way that the value of the contract at initiation is nil.

Recall the marking to market mechanism discussed in Section 10.5. At time 1 the cash flow resulting from the marking to market (of a futures contract initiated at time 0 for delivery at time 2) at node (i, j) is $F0 - Fu_{i,j}$. At time zero, $F0$ will be fixed such that the value of the contract at that time will be zero. Hence $F0$ must satisfy equation[10] (14.43).

$$du\,(F0 - Fu_{1,0}) + dd\,(F0 - Fu_{1,1}) = 0. \qquad (14.43)$$

Equation (14.43) is submitted to MAPLE to be solved for $F0$, but we delay the substitution of Fu in terms of S and $R(1, j)$ as in equation (14.38). This is done by enclosing it in ' ', i.e., as '$Fu[1, 1]$'.

```
> solve(du*(F0-'Fu[1,1]')+dd*(F0-'Fu[1,0]')=0,F0);
```

$$\frac{du\,Fu_{1,1} + dd\,Fu_{1,0}}{du + dd}$$

Recognizing that $\frac{du}{dd+du} = qu$ and $\frac{dd}{dd+du} = 1 - qu$, the solution presented by MAPLE can be written as in equation (14.44).

$$qu\;F_{1,1} + (1 - qu)\,F_{1,0} = F0 \qquad (14.44)$$

Equation (14.44) conveys a general property of futures prices. Under the risk-neutral probability, the expected value of the futures price (at some future time) is the current futures price. Put differently, the futures price is a Martingale. The process of a forward price in a continuous-time setting is discussed in Chapter 15.

The next MAPLE commands substitute for du and dd in terms of $R0$, Up, and Do in the solution to equation (14.43) and evaluate $Fu_{1,j}$ in terms of $R(1, j)$. This produces the expression for $F0$, the futures price of the contract at time zero.

```
> simplify(subs(du=(R0-Do)/(R0*(Up-Do)),\
> dd=(Up-R0)/(R0*(Up-Do)),%));
```

$$-\frac{S\,(R0\,\mathrm{R}(1,\,1)\,Up - \mathrm{R}(1,\,1)\,Up\,Do + \mathrm{R}(1,\,0)\,Do\,Up - R0\,\mathrm{R}(1,\,0)\,Do)}{-Up + Do}$$

[10] Indeed, in this environment of stochastic interest rates, du and dd are functions of the one-period interest rate, $R0$ in equation (14.43). However, for simplicity we suppress this dependency.

Hence the futures price is given by

$$F0 = S\frac{UpR\,(1,1)\,(R0 - Do) + DoR\,(1,0)\,(Up - R0)}{Up - Do} \tag{14.45}$$

while the forward price is the future value of S, i.e., $SR2$, which based on equation (14.37) is given by equation (14.46).

$$\frac{R0\,(Up - Do)\,R\,(1,1)\,R\,(1,0)\,S}{R\,(1,0)\,(R0 - Do) + R\,(1,1)\,(Up - R0)} \tag{14.46}$$

Consequently, the difference between $F0$ and $R2S$ is given by the variable $F0\_SR2$,

```
> FO_SR2:=S*(-Up*R(1,1)*RO+Up*R(1,1)*Do-Do*\
> R(1,0)*Up+Do*R(1,0)*RO)/(-Up+Do)- \
> S*(RO*(-Up+Do)*R(1,1)*R(1,0)/(-R(1,0)*RO+\
> R(1,0)*Do-R(1,1)*Up+R(1,1)*RO));
```

$$F0\_R2S := \frac{S\,(-R0\,\mathrm{R}(1,\,1)\,Up + \mathrm{R}(1,\,1)\,Up\,Do - \mathrm{R}(1,\,0)\,Do\,Up + R0\,\mathrm{R}(1,\,0)\,Do)}{-Up + Do}$$
$$- \frac{R0\,(-Up + Do)\,\mathrm{R}(1,\,1)\,\mathrm{R}(1,\,0)\,S}{-\mathrm{R}(1,\,0)\,R0 + \mathrm{R}(1,\,0)\,Do - \mathrm{R}(1,\,1)\,Up + \mathrm{R}(1,\,1)\,R0}$$

The next command calculates the algebraic expressions for $F0\_SR2$, the difference between the forward and futures prices, under the assumption that $R0 = R(1,0) = R(1,1)$. This should confirm the equality between futures and forward prices in a deterministic term structure environment. Indeed, this is the case: $F0 - SR2 = 0$, as shown below.

```
> algsubs(R(1,0)=RO,FO_SR2);
```

$$\frac{S\,(-R0\,\mathrm{R}(1,\,1)\,Up + \mathrm{R}(1,\,1)\,Up\,Do + Do\,R0^2 - Up\,Do\,R0)}{-Up + Do}$$
$$- \frac{R0^2\,(-Up + Do)\,\mathrm{R}(1,\,1)\,S}{Do\,R0 - R0^2 - \mathrm{R}(1,\,1)\,Up + \mathrm{R}(1,\,1)\,R0}$$

```
> simplify(algsubs(R(1,1)=RO,%));
```

$$0$$

Let us substitute the following values for Up, Do, and S in the expression for $F0 - SR2$:

```
> Up=evalf(exp(1/5*sqrt(1/2)));
> Do=evalf(exp(-1/5*sqrt(1/2)));S=100;
```

$$Up = 1.151909910$$

$$Do = .8681234454$$
$$S = 100$$

As long as $R(1,1)$, $R0$, and $R(1,0)$ are smaller than Up and larger than Do, the no-arbitrage condition is satisfied. We present below a numerical example that demonstrates the difference between futures and forward prices when interest rates are stochastic.

```
> subs(Up=evalf(exp(1/5*sqrt(1/2))),\
> Do=evalf(exp(-1/5*sqrt(1/2))),S=100, \
> R0=1.07,R(1,1)=1.12,R(1,0)=1.09,S*(-Up*R(1,1)*\
> R0+Up*R(1,1)*Do-Do*R(1,0)*Up+Do*R(1,0)*R0)/(-Up+Do));
```

$$119.0882949$$

```
> subs(Up=evalf(exp(1/5*sqrt(1/2))),\
> Do=evalf(exp(-1/5*sqrt(1/2))),S=100, \
> R0=1.07,R(1,1)=1.12,R(1,0)=1.09,R0*(-Up+Do)*R(1,1)*\
> R(1,0)*S/(-R(1,0)*R0+R(1,0)*Do-R(1,1)*Up+R(1,1)*R0));
```

$$118.8954936$$

As we can see in this environment futures and forward prices are no longer the same.

14.4 Brownian Motion

The continuously compounded rate of return for a risky asset was first discussed in Section 5.5. We assumed there that over a period $[0,T]$, $Y(T)$ follows a normal distribution with an expected value of μT and a standard deviation of $\sigma\sqrt{T}$, for some constants μ and σ. In fact, the assumption of a normal distribution is redundant.

The assumption[11] that over periods of equal length, the rates of return are i.i.d. distributed random variables already implies that $Y(T)$ follows the normal distribution. Dividing a given time period into N equal periods so that the rate of return over the original period becomes the sum of the returns over these small periods produces this conclusion. Letting N approach infinity and applying to the Central Limit Theorem provides the required result. We have already stipulated a way of choosing the parameters in the binomial model so that in the limit $Y(T)$ follows the normal distribution. Hence, the price process, $S\,e^{Y(T)}$, follows the lognormal distribution.

[11] See the assumption at the end of Section 5.5 and equations (5.30) and (5.31).

Equations (14.6) and (14.7) present a sufficient condition under which the limiting distribution produces an expected value of μT and a standard deviation of $\sigma\sqrt{T}$. As we can see in the MAPLE commands below, the general solution, u, v, and qu, to these equations does not uniquely determine the value of the parameters.

```
> u:='u':v:='v':N:='N':qu:='qu'::
> solve({mu*T=N*(qu*u+(1-qu)*v),sigma^2*T=\
> N*((qu-qu^2)*u^2-2*(1-qu)*v*qu*u\
> +(1-qu)*v^2-(1-qu)^2*v^2)},{u,v,qu});
```

$$\{qu = \frac{-2\,\mu\,T\,N\,v + N^2\,v^2 + \mu^2\,T^2}{\sigma^2\,T\,N - 2\,\mu\,T\,N\,v + N^2\,v^2 + \mu^2\,T^2},$$
$$u = -\frac{(-\mu\,N\,v + \sigma^2\,N + T\,\mu^2)\,T}{(-\mu\,T + N\,v)\,N},\ v = v\}$$

Moreover, these conditions are too strong. There is no need for the equations to be satisfied for each N. Rather they should be satisfied only in the limit as N approaches infinity. A solution for which equation (14.7) is satisfied only in the limit is presented below. It makes it easier to motivate some properties of the Brownian motion.

Consider a solution[12] where $u = -v$ and $u = \sigma\sqrt{\frac{T}{N}}$, i.e.,

```
> u:=sigma*sqrt(T/N);
```

$$u := \sigma\,\sqrt{\frac{T}{N}}$$

Note that a property of this solution is that the absolute value of the change in the continuously compounded rate of return is $\sigma\sqrt{\frac{T}{N}}$ with certainty, and that it depends on the time interval via the square root of its length. These properties highlight some of the properties of the Brownian motion as we shall soon see. The expected value of Y_N is given by

```
> Ex:=simplify(N*(qu*u+(1-qu)*v));
```

$$Ex := N\,\sigma\,\sqrt{\frac{T}{N}}\,(2\,qu - 1)$$

[12] The solution presented here follows [20]. The reader is asked, in the exercises, to utilize the solution presented in Section 14.1.1 in order to motivate the properties of the Brownian motion that are discussed in the rest of this section.

and its variance is given by

```
> Var:=collect(N*((qu-qu^2)*u^2-2*(1-qu)*\
> v*qu*u+(1-qu)*v^2-(1-qu)^2*v^2),qu);
```

$$Var := -4\,\sigma^2\,T\,qu^2 + 4\,\sigma^2\,T\,qu$$

Let us assume that qu is given by equation (14.47),

$$qu = \frac{1 + \frac{\mu\sqrt{\frac{T}{N}}}{\sigma}}{2}, \tag{14.47}$$

and assign qu its value so we can calculate the induced expected value (denoted Ex) and variance (denoted Var) of Y_N. We can thereby we can check if equations (14.6) and (14.7) are satisfied.

```
> qu:=(1+mu*sqrt(T/N)/sigma)/2;
```

$$qu := \frac{1}{2} + \frac{1}{2}\,\frac{\mu\sqrt{\frac{T}{N}}}{\sigma}$$

```
> simplify(Ex);
```

$$\mu\,T$$

```
> expand(Var);
```

$$\sigma^2\,T - \frac{T^2\,\mu^2}{N}$$

We therefore see that equation (14.6) is satisfied but equation (14.7) is not. However, as N goes to infinity and the length of the time interval approaches zero, indeed the variance approaches $\sigma^2\,T$ as desired. This is demonstrated in the MAPLE command below.

```
> limit(Var,N=infinity);
```

$$\sigma^2\,T$$

In both solutions, the one presented here and the one presented in Section 14.1.1, the possible realization of the change in Y, u, and v, over a period of length $\frac{T}{N}$, depends on $\sqrt{\frac{T}{N}}$. It ensures that the variance over the interval $[0, T]$, as N approaches infinity, depends on T and not on N. It therefore explains why Assumption II in Section 5.4 stipulates that Y depends, via its standard deviation, on $\sqrt{T}$. Furthermore, with the specifications as in equation (14.47), the absolute value of the change in Y over a period, in the binomial mode, is with certainty $\sigma\sqrt{\frac{T}{N}}$ since $u = |v|$. Hence the expected value of the absolute change over a period in the binomial model is $\sigma\sqrt{\frac{T}{N}}$.

Consequently, the total expected change of Y over $[0, T]$ is $N\sigma\sqrt{\frac{T}{N}}$, which approaches infinity as N goes to infinity.

```
> assume(T>0,sigma>0);
> limit(N*sigma*sqrt(T/N),N=infinity);
```

$$\infty$$

This explains why the graph of a realization of a Brownian motion $Y(t)$ versus t "travels" an infinite distance over a finite time[13] interval and therefore looks very jagged. Furthermore, the absolute value of the rate of change in Y over a period in the binomial model is given by $\frac{\sigma\sqrt{\frac{T}{N}}}{\frac{T}{N}}$, which as N goes to infinity, as shown below, explodes.

```
> limit(sigma*sqrt(T/N)/(T/N),N=infinity);
```

$$\infty$$

Hence it explains why the (calculus) derivative of the function $Y(t)$ with respect to t does not exist.

Readers who opt to skip the next Chapter may wish to execute the procedure **Simudif** below now. It is a simulation of the realization of a Brownian motion as shown in Figure 14.7. The parameters for this procedure are given below:

- $Y0$, the current value of Y, 0.20 in the example below
- $t0$ and $t1$ the left and right end points of the time interval, respectively; 0 and 1 in our example
- N, μ, and σ as defined above, which are set to 1000, 0.20, and 0.18, respectively, in the example
- The last parameter "plot", if omitted, causes the procedure to produce an animation instead of a static graph.

```
> Simudif(.20,0,1,1000,.20,.18,"plot");
```

The linear graph in Figure 14.7 is of the expected value, μt, while the jagged graph is a realization of the path taken[14] by the Brownian motion. The reader may want to change the parameter values and rerun the procedure to gain a better appreciation of the process. Figure 14.7 makes visual

[13] See our discussion of total variation in the next Chapter.

[14] This animation is produced by sampling a normal random variable, instead of a binomial variable, at the beginning of each period, the length of which is $\frac{1}{1000}$. A very similar graph would have been produced if the sampling were done with a binomial variable adopting one of the solutions presented in this Chapter.

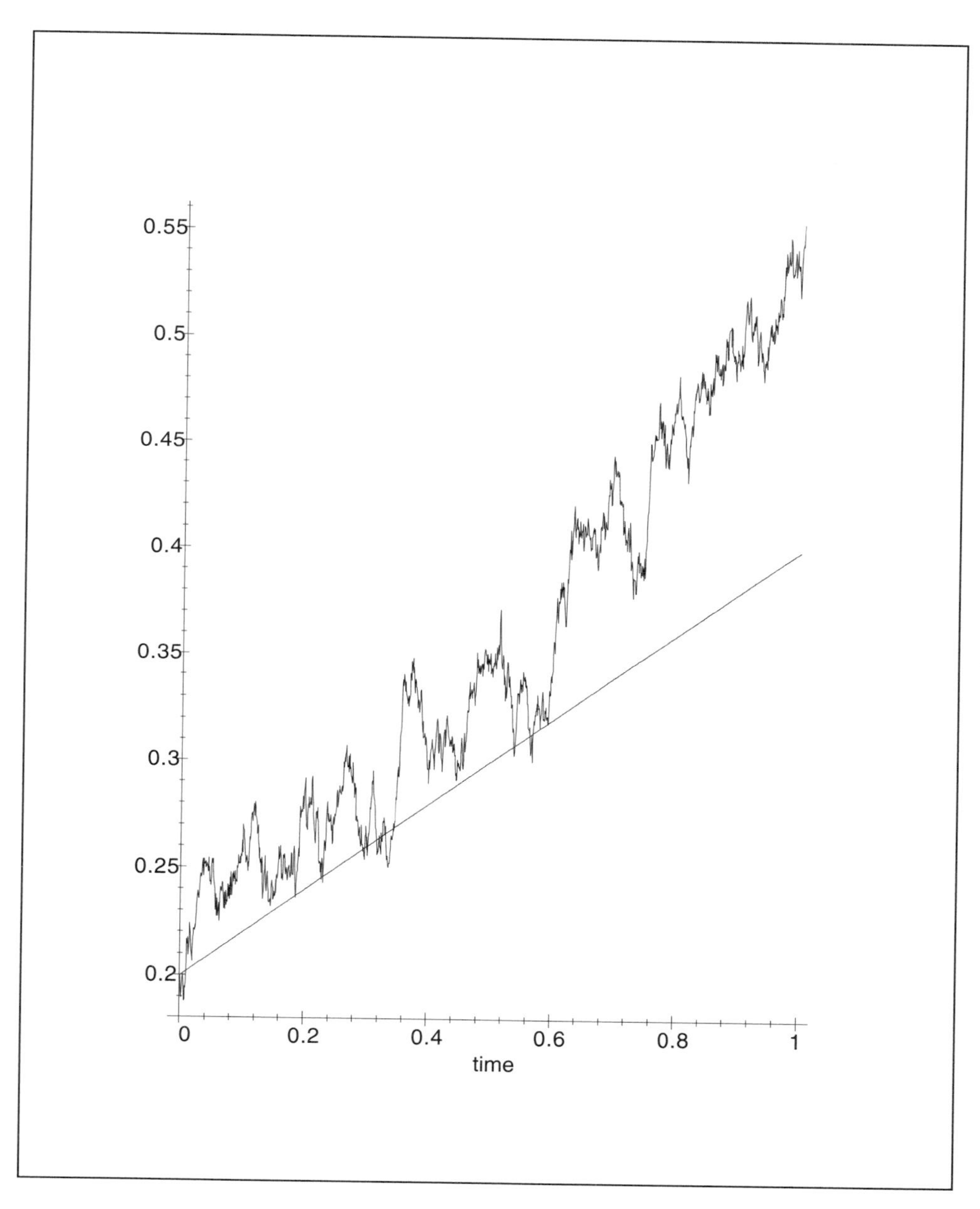

Figure 14.7: A Realization of a Brownian Motion, $Y(t)$ versus t

the reasons why the tools used in deterministic calculus are not able to handle such circumstances. The next Chapter introduces the reader to methods by which the Black–Scholes formula can be derived in this stochastic environment.

14.5 Concluding Remarks

This Chapter completes our treatment of the binomial model. We utilized the binomial model for two primary goals, as a practical tool for numerical valuations and as a simplified environment of a continuous-time model, to explore some properties that hold in general. This Chapter also addresses how dividends can be incorporated into the binomial model and some difficulties that arise when the dividends are modeled as absolute payments rather than as a dividend yields.

Chapter 13 demonstrated valuing American options with the aid of the binomial model. Here we justified the choice of the parameters in the binomial model, the guideline being the requirement that, as the number of periods in the binomial model increases to infinity, and the length of each period decreases to zero, the distribution of the underlying asset approaches the lognormal distribution. This result is justified in the text by applying to a property known in probability theory. Appendix 14.7 derives it from first principles applying to the Central Limit Theorem.

The binomial model facilitates an easy introduction to the Martingale concept and its connections to the risk-neutral probability. In a one-period model, the existence of a risk-neutral probability is equivalent to the satisfaction of the no-arbitrage condition. In a multiplied model where there are no arbitrage opportunities, Theorem 11, in Section 13.2.1, implies that there exists a risk-neutral probability governing the changes in the underlying asset from each period to the next. Therefore, the no-arbitrage condition in this setting is equivalent to the stochastic process of the relative price (as defined in equation (14.20)) being a Martingale. This same setting also provides for a "friendly environment" in which the difference between futures and forward prices is explored.

As N, the number of periods, approaches infinity and Δt, the length of each period, approaches zero, the binomial model converges to a continuous-time model. This continuous-time model is explored in the next Chapter. The reader might use this limiting process as a point of reference when reading the next Chapter. Indeed, the next Chapter will apply again to the replication arguments, deriving the Black–Scholes formula in a continuous-time setting. In order to prepare the reader for the replication argument in this setting we offer here another interpretation of the Black–Scholes formula. The Black–Scholes formula, equation (6.2), was developed in Chapter 6 and is repeated below.

$$PN\left(\frac{\ln\left(\frac{P}{K}\right)+\left(r+\frac{\sigma^2}{2}\right)t}{\sigma\sqrt{t}}\right)-Ke^{-rt}N\left(\frac{\ln\left(\frac{P}{K}\right)+\left(r-\frac{\sigma^2}{2}\right)t}{\sigma\sqrt{t}}\right)$$

In the binomial setting Xs represents the holding (number of units) of the stock in the replicating portfolio and Xb stands for the amount (borrowed) invested in the risk-free rate. The value of the replicating portfolio is thus $XsP + Xb$, which equals the value of the option. Comparing it to the above equation reveals that Xs and Xb are the counterparts of $N\left(\frac{\ln\left(\frac{P}{K}\right)+\left(r+\frac{\sigma^2}{2}\right)t}{\sigma\sqrt{t}}\right)$ and $-Ke^{-rt}N\left(\frac{\ln\left(\frac{P}{K}\right)+\left(r-\frac{\sigma^2}{2}\right)t}{\sigma\sqrt{t}}\right)$, respectively. Hence, the number of units of the stock in the replicating portfolio is

$$N\left(\frac{\ln\left(\frac{P}{K}\right)+\left(r+\frac{\sigma^2}{2}\right)t}{\sigma\sqrt{t}}\right)$$

while

$$-Ke^{-rt}N\left(\frac{\ln\left(\frac{P}{K}\right)+\left(r-\frac{\sigma^2}{2}\right)t}{\sigma\sqrt{t}}\right)$$

can be further interpreted as the product of

1. $-N\left(\frac{\ln\left(\frac{P}{K}\right)+\left(r-\frac{\sigma^2}{2}\right)t}{\sigma\sqrt{t}}\right)$, the number of units of zero-coupon bonds being held short, and
2. Ke^{-rt}, the price of a zero-coupon bonds which matures, when the option does, at time t.

However, in contrast to the binomial model, the replication portfolio in a continuous-time setting must be revised every instant rather than every Δt. Referring to the binomial model may help the reader in clarifying the conceptual process and further enhance the connection between Xs, as defined in Section 13.3.2, and the sensitivity measure Δ, defined in Section 7.2. Thinking about Δ as the counterpart of Xs, as interpreted above, enhances the intuition behind the arguments explored in the next Chapter.

14.6 Questions and Problems

Problem 1. In the context of a binomial model, prove for an American put option that if at time t it is optimal to exercise it when the price of the underlying asset is S, then it will be optimal to exercise it

1. at t when the underlying price is smaller than S, and
2. at any time after t if the underlying price is smaller than or equal to S.

Problem 2. Consider the binomial tree in Table 13.2 and assume that the price evolution described there is of the stock only (not including dividends). Assume that the stock pays a one-time dividend $Di = \$3$ at time 3. The price of the stock, at each node (i, j), such that $i < 3$, is then $SUp^j Do^{i-j}$ plus the present value of the dividend.

1. Calculate the price of a European call option with an exercise price of $100.
2. Calculate the price of an American option with the same exercise price; identify if and when it is optimal to exercise the option.

Problem 3. Consider the binomial tree in Figure 14.6 Section and verify that in this symbolic example, equation (14.23) is satisfied in spite of the stochastic interest rate[15] environment. This is done noticing that at time $k-1$, $S(k)$ is a random variable but the interest rate, $r(k-1,k)$, spanning the time interval $[(k-1)\,\Delta t, k\Delta t]$, will be realized at time $k-1$, and thus

$$E_{k-1}\left(\frac{S\,(k)}{e^{r(k-1,k)}}\;\middle|\; S\,(k-1)\right) = \frac{1}{e^{r(k-1,k)}} E_{k-1}\left(S(k)\mid S\,(k-1)\right).$$

Problem 4. Prove Theorem 12. Namely, prove that in the context of the binomial model the no-arbitrage condition is satisfied if, and only if, the relative price process is a Martingale.

[15] At time $k-1$, the interest spanning the time interval $[(k-1)\,\Delta t, k\Delta t]$ will be realized. In the parlance of stochastic processes, such a process is said to be *predictable*. That is, given the set of information at time $k-1$, r is a deterministic constant while the price process $S(k)$ (or $S(i)$, $i = k, k+1, ...$) is a random variable, the distribution of which is known (as we mentioned before, $S(k)$ is measurable with respect to the information at time $k-1$). In view of this footnote it might be more appropriate to interpret the notation $E(S(m)|\,S(n))$ as the expectation of $S(m)$ given the information set at time n, $n < m$.

Problem 5. Prove Theorem 12 when interest rates are stochastic.

Problem 6. Consider the binomial tree in Table 13.2. Price a European option that pays the maximum between zero and the difference of the price of the stock and its arithmetic average price. The arithmetic price is calculated as a simple average of the price history of the stock in each node realized. Such an option is termed a path dependent option as it does depend on the path of the price history. The valuation procedure is still done in the usual manner; one should simply consider carefully all the price evolutions as the tree (with resect to the price of the option) is not longer a recombining tree. Could the value of an American option, as described above, be calculated using the binomial tree?

14.7 Appendix

14.7.1 The Black–Scholes Formula as a Limit of the Binomial Formula

The binomial distribution, $B(N, qu)$, can be interpreted as a sum of N independent Bernoulli random variables. A Bernoulli random variable is a variable that can take only two values, e.g., 0 and 1, that are associated with the states **Up** and **Down** in our case. Hence, $B(N, qu)$ equals the sum of N independent Bernoulli random variables denoted as $B(1, qu)$. It is easy to verify, following the definition, that the expected value of a Bernoulli variable is $qu \times 1 + (1 - qu) \times 0 = qu$. Hence, if $B_i \sim B(1, qu)$ and $B_N = \sum_{i=0}^{N-1} B_i$, then $B_N \sim B(N, qu)$ and

$$E(B_N) = \sum_{i=0}^{N-1} E(B_i) = Nqu.$$

Since the B_i's are independent random variables, the variance of B_N satisfies equation (14.48).

$$Var(B_N) = \sum_{i=0}^{N-1} Var(B_i). \tag{14.48}$$

The variance of each B_i is given by $Var(B_i) = E\left(B_i^2\right) - (E(B_i))^2$. However, $E\left(B_i^2\right) = 1^2 \times qu + 0^2 \times (1 - qu) = qu$ and, as shown above, $E(B) = qu$. Therefore, $E(B_i)^2 = qu^2$. Thus,

$$Var(B_i) = qu - qu^2 = qu(1 - qu),$$

and hence, by equation (14.48),

$$\sum_{i=0}^{N-1} Var(B_i) = N\, qu\; (1 - qu)\,. \tag{14.49}$$

The number of time periods and movements in this model are intimately related to the continuously compounded rate of return. This relation will prove useful very shortly. Assume that the current price of the stock is $S(0)$, and that its price distribution in k periods, or in $k\Delta t$ units of time, is given by equation (14.50),

$$\Pr\left(S\,(k) = S\,(0)\,Up^j\,Do^{k-j}\right) = \frac{k!}{j!\,(k-j)!} qu^j\,(1-qu)^{k-j}\,, \tag{14.50}$$

where $j = 0, ..., k \leq N$. The continuously compounded rate of return over the period is given by the natural logarithm of the price ratio $\frac{S(k)}{S(0)}$, and hence it is

$$\ln\left(\frac{S\,(k)}{S\,(0)}\right) = \ln\left(Up^j\,Do^{k-j}\right), \tag{14.51}$$

where j is a binomial random variable, i.e., $j \sim B\,(k, qu)$.

In some of the models, see equation (13.4), the factors Up and Do are specified as $e^{V_{\Delta t}}$, where $V_{\Delta t}$ is a random variable taking a positive value u if state **Up** occurs and a negative value v if state **Down** occurs. For this specification the random variable delineating the continuously compounded rate of return over the period $[0, k\Delta t]$ is given by $uj + v\,(k - j)$, where $j \sim B\,(k, qu)$. Under such a specification the calculation of the variance and expected value of the rate of return over $[0, k\Delta t]$, based on equation (14.51), is an easy task, and is left for the reader.

Consider a European call option on a non-dividend-paying stock. Assume that the price of the stock now is S and that the option matures at time T and has an exercise price K. Given that there are N periods in the binomial model we know that there are $N + 1$ possible states of nature at maturity. The stock will be worth $S\,Up^j\; Do^{N-j}$ at time N if j up movements have occurred by that time, and the corresponding value of the European call option in state j will be $\max\left(0,\; S\,Up^j\,Do^{N-j} - K\right)$.

Applying the risk-neutral valuation concept, the value of this option now can be determined as the expected value of its payoff at maturity, discounted at the risk-free rate. In other words, the value of the call option can be written as

$$C(S, K, T, N) = e^{-r_f\,T} \sum_{j=0}^{N} \binom{N}{j} qu^j\,(1-qu)^{N-j} \max\left(0, SUp^j\,Do^{N-j} - K\right),$$

where qu is the risk-neutral probability of an up movement. We know that at all states of nature at which $S\,Up^j Do^{N-j} < K$ the option will pay nothing. Define m to be the smallest integer such that $K < S\,Up^m Do^{N-m}$. Then

$$S\,Up^{m-1} Do^{N-(m-1)} \leq K$$

and

$$K < S\,Up^m\, Do^{N-m},$$

or equivalently,

$$\frac{\ln\left(\frac{K}{S\,Do^N}\right)}{\ln\left(\frac{Up}{Do}\right)} \leq m$$

and

$$m < \frac{\ln\left(\frac{K}{S\,Do^N}\right)}{\ln\left(\frac{Up}{Do}\right)} + 1.$$

Taking this definition of m into account the price of the option can now be written as

$$C(S,\,K,\,T,\,N) = e^{-r_f T} \sum_{j=m}^{N} \binom{N}{j} qu^j\,(1-qu)^{N-j}\,\left(S\,Up^j\,Do^{N-j} - K\right).$$

We can further rewrite the price of the option $C(S,\,K,\,T,\,N)$ as

$$S \sum_{j=m}^{N} \binom{N}{j} qu^j\,(1-qu)^{N-j} \left(\frac{Up}{e^{\frac{r_f T}{N}}}\right)^j \left(\frac{Do}{e^{\frac{r_f T}{N}}}\right)^{N-j} - \\ K\,e^{-r_f T} \sum_{j=m}^{N} \binom{N}{j} qu^j\,(1-qu)^{N-j}$$

or as

$$S \sum_{j=m}^{N} \binom{N}{j} \left(\frac{qu\,Up}{R}\right)^j \left(\frac{(1-qu)\,Do}{R}\right)^{N-j} - \\ K\,e^{-r_f T} \sum_{j=m}^{N} \binom{N}{j} qu^j\,(1-qu)^{N-j},$$

where $R = e^{\left(\frac{r_f T}{N}\right)}$, as defined before. One can check that given the definition of the risk-neutral probability qu as in equation (14.16),

$$qu\frac{Up}{R} + (1-qu)\,\frac{Do}{R} = 1.$$

This together with the fact that Up, Do, and R are positive allows us to interpret $\frac{quUp}{R}$ and $\frac{(1-qu)\,Do}{R}$ as probabilities.

Utilizing this fact, the first summation

$$\sum_{j=m}^{N} \binom{N}{j} \left(\frac{qu\,Up}{R}\right)^{j} \left(\frac{(1-qu)\;Do}{R}\right)^{N-j}$$

can simply be thought of as a probability of a random variable having a binomial distribution, $B\left(N,\, \frac{qu\,Up}{R}\right)$, being greater than or equal to m. The second summation

$$\sum_{j=m}^{N} \binom{N}{j} qu^{j}\;(1-qu)^{N-j}$$

is simply a probability of the random variable having a binomial distribution, $B\,(N,\, qu)$, being greater than or equal to m.

The value of the option can therefore be written as

$$\begin{aligned} C(S,\, K,\, T,\, N) \;&=\; S\left(1 - B\left(m,\, N,\, \frac{qu\,Up}{R}\right)\right) \\ &\quad -K\,e^{-r_f T}\,(1 - B\,(m,\, N,\, qu))\,. \end{aligned} \tag{14.52}$$

One may already notice the similarity of this expression to the Black–Scholes formula. Let us now see what happens to this expression when N converges to infinity. As was explained before, the binomial distribution $B\,(N,\, p)$ can be thought of as the sum of N independent identically distributed Bernoulli random variables with $B\,(1,\, p)$. This means that by the Central Limit Theorem, as N converges to infinity, the binomial distribution converges to the normal distribution. Taking this into account, the $\Pr\,(m \le j)$ can now be approximated by

$$\begin{aligned} \Pr\,(m \le j) \;&=\; 1 - \Pr\left(\frac{j - N\,p}{\sqrt{N\,p\,(1-p)}} \le \frac{m - N\,p}{\sqrt{N\,p\;(1-p)}}\right) \\ &=\; 1 - \mathrm{N}\left(\frac{m - N\,p}{\sqrt{N\,p\;(1-p)}}\right) \\ &=\; \mathrm{N}\left(\frac{N\,p - m}{\sqrt{N\,p\;(1-p)}}\right) \end{aligned}$$

and

$$\Pr\,(j \le m) = \mathrm{N}\left(\frac{m - N\,p}{\sqrt{N\,p\,(1-p)}}\right),$$

where N is a cumulative distribution function of the standard normal distribution. Hence, as N converges to infinity

$$1 - B\left(m,\, N,\, \frac{qu\,Up}{R}\right) = \mathrm{N}\left(\frac{N\frac{qu\,Up}{R} - m}{\sqrt{N\frac{qu\,Up}{R}\left(1 - \frac{qu\,Up}{R}\right)}}\right)$$

$$1 - B\left(m,\, N,\, qu\right) = \mathrm{N}\left(\frac{N\,qu - m}{\sqrt{N\,qu\,\left(1 - qu\right)}}\right)$$

and the binomial formula $C(S, K, T, N)$ in equation (14.52) which resembles the Black–Scholes formula indeed converges to it. However, we should point out that the "classical" Central Limit Theorem claims that the summation of i.i.d. random variables as N approaches infinity approaches the normal distribution. In the classical case as N increases say, from N to $N + 1$, the $N + 1$ random variables have the same distribution as that of that of the N random variables. In our case, as N increases the random variables that are summed are different for every value of N. This is because the expected value and variance of the summed variables are functions of N. Nevertheless, the Central Limit Theorem applies in this case as well, provided certain regularity conditions are satisfied. The interested reader is referred to [17] for the rigorous investigation of the theorem.

Chapter 15

A Second Look at the Black–Scholes Formula

15.1 An Overview

The presentation order of the material in this book has been induced largely by a natural progression from the simple model introduced in Chapter 1. We now arrive at the most general version of the model that will be presented in this book. We started with a one-period model in which there were a finite number of possible states of nature. This model was generalized to a one-period model with a continuum of states of nature. In Chapter 6 we derived the Black–Scholes formula in that latter setting. That derivation relied on the equivalency of the no-arbitrage condition to the existence of a stochastic discount factor and the risk-neutral distribution function.

We assumed that the real-life and risk-neutral distributions belong to the same family of distributions, the lognormal distribution, and have the same standard deviation. Consequently, equation (5.41) implies that the expected value of the continuously compounded rate of return, under the risk-neutral probability, must be $r - \frac{\sigma^2}{2}$, where r is the risk-free rate of interest. The Black–Scholes formula for valuing options then was derived by equating the value of the option to the discounted value of its expected payoff.

We also looked at a multiperiod model with a finite number of time periods and states of nature, e.g., the binomial model of Chapter 13 and Chapter 14. The Black–Scholes formula was derived in this setting through use of limiting arguments. The binomial model can be understood as a sequence of N one-period, two-states-of-nature models. Recursively applying the replication argument to that model structure resulted in the price of the derivative security in the N-period model setting. In Section 14.7 the Black–Scholes formula was shown to be the limit as N approaches infinity of the price of the derivative security in the N-period binomial model.

In Chapters 9 and 10 we analyzed forwards and swaps in the setting of a model with finitely many time periods and a continuum of states of nature. This Chapter introduces a model in which both the states of nature and time are continuous. This means that trading can take place at every instant of time and that possible realizations of the price of the underlying asset may take any value in the interval $[0, \infty)$. In this newest context we now take a second (or perhaps a third) look at the Black–Scholes formula.

The key idea for deriving the Black–Scholes formula in this setting can be presented in several ways. The formula can be derived via replication pricing arguments, "adjusted" for the current environment. It can also be uncovered from the construction of a strategically designed portfolio. This portfolio is risk-free over an instant of time. It is this latter derivation on which we focus in this overview, attempting to explain the philosophy behind it. Section 15.6 of this Chapter derives the Black–Scholes formula both ways, beginning with the former.

The price of a risk-free portfolio was discussed in Chapter 1 in the discussion preceding equation (1.8). We have also shown in Section 2.1 that the return on such a portfolio, a risk-free portfolio, must be the risk-free rate of interest, or else arbitrage opportunities would exist. A risk-free portfolio composed of an option and a bond was also referred to in the introduction of Chapter 8 when we investigated hedging strategies. Similarly, invoking arbitrage arguments of the same sort here, we will conclude that the rate of return on a portfolio that is risk-free, even only for an instant in time, must be equal to the continuously compounded risk-free rate.

The source of uncertainty regarding the price of the derivative is the price of the underlying asset. Intuition suggests that we might be able to combine the derivative and the underlying asset in a portfolio in such a way that the combination does not require the portfolio holder to bear risk. We must keep in mind that we are speaking about zero risk only for an instant of time. This is the same old concept of instantaneous hedging. As anticipated, it is related to the sensitivity measures of Chapter 7 and to hedging with the Greeks of Chapter 8. Accordingly, we will derive the Black-Scholes formula using instantaneous hedging arguments, as in Chapter 8, and using a version of the no-arbitrage condition appropriate for this type of an environment.

The environment now is different than that encountered in the model of Chapter 1. The length of the time period we are looking at is only an instant. We thus have to ensure that for every possible realization of the price of the underlying asset in the next instant of time, a portfolio composed of a position in the underlying security and in the derivative is risk-free. How do we go about it? Given the explanation of the sensitivity measures in Chapter 7, the reader already might have anticipated that we will make use of the Taylor approximation.

Consider a portfolio composed of the derivative security and the underlying asset. We will use a Taylor series expansion to find an expression for the sensitivity of the portfolio value (or its incremental change) to changes in the price of the underlying asset and to the passage of time. Once the sensitivity is understood, we will know how to construct a portfolio that is insensitive to changes in the price of the underlying asset. Such a portfolio is not risky. Arbitrage arguments imply that its rate of return, over the next instant, must be equal to the risk-free rate of interest. Since we are dealing with an infinitesimally short time period, such arguments will induce a differential equation for the price of the derivative security.

There are some issues that need to be addressed before the outlined plan for valuation can be executed. The rate of change of a function over the next instant is its derivative with respect to time. Appendix 15.11.1 of this Chapter is dedicated to reviewing the meaning of the change over the next instant. Some readers may opt to read this appendix before they delve into the sections of this Chapter. Others[1] may feel it is not needed. Returning to the binomial model and the explanation of incremental changes in Section 14.2 may also help. The concepts are similar in nature albeit here the time is continuous.

As time passes, the price of the underlying asset changes, but in a random way. As a result, the rate of change of the value of the derivative security over the next instant of time will involve "differentiation with respect to a random variable". This in turn requires a second look at the stochastic price process followed by the underlying security and its induced instantaneous increments. That is, we would like to investigate the properties of

$$\lim_{\Delta t \to 0} S_{t+\Delta t} - S_t,$$

where S_t is the value of the underlying asset at time t.

The "change over the next instant" is defined by a Taylor series approximation of appropriately chosen order. Both of these concepts depend on limiting arguments applied to random variables. The text will explain

[1] Readers who are comfortable with the limit concept and the $o(h)$ notation may skip Appendix 15.11.1 without discontinuity. Others may like to pause and read it before proceeding to the next section.

these concepts in an intuitive way. For readers who are interested in a more in-depth analysis, we dedicate an appendix, Appendix 15.11.2, to the explanation of limits in this context.

Having completed the explanations of the required concepts we can attend to the main question: Given the instantaneous increment of the price process of the underlying asset, what is the instantaneous increment of a derivative security? That is, what is $\lim_{\Delta t \to 0} V(S_{t+\Delta t}) - V(S_t)$, where V denotes the value of the derivative security.

The answer to this question is found in a result known as Ito's lemma. With the aid of this powerful mathematical tool we can proceed, as we have outlined, to produce the differential equation satisfied by the price of the derivative security. Our treatment of this topic is done on an intuitive and heuristic level. Reading the appendixes will help the reader to acquire a more in-depth understanding. The text is sufficient to convey an appreciation of the material and the ability to apply it to practical problems.

Readers who are interested in a more in-depth mathematical analysis of the topic should consult[2] other books. Conversely, readers who are less mathematically inclined may opt to skip this Chapter, at least on the first reading. The representation of the Black–Scholes formula as a solution to a partial differential equation has some advantages. We therefore recommend reading on and have attempted to make this experience not excruciating. The on-line copy of the book exemplifies the concepts with live animation when possible. This, presumably, makes the material easier to follow and perhaps even entertaining.

15.2 The Price Process: A Second Look

As pointed out in the text and in several footnotes, Assumptions I, II, and III of Section 5.4 are intimately related to the instantaneous behavior of the price process. The next few sections are devoted to a second look at this price process. This section starts with a generalization of equations (5.17) and (5.18). In developing these equations we have assumed that the current time is 0 and that we are investigating the price of a stock t years from now. The stock is the asset underlying the derivative. If we denote the current time by t_0 and the current stock price by S_{t_0}, then $E[S_T]$, the expected value of the stock price at time T, $T - t_0$ years from now, satisfies

$$E_f(S_T) = S_{t_0}\, e^{r(T-t_0)} \tag{15.1}$$

[2] Some suggestions listed in increasing level of difficulty are [34], [2], [21], and [35].

since

$$S_{t_0}\, e^{r(T-t_0)} = \int_0^{\infty} s\, f(s)\, ds, \tag{15.2}$$

where f is the risk-neutral density function of the stock price at time T. The risk-neutral density function f should have been indexed by t_0, i.e., as f_{t_0}, but for simplicity it was omitted. Equations (15.2) and (15.1) describe the following phenomenon. Given that the current price of the stock is S_{t_0}, its expected value[3] at time T is the future value of its current price, based on the risk-free rate of interest. Hence, under the risk-neutral probability, one does not expect a change in the value of the stock but one does recognize the effect of the time value of money. Since S_T is random, its actual future value may differ from its expected future value.

We are interested in the random evolution of the stock price, or equivalently, in the continuously compounded rate of return Y, from its value at time t_0 to its value at time T. (A realization of this random evolution is also referred to as a sample path.) Our study of the instantaneous random evolution is built upon a discretization of the time space and its comparison to the deterministic evolution governed by the risk-free rate of interest. According to equation (5.32)

$$S_T = S_{t_0}\, e^{\mu_Y\,(T-t_0)+Z\,\sigma_Y\,\sqrt{T-t_0}}, \tag{15.3}$$

where $Z \sim \mathrm{N}(0,\, 1)$ and μ_Y and σ_Y are the expected value and standard deviation of Y, respectively. The exponent in equation (15.3),

$$\mu_Y\,(T-t_0) + Z\,\sigma_Y\,\sqrt{T-t_0},$$

is the (random) continuously compounded rate of return (over a period $[t_0, T]$) on the stock, Y_{T-t_0}. The continuously compounded rate of return on the stock is a normally distributed random variable,

$$Y_{T-t_0} \sim \mathrm{N}\left(\mu_Y(T-t_0),\, \sigma_Y\sqrt{T-t_0}\right).$$

It was defined in equation (5.24) as $\ln\left(\frac{S_T}{S_{t_0}}\right)$. Once again, we emphasize that it is the evolution according to the "real-life" distribution that is being

[3]Those readers familiar with the Martingale property of a stochastic process (or those who did not skip over the explanation in Section 14.2) will recognize that this last equation expresses a similar property. When $r = 0$, $e^{-r\,t} = 1$ and thus $S_{t_0} = E_f\,(S_T|\; S_{t_0})$, i.e., the conditional expectation of S_T given S_{t_0} is S_{t_0}, which is the Martingale property. When $r \neq 0$ the process of the "relative price", as defined in Section 14.20, satisfies equation 14.20 and thus is a Martingale.

analyzed here. Thus, μ_Y, the expected value (over a unit of time), is not replaced with $r - \frac{\sigma^2}{2}$ for some constant σ, as was done in equation (5.42). Rather μ_Y is the **true** continuously compounded annual expected return on the stock.

A word of caution is probably appropriate here. As we have seen before, in equation (5.38), if $\mu_Y = \mu$ and $\sigma_Y = \sigma$ for some constants σ and μ, then the expected value of the stock price at time T as of time t_0 is $S_{t_0} e^{\mu\,(T-t_0)+\frac{\sigma^2}{2}\,(T-t_0)}$. This is because the stock price S_T is not a linear function of Y_T and thus $E\left(S_{t_0} e^{Y_{T-t_0}}\right)$ does not equal[4] $S_{t_0} e^{E(Y_{T-t_0})} = S_{t_0} e^{\mu\,(T-t_0)}$.

The expected value of the stock price S_T will be $S_{t_0} e^{\mu\,(T-t_0)}$ if we stipulate that

$$\mu_Y = \mu - \frac{\sigma^2}{2} \tag{15.4}$$

and that

$$\sigma_Y = \sigma \tag{15.5}$$

for some constants μ and σ. An exercise at the end of this Chapter asks the reader to verify this. The reader should keep this point in mind since it is different from the deterministic case to which we refer next in our explanation.

In the deterministic case, if the continuously compounded rate of return is r, an investment V will grow to $V\,e^{rt}$ in t units of time. As we progress through the exposition of the model of this Chapter we will get another look at the source of the difference between the two cases. The difference will be related to the different rules that apply to "differentiating" (as in calculus) deterministic and stochastic functions.

Our starting point is the similarity of equation (15.3) to its counterpart in the risk-free case. Recall the continuous compounding discussion of Section 3.4.1. If at time t_0 the amount $V(t_0)$ is invested at the risk-free rate r, then at time T its value will be

$$V(T) = V(t_0)e^{r\,(T-t_0)}. \tag{15.6}$$

The instantaneous evolution of $V(t)$ through time, as described by $\frac{dV(t)}{dt} = V(t)r$, and its relation to equation (15.3) was discussed in Section 5.6.

One immediately notices that $V(T)$ and the exponent $r(T - t_0)$ in equation (15.6) play the same role as S_T and the part of the exponent $\mu(T - t_0)$

[4] Given a convex function g and a random variable x, the relation known as Jensen's inequality stipulates that $g\,(E(x)) \leq E\,(g(x))$. The inequality noted in the text is explained via Jensen's inequality.

in equation (15.3). In both cases the exponent, when $T - t_0$ is equal to one, represents the expected[5] instantaneous incremental change of V, or of S, per $1 value of the original asset. Given r, $V(t)$ is a function only of time while S_t depends on the realization of the random variable Z in equation (15.3).

As it turns out, the relation between equation (15.3) and the instantaneous changes in S_t are similar in spirit to the relation between $V(t)$ and $\frac{dV(t)}{dt} = V(t)r$. By the "instantaneous changes in S_t" we mean the evolution of S_t as time advances instant by instant. This process was visualized by using a discretization in Figure 14.7. To emphasize the similarities, the notation $S(t)$ instead of S_t will be used in some parts of the subsequent analysis.

The instantaneous change in the price process $S(t)$ can also be thought of, roughly, as "$\frac{dS(t)}{dt}$". It is the random part, $Z\sigma\sqrt{t}$, that requires extra care when analyzing the stochastic instantaneous increments of the path of $S(t)$ as a function of time. Put differently, there are two ways of describing a function. A function can be described directly using an expression like $V(t) = V_0e^{rt}$ or it can be described using its derivative. For the latter description, an expression such as $\frac{dV(t)}{dt} = V(t)r$ is used. This expression specifies the instantaneous evolution of $V(t)$ and thereby the function, but only up to a constant of integration (knowing the value of V at the initial time, known as an initial condition, one can recover the constant of integration). This is where discretization enters the explanation.

15.2.1 Stochastic Evolution: The Discrete Case

The changes in $S(t)$ and in $V(t)$ occur in a continuous manner, every instant of time. Let us approximate those changes by assuming changes that occur at discrete time points. If at time 0 the stock price is $S(0)$, its value at some future time T, $S(T)$, can be approximated by dividing the interval $[0, T]$ into n subintervals. The same holds true for $V(t)$. Defining

$$\Delta t = \frac{T}{n}, \quad t_i = 0 + i\,\Delta t, \quad i = 1, ..., n, \tag{15.7}$$

and assuming that interest is paid every Δt, i.e., at every t_i, $i = 1, ..., n$, (rather than at every instant) yields

$$\frac{V(t + \Delta t) - V(t)}{\Delta t} = V(t)r, \tag{15.8}$$

[5]In the case of deterministic constant r, the expected return equals the actual return since the expected value of a constant is the constant itself.

or equivalently that

$$\Delta V(t) = V(t) r \Delta t, \tag{15.9}$$

where

$$\Delta V(t) = V(t + \Delta t) - V(t).$$

The change in V, $\Delta V(t)$, from time t to time $t + \Delta t$ is the increment of V at t. The value of $V(t_i)$ will thus be $V(0)$ plus the aggregate of all the increments until time t_i.

$$\begin{aligned} V(t_i) &= V(0) + \Delta V(t_0) + \cdots + \Delta V(t_{i-1}) \\ &= V(t_0) + \sum_{j=0}^{i-1} V(t_j)\, r\, \Delta t. \end{aligned} \tag{15.10}$$

Suppose that σ is equal to zero. In this case $S(t)$ has no random part and it behaves, in principle, like $V(t)$, albeit with instantaneous rate μ instead of r. For $S(t)$, rather than assuming that interest is paid every Δt, it is equivalent to assuming that the price appreciates every Δt. When σ is not equal to zero, the increments of $S(t)$ include the random part $\sigma \sqrt{\Delta t}\, Z$. At each point t_i, $S(t_i)$ is a known deterministic constant: at that point in time, the value of S is known with certainty.

The increment of $S(t_i)$ is composed of the deterministic part $S(t_i)\, \mu\, \Delta t$ plus the stochastic part $S(t_i)\, \sigma \sqrt{\Delta t}\, Z_i$, where $Z_i \sim \mathrm{N}(0,\, 1)$ and are i.i.d. Consequently, the counterparts of equations (15.9) and (15.10) are equation (15.11),

$$\Delta S(t) = S(t) \mu \Delta t + S(t) \sigma \sqrt{\Delta t} Z, \tag{15.11}$$

where

$$\Delta S(t) = S(t + \Delta t) - S(t), \tag{15.12}$$

and equation[6] (15.13),

$$S(t_i) = S(t_0) + \Delta S(t_0) + \cdots + \Delta S(t_{i-1})$$

[6]Note that $S(T)$ can be modeled in a similar way by using discretization and continuous compounding. Assume that $S(t_i + \Delta t) = S(t_i) e^{W_{\Delta\, t_i}}$, where $W_{\Delta\, t_i}$ is a random variable denoting the continuously compounded return over the time interval $\Delta\, t$, satisfies Assumptions I through III of Section 5.4. Thus, $E(W_{\Delta\, t_i}) = \mu \Delta t$ and $Var(W_{\Delta\, t_i}) = \sigma^2 \Delta t$. No assumptions are made regarding the distribution of $W_{\Delta\, t_i}$. As $n \to \infty$, applying the Central Limit Theorem yields that $Y_{T-t_0} = \sum_i W_{\Delta\, t_i} \sim N(\mu(T - t_0), \sigma\sqrt{T - t_0})$, and that $\frac{S(T)}{S(t_0)} = e^{Y_{T-t_0}}$ is a lognormally distributed random variable with an expected value of $\mu(T - t_0)$ and a standard deviation of $\sigma\sqrt{T - t_0}$. This approach may have the seeming advantage of not assuming directly that $W_{\Delta\, t}$ is normally distributed, but rather justifying the lognormal distribution assumption as a consequence of the Central Limit Theorem.

$$= \quad S(t_0) + \sum_{j=0}^{i-1} \left(S(t_j)\mu\Delta t + S(t_j)\sigma\sqrt{\Delta t} Z_j \right), \qquad (15.13)$$

respectively.

At time t, $S(t)$ has been realized and is no longer a random variable, but $S(t+\Delta t)$ is a random variable. Its value is described by

$$S(t+\Delta t) = S(t) + S(t)\mu\Delta t + S(t)\sigma\sqrt{\Delta t} Z, \qquad (15.14)$$

which depends on the realization of Z. Equation (15.14) is just another way of writing equation (15.11). It emphasizes the fact that by knowing $S(t)$ and calling a random number generator to simulate a value for Z, $S(t+\Delta t)$ may be calculated. Proceeding thus from time t_0 on, based on equation (15.13), the values $S(t_i)$, for $i = 1, ..., n$, are calculated. The procedure **Simudif** introduced in Section 14.4 calculates $S(t_i)$ for $i = 1, ..., n$, in the manner suggested. It draws or animates a sample path of $S(t)$ using linear interpolation.

The stochastic evolution of $S(t)$, as described by equation (15.14), can be generalized. We can allow the mean and standard deviation of the incremental change to depend on time and the current value of the stock, $S(t)$. Consider a stochastic evolution described by

$$S(t+\Delta t) = S(t) + a(S,t)\Delta t + b(S,t) Z\sqrt{\Delta t}, \qquad (15.15)$$

where $a(S,t)$ and $b(S,t)$ are[7] functions that depend on the time, t, and on the realization of S at time t.

The reader should immediately recognize that $a(S,t)$ and $b(S,t)$ in equation (15.15) are the $S(t)\mu$ and $\sigma\, S(t)$ of equation (15.14), respectively. Based on equation (15.15) and using the Δ notation, the stochastic increments of S can be written as

$$\Delta S = a(S,t)\Delta t + b(S,t)\Delta Z. \qquad (15.16)$$

In equation (15.16) instead of $Z\sqrt{\Delta t}$, where $Z \sim \mathrm{N}(0,1)$, we write ΔZ, which is a normally distributed random variable with an expected value of zero and a standard deviation of $\sqrt{\Delta t}$. Using this notation, $Z\sqrt{\Delta t}$ is the same random variable as ΔZ. ΔZ can therefore be interpreted as the stochastic increments of $Z(t)$ which is normally distributed with a mean of zero and a standard deviation of $\sigma\sqrt{t}$, i.e., $Z(t) \sim \mathrm{N}(0, \sqrt{t})$. We shall return to the notion of $\Delta\, Z$ in the near future but first let us run a few animations.

[7]In fact, a forthcoming analysis requires some further regularity conditions to be imposed on these functions. Formally speaking, we should have used the notation $a(S(t),t)$ and $b(S(t),t)$ instead of $a(S,t)$ and $b(S,t)$, respectively. These notations will be used interchangeably in the text.

15.3 Simulation of Stochastic Evolution

The procedure **Simudif** simulates the process[8] followed by $S(t)$ as it evolves over time as specified in equation (15.16). The parameters of the procedure are as outlined in Section 14.4. The output from the procedure **Simudif** is an animation (or a graph) of a sample path of the price process as well as the path when the stochastic element is not present, i.e., when $\sigma = 0$.

The reader should run the procedure several times without changing the value of the parameters. Each time the procedure is run the resultant sample path of $S(t)$ will be different. Each time the procedure is run, a random number generator is summoned to supply a value, and the procedure draws the sample path accordingly.

Let us start with a stock for which the price at time zero is \$100 with a μ parameter of 0.20 (expected return of 20 percent) and a σ parameter of 0.18 (volatility of 18 percent). (Recall that our units of time are measured in years and μ and σ are quoted as annual figures.)

```
> Simudif(100,0,1,50,.20*s,.18*s);
```

Hence, the on-line version of the book animates the process

$$\Delta S = S\mu\Delta t + S\sigma\Delta Z, \tag{15.17}$$

where $a(S,t) = S\mu = 0.20S$ and $b(S,t) = S\sigma = 0.18S$. A static version of the animation is presented in Figure 15.1. In this case a and b are functions only of the realization of S at time t, and are not functions of t. This process may also be expressed in the form of

$$\frac{\Delta S}{S} = \mu\Delta t + \sigma\Delta Z. \tag{15.18}$$

Let us rerun the above procedure with a larger N, say, $N = 100$, and display a static version of it in Figure 15.2.

```
> Simudif(100,0,1,100,.20*s,.18*s);
```

It should be obvious that the path for $\sigma = 0$ produces the same path each run: the expected value of the path. Since $E(\Delta Z) = 0$, and $S(t_0)$ is a deterministic constant, the value of $E(S(t_i))$ is obtained by substituting[9]

[8] Since we use discretization, the value of $S(t)$ is calculated only at the discrete points. The larger N is, the better the approximation.

[9] When $\sigma = 0$, $S(t+\Delta t) - S(t) = S(t)\mu\Delta t$, and thus $S(t+\Delta t) = S(t) + S(t)\mu\Delta t = S(t)(1+\mu\Delta t)$, and $S(T) = S(t_0+n\Delta t) = S(t_0)(1+\mu\Delta t)^n$. $S(T)$ is therefore an element in a geometric sequence, and perhaps this is the reason the process, when σ is different from zero and Δt approaches zero, is called geometric Brownian motion. Of course when σ is different from zero then $S(T) = S(t_0)\prod_{i=1}^{n}(1+\mu\Delta t + \Delta Z_i)$, where $\Delta Z_i \sim N(0, \sigma\sqrt{\Delta t})$ yields the same expression as in equation (15.13).

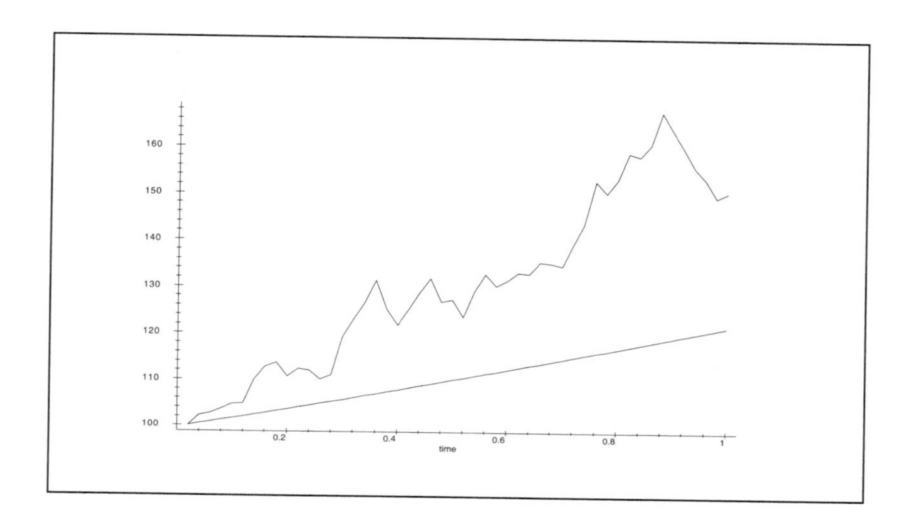

Figure 15.1: Simulation of $\frac{\Delta S}{S} = 0.2\Delta t + 0.18\,\Delta Z$ with $N = 50$

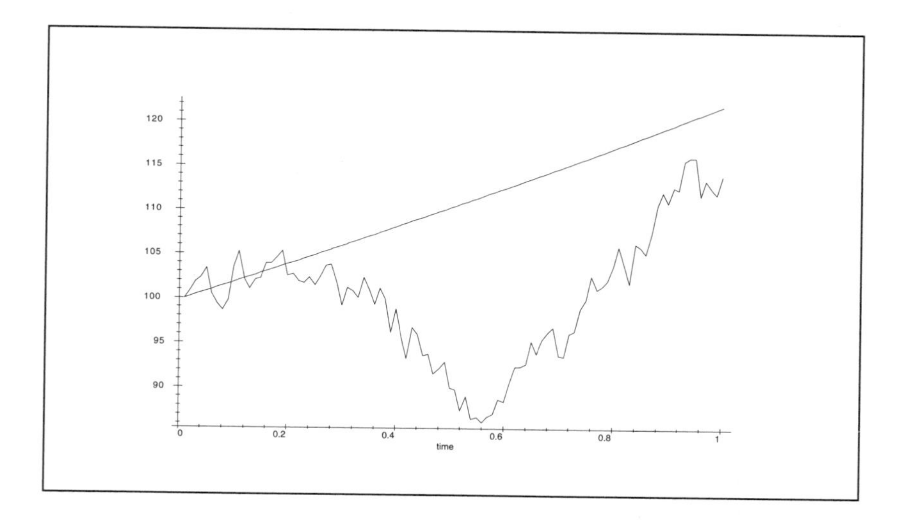

Figure 15.2: Simulation of $\frac{\Delta S}{S} = 0.2\Delta t + 0.18\Delta Z$ with $N = 100$

$\sigma = 0$ in (15.13) and is given by

$$E(S(t_i)) = S(t_0) + \sum_{i=1}^{n} S(t_i)\mu\Delta t.$$

Before looking at some other animations we would like to justify the notion that the increments of the variable Z, ΔZ, are normally distributed random variables with a mean of zero and a standard deviation of $\sigma\sqrt{\Delta t}$. That is, the increments satisfy equation (15.19) in the sense that the random variables on the left- and right-hand sides of the equation are indeed equal.

$$\Delta Z = Z(t + \Delta t) - Z(t) \tag{15.19}$$

This follows simply from the fact that Z is a normal random variable with independent increments (Assumption II of Section 5.4). As a consequence of this assumption we have that

$$Z(t) + Z(\Delta t) = Z(t + \Delta t), \tag{15.20}$$

where $Z(t) \sim \mathrm{N}(0, \sqrt{t})$ and $Z(\Delta\, t) \sim \mathrm{N}(0, \sqrt{\Delta t})$. Recall that the variance of the sum of two independent random variables is the sum of their individual variances, i.e., $Var\,(Z(t) + Z(\Delta t)) = Var\,(Z(t)) + Var\,(Z\Delta t) = t + \Delta t$, which is indeed the variance of $Z(t{+}\Delta t)$. The normality assumption assures that $Z(t{+}\Delta t)$ is a normally distributed random variable since it is the sum of normally distributed random variables. $Z(t{+}\Delta t)$ can therefore be written as $Z(t){+}Z(\Delta t)$, the way it is presented in equation (15.20). Denoting $Z(\Delta t)$ as ΔZ and presenting the evolution as $\Delta S = a(S,\, t)\Delta t{+}b(S,\, t)\Delta Z$ strengthens the notion of stochastic increments modeled by the random variable ΔZ and deterministic increments modeled, as usual, in deterministic calculus by Δt.

The reader can observe the phenomenon of the stochastic increments as related to the deterministic ones by running the procedure for a large value of N, say, $N = 1000$, for a time interval of one year. This is shown in Figure 15.3. (To save time we run it here as a plot. Omitting the parameter "plot" in the command line will generate an animation.)

```
>  Simudif(100,0,1,1000,.20*s,.18*s,plot);
```

Observe the relation between a realization (for a particular random number) of the sample path and the deterministic path. This also demonstrates what happens to the graph as N approaches infinity. It becomes very "pointy" but still forms a continuous path. The reader is invited and encouraged to run further animations making use of the **Simudif** procedure.

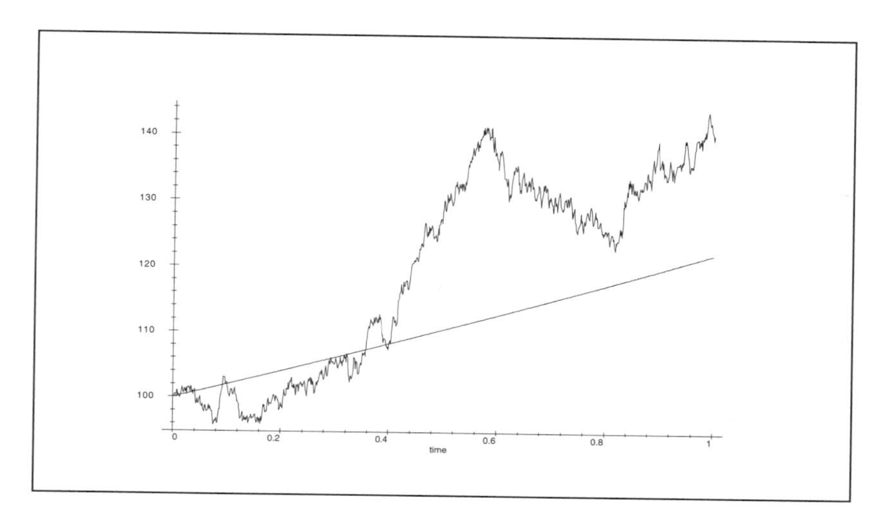

Figure 15.3: Simulation of $\frac{\Delta S}{S} = 0.2\Delta t + 0.18\Delta Z$ with $N = 1000$

The procedure **Simudif** can also simulate the instantaneous increments of the process of the continuously compounded rate of return. Recall that the continuously compounded rate of return over the period $[0,\, T]$, Y_T, was assumed to follow the distribution $\mathrm{N}(\mu\, T,\, \sigma\sqrt{T})$ with independent increments. Hence, given equally spaced knot points t_i, as defined in (15.7), implies equation (15.21).

$$Y_T = Y_0 + \sum_{i=1}^{n} \left(Y_{t_i} - Y_{t_{i-1}}\right) \tag{15.21}$$

In equation (15.21) Y_0 is the realization of the continuously compounded rate of return at time zero (spanning the time interval $[0,\, \Delta t]$). The rate of return from time zero to time t_i is $Y_{t_i} \sim N(\mu t_i,\, \sigma\sqrt{t_i})$. Denoting, as before,

$$\Delta Y_i = Y_{t_i} - Y_{t_{i-1}}, \quad i = 1, ..., n, \tag{15.22}$$

Y_{t_i} can be represented as

$$Y_{t_i} = Y_0 + \sum_{j=1}^{i-1} \Delta Y_{t_j}, \tag{15.23}$$

where the terms ΔY_{t_j} are i.i.d. random variables satisfying

$$\Delta Y_{t_j} \sim N(\mu\Delta t,\, \sigma\sqrt{\Delta t}).$$

Furthermore, ΔY_{t_i} as we have already seen, can be written as

$$\Delta Y_t = \mu \Delta t + \sigma \Delta Z_t, \tag{15.24}$$

or as

$$Y_{t_i} = Y_0 + \sum_{j=1}^{i-1} \mu \Delta t + \sum_{j=1}^{i-1} \sigma \Delta Z_{t_j}, \tag{15.25}$$

where $\Delta Z_{t_i} \sim N(0, \sqrt{\Delta t})$ and, for simplicity, the subindex i is omitted from equation (15.24).

We shall revisit equation (15.25) after we run the next few animations. Equation (15.24) is a special case of the general equation (15.16) since $a(S,\, t) = \mu$ and $b(S,\, t) = \sigma$. We can therefore use the **Simudif** procedure to simulate the stochastic evolution of the rate of return. This is done in the command below, where $Y_0 = 0.15$, $\mu = 0.15$, and $\sigma = 0.18$, and the output is presented in Figure 15.4.

```
> Simudif(.15,0,1,1000,.15,.18,plot);
```

One can also simulate the behavior of a standard Brownian motion, i.e., a process Z_t such that $Z_0 = 0$ and its evolution is given by $\Delta\, Z_{t_i}$. In this case $a(S(t),\, t) = 0$, $b(S(t),\, t) = 1$, and

$$Z(t_i) = \sum_{j=1}^{i-1} \Delta\, Z_{t_j}. \tag{15.26}$$

The deterministic path of $Z(t)$ coincides with the x-axis since $E(\Delta\, Z) = 0$. To see this, change the width of the line in the figure on-line. Place your cursor in the figure, click the right button, choose "style", and then from the submenu choose "line width" and then "medium". Note how the random shocks in Figure 15.5 cause the realization to move around the horizontal axis. Rerun the simulation for larger and for smaller values of sigma in order to appreciate the affect.

```
> Simudif(0,0,1,1000,0,1,plot);
```

To illustrate that this model of stochastic increments is not as restrictive as it may seem, we display a process called a "mean reverting process" in Figure 15.6. In such a process, denoted here as r, $a(r(t), t)$ is of the form $(r_l - r)\,\beta$, where r_l is believed to be the long-term value of the process. Hence when r deviates from its long-term value of r_l, the deterministic increment (which is also the mean of the increment) "corrects" the value of

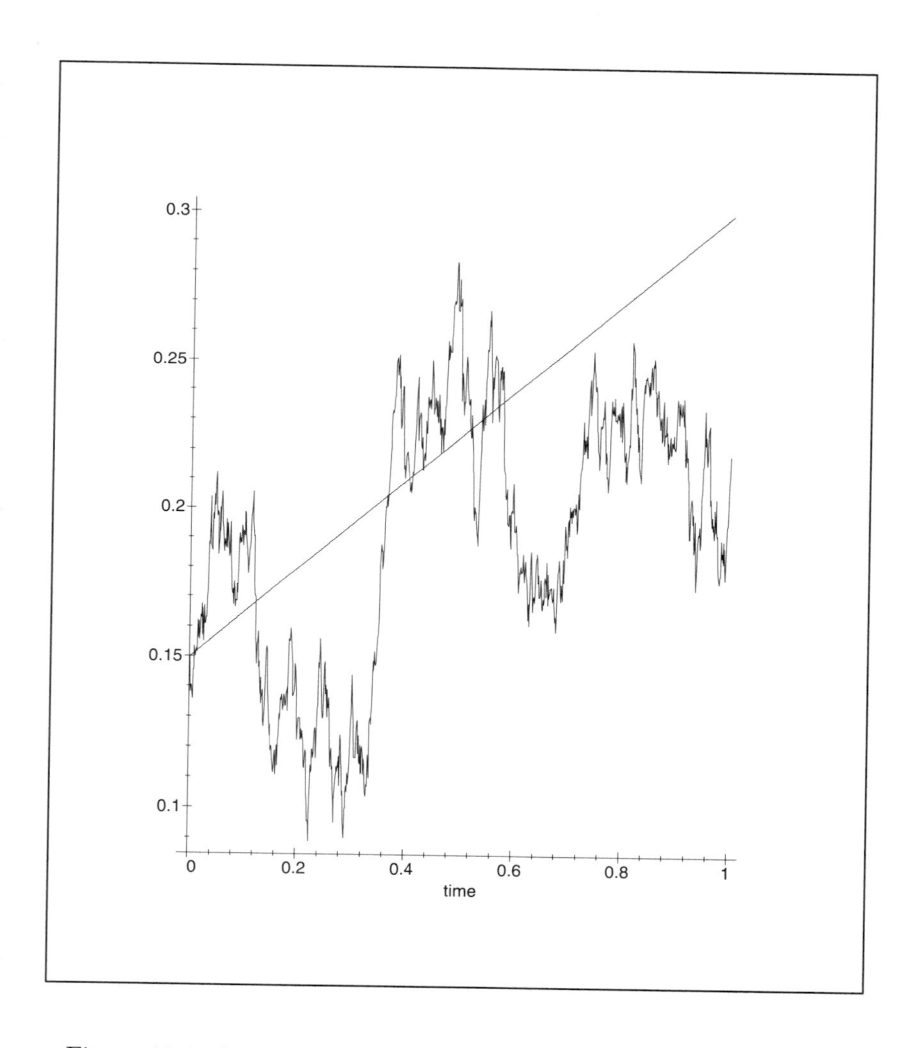

Figure 15.4: Simulation of $\Delta Y_t = 0.15\,\Delta t + 0.18\,\Delta Z_t$ for $N = 1000$

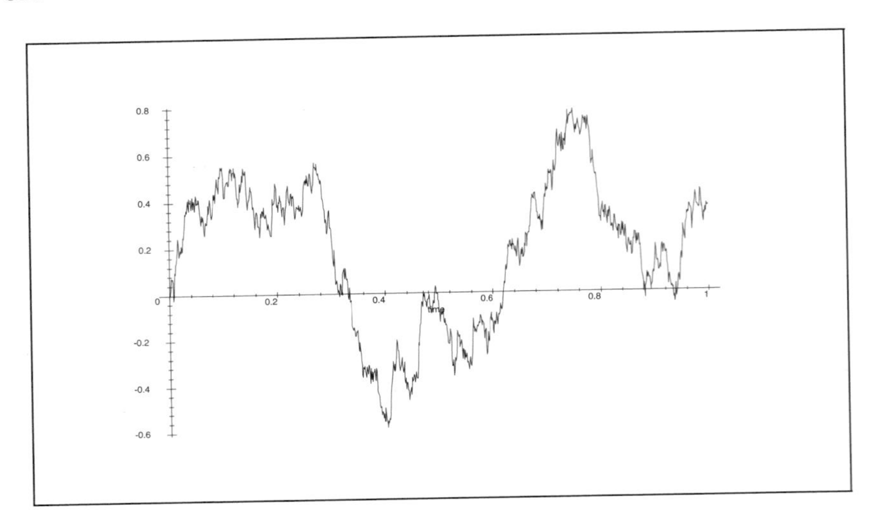

Figure 15.5: Simulation of $\Delta Z_t = 0\,\Delta t + \Delta Z_t$ for $N = 1000$

the process. It corrects it by a factor of β multiplied by the deviation from the long-term value r_l, $r_l - r$. Figure 15.6 displays the evolution of a process with increments $\Delta\, r$ modeled by

$$\Delta r = (r_l - r)\beta\Delta t + \sqrt{r}\sigma\Delta Z. \tag{15.27}$$

It is generated by the following command:

```
> Simudif(0.05,0,5,500,.07-s,.05*sqrt(s),plot);
```

Note, that if the input parameter to **Simudif** is a function, its argument must be denoted by s. Figure 15.6 is worth a thousand words of explanation as to the nature of the name of this process. These types of processes are very useful in modeling the stochastic evolution of the interest rate.

Watch the behavior of the deterministic part (the expected value) of the sample path. Try to run the simulation with different starting points, above and below the long-term value r_l, for different values of r_l, for different values of β and for different values of $b(r, t)$. Note that in the specification of equation (15.27), the standard deviation of the increment (volatility) of the process is an increasing function of the value of the process.

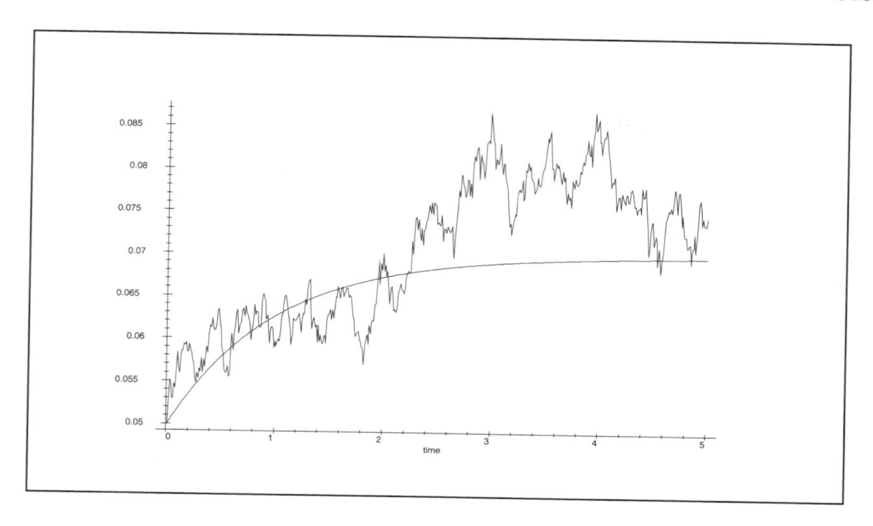

Figure 15.6: Simulation of $\Delta r = (r_l - r)\beta\Delta t + \sqrt{r}\sigma\Delta Z$

15.4 Stochastic Evolution: Toward a Continuous Model

As Δt approaches zero or as n approaches infinity our approximation becomes more and more accurate. In the case of a deterministic evolution such as that of V, $\frac{V(t+\Delta t)-V(t)}{\Delta t}$ approaches $\frac{dV(t)}{dt}$ and equation (15.8) becomes

$$\frac{dV(t)}{dt} = V(t)\,r. \tag{15.28}$$

Chapter 3 demonstrated that if the instantaneous evolution of $V(t)$ is as in (15.28), $V(T)$ is given[10] by (15.6). This equation is commonly written (by abuse of notation) as

$$dV(t) = V(t)rdt \tag{15.29}$$

or sometimes as

$$\frac{dV(t)}{V(t)} = rdt. \tag{15.30}$$

These equations convey the notion that the change in $V(t)$ over an instant of time is the limit of the change in $V(t)$ over a length of time Δt

[10] If you skipped the discussion of continuous compounding in Section 3.4.1 you can easily verify this result now. Simply take the derivative of $V(t) = e^{r\,t}$ with respect to t.

when Δt approaches zero. Also recall that the change over an instant of time is discussed in Appendix 15.11.1. Equation (15.29) therefore describes the limit, as Δt approaches zero, of equation (15.9), which is repeated here for convenience and is labeled as equation (15.31).

$$\Delta V(t) = V(t)r\Delta t \tag{15.31}$$

As Δt approaches zero and n approaches infinity, equation (15.10), repeated here for convenience and labeled equation (15.32),

$$\begin{aligned} V(t_i) &= V(t_0) + \Delta V(t_1) + \cdots + \Delta V(t_{i-1}) \\ &= V(t_0) + \sum_{j=1}^{i-1} V(t_j)r\Delta t, \end{aligned} \tag{15.32}$$

becomes

$$V(T) = V(t_0) + \int_{t_0}^{T} V(t)rdt. \tag{15.33}$$

This is the precise meaning of equation (15.29). In other words, equation (15.29) is really an integral equation but it uses a form of a shorthand notation and it is written as a differential equation.

What is the parallel limiting process by which (15.32) passes into (15.29) for a stochastic evolution? To attempt an answer to this question we focus first on the simplest stochastic evolution process we have described thus far. Consider the evolution through time of a Brownian motion, like Y_t, describing the continuously compounded rate of return on a certain stock. The models discussed above are, in a way, functions of this simple stochastic variable. We consider an evolution of the type described in equations (15.24) and (15.25), renumbered here as (15.34) and (15.35), respectively.

$$\Delta Y_{t_i} = \mu\Delta t + \sigma\Delta Z_{t_i} \tag{15.34}$$

$$Y_T = Y_0 + \sum_{i=0}^{n-1} \mu\Delta t + \sum_{i=0}^{n-1} \sigma\Delta Z_{t_i} \tag{15.35}$$

A full explanation of this limiting process is beyond the scope of this book and the reader is referred to the stochastic calculus books of footnote 2. Heuristic explanations (on two levels) are, however, offered herein.

The increments of Y_t are composed of a deterministic piece, $\mu\Delta t$, and a stochastic piece, $\sigma\Delta tZ$. Consider first the deterministic part. As Δt approaches zero, the same abuse of notation as was found in equation (15.29)

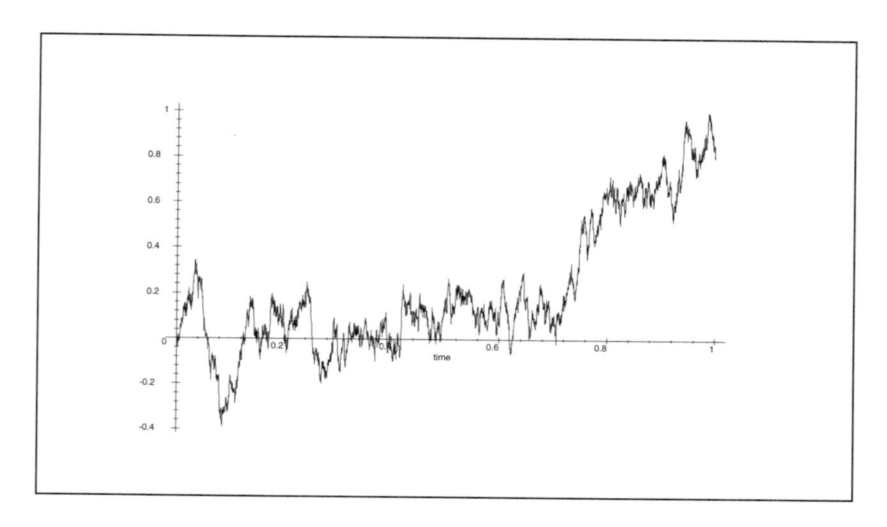

Figure 15.7: Simulation of $\Delta Z_t = 0\Delta t + \Delta Z_t$ for $N = 5000$

yields the relation $dY(t) = \mu dt$, with the same interpretation. The derivative of $Y(t)$ with respect to t is μ. Alternatively, as an integral equation, this can be written in the form $Y(t) = Y_0 + \int_0^t \mu dt$. When Δt approaches zero, $\lim_{\Delta t \to 0} \frac{\Delta Y_t}{\Delta t} = \frac{\partial Y_t}{\partial t}$ and $\lim_{\Delta t \to 0} \sum_{i=1}^{\frac{t}{\Delta t}} \mu \Delta t = \int_0^t \mu dt$.

As Δt approaches zero the increments of the stochastic part of Y_t, $\sigma \Delta Z$, also become instantaneous increments. These are realizations of a normally distributed random variable with an expected value of zero and a variance of Δt, i.e., $\Delta Z \sim N(0, \sqrt{\Delta t})$. As Δt approaches zero, the variance of ΔZ approaches zero also, and the stochastic increments are realizations of a random variable with a "very small" variance.

We cannot keep the same interpretation of the derivative of Y_t with respect to t as we had in the deterministic case. Let us take a look again, Figure 15.7, at a plot of the realization of the stochastic part, i.e., of $\sum_{i=1}^{n} \sigma \Delta Z_{t_i}$, where the interval $[0, 1]$ is divided into subintervals, each of length $\Delta t = \frac{1}{5000}$. This should give us a good appreciation of the limiting process.

```
>  Simudif(0,0,1,5000,0,1,plot);
```

One notices immediately that the sample path is not smooth at all; rather it is very "pointy". This does not look like a curve that is differentiable.

Recall that the function $|x|$ is not differentiable at zero because[11] of it being "pointy" there. The green sample path in the preceding diagram has many such "pointy" locations — in fact every point on the graph is "pointy". Consequently, it is possible to prove mathematically that the sample path is nowhere differentiable.

Furthermore, notice how the graph becomes thick at certain subintervals. This is because the graph jumps up and down with high frequency in very small intervals. It is like "up and down motions" next to each other generating a thickness of the graph. Put differently, the distance that this graph travels in a finite interval $[0, t]$ is not finite. Mathematically, this distance is referred to as the total variation of the graph, and is defined in equation (15.36). Denoting by $Z(t_i)$ the value of Z at time t_i, the total variation is

$$\lim_{n\to\infty}\sum_{i=1}^{n}\left|Z_{t_i}-Z_{t_{i-1}}\right| = \lim_{n\to\infty}\sum_{i=1}^{n}\left|\Delta Z_{t_i}\right|. \tag{15.36}$$

The total variation of the sample path will not be bounded, i.e., it will explode to infinity. (The precise claim is that the probability of the explosion happening is one.) A classical example of unbounded variation for a deterministic function is $x sin(\frac{\pi}{x})$ on, for example, the interval $(0, 1]$. You can watch this phenomenon (the explosion to infinity) in Figure 15.8 by executing the command below.

```
> plot(x*sin(Pi/x),x=0...1,numpoints=500,resolution=400);
```

The unbounded variation suggests that the sample path is not differentiable. Moreover, the "slope" of the sample path changes so frequently as it travels an infinite distance over a finite interval. Consequently, the slope, defined as a limit in the usual sense, does not exist. Yet, the probability of the sample path being continuous is equal to one.

As Δt approaches zero, the density function of ΔZ becomes more and more concentrated around zero. See Figure 5.4 in Chapter 5. This means that the realization of ΔZ, as Δt approaches zero, which is denoted by dZ, is almost surely going to be zero. We conclude then that

$$\lim_{\Delta t\to 0}\Pr\left(Y(t+\Delta t)\neq Y(t)\right)=0.$$

[11] There could be many lines tangent to the curve at the point zero. The derivative is the slope of such a line and hence cannot be uniquely defined. You can execute the command below, in the on-line version of the book, to observe this.

```
> plot(abs(x),x=-10..10);
```

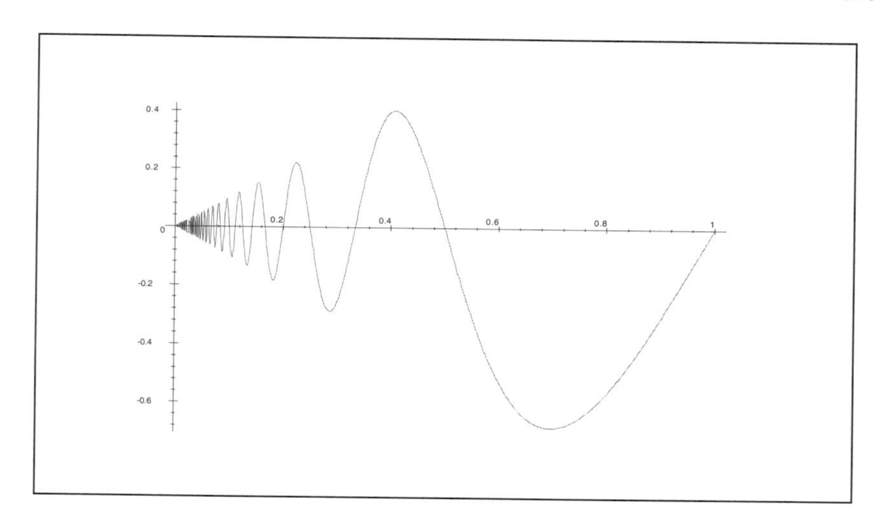

Figure 15.8: A Graph of the Function $x \sin\left(\frac{\pi}{x}\right)$

This is the reason that, as Δt approaches zero, the probability that the sample path will not be continuous is equal to zero. We therefore have a graph that is continuous with unbounded variation and very pointy. Even if we try to define[12] a derivative for this process in a probabilistic sense, we fail.

There is a way of defining an integral in this environment and keeping the spirit of the interpretation of the integral equation (15.33) from the deterministic case. This requires, as mentioned at the beginning of the Chapter, the use of a limit concept for random variables. In this environment we do not have a differentiation theory but we do have an integration theory. With the notation of dZ and dt, the instantaneous increments of $Y(t)$ are written (by abuse of notation) as

$$dY(t) = \mu dt + \sigma dZ. \tag{15.37}$$

This equation is the counterpart of equation (15.29). Here, though, the increments are stochastic and for this reason equation (15.37) is termed a stochastic differential equation.

The multiplier of dt is referred to as the drift (or expected value) coefficient, since it represents the constant drift of the process. It is also the

[12] Such a definition might be the expected value of the deterministic definition.

expected value of the instantaneous increment. The multiplier of dZ is referred to as the diffusion coefficient and it is the standard deviation of the instantaneous increment. It is the coefficient that accounts for the volatility of the stochastic increments of the process $Y(t)$.

We interpret equation (15.37) as an integral equation, i.e., as

$$Y(t) = Y_0 + \int_0^t \mu dt + \int_0^t \sigma dZ, \tag{15.38}$$

but we still have to define $\int_0^t \sigma dZ$. For a general process, $dS(t) = a(S(t),t)dt + b(S(t),t)dZ$, we have to define $\int_0^t b(S(u),u)dZ$. Intuition suggests that the integral should be defined as the expression in equation (15.39).

$$\lim_{n\to\infty} \sum_{i=1}^{n} \sigma \Delta Z_{t_i} \tag{15.39}$$

However, the sample path is of unbounded variation, notwithstanding the fact that it is continuous. This limit, if interpreted as the limit over the sample path, does not exist. However, the limit in equation (15.39) does exist when it is understood in the sense explained in Appendix 15.11.2.

The integral $\int_0^t \sigma dZ$ is a random variable by virtue of it "being the summation" of random increments. It is therefore termed a stochastic integral and is called an Ito integral in order that it be distinguished from any other definition of a stochastic integral. We shall refer to it simply as a stochastic integral since we do not deal with the other type of stochastic integral.

The integral $\int_0^t \sigma dZ$ is probably the simplest stochastic integral. dZ is the limit of $\Delta Z \sim \mathrm{N}(0, \sqrt{\Delta t})$ as Δt goes to zero. Hence one can think about a random variable $Z(T) \sim \mathrm{N}(0, \sqrt{T})$, where Δt are subintervals of $[0, T]$, as being the aggregate of infinitely many random variables dZ. Z was sliced into infinitely many independent components dZ. These components are i.i.d. with a distribution in the same family as Z. Thus Z is said to be infinitely divisible.

The integral operation reverses the slicing of Z into infinitely many dZ pieces. It aggregates (sums) all the dZ pieces back together to form the random variable Z. As a result $\int_0^T \sigma dZ = \sigma \int_0^T 1 dZ$, which in turn is equal to $\sigma Z(T) + \sigma Z(0)$. For a generalized process, one for which σ is replaced with a function $b(S(t), t)$, the integral is defined to be as in equation (15.40),

$$\lim_{n\to\infty} \sum_{i=1}^{n} b(S(t_i), t_i)\Delta Z_{t_i}, \tag{15.40}$$

where the limit is defined in the mean-square sense as explained in Appendix 15.11.2.

The right-hand side of equation (15.38) is composed of two integrals. The second is the one discussed above, the stochastic integral. The first is a regular integral over a deterministic function and it does not present a problem. However, for a general process such as $dS = a(S,t)\,dt + b(S,t)\,dZ$, the first integral, $\int a(S,t)\,dt$, is also a random variable. It is defined to be the limit of

$$\sum_{i=1}^{n} a(S(t_i),\, t_i)\Delta t_i \tag{15.41}$$

as n approaches infinity. For every n, the sum is a linear combination of random variables since $S(t_i)$ is a random variable. Nevertheless the same limiting concept should be adapted to this stochastic environment. This issue is the subject of Appendix 15.11.2.

We would like to emphasize here that, unlike in the deterministic case, it does matter if $S(t_i)$ is the value of S at the left or the right end point of the interval Δt_i. Our explanation here follows the Ito integral based on t_i being the left end point of the interval. That is, when applied to a deterministic function, for example, t^2, the sum is the area under the rectangles in Figure 15.9.

```
> student[leftbox](t^2,t=-3..3, 10,colour=BLUE);
```

The explanation of the limit concept in Appendix 15.11.2, used in deriving the Ito integral, sheds some light on the source of this difference. Readers who are not interested in this level of detail may continue to the next section, "Ito's Lemma". Ito's lemma is the main tool that allows us to derive the differential equation satisfied by a derivative security.

The next section starts with an intuitive insight into Ito's lemma and can be followed without impairing the continuity of the rest of the book by readers who opt not to read Appendix 15.11.2. However, for readers who are interested in a more in-depth analysis we suggest the following reading order: Appendix 15.11.2, followed by the next section, Section 15.5, followed by Appendix 15.11.3. The last appendix explains Ito's lemma, more rigorously, with a reference to the concept of a limit, as introduced in Appendix 15.11.2.

15.5 Ito's Lemma

The process followed by the price of a stock over time, as we discovered in the previous section, is a function of a variable that follows a Brownian

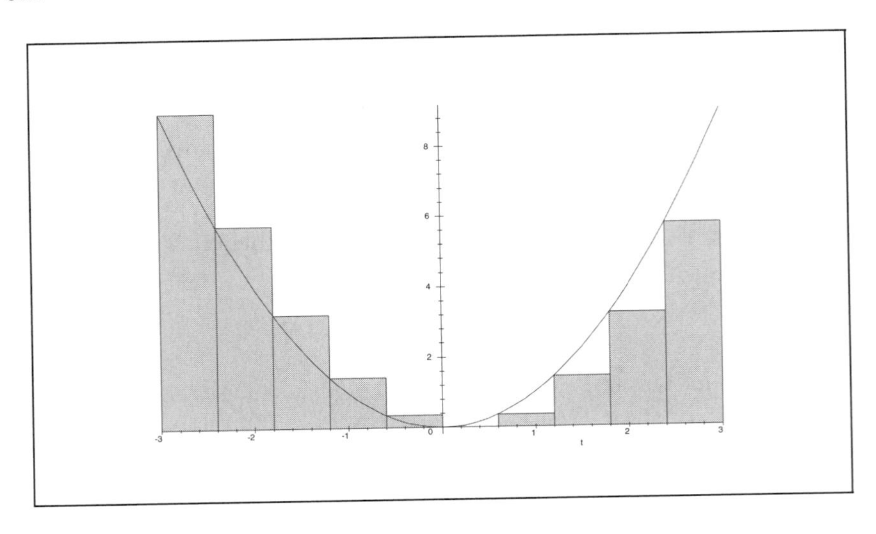

Figure 15.9: The Sum $\sum_{i=1}^{n} \Delta t_i t_i^2$, Based on the Left End Point of Δt_i

motion: the continuously compounded rate of return. We also heuristically got ourselves to the conclusion, see footnote 9, that the price process itself is a geometric Brownian motion. Brownian processes are also referred to as Ito processes and we will use either of these terminologies interchangeably. Ito's lemma addresses the instantaneous behavior of a function of an Ito process.

Ito's lemma claims that, given an Ito process, $x(t)$, for example, and given a deterministic function, for example, g of both x and t (where g is twice continuously differentiable), then $g(x(t), t)$ is again an Ito process. Ito's lemma specifies the stochastic differential equation satisfied by the new process.

Since a derivative security is a function of the underlying asset, applying Ito's lemma to this function will yield the instantaneous behavior of the derivative security. In what follows we will state Ito's lemma formally, and follow that statement with intuitive sketches of the proof of the lemma.

Lemma 13 (Ito's Lemma) *Let $x(t)$ be an Ito process given by*

$$dx = a(x,t)dt + b(x,t)dZ$$

and let g be a function that is twice continuously differentiable. Then $g(x(t),t)$

is again an Ito process and

$$dg = \left(ag_x + g_t + \frac{1}{2}b^2 g_{xx}\right) dt + bg_x dZ, \tag{15.42}$$

where $g_x = \frac{\partial g}{\partial x}$, $g_{xx} = \frac{\partial^2 g}{\partial x^2}$, *and* $g_t = \frac{\partial g}{\partial t}$. *For simplicity, the arguments of* a, b, x, *and* g *are omitted, e.g., we use* g_x *instead of* $\left.\frac{\partial g}{\partial x}\right|_{(x(t),t)}$ *and* x *instead of* $x(t)$.

15.5.1 Heuristic Proofs of Ito's Lemma

As promised, the explanation of the results contained in Ito's lemma will be provided on two levels. The explanation in the text is provided for readers who opted not to read Appendix 15.11.2, which deals with the limit of a random variable – that is, for those who are reading this section immediately after having read Section 15.4.

Both levels of explanation begin by applying to the Taylor series approximation. In the absence of a theory of differentiation in this stochastic environment, we turn to a Taylor series approximation in order to uncover the instantaneous increment of the process. However, this time, and not as in Chapter 7, footnote 6, we deal with a function of two variables, x and t.

Assuming that the current time is t, we would like to approximate the increment in the function g from time t to time $t+h$. To this end, we perform a Taylor series expansion on the function g. (Recall that g is a deterministic and twice continuously differentiable function). As time passes, the change in x from $x(t)$ to $x(t+h)$ is a random variable and we shall soon see how this affects our approximation.

To facilitate some of the MAPLE calculations we use dt to represent $\Delta t = t + h - t$, a nonzero step in time, and dx to represent $\Delta x = x(t+h) - x(t)$. In other words, we abuse the notation and use dx and dt to represent nonzero increments in x and t, respectively, although such notation is properly used only when Δt approaches zero. Consequently, calculating a Taylor series approximation of the second order to the increment of $g(x(t), t)$, $g(x(t+h), t+h) - g(x(t), t)$, yields equation (15.43).

$$\begin{aligned} g(x(t+h), t+h) - g(x(t), t) &= g_x dx + g_t dt + \\ &\quad g_{xt} dx dt + \frac{1}{2} g_{xx} (dx)^2 + \frac{1}{2} g_{tt} (dt)^2 + R \end{aligned} \tag{15.43}$$

where R is the remainder of the approximation involving higher order derivatives.

To understand this equation the reader may want to recall the discussion of Chapter 7, footnote 6, and read the footnote[13] below. We continue now with an intuitive explanation of Ito's lemma. A sketch of the more rigorous proof of the lemma is offered in Appendix 15.11.3. We recommend that the reader who is interested in that level of exposition read it following the end of this section.

Applying the guidelines of the deterministic case, as explained in Appendix 15.11.1, we ignore[14] elements in the expansion sequence that are of order $o(h)$. This explains why, as dt approaches zero, elements that involve powers of h that are greater than one are ignored. Consequently, the element $x_{tt}\,(dt)^2$ should be dropped from the approximation. Moreover, if we improve the approximation via Taylor series expansion by including higher order derivatives, each element of the expansion will involve a power of dt greater than one. Thus these elements would be ignored in the limiting process.

The reader may wish to execute the commands below to see what the approximation would look like if higher order partial derivatives were included in the Taylor series expansion. The command below generates the approximation including derivatives of up to the fourth order. Recall that $D_{i,j}(g)$ in MAPLE stands for the mixed derivative of g with respect to the ith and j^{th} arguments. The number of times you take the derivative with respect to the ith variable is represented by the number of times i appears in the subscript, e.g., $g_{xxt} = D_{1,1,2}(g)$ and $g_{xtt} = D_{1,2,2}(g)$.

```
> mtaylor(g(x+dx,t+dt),[dx,dt],5);
```

$$\begin{aligned}
&\mathrm{g}(x,\,t) + \mathrm{D}_1(g)(x,\,t)\,dx + \mathrm{D}_2(g)(x,\,t)\,dt + \frac{1}{2}\,\mathrm{D}_{1,\,1}(g)(x,\,t)\,dx^2 \\
&\quad + dx\,\mathrm{D}_{1,\,2}(g)(x,\,t)\,dt + \frac{1}{2}\,\mathrm{D}_{2,\,2}(g)(x,\,t)\,dt^2 \\
&\quad + \frac{1}{6}\,dx^3\,\mathrm{D}_{1,\,1,\,1}(g)(x,\,t) + \frac{1}{2}\,dx^2\,\mathrm{D}_{1,\,1,\,2}(g)(x,\,t)\,dt \\
&\quad + \frac{1}{2}\,dx\,dt^2\,\mathrm{D}_{1,\,2,\,2}(g)(x,\,t) + \frac{1}{6}\,dt^3\,\mathrm{D}_{2,\,2,\,2}(g)(x,\,t)
\end{aligned}$$

[13] Recall footnote 6 in Chapter 7 where a Taylor expansion of a function of one variable was presented. Bearing that example in mind, fix x thinking about g as a function of only one variable, t. This thought process explains the elements in equation (15.43) that involve derivatives with respect to t alone. In a symmetric way, fixing t explains the elements involving derivatives with respect to x alone. However, the movements in both x and t together also affect the value of the function and these are captured by the mixed derivative g_{xt}.

[14] Readers who are not familiar with this notation may like to pause and review it in Appendix 15.11.1.

$$+\frac{1}{24}\,dx^4\,\mathrm{D}_{1,1,1,1}(g)(x,\,t)+\frac{1}{6}\,dx^3\,\mathrm{D}_{1,1,1,2}(g)(x,\,t)\,dt$$
$$+\frac{1}{4}\,dx^2\,dt^2\,\mathrm{D}_{1,1,2,2}(g)(x,\,t)+\frac{1}{6}\,dx\,dt^3\,\mathrm{D}_{1,2,2,2}(g)(x,\,t)$$
$$+\frac{1}{24}\,dt^4\,\mathrm{D}_{2,2,2,2}(g)(x,\,t)$$

For reasons explained earlier, we limit ourselves to a second-order approximation via Taylor series expansion. This is done in **step0** below.

```
> step0:=g(x+dx,t+dt)=mtaylor(g(x+dx,t+dt), [dx,dt], 3)-\
> 1/2*D[2,2](g)(x,t)*dt^2;
```

$$\mathit{step0} := \mathrm{g}(x+dx,\,t+dt) = \mathrm{g}(x,\,t)+\mathrm{D}_1(g)(x,\,t)\,dx+\mathrm{D}_2(g)(x,\,t)\,dt$$
$$+\frac{1}{2}\,\mathrm{D}_{1,1}(g)(x,\,t)\,dx^2+dx\,\mathrm{D}_{1,2}(g)(x,\,t)\,dt$$

Denoting $g(x+dx,\,t+dt)-g(x,\,t)$ by dg, we have

```
> step1:=dg=D[1](g)(x,t)*dx+D[2](g)(x,t)*dt+\
> 1/2*D[1,1](g)(x,t)*dx^2+dx*D[1,2](g)(x,t)*dt;
```

$$\mathit{step1} := dg = \mathrm{D}_1(g)(x,\,t)\,dx+\mathrm{D}_2(g)(x,\,t)\,dt+\frac{1}{2}\,\mathrm{D}_{1,1}(g)(x,\,t)\,dx^2$$
$$+\,dx\,\mathrm{D}_{1,2}(g)(x,\,t)\,dt$$

Recall that

```
> dx=a*dt+b*dZ;
```

$$dx = a\,dt+b\,dZ$$

Substituting it in **step1** and expanding dx, we obtain

```
> step2:=expand(subs(dx=a*dt+b*dZ,step1));
```

$$\mathit{step2} := dg = \mathrm{D}_1(g)(x,\,t)\,a\,dt+\mathrm{D}_1(g)(x,\,t)\,b\,dZ+\mathrm{D}_2(g)(x,\,t)\,dt$$
$$+\frac{1}{2}\,\mathrm{D}_{1,1}(g)(x,\,t)\,a^2\,dt^2+\mathrm{D}_{1,1}(g)(x,\,t)\,a\,dt\,b\,dZ$$
$$+\frac{1}{2}\,\mathrm{D}_{1,1}(g)(x,\,t)\,b^2\,dZ^2+\mathrm{D}_{1,2}(g)(x,\,t)\,dt^2\,a$$
$$+\,\mathrm{D}_{1,2}(g)(x,\,t)\,dt\,b\,dZ$$

and since $(dt)^2$ is to be eliminated we arrive at the expression below, **step3**, for dg.

```
> step3:=subs(dt^2=0,step2);
```

$$\mathit{step3} := dg = \mathrm{D}_1(g)(x,\,t)\,a\,dt+\mathrm{D}_1(g)(x,\,t)\,b\,dZ+\mathrm{D}_2(g)(x,\,t)\,dt$$
$$+\,\mathrm{D}_{1,1}(g)(x,\,t)\,a\,dt\,b\,dZ+\frac{1}{2}\,\mathrm{D}_{1,1}(g)(x,\,t)\,b^2\,dZ^2$$
$$+\,\mathrm{D}_{1,2}(g)(x,\,t)\,dt\,b\,dZ$$

Consider the term $D_{1,2}(g)(x,t)dt\ b\ dZ$, or in a conventional notation,

$$bg_{xt}\Delta t\Delta Z.$$

The random variable $\Delta Z \sim \mathrm{N}(0, \sqrt{\Delta t})$ can also be expressed as[15] $\sqrt{\Delta t}Z$ for $Z \sim N(0,1)$. Employing this representation for ΔZ, we have that

$$\Delta t\Delta Z = \Delta t\sqrt{\Delta t}Z, \tag{15.44}$$

and hence the power of Δt in the expression $bg_{xt}\Delta t\Delta Z$ is larger than one. Consequently, this expression is of order $o(\Delta t)$ and can be ignored when Δt approaches zero. We therefore have to eliminate it from the expression dg in **step3**.

```
> step4:=algsubs(dt*dZ=0,step3);
```

$$step4 := dg = \frac{1}{2}\mathrm{D}_{1,1}(g)(x,\,t)\,b^2\,dZ^2 + (\mathrm{D}_1(g)(x,\,t)\,a + \mathrm{D}_2(g)(x,\,t))\,dt + \mathrm{D}_1(g)(x,\,t)\,b\,dZ$$

It remains for us to consider the expression involving $(dZ)^2$.

This is probably the most puzzling result for readers who encounter it for the first time. Making use of the same arguments as above, $(\Delta Z)^2 = \left(\sqrt{\Delta t}Z\right)^2 = \sqrt{\Delta t}\sqrt{\Delta t}Z^2 = \Delta t Z^2$, and hence it is not of order $o(\Delta\, t)$ and cannot be ignored. On the other hand, note that $E\left((\Delta\, Z)^2\right) = \Delta\, t$ and $Var((\Delta Z)^2) = 2(\Delta t)^2$. The reader may want to utilize the MAPLE procedure **MgmN** to calculate[16] the expected value of Z^2, where $Z \sim N(\mu,\,\sigma)$. We use dZ for $\Delta\, Z$ and dt for $\Delta\, t$ in the MAPLE calculations.

```
> Var((dZ)^2)=MgmN(4,0,sqrt(dt))-MgmN(2,0,sqrt(dt))^2;
```

$$\mathrm{Var}(dZ^2) = 2\,dt^2$$

```
> E((dZ)^2)=MgmN(2,0,sqrt(dt));
```

$$\mathrm{E}(dZ^2) = dt$$

[15] The latter follows from the fact that if Z is a normally distributed random variable with an expected value of zero and a standard deviation equal to one then $\sqrt{\Delta t}Z$ is a normally distributed random variable with an expected value equal to $E(\sqrt{\Delta t}Z) = \sqrt{\Delta t}E(Z) = 0$, and the variance of $\sqrt{\Delta t}Z$ is equal to $Var(\sqrt{\Delta t}Z) = \Delta t\, Var(Z) = \Delta t$ and hence a standard deviation of $\sqrt{\Delta t}$. Therefore ΔZ is equivalent to $Z\sqrt{\Delta t}$.

[16] The reader may want to utilize the MAPLE procedure **MgmN** to calculate the expected value of Z^n, where $Z \sim N(\mu,\,\sigma)$. This procedure is introduced in Appendix 15.11.3 and uses the input parameters n, μ, and σ (in that order).

We see that as Δt approaches zero, the variance of $(\Delta Z)^2$ approaches zero, as does its expected value of $(\Delta Z)^2$. However, the variance is of order $o(\Delta t)$, but not the expected value. Roughly speaking, as Δt approaches zero, the variance "reaches zero" faster than the expected value since the expected value is Δt and the variance is $2(\Delta t)^2$. However, when the variance "reaches zero", $(\Delta Z)^2$ ceases to be a random variable. A random variable with a variance of zero is simply a deterministic constant. The expected value of $(\Delta Z)^2$ is Δt and so it becomes the deterministic constant Δt.

It is also the case that as Δt approaches zero this deterministic constant becomes an instant of time. That is, it becomes dt. Put differently, the variance goes to zero faster than the expected value does and so the random variable becomes a constant before it becomes a (degenerate) random variable of zero. Thus it cannot be ignored and it becomes an instant of time. We encounter a puzzling result (which we hope will not be too puzzling after our explanation and the animation):

$$(dZ)^2 = dt. \tag{15.45}$$

Readers who are interested in a more rigorous argument can read Appendix 15.11.3. We, however, recommend that all readers run the animation below (not produced in the hard copy of the book) as a means of more fully appreciating the differences between the behavior of ΔZ and that of $(\Delta Z)^2$ as Δt approaches zero. Observe the animations produced in the on-line version of the book and the behavior of ΔZ as Δt approaches zero (the red graph) versus the behavior of $(\Delta Z)^2$ as Δt approaches zero (the blue graph). This may help shed some light as to why ΔZ remains a random variable (albeit with very small variance) as Δt approaches zero, while $(\Delta Z)^2$ becomes a deterministic "instant".

The density function of ΔZ as Δt approaches zero looks like a spike at zero. A "spike" is a function whose integral is equal to one and whose value is zero, except at the point zero. Note that it is important to run the animation in the correct direction. Make sure that you push the arrow in the lower part of the upper bar that points to the left (<—) before you run the animation. (The arrow will be visible when the cursor is in the graph area.)

```
> plotNormal1:=plots[animate](Normalpdf(y,0,Delta),\
> y=-10..10,Delta=0..1,thickness=2,color=red,frames=50):

> plotNormal2:=plots[animate](Normalpdf(y,Delta,2*Delta^2)\
> ,y=-10..10,Delta=0..1,thickness=2,color=blue,frames=50):
```

```
> plots[display](plotNormal1,plotNormal2):
```

On the other hand, the graph showing the behavior of $(\Delta Z)^2$ as Δt approaches zero in contrast to ΔZ (the blue graph, on its own below) eventually settles at zero. Its variance approaches zero, making it a nonrandom variable. Its expected value also goes to zero but at a faster rate.

```
> plots[animate](Normalpdf(y,Delta,2*Delta^2),y=-10..10,\
> Delta=0..1,thickness=2,color=blue,frames=50):
```

Operating these two animations develops some insight regarding why the Brownian motion is defined in such a way that its standard deviation depends on the square root of the time interval. Any other construction would either be trivial or lose the randomness over an instant dt.

The substitution of equation (15.45) in our expression for dg in **step4** yields **step5**.

```
> step5:=subs(dz*dz=dt,step4);
```

$$step5 := dg = (\mathrm{D}_1(g)(x,\,t)\,a + \mathrm{D}_2(g)(x,\,t))\,dt + \mathrm{D}_1(g)(x,\,t)\,b\,dz + \frac{1}{2}\,\mathrm{D}_{1,\,1}(g)(x,\,t)\,b^2\,dt$$

Collecting terms based on dt will generate the required result.

```
> step6:=collect(step5,dt);
```

$$step6 := dg = (\mathrm{D}_1(g)(x,\,t)\,a + \mathrm{D}_2(g)(x,\,t) + \frac{1}{2}\,\mathrm{D}_{1,\,1}(g)(x,\,t)\,b^2)\,dt + \mathrm{D}_1(g)(x,\,t)\,b\,dz$$

This is the result claimed[17] in equation (15.42).

15.5.2 Examples Utilizing Ito's Lemma

Armed with Ito's lemma we can calculate the instantaneous increment of a function of an Ito process. This is our main goal. After all, the price of a derivative security is a function of both the underlying asset price and time. We shall see in the next section how this result is utilized to derive the differential equation satisfied by a derivative security.

To ease the burden of calculation we have programmed a procedure in MAPLE and named it **ItosLemma**. The inputs to the procedure are, in order, the μ and σ parameters of an Ito process. These could be either

[17] Readers who are interested in a more rigorous explanation of Ito's lemma may want to read Appendix 15.11.3 now, followed by the subsequent subsections, proceeding finally to reading the subsection containing examples.

numerical values or functions of the process and of time. The process must be denoted by x and the time parameter by t. The next input parameter is the function of the process and again it must be an expression of x and t. The last two parameters are names, chosen by the user, to which the procedure will assign the drift coefficient (the multiplier of dt) and the volatility or diffusion coefficient (the multiplier of dZ), respectively, of the resulting process. (The procedure assumes that the process is given in the form $dx = \mu(x,t)dt + \sigma(x,t)dZ$.) The printed output of the procedure is the stochastic differential equation of the new process.

Let us consider a few examples that will help us adapt to the new rules of calculating the instantaneous increment of a function of an Ito process. There is no differentiation theory in this environment and we focus solely on functions of Brownian motions. Consequently, the instantaneous increments are found by an application of Ito's lemma. Moreover, the solution to a differential equation is a random variable whose increments are specified by the differential equation. Thus, finding the solution to a given differential equation amounts to finding a function (of a Brownian motion and of time) to which the application of Ito's lemma will result in the specified differential equation.

The Price Process of a Stock

At the beginning of the Chapter we provided an intuitive argument for concluding that if the price of the stock at time t satisfies $S(t) = S0e^{Y_t}$, where $Y_t \sim \mathrm{N}(\mu_Y t,\, \sigma_Y \sqrt{t})$, then its increments are given by

$$\Delta S(t) = S(t)\mu\Delta t + S(t)\sigma\Delta Z,$$

where $\mu = \mu_Y + \frac{\sigma^2}{2}$ and $\sigma = \sigma_Y$. In this last expression $S(t)\mu$ and $S(t)\sigma$ are the drift and diffusion coefficients of (the instantaneous increments of) $S(t)$, respectively.

We also remind the reader of the words of caution offered at the beginning of the Chapter. The caution concerned the relation of μ_Y and σ_Y to μ and σ in equations (15.4) and (15.5). This relation was discussed first in Chapter 5, footnote 21, which alerted the reader to the fact that $E(S(t) \mid S0) = E\left(S0e^{Y_t}\right)$, which equals $S0e^{\mu_{Y_t} + \frac{1}{2}\sigma_{Y_t}^2}$. Thus, for the expected value of the increments of $S(t)$ to be $S(t)\mu$, μ_Y should be equal to $\mu - \frac{\sigma^2}{2}$. Hence, on an intuitive level we arrived at the conclusion that the solution to the stochastic differential equation

$$dS(t) = S(t)\,\mu\,dt + S(t)\,\sigma\,d\,Z \tag{15.46}$$

is $S(t) = S0e^{X_t}$, where $X_t \sim \mathrm{N}\left(\left(\mu - \frac{\sigma^2}{2}\right)t, \sigma\sqrt{t}\right)$. This in turns implies that X_t is a Brownian motion with a drift parameter of $\mu - \frac{\sigma^2}{2}$ and a diffusion parameter σ. Note that in equation (15.46) the drift and diffusion coefficients are functions of both $S(t)$ and t.

Let us see how the solution to equation (15.46) can be verified with the aid of Ito's lemma. As we have mentioned before, in the absence of a differentiation theory, given a stochastic differential equation, we have to "guess" a solution that is a function of a Brownian motion and then apply Ito's lemma to that solution. This will result in the stochastic differential equation that is satisfied by our "guess" which, we hope, coincides with the given stochastic differential equation.

Assume therefore that we are presented with the stochastic differential equation (15.46) and we "guess" the solution $S0e^{X_t}$. To verify our guess we make use of the procedure **ItosLemma** and derive the stochastic increments of $S0e^{X_t}$, where $X_t \sim \mathrm{N}\left(\left(\mu - \frac{\sigma^2}{2}\right)t, \sigma\sqrt{t}\right)$.

```
> ItosLemma(mu-(sigma^2)/2,sigma,S0*exp(x),GBMmu,GBMsigma);
```

$$\mathrm{d}(S0\ e^x) = S0\ e^x\ \mu\ dt + S0\ e^x\ \sigma\ dZ$$

Denoting $S0e^{X_t}$ as $S(t)$ confirms that the same stochastic differential equation is obtained. Thus we know that the solution to the differential equation (15.46) is $S0\,e^{X_t}$. Recall that the procedure assigns to the parameters **GBMmu** and **GBMsigma** the expected value and standard deviation of the increments of $S0e^{X_t}$, respectively. This is verified by the MAPLE commands below.

```
> GBMmu;
```

$$S0\ e^x\ \mu$$

```
> GBMsigma;
```

$$S0\ e^x\ \sigma$$

The Process of a Forward Price

Let us look at another example. Consider the forward price of a stock. In equation (2.9) we explained that the forward price of a stock at time t is $S(t)e^{r(T-t)}$, where T is the delivery time of the forward contract. As the price of the stock evolves through time so will the forward price.

Assume, as usual, that the price of the stock follows a geometric Brownian motion as in equation (15.46). Thus, we can employ Ito's lemma to find the stochastic differential equation followed by the forward price. To this end we execute the command below, where μx and σx are the drift and diffusion parameters of $S(t)$, respectively, and **Newmu** and **Newsigma** are the parameters of the process $F(t) = S(t)e^{r(T-t)}$.

```
> ItosLemma(mu*x,sigma*x,x*exp(r*(T-t)),Newmu,Newsigma);
```

$$\mathrm{d}(x\,e^{(r\,(T-t))}) = (e^{(-r\,(-T+t))}\,\mu\,x - x\,r\,e^{(-r\,(-T+t))})\,dt + e^{(-r\,(-T+t))}\,\sigma\,x\,dZ$$

Denoting $xe^{r\,(T-t)}$ by F we have

$$dF = F(\mu - r)dt + F\sigma dZ, \tag{15.47}$$

which means that the forward price F follows a geometric Brownian motion with a drift of $F(\mu - r)$. Thus, under the risk-neutral probability, when μ, the drift parameter of x, is replaced with r, F follows

$$\frac{dF}{F} = \sigma dZ.$$

This is verified below by executing **ItosLemma** where μx is replaced with rx.

```
> ItosLemma(r*x,sigma*x,x*exp(r*(T-t)),Newmu,Newsigma);
```

$$\mathrm{d}(x\,e^{(r\,(T-t))}) = e^{(-r\,(-T+t))}\,\sigma\,x\,dZ$$

Thus, under the risk-neutral probability the expected value of F is its current value and consequently $F(t)$ is a Martingale. An exercise at the end of the Chapter asks the reader to justify this conclusion, modeling F as a function of the continuously compounded rate of return on the stock.

Solving a Stochastic Differential Equation

Since we do not have a differentiation theory in this environment, finding a solution to a stochastic differential equation might involve some "guessing". We have alluded to this method in the former section. Let us now explain this with an example in which we begin with an equation which has not been discussed before. Consider the stochastic differential equation (15.48) below.

$$dw = \frac{2\mu e^{w} - \sigma^2}{2e^{2w}}dt + \frac{\sigma}{e^{w}}dZ \tag{15.48}$$

Solving equation (15.48) means finding a function g of a Brownian motion with increments $dx = \mu dt + \sigma dZ$ such that the increments of the function g, dg, will behave like the given stochastic differential equation. We therefore have to make a "guess" at the function and this may not always be an easy task.

Let us try to find the increments of $g(x) = \ln x$ with the aid of the procedure **ItosLemma**. (Recall that S should be replaced with x in the MAPLE procedure.)

```
> ItosLemma(mu,sigma,ln(x),Wmu,Wsigma);
```

$$\mathrm{d}(\ln(x)) = \frac{1}{2}\,\frac{(2\,\mu\,x - \sigma^2)\,dt}{x^2} + \frac{\sigma\,dZ}{x}$$

Applying Ito's lemma to the function $\ln x$ confirms that $x = e^w$ is the solution to equation (15.48).

Having acquired some insight into Ito's lemma we are ready to put it to work. Ito's lemma provides the tool that was previously missing in order to develop the Black–Scholes formula. The next section utilizes this lemma to arrive at the differential equation satisfied by every derivative security.

15.6 The Black–Scholes Differential Equation

We are finally at the point where we are ready to assemble the pieces introduced in this Chapter and to use them to derive the Black–Scholes formula. Consider a derivative security that expires at time T. Let us denote, for the purpose of this derivation, the value of a derivative security at time t, i.e., $T-t$ units of time prior to expiration, by the function $V(S(t), t)$, where $S(t)$ is the price of the underlying asset at time t. At time T, if the derivative security is a European call option then $V(S(T), T) = \max(0, S(T) - K)$. We would like to see what can be said about the value of the derivative security, $V(S(t), t)$, at some time $t < T$, prior to its expiration.

For the rest of this derivation we will not specify whether the derivative security is a call or a put option, or any other type of a derivative security. This derivation is independent of the type of the derivative security. What matters is that the payoff from the derivative security is contingent only on the price of the underlying asset, $S(t)$, and on time t. In turn, the price of the underlying asset depends on the realization of the continuously compounded rate of return. The uncertainty regarding this rate of return is driven by an Ito process. Hence, as we already alluded to in Section 5.1, the states of nature at time t can be thought of as the possible realizations of an Ito process dZ. This will become clearer as we proceed with the specifications of this new environment and its analogy to the models that we have discussed in past Chapters.

There is a new continuous-time stochastic environment in which we are now operating. It has some different (perhaps even strange) rules, but the core ideas used in deriving the formula are unchanged. We are simply going to use the no-arbitrage condition and pricing by replication introduced in Chapter 1, with some modifications for use in this new environment. We will demonstrate two ways (that are essentially the same) of deriving the differential equation. The analogy between this derivation and the no-arbitrage condition of Chapter 1 is introduced with the following way of reflecting on the concepts.

We have a sequence of one-period models. In each, the length of a time period is an instant, in the sense explained in Appendix 15.11.1. In the model of Chapter 1 there were a finite number of states of nature represented by the columns of the payoff matrix. The value of each security in each state was represented by the element under the column for each security. This model was later generalized to a continuum of states of nature represented by the price of the underlying asset, S. In that case we represented the payoff or value of each security as a function of S which acted as the state of nature.

We are also well versed in the binomial model of Chapters 13 and 14. The binomial model involves two assets, one of which is risky. The binomial model was portrayed as a sequence of one-period models, each of which can have only two possible states of nature. Each of the models in the sequence specified the value of the underlying asset at the beginning of the period and the two possible values of that same asset at the end of the period. In that setting, this was sufficient to replicate the derivative security. Operating recursively, beginning with the time of maturity of the derivative and working backward, we calculated the value of the derivative security given the price of the underlying asset at the beginning of each time period.

Conceptually, the following development, in this new environment, outlines a pricing model for a derivative security that uses the same basic ingredients of the binomial model. We now have a sequence of one-period models, the length of each time period being an instant. There is a continuum of states of nature and they are represented as the realization of a random variable dZ. At the beginning of each such period we only need to know the value of the underlying asset and, with the aid of the risk-free asset, we can replicate the payoff from the derivative security. Let us see how this works.

We focus on three assets in this market: a stock the price of which is denoted by x, a derivative security, V, that is contingent on it, and a risk-free asset. We assume a stock that pays a continuous dividend yield at a rate of y (a deterministic constant). Thus the dividend payment at each instant of time is $yxdt$. We shall see shortly how this assumption affects our calculation. The increments of the stock price are specified by equation

(15.49).

$$dx = \mu x dt + \sigma x dZ \tag{15.49}$$

```
> dx:=mu*x*dt+sigma*x*dZ;
```

$$dX := \mu\, x\, dt + \sigma\, x\, dZ$$

Holding the stock produces the continuous stream of dividends, denoted by dDx in equation (15.50) and defined in MAPLE below. (See also the explanation in Sections 6.4 and 10.3.2.)

$$dDx = yxdt \tag{15.50}$$

```
> dDx:=y*x*dt;
```

$$dDx := y\, x\, dt$$

The states of nature in this environment are, rather than the price of the underlying asset, the increments dZ of Z. Thus the possible states of nature are represented by the dZ. dZ can be thought of as the "generator" of the randomness (states of nature) in this environment. Each realization of dZ corresponds to a column in the payoff matrix of Chapter 1. The dZ in this model plays the role of the two possible values of the underlying asset in the binomial model.

The increments, dx, which represent the infinitesimal change in the price of the stock over the next instant, are random variables. They depend on the price of the underlying asset at the beginning of the period and on the realization of dZ. Thus we use the same method of representation as in the case with a continuum of states of nature. The expression dx specified in equation (15.49) is a random variable specifying the increments of x for every state of nature dZ.

The second risky asset is the derivative security with increments that are dependent on the realization of x, the underlying asset, and on time. Since the realization of x depends on dZ, so does the derivative security that is contingent on x. The increment of the derivative security is denoted by dV and is specified with the aid of Ito's lemma in equation (15.51). (Note: to operate **ItosLemma** the variable must be defined as small x.)

```
> ItosLemma(mu*x,sigma*x,V(x,t),muV,sigmaV);
```

$$\mathrm{d}(\mathrm{V}(x,\, t)) = ((\frac{\partial}{\partial x}\,\mathrm{V}(x,\, t))\,\mu\, x + (\frac{\partial}{\partial t}\,\mathrm{V}(x,\, t)) + \frac{1}{2}\,(\frac{\partial^2}{\partial x^2}\,\mathrm{V}(x,\, t))\,\sigma^2\, x^2)\; dt$$
$$+ (\frac{\partial}{\partial x}\,\mathrm{V}(x,\, t))\,\sigma\, x\, dZ$$

To simplify matters, we use a more compact notation than that of MAPLE. In equation (15.51) we omit the arguments of the functions involved and denote derivatives by a subindex, i.e., $\frac{\partial V}{\partial x}(x,t) = V_x$, $\frac{\partial^2 V}{\partial x^2}(x,t) = V_{xx}$ etc.

$$dV = \left(\mu x V_x + V_t + \frac{1}{2}\sigma^2 x^2 V_{xx}\right) dt + V_x \sigma x dZ \tag{15.51}$$

In MAPLE, we also define the variable dV below. (Note that **muV** and **sigmaV** were defined by the **ItosLemma** procedure above.)

```
> dV:=muV*dt+sigmaV*dZ;
```

$$dV := \left(\left(\frac{\partial}{\partial x}\,\mathrm{V}(x,\,t)\right)\mu\,x + \left(\frac{\partial}{\partial t}\,\mathrm{V}(x,\,t)\right) + \frac{1}{2}\left(\frac{\partial^2}{\partial x^2}\,\mathrm{V}(x,\,t)\right)\sigma^2\,x^2\right) dt + \left(\frac{\partial}{\partial x}\,\mathrm{V}(x,\,t)\right)\sigma\,x\,dZ$$

The third asset is a bond (with a face value of $1) that pays interest continuously. It can also be thought of as a money market account. The bond has (as it should) the same return for every state of nature. The bond value in each possible state of nature is not dependent on dZ. It is analogous to the risk-free rate of interest in the model of Chapter 1. In that model, the risk-free asset had the same number appearing in each column of the payoff matrix. Here, regardless of the value of dZ, the same return is realized. Thus the increments of the bond are specified by equation (15.52).

$$dB = rdt, \tag{15.52}$$

where r is the continuously compounded risk-free rate. We also define in MAPLE the variable dB.

```
> dB:=r*dt;
```

$$dB := r\,dt$$

In order to price the derivative security by replication we operate in exactly the same manner that we have done before. We try to find a portfolio that replicates the return on the derivative security. If we succeed, by arbitrage arguments the price of the portfolio must equal the price of the replicated asset. Thus we start by solving for a portfolio composed of β units of the bond and δ units of the stock.

The letter δ is the lower case delta, and we have chosen this Greek symbol for a reason. As we shall soon see it is the Delta (Δ) that we already know from hedging with the Greeks (Chapter 8). In order not to confuse it with a Δ step in time we will use here the lower case δ.

At this stage we have to account for dividends. The stock pays a continuous dividend stream at a rate of y. Thus, holding one unit of the stock over dt units of time provides the holder with the (deterministic) dividend stream paid during that time, namely, $xydt$. Consequently, holding δ units of the stock over dt entitles the holder to the dividends payments of $\delta xydt$. Thus the increments of the portfolio (consisting of δ units of stock and β units of the bond), which we denote by $dP = \delta dx + \delta xydt + \beta dB$, are given by equation (15.53).

$$dP = (\beta r + \delta(\mu + y)x)\, dt + \delta\sigma x dZ \tag{15.53}$$

We also define the variable dP in MAPLE.

```
> dP:=beta*r*dt+delta*((mu+y)*x*dt+sigma*x*dZ);
```

$$dP := \beta\, r\, dt + \delta\,((\mu + y)\, x\, dt + \sigma\, x\, dZ)$$

The next step is the counterpart of solving for a replicating portfolio, i.e., for δ and β. In the binomial model this amounted to solving two equations, the system of equation in (13.16), with two unknowns: the holding of the underlying asset and the bond. Each equation constrains the payoff from the portfolio to be equal to that of the replicated security in one of the two states of nature.

In the current environment the states of nature are a continuum generated by dZ. However, there is the same uncertainty source dZ in both the underlying asset and the derivative security. Thus, ensuring that the multipliers of dt (the drift) and of dZ (the diffusion coefficient) in dP and dV are the same[18] guarantees that for every realization of Z, $dV = dP$. That is, the portfolio P replicates the derivative security V. To solve for δ and β we first collect the increments of dP based on dt and dZ.

```
> dP:=collect(dP,dt);
```

$$dP := (\beta\, r + \delta\,(\mu + y)\, x)\, dt + \delta\, \sigma\, x\, dZ$$

Equating the multipliers of dZ in both dP in equation (15.53) and dV in equation (15.51) yields equation (15.54).

$$\delta\sigma x = \frac{\partial V}{\partial x}\sigma x \tag{15.54}$$

Thus,

$$\delta = \frac{\partial V}{\partial x}, \tag{15.55}$$

[18] Equating the multipliers of dt and dZ is not only sufficient, but also necessary for the replication. This is due to the unique decomposition of an Ito process.

which is confirmed by MAPLE below.

```
> solve(delta*sigma*x=sigmaV,delta);
```

$$\frac{\partial}{\partial x}\mathrm{V}(x,\,t)$$

The reader may note that δ is indeed the partial derivative of the value of the derivative security with respect to the price of the underlying asset. This is the way it was defined[19] in equation (7.6). The solution for β is obtained by equating the multipliers of dt in dP and dV.

```
> solve(beta*r+delta*(mu+y)*x=muV,beta);
```

$$\frac{1}{2}\,\frac{-2\,\delta\,x\,\mu-2\,\delta\,x\,y+2\,(\frac{\partial}{\partial x}\mathrm{V}(x,\,t))\,\mu\,x+2\,(\frac{\partial}{\partial t}\mathrm{V}(x,\,t))+(\frac{\partial^2}{\partial x^2}\mathrm{V}(x,\,t))\,\sigma^2\,x^2}{r}$$

The command below substitutes the solution of δ in the expression for β.

```
> subs(delta=diff(V(x,t),x),\
> solve(beta*r+delta*(mu+y)*x=muV,beta));
```

$$\frac{1}{2}\,\frac{-2\,(\frac{\partial}{\partial x}\mathrm{V}(x,\,t))\,x\,y+2\,(\frac{\partial}{\partial t}\mathrm{V}(x,\,t))+(\frac{\partial^2}{\partial x^2}\mathrm{V}(x,\,t))\,\sigma^2\,x^2}{r}$$

Hence, the solution for β is given in equation (15.56).

$$\beta=-\frac{2xyV_x-2V_t-\sigma^2x^2V_{xx}}{2\,r} \tag{15.56}$$

We now proceed exactly as we did in every replication argument used thus far. We equate the price of the portfolio to the price of the replicated asset to arrive at equation (15.57). (Since one unit of the bond costs \$1, buying β units of the bond costs \$$\beta$.)

$$V(x,t)=\delta x+\beta \tag{15.57}$$

The equation is produced in MAPLE below.

```
> diff_BS1:=V(t,x)=solve(delta*sigma*x=sigmaV,delta)*x+\
> subs(delta=diff(V(x,t),x),solve(beta*r+delta*(mu+y)*
> x=muV,beta));
```

$$\mathit{diff\_BS1}:=\mathrm{V}(t,\,x)=(\frac{\partial}{\partial x}\mathrm{V}(x,\,t))\,x-\frac{1}{2}\,\frac{2\,(\frac{\partial}{\partial x}\mathrm{V}(x,\,t))\,x\,y-2\,(\frac{\partial}{\partial t}\mathrm{V}(x,\,t))-(\frac{\partial^2}{\partial x^2}\mathrm{V}(x,\,t))\,\sigma^2\,x^2}{r}$$

[19] Granted, this derivative is shown with a different sign here. This is a result of the way that t is defined in each case.

Multiplying both sides by r and rearranging generates the partial differential equation (15.58) as confirmed by MAPLE.

```
> collect(simplify(diff_BS1*r),diff(V(x,t),x));
```

$$r\,\mathrm{V}(t,\,x) = (x\,r - x\,y)\,(\frac{\partial}{\partial x}\,\mathrm{V}(x,\,t)) + (\frac{\partial}{\partial t}\,\mathrm{V}(x,\,t)) + \frac{1}{2}\,(\frac{\partial^2}{\partial x^2}\,\mathrm{V}(x,\,t))\,\sigma^2\,x^2$$

$$rV = x(r-y)V_x + V_t + \frac{1}{2}\sigma^2 x^2 V_{xx} \tag{15.58}$$

If no dividends are paid by the underlying asset over the life of the derivative security then we can substitute $y = 0$ in equation (15.58). Consequently, the partial differential equation satisfied by the derivative security is given by equation (15.59).

$$rV = xrV_x + V_t + \frac{1}{2}\sigma^2 x^2 V_{xx} \tag{15.59}$$

At first glance, the reader might be puzzled by the fact that only σ, and not μ, appears in the differential equation. One must remember, however, that the differential equation was derived based on arbitrage arguments which are independent of risk preferences. All investors in the economy regardless of their risk preference must agree that this equation must be satisfied, or else arbitrage opportunities exist. We have encountered this result, albeit from a different angle, at the end of Section 5.9.

The expected value of a stock compensates for its risk. Therefore, investors with different risk "appetites" (preferences) may disagree on what should be the expected value of a stock for it to be "attractive". Investors who find the expected value to compensate for the risk implicit in the stock hold the stock, and others do not. Hence, a pricing relation for the derivative security that is dependent on the expected value of the underlying asset, μ, might not be valid for all investors.

The lack of μ in the differential equation is consistent with the arbitrage arguments that induce this relation. In fact we should not be surprised by the lack of μ, but rather should expect the equation not to be dependent on μ. In Section 15.6.1 we derive the differential equation utilizing risk-neutral valuation arguments. This will uncover the usual duality between the two methods of pricing mentioned first in Chapter 1. Thereby it will shed some more light on the lack of μ in equation (15.58).

Throughout this development we suppressed the fact that δ and β are actually time dependent. The price of the derivative security is calculated in this continuous-time model very much like in the binomial model. At the beginning of each time period, in both models, we know the value of the underlying asset and of the derivative security. Given that information, we solve for the position in the bond and in the underlying asset that replicates the holding of the derivative security over the next period. In the binomial case the period length is some positive Δt but here it is an instant. While the tools we used for these solutions are different, the concepts are much the same. In fact, what we have accomplished above is exactly the same as what we have done in the binomial model.

We have solved symbolically for the replicating portfolio δ and β in terms of the value of the underlying asset and the value of the derivative security at the beginning of the period. The δ and β are the counterparts of Xs and Xb in equation (13.16) of Section 13.3.2, respectively. Thus, in principle, we are solving for the value of the derivative security by recursion from the maturity time, T, backward to the initial time instant by instant. This is exactly the same way it was done in the binomial model in Section 13.3. However, the binomial model is in a setting of discrete time. The realizations of the states of nature are only **Up** an **Down** while here they are represented by dZ.

In the binomial model we specified the payoffs from the derivative security in each possible state of nature at its maturity. This knowledge enabled us to solve for the value of the derivative security one period prior to maturity and to continue in a recursive manner to its value at time zero. We are missing the analogue of that information here. The missing information is called a boundary condition and is needed to solve the differential equation (15.58).

Note also that we have not specified the nature of the derivative security, e.g., put, call, American, or European. Hence the arguments presented above hold for every derivative security. What distinguishes one from the other are the boundary conditions. In order to solve for the price of a European call option we have to use the boundary condition in equation (15.60).

$$V(x,T) = max(x-K,0) \tag{15.60}$$

The reader is asked, in the exercises of this Chapter, to verify that the expression in equation (6.16), specifying the price of a call option, satisfies equations (15.59) and (15.60).

Equation (15.59) also demonstrates a relation that exists among three of the "Greeks" introduced in Chapter 7. Recall that Θ was defined, in equation (7.1), as the partial derivative of the call value with respect to the variable "time to maturity". In the notation used in this section, the time to maturity is $T-t$, where T is the maturity of the option and t is the current time. Hence $V_t = -\Theta$. A careful examination of equation (15.59) reveals the relation in equation (15.61), where we use δ for the Δ of Chapter 7.

$$rV = \delta x r - \Theta + \frac{1}{2}\Gamma\sigma^2 x^2 \tag{15.61}$$

15.6.1 A Second Derivation

As we have encountered many times in the past, arbitrage arguments are based on the creation of a riskless position. Such a position is the result of a portfolio that produces the same return for every state of nature. To produce an arbitrage portfolio in this setting will require a statement that will hold true for every realization of dZ.

One way of doing it is finding a portfolio with increments that are not dependent on dZ. This can be accomplished by combining two assets in such a way that the multiplier of dZ (the drift coefficient) in the portfolio is zero. In other words, we would like to find two assets for which in every state of nature the stochastic increment of one is the negative of that of the other.

Combining these assets in a portfolio will offset the effect of dZ on the increments and hence all the risk will be eliminated from this portfolio. The increments of such a portfolio will thus be the risk-free rate, or else arbitrage opportunities exist. For simplicity we derive this second alternative for a derivative written on an underlying asset that pays no dividends over the life of the derivative.

The increments of the derivative security, as we have shown, are given below.

```
> ItosLemma(mu*x,sigma*x,V(x,t),muV,sigmaV);
```

$$\mathrm{d}(\mathrm{V}(x, t)) = ((\frac{\partial}{\partial x}\mathrm{V}(x, t))\,\mu\, x + (\frac{\partial}{\partial t}\mathrm{V}(x, t)) + \frac{1}{2}(\frac{\partial^2}{\partial x^2}\mathrm{V}(x, t))\,\sigma^2\, x^2)\,dt$$
$$+ (\frac{\partial}{\partial x}\mathrm{V}(x, t))\,\sigma\, x\, dZ$$

It is apparent that the stochastic "driver" of both the derivative security and the underlying asset is the same process. This is not surprising since the payoff from the derivative security is a function of the price of the underlying asset. The movements in the price of the underlying asset and the value of the derivative security are therefore correlated. Moreover, Ito's lemma clearly shows the multiplier of dZ in the equation for the increments of the derivative security as being $\frac{\partial V}{\partial x}\sigma x$.

Thus it is easy to see how many units of the underlying asset one needs to hold together with a derivative security to make the multiplier of dZ

zero. Nevertheless, we will denote it η and employ Ito's lemma again to see the multiplier of dZ as a function of η. Let us see what the instantaneous increments of a portfolio of δ units of the stock and a long position in the derivative security, $V(x,t)$, will be. To this end we can simply employ Ito's lemma on the expression describing the portfolio so constructed, $\eta x+V(x,t)$.

```
> diff_BS:=ItosLemma(mu*x,sigma*x,V(x,t)+eta*x\
> muPor,sigmaPor);
```

$$diff\_BS := \mathrm{d}(\mathrm{V}(x,\, t) + \eta\, x) = \\ ((\frac{\partial}{\partial x}\,\mathrm{V}(x,\, t))\,\mu\, x + \mu\, x\, \eta + (\frac{\partial}{\partial t}\,\mathrm{V}(x,\, t)) + \frac{1}{2}\,(\frac{\partial^2}{\partial x^2}\,\mathrm{V}(x,\, t))\,\sigma^2\, x^2) \\ dt + ((\frac{\partial}{\partial x}\,\mathrm{V}(x,\, t)) + \eta)\,\sigma\, x\, dZ$$

As anticipated, for the diffusion coefficient to be zero, η should be chosen to satisfy (15.62).

$$\eta = -V_x \tag{15.62}$$

We confirm this solution using MAPLE by asking it to solve the equation below.

```
> solve(sigmaPor=0,eta);
```

$$-(\frac{\partial}{\partial x}\,\mathrm{V}(x,\, t))$$

Let us substitute this solution into the right-hand side of the stochastic differential equation **diff_BS** produced by MAPLE above.

```
> diff_BS:=subs(eta=-diff(V(x,t),x),rhs(diff_BS));
```

$$diff\_BS := ((\frac{\partial}{\partial t}\,\mathrm{V}(x,\, t)) + \frac{1}{2}\,(\frac{\partial^2}{\partial x^2}\,\mathrm{V}(x,\, t))\,\sigma^2\, x^2)\; dt$$

Remember, though, that as we explained at the very beginning of this Chapter, the instantaneous increments of this portfolio are no longer random. Hence its increment dP must be equal to the instantaneous increment paid on the value of the portfolio at the risk-free rate of interest. (To avoid the existence of arbitrage opportunities, this must be the case. Otherwise, by taking appropriate short and long positions in the two assets, the investor can generate riskless arbitrage profit over an instant of time.)

The portfolio value is thus $-V_x x + V$ and hence its instantaneous increments must be $(-V_x x + V)rdt$. We therefore arrive at the differential equation below.

```
> diff_BS:=diff_BS=(-diff(V(x,t),x)*x+V(x,t))*r*dt;
```

$$diff\_BS := ((\frac{\partial}{\partial t} \mathrm{V}(x, t)) + \frac{1}{2}(\frac{\partial^2}{\partial x^2} \mathrm{V}(x, t))\, \sigma^2 x^2)\, dt =$$
$$(-(\frac{\partial}{\partial x} \mathrm{V}(x, t))\, x + \mathrm{V}(x, t))\, r\, dt$$

This equation can be rearranged as follows, and it is of course the main equation obtained in the previous section, i.e., equation (15.59).

```
> diff_BS:=diff(V(x,t),t)+(diff(diff(V(x,t),x),x)/2)* \
> sigma^2*x^2= (-diff(V(x,t),x)*x+V(t,x))*r;
```

$$diff\_BS := (\frac{\partial}{\partial t} \mathrm{V}(x, t)) + \frac{1}{2}(\frac{\partial^2}{\partial x^2} \mathrm{V}(x, t))\, \sigma^2 x^2 =$$
$$(-(\frac{\partial}{\partial x} \mathrm{V}(x, t))\, x + \mathrm{V}(t, x))\, r$$

```
> diff_BS:=expand(diff_BS);
```

$$diff\_BS := (\frac{\partial}{\partial t} \mathrm{V}(x, t)) + \frac{1}{2}(\frac{\partial^2}{\partial x^2} \mathrm{V}(x, t))\, \sigma^2 x^2 =$$
$$-r\,(\frac{\partial}{\partial x} \mathrm{V}(x, t))\, x + r\, \mathrm{V}(t, x)$$

15.7 Reconciliation with Risk-Neutral Valuation

The derivation of the Black–Scholes differential equation, as in Section 15.6, was basically a "pricing by replication" argument. The reader may be wondering about the connection between that derivation and pricing utilizing the stochastic discount factors. The latter, as we have shown numerous times, is equivalent to pricing by discounting the expected value (under the risk-neutral probability) of the payoff.

This section demonstrates how the differential equation can be derived using the expectation under the risk-neutral probability. We will thereby provide the link between these two methods, showing that indeed the same differential equation is obtained. Additionally, we will demonstrate why μ does not appear in the differential equation.

Recall that under the risk-neutral distribution the expected value of every asset appreciates, as does the risk-free asset. Hence, as we reported before, $E(S(t)) = E(S0e^{Y_t}) = S0e^{r\,t}$. Thus, as stipulated by equation (15.4) and our word of caution following that equation, under the risk-neutral probability, $Y_t \sim \mathrm{N}\left(\left(r - \frac{\sigma^2}{2}\right) t, \sigma\sqrt{t}\right)$. The stochastic evolution of the stock price under the risk-neutral probability is thus found by applying

Ito's lemma to the function $g(Y) = S0e^Y$, as below. (Recall that we must use x as the parameter in this procedure.)

```
> ItosLemma(r-(sigma^2)/2,sigma,S0*exp(x),Newmu,Newsigma);
```

$$\mathrm{d}(S0\, e^x) = S0\, e^x\, r\, dt + S0\, e^x\, \sigma\, dZ$$

It follows that under the risk-neutral probability the stock price, $S(t)$, follows a geometric Brownian motion[20] with a drift parameter rS and a diffusion parameter $S\sigma$. This is outlined in equation (15.63),

$$dS = Srdt + S\sigma dZ, \tag{15.63}$$

and is consistent with our explanation in Section 5.7. Under the risk-neutral probability, every asset in the market must have an expected value that will be equal to the appreciation of the asset based on the risk-free rate. Hence, over an instant, the expected value of the price appreciation of the stock must be Sr.

This is verified by taking the expectation of the right-hand side of equation (15.63). By the same argument, the expected value of the price appreciation of a derivative security must also behave in exactly the same way. That is, the expected value of the price appreciation, under the risk-neutral probability, of a derivative security whose current value is $V(S(t), t)$ must be $V(S(t), t)r$.

We can calculate the increments of the derivative security (under the risk-neutral probability) by applying Ito's lemma to $V(S(t), t)$, where S follows the process as in (15.63). (Recall again that we have to use x in the procedure and thus we execute the command below.)

```
> subs(V(x,t)=V(S,t),x=S,\
> ItosLemma(r*x,sigma*x,V(x,t),RNmu,RNsigma));
```

$$\mathrm{d}(\mathrm{V}(S,\, t)) = ((\frac{\partial}{\partial S}\,\mathrm{V}(S,\, t))\, r\, S + (\frac{\partial}{\partial t}\,\mathrm{V}(S,\, t)) + \frac{1}{2}\,(\frac{\partial^2}{\partial S^2}\,\mathrm{V}(S,\, t))\,\sigma^2\, S^2)\, dt + (\frac{\partial}{\partial S}\,\mathrm{V}(S,\, t))\,\sigma\, S\, dZ$$

[20] The process dZ that appears in equation (15.49) describing the increments of the stock is not the dZ that appears in equation (15.63). They both have the same parameters, but the risk-neutral process is a different random variable.

The expected value of the increments of $V(S(t), t)$, based on the risk-neutral distribution, is **RNmu** (since the expected value of dZ is zero). Based on the discussion above, **RNmu** must equal $V(S(t), t)\, r$. Consequently, we obtain the differential equation shown below.

```
> subs(V(x,t)=V(S,t),RNmu)=V(S,t)*r;
```

$$\left(\frac{\partial}{\partial x} \mathrm{V}(S,\, t)\right) r\, x + \left(\frac{\partial}{\partial t} \mathrm{V}(S,\, t)\right) + \frac{1}{2} \left(\frac{\partial^2}{\partial x^2} \mathrm{V}(S,\, t)\right) \sigma^2\, x^2 = \mathrm{V}(S,\, t)\, r$$

This is the same equation as equation (15.59), derived based on the replication argument in the former section. The derivation in this section is based on the risk-neutral valuation and as anticipated yields the same results. Thus, we observe that the pricing guidelines developed in the very first model of Chapter 1 extend to the more complex model of continuous time.

15.8 American vs. European

We first encountered the reason an analytical solution does not exist for an American option in the Concluding Remarks to Chapter 12. The preceding Chapter further investigated this reason within the realm of the binomial model. Since early exercise is possible with an American option, to value the option one should know the set of pairs of possible stock price and time $(S(t), t)$ at which it will be optimal to exercise the option. In other words, one should know the boundary determining optimal early exercise.

However, a prerequisite for determining where the boundary lies is the price of the option itself. The essence of the difficulty for producing an analytical solution is the fact that we do not know where the boundary lies *a priori* (before knowing the price of the option). A similar difficulty exists in many physics problems for which the solution is classified by a partial differential equation like equation (15.58). These types of problems are referred to by the general[21] name, "free boundary problems", perhaps because the boundary is not explicitly specified.

In the binomial model, time is replaced by a discrete grid of points in time. The discrete grid for the stock price is induced by the time grid and by our assumptions[22] regarding the price process. Figure 14.5 for a put option

[21] See [47] for a derivation of option pricing theory from a point of view that utilizes the vast literature on partial differential equations.

[22] The *Up* and *Do* factors are chosen to ensure convergence of the price process to the lognormal distribution with the appropriate parameters. See equations (14.9) and (14.10) in Section 14.1.

and Figure 16.1 in for a call option demonstrate the existence of this "free boundary". It is the upper and lower envelope of the area filled with squares in Figures 16.1 and 14.5, respectively.

In a binomial tree, as in either of the figures just mentioned, the time is marked on the "Time" axis while the "Up" axis corresponds to the price of the stock. The boundary divides the time–price plane into two parts: where it is optimal to exercise the (American) option (early) and where it is not. The boundary was generated numerically, as explained in Section 14.1.2, by comparing the value of the American option if exercised to its value if kept alive at each node.

Recall the development of the partial differential equation in Section 15.6. The driving force was an arbitrage argument claiming that if a portfolio composed of stocks and bonds replicates the return on the derivative security (if it is not exercised), then the price of the replicating portfolio must equal the price of the derivative security. The price of the replicating portfolio, after solving for β and δ as in equations (15.55) and (15.56), is

$$V_S S - \frac{ySV_S - V_t}{r} - \frac{\sigma^2 S^2 V_{SS}}{2r}. \tag{15.64}$$

Hence, if the option is not exercised we have equality between the price of the replicating portfolio and the price of derivative security. If, however, it is optimal to exercise, then the option value if exercised must be above its value if not exercised. Its value if not exercised equals the value of the replicating portfolio, which therefore should be, in general, less than or equal to the value of the option. This is stipulated in equation (15.65) below.

$$V \geq SV_S - \frac{ySV_S}{r} - \frac{V_t}{r} - \frac{\sigma^2 S^2 V_{SS}}{2r} \tag{15.65}$$

Consequently, for an American option, equality holds in equation (15.65) only if it is not optimal to exercise. If it is optimal to exercise, an inequality, conveying the fact that the value of the derivative security exceeds that of the replicating portfolio, is valid. After some algebraic manipulation, assuming that the dividend yield is zero, equation (15.65) can be presented as

$$rSV_S - V_t - \frac{\sigma^2 S^2 V_{SS}}{2} \leq rV. \tag{15.66}$$

Equation (15.66) is the familiar Black–Scholes equation, provided the inequality is replaced with an equality.

The inequality obtained in this case should not come as a surprise. After all, we have experienced, even in the simplest model investigated in this book, that we are unable to price a cash flow if an exact replicating strategy does not exist. In such cases only bounds (inequalities) on the price of the cash flow can be derived.

The fact that for an American option we have a differential inequality rather than a differential equation can also be viewed using the risk-neutral arguments as in Section 15.7. Under the risk-neutral probability, assets appreciate in value at the risk-free rate. It follows that since it is optimal to exercise, the risk-free rate of return is achieved if the option is exercised.

Thus if the option is kept alive, the return obtained is smaller than the risk-free rate. For this reason it is optimal to exercise and deposit the proceeds in a money market account or buy the bond. Therefore, there exists a boundary in the time–price plane, such as in the binomial model, that divides the plane in two: on one side of the boundary, relation (15.66) holds as an equality, and on the other side of the boundary, this relation holds as an inequality.

Proceeding to solve for the value of an American option using the point of view induced by the differential equation is similar in some aspects to the "binomial philosophy". It starts again with a discretization. The time–price plane is discretized and the value of the option on the boundary is calculated based on the boundary conditions. The discretization employed in the current context is different from the one employed by the binomial model.

The range of the time variable is chosen (obviously) to correspond to the maturity of the option. Hence we deal in the range $[0, T]$ that is divided into steps Δt, as we have seen before. The stock price range is chosen to be $[0, S(\infty)]$, where $S(\infty)$ plays the numerical value of infinity and is also divided into steps of size ΔS. Thus, if we superimpose this discretization, denoted by squares, on a binomial tree with, for example, 12 nodes, the result[23] will be as shown in Figure 15.10.

```
>  BinTree(12,[seq(seq([j,i],i=0..12),j=0..12)]);
```

[23]This superimposition is not precise but it demonstrates the idea. In the binomial model the discretization of stock prices is done based on the factors Up and Do, i.e., j on the Up axis corresponds to $S(0)Up^j$, $j = 0, ..., 12$, where $S(0)$ is the stock price at time zero, and $S(\infty) = S(0)\,Up^{12}$. Approaching the solution from the differential equation point of view, the discretization is done by dividing the interval $[0,\ S(\infty)]$ in to subintervals of the same length as is done for the time dimension (in both models). Solving for the differential equation, the number of subintervals along the time dimension and along the price dimension need not be equal as is shown in Figure 15.10. However, one of the knot points on $[0,\ S(\infty)]$ should be the current stock price.

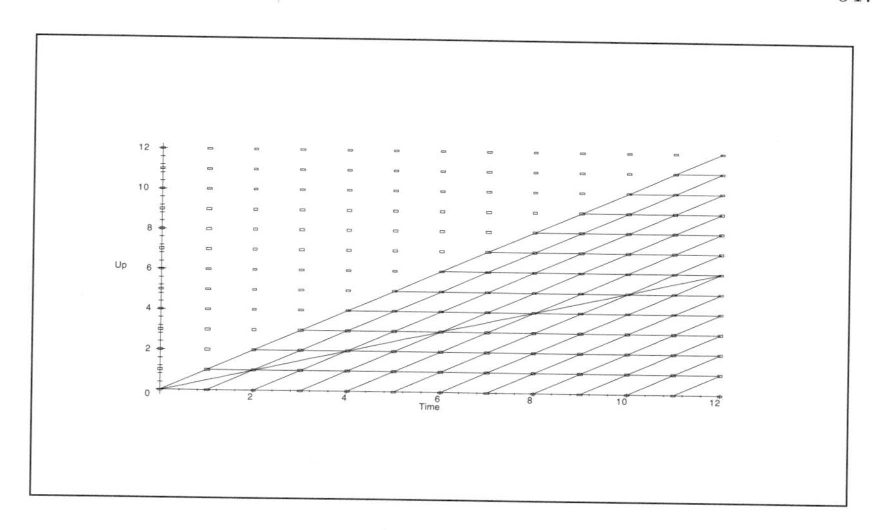

Figure 15.10: A Discrete Grid of (t, S) for a Numerical Solution of the Differential Equation

In both the binomial model and the differential equation approach to the valuation of the derivatives, at $(T,\ S(0)Up^j)$, $j = 0, ..., 12$, the value of the derivative security is set equal to the payoff at maturity. This is one boundary condition. However, while this is sufficient to start the traversing backward in the binomial setting, in the current setting it is not. In order to solve the differential equation numerically we need further information: the first-order derivative of the value of the derivative security with respect to t, and the first- and the second-order derivatives of the value of the derivative security with respect to S. Hence, we need to apply another boundary condition.

The value of the derivative security along the boundaries $(t, S(\infty))$ and $(t, 0)$, $t = 0, ..., 12$, will supply the missing boundary conditions. For a call option the value along $(t, S(\infty))$ is set equal to $S(\infty)$, since the value of the call as S approaches infinity approaches the value of the stock. For a put option, the value along $(t, S(\infty))$ is set equal to 0 since when S approaches infinity the value of the put approaches zero.

The value along $(t, 0)$ is set to zero for a call option and to the exercise price for a put option. When the value of the stock hits zero[24] the value

[24]When the value of the stock hits zero, it stays zero at any time afterward. This is

of the call option vanishes and stays zero at any time afterward. Once the price of the stock vanishes, the value of the put option does not change. The put option (if it is American) had better be exercised on the spot since it has reached its maximal value.

Other types of derivative securities induce different boundary conditions. Indeed, the differential equation, as we noted in Section 15.6 prior to equation (15.60), is generic and the boundary conditions distinguish between the various derivative securities. We also need to take account of the free boundary condition induced by the fact that the derivative is American. Namely, at each time and stock price the value of the security must be greater than or equal to its value if exercised. Thus the free boundary condition for an American call option states that $V(S(t),t) \geq \max(0, S(t) - K)$ and that only on the boundary, $\{(S(t),t)|V(S(t),t) = \max(0, S(t) - K)\}$, equality holds in relation (15.66).

Let us denote the value of the derivative security at the discrete nodes by $V(j,i)$, where $j = 0, ..., m$, is the index of the discrete stock price and $i = 0, ..., n$, is the index of the discrete time. Figure 15.10 displays a case where $m = n = 12$. The value of $V(j,n)$ is known — this is one of the boundary conditions implied by the value of the derivative at its maturity. Furthermore, due to the other two boundary conditions, the values $V(0, n-1)$ and $V(m, n-1)$ are also known. Hence the values of $V(j, n-1)$, $j = 1, ..., m-1$, should be calculated.

The differential equation induces $m - 1$ equations with $m - 1$ unknowns from which the values $V(j, n-1)$, $j = 1, ..., m-1$, are determined. In order to solve these equations we need to estimate the first- and second-order derivatives of V with respect to S, and the first derivative of V with respect to t. There are a few ways of doing this. We will give only a sketch of one such method here.

The first-order derivatives of V are estimated by calculating the difference in the value of the function between two adjacent points and dividing that difference by the step size. Hence, for example, we can estimate the first-order derivative of V with respect to t at the point $(j, n-1)$, as is done

---

apparent from the stochastic differential equation, equation (15.46), describing the evolution of the price process, since when $S(\tau) = 0$, both the drift and the diffusion coefficient are zero at τ, and equation (15.46) thus implies that $S(t) = 0$ for every $t > \tau$. In the terminology of stochastic processes, zero is an absorbing state of the price process of the stock. We mentioned this in Section 12.2.2, pointing out an obvious state of nature in which an American put option should be exercised on the spot.

in equation (15.67).

$$V_t(j, n-1) = \frac{V(j, n) - V(j, n-1)}{\Delta t} \tag{15.67}$$

The first-order derivative of V with respect to S is similarly estimated, i.e., as

$$V_S(m-1, i) = \frac{V(m, i) - V(m-1, i)}{\Delta S}. \tag{15.68}$$

The second-order derivative of V with respect to S, V_{SS}, can be estimated by applying the same difference notion to V_S. Hence we obtain

$$V_{SS}(m-1, i) = \frac{V_S(m, i) - V_S(m-1, i)}{\Delta S}. \tag{15.69}$$

Once this is done, we substitute the estimated values of the first-and second-order derivatives, V_t, V_S, and V_{SS} from equations (15.67), (15.68), and (15.69), respectively, into the differential equation

$$rSV_S + V_t + \frac{1}{2}\sigma^2 S^2 V_{SS} = rV,$$

where r is assumed (as usual) to be fixed and deterministic and $S = j\Delta S$.

The differential equation thus induced $m-1$ difference equations for $j = 1, ..., m-1$, and for $i = n-1$, with $m-1$ variables $V(j, n-1)$. The so-generated system of equations is solved, and the free boundary conditions should be checked at this stage. This is done in the same way as the possibility of early exercise is handled in the binomial model. $V(j, n-1)$ is set to the maximum of the value solved for, and the value of the derivative security if exercised at this price in this time period.

Recovering the values of $V(j, n-1)$ for $j = 1, ..., m-1$, we proceed to solve for $V(j, n-2)$ in exactly the same manner. When $V(j, n-1)$ for $j = 1, ..., m-1$, is known, since $V(0, n-2)$ and $V(m, n-2)$ are known from the boundary conditions, solving for $V(j, n-2)$ for $j = 1, ..., m-1$, generates a similar system of equations, but with different indexes. Proceeding recursively in the same manner will recover $V(j, 0)$ for $j = 1, ..., m-1$, and since the knot points are chosen such that for a particular j, say j_1, $S = j_1 \Delta S$ is the current stock price, $V(j_1, 0)$ is the current price of the derivative security.

15.9 Concluding Remarks

This Chapter has described the most general model that is discussed in this book. It is a model for which both the time space and the states of nature space are continuous. In an environment such as this, the dynamic evolution of the price of a financial asset is described by a stochastic differential equation. This Chapter introduced the concepts needed to understand such equations in a heuristic manner.

Ito's lemma was used to derive the stochastic differential equation that is satisfied by a derivative security. The lemma was explained at two levels of difficulty. (The more advanced explanation is contained in Appendix 15.11.3.) An arbitrage argument facilitates the derivation of the deterministic partial differential equation that a derivative security must satisfy. In particular, this differential equation, termed the Black–Scholes differential equation, falls into the category of parabolic equations.

The literature concerning numerical solutions to partial differential equations is very extensive. Presenting the option pricing formula as a solution to a differential equation allows us to tap this vast literature. As such, we can price derivatives even though the Black–Scholes differential equation has no analytical solution. The last part of Section 15.8 gives an overview of the guidelines of numerical solutions that are induced by the differential equation.

The Chapter has also highlighted the connection between the sensitivity measures developed in Chapter 7 and the relationship between those measures and the hedging arguments developed in Chapter 8. Deriving the differential equation (15.59) via the replication argument, as is done in Section 15.6.1, is in the spirit of Section 2.1. There, in the context of a one-period model, the risk-free rate was shown to be implicit in the prices of the securities in the market.

The various derivations of the differential equation (15.59) in Section 15.7 make use of risk-neutral valuation and replication arguments. Thereby Section 15.7 closes a circle. The circle connects the guidelines and duality that exist in the very simple model of Chapter 1 to the more complex model presented in this Chapter. It demonstrates yet again that the conceptual framework stays intact, even in an environment that requires the use of sophisticated techniques. As such, this Chapter concludes the adventures on which the reader embarked in the first Chapter when this framework was introduced.

It remains to demonstrate that these guidelines can be applied to different types of options, e.g., options for which the underlying asset is not a stock. Indeed, this is the topic of the next Chapter.

15.10 Questions and Problems

Problem 1. Verify that if the price of a stock at time T, $S(T)$, is given by $S_{t_0}e^{Y_{T-t_0}}$, where

$$Y_{T-t_0} \sim \mathrm{N}\left(\mu_Y(T-t_0), \sigma_Y\sqrt{T-t_0}\right)$$

and t_0 is the current time, then $E(S(T))$ under the risk-neutral distribution is $S_{t_0}e^{r(T-t_0)}$, where r is the risk-free rate.

Problem 2. Under the risk-neutral distribution,

$$Y_{T-t_0} \sim \mathrm{N}\left(r - \frac{\sigma_Y^2}{2}, \sigma_Y\sqrt{T-t_0}\right),$$

where the notation is as defined in Problem 1 and the stochastic differential equation satisfied by Y is

$$dY = \left(r - \frac{\sigma_Y^2}{2}\right)dt + \sigma_Y dZ,$$

where Z is a standard Brownian motion, i.e., $Z(t) \sim \mathrm{N}\left(0, \sqrt{t-t_0}\right)$. Apply Ito's lemma to $S(T) = S_{t_0}e^{Y_{T-t_0}}$ and determine the stochastic differential equation satisfied by $S(T)$ under the risk-neutral distribution.

Problem 3. Under the real-life distribution

$$Y_{T-t_0} \sim \mathrm{N}\left(\mu_Y(T-t_0), \sigma_Y\sqrt{T-t_0}\right),$$

where the notation is as defined in Problem 1, assume a stock that pays a deterministic dividend yield of δ. What is the stochastic differential equation satisfied by $S(T)$? (Assume that Y models the price appreciation of the stock, not including the dividend effect, as in the explanation of Section 14.1.3.) Use Ito's lemma to solve the stochastic differential equation you have derived. Compare your result to equation (6.15).

Problem 4. Use Ito's lemma to solve the stochastic differential equation

$$dx = -\frac{\sigma^2}{2}dt + \sigma dZ$$

and determine the process $x(t)$, where Z is a standard Brownian motion, i.e., $Z(t) \sim \mathrm{N}\left(0, \sqrt{t-t_0}\right)$. Compare your results to the statement at the end of Section 15.5.2.

Problem 5. Section 15.6 develops the differential equation satisfied by a derivative security, equation (15.51), by applying Ito's lemma to $V(S,t)$, where $S(t)$ follows a geometric Brownian motion. Develop the same equation using Ito's lemma when V is written as $V(S0e^{Y_t},t)$, where Y_t follows a Brownian motion.

Problem 6. Verify that $V(S,t)$, as defined in Chapter 6, in equations (6.3), (6.4), (6.5), and (6.6), solves the differential equation (15.59) with the boundary condition of a call option.

Problem 7. The differential equation (15.59) induces $m-1$ equations with $m-1$ unknowns from which the values $V(j,n-1)$, $j=1,...,m-1$, as defined on page 648, are determined. State the estimates of the first- and second-order derivatives of V with respect to S, and the first derivative of V with respect to t, and set up these equations.

Problem 8. Show that a self-financing portfolio which is Delta and Gamma neutral is also Theta neutral.

Problem 9. Utilize Ito's lemma, modeling the forward price, F, of a stock as a function of the continuously compounded rate of return, to show that F is a Martingale.

15.11 Appendix

15.11.1 A Change over an Instant

The change "over the next instant" has a precise meaning in terms of a limit. An instant in time later than t is defined to be $\lim_{h\to 0^+} t+h$, i.e., when h approaches zero from above (the positive side). One should recall that the limit does not mean substitution of zero for h but rather what happens to $t+h$ as h gets smaller. For a function f which is discontinuous at t there exists at least one direction for which $\lim_{h\to 0} f(t+h)$ does not coincide with $f(t)$. For a continuous function, the limit from above coincides with the limit from below and both equal the value of the function at that point. We will be dealing with such functions.

It is customary to denote the time increment by Δt. However, for ease of use in the MAPLE calculations, we will use h instead of Δt. Intuitively, the (average) rate of change of a function f, at some time t_0 over some time h, is $\frac{f(t_0+h)-f(t_0)}{h}$. Using the definition of an instant, the rate of change of

f over the next instant will be

$$\frac{\partial f}{\partial t}(t_0) = \lim_{h \to 0} \frac{f(t_0 + h) - f(t_0)}{h}. \tag{15.70}$$

Thus the **rate of change,** at time t_0 over the next instant of time, of a continuous function f is the function $\frac{\partial f}{\partial t}$ whose value at t_0 is defined by equation (15.70). The rate of change is $\frac{\partial f}{\partial t}(t_0)$ — the derivative of the function f with respect to t evaluated at t_0. Roughly speaking, the change over one instant of time, at time t_0, is therefore

$$\frac{\partial f}{\partial t}(t_0) \text{ multiplied by an "instant"}, \tag{15.71}$$

where the meaning of an "instant" is the limit of h when h goes to zero.

When we substitute a numerical value for the "instant", even a very small numerical value, we get an approximation to the change. The mathematical notation for an instant is dt. Thus, we use the notation

$$df = \frac{df}{dt} dt \tag{15.72}$$

to say that the df, the change in f over an instant, is the derivative of f, $\frac{df}{dt}$ ("the rate of change over an instant"), multiplied by the length of the interval, dt ("an instant of time"). Given the (constant) rate of change of a function over an interval, the total change is the rate of change multiplied by the length of the interval. The same notion remains correct when the interval length is a mere instant. There is (roughly speaking) only one point in an interval. The length of the interval is an instant of time. Therefore, the rate of change over such an interval is a constant. This idea is only correct in the limiting[25] sense.

Let us consider a simple example which will illustrate what is meant by "only in the limiting sense". Consider the function $f(t) = t^2$. The rate of change of this function at the point $t_0 = 1$ is 2. This can be verified by calculating the first derivative of $f(t) = t^2$ and substituting $t = 1$ in the expression for the first derivative.

```
>  subs(t=1,diff(t^2,t));
```

2

[25] More precisely, equation (15.72) is a shorthand notation for $f(t) = f(0) + \int_0^t f(x)\, dx$. When investigating changes with respect to random variables, some modifications are needed.

Let us take another look at the first derivative of this function, calculated this time from the formal definition. The incremental change of this function over the time interval h is

```
> Incf:=((t+h)^2-t^2);
```

$$Incf := (t+h)^2 - t^2$$

```
> expand(Incf);
```

$$2\,t\,h + h^2$$

The incremental change of this function when t changes from t to $t+h$ is given above. Thus, the average rate of change over the interval $[t, t+h]$ is found by dividing the expression above by h.

```
> simplify(Incf/h);
```

$$2\,t + h$$

The incremental change in the function for an infinitesimal movement in t is the limit of the above expression, taken as h approaches zero. It is straightforward to verify that the limit of the above expression as h goes to zero is $2t$. (The MAPLE syntax for finding the limit as h approaches zero is $h = 0$ but indeed, as previously mentioned, the limit is not calculated simply by substituting $h = 0$ in the expression.)

```
> limit(Incf/h,h=0);
```

$$2\,t$$

To recap, at the point t over an instant of time the incremental change will be $2t$ multiplied by the instant. Using our notation, this is $2t\,dt$, where dt is equal to the change in t which is h. However, the incremental change of f over the interval $[t, t+h]$ is actually $(2t+h)h = 2t\,h + h^2$, for which $2th$ is only an approximation. The approximation is better the closer h is to zero.

We see that in generating this approximation to the change we ignore the h^2 term in the expression $2t\,h + h^2$ because $\lim_{h\to 0} \frac{h^2}{h} = 0$. Furthermore, when we refer to the instantaneous (total) change, df (also referred to as the differential of the function), we replace h by dt and say that $df = 2t\,dt$. In general, to arrive at the approximation, we ignore (additive) parts of the expression which are dependent on h and satisfy a certain condition.

Consider an expression $g(h)$. If

$$\lim_{h\to 0} \frac{g(h)}{h} = 0 \tag{15.73}$$

we neglect this term in order to generate the approximation, since when calculating the derivative it will vanish anyway. For this reason the h^2 term above is eliminated to arrive at the approximation. The (additive) parts that we ignore are functions of h which approach zero faster than h approaches zero.

An expression which is a part of the approximation may approach zero as h does and yet may not be eliminated. The $g(h)$ in equation (15.73) will be eliminated since it approaches zero faster than h does. As h becomes smaller, $g(h)$ becomes smaller, but if $g(h)$ does not approach zero fast enough, the limit in equation (15.73) will not approach zero. For example, if $g(h)$ is $\frac{h}{2}$ the limit of $g(h)$ over h will be $\frac{1}{2}$. Indeed, $g(h) = \frac{h}{2}$ is smaller than h, for every positive h, but not small enough to cause the limit to be zero. $g(h) = \frac{h}{2}$ does approach zero, but not fast enough to cause the limit of $\frac{g(h)}{h}$ to be zero.

Consider now the function $g(h) = \sqrt{h}$. Both h and $g(h) = \sqrt{h}$ approach zero as h approaches zero. Moreover, in this case h is approaching zero faster than $\sqrt{h}$, since in the vicinity of zero, $\sqrt{h}$ is greater than h. In this case $\sqrt{h}$ is not approaching zero fast enough to cause the quotient $\frac{\sqrt{h}}{h}$ to be zero. This becomes apparent in Figure 15.11, where it can be seen that in the vicinity of zero, the linear graph (green) is always smaller than the curve describing the square-root function, but the graph of the quotient approaches infinity.

```
> plot([h,sqrt(h),sqrt(h)/h],h=0..2,\
> color=[green,red,blue],thickness=2);
```

Thus, the parts which we ignore in this approximation are those parts, $g(h)$, that approach zero fast enough to make $\frac{g(h)}{h}$ approach zero. The mathematical notation for those $g(h)$ that satisfy equation (15.73) is $o(h)$, which is read as "little o of h". For example, h^2, as we saw above, approaches zero faster than h; fast enough to make $\frac{h^2}{h}$ approach zero as h approaches zero. Hence we say that h^2 is of order little o of h.

In summary, therefore, given a function of t, its rate of change over the next instant of time is $\frac{\partial f}{\partial t}$ and the change over the next instant is $\frac{\partial f}{\partial t}dt$. Saying that we are ignoring elements of the function that are of order $o(h)$ in calculating the rate of change over the next instant, or that $\frac{\partial f}{\partial t}$ is the rate of change over the next instant, is nothing but a tautology. It is simply a consequence of the definition of the derivative as the limit of $\frac{f(t+h)-f(t)}{h}$ as h goes to zero. Obviously if $f(t+h) - f(h)$ includes a few expressions, labeled here $w(h)+k(h)$, that are of order $o(h)$, i.e., $f(t+h)-f(t) = w(h)+k(h)+\cdots$, then these expressions, when divided by h, will vanish as h approaches zero and will not be present in the expression for the derivative.

Things are, however, somewhat different when calculating the rate of change of a function that involves a random variable. Nevertheless, the

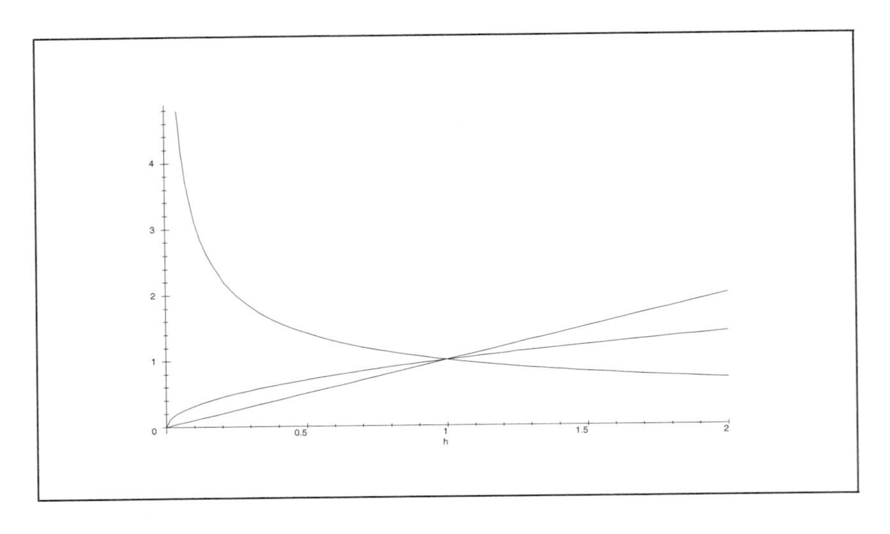

Figure 15.11: The Graphs of h, $\sqrt{h}$, and $\frac{\sqrt{h}}{h}$

guidelines stay the same: ignore expressions that are of order $o(h)$.

Our interest in the expression for the rate of change over an instant stemmed from our desire to compose a hedged portfolio. Our goal is to compose a portfolio such that its rate of change over the next instant will not be affected by the change in the price of the underlying asset. Such a portfolio will be riskless over the next instant of time. Let us see how this idea will be applied in our context. We are interested in the rate of change (or rather change over the next instant) of specific random variables (or price processes) — namely, those that model the price process of the assets underlying the derivatives.

15.11.2 The Limit of a Random Variable

As the discussion above indicated, to arrive at the change over the next instant we have to apply to a concept of a limit. However, since we are interested in the instantaneous change of a random variable we have to define a limit for a random variable. There exist a few such definitions. We will only define the one that is useful in our context.

Given a sequence of real numbers $\{a_n\}_{n=1}^{\infty}$ we know that the sequence converges to a if the distance between a_n and a goes to zero as n goes to infinity. The distance in such a case is simply the absolute value of the difference between a_n and a, i.e., $|a_n - a|$. Hence we say that a_n converges to a if $\lim_{n\to\infty} |a_n - a| = 0$. How do we judge a distance when the elements of a sequence are random variables? In order to answer this question we first address the concept of a sequence of functions[26] that converge to another function.

What we are missing, however, is a measure of a distance between two functions. Once this is supplied, the definition will simply follow the guidelines of the counterpart for a sequence of numbers. Consider two functions, f and g, defined over the interval $[0, T]$ and let us define the function $f - g$ over the same interval. If f is close to g, then $f - g$ should be close to the constant function zero. That is, the function $(f - g)(x)$ should be close to zero for every x, albeit $f - g$ can be both positive and negative. Hence, a better measure will be obtained by considering $(f-g)^2$, which should also be close to zero for every x. In that case the graph of $(f-g)^2$ should be close to the x-axis and hence the area under the graph of $(f - g)^2$, $\int_0^T (f-g)^2(x)dx$, should be close to zero.

One natural measure of a distance between two functions is[27]

$$\int_0^T (f-g)^2(x)\, dx. \tag{15.74}$$

Consider the sequence of functions f_n and a function f defined over the interval $[0, T]$. Armed with a measure for the distance between two functions, we say that f_n converges to f if the sequence of numbers $\int_0^T (f_n - f)^2(x)dx$ converges to zero, i.e., if $\lim_{n\to\infty} \int_0^T (f_n - f)^2(x)\, dx = 0$. For simplicity, and since this is the case in most of our applications, we use functions that are defined over the interval $[0, T]$. In general that interval might be $(-\infty, \infty)$ or some other interval. The modification in the definitions is self-explanatory, as demonstrated by the examples below.

Consider the function

$$G1(n, x) = \begin{cases} n & if \quad 0 < x < \frac{1}{n} \\ 0 & otherwise \end{cases}$$

[26] After all, the precise definition of a random variable is a function from the sample space to the real line.

[27] In order to measure the distance in the original units one should actually look at $\sqrt{\int_0^T (f-g)^2(x)\, dx}$. Note that this is a natural extension of the distance between two vectors in R^n. If one replaces the integral with a summation and thinks about a vector $(x_1, ..., x_n)$ as a function defined over $1, ..., n$, so x_1 is $x(1)$, then the distance between a vector x and a vector y is as defined for two functions, i.e., $\sqrt{\sum_{i=1}^n (x_i - y_i)^2}$.

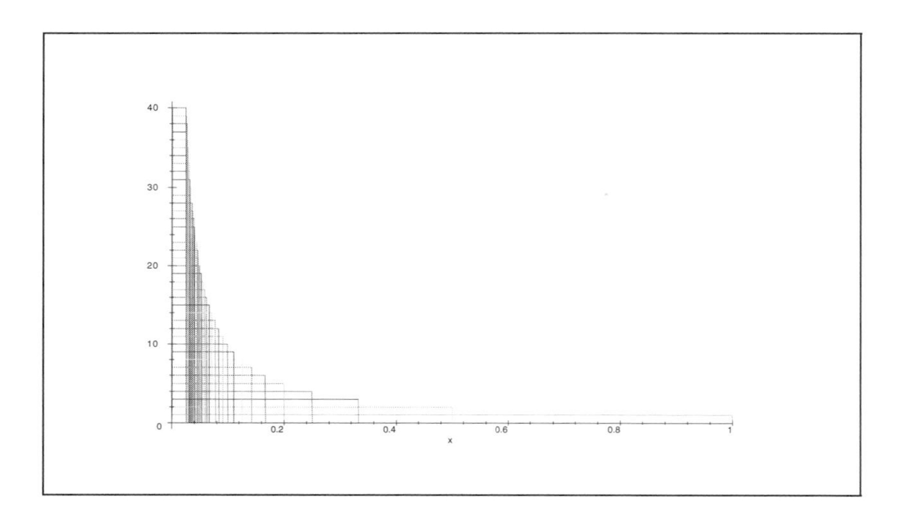

Figure 15.12: The Graphs of $G1$ for $n = 1, ..., 40$

defined in MAPLE below.

```
> G1:=(n,x)->piecewise(x>0 and x<(1/n),n,0);
```

$$G1 := (n, x) \to \text{piecewise}(0 < x \textbf{ and } x < \frac{1}{n}, n, 0)$$

Let us first look at an animation of the sequence of the functions $G1$ for $n = 1, ..., 40$, to appreciate what is happening here. A static version of the animation is displayed in Figure 15.12.

```
> plots[animate](G1(n,x),x=0..1,n=1..40,\
> thickness=3,colour=red);
```

As n gets larger, the interval $\left(0, \frac{1}{n}\right)$, where the function's value is different than zero, shrinks. However, the function's value at that interval, n, gets larger and larger. The question we pose is whether the sequence of functions $G1(n, x)$ so defined converges to zero. You may want to (and should) think about this zero as a function. It is the constant function zero, i.e., that assigns the value zero for every x.

In order to reply to this question we need to determine what happens to the distance between $G1(n, x)$ and the function $G(x) = 0$, for every x, as n gets larger. Since $G(n, x)$ differs from zero on the interval $\left(0, \frac{1}{n}\right)$ and its value there is n, the measure of the distance will be $\int_0^{\frac{1}{n}} n^2 \, dx$, as calculated below by MAPLE.

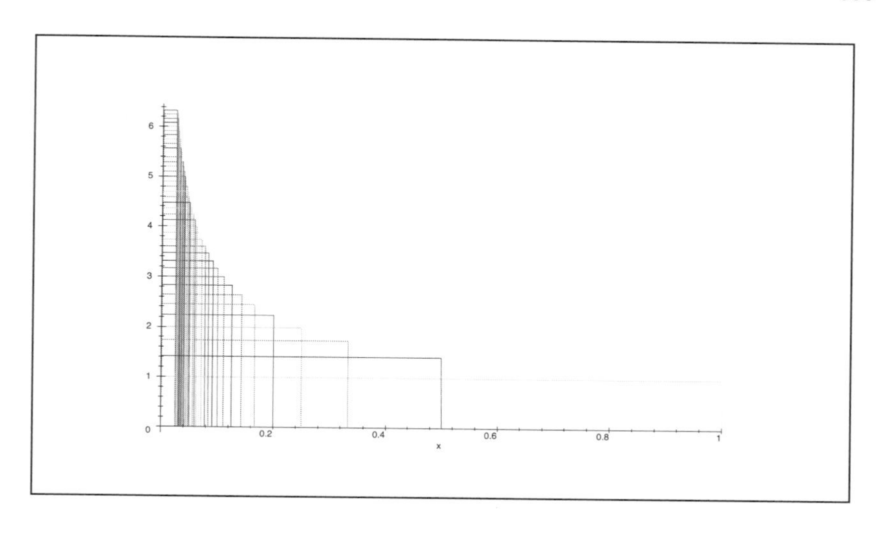

Figure 15.13: The Graphs of $G2$ for $n = 1, ..., 40$

```
> int(n^2,x=0..1/n);
```

$$n$$

Clearly, as n approaches infinity, the distance increases to infinity and hence we cannot claim that $G1(n,x)$ converges to the function $G(x) = 0$ for every x.

Let us now consider another function $G2(n,x)$,

$$G2(n,x) = \begin{cases} n^{\frac{1}{2}} & if \quad 0 < x < \frac{1}{n} \\ 0 & otherwise, \end{cases}$$

defined in MAPLE below.

```
> G2:=(n,x)->piecewise(x>0 and x<(1/n),n^(1/2),0);
```

$$G2 := (n, x) \to \text{piecewise}(0 < x \textbf{ and } x < \frac{1}{n}, \sqrt{n}, 0)$$

Take a look at the animation of this function as n gets larger. A static version of the animation is displayed in Figure 15.13.

```
> plots[animate](G2(n,x),x=0..1,n=1..40,\
> thickness=3,colour=blue);
```

We observe roughly the same phenomenon. Now, however, the height of the function on the interval $(0, \frac{1}{n})$ is only $\sqrt{n}$. As n gets closer to infinity the height of $G2$ indeed increases, but less than the height of $G1$. Will the sequence of the functions $G2(n,x)$ converge to zero as n approaches infinity?

Again, this question is similar to the speed of convergence problem we dealt with in the definition of little o. The interval $(0, \frac{1}{n})$ shrinks to zero, but the level of the function on that interval increases. Which process is faster ("wins") is actually the question we asked. What happens to the distance between $G2(n,x)$ and $G(x) = 0$ as n gets larger? This is determined by calculating the integral below.

```
> int((n^(1/2))^2,x=0..1/n);
```

$$1$$

Since the distance between each $G2(n,x)$ and $G(x) = 0$ is 1 for every n, the limit of the distance, as n approaches infinity, will be one as well. Thus, we see that indeed the functions in this sequence are closer to zero than the functions in the sequence $G1(n,x)$. However, we still cannot claim that the sequence $G2(n,x)$ converges to zero. Furthermore, the distance between $G2(n,x)$ and $G(x)$ is 1 for every n.

Finally, we introduce a third function

$$G3(n,x) = \begin{cases} n^{\frac{1}{3}} & if \quad 0 < x < \frac{1}{n} \\ 0 & otherwise \end{cases}$$

defined in MAPLE below.

```
> G3:=(n,x)->piecewise(x>0 and x<(1/n),n^(1/3),0);
```

$$G3 := (n, x) \rightarrow \text{piecewise}(0 < x \textbf{ and } x < \frac{1}{n}, n^{1/3}, 0)$$

Let us first calculate the limit of the distance between $G3(n,x)$ and $G(x) = 0$.

```
> int((n^(1/3))^2,x=0..1/n);
```

$$\frac{1}{n^{1/3}}$$

```
> limit(int((n^(1/3))^2,x=0..1/n),n=infinity);
```

$$0$$

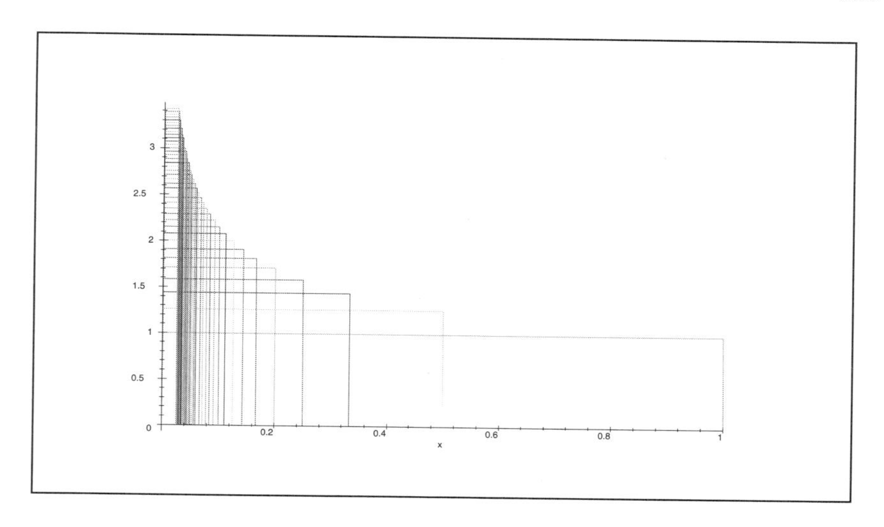

Figure 15.14: The Graphs of $G3$ for $n = 1, ..., 40$

The sequence $G3(n, x)$ converges to zero as n approaches infinity. This is the result of the height of $G3(n, x)$ on $\left(0, \frac{1}{n}\right)$ not going to infinity as fast as both the heights of $G1$ and $G2$ on that interval.

The animation below presents the behavior of the three functions as n approaches infinity. The upper red graph is the graph of $G1$, the middle blue graph is the graph of $G2$, and the lower green graph is the graph of $G3$. A static version of only $G3$ is displayed in Figure 15.14.

```
> aniG1:=plots[animate](G1(n,x),x=0..1,n=1..40, thickness=
> 3,colour=red):

> aniG2:=plots[animate](G2(n,x),x=0..1,n=1..40, thickness=
> 3,colour=blue):

> aniG3:=plots[animate](G3(n,x),x=0..1,n=1..40, thickness=
> 3,colour=green):

> plots[display](aniG1,aniG2,aniG3);
```

One should note the progression of the limit of the distance between G and $G1$, $G2$,and $G3$: infinity, 1, and 0 respectively. The functions $G1(n, x)$ do not get close to G as n approaches infinity. In fact they get farther away from G. We see that the distance measure diverges as n approaches infinity. For $G2$, the limit is a finite number 1, reflecting the fact that as n approaches infinity, the distance between $G2$ and G does not change. As $G3$ is getting closer and closer to G, the limit of the distance is zero.

Our ultimate goal is the definition of distance between two random variables. We would like to look at an example that will clarify the forthcoming discussion. Consider the function

$$G4(n,x) = \begin{cases} 1 & if \quad x = 0 \\ \frac{1}{n} & if \quad 1 < x < 2 \\ 0 & otherwise \end{cases}$$

defined in MAPLE below.

```
> G4:=(n,x)->piecewise(x=0,1,x>1 and x<2,1/n,0);
```

$$G4 := (n, x) \to \text{piecewise}(x = 0,\ 1,\ 1 < x \textbf{ and } x < 2,\ \frac{1}{n},\ 0)$$

Let us take a look at the animation of this function as n gets bigger. The function gets closer to zero for the most part, although at $x = 0$, for every n, the value of the function is 1. Nevertheless, our distance criteria will tell us that $\lim_{n\to\infty} \int_0^\infty (G4(n,x) - 0)^2\, dx = 0$. This is because there really is no area under $G4(n,x)$ at $x = 0$, and hence we can calculate the integral of $G4$ ignoring the point $x = 0$. The animation below exemplifies this matter. A static version of it is displayed in Figure 15.15.

```
> plots[animate](G4(n,x),x=0..3,n=0..200,
> thickness=3,frames=30,color=yellow);
```

Suppose now that x is a random variable that takes the value 0 with a probability of $\frac{1}{2}$ and any point in the interval $(1,2)$ with equal probability. In such a case $G4(n,x)$ is a random variable. Does it still make sense to claim that $G4$ approaches zero as n approaches infinity?

This time when we say zero we mean a degenerate random variable that takes the value zero with a probability one. Indeed, as n approaches infinity, $G4(n,x)$ is approaching zero at nearly every point but one, $x = 0$, at which $G4(n,0) = 1$ for every n. Thus, there is a probability of $\frac{1}{2}$ that $G4(n,x)$ will not be zero for every n.

It would not make sense to have a criterion that still says that $G4(n,x)$ approaches zero in this stochastic environment. In a deterministic environment the importance of every point is the same, but in a stochastic environment we must somehow reflect the importance of each point based on its probability. How do we then judge a distance between two functions when the functions are no longer deterministic but rather random variables?

To further illustrate the idea we remind the reader of footnote 8 in Section 5.2 that refreshed the definition of an integral. In what follows, we will

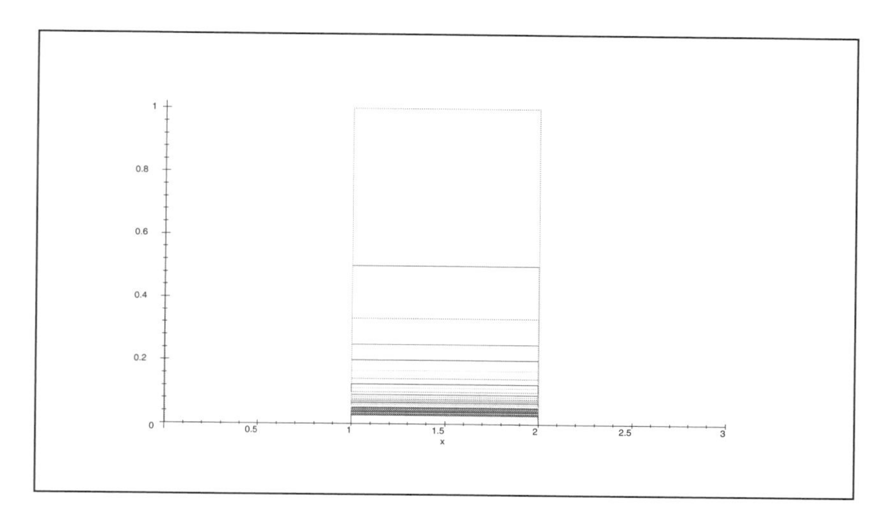

Figure 15.15: The Graphs of $G4$ for $n = 1, ..., 40$

somewhat extend the explanation provided there. This will also help us in the forthcoming analysis when we will deal with stochastic integrals.

Recall that if the (Riemann) integral of a function exists, it is given in terms of the limit of a sum. Consider for a moment the interval $[0, S]$ and let us divide it into N equal subintervals, $[0, s_1], [s_1, s_2], ..., [s_{N-1}, S]$, each of which has width of $\frac{S}{N} = \Delta s$. We have explained in footnote 8 of Section 5.2 the intuition behind the integral $\int_0^S h(s)\, ds$ being defined as

$$\lim_{N \to \infty} \sum_{j=0}^{N-1} h(s_j)\Delta s. \tag{15.75}$$

Note, however, that the index j goes from 0 to $N-1$. This means that we chose the value of the function at each left point of the subinterval. While it has no effect in the deterministic case, it has a marked effect in the stochastic case (although it may not be apparent at the level of our explanation).

In a generalized setting we may want to replace Δs with some other expression. Δs is the length of the interval and thus $h(s_j)\Delta s$ is the area of the rectangle that is built on the base Δs, where the height of the rectangle is the value of the function h at the left end point. Sometimes we are interested not in the physical length of the interval but in a function of it. For example, suppose S is a random variable, say, a price of a stock that can take values in $[0, \infty)$.

Assume we have another two random variables that are defined as functions of the initial random variable, say $g(S)$ and $v(S)$. How would we define the distance between $g(S)$ and $v(S)$? S is a random variable, and if we keep the same definition of the distance, i.e., $\int_0^\infty (v(s) - g(s))^2 \, ds$, we may not measure a relevant expression.

Consider a partition into subintervals of the interval $[0, \infty)$, where the knot points are $s_i = i\Delta s$ for $i = 0, 1, 2, ...$, for some fixed value of Δs. There might be a big difference between the two functions $g(s)$ and $v(s)$ when, for example, S is between s_1 and s_2. However, if the likelihood of S being in such a neighborhood is very slim, then the functions could still be very likely to be close to each other.

"Close" in this context must be interpreted in the sense that the likelihood of it being close is "large". Somehow this stochastic environment should be reflected in our measure.

One possibility is to maintain the function $(v(s_1) - g(s_1))^2$ representing the distance between the two functions over the interval $[s_1, s_2]$. To reflect the uncertainty environment we modify our measure by multiplying $(v(s_1) - g(s_1))^2$ by the probability of S being in the interval $[s_1, s_2]$ rather than by the length of $[s_1, s_2]$. When Δs approaches zero, $(v(s_1) - g(s_1))^2$ will be a good approximation for the difference between the two functions over the interval $[s_1, s_2]$.

Adopting this approach means that what matters is not the actual physical length of the intervals $\Delta s = s_2 - s_1$, but the probability of the interval $[s_1, s_2]$, that is, the probability that $s1 \leq S \leq s2$. Let us denote by F_S the cumulative distribution function of S. In terms of F_S, the likelihood of the interval $[s_1, s_2]$ is $F_S(s_2) - F_S(s_1)$. So a sensible measure of the distance between two random variables will thus be the summation of the probability of $[s_{i-1}, s_i)$ multiplied by the squared difference of the functions.

$$\sum_{i=1}^{\infty} (v(s_i) - g(s_i))^2 \left(F_S(s_i) - F_S(s_{i-1})\right) \tag{15.76}$$

Note that equation (15.76) can be thought of as a generalization of the summation in equation (15.75). If F is replaced with the identity function $I(s) = s$ then $I(s_2) - I(s_1)$ is indeed Δs. Furthermore, take a look at equation (15.76), where the length of each interval is replaced with its probability. When we multiply the probability by the squared difference of the functions and sum it up we obtain a weighted average.

When Δs goes to zero we will actually have the expected value (if it exists) of the squared differences. The integral defined as

$$\lim_{\Delta s \to 0} \sum_{i=1}^{\infty} (v(s_i) - g(s_i))^2 (F_S(s_i) - F_S(s_{i-1})) \tag{15.77}$$

is the expected value of the squared differences. This integral was defined with respect to the increment of the function F_S, usually denoted as $\Delta F_{s_i} = F_S(s_i) - F_S(s_{i-1})$, instead of with respect to the increments of S. Hence the notation reserved for such an integral is $\int_0^\infty (v(s) - g(s))^2 \, dF$ and we have that[28]

$$E\left((v(s) - g(s))^2\right) = \int_0^\infty (v(s) - g(s))^2 dF. \tag{15.78}$$

We have justified the distance measure as being the mean (expected value) of the squared differences. With this measure at hand, we say that the sequence of random variables $v_n(s)$ converges to another random variable $v(s)$ if $\lim_{n \to \infty} E\left((v_n(s) - v(s))^2\right) = 0$. Not surprisingly, such a convergence is termed convergence in the mean-square sense.

Consider a sequence of random variables $Z_{\Delta t} \sim N(0, \sqrt{\Delta t})$. $Z_{\Delta t}$ so defined is actually a Brownian motion (recall the definition in Section 5.4 and Section 14.4). In our analysis we will be interested to know if the mean-square limit of certain random quotients, e.g.,

$$\lim_{\Delta t \to 0} \frac{Z_{\Delta t}}{\Delta t}, \tag{15.79}$$

vanishes. According to the definition just presented, the limit of this quotient is zero if $\lim_{\Delta t \to 0} E\left(\left(\frac{Z_{\Delta t}}{\Delta t} - 0\right)^2\right) = 0$. However, $E\left(\left(\frac{Z_{\Delta t}}{\Delta t} - 0\right)^2\right) = \frac{E\left((Z_{\Delta t})^2\right)}{(\Delta t)^2}$, so we need to calculate $E\left((Z_{\Delta t})^2\right)$. To ease the calculation we have introduced the procedure[29] **MgmN**.

The procedure **MgmN** is a function that calculates the expected value of Z^n for a normally distributed random variable that has an expected value of μ and a standard deviation of σ. The input parameters to the procedure are n, μ, and σ. To calculate the expected value of $(Z_{\Delta t})^2$ we execute the command below. (We use Δ instead of Δt for the MAPLE calculations.)

---

[28] When F is differentiable then it can be shown that $\int_0^\infty (v(s) - g(s))^2 \, dF = \int_0^\infty (v(s) - g(s))^2 f(s) ds$, where $f = \frac{\partial F}{\partial s}$ is the density function of the random variable S. This integral is called the Stiltitz integral.

[29] The procedure simply uses the moment generating function in order to calculate the required expected value.

```
> MgmN(2,0,sqrt(Delta));
```

$$\Delta$$

Hence, $E\left((Z_{\Delta t})^2\right) = \Delta\, t$ and thus $\frac{E\left((Z_{\Delta t})^2\right)}{(\Delta t)^2} = \frac{1}{\Delta\, t}$, which obviously does not go to zero as Δt approach zero. In fact it diverges, and since

$$\lim_{\Delta t \to 0} E\left(\left(\frac{Z_{\Delta t}}{\Delta t}\right)^2\right) \tag{15.80}$$

does not exist, it means that in the mean-square sense, the limit in equation (15.79) does not converge to zero as Δt approaches zero.

Consider the mean-square limit of the quotient in equation (15.81).

$$\lim_{\Delta t \to 0} \frac{(Z_{\Delta t})^2}{\Delta t} \tag{15.81}$$

This time we have to calculate the expected value $E\left(\left(\frac{(Z_{\Delta t})^2}{\Delta t} - 0\right)^2\right) = \frac{E\left((Z_{\Delta t})^4\right)}{(\Delta t)^2}$. Let us first calculate the expected value of $(Z_{\Delta t})^4$.

```
> MgmN(4,0,sqrt(Delta));
```

$$3\, \Delta^2$$

Hence $\lim_{\Delta\, t \to 0} E\left(\left(\frac{(Z_{\Delta t})^2}{\Delta t}\right)^2\right) = 3$ and thus again the limit is not zero. This time, unlike the limit in (15.80), the limit exists, which means that as $\Delta\, t$ goes to zero the distance between zero and $\frac{(Z_{\Delta t})^2}{\Delta t}$ is a finite number.

Recall again that when we say zero, we mean a random variable such that the probability of it taking the value zero is one. Thus it is a random variable that assigns the value zero to all outcomes in the sample space. In our case the sample space is $[0, \infty)$ and thus this random variable is like the zero function defined over that interval.

Finally, we take a look at the limit of $Z_{\Delta t}$ as Δt goes to zero. Thus we need to calculate $E\left((Z_{\Delta t})^2\right)$, which, as was shown above, is Δt. Hence, it is clear that as Δt goes to zero the limit is zero. This means that while $Z_{\Delta t}$ is a random variable, the limit of it as Δt goes to zero will be a degenerate random variable, i.e., one that takes the value zero with probability one.

15.11.3 A More Rigorous Insight into Ito's Lemma

The material contained herein assumes that the reader has read or is familiar with the material from Appendix 15.11.1 and Appendix 15.11.2. As was mentioned above, the classical notation that is used in the statement of Ito's lemma, equation (15.42), is only a shorthand notation. In fact, the meaning of this notation is an integral equation as in equation (15.82) (where u is the integration variable for the second argument of g),

$$\begin{aligned} g(x,t) &= g(x_0,0) + \\ & \int_0^t \left(a(x,u)\, g_x(x,u) + g_u(x,u) + \frac{1}{2}\, b(x,\, u)^2\, g_{xx}(x,\, u) \right) du \\ & + \int_0^t b(x,u)\, g_x(x,u) dZ, \end{aligned} \tag{15.82}$$

where x is an Ito process and is given (using the integral notation) by

$$x(t) = x(0) + \int_0^t a(x,u)dt + \int_0^t b(x,u)dZ. \tag{15.83}$$

Utilizing the same approach as before, the interval $[0,t]$ is divided into n subintervals with the knot points t_i, $i = 0, ..., n-1$, where $t_0 = 0$. Hence,

$$g(x,t) = g(x_0,0) + \sum_{i=0}^{n-1} \Delta g(x_i,t_i), \tag{15.84}$$

where x_i is the value of x at time t_i and $\Delta g(x_i,t_i) = g(x_{i+1},t_{i+1}) - g(x_i,t_i)$ for $i = 0, ..., n-1$. Utilizing Taylor approximation, each $g(x_{i+1},\, t_{i+1})$ is approximated around the point (x_i,t_i) and $\Delta g(x_i,t_i)$ can be written as

$$\begin{aligned} \Delta\, g(x_i,t_i) &= g_x(x_i,t_i)\Delta x_i + g_t(x_i,t_i)\Delta t_i + \frac{1}{2} g_{xx}(x_i,t_i)\Delta x_i^2 \\ & \quad + g_{xt}(x_i,t_i)\Delta x_i \Delta t_i + \frac{1}{2} g_{tt}(x_i,t_i)\Delta t_i^2 + R_i, \end{aligned}$$

where $\Delta t_i = t_{i+1} - t_i$, $\Delta x_i = x_{i+1} - x_i$, and R_i is a remainder term. Substituting these results in equation (15.84), $g(x,t) - g(x_0,0)$ can be written as in equation (15.85).

$$\begin{aligned} & \sum_{i=0}^{n-1} g_x(x_i,t_i)\Delta x_i + \sum_{i=0}^{n-1} g_t(x_i,t_i)\Delta t_i + \\ & \frac{1}{2}\sum_{i=0}^{n-1} g_{xx}(x_i,t_i)\,(\Delta x_i)^2 + \sum_{i=1}^{n-1} g_{xt}(x_i,t_i)\Delta x_i \Delta t_i + \\ & \frac{1}{2}\sum_{i=0}^{n-1} g_{xt}(x_i,t_i)\,(\Delta t_i)^2 \end{aligned} \tag{15.85}$$

As Δt_i approaches zero, or as n approaches infinity, we have (by definition) that the first two elements in equation (15.85) converge to the expressions in equations (15.86) and (15.87), respectively. Equation (15.86) is an Ito integral, while the integral in equation (15.87) is with respect to t, but as explained prior to equation (15.41), it is also a random variable.

$$\lim_{n\to\infty} \sum_{i=0}^{n-1} g_x(x_i, t_i)\Delta x_i = \int_0^t g_x(x, u)dx \qquad (15.86)$$

$$\lim_{n\to\infty} \sum_{i=0}^{n-1} g_t(x_i, t_i)\Delta t_i = \int_0^t g_u(x, u)du. \qquad (15.87)$$

The third element in equation (15.85),

$$\sum_{i=0}^{n-1} g_{xx}(x_i, t_i)\Delta x_i^2, \qquad (15.88)$$

is "responsible" for the puzzling results of $dZ\,dZ = dt$. See the discussion prior to equation (15.45).

Let us look first at what happens to (15.88) as n approaches infinity. Substituting $\Delta x_i = a_i \Delta t + b_i \Delta Z_i$ in equation (15.88), where for simplicity a_i and b_i denote $a(x_i, t_i)$ and $b(x_i, t_i)$, respectively, we obtain

$$\begin{aligned} \sum_{i=0}^{n-1} g_{xx}(x_i, t_i)(\Delta x_i)^2 &= \sum_{i=0}^{n-1} g_{xx}(x_i, t_i)a_i^2\,(\Delta t_i)^2 + \\ &\sum_{i=0}^{n-1} 2g_{xx}(x_i, t_i)a_i b_i \Delta t_i \Delta Z_i + \\ &\sum_{i=0}^{n-1} g_{xx}(x_i, t_i)b_i^2\,(\Delta Z_i)^2 . \end{aligned} \qquad (15.89)$$

The first two sums in the above expression are of order $o(\Delta t)$. This will also explain why the last two elements in equation (15.85) are eliminated from the instantaneous increments. Let us see why this is the case. Consider, for example, the second sum in equation (15.89) (the first is obviously of order $o(\Delta t)$),

$$\sum_{i=0}^{n-1} g_{xx}(x_i, t_i)a_i b_i \Delta t_i \Delta Z_i. \qquad (15.90)$$

To show that the expression in equation (15.90) approaches zero in the mean-square sense we have to examine

$$E\left(\left(\sum_{i=0}^{n-1} g_{x\,x}(x_i,t_i)a_ib_i\Delta t_i\Delta Z_i\right)^2\right). \tag{15.91}$$

Since Z has independent increments and $E(\Delta Z_i)=0$, the expected value of products of the form

$$\left(g_{xx}(x_i,t_i)a_ib_i\Delta t_i\Delta Z_i\right)\left(g_{xx}(x_j,t_j)a_jb_j\Delta t_j\Delta Z_j\right),$$

for i different from j, will vanish. The independent increments also imply that ΔZ_i and $g_{xx}(x_i,t_i)$ are independent. Hence, the expected value in equation (15.91) is equal to

$$\sum_{i=0}^{n-1} E\left(g_{xx}(x_i,t_i)a_ib_i\right)(\Delta t_i)^2\, E\left((\Delta Z_i)^2\right). \tag{15.92}$$

As we have already seen, $E\left((\Delta Z_i)^2\right)=\Delta t_i$ and thus the expression in equation (15.92) reduces to

$$\sum_{i=0}^{n-1} E\left(g_{xx}(x_i,t_i)a_ib_i\right)(\Delta t_i)^3,$$

which is of order $o(\Delta t)$. (See footnote 31.)

Finally, we arrive to the term behind the puzzling results of $dZ\,dZ=dt$, that is, to

$$\sum_{i=0}^{n-1} g_{xx}(x_i,t_i)b_i^2(\Delta Z_i)^2.$$

The meaning of $dZ\,dZ=dt$, as we have alluded to before, is actually a claim about a certain stochastic integral converging to a deterministic integral. More precisely the claim is

$$\lim_{n\to\infty}\sum_{i=0}^{n-1} g_{xx}(x_i,t_i)b_i^2(\Delta Z_i)^2=\int_0^t g_{xx}(x,u)b(x,u)^2du. \tag{15.93}$$

One can see why the shorthand notation for this claim is $dZ\,dZ=dt$, since if $(\Delta Z_i)^2$ in equation (15.93) is replaced with Δt then (15.93) stands true.

To gain an insight into this claim we proceed in the following way. The $\int_0^t g_{xx}(x,u)b(x,u)^2du$ can also be written as

$$\lim_{n\to\infty}\sum_{i=1}^{n-1} g_{xx}(x_i,t_i)b_i^2\Delta t_i.$$

Hence we need to show that

$$\lim_{n\to\infty}\sum_{i=0}^{n-1} v_i\left(\Delta Z_i\right)^2 = \lim_{n\to\infty}\sum_{i=0}^{n-1} v_i\Delta t_i,$$

where for simplicity $g_{xx}(x_i,t_i)b_i^2$ is denoted by v_i. Since in this environment convergence is considered in the mean-square sense, to show that the above equality holds we have to show that

$$\lim_{n\to\infty} E\left(\left(\sum_{i=0}^{n-1} v_i\left(\Delta Z_i\right)^2 - \sum_{i=0}^{n-1} v_i\Delta t_i\right)^2\right) = 0. \tag{15.94}$$

A careful inspection reveals that

$$\begin{aligned} & E\left(\left(\sum_{i=0}^{n-1} v_i\left(\Delta Z_i\right)^2 - \sum_{i=0}^{n-1} v_i\Delta t_i\right)^2\right) & (15.95)\\ = \; & E\left(\sum_{j=0}^{n-1}\sum_{i=0}^{n-1} v_i v_j\left(\left(\Delta Z_i\right)^2 - \Delta t_i\right)\left(\left(\Delta Z_j\right)^2 - \Delta t_j\right)\right) \end{aligned}$$

and that due to the independence of the increments of Z, the left-hand side of equation (15.95) equals the left-hand side of equation (15.96) and thus

$$\begin{aligned} & E\left(\sum_{i=0}^{n-1} v_i^2\left(\left(\Delta Z_i\right)^2 - \Delta t_i\right)^2\right) & (15.96)\\ = \; & \sum_{i=0}^{n-1} E\left(v_i^2\right) E\left(\left(\Delta Z_i\right)^4 - 2\left(\Delta Z_i\right)^2\Delta t_i + \left(\Delta t_i\right)^2\right). \end{aligned}$$

As can be verified,[30] the expected value of $\left(\Delta Z_i\right)^4$ is $3\left(\Delta t_i\right)^2$, and the ex-

[30] Utilizing the procedure **MmgN** as in the previous Appendix, this can be done by executing the following commands:

```
>MgmN(2,0,sqrt(Delta));
>MgmN(4,0,sqrt(Delta));
```

pected value of $(\Delta Z_i)^2$ is Δt_i. Therefore, we have that

$$\begin{aligned}
& E\left(\sum_{i=0}^{n-1} v_i^2 \left((\Delta Z_i)^2 - \Delta t_i\right)^2\right) \\
= & \sum_{i=0}^{n-1} \left(3\,(\Delta t_i)^2 - 2\,(\Delta t_i)^2 + (\Delta t_i)^2\right) E\left(v_i^2\right) \\
= & 2\sum_{i=0}^{n-1} E\left(v_i^2\right)(\Delta t_i)^2,
\end{aligned}$$

which converges to zero as Δt converges to zero (or equivalently, as n converges to infinity) and thereby[31] proves equation (15.94). Putting all of this together produces Ito's lemma.

[31] Recall that if $\{y_n\}$ and $\{w_n\}$ are two converging sequences, the sequence $\{y_n w_n\}$ also converges. Moreover, $\lim_{n\to\infty} y_n w_n = (\lim_{n\to\infty} y_n)(\lim_{n\to\infty} w_n)$. This in turn means that $\lim_{n\to\infty} 2\sum_{i=0}^{n-1} E\left(v_i^2\right)(\Delta t_i)^2 = 2\left(\lim_{n\to\infty}\sum_{i=0}^{n-1} E\left(v_i^2\right)\right)\left(\lim_{n\to\infty}(\Delta t_i)^2\right)$, assuming of course that Δt_i is the same for all i. The first limit in the right hand side of this expression is finite and the second one is zero. Hence $2\sum_{i=0}^{n-1} E\left(v_i^2\right)(\Delta t_i)^2$converges to zero as n goes to infinity.

Chapter 16

Other Types of Options

Until now we have only implicitly addressed options written on an underlying asset different than a stock. Occasionally, we did make references to such options, noting that the theory provided so far, under certain conditions, will be equally applicable to other types of options. As well, we hardly touched options that possess features different than the simple, named "plain vanilla," American or European put or call options. The purpose of this Chapter is to explicitly investigate options written on underlying assets different than a stock and those with special features, referred to as exotic options. We will also touch upon some interest-rate-contingent-claims, although a full investigation of interest rate processes is beyond the scope of this book. The idea of arbitrage pricing and replication is of course generic, and remains the tool by which even exotic options are being priced.

We focus our attention firstly on plain vanilla options where the underlying asset is not a stock. As long as the underlying price process is assumed to follow a geometric Brownian motion[1] the modifications needed to value such options are simple in nature. Options on indexes, foreign currencies, and futures fall into this category.

The valuations of options on indexes, foreign currency, and futures are like the valuation of an option on a dividend-paying stock. Hence, before justifying these similarities and exploring these type of options, we review

[1] A weaker assumption is that the risk-neutral price distribution at maturity is lognormal. Such an assumption is weaker as it does not constrain the behavior of the price process in the interim. It will, however, allow the derivation henceforth when applied only to European-type options. Readers who opted not to skip Chapter 15 should feel comfortable with the notion of the lognormal distribution being a solution to the stochastic differential equation describing a geometric Brownian motion. See the example in Section 15.5.2.

the valuation of options on a dividend-paying stock.

16.1 American Options, Dividend-Paying Stocks, and Binomial Models

Section 6.4 investigated the valuation of a European option on dividend-paying stocks. It concluded that for a dividend-paying security the option pricing formula should be modified to use an initial price of $S\,e^{-Di\,t}$ instead of the (true) initial price of S, where the time to maturity is t and Di is the dividend yield. However, an American option (even a call) on a dividend-paying stock requires a numerical solution, which is reviewed below.

An American option on a dividend-paying stock can be valued utilizing the binomial model. The effect of dividends on the binomial model was discussed in Section 14.1.3. Dividends affect the expected value of the risk-neutral distribution of the binomial model in the manner discussed in Section 6.4. Consequently, the parameters of the Binomial model, qu, u, and v, should satisfy equations (16.1) and (16.2) below.

$$\left(r - Di - \frac{\sigma^2}{2}\right) T = N\left(qu\,u + (1 - qu)\,v\right) \tag{16.1}$$

$$\sigma^2 T = N qu\,(1 - qu)\,(u - v)^2 \tag{16.2}$$

Equations (16.1) and (16.2) are the counterparts of equations (14.6) and (14.7) in Section 14.1.1. The only difference is that the left-hand side of equation (16.1) is $(r - Di - \frac{\sigma^2}{2})T$ instead of μT as in equation (14.6). In the absence of dividends, for the binomial model to converge to the risk-neutral distribution of the continuously compounded rate of return, μ should be replaced with $r - \frac{\sigma^2}{2}$. This replacement is based on Proposition 9 (page 210) and equation (5.42). It was also explained utilizing Ito's Lemma in Section 15.5.2.

When the underlying asset pays a constant dividend yield Di, the expected value $r - \frac{\sigma^2}{2}$ of the risk-neutral distribution, as was explained in Section 6.4, should be further reduced by Di. Consequently, u, v, and qu can be chosen to satisfy equations (14.9) and (14.10) in Section 14.1.1 when μ is replaced with $r - Di - \frac{\sigma^2}{2}$, namely,

$$\begin{aligned} u &= \left(r - Di - \frac{\sigma^2}{2}\right) \Delta t + \sigma\sqrt{\Delta t} \\ v &= \left(r - Di - \frac{\sigma^2}{2}\right) \Delta t - \sigma\sqrt{\Delta t} \end{aligned} \tag{16.3}$$

$$qu \quad = \quad \frac{1}{2}.$$

That u, v, and qu, as stipulated in equation (16.3), satisfy equations (16.1) and (16.2) is demonstrated by MAPLE below. We first assign the solutions.

```
> u:=(r-Di-sigma^2/2)*T/N+sigma*sqrt(T/N);
```

$$u := \frac{(r - Di - \frac{1}{2}\sigma^2)\,T}{N} + \sigma\sqrt{\frac{T}{N}}$$

```
> v:=(r-Di-sigma^2/2)*T/N-sigma*sqrt(T/N);
```

$$v := \frac{(r - Di - \frac{1}{2}\sigma^2)\,T}{N} - \sigma\sqrt{\frac{T}{N}}$$

```
> qu:=1/2;
```

$$qu := \frac{1}{2}$$

The expected values of the rate of return and of the rate of return squared, over a period in the binomial model, are given respectively by the MAPLE groups below.

```
> Ex:=(qu*u+(1-qu)*v);
```

$$Ex := \frac{(r - Di - \frac{1}{2}\sigma^2)\,T}{N}$$

```
> Ex2:=simplify(qu*u^2+(1-qu)*v^2);
```

$$Ex2 := \frac{1}{4}T(4\,T\,r^2 - 8\,T\,r\,Di - 4\,T\,r\,\sigma^2 + 4\,T\,Di^2 + 4\,T\,Di\,\sigma^2 + T\,\sigma^4 + 4\,\sigma^2\,N)/N^2$$

Hence, the variance of the rate of return over a period in the binomial model is given by

```
> Var:=simplify(Ex2-Ex^2);
```

$$Var := \frac{T\,\sigma^2}{N}$$

Thus, the variance over the complete period, due to the independent increments property, is $N\frac{\sigma^2 T}{N} = \sigma^2 T$, which is the required result.

The procedure **AmerOpt** utilizes the binomial model, based on equation (16.3), to value American options on a dividend-paying stock. It is very similar to the procedures **AmerPut** and **AmerCall**. However, it uses a different approximation of the binomial model and allows the user to specify the payoff function at maturity in the same way that the procedure **Valbin** does. Consequently, the procedure **AmerOpt** allows the flexibility of valuing all types of American options as long as the payoff depends only on the value of the underlying asset at maturity. **AmerOpt,** as **AmerPut,** has an optional parameter to store the coordinates at which it was optimal to exercise the option. Hence, these coordinates can be visualized in a plot with the aid of the **BinTree** procedure. The input parameters to **AmerOpt** in order are:

- N, the number of nodes in the binomial model
- f, a function of the payoff at maturity, where f is explicitly defined in the input field (see the example below)
- T, time to maturity in years
- S, the current price of the stock
- σ, the standard deviation of the stock price (in percentage per year)
- r, the (continuously compounded) risk-free rate (per year)
- D, the dividend yield
- Path, an (optional) parameter used to store the coordinates at which it was optimal to exercise the option

Thus, for example, the value of an American call with an exercise price of \$80, where $N = 30$, $S = 100$, $T = \frac{3}{12}$, $\sigma = 0.26$, $r = .09$, and $Di = 0$, is given by

```
> AmerOpt(30, x->max(0,x-80), 3/12, 100, 0.25, 0.09,0);
                        21.87734734
```

Note that the price obtained is slightly larger that the price of 21.87734722 calculated for this option when the procedure **Valbin** was applied to the same example in Section 14.1.2. The "inconsistency" is due to the different approximation[2] used in these two procedures.

Before we move on, we would like to present an example of a call option where the dividend yield is not zero. The price of the call option above, but with $Di = 0.10$, is presented below.

```
> AmerOpt(30, x->max(0,x-80), 3/12, 100, 0.25, 0.09,0.10,
```

[2]For this reason, applying the procedure **AmerOpt** to the valuation of an American put may result in a slightly different value than the procedure **AmerPut** will yield for the same put.

```
> optc);
```

20.01535153

The **BinTree** procedure is utilized to visualize, Figure 16.1, the coordinates where it is optimal to exercise the call option as per our discussion in Chapter 12.

```
> BinTree(30,optc):
```

The reader is asked in one of the end-of-Chapter exercises to justify the shape of this set in comparison to the set depicted in Figure 14.5 that pertains to a put option. Some institutional details concerning options on indexes, foreign currency, and futures are needed to show that the theory developed for plain vanilla options can be applied to these types of options. This is done in the next section.

16.2 Options on Indexes, Foreign Currency, and Futures

16.2.1 Stock Index Options

Indexes are portfolios composed of certain stocks that are traded in a certain market or markets. Usually the indexes are formed so as to represent a certain segment of the market, e.g., blue chips or the performance of the market as a whole. For example, the S&P 500 index includes common stocks of industries, utilities, transportation, and financial institutions.

Options on the indexes are traded in various exchanges around the world. Some are European options like the options on the S&P 500, and some are American, like those traded on the S&P 100 and on the Major Market Index, which is composed of 20 blue chip stocks traded on the New York stock exchange. As for some individual stocks, options with longer maturities (up to three years) are also available on indexes. The acronym for these long-term options is LEAPS (Long Term Equity Anticipation Shares).

A call option on the index thus gives the holder the right to buy the index for the strike price. There is no difference between options on a stock index and an option on a stock, except for the fact that the former is settled in cash (not by delivering the portfolio) and that one contract is 100 times the value of the index. Such is the case for contracts written on both the S&P 500 and the S&P 100.

Hence, if the risk-neutral distribution of the value of the index is assumed

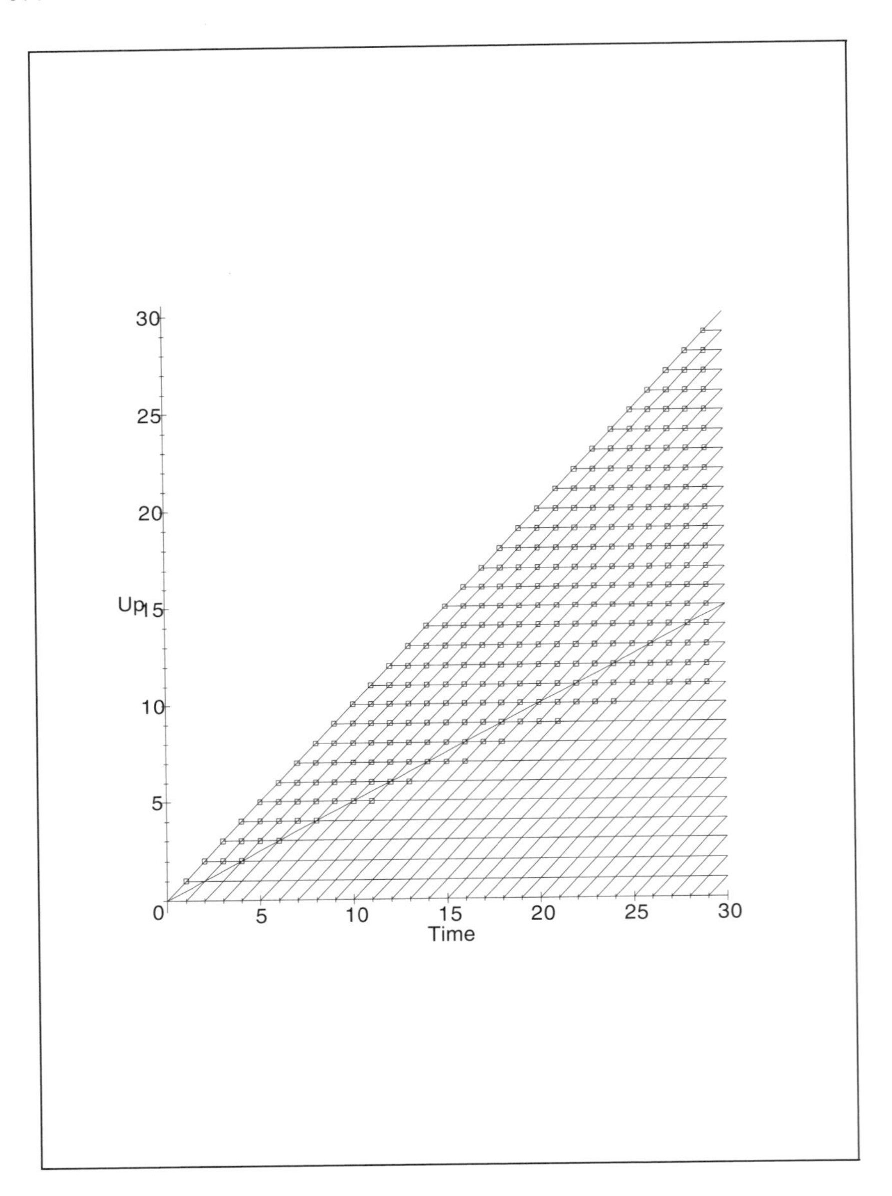

Figure 16.1: The Collections of (T, Up) Where It Is Optimal to Exercise Early an American Call on a Dividend-Paying Stock

to be lognormal,[3] one needs only to take into account the dividend yield of the index (portfolio) and apply equation (6.16) to it. The value of a call on the index will be 100 times the value obtained by equation (6.16). In the case of an American option, the binomial tree can be utilized in order to solve for the numerical value of the option.

Consider a call option on an index with a volatility of 0.20 where the current level of the index is 250, the dividend yield is 0.058, and the option matures in 2 months. The risk-free rate is 0.10 and the exercise price is 240. If the option is European, we can find its value by using the **Bs** procedure where instead of the current level of the index, we substitute $250e^{-0.058(\frac{2}{12})}$. According to the order of parameters in the **Bs** procedure, we should execute the command below.

```
> Bs(240,2/12,250*exp(-0.058*2/12),0.20,0.10);
```

$$14.96773756$$

Hence one contract would cost

```
> 100*Bs(240,2/12,250*exp(-0.058*2/12),0.20,0.10);
```

$$1496.773756$$

On the other hand, if the option is American we should use the procedure **AmerOpt** and execute the following command:

```
> 100*AmerOpt(40,x->max(0,x-240),2/12,250,0.2,0.10,0.058);
```

$$149.9097918$$

The reader may like to run this last command with the optional parameter in order to see if it would be optimal to exercise this option early.

16.2.2 Currency Options

A foreign currency option also lends itself to the use of equation (6.16) for the value of an option on a dividend-paying stock. The underlying asset in this case, the foreign currency, produces an income stream — the interest. The foreign risk-free rate earned on holding the foreign currency plays the same role as the dividend yield paid by a stock.

Let us specify the following to allow us to draw the parallel to a dividend-paying stock more accurately. The price of the underlying asset is the spot

[3] Theoretically, this assumption is inconsistent with the assumption that the price process of each stock follows a lognormal distribution, since the sum of lognormal random variables is not a lognormal variable. In the marketplace, however, this is the common practice, which seems to produce a reasonable approximation.

rate, i.e., the value of one unit of the foreign currency in terms of the local currency. The spot rate is assumed to follow a lognormal distribution. The risk-free rates in the domestic and foreign markets are assumed to be constant during the life of the option. A foreign currency call option gives the holder the right to buy an agreed-upon principal (denominated in the foreign currency), paying the exercise price specified on the option per one unit of the foreign currency.

The exercise price fixes the exchange rate; hence foreign currency options can be used by corporations to hedge the risk induced by fluctuations in the exchange rate. Foreign currency options are settled by delivering the principal (in foreign currency). Therefore, the value of a foreign currency call option is given by the principal times the value of equation (6.16) where the dividend yield is replaced by the foreign interest rate. Another feature of foreign currency options is a symmetry that exists between a foreign currency call and a put under a certain condition.

Assume two markets, **A** and **B,** and denote their currencies by A and B, respectively. Consider a (foreign currency) call option in market A to buy P units of currency B with a strike price of K. Now consider a (foreign currency) put option in market B to sell PK of currency A for a strike price of $\frac{1}{K}$ with the same maturity as the call option. A careful examination of the consequent cash flows involved reveals that these put and call options are nothing but the same asset.

We analyze these consequent cash flows with the help of MAPLE, starting with our assumptions on S and K being positive and the case where $S > K$. (We interpret S as being the spot exchange rate of $\frac{local\ currency}{foreign\ currency}$.)

```
> assume(S>0,K>0,S>K);
```

The call option, in this case, will be in-the-money and each unit of the principle (which is expressed in the foreign currency) contributes $S - K$ of the foreign currency to the cash flow. Hence, in the local currency the cash flow at maturity will be as below.

```
> P*max(0,S-K)/S;
```

$$\frac{P\,(S-K)}{S}$$

We can also express it using the **Call** function:

```
> P*Call(S,K)/S;
```

$$\frac{P\,(S-K)}{S}$$

The put option, in that case, will also be in-the-money. The exercise price is $\frac{1}{K}$ and the exchange rate (from the point of view of the foreign market) will be $\frac{1}{S}$, but since $S > K$, we have $\frac{1}{S} < \frac{1}{K}$. Each unit of the principle (which is now presented in the units of the local market, as the put is written in the foreign market) contributes $\frac{1}{K} - \frac{1}{S}$ to the cash flow at maturity. Thus the cash flow at maturity is

```
> P*K*max(0,1/K-1/S);
```

$$P\,K\,(-\frac{1}{S}+\frac{1}{K})$$

or in terms of the **Put** function,

```
> P*K*Put(1/S,1/K);
```

$$P\,K\,(-\frac{1}{S}+\frac{1}{K})$$

which verifies the conjecture above, for the case $S > K$. If, on the other hand, $K \geq S$ then both the put and the call are expired worthless. This is also expressed by the MAPLE commands below.

```
> assume(S>0,K>0,K>=S);
> P*max(0,S-K)/S;
```

$$0$$

```
> P*Call(S,K)/S;
```

$$0$$

```
> P*K*max(0,1/K-1/S);
```

$$0$$

```
> P*K*Put(1/S,1/K);
```

$$0$$

Let us look at an example of a Canadian dollar (American type) put option traded on the Philadelphia exchange. The contract size for such options is 50,000 Canadian dollars. Assume an expiration time of three months where the spot rate is $0.63\frac{US}{Can}$, the risk-free rate is 5% and 8% in the Canadian and American markets, respectively, the volatility of the spot rate is 18%, and the strike price is $0.67\frac{US}{Can}$. Utilizing the binomial model, the price of such an option can be calculated as below.

```
> AmerOpt(40,x->max(0,.67-x),3/12,.63,0.18,0.08,0.05,FC);
```

$$.04587884706$$

The price of such a contract will thus be

```
> 50000*AmerOpt(40,x->max(0,.67-x),\
> 3/12,.63,0.18,0.08,0.05,FC);
```

$$2293.942353$$

The nodes at which the put option should be exercised are displayed (only in the on-line version of the book) below.

```
> BinTree(40,FC);
```

16.2.3 Options on Futures Contracts

A call option on a futures contract gives its holder the right to assume a long position in a futures contract with a delivery price of K — the exercise price of the option. The delivery time of the futures contract, T_f, is larger than or equals the maturity time of the option T. When the option is exercised, at, say, time t, the futures contract is immediately marked to market. Hence, the holder of the option is being paid $F(t,T_f)-K$, where $F(t,T_f)$ is the futures price at time t for delivery at time T_f. The holder of the option also assumes a long position in the futures contract, the value of which (after the marking to market) is zero. Thus, the position in the futures contract can be closed at no cost. It follows that the payoff from the option at maturity, time T, is

$$\max\left(F\left(T,T_f\right)-K,0\right). \tag{16.4}$$

Options on a futures contract exist for financial futures as well as commodities which are sometimes referred to as futures options. A put option on a futures contract grants its holder the right to sell a futures contract. Hence, when the put option is exercised, the holder assumes a short position in a futures contract which is again immediately marked to market. Therefore, the payoff at maturity from such a put option is

$$\max\left(K-F\left(T,T_f\right),0\right). \tag{16.5}$$

We start by addressing the valuation of European-type options. Assuming, as we always do for the Black–Scholes analysis, that interest rates are deterministic, we know from Section 10.6 that futures[4] prices are the same as forward prices. From our discussion of forward prices in Section 2.4.1 we know that $F\left(T,T_f\right)=e^{r\left(T_f-T\right)}S(T)$, where $S(T)$ is the price at time T of the underlying asset on which the futures contract is written. Thus, the payoff at maturity of the option is

$$\max(e^{r\left(T_f-T\right)}S(T)-K,0), \tag{16.6}$$

[4]We will use the notation F for both futures and forward prices since indeed in a deterministic interest rate environment they are the same.

which can be written as in equation (16.7).

$$e^{r(T_f-T)} \max\left(S(T) - e^{-r(T_f-T)}K, 0\right) \tag{16.7}$$

The expression in equation (16.7) is $e^{r(T_f-T)}$ times the payoff from a call option where the underlying asset is S and the exercise price is $e^{-r(T_f-T)}K$. When S is assumed to follow the lognormal distribution and all the other assumptions of the Black–Scholes formula are satisfied, valuing such an option is a straightforward matter. The value of a European call option with an exercise price K on a futures contract is $e^{r(T_f-T)}$ times the value of a European call option, with an exercise price of $e^{-r(T_f-T)}K$. Hence, based on equation (6.1) the value of a call option on a futures contract is given[5] by

$$e^{r(T_f-T)}\left(S(t_0)N(d_1) - e^{-r(T-t_0)}e^{-r(T_f-T)}KN(d_2)\right), \tag{16.8}$$

where t_0 is the current time, N is the standard cumulative normal distribution,

$$d_1 = \frac{\ln\left(\frac{S(t_0)}{Ke^{-r(T_f-T)}}\right) + \left(r + \frac{\sigma^2}{2}\right)(T - t_0)}{\sigma\sqrt{T - t_0}}, \tag{16.9}$$

and

$$d_2 = d_1 - \sigma\sqrt{T - t_0}. \tag{16.10}$$

If S is an asset that generates an income stream at a constant yield then the equation above will be modified based on the guidelines explained in Section 6.4. See also one of the exercises at the end of this Chapter.

By equation (10.5), $F(t_0, T_f)$, the futures price as of time t_0 for delivery at time T_f, is given by $S(t_0)e^{r(T_f-t_0)}$. Thus substituting $e^{-r(T_f-t_0)}F(t_0, T_f)$ in equations (16.8) and (16.9) for $S(t_0)$ yields that the value of the call option can be expressed as

$$e^{-r(T-t_0)}F(t_0, T_f)N(d_1) - N(d_2)e^{-r(T_f-t_0)}K, \tag{16.11}$$

where

$$d_1 = \frac{\ln\left(\frac{F(t_0, T_f)e^{-r(T-t_0)}}{K}\right) + \left(r + \frac{\sigma^2}{2}\right)(T - t_0)}{\sigma\sqrt{T - t_0}} \tag{16.12}$$

and d_2 is given as in equation (16.10).

[5] This derivation is shown in [6].

As we mentioned before, $T_f \geq T$, but in many cases T_f is very close or even equal to T. Such is the case for options where the underlying asset is a futures contract on the NYSE index or the S&P. Section 6.4 investigated the valuation of option on a stock paying a constant dividend yield of Di. It concludes that such options can be valued applying the Black–Scholes formula, equation (6.16), but substituting $S(t_0)e^{-Di\,t}$ for $S(t_0)$, the "true" initial price of the stock. Comparing equations (16.11) and (16.12) to equations (6.16) and (6.18) reveals[6] that options on futures are priced as if $F(t_0, T)$ is the current price of an underlying stock that pays a constant dividend yield at the rate of r — the risk-free rate.

Of course this is just a tautology since when $T_f = T$ a European option contingent on $S(T)$ has the same payoff at maturity as a European option contingent on F. This can easily be verified by substituting $T_f = T$ in equation (16.6). It is just another way of saying that at maturity the futures price is the spot price, i.e., $F(T, T) = S(T)$. Hence, if the initial price of the underlying asset is written as $F(t_0, T_f)\, e^{-r(T-t_0)}$ or equivalently as $S(t_0)$ the same result would be obtained. Either way we look at this option, it is like a European contingent claim on the stock with an exercise price of K.

To value an American option on a futures contract we should proceed with care. First, let us look at the case $T_f = T$, i.e., the futures contract matures when the option matures. Hence, exercising such an option at time t grants its holder the cash flow $\max(F(t, T) - K, 0)$. Since we assume that the interest rate is deterministic, we can again replace the futures price with a forward price and write the cash flow upon exercising as

$$\max(e^{r(T-t)}S(t) - K, 0). \tag{16.13}$$

Presenting it in this manner we see that the cash flow is very similar to a cash flow of an American option on $S(t)$. The uncertainty is induced by $S(t)$, albeit the cash flow upon exercising is also a function of the time to maturity. Therefore, it is possible to use the binomial model where the parameters are chosen in the usual manner, such that the distribution of $S(t)$ will converge to the lognormal distribution. However, the value if exercised is calculated at each node based on equation (16.13). Following the usual

[6] Readers that did not skip Chapter 15 may recall that in Section 15.5.2 we demonstrated that F and S share the same volatility parameter. The σ that appears in (16.12) applies to both F and S. Assuming that the stochastic differential equation followed by F is $dF = F\,\mu\,dt + F\,\sigma\,dZ$ and applying the hedging arguments, in a continuous-time setting, induces the deterministic partial differential equation satisfied by an option written on a futures contract. One of the end-of-Chapter exercises asks the reader to prove equation (16.11) in this manner.

recursive procedure will generate the price of the option. Only a small modification is required for the above procedure when $T_f \neq T$, i.e., upon exercising, the cash flow will be

$$\max\left(e^{r(T_f-t)}S(t)-K,0\right). \tag{16.14}$$

The procedure **AmerOpt** is capable of using the binomial model in order to value options on futures. The parameters are as specified in the former subsection. However, when the f parameter (the function specifying the payoff from the option) is defined to have two arguments, the first is interpreted as the current price of the underlying asset S, and the second as the time to maturity. Hence, it allows the valuation of options where the cash flow upon exercising (or at maturity) is also a function of the time.

Thus, to value an option on a futures contract where $T_f = T$, f should be defined in our example as $(x,t) \to \max\left(0, e^{0.08t}x - 0.67\right)$. Assume that we would like to value a call option on a futures contract on the exchange rate. Let the value of the parameters be as in the former subsection, where we value a put option on the exchange rate. We therefore execute the command below.

```
>  AmerOpt(40,(x,t)->max(0,exp(.08*t)*x-.67),\
>  3/12,.63,0.18,0.08,0.05,FC);
                              .01146137150
```

We can also view the nodes at which it is optimal to exercise the option early. This is shown in Figure 16.2.

```
>  BinTree(40,FC);
```

Assume now that the futures contract matures a month after the option matures. This will require us to change the definition of the f parameter to be $(x,t) \to \max\left(0, e^{0.08\left(t+\frac{1}{12}\right)}x - 0.67\right)$. Valuing such an option is demonstrated in the next MAPLE command.

```
>  AmerOpt(40,(x,t)->max(0,exp(.08*(t+1/12))*x-.67),\
>  3/12,.63,0.18,0.08,0.05,FC);
                              .01295170256
```

Options on Forward Contracts

Options on futures contracts are different than options on forward contracts, which are traded over the counter. The differences arise for the following reason. Consider a call option written on a forward contract where we utilize

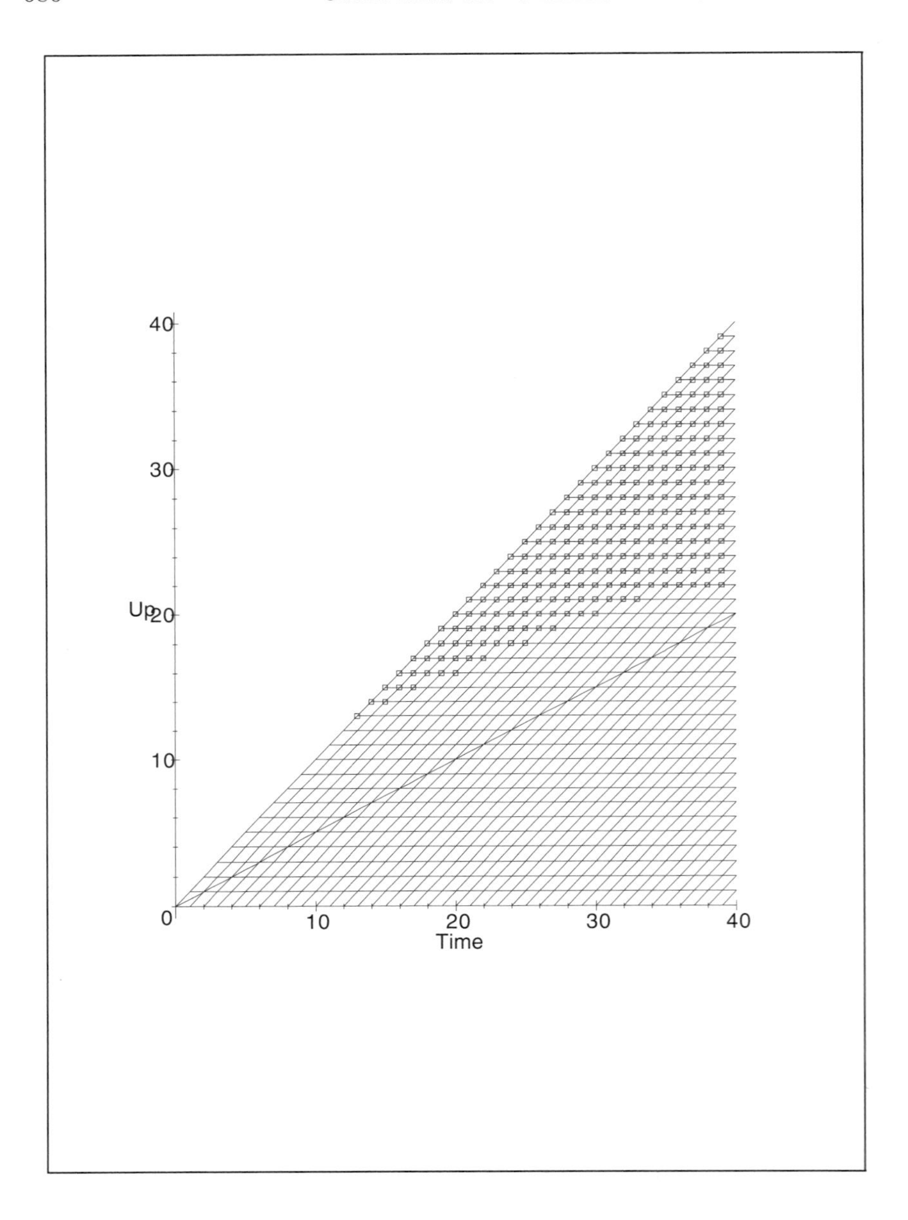

Figure 16.2: The Optimality of Exercising Early a Call Option on a Futures Contract

the same notations as for a futures contract. At the maturity of the option T, its value is the maximum between zero and the value of a forward contract with a delivery time T_f ($T_f > T$) and a delivery price of K (the exercise price of the option). Let $F(T, T_f)$ be the forward price of a contract written at time T specifying delivery at T_f. We know from equation (10.2) that the value of such a forward contract at time T is $e^{-r(T_f - T)}(F(T, T_f) - K)$. Thus, the value of the option at time T is

$$\max\left(e^{-r(T_f-T)}\left(F(T,T_f)-K\right),0\right), \tag{16.15}$$

which equals

$$e^{-r(T_f-T)}\max\left(F(T,T_f)-K,0\right). \tag{16.16}$$

Since we assume that interest rates are deterministic, the futures and forward prices are the same. Consequently, the value of the expression in equation (16.16) is simply $e^{-r(T_f-T)}$ times the value of an option on a futures contract, i.e., the expression in equation (16.6). Thus, the difference between the payoff of a European option on a forward contract and that of a European option on a futures contract is just the factor multiplication of $e^{-r(T_f-T)}$. The value of a European option on a forward contract is thus also $e^{-r(T_f-T)}$ times the expression in equation (16.8), i.e.,

$$S(t_0)N(d_1) - e^{-r(T_f-t_0)}K\,N(d_2). \tag{16.17}$$

If the option is American and it is being exercised early, say at time t, its payoff is

$$\max\left(e^{-r(T_f-t)}\left(F(t,T_f)-K\right),0\right) = \max\left(S(t) - e^{-r(T_f-t)}K, 0\right) \tag{16.18}$$

and the above conclusion is no longer valid. We can no longer rewrite the payoff as in equation (16.16) since $e^{-r(T_f-t)}$ depends on the exercising time t, and cannot be taken outside the maximum operator. The payoff from an American option on a forward contract is therefore not only a constant multiplication of the payoff to an American option on a futures contract. We can, however, still utilize the procedure **AmerOpt** to value an option on a forward contract. This will be accomplished by changing the f parameter to be $(x,t) \rightarrow \max\left(x - e^{0.08\left(t+\frac{1}{12}\right)}0.67, 0\right)$. Consider the option on the futures contract we just valued. If this would have been an option on a forward contract its value would have been given by executing the MAPLE command below.

```
> AmerOpt(40,(x,t)->max(0,x-exp(-.08*(t+1/2))*.67),\
> 3/12,.63,0.18,0.08,0.05,FC);
                              .02109871013
```

Moreover, this difference affects the decision of exercising early an American option on a forward contract. Consequently, it might be optimal to exercise early an option on a forward contract, but not on a futures contract. This issue and the put–call parity for these two types of options are elaborated on in the exercises at the end of the Chapter.

16.3 Examples of Exotic Options

"Exotic options" is the name reserved for options that do not have a payoff defined by $\max(S-K,0)$ or $\max(K-S,0)$ (i.e., the plain vanilla put and call options). In most cases, however, even for exotic options, the basic assumption is that the underlying asset follows a geometric Brownian motion. Consequently, European-type options can be valued simply by applying the guidelines as in equation (5.19) or as in equation (15.59).

The former values the option as the discounted expected payoff, based on the risk-neutral distribution. The latter values it as the solution to a partial differential equation satisfied by every derivative, with boundary conditions specific to the derivative at hand. American-type options will be valued numerically either utilizing the binomial model or by numerically solving the differential equation. The landscape of options includes many varieties of options, and new instruments are being developed regularly. In some cases the payoff is even simpler in nature than that of the plain vanilla options. In these cases, calculating the expected value, the approach adopted herein, is a straightforward task.

In other cases, the payoff is more complex and even finding its risk-neutral distribution is not trivial. Nevertheless, once it has been identified (or approximated) the same mechanism is applied in order to proceed with the valuation. In this section we survey some of the less-complicated exotic options, starting with an option that we have already been introduced to in Chapter 4. Moreover, we will demonstrate that we can also make use of the building blocks approach in the context of exotic options.

Once we introduce exotic options that we know how to value, our universe has been expanded. We can now generate new instruments by combining the plain vanilla options with the new exotic options. The price of the so generated instrument will be easily valued applying the replication argument.

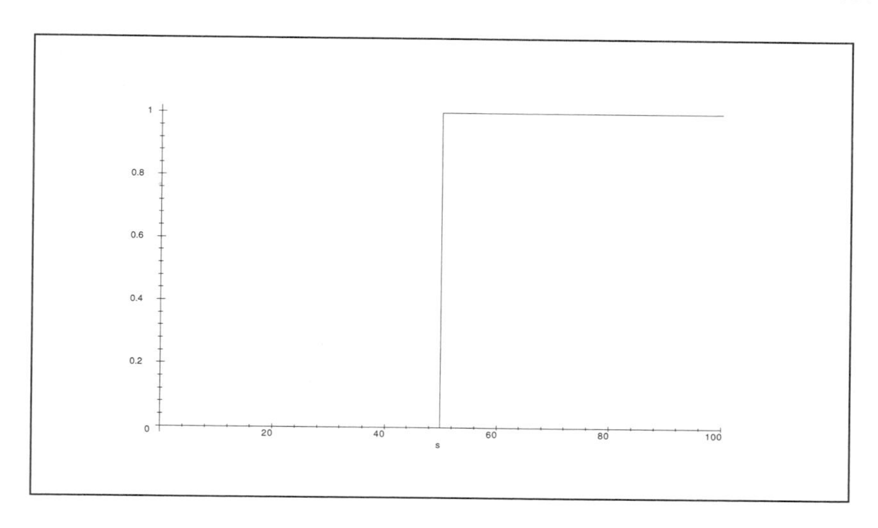

Figure 16.3: A Payoff from a Digital Option with Exercise Price $K = 50$

16.3.1 Binary (Digital) Options

In Section 4.5 we introduced the binary call option: An option that pays \$1 if the price of the stock is above the exercise price and zero otherwise. Its payoff is therefore given by the function

$$Dig(S,\, K) = \begin{cases} 0 \text{ if } S < k \\ 1 \text{ if } S \geq K. \end{cases} \tag{16.19}$$

We defined this function in MAPLE and display its graph in Figure 16.3.

```
> plot(Dig(s,50),s=0..100, title="Figure 16.3:  A\
> Payoff from a Digital Option with Exercise Price K=50");
```

Furthermore, we explained in Section 4.5 why its payoff cannot be replicated with the payoffs of plain vanilla options. The payoffs of plain vanilla options are piecewise-linear continuous functions. Portfolios are linear combinations of the asset included in the portfolio. The payoffs of portfolios are linear combinations of the payoff functions of the asset that are included in the portfolio. Linear combinations of continuous functions, however, cannot be combined to create discontinuous payoffs. The payoff of Digital options is not continuous, and hence cannot be generated with the plain vanilla options. In Chapter 4 we exemplified, with the aid of Figure 4.39, how it can be approximated by a portfolio of plain vanilla options.

Having the knowledge acquired since Chapter 4, we can actually value a digital option as the discounted expected value of its payoff. In fact, its valuation is even simpler than the valuation of plain vanilla options. This should not be such a surprise, since the payoff of a digital option is simpler than those of plain vanilla options. In fact, utilizing our guideline we simply need to calculate

$$e^{-r\,t}\int_0^\infty Dig(s,\,K)Lnrn(s,t,P,r,\sigma)ds, \tag{16.20}$$

where **Lnrn**, as defined in Section 6.1, is the risk-neutral lognormal distribution given by

```
> Lnrn(s,t,P,r,sigma);
```

$$\frac{1}{2}\,\frac{\sqrt{2}\,e^{\left(-1/2\,\frac{(\ln(s)-\ln(P)-(r-1/2\,\sigma^2)\,t)^2}{\sigma^2\,t}\right)}}{s\,\sigma\,\sqrt{\pi\,t}}$$

In the above definition s is the argument of the function, t the time to maturity, P the current price of the stock, r the risk-free rate, and σ the standard deviation.

In view of the structure of $Dig(s, K)$, as stipulated in equation (16.19), the integral in equation (16.20) is simply the probability of the underlying asset having a value larger than the exercise price at the maturity time. To facilitate the valuation with MAPLE we assume the following

```
> assume (t>0,r>0,K>0,sigma>0);
```

and ask MAPLE to value the expression in equation (16.20) and assign the result to **CprDig**.

```
> CprDig:=exp(-r*t)*int(Dig(s,K)*Lnrn(s,t,P,r,sigma),\
> s=0..infinity);
```

$$CprDig := \frac{1}{2}\,\frac{e^{(-r\,t)}\,\sqrt{\pi}\,\sqrt{t}\,(1+\mathrm{erf}(\frac{1}{4}\,\frac{\sqrt{2}\,(-2\ln(K)+2\ln(P)+2\,r\,t-t\,\sigma^2)}{\sigma\,\sqrt{t}}))}{\sqrt{\pi\,t}}$$

Proceeding as we did in Chapter 6 we ask MAPLE to express the pricing formula in terms of the cumulative standard normal distribution denoted here by N. To this end we again define the function "newerf," which is simply the **erf** function in terms of the N function. Recall from Chapter 6 that this relationship is

```
> newerf:= t -> 2*N(t*sqrt(2))-1;
```

$$newerf := t \to 2\,\mathrm{N}(t\,\sqrt{2})-1$$

We can now substitute **newerf** for **erf** in the **CprDig** expression. The pricing formula is now expressed in terms of the N function

```
> CprDig:= simplify(subs(erf=newerf,CprDig));
```

$$CprDig := e^{(-r\,t)}\,\mathrm{N}(\frac{1}{2}\,\frac{-2\ln(K)+2\ln(P)+2\,r\,t-t\,\sigma^2}{\sigma\,\sqrt{t}})$$

It can thus be written as

$$e^{-r\,t}N\left(\frac{\ln\left(\frac{P}{K}\right)+\left(r-\frac{\sigma^2}{2}\right)t}{\sigma\sqrt{t}}\right) \tag{16.21}$$

or as

$$e^{-r\,t}N(d_2), \tag{16.22}$$

where $d_2 = \frac{\ln\left(\frac{P}{K}\right)+\left(r-\frac{\sigma^2}{2}\right)t}{\sigma\,\sqrt{t}}$.

The procedure **CallPriceDig** was defined in MAPLE to value a European digital option. The input parameters are, in order, exercise price, time to maturity, current stock price, standard deviation, and the risk-free rate. Hence a digital option where $K = 110$, $t = \frac{3}{12}$, the current price of the stock is 100, $\sigma = 0.18$, and $r = 0.09$ is valued by the MAPLE command below.

```
> CallPriceDig(110,3/12,100,0.18,0.09);
```

$$.1921789031$$

A digital American option can be valued numerically by the procedure **AmerOpt**. One needs simply to specify the payoff function accordingly. For example, consider an American digital option with the same parameters as the European option just valued above. Its payoff function at maturity can thus be written in MAPLE as $piecewise(110 \le x,\ 1,\ 0)$. Consequently, its numerical value will be given by executing the procedure **AmerOpt** with parameters as specified below.

```
> AmerOpt(30,x->piecewise(x>=110,1,0),\
> 3/12,100,0.18,0.09,0,node0);
```

$$.3158523708$$

Let us see at which nodes it was optimal to exercise this option. First, however, we would like to point out that six up steps should be realized in the binomial model for the price of the stock to start from \$100 at time 0 and exceed \$110 at time 6. Thus, it should be optimal to exercise if six up steps were realized. One of the exercises at the end of this chapter guides the reader in verifying this point as well as verifying the nodes at which it is optimal to exercise this option. These nodes can be viewed, Figure 16.4, by executing the command below.

```
> BinTree(30,node0);
```

Utilizing the method explained in the exercise mentioned above one could confirm that the (blue) squares are superimposed on every coordinate where the price of the underlying asset is at least \$110. Figure 16.4 therefore suggests that it is optimal to exercise this option as soon as the price of the underlying asset hits \$110. The coordinates of these nodes are stored in the list **node0** and could be displayed by executing the command below.

```
> node0;
```

Let us further enhance this phenomenon by calculating the value of this option when the price of the stock is \$110.

```
> AmerOpt(30,x->piecewise(x>=110,1,0),\
> 3/12,110,0.25,0.09,0,node1);
```

$$1$$

It is not surprising that the value of the option is \$1 as it is worth exercising it *immediately*. It will never yield more than \$1, so there is no possible gain to be made by waiting for the price of the stock to increase above \$110. The on-line version of the book displays below the nodes at which it is optimal to exercise this option.

```
> BinTree(30,node1);
```

We realize again that the option should be exercised on the spot, and if it is left unexercised, it is exercised at any nodes at which the underlying asset price exceeds \$110.

Figure 16.4 visualizes the fact that it is optimal to exercise the option as soon as the price of the underlying asset reaches \$110. This is also the reason an analytical solution for the value of an American digital option exists. While the details of the solution are beyond the scope of this book, it is important to understand why such a solution exists. Recall our discussion of the vicious circle in Sections 12.5 and 14.1.2. It was explained why this vicious circle is the cause for the absence of an analytical solution to a plain vanilla American option. We mentioned that not knowing *a priori* when the option should be exercised is the reason for a lack of an analytical solution. In the case of an American digital option, in contrast to a plain vanilla American option, it is clear that once the value of the stock reaches the exercise price it is optimal to exercise it.

The reason is quite obvious: there is no advantage to holding the option as its value never appreciates above its value when it is in-the-money.

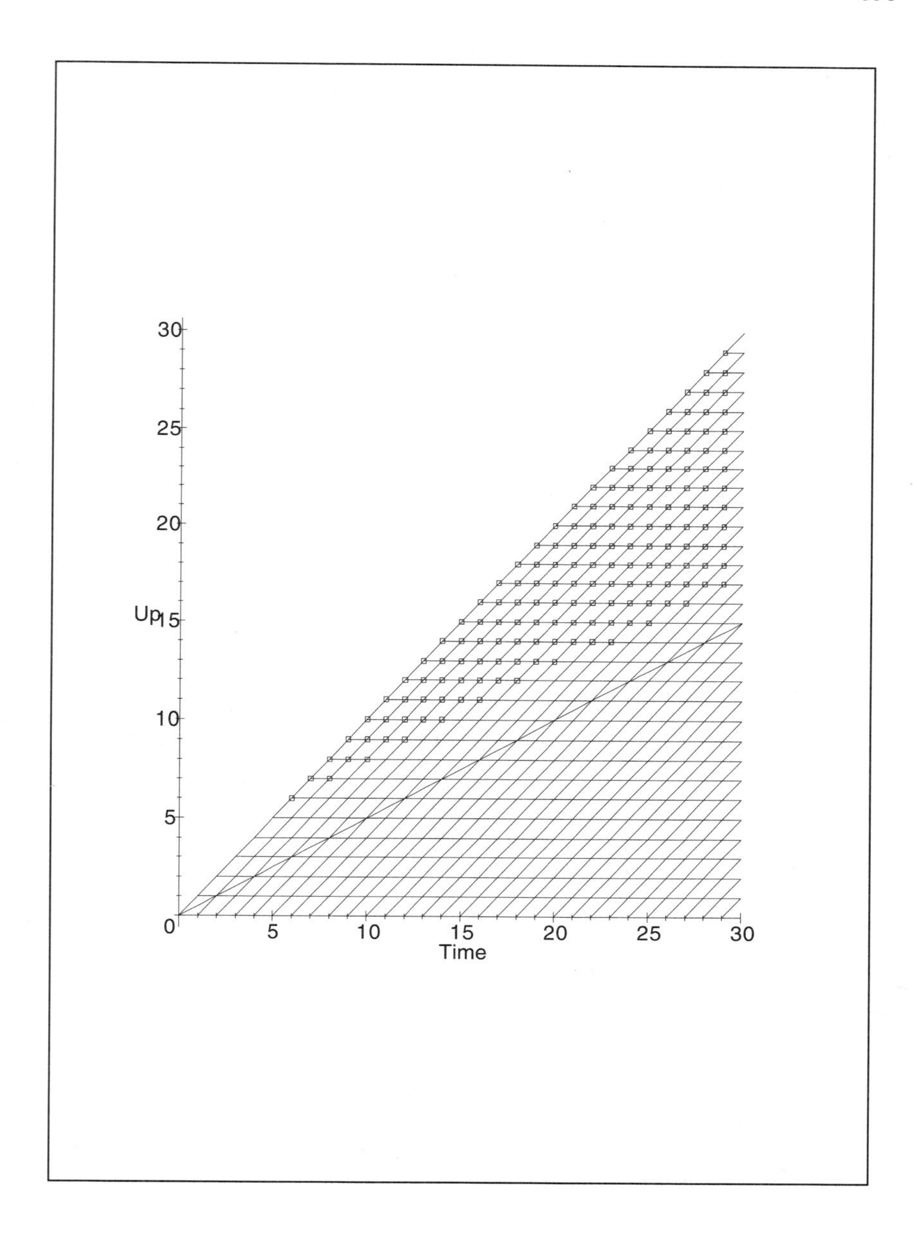

Figure 16.4: Digital Option, Optimal Exercising

Hence waiting will only cause losing the time value of money. Therefore the problem is no longer of the free boundary type since the optimal exercising strategy is known *a priori*. To analytically value this option the discounted expected value (under the risk-neutral distribution) of its payoff should be calculated. However, in order to proceed in this manner we should find the distribution of τ, the first time[7] the underlying asset value reaches the exercise price. Given τ, the expected value of the option is $g(\tau) = 1e^{-r\tau}$ if τ is smaller than the maturity time, and zero otherwise. Utilizing the technique of conditional expectation, the value of the option is the expected value of $e^{-r\tau}$ (under the risk-neutral distribution of τ).

A digital put option is valued analogously and we leave the details of this valuation as an exercise for the reader. One can imagine that the Delta of a digital option is going to be very unstable in the vicinity of the exercise price. This is exemplified in Figure 16.5, where the value of the digital option with its Delta is presented in the same plane.

```
> plot([CallPriceDig(40,2/12,S,.18,.05),\
> diff(CallPriceDig(40,2/12,S,.18,.05),S)],\
> S=0..80,colour=[red,blue],thickness=3,\
> title="Figure 16.5:  The Value and Delta of\
> a Digital Option as Functions of the Stock Price");
```

Delta hedging a digital option can thus be very costly. Instead, it may be more cost efficient to use a static hedging. This can be done utilizing the same technique employed in Chapter 4 to produce bounds on the price of a digital option using plain vanilla options. See the exercises at the end of the Chapter for a demonstration of this issue.

Having valued the digital option we can add it to our universe (arsenal) of building blocks and explore other types of exotic options we can create combining the digital option with the plain vanilla options. We will proceed in doing that in the next subsection where we employ the technique utilized in Chapter 4.

16.3.2 Combinations of Binary and Plain Vanilla Options

Now that we know how to value a digital option we can generate non-continuous payoffs by combining plain options, the stock, and the Digital option. The digital option, as displayed in Figure 16.3, allows us to have a "jump" in the payoff. For example, consider a slight modification of the

[7] Readers that are familiar with the concept of stopping time will recognize τ as such time.

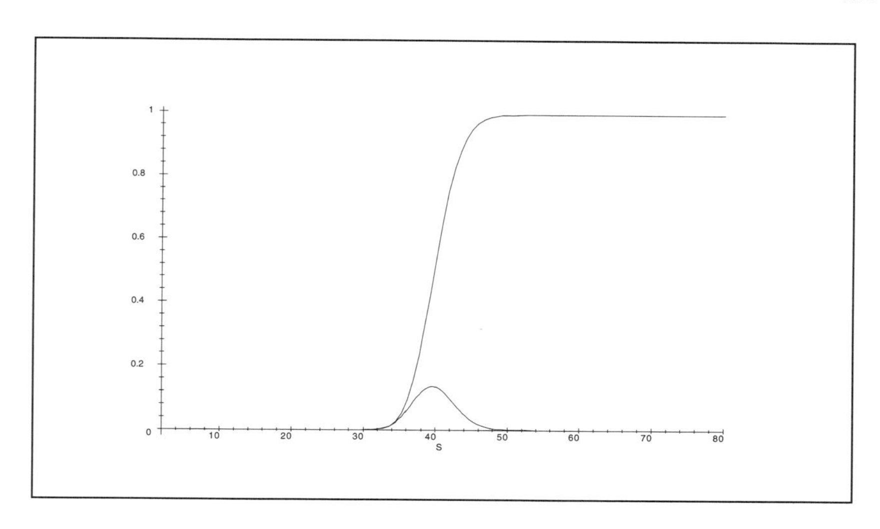

Figure 16.5: The Value and Delta of a Digital Option as Functions of the Stock Price

payoff from a plain vanilla call option with an exercise price of \$4. The modification we make is that the option pays $s - (K - 1)$ if it is in-the-money, instead of $s - K$. This payoff is simply the sum of a digital option with $K = 4$ and a call option with $K = 4$ as exemplified in Figure 16.6.

```
>  plot(Call(s,4)+Dig(s,4),s=0..10,thickness=2,\
>  title="Figure 16.6:  A Payoff from a Digital Option \
>  Plus a Call Option, Both with an Exercise Price K=4",\
>  titlefont=[TIMES,BOLD,8]);
```

In what follows we utilize the digital option to generate a few modifications of plain vanilla options.

16.3.3 Gap Options

Consider an option that pays upon maturity according to the function

$$GapC(s,\ K, G) = \begin{cases} 0 & \text{if} \quad s < K \\ s - G & \text{if} \quad s \geq K, \end{cases} \tag{16.23}$$

which we also defined in MAPLE.

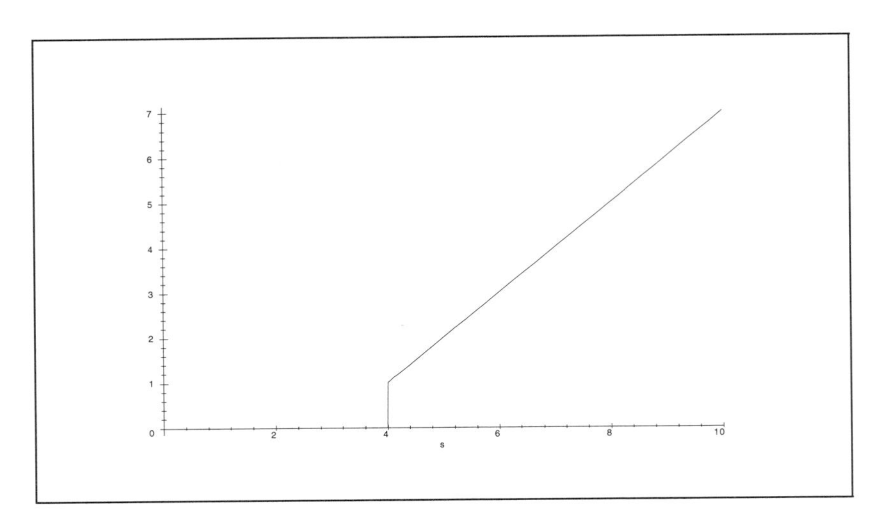

Figure 16.6: A Payoff from a Digital Option Plus a Call Option, Both with an Exercise Price $K = 4$

This payoff is the general case of the one shown in Figure 16.6, where K was 4 and G was 1. These types of options are called gap options, as there is a gap, or a jump, in their payoffs.

The function **GapC** has been defined in MAPLE and we can utilize it to generate a payoff where the jump at K is not necessarily 1 as in Figure 16.6. Note that G might be smaller or larger than K, which means that the option may have a negative cash flow in certain states of nature, namely, for s such that $K < s < G$.

Consider, for example, the payoff $GapC(s, 4, 6)$. A careful examination of figure 16.6 reveals that the payoff there differs from $GapC(s, 4, 6)$ only in that the jump at $s = 4$ is of two units instead of one unit. Therefore, $GapC(s, 4, 6)$ can be written as $GapC(s, 4, 6) = Call(s, 4) + 2Dig(s, 4)$, as is visualized in Figure 16.7.

```
> plot(GapC(s,4,6),s=0..10,thickness=2,title="Figure\
> 16.7:  A Payoff from a Gap Option with K=4 and\
> G=6",titlefont=[TIMES,BOLD,8]);
```

In general we thus have the relation

$$GapC(s, K, G) = Call(s, K) + (G - K)Dig(s, K), \tag{16.24}$$

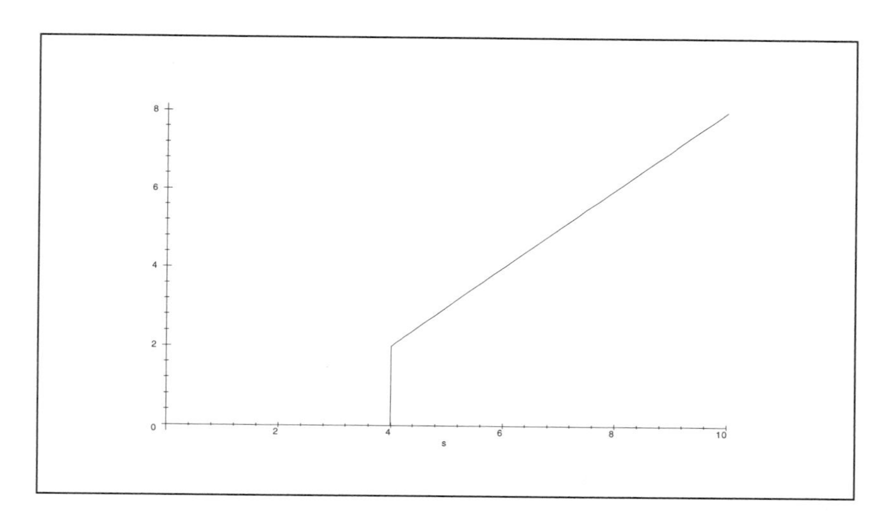

Figure 16.7: A Payoff from a Gap Option with $K = 4$ and $G = 6$

which points out a way of evaluating a gap call option. Its price must equal that of a call option plus $(G - K)$ units of a digital option, or else arbitrage opportunities would exist. Hence, substituting in equation (16.24) the expression of the value of a digital option, equation (16.21), for $Dig(s, K)$ and the expression of the value of a plain vanilla call option, equation (6.3), for $Call(s, K)$ yields the value of the Gap option,

$$\begin{aligned} & N(d_1) - e^{-rt}KN(d_2) + (G - K)e^{-rt}N(d_2) \\ = \; & N(d_1) + (G - 2K)e^{-rt}N(d_2). \end{aligned} \tag{16.25}$$

The procedure **CallPriceGap** was defined in MAPLE to value a European gap option. The input parameters are, in order, G, exercise price, time to maturity, current stock price, standard deviation, and the risk-free rate. Hence a gap option where $G = 100, K = 110, t = \frac{3}{12}$, the current price of the stock is 100, $\sigma = 0.18$, and $r = 0.09$ is valued by the MAPLE command below.

```
> CallPriceGap(100,110,3/12,100,0.18,0.09);
                    3.025412201
```

When $G = 0$ the payoff function is of the form

$$GapC(s, K, 0) = \begin{cases} 0 & \text{if} \quad s < K \\ s & \text{if} \quad s \geq K \end{cases}$$

and the option is sometimes referred to as an "asset or nothing" option. The reason for this name is self-explanatory and the valuation formula is obtained by substituting $G = 0$ in equation (16.25). Thus it will be valued as below.

```
> CallPriceGap(0,110,3/12,100,0.18,0.09);
```

$$22.24330251$$

Consider the European call gap option valued above, and assume that it is of an American type. Such an American gap option can be valued numerically utilizing the **AmerOpt** procedure. Its value is given below.

```
> AmerOpt(30,x->piecewise(x>=110,x-100,x<=110,0),\
> 3/12,120,0.25,0.09,0,node);
```

$$22.42199536$$

We anticipate, of course, its value to be at least the value of its European counterpart. There is, however, a point we would like to highlight here. We know that it is not optimal to exercise an American call (on a non-dividend-paying stock) and that it is optimal to exercise a digital option as soon as it is in-the-money. A gap option is a combination of a digital and call option, which in certain regions affect the decision of early exercising in an offsetting manner. When, therefore, will it be optimal to exercise the above gap option? What would be the shape of the set of coordinates at which exercising should take place?

Figure 16.8 displays the set of coordinates at which early exercising should take place. These coordinates are stored in the list node, as shown below. We leave, however, as an exercise to the reader the explanation for the form of the set presented in Figure 16.8.

```
> node;
```

$$[[24, 10], [26, 11], [28, 12]]$$

```
> BinTree(30,node);
```

Similarly, it is possible to define a gap option which is of a put type. The payoff function of such an option is defined in equation (16.26).

$$GapP(s,\ K, G) = \begin{cases} G - s & \text{if} \quad s < K \\ 0 & \text{if} \quad s \geq K \end{cases} \tag{16.26}$$

It is again a combination of a Digital option and a plain vanilla put option. The valuation of such an option is left as an exercise for the reader. However, before we proceed to the next combination we would like to display the counterpart of Figure 16.8 for a put gap option. Consider a put gap option

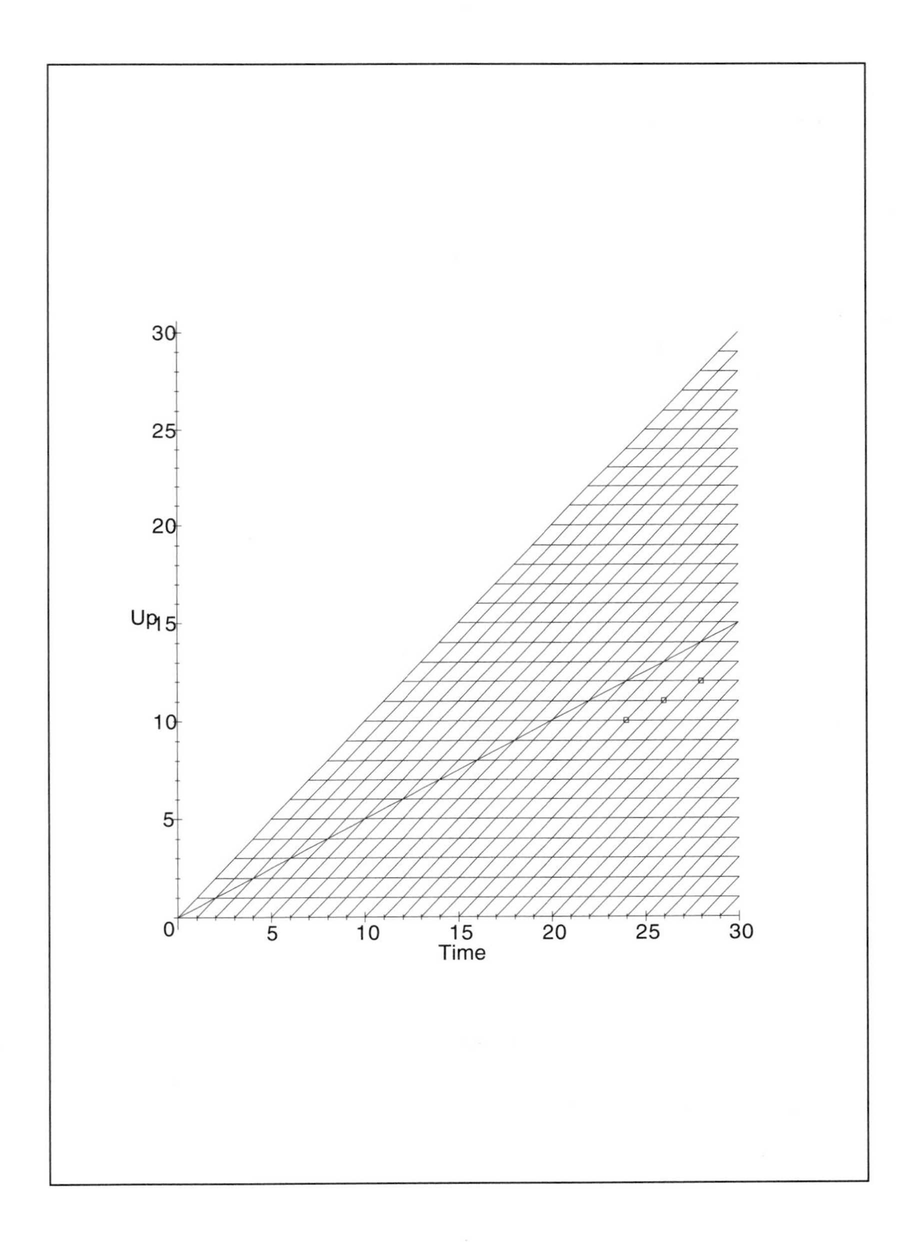

Figure 16.8: The Optimality of Early Exercise of a Call Gap Option

where $G = 100$, $K = 110$, and the rest of the parameter are as above. To value this option we should operate the **AmerOpt** procedure when the f parameter is $x- > piecewise(x <= 110, 100 - x, x >= 110, 0)$.

```
> AmerOpt(30,x->piecewise(x<=110,100-x,x>=110,0),3/12,\
> 100,0.25,0.09,0,node);
```

$$3.672693649$$

We ask the reader to try and explain the shape of the set of coordinates at which early exercise should take place. It is displayed in Figure 16.9. The fact that when $s \geq G$ the cash flow from the option is negative may shed some light on the shape of this set.

```
> BinTree(30,node);
```

16.3.4 Paylater (Cash on Delivery) Options

The paylater option is, as the name suggests, an option where at initiation no cash changes hands. It is a zero cost position and in some ways resembles a forward contract. The holder of such an option is required to exercise it, if it is in-the-money, and pay the premium at that time. The payoff from a Paylater call option can be negative. This occurs when the premium exceeds the difference between the exercise price and the value of the underlying asset.

The payoff at maturity of a paylater option is given by the function

$$PaylC(s, K, C) = \begin{cases} 0 & \text{if} \quad s < K \\ s - K - C & \text{if} \quad s \geq K, \end{cases} \tag{16.27}$$

where s and K are defined as usual, and C is the premium which is paid at maturity. We defined this function in MAPLE and display the payoff function in Figure 16.10 for $K = 10$ and $C = 1$.

```
> plot(PaylC(s,10,1),s=0..15,thickness=3,title="Figure \
> 16.10:  The Function PaylC(s,10,1)");
```

An examination of the payoff in Figure 16.10 reveals that given the premium C, the payoff function is a linear combination of a payoff of a call option and $-C$ units of a digital option with the same exercise price. Thus the payoff can be written as

$$Call(s, K) - CDig(s, K).$$

Hence given C, the price of the paylater option is the price of a call option minus C times the price of a digital option. Thus, given C, the price

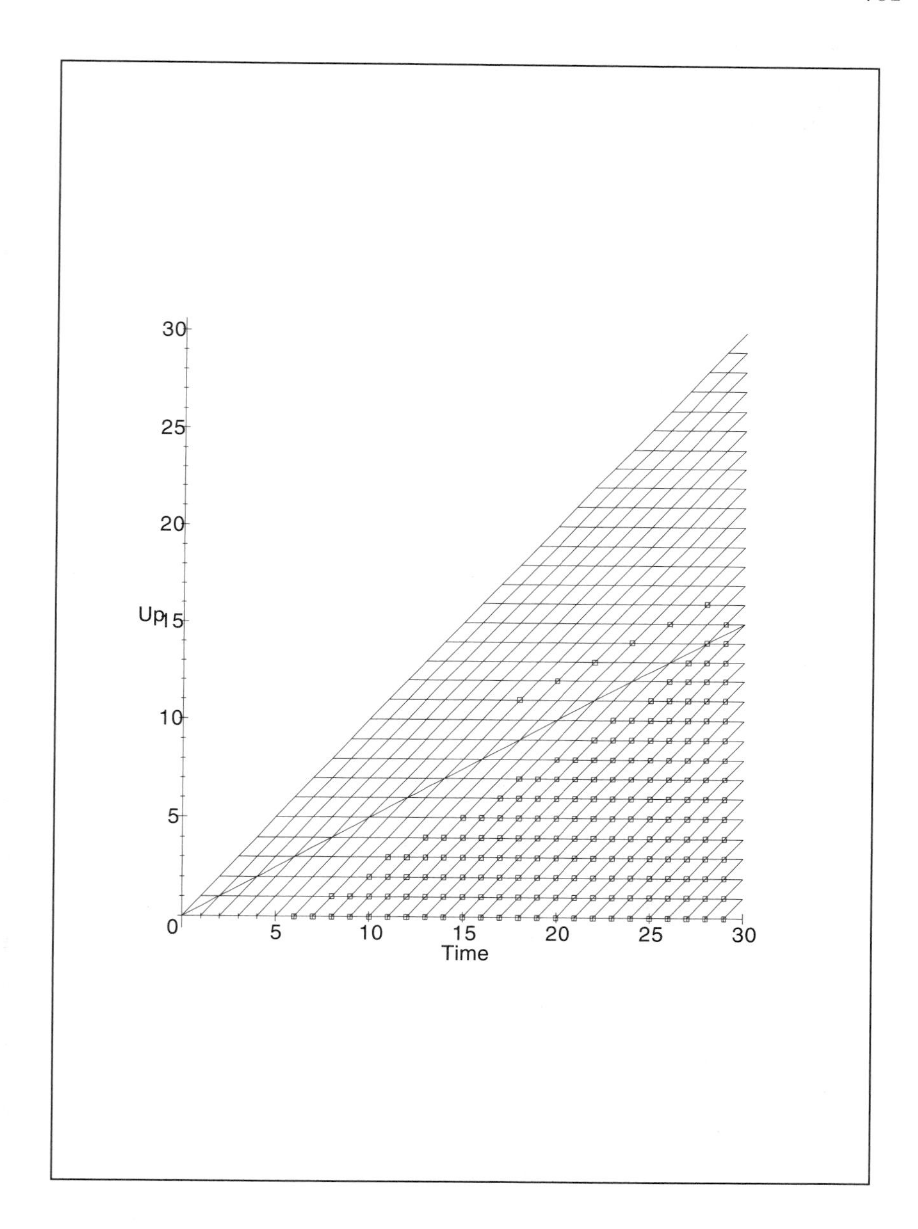

Figure 16.9: The Optimality of Early Exercis a Put Gap Option

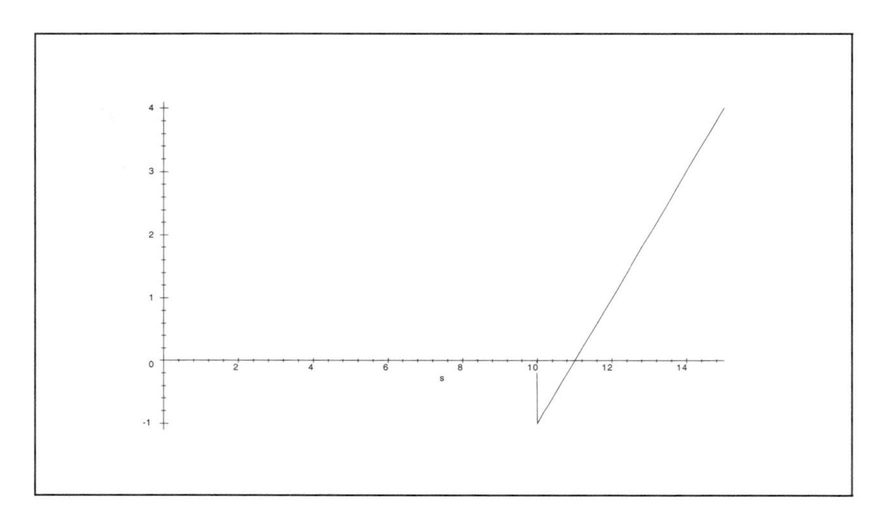

Figure 16.10: The Function PaylC(s,10,1)

of the paylater option is given by equation (16.28).

$$N(d_1) - e^{-rt}\,(K + C)\,N(d_2) \tag{16.28}$$

Since it costs nothing to buy a paylater option its value at initiation must be zero. Indeed the exercise price of such an option is determined, similarly to the way a forward price is determined, such that the value of the option will be nil.

The MAPLE procedure **CallPricePL** calculates the expression in equation (16.28). Its parameters are, in order,

- C, the premium paid when the option is exercised
- K, the exercise price
- t, the time to maturity
- P, the current price of the underlying asset
- σ, the standard deviation of the underlying asset
- r, the risk-free rate

The procedure **CallPricePL** is capable of valuing a paylater call option and determining the value of the premium at initiation. At initiation, the value of the option as a function of the premium C is determined by issuing the command below.

```
> CallPricePL(C,30,4/12,40,.18,.05);
```

$$10.49760615 - .9814949737\,C$$

Hence, C is the solution to the equation

$$10.49760615 - 0.9814949737C = 0.$$

The MAPLE command below solves for C and assigns its value to **PrPL**.

```
> PrPL:=solve(CallPricePL(C,30,4/12,40,.18,.05)=0);
```

$$PrPL := 10.69552716$$

Once C is determined at the initiation time, it is treated as a parameter of the option.

Consider the option we just valued, and assume that a month passed since the initiation time. Therefore, the time to expiration is now $\frac{3}{12}$ years and the premium is **PrPL** as determined above. Thus, to solve for the value of the option now, we execute the command below.

```
> CallPricePL(PrPL,30,3/12,40,.18,.05);
```

$$-.18437042$$

We have mentioned before that the value of a paylater option may be negative. This is due to the fact that the holder of the option is obligated to exercise it even if the net cash flow is negative. We visualize this phenomenon in Figure 16.11. It displays the value of the option as a function of time to maturity and the price of the underlying asset.[8]

```
> plot3d(CallPricePL(PrPL,30,t/12,S,.18,.05),S=0..80,\
> t=0..11,axes=FRAME, orientation=[132,55],shading=ZHUE,\
> title="Figure 16.11:  The Value of a Paylater Option as a\
> Function of the Stock Price and\
> the Time to Maturity" ,titlefont=[TIMES,BOLD,8]);
```

A paylater put option is defined and valued analogously, but we leave the details of this valuation as an exercise for the reader.

[8] We do not demonstrate an American paylater option as our **AmerOpt** procedure is not designed to force exercising, which is the case in these types of options. The interested reader can however utilize the code in Section 13.4.4 and modify it slightly to accommodate an American paylater option. See also the exercises at the end of this Chapter.

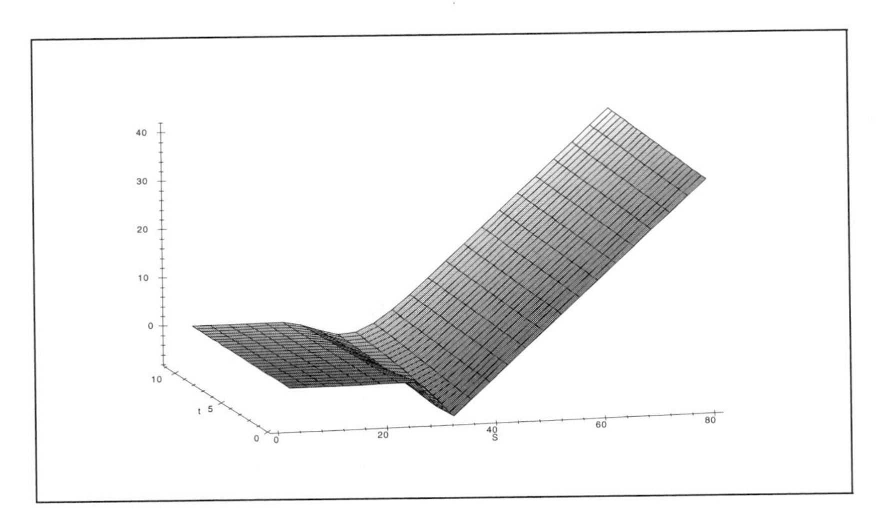

Figure 16.11: The Value of a Paylater Option as a Function of the Stock Price and the Time to Maturity

16.4 Interest Rate Derivatives

A complete investigation of interest rate derivatives and the stochastic process governing the interest rate process is beyond the scope of this book. Modeling interest rate processes is a more intricate issue than modeling the price process of a stock. This is the case for a number of reasons.

The modeling of interest rate processes actually pertains to the movements of the term structure of interest rates. At each future date t, we have a realization of "many" random variables. These random variables are the spot rates for different maturities, i.e., $r(t, T)$ for different $T \geq t$. In contrast, modeling the stock price involves only one random variable, the price of the stock at some future time, t.

Furthermore, interest rate processes must ensure that the evolution of the curve $r(t, T)$ is arbitrage-free. Hence, special attention must be given to the comovements of the $r(t, T)$ for all $T \geq t$. In addition, the parameters of the process must be chosen so that the process is consistent with the observed (current) bond prices (or equivalently with the observed term structure of interest rates) and that the volatility of a bond price is zero at the bond's maturity. After all, at maturity the value of a bond is a deterministic constant: its face value. In fact, these issues are related to the assumption that a constant volatility, used in modeling the stock price, is not reasonable for modeling interest rate processes. Hence, the issue of the term structure of volatility will also have to be dealt with, i.e., the structure of σ_{T-t}, the volatility of $r(t,T)$. Notwithstanding the comments above, in the industry there is a widely used model that does not adhere to the points just raised. This model which is presented next, is due to Black, and is based on the derivation in [6].

16.4.1 Black's Model

Our presentation of Black's model, for pricing European bond options, will be in a somewhat unconventional (although equivalent) fashion. We will thereby emphasize it as an extension of the Black–Scholes formula and demonstrate at least one of its deficiencies when applied to interest-rate-contingent claims. Commonly, Black's model is presented as if the payoff of the option at maturity is a function of the futures price of the asset, rather then the spot price $S(T)$.

In dealing with European-type options, these two presentations are equivalent, as the futures price $F(t,T)$, for a contract written at time t to be delivered at some future time T, satisfies $F(T,T) = S(T)$. The payoff at maturity of a call option can thus be written as $\max(F(T,T) - K, 0)$ or equivalently as $\max(S(T) - K, 0)$. In the first expression the underlying asset is the future price, while in the second is it is the spot price. As we have already seen in equation (16.8) and equation (16.9), both presentations lead to the same value of the option. Rewriting these equations in terms of the current notation, we obtain that the value of an option with a payoff $\max(F(T,T) - K, 0)$, where the underlying asset is $F(T,T)$, is given by

$$e^{-r(T-t)}\left(F\left(t,T\right)N(d_1) - KN(d_2)\right), \tag{16.29}$$

where K is the exercise price,

$$d_1 = \frac{\ln\left(\frac{F(t,T)}{K}\right) + \frac{\sigma^2}{2}(T-t)}{\sigma\sqrt{T-t}}, \tag{16.30}$$

and

$$d_2 = d_1 - \sigma\sqrt{T-t}. \tag{16.31}$$

Hence, when $F(t,T) = S(t)e^{r(T-t)}$ the formulas (16.8) and (16.29) coincide. This is indeed the case when the Black model is used to price European bond options, but not for pricing interest rate options.

The core assumption of Black's model is that the price of the underlying asset, the futures price $F(t,T)$, at the maturity of the option, follows the lognormal distribution. This facilitates, albeit with some minor modification, its application in a straightforward manner to various interest-rate-contingent claims. Perhaps even more troubling than the deficiencies raised above is the inconsistency inherent in such applications. The Black–Scholes formula assumes that the term structure of interest rate is flat and constant, yet we apply it to a situation where interest rates are assumed to be stochastic. Nevertheless, due to the popularity of this model we will use it to price European options on bonds, interest rate options, and options on swaps (also referred to as swaptions).

Although the material covered until now, particularly Chapter 15, presents a reasonable background to delve into the study of the interest rate process, only one model (Black, Derman, and Toy's [7]) will be covered here. This model, presented in the next section, lends itself to numerical solutions based on recombining binomial trees.

European Bond Options

Consider a European call option on a zero-coupon bond. Assume that the current time is t, the option matures at time T, and the bond matures at some time after T. Thus the payoff of the call is

$$\max\left(B_T - K, 0\right),$$

where K is the exercise price and B_T is the price of the zero-coupon bond at time T. If we are willing to assume further that B_T follows the lognormal distribution, we can apply the Black–Scholes formula, equation (6.3), to price the option as

$$B_t N(d_1) - e^{-r(T-t)} K N(d_2), \tag{16.32}$$

where N is the standard cumulative normal distribution,

$$d_1 = \frac{\ln\left(\frac{B_t}{K}\right) + \left(r + \frac{\sigma^2}{2}\right)(T-t)}{\sigma\sqrt{T-t}}, \tag{16.33}$$

and

$$d_2 = d_1 - \sigma\sqrt{T-t}. \tag{16.34}$$

In the above equations σ is the standard deviation of $\ln(B_T)$ and r is the continuously compounded interest rate spanning the time interval $[t,T]$. The price of a put option with a payoff given by $\max(K - B_T, 0)$ is obtained based on equation (6.7), and is given by

$$e^{-r(T-t)}KN(-d_2) - B_t N(-d_1), \tag{16.35}$$

where d_1 and d_2 are defined in equation (16.33) and equation (16.34), respectively.

If the option is written on a coupon-paying bond, the situation is similar to an option on a dividend-paying stock. In Section 6.4 we provided an explanation for the effects of dividends on the option pricing formula. Namely, substituting in equation (6.3) the current price of the stock discounted by the dividend yield for the current price.

The explanation offered here is equivalent to the former, but is presented from a slightly different point of view. The application of the Black–Scholes formula requires the substitution of the current price of the asset, the future value of which is assumed to follow the lognormal distribution. However, if the bond is paying coupons during the life of the option, the current price of the bond is not the present value of the price of the bond at T (which is assumed to follow the lognormal distribution). This is because the value of the bond today is the present value of all the future cash flows from the bond. This includes the coupons that will be received during the life of the option. The value of the bond at time T, however, is the present value, as of time T, of the cash flows obtained from the bond only after time T.

It follows that the present value of the price of the bond at time T is the current price of the bond less the present value of the coupons that will be received during the life of the option. However, before proceeding to an example of pricing a call option we have to be alerted to the issue raised in footnote 2 at the beginning of Section 3.2. When we use the term "the current price of the bond" we mean the aggregate amount, referred to also as the cash price, one pays to purchase a bond. It includes the quoted price plus the accrued interest. Similarly, the exercise price refers to the cash price to be paid upon exercising the option. If this is not the case, the valuation formula should be adjusted in a straightforward manner.

Consider the bond market that was specified in Table 3.1 and is repeated here for convenience as Table 16.1.

It includes three bonds paying coupons semiannually, on the same day. For simplicity, let us assume that the current time is immediately after a coupon payment, so that the time until the next coupon payment is six months. We verify below that there are no arbitrage opportunities presented by the current term structure of interest rates and solve for the discount

Price/Time	1	2	3	Security
$94.5	$105	$0	$0	$B1$
$97.0	$10	$110	$0	$B2$
$89.2	$8	$8	$108	$B3$

Table 16.1: A Simple Bond Market Specification

factors $d1$, $d2$, and $d3$ for times 1, 2, and 3, respectively. This is done by executing **NarbitB** as below, where we also solve for the continuous approximation of the term structure and assign it to the function *dd*.

```
> NarbitB([[105,0,0],[10,110,0],[8,8,108]],[94.5,97,89],\
> 4,dd,0);
```

The no − arbitrage condition is satisfied.

The discount factor for time, 1, *is given by*, .9000000000

The interest rate spanning the time interval, [0, 1], *is given by*, .111

The discount factor for time, 2, *is given by*, .8000000000

The interest rate spanning the time interval, [0, 2], *is given by*, .250

The discount factor for time, 3, *is given by*, .6981481481

The interest rate spanning the time interval, [0, 3], *is given by*, .432

The function Vdis([c1, c2, ..]), values the cashflow [c1, c2, ..]

The continuous discount factor is given by the function, '*dd*', (.)

We have an exact fit of the continuous approximation to the three discount factors estimated from market prices. This can be verified by noticing that the value of **SumAbsDiv** is zero (some readers may want to refer back to Chapters 3 and 9).

```
> SumAbsDiv;
```

0

Consider a European call option that matures in eight months, with an exercise price of $90, written on the 10% coupon bond (that matures in a year). The face value of the bond is $100, it pays a coupon in six months, and its cash value is $97. The present value of the coupon payment can be calculated as follow:

```
> 10*dd(1);
```

9.000000000

Thus, the price of the bond minus the present value of the coupon payment is $97-9=88$. Assume further that volatility of the bond price in eight months is 9% per annum. The eight-month, continuously compounded, risk-free rate per annum is solved for in the next MAPLE command.

```
> solve(dd(8/6)=exp(-r*8/12),r);
                              .2026862736
```

Hence the value of this option will be given by equation (16.32) where instead of the price of the bond we substitute $97-9=88$. Thus, the price of the call option is given by executing the **Bs** procedure as below:

```
> Bs(90,8/12,88,0.09,0.2026862736);
                              9.54133989
```

Interest Rate Options

Interest rate options are contingent claims where the underlying asset is the interest rate. They are typically related to floating rate loans and used to protect the borrower (call option) or the lender (put option) from unfavorable movement of the interest rates. Essentially, these options are (or can be) used in the same manner as their equity counterparts. They bound the potential losses due to the uncertainty associated with the movements of the interest rate. Let us explain by way of an example.

Consider a one-year (variable) floating rate loan commencing now at time $t=0$, and requiring payments m times a year. We assume, as is the common practice in the marketplace, that the rates are quoted based on compounding m times a year as well. For simplicity we set the face value of the loan to \$1 and m to $\frac{1}{2}$, i.e., compounding is done semiannually. At $t=0$ the first payment of the loan at $t=0.5$ years is known. This is because at $t=0$ the interest rate spanning the time period $[0,0.5]$ is known. However, the (loan) interest rate $r_{0.5}(0.5,1)$ that at time $t=0.5$ will span the time period $[0.5,1]$ is not known at time $t=0$. The reader may like to recall our discussion of a floating rate loan in Section 9.4. Further clarifications for the timing at which the rates are realized are offered in Sections 14.2 and 14.3, and in particular, in footnote 9 on page 575.

At time $t=1$ the borrower will have to pay the lender the last interest payment of $\frac{1}{2}r_{0.5}(0.5,1)$. This rate, however, is not known at time zero; it will be known only at time $t=0.5$. The borrower, fearing an increase in interest rates, may like to be protected and buy a call option written on $\frac{1}{2}r_{0.5}(0.5,1)$ with an exercise price of $\frac{1}{2}K$. It will ensure the borrower that the interest payment due at time $t = 1$ will not exceed $\$\frac{K}{2}$. The payoff of such a call option is

$$\max\left(\frac{r_{0.5}(0.5,1)}{2} - \frac{K}{2}, 0\right) = \frac{1}{2}\max(r_{0.5}(0.5,1) - K, 0). \qquad (16.36)$$

Hence, if $r_{0.5}(0.5,1) > K$ the payoff from the call option will be $\frac{r_{0.5}(0.5,1)-K}{2}$, and since the borrower is committed to pay $\frac{r_{0.5}(0.5,1)}{2}$, the net payment will be $\frac{r_{0.5}(0.5,1)}{2} - \frac{r_{0.5}(0.5,1)-K}{2} = \frac{K}{2}$. If, on the other hand, $r_{0.5}(0.5,1) \leq K$, the payoff from the option is nil and the borrower will pay an amount less than $\frac{K}{2}$.

The rate $\frac{r_{0.5}(0.5,1)}{2}$ is being realized at time $t = 0.5$, at which point the holder of the call option should decide whether to exercise it or not. However, unlike a regular call option, the payment will not be commencing until time $t = 1$. We shall attend to the effect of this timing issue on the valuation formula very shortly. We see, therefore, that buying such an option caps the rate the borrower will have to pay at time $t = 0.5$ at K. Indeed, such interest rate options are also refereed to as *caplets*. A loan in which each interest rate payment is protected in such a way is referred to as a *cap*. It is a loan together with a portfolio of options on each interest payment, each of which is called a caplet. A cap is thus a floating rate loan, with a guarantee that the interest rate payments will not exceed $\frac{K}{2}$.

These options that are attached to the loan (the caplets) are over-the-counter instruments and they are not traded in the market. They are usually offered to the clients by the financial institution originating the loan. In a symmetric manner the lender may fear a decrease in interest rates and may like to ensure that the payment will not be less than, say, $\frac{K}{2}$. This will be accomplished by buying a put option on $r_{0.5}(0.5,1)$. The payoff of such a put option will be

$$\frac{1}{2}\max(K - r_{0.5}(0.5,1), 0), \qquad (16.37)$$

and again the timing of the realization of $r_{0.5}(0.5,1)$ and of exercising will be as for the call option. A floating rate loan in which the rates are protected from being below a certain level, by a portfolio of put options is called a *floor*. Again, these options are "tailor-made", usually by one institution for another, and are not traded in the marketplace.

Let us now see how Black's model is applied to value the caplet in equation (16.36). If $r_{0.5}(0.5,1)$ is assumed to satisfy all the assumptions required by the Black–Scholes formula, only the timing issue should be resolved.

What is the effect induced by the resultant payment from exercising at time $t = 0.5$ being obtained at time $t = 1$?

One may intuitively reach the conclusion that the price of such an option is simply the price of the same option with the payment at $t = 0.5$ discounted, to time $t = 0.5$, by the forward rate spanning the tine interval $[0.5, 1]$. This is indeed the case, and it could be proven applying arbitrage arguments of the type we have seen before. In fact, the question can be posed in a more general way, which is not necessarily related to option pricing. We assume that the term structure, at time $t = 0$, is known, is reported in a continuous compounding manner (where the unit of time is a year), and is denoted by R. The rate of the loan[9] is reported based on semiannual compounding and is denoted by r, as above.

Consider a risky amount to be obtained at time $t = 0.5$, the value of which at time $t = 0$ is know to be X. What should be the value of the same risky amount obtained at time $t = 1$? One way of thinking about this is related to the concept of the certainty equivalent discussed in Section 1.4.1, equation (1.10). The value of the risky amount to be obtained at time $t = 0.5$ is the same as the sure amount $Xe^{\frac{1}{2}R_0(0,0.5)}$ obtained at $t = 0.5$. Thus, the question can be rephrased as, what would be the value, at $t = 0$, of the sure amount $Xe^{\frac{1}{2}R_0(0,0.5)}$ if it would have been received at time $t = 1$? The answer to this question is simply $Xe^{\frac{1}{2}R_0(0,0.5)}e^{-R_0(0,1)}$, which equals $Xe^{-\frac{1}{2}R_0(0.5,1)}$, where $R_0(0.5, 1)$ is the continuously compounded forward rate spanning the time interval $[0.5, 1]$.

A caplet with a payoff as in equation (16.36) could therefore be valued based on equation (16.29). In equation (16.29), $F(t, T)$ is playing the role of the futures price of the asset at time t. In our context, it is the rate one can secure now, for a loan to be commencing at time T and agreed upon now. That is, it is the forward rate $r_{0.}(0.5, 1)$ that would apply as of the current time for a loan to be starting at T (where the rates are to be reported in the same manner as the rate on the loan). Hence, we need to discount the value obtained by equation (16.29), $\frac{1}{2}e^{-R}\left(r_0(0.5, 1)N(d_1) - KN(d_2)\right)$, to arrive at the value of the caplet as in equation (16.38).

$$e^{-\frac{1}{2}R_0(0.5,1)}\frac{1}{2}e^{-R}\left(r_0(0.5, 1)N(d_1) - KN(d_2)\right) \tag{16.38}$$

In general, therefore, where the compounding is done m times a year, the principal is N, and the caplet is for an interest payment commencing at time

[9] Due to the risk of default the rate of the loan may be different than the rate of the term structure. This risk is ignored in the analysis henceforth.

$T + \frac{1}{m}$, equation (16.38) is replaced with equation (16.39),

$$\frac{1}{m} e^{-\frac{1}{m} R_0 (T, T+\frac{1}{m})} e^{-RT} \left(r_0 \left(T, T + \frac{1}{m} \right) N(d_1) - K N(d_2) \right), \qquad (16.39)$$

where d_1 and d_2 are as defined in equation (16.30) and equation (16.31), respectively.

Floors and caps can also be represented by a portfolio of options on discount bonds rather than on the interest rates. Sometimes a floating rate loan is combined with lower and upper bounds on the rates. Such an instrument is called a *collar*. We would like to emphasize (again) that assuming that the option is written on the futures price of the underlying asset or on the spot price does not make a difference as long as $F(t, T) = S(t)e^{r(T-t)}$. However, here $F(t, T)$ is replaced with a forward rate, and $S(t)$ with a spot rate, and thus the above relation is not satisfied. Some of the exercises at the end of this Chapter pertain to the valuation of these instruments and to the effect of the above assumption, which is done in the same vein as shown in the text.

Swaptions

Another instrument to which Black's model is applied for valuation is an option on a swap, also referred to as a *swaption*. A call option on a swap gives the holder the right to enter into a fixed-for-float swap (receives fixed and pays float), with a prespecified fixed rate. These types of call options specify the details of the swap, i.e., the notional principal, the payment period, the duration of the agreement, and the starting time.

Such an option might be purchased by a corporation that anticipates getting into a swap agreement (paying float and receiving fixed) in the future, and would like to ensure that the fixed rate it will receive will not be below a certain level. Assume that the notional principal of the agreement is N and payments are to be swapped m times a year. The swap rates, as before, are reported based on compounding m times a year, while the term structure is reported based on continuous compounding.

Let us apply Black's model to value a call option on a swap. Assume that the current time is $t = 0$ and that the maturity of the options is T. The exercise price K is the fixed rate that would apply to the swap if the option is exercised. For simplicity let us assume that the swap requires exchanging cash flows every year so that the first exchange will occur at time $T + 1$. At time T the holder of the option may decide to exercise it or not, depending on the conditions in the market at that time. If at time T the fixed rate of a swap, with the above specification, is larger than K, the holder will exercise the option. The resultant payoff will be obtained only at time $T+1$. We will therefore have to take care of this lag, as we did in the case of caps. At time T the value of the fixed rate r for such a swap will be realized and the option will be exercised if $r > K$ or if $r - K > 0$. The payoff from this call option is thus

$$N \max(r - K, 0) \tag{16.40}$$

and is obtained at time $T+1$.

In order to apply Black's model as in equation (16.29) we have to stipulate what replaces $F(0,T)$ in this context, and assume that it follows the lognormal distribution. $F(0,T)$ is the futures price of the asset and, since for this purpose the interest rate is assumed to be deterministic and fixed, it is the forward price. Hence, $F(0,T)$ is the price at which a commitment to buy the asset a time T can be agreed upon at time 0. In our context therefore, it is the fixed rate of a fixed-for-float swap commencing at time T to which the parties have committed at the present time. Applying the same arguments as in Section 11.1, it is easy to determine this rate, which we denote here by r_s. We therefore leave it as an exercise for the reader. In equation (16.40) r is a random variable which the approximation by Black's model replaces with the forward rate of a swap, as explained above. This forward rate is assumed to follow the lognormal distribution. Hence, the value of such a swaption based on equation (16.29) is given by equation (16.41).

$$Ne^{-R_0(T,T+1)}e^{-RT}\left(r_s N(d_1) - KN(d_2)\right) \tag{16.41}$$

In general the swap duration can be for n years, specifying an exchange of cash flows m times a year. The value of such a swap will thus be the sum of the discounted values of the payoffs

$$\frac{N}{m} \max(r - K, 0) \tag{16.42}$$

occurring at time $t = T + \frac{1}{m}, T + \frac{2}{m}, ..., T + \frac{mn}{m}$. It is therefore given by equation (16.43),

$$\sum_{i=1}^{mn} \frac{N}{m} e^{-R_0\left(T,T+\frac{i}{m}\right)} e^{-RT}\left(r_s N(d_1) - KN(d_2)\right), \tag{16.43}$$

where d_1 and d_2 are as defined in equation (16.30) and equation (16.31), respectively.

16.4.2 The Black, Derman, and Toy Model

As we have explained before, a complete treatment of interest rate derivatives and stochastic process that govern the term structure is beyond the scope of this book. Readers who are interested in this topic should consult other books such as [37]. We hope that the level used in [37] will not present a difficult challenge for readers who opted not to skip any of the chapters in this book.

Given the drawbacks of Black's model, we would like to present an alternative model. We choose to present a model suggested by Black, Derman, and Toy (BDT) in [7], that belongs to the so-called group of no-arbitrage models[10] of the term structure. This model, referred to henceforth as the BDT model, is a natural choice, given the no-arbitrage spirit of the book and the use of symbolic computation that avoids the trial and error approach commonly used to calibrate this model. The BDT model assumes that the source of uncertainty in the term structure[11] is the short rate, which is assumed to follow the lognormal distribution. Thus, in a sense, the short-term interest rate is treated as a risky "asset" similar to the way the price of a stock is modeled.

To accommodate the need for bond prices to be consistent with the evolution of the term structure, the model allows for more degrees of freedom, with respect to the model of the stock price, in its parameters. These parameters, soon to be specified, are determined in an implied manner, much like the implied volatility approach in Chapter 6. The parameters are chosen so that the prices of outstanding bonds, based on the evolution of the term structure, coincide with the observed prices.

These parameters are being "fitted" to the observed prices by approximating the evolution of the term structure using a recombining binomial tree. The tree is being solved for numerically, step-by-step, ensuring the satisfaction of the no-arbitrage condition in each of the one-period binomial

[10] There are different approaches to the modeling of interest rate processes. A complete coverage of this subject is beyond the scope of this book. The BDT model falls into the category of the no-arbitrage approach and is very popular among practitioners. We therefore choose to explore the BDT model as it suits the no-arbitrage spirit of this book. As well, it cements the ties between the binomial models as applied to equality derivatives and the estimation of the term structure investigated in Chapter 9.

[11] Models of the term structure of interest rates are also categorized by the number of sources of uncertainty incorporated into the model. In the BDT model there is only one source of uncertainty, the short rate. It belongs to a group called "one-factor models". In a continuous-time framework, the short rate is the instantaneous rate. The BDT model, as we shall soon see, is a discrete model. "Short" in this context thus refers to the length of a period.

trees,[12] the combination of which composes the full tree. Once the evolution of the term structure is solved for, the tree can be utilized to value interest rate derivatives in the same manner as we used it in Chapters 13 and 14 to value equity derivatives.

In inputting the parameters of the tree we make use of an approximation to the lognormal distribution, essentially the one used in Section 14.1.1, equations (14.9) and (14.10). The free parameters are constrained so that the tree recombines. The volatility is assumed to be a known deterministic function of time, not a constant as in the price process of a stock. This still leaves enough degrees of freedom to set the parameters so that the evolution of the short rate is consistent with the observed prices of the bonds.

The BDT model stipulates that the instantaneous short rate at time t is given by

$$r(t) = M(t)e^{\sigma(t)Z}, \tag{16.44}$$

where $M(t)$ is a deterministic function of time, Z is a normal random variable with an expected value of zero and a variance of t, and $\sigma(t)$ is a deterministic function of time. Thus, $\sigma(t)Z$ is a normal random variable with an expected value of zero and a variance of $\sigma^2(t)t$. Therefore, $r(t)$ follows the lognormal distribution, while $M(t)$ is a deterministic function. It is the degrees of freedom introduced by $M(t)$ that allow the fitting of the evolution of the term structure to the observed prices of the bonds. We shall solve for $M(t)$ by designing a binomial tree that approximates the distribution of $\ln r(t)$, which is a normal distribution. To begin our approximation (although, as we shall see, the parameters need not be specified in a direct way) we first call upon Ito's lemma in order to uncover the (stochastic) instantaneous increments of $\ln r(t)$. At this point, we would like to emphasize that it is the evolution under the risk-neutral probability that is being dealt with here. This is the reason a security in this model is valued simply as the discounted expected value of its cash flow.

To calculate the stochastic instantaneous increments of $r(t)$ in equation (16.44) we operate the procedure **ItosLemma**. Recall that Z must be renamed as x for the procedure to work, which is assumed to have a standard deviation of $\sqrt{t}$ and an expected value of zero. The first two parameters in **ItosLemma** allow us to change the expected value of x and its standard deviation (see the Help on **ItosLemma**). However, since we would like to have a standard deviation of x that is a function of time we must stipulate it by multiplying x by $\sigma(t)$. This ensures that the standard deviation of

[12]This in turn guarantees, based on Theorem 11 in Section 13.2.1, that the no-arbitrage condition is satisfied for the complete period modeled.

$x\sigma(t)$ is $\sigma(t)\sqrt{t}$ while facilitating the treatment of $\sigma(t)$ as a symbolic function. Executing the procedure below with the third parameter[13] being the function applied to x results in the stochastic differential equation specifying the process followed by $\ln r(t)$.

```
> ItosLemma(0,1,ln(M(t)*exp(sigma(t)*x)),muBDT,sigmaBDT);
```

$$\mathrm{d}(\ln(\mathrm{M}(t)\, e^{(\sigma(t)\, x)})) = \frac{((\frac{\partial}{\partial t}\,\mathrm{M}(t)) + \mathrm{M}(t)\,(\frac{\partial}{\partial t}\,\sigma(t))\,x)\,dt}{\mathrm{M}(t)} + \sigma(t)\,dZ$$

Consider the drift parameter of $\ln r(t)$ which is now assigned to **muBDT**.

```
> muBDT;
```

$$\frac{(\frac{\partial}{\partial t}\,\mathrm{M}(t)) + \mathrm{M}(t)\,(\frac{\partial}{\partial t}\,\sigma(t))\,x}{\mathrm{M}(t)}$$

Our first step is to solve for x in terms of $r(t)$ and $M(t)$, based on equation (16.44). The next step is to substitute the result back in **muBDT** and expand the expression. This is done below after we rewrite in terms of r, instead of x, the stochastic differential equation satisfied by $\ln r(t)$.

```
> muBDT:=expand(subs(x=solve(ln(r)=ln(M(t)*\
> exp(sigma(t)*x)),x),muBDT));
```

$$muBDT := \frac{\frac{\partial}{\partial t}\,\mathrm{M}(t)}{\mathrm{M}(t)} + \frac{(\frac{\partial}{\partial t}\,\sigma(t))\ln(r)}{\sigma(t)} - \frac{(\frac{\partial}{\partial t}\,\sigma(t))\ln(\mathrm{M}(t))}{\sigma(t)}$$

Recognizing that $\frac{\frac{\partial}{\partial t}M(t)}{M(t)} = \frac{\partial}{\partial t}\ln M(t)$, **muBDT** can be written as

$$\frac{\partial}{\partial t}\ln M(t) + \frac{\partial}{\partial t}\ln\sigma(t)\,(\ln r - \ln M(t))$$

and the stochastic differential equation stipulating the stochastic increments of $\ln(r)$ is thus

$$d\ln r = \left(\frac{\partial}{\partial t}\ln M(t) - \frac{\partial}{\partial t}\ln\sigma(t)\,(\ln M(t) - \ln r)\right)dt + \sigma(t)dZ. \qquad (16.45)$$

A few implications are deduced directly from equation (16.45). If σ is a constant, then $\frac{\partial}{\partial t}\ln\sigma(t) = 0$, and the process of $d\ln r$ is a Brownian motion but with a drift that is a function of $M(t)$ only. If, on the other hand, $\sigma(t)$ is a decreasing function of time, then $-\frac{\partial}{\partial t}\ln\sigma(t)$ is positive, which induces a reversion of $\ln(r)$ to $\ln M(t)$. This property of mean-reversion was briefly discussed in Chapter 15 for a simpler case, equation (15.27), and was demonstrated by a simulation in Figure 15.6 in Section 15.3.

[13]The last two parameters are names only, specified by the user, to which the drift and diffusion parameters of the new process are assigned.

The BDT model is actually a discrete model which approximates a continuous process described in equation (16.45). It does so by a recombining binomial tree in the spirit of the approximation introduced in Section 14.1.1, equation (14.8). It starts with the specification of the length of a period in each of the one-period binomial trees composing the total period. In what follows we use the notation introduced in Chapters 13 and 14, refreshing some of it as we go along.

Denoting the length of time period in each of the one-period binomial models by Δt, and applying equation (14.8) to the process in equation (16.45), we can approximate $\ln r$ with a binomial tree. Hence, $\ln r(t+\Delta t) - \ln r(t)$ is a random variable that takes on two values, u in state **Up** and v in state **Down,** each with a probability of 0.5. Here $u = \mu(t)\Delta t + \sigma(t)\sqrt{\Delta t}$ and $v = \mu(t)\Delta t - \sigma(t)\sqrt{\Delta t}$, where $\mu(t) = \frac{\partial}{\partial t}\ln M(t) - \frac{\partial}{\partial t}\ln \sigma(t)\,(\ln M(t) - \ln r)$. Since $\sigma(t)$ is positive, the **Up** state is a state where $\ln r(t+\Delta t) > \ln r(t)$ and the **Down** state is a state where $\ln r(t+\Delta t) < \ln r(t)$. It is also easy to show[14] that the standard deviation of the random variable $\ln r(t+\Delta t)$ is $\sigma(t)\sqrt{\Delta t}$.

The function $\sigma(t)$ specifying the volatility of the short interest rate at time t, referred to as the term structure of volatility, is assumed to be known. The value of $\mu(t)$ is constrained by the requirements that the tree will be recombining, that the no-arbitrage condition will be satisfied, and that the observed prices will be consistent with the evolution of the term structure. Let us see what these requirements mean for $\mu(t)$.

As in Chapter 13, the investigation below uses the notation of i (replacing t) for the discrete time and j for the number of up movements since time zero. Hence, in this notation $r(i,j)$ stands for the short interest rate prevailing in the market at time i if there were j up movements since time zero. Assume that at time $i-1$ the realization of the state of nature was j (recall again the sequence of events and the timing of the realization of r discussed in Section 16.4.1). At time i the short interest rate will be $r(i,j+1)$ if an up movement is realized or $r(i,j)$ if a down movement is realized, i.e.,

$$\ln r(i,j+1) = \ln r(i-1,j) + \mu(i)\Delta t + \sigma(i)\sqrt{\Delta t}$$

or

$$\ln r(i,j) = \ln r(i-1,j) + \mu(i)\Delta t - \sigma(i)\sqrt{\Delta t}, \tag{16.46}$$

[14] Consider a random variable that takes only two values, u and v $(u > v)$, each with a probability of $\frac{1}{2}$. The expected value of such a variable is $\frac{u+v}{2}$. The variance of this variable is $\frac{1}{2}\left(u - \frac{u+v}{2}\right)^2 + \frac{1}{2}\left(v - \frac{u+v}{2}\right)^2$, which equals $\frac{1}{2}\left(\frac{u-v}{2}\right)^2 + \frac{1}{2}\left(\frac{v-u}{2}\right)^2 = \left(\frac{u-v}{2}\right)^2$, and hence its standard deviation is $\frac{u-v}{2}$.

which implies, see also footnote 14, that

$$\frac{\ln r(i,j+1) - \ln r(i,j)}{2} = \sigma(i)\sqrt{\Delta t}. \tag{16.47}$$

The evolution of the interest rate can therefore be displayed as in Figure 13.2. At time i, there are $i+1$ possible states of nature corresponding to the number of up movements, $0, 1, 2, ..., i$, that occurred from time 0 to time i. The structure adopted by the binomial approximation imposes a relation on the different realization of the short rate at time i and, thereby, on $\mu(t)$. Specifically, the rate $r(i,0)$ is related to $r(i,1)$ via equation (16.47), and hence

$$\frac{\ln r(i,1) - \ln r(i,0)}{2} = \sigma(i)\sqrt{\Delta t},$$

which implies that

$$r(i,1) = r(i,0)e^{2\sigma(t)\sqrt{\Delta t}}. \tag{16.48}$$

Let us look at an example where we model a total period of three years and each period corresponds to a year, i.e., $\Delta t = 1$. Assume that at time $i = 2$ the realization of the state of nature was $j = 0$. In this case equation (16.48) relates the short interest rates at nodes $(3,0)$ and $(3,1)$, which are designated with squares in Figure 16.12.

```
> BinTree(3,[[3,0],[3,1]]);
```

Applying the same argument as above, we can relate the short rates $r(i,2)$ and $r(i,1)$ by equation (16.49):

$$r(i,2) = r(i,1)e^{2\sigma(t)}. \tag{16.49}$$

If at time $i = 2$ the realization was of $j = 1$, we can relate the interest rate in the nodes $(3,1)$ and $(3,2)$, which are emphasized with a square in Figure 16.13.

```
> BinTree(3,[[3,1],[3,2]]);
```

Since the binomial tree is required to be recombining, an up movement from $(2,0)$ and a down movement from $(2,1)$ end up at the same state, namely, at $(3,1)$. Equation (16.48) and equation (16.49) can, therefore, be applied at each node of the tree and $r(i,1)$ in equation (16.49) can be replaced by its value in equation (16.48). We therefore arrive at equation (16.50):

$$r(i,2) = r(i,0)e^{(2+2)\sigma(t)}. \tag{16.50}$$

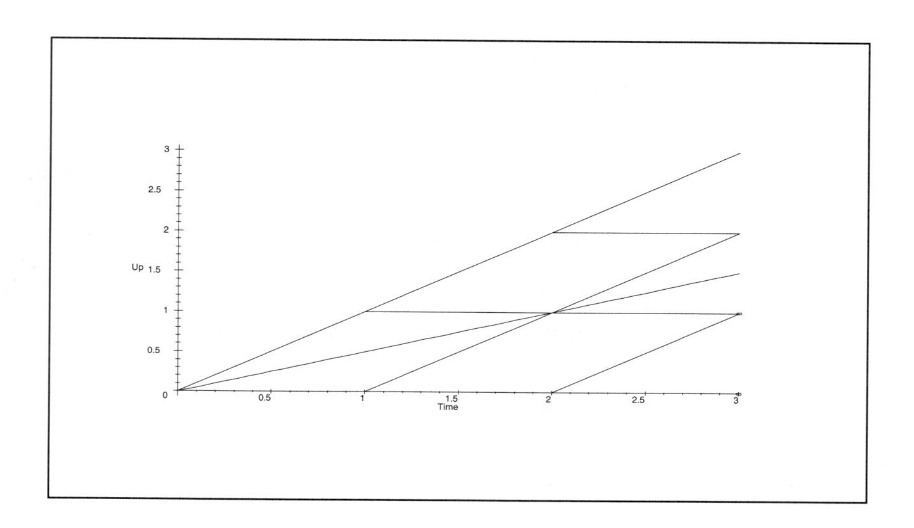

Figure 16.12: Solving for the Short Rates in Nodes $(3,0)$ and $(3,1)$

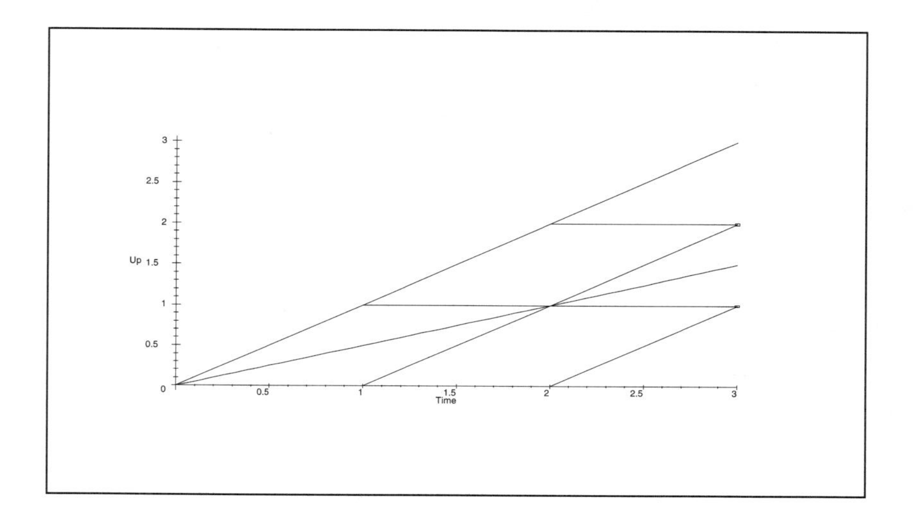

Figure 16.13: Solving for the Short Rates in Nodes $(3,1)$ and $(3,2)$

Following the same logic, the general relation

$$r(i,j) = r(i,0)e^{2j\sigma(i)\sqrt{\Delta t}} \tag{16.51}$$

for every i and $j = 0, ..., i$, is obtained. Consequently, when the function $\sigma(t)$ is specified, the realization of the short interest rate at time i is a function only of $r(i,0)$. Given the current term structure of interest rates, we can solve for the numerical value of $r(i,0)$, while ensuring the satisfaction of the no-arbitrage condition and consistency with the observed bonds prices.

Let us demonstrate how this is done by enhancing the structure of our example. To this end we enrich the example with the bond market specified in Table 16.1. We already verified that no arbitrage opportunities are presented by the current term structure of interest rate. This time, however, let us assume that these three bonds pay coupons **annually**, on the same day. Assume that the current time is immediately after a coupon payment, so that the time until the next coupon payment is a year.

To simplify matters further, we replace the original market with an equivalent one that is composed only of zero-coupon bonds with a face value of \$1. The equivalent market is specified below, where the price of a zero-coupon bond maturing at time t is given by $dd(t)$. The reader may want to refer to Section 2.1 for an example of equivalent markets.

```
> NarbitB([[1,0,0],[0,1,0],[0,0,1]],[dd(1),dd(2),dd(3)],\
> 4,d,0);
```

The no − arbitrage condition is satisfied.

The discount factor for time, 1, *is given by*, .9000000000

The interest rate spanning the time interval, [0, 1], *is given by*, .111

The discount factor for time, 2, *is given by*, .8000000000

The interest rate spanning the time interval, [0, 2], *is given by*, .250

The discount factor for time, 3, *is given by*, .6981481481

The interest rate spanning the time interval, [0, 3], *is given by*, .432

The function Vdis([c1, c2, ..]), values the cashflow [c1, c2, ..]

The continuous discount factor is given by the function, 'd', (.)

That indeed these markets are equivalent can be verified by noting that the discount factors in these two markets are the same. Since $\Delta t = 1$, the short rate at time zero, using continuous compounding, is simply the solution to

```
> solve(d(1)=exp(-r[0]),r[0]);
```

.1053605157

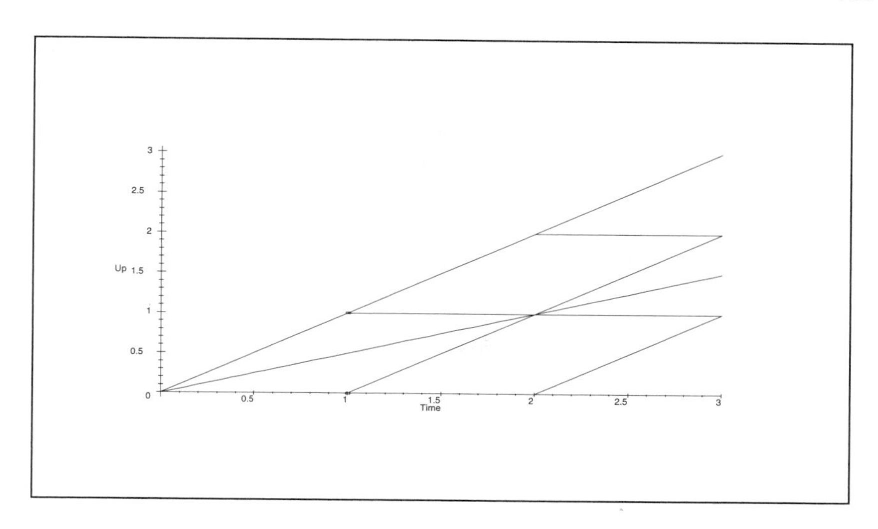

Figure 16.14: The Realization of the Short Rates $r(1,0)$ and $r(1,1)$ at Time $i = 1$

Note that the procedure used discrete compounding in calculating the rates reported in the output.

At time $i = 1$ there are two possible realizations of the short rate (refer to Figure 16.14), $r(1,0)$ and $r(1,1)$, that are related to each other by equation (16.48), i.e., $r(1,1) = r(1,0)e^{2\sigma(1)}$.

```
> BinTree(3,[[1,0],[1,1]]);
```

The bond maturing at time $i = 2$ pays a dollar at that time regardless of the state of nature. Hence, if at time $i = 1$ state one is realized, i.e., the process is at node $(1,1)$, the price of this bond will be $1 \times e^{-r(1,1)}$. If at time 1 state 0 is realized, i.e., the process is at node $(1,0)$, the bond maturing at time 1 will have a price of $1 \times e^{-r(1,0)}$. At time 0 the observed price of bond 2 is $d(2)$.

```
> d(2);
```

$$.8000000000$$

To avoid arbitrage, the price of bond 2 at time 0 should be the discounted expected value of its price at time 1. Hence, we obtain equation (16.52).

$$d(2) = d(1)\left(\frac{e^{-r(1,1)}}{2} + \frac{e^{-r(1,0)}}{2}\right) \tag{16.52}$$

Substituting for $r(1,1)$ in terms of $r(1,0)$, based on equation (16.48), yields equation (16.53), where the only unknown is $r(1,0)$.

$$d(2) = d(1)\left(\frac{e^{-r(1,0)e^{2\sigma(1)}}}{2} + \frac{e^{-r(1,0)}}{2}\right). \tag{16.53}$$

Solving equation (16.53) for $r(1,0)$, we can recover $r(1,1)$ in a way that the evolution of the short rate complies with the no-arbitrage conditions and is consistent with the observed bond prices.

Knowing $r(0,1)$ and $r(1,0)$ allows us to solve for $\mu(1)$ and $M(1)$. However, knowing the values of $\mu(1)$ and $M(1)$ is not needed in order to value derivative securities based on the so generated binomial tree. Once the evolution of the term structure is determined, the tree can be used to value interest rate derivatives in the usual way.

We have, therefore, recovered the evolution of the term structure up to time 2. Our next step is to show how the evolution of the term structure to time 3 is recovered, and then to generalize the method. However, before addressing this task we would like to point out a relation that is a consequence of equation (16.53) and, as we shall soon see, holds in general. Dividing equation (16.52) by $d(1)$ and realizing that $\frac{d(2)}{d(1)}$ is $e^{-r_0(1,2)}$, where $r_0(1,2)$ is the forward rate (as of time zero) spanning the time interval $[1,2]$, we obtain equation (16.54).

$$e^{-r_0(1,2)} = \frac{e^{-r(1,1)}}{2} + \frac{e^{-r(1,0)}}{2} \tag{16.54}$$

Equation (16.54) states that the procedure of solving for the evolution of the term structure ensures a certain property of the forward rates in this market. It fixes the evolution such that the expected value, under the risk-neutral distribution, of the discount factor from time 1 to time 2 equals the discount factors that are based on the forward rate.

We proceed now to solve for the rates $r(2,0)$, $r(2,1)$, and $r(2,2)$. We know the evolution of the short rate for the nodes on which the squares were superimposed in Figure 16.14, and we would like to find out the rates for the nodes $(2,0)$, $(2,1)$, and $(2,2)$ that are superimposed with squares in Figure 16.15.

```
> BinTree(3,[[2,0],[2,1],[2,2]]);
```

Accomplishing this task provides us with the evolution of the short rate over the 3-year period. Although we want to solve for the three rates $r(2,0)$, $r(2,1)$, and $r(2,2)$, we know that these rates are related to each

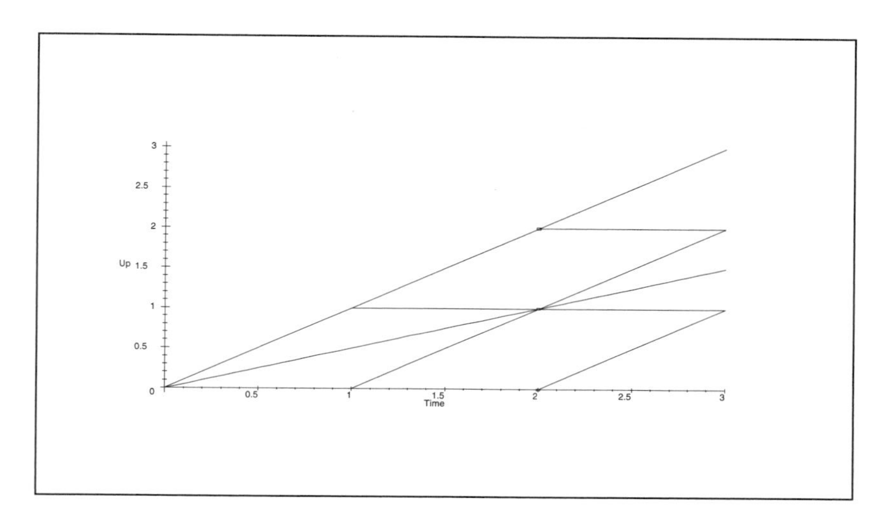

Figure 16.15: Solving for the Rates $r(2,0)$, $r(2,1)$, and $r(2,2)$ at the Nodes Superimposed with Squares

other through equation (16.51). Hence, we only have one unknown, $r(2,0)$, as equation (16.51) implies that

$$\begin{aligned} r(2,1) &= r(2,0)e^{(1\times 2)\sigma(2)} \\ r(2,2) &= r(2,0)e^{(2\times 2)\sigma(2)}. \end{aligned}$$

The value of $r(2,0)$ will be recovered utilizing the requirement that the price of a zero-coupon bond maturing at time 3, based on the evolution of the short rate, should equal its observed price. At node $(2,j)$ the price of this bond will be $e^{-r(2,j)}$. Hence, the price of it at time zero can be calculated discounting by $d(2)$ the expected value of the price of the bond as of time 2. The probability of arriving to node j at time 2 is $\binom{2}{j}\left(\frac{1}{2}\right)^2$. Thus, the discounted expected value of the bond is $d(2)\sum_{j=0}^{2}\binom{2}{j}\left(\frac{1}{2}\right)^2 e^{-r(2,j)}$ and must satisfy equation (16.55).

$$d(2)\sum_{j=0}^{2}\binom{2}{j}\left(\frac{1}{2}\right)^2 e^{-r(2,j)} = d(3). \tag{16.55}$$

If we substitute for $e^{-r(2,j)}$ in terms of $r(2,0)$ we obtain equation (16.56),

$$d(2)\sum_{j=0}^{2}\binom{2}{j}\left(\frac{1}{2}\right)^{2} e^{-r(2,0)e^{2j\sigma(2)}} = d(3), \tag{16.56}$$

which can be solved to recover the value of $r(2,0)$. Dividing equation (16.55) by $d(2)$ yields equation (16.57),

$$e^{-r_0(2,3)} = \sum_{j=0}^{2}\binom{2}{j}\left(\frac{1}{2}\right)^{2} e^{-r(2,j)}, \tag{16.57}$$

which conveys the same intuition as that of equation (16.54). This intuition can be expressed in terms of the prices of the bonds, rather than in terms of discount factors. Stating it in the latter manner, the right-hand side of equation (16.57) is the expected value at time two, as of time zero, of a zero-coupon bond issued at time two and maturing at time three. The left-hand side of the equation is the forward price, at time zero, of a zero-coupon bond maturing at time three and issued at time two.

Let us look at a numerical example utilizing the equivalent market of this section to demonstrate the solution for an evolution of the short rate. To this end we need estimates of the volatility of the short rate in a year, two years, and three years. These are given in the list sd, where sd_i is the volatility of the short rate at time i.

```
> r:=evaln(r): sd:=[0.15,0.21,0.18];
```

$$sd := [.15,\ .21,\ .18]$$

We will use a table (a form of an array in MAPLE) to store the values of the short rate. The table will be denoted by r and will have two indexes. As per our usual convention the first index, i, will denote the time, and the second, the number of up movements realized from time 0 to time i. The short rate at time zero is implicit in the price of the bond maturing in a year, or equivalently in the discount factor $d(1)$. The rate $r_{0,0}$ is, as explained above, given by

```
> r[0,0]:=-ln(evalf(d(1)));
```

$$r_{0,0} := .1053605157$$

To proceed and solve for the short rate $r_{1,0}$ (at time 1) we need to solve equation (16.53). Below we ask MAPLE to solve this equation numerically.

```
> fsolve(d((2))/d(1)=sum('(1/2)^1*exp(-r[1,0]*\
> exp(j*2*sigma[1]))','j'= 0..1),r[1,0],0..1);
```

$$.1003781062$$

Next, the obtained solution is assigned to $r_{1,0}$.

```
> r[1,0]:=%;
```

$$r_{1,0} := .1003781062$$

The value of $r_{1,1}$ is recovered based on equation (16.48).

```
> r[1,1]:= r[1,0]*exp(2*1*sigma[1]);
```

$$r_{1,1} := .1354962708$$

The solution of equation (16.56) determines the value of $r_{2,0}$. Equation (16.56) is solved numerically below, and the solution is assigned to $r_{2,0}$.

```
> fsolve(d(3)/d(2)=sum('(binomial(2,j)*(1/2)^2)*\
> exp(-r[2,0]*exp(j*2*sigma[2]))','j'=0..2));
```

$$.08615753588$$

```
> r[2,0]:=%;
```

$$r_{2,0} := .08615753588$$

The other short rates at time two, $r_{2,1}$ and $r_{2,2}$, are obtained by equation (16.50). The next two MAPLE commands assign the values of these rates based on equation (16.50).

```
> r[2,1]:= r[2,0]*exp(2*sigma[2]);
```

$$r_{2,1} := .1311284574$$

```
> r[2,2]:= r[2,0]*exp(2*2*sigma[2]);
```

$$r_{2,2} := .1995724709$$

We can now ask MAPLE to display the values in the table r. This is done below, where, for example, "(2, 2) = 0.1995724709" is MAPLE's way of reporting that $r_{2,2} = 0.1995724709$.

```
> eval(r);
```

table([
(2, 1) = .1311284574
(2, 2) = .1995724709
(1, 0) = .1003781062
(0, 0) = .1053605157
(2, 0) = .08615753588
(1, 1) = .1354962708
])

Having solved for the evolution of the short rate we can solve for the evolution of the prices of the bonds. Take, for example, the zero-coupon bond (with a $1 face value) maturing at time 3. Its value at time 2 in state j is given by $e^{-r_{2,j}}$. We use the table **Bond3** to store the evolution of the price of the bond maturing at time 3, where the indexing is used in the same manner as above. The price of this bond at time 2, in each state, is given below.

```
> i:='i': j:='j':
> for j from 0 to 2 do;
> Bond3[2,j]:=1*exp(-r[2,j]);
> od;
```

$$Bond3_{2,0} := .9174496886$$
$$Bond3_{2,1} := .8771050965$$
$$Bond3_{2,2} := .8190808591$$

```
> i:='i': j:='j':
```

In evaluating the price of the bond at time i ($i = 0, 1$) in state j, we approach the bond as if it is a risky security. Hence, we calculate its value by discounting (with $r_{i,j}$) its expected value at time $i + 1$, as of time i. This is done in exactly the same manner as an equity derivative security was valued in Chapter 13. The value of the bond at time 1 and 0 is calculated by the double "do-loop" as below.

```
> for i from 1 to 0 by -1 do;
> for j from 0 to i do;
> Bond3[i,j]:=exp(-r[i,j])*\
> (Bond3[i+1,j]/2+Bond3[i+1,j+1]/2);
> od;
> od;
```

We are now in a position to verify that indeed the value of the bond maturing at time 3, as evaluated above based on the evolution of the short rate, coincides (**numerically**) with its observed price. We thus expect, that the price of this bond at time zero would coincide (but with roundoff errors, as we solved it numerically) with $d(3)$. This is confirmed by the next MAPLE commands.

```
> Bond3[0,0];
```

$$.6984935660$$

```
> d(3);
```

$$.6981481481$$

The reader is encouraged to confirm that the prices of the other bonds, as calculated based on the evolution of the term structure, coincide with the observed prices. We unassign below the variables used in this example. The reader may like to construct other examples or change some of the parameters in this example and resolve it.

```
> i:='i': j:='j': evaln(Bond3): evaln(r):
```

The prices of the zero-coupon bonds, in the artificial equivalent market, are consistent with the evolution of the short rate in the BDT model. Hence, so should be the observed prices of the bonds in the original market (Table 16.1). We leave the proof of this conjecture to the reader. The exercises of this Chapter point out prices of other financial assets that should be consistent with the evolution of the short rate in the BDT model.

In general, given the current discount factor function $d(\cdot)$, the evolution of the term structure from time i to time $i+1$ is given by equation (16.58),

$$d(i)\sum_{j=0}^{i}\binom{i}{j}\left(\frac{1}{2}\right)^{i}e^{-r_{i,0}e^{2j\sigma_i\sqrt{\Delta t}}}=d(i+1), \tag{16.58}$$

for $i=1,...,N-1$, where the length of a time period is Δt and the number of periods in the model is N. Equation (16.58) stipulates the value of $r(i,0)$ since it is the only unknown variable in the equation. The value of $r(i,j)$ for $j=1,...,i$, was given by equation (16.51) and is repeated below in equation (16.59).

$$r_{i,j}=r_{i,0}e^{2j\sigma_i\sqrt{\Delta t}} \tag{16.59}$$

Working with the BDT model, for a general case, follows the same guidelines demonstrated in the above example. The procedure **BDTree** builds an implied (by the set of discount factors) binomial tree based on the BDT model. Its input parameters, in order, are:

- N, number of nodes in the binomial tree
- T, the length of the complete period modeled, so that the length of each subperiod is $\Delta t=\frac{T}{n}$ (that is, the short rate is the rate spanning a period of length Δt)
- d, the discount factor function which can be generated by executing **NarbitB**
- σ, an array of the volatilities of the short rate, where σ_i is the volatility of the short rate at time i, $i=1,...,N$

- *Name*, a name supplied by the user to which the table of short rates will be assigned ($Name_{i,j}$ is the short rate at time i in state j)

Hence, in order to solve with **BDTree** the example we just solve manually, we should execute the command below.

```
> BDTree(3,3,d,sd,BDT);
```

To verify that the same results are obtained, we ask MAPLE to print the table BDT generated by the **BDTree** procedure.

```
> eval(BDT);
```

table([
(2, 0) = .08615753588
(0, 0) = .1053605157
(2, 1) = .1311284574
(2, 2) = .1995724709
(1, 1) = .1354962708
(1, 0) = .1003781062
])

This table and the table r we calculated above are indeed the same.

We conclude this section by demonstrating a run of the **BDTree** procedure. We solve, with **BDTree**, for the evolution of the short rate in the example above, but where the length of a period is one month rather than a year. To this end we need to have zero-coupon bonds maturing in each month for the next three years. The prices of these bonds are not observed directly in the given market (Table 16.1). Rather we have to use our estimate of the continuous discount function, since the meaning of $d(\frac{i}{36})$, $i = 1, ..., 36$, is the price of a zero-coupon bond maturing at time i. We just need, therefore, to change the definition of sd, the term structure of volatility, as it should have estimates of 36 volatilities. Assume just for simplicity, of course, that the sequence of 36 volatilities repeats itself and is given as below.

```
> sd:=[seq(op(sd),i=1..12)];
```

sd := [.15, .21, .18, .15, .21, .18, .15, .21, .18, .15, .21, .18, .15, .21, .18, .15, .21, .18, .15, .21, .18, .15, .21, .18, .15, .21, .18, .15, .21, .18, .15, .21, .18, .15, .21, .18]

We are now ready to run the procedure which will redefine the table BDT such that $BDT_{i,j}$ is the short rate at time i, $i = 1, ..., 36$, in state j $(j = 0, ..., i)$.

```
>  BDTree(36,3,d,sd,BDT);
```

We can display the result as before issuing the command **eval**(**BDT**), which will generate a long column of all the short rates. We can display some of the results by issuing commands like

```
>  seq(BDT[8,j],j=0..8);
```

.004491440914, .005070386884, .005723958891, .006461776217,
.007294698071, .008234983409, .009296471372,
.01049478495, .01184756094

or like

```
>  seq(BDT[i,8],i=8..24);
```

.01184756094, .01026217725, .009171511843, .009504268985,
.008824302503, .008444044758, .008605719847,
.008487562695, .008572370839, .008439539930,
.008669143578, .009058634378, .008452170704,
.008884206010, .009451700609, .008236955263,
.008739507798

Once the evolution of the term structure has been identified, the tree can be used to value different types of interest rate derivatives. This can be done in a recursive manner for an American option, or simply by calculating the discounted expected value at time N of a European option. The exercises at the end of this Chapter guide the reader in practicing valuation of interest rate derivatives, like those valued with Black's model in the former section, with the BDT model.

16.5 Concluding Remarks

As suggested by its title, this Chapter it is devoted to "other types of options". In Chapter 5, we introduced the (continuous) risk-neutral distribution, and extended the concept of valuation as the discounted expected value, under this distribution, to a continuous-time setting. We stated (on page 190), that while our analysis was done in terms of stock options, it applies to other contingent claims as well. This Chapter demonstrates how contingent claims, where the underlying asset is not a stock, can be priced utilizing the guidelines developed for stock options.

This Chapter began by looking at contingent claims similar (from a computation point of view) to options on a stock that pays dividends. Included in this group are options on indexes, foreign currency, and futures, the pricing of which is a direct application of pricing options on a dividend-paying stock. The Chapter then continued with examples of some exotic options. Yet again, we notice that the guidelines developed throughout the book are applicable to pricing of "nonstandard" options. The exercises at the end of this Chapter will introduce the reader to some more exotic options, asking the reader to price these options. The prices of many exotic options can be determined if one notices that the exotic option at hand can actually be written as a combination of options with known prices. This is just another application of the building blocks approach used extensively in Chapter 4.

The Chapter offered a brief investigation of interest rate derivatives. This investigation was initiated with Black's model, which is essentially a direct application of the Black–Scholes formula to interest rate derivatives. While this model is commonly used in the industry, it suffers from some deficiencies. These are due to the characteristics of interest rate processes, which differ from the process of the price of stocks. The Chapter concluded with the introduction of the BDT model for interest rate processes. This model does not suffer from some of the deficiencies of the Black model. Furthermore, it is essentially a direct application of the binomial model investigated in Chapters 4 and 14 for stock options. The BDT model can be thought of as an implied binomial tree (similar in philosophy to the concept underlying the implied volatility investigated in Chapter 6). In the BDT model the probability of the realization in each one-period model is assumed to be one-half and the volatility of the short rate is given. The model then solves numerically for the realization of the short rate, in each node, so that the observed prices of the bonds will be consistent with the evolutions of the short rate.

Pricing interest rate derivative securities with this model is very similar to the way stock options were priced in the binomial model. The exercises at the end of the Chapter provide some examples for this valuation. This Chapter is essentially the last one in the book; it marks the end of the journey started with the simple model of Chapter 1. But is it the end, or essentially the beginning of new journeys the reader is now hopefully qualified to take: journeys into the land of new continuously evolving derivatives. The next Chapter provides a perspective on the material covered in the book, and thereby attempts a reply to the above question.

16.6 Questions and Problems

Problem 1. The **BinTree** procedure was utilized to visualize, Figure 16.1, the coordinates where it is optimal to exercise a call option. Explain the shape of this set in comparison to the set depicted in Figure 14.5 that pertains to a put option.

Problem 2. The value of a call option on a futures contract was given in the text (equation (16.8)) as

$$e^{r(T_f-T)}\left(S(t_0)N(d_1)-e^{-r\,(T-t_0)}e^{-r(T_f-T)}KN(d_2)\right).$$

How would the equation above be modified if S is an asset that generates an income stream at a constant yield?

Problem 3. Would it be optimal to exercise early an option on a futures contract? Would you change your answer if the option would be on a forward contract and why?

Problem 4. State the put–call parity for options on a forward contract.

Problem 5. The **AmerOpt** procedure uses a binomial model to solve for the value of American options. It uses the following approximation of the binomial distribution to the price of the underlying asset (compare to equations (14.9) and (14.10) in Section 14.1.1). Assume that the price of the asset at time zero is S. Then its price in node N, if there are j up movements, is $Se^{ju+(N-j)d}$, where $u = \left(r - \frac{\sigma^2}{2}\right)\Delta t + \sigma\sqrt{\Delta T}$ and $v = \left(r - \frac{\sigma^2}{2}\right)\Delta t - \sigma\sqrt{\Delta t}$ are possible realizations of a random variable, the probability of each being 0.5. Consider the example of a digital American option solved in the text. Show that six up steps should be realized in the binomial model for the price of a stock starting from \$100 at time 0 exceeding \$110 at time 6, and thus it should be optimal to exercise the option as soon as six up steps are realized.

Problem 6. Delta hedging of a digital option can be very costly. Develop a hedging environment like the one in Section 8.2.1 (Table 8.3–a spreadsheet in the on-line version of the book) and try to hedge a digital option. Compare your results to a static hedging utilizing the same technique employed in Section 4.5 to produce bounds on the price of a digital option using plain vanilla options.

Problem 7. Consider a European option written on a stock with a maturity of T (current time is assumed to be zero) and an exercise price of K. Assume that at some future time $t < T$ the holder of the option can decide if the option is a put or a call. Such an option is named a "chooser option" or "as you like it option". What will be the value of such an option? (Hint: use the put–call parity.)

Problem 8. Consider a European option to exchange one asset with another, sometimes referred to as an "exchange option". Let S_1 be one asset and S_2 be another asset; then the payoff of an option to exchange S_1 with S_2 is $\max(S_2 - S_1, 0)$. Value such an option. (Hint: assume that the correlation between the value of S_1 and S_2 at maturity is ρ and take advantage of the fact that if X and Y are two normal random variables, with expected values of μ_X and μ_Y, respectively, standard deviations of σ_x and σ_Y, respectively, and a correlation coefficient of ρ, then $X - Y$ is a normal random variable with an expected value of $\mu_X - \mu_Y$ and a standard deviation of $\sqrt{\sigma_X^2 - \sigma_Y^2 - 2\rho\sigma_X\sigma_Y}$.)

Problem 9. Consider a European option to obtain minimum between two assets, S_1 and S_2. The payoff of such an option is $\min(S_1, S_2)$. A similar type of option grants the holder the payoff $\max(S_1, S_2)$. Find the value of these options. (Hint: $\min(S_1, S_2)$ can be written as $S_2 - \max(S_2 - S_1, 0)$, and $\max(S_1, S_2)$ can be written as $S_1 + \max(S_2 - S_1, 0)$.)

Problem 10. Almost exclusively, the derivative securities we dealt with depended on the price of the underlying asset and not on its history. That is, the payoff was contingent only on the price of the underlying asset at the exercising time. In some cases the payoff is contingent on the path of the price of the underlying asset. Such derivative securities are termed path-dependent. Consider a derivative, the payoff of which is the amount the current stock price exceeds the maximum stock price until the current time. Assume a binomial model with the price evolution as described in Table 13.1 of Section 13.1 and value the derivative security. Note that this requires calculating the maximum price achieved along each path and thus the tree is essentially a non-recombining. The particular path-dependent option illustrated here is named a look back option. The binomial tree can also be utilized to value an American look back option. Value an American look back option as above. Think carefully as you "roll back" to get its value at time zero.

Problem 11. The prices of the zero-coupon bonds, in the artificial equivalent market in the text, are consistent with the evolution of the short rate in the BDT model. This implies that the observed prices of the (coupon) bonds in the original market (Table 16.1) are also consistent with the evolution of the short rate. Prove this statement and use the same argument to determine if the prices of the financial assets listed below are consistent with the evolution of the short rate in the BDT model.

1. FRA
2. Swaps
3. Caplet and caps
4. Swaptions

Problem 12. Use the BDT model to price all the instruments that were priced in the text with Black's model.

Chapter 17

The End or the Beginning?

Although it may not seem so, this book has opened only a window to a new world. It is almost an impossible task to have one book that introduces the subject, and encompasses all the various existing derivatives and their regulations while at the same time putting the reader at the frontier of the literature. The market of derivative securities is vibrant and changing continuously. New models and products are being introduced to the market regularly. As in all subjects, a textbook can never cover all the situations that might arise in reality and this book is no exception. It does, however, introduce the reader to the main concepts and ideas in the area of contingent claims.

The book attempted to equip the reader with a "way of thinking", or rather an analytical framework, with which topics in derivative securities can be analyzed. To achieve this goal the book tried to convey the "feeling" of what arbitrage pricing is all about. Indeed, as mentioned in the preface, the motto of this book is the satisfaction of the no-arbitrage condition by prices in the market. Consequently, the order of coverage of topics was dictated by the complexity of the no-arbitrage condition needed for valuation.

Chapter 1 began by defining the no-arbitrage condition in the simplest one-period model. In Chapter 2 it is used to value, and thereby familiarize the reader with, a variety of contingent claims. Gradually, the book extended the simple model to a more realistic situation, permitting the valuation of more complicated securities. The message is that arbitrage pricing, while being a simple concept, can go a long way. Indeed, as we have realized, it was applicable to both the first, very simple environment introduced in the book and the most complex setting investigated. Granted the tools used in exploiting these arguments might be different, in the various environments considered, but the concept stays intact.

One may like to keep this principle in mind: Even when faced with a new instrument that requires different and complex calculations, resorting to the pure and simple arbitrage arguments might help. This may lead to solving an unfamiliar differential equation or calculating an integral that might not be easy. These, however, are technical points. Notwithstanding their importance, once the techniques needed are identified, a solution is much closer.

The interpretation of the one-period model and of the no-arbitrage condition was slightly modified in Chapter 3. This allowed the investigation of the bond market and the estimation of the term structure. Chapter 4 generalized the one-period model so that it can accommodate a continuum of states of nature. This last model is utilized throughout Chapters 5 to 8 to introduce the reader to plain vanilla European options and their properties and valuation.

Chapters 9, 10, and 11 expanded on the one-period model in yet another direction, resulting in a discrete multiperiod model. In this environment the different varieties of swaps, forwards, and futures agreements were investigated. Chapter 12 defined American options and explored bounds on the pricing of such options. Numerical valuations of these options in the framework of the binomial model were explored in Chapters 13 and 14.

Chapter 15 dealt with the most complex model explored in the book: a model of continuous time and continuous states of nature. This facilitated the linkage between option pricing and numerical solutions of differential equations. The Chapter introduced the tools used in this setting, Ito's Lemma, to explore arbitrage pricing. It also presented a rigorous justification of the Black–Scholes formula that was developed in Chapter 6, intuitively. Finally, Chapter 16 demonstrated that the concepts studied thus far could be utilized to value different kinds of contingent claims, e.g., claims where the underlying asset is not a stock and some exotic options. The Chapter also suggested, by considering the BDT model, that the no-arbitrage arguments and methodologies utilized in the equity market can be applied to the valuation of interest-rate-contingent claims.

While arbitrage pricing is a cornerstone of modern finance, it can lead us to a certain point only. We should now appreciate that a risk has a price and so does the comfort of pricing by arbitrage. In certain instances arbitrage pricing cannot generate a unique price or even a significant price. Such cases arise, for example, when the assumption of a fixed and deterministic volatility is relaxed. Other assumptions on "market price of risk" or on the attitude toward risk in these cases must be employed. Yet it is always important to know what part of the result is dependent on these added assumptions, and how far we can "push" the result without them. After all, when pricing a customized position, if arbitrage pricing is not sufficient, we should employ the risk attitude of the "client". The price obtained should reflect the client's appetite for risk, or what the correct price is for this client.

The field of derivative securities continues to grow and expand. Assumptions that were taken for granted in the development stage of the basic theory are now relaxed and new boundaries of knowledge are being defined. Yet, we believe that the conceptual foundation of pricing by arbitrage and the analysis of derivative securities has been laid down in this book. We hope that after reading the book readers will be able to see a clear way to the analysis of new instruments not discussed here.

The boundaries between the classical, or perhaps by now, ancient, separation of corporate finance and investments are gradually becoming fuzzy. This is mostly attributed to the popularity of derivative securities with which one can transform one type of risk to another. Contingent claims increased the degrees of freedom in carving out the risk profile one wishes to sustain. At the same time, of course, it allows more creativity in executing speculative strategies. Derivative securities enlarged (or introduced) a market for risk in which investors can buy and sell "portions of risk" implicit in their portfolio holdings. Hence, financial managers use options for both speculation and hedging. Some subjects that were seemingly not related to contingent claims, and classically were not analyzed in this way, are now being viewed with the aids of such claims. Examples of this new trend exist in the analysis of real options, mining, energy derivatives, investment timing, the decision to invest, and valuation of projects. The topic of risk management is intimately related to contingent claims, as is the recently coined concept of value at risk (VAR).

The reader may like to take advantage of the dynamic features of this book in pursuing further interest in contingent claims and the related issues mentioned above. Perhaps the best way to end the journey, started at the preface of the book, is with essentially the same phrase mentioned there. Learning is enhanced by altering the commands on-line. They can be varied at will in order to experiment with applications of the concepts and different (reader-generated) examples, in addition to those that are already in the book. It is this interaction and experimentation, making use of MATLAB and MAPLE, together with the ability to bring to life the theoretical material on the screen, that provide a unique, powerful, and entertaining way to learn about derivatives.

References

[1] K. J. Arrow. The role of securities in the optimal allocation of risk bearing. *Review of Economic Studies 31*, pages 91–96, 1964.

[2] M. Baxter and A. Rennie. *Financial Calculus: An Introduction to Derivative Pricing.* Cambridge University Press, 1996.

[3] S. Beckers. Standard deviations implied in option prices as predictors of future stock price variability. *Journal of Banking and Finance 5*, pages 363–381, 1981.

[4] F. Black. Interest rates as options. *Journal of Finance 50*, pages 1371–1376, 1995.

[5] F. Black. Fact and fantasy in the use of options and corporate liabilities. *Financial Analysts Journal*, pages 36–41, July-August 1975.

[6] F. Black. The pricing of commodity contracts. *Journal of Financial Economics 3*, pages 167–179, March 1976.

[7] F. Black, E. Derman, and W. Toy. A one-factor model of interest rates and its application to treasury bond options. *Financial Analysis Journal*, pages 33–39, January-February 1990.

[8] F. Black and M. Scholes. The pricing of options and corporate liabilities. *Journal of Political Economy 81*, pages 637–659, 1973.

[9] J. Bowie and P. Carr. Static simplicity. *Risk*, pages 45–49, 1994.

[10] P. Boyle and D. Emanuel. Discretely adjusted option hedges. *Journal of Financial Economics 8*, pages 259–282, 1980.

[11] P. Buchen and M. Kelly. The maximum entropy distribution of an asset inferred from option prices. *Journal of Financial and Quantitative Analysis 31*, pages 143–159, 1996.

[12] C. J. Corrado and T. W. Miller. A note on a simple, accurate formula to compute implied standard deviations. *Journal of Banking and Finance 20*, pages 595–603, 1996.

[13] D. R. Cox and H. D. Miller. *The Theory of Stochastic Processes.* Chapman and Hall, 1965.

[14] J. C. Cox, J. Ingersoll, and S. Ross. The relation between forward prices and futures prices. *Journal of Financial Economics 9*, pages 321–346, 1981.

[15] J. C. Cox, S. Ross, and M. Rubenstein. Option pricing: A simplified approach. *Journal of Financial Economics 7*, pages 229–264, 1979.

[16] E. Crow and K. E. Shimizu. *Lognormal Distributions, Theory and Applications.* Marcel Dekker Inc., 1988.

[17] J. Davidson. *Stochastic Limit Theory.* Oxford University Press, 1994.

[18] E. Derman, D. Ergener, and I. Kani. Static option replication. *Journal of Derivatives*, 1995.

[19] J. C. Dermody and E. Z. Prisman. No-arbitrage and valuation in markets with realistic transaction costs. *Journal of Financial and Quantitative Analysis 28*, pages 65–80, 1993.

[20] A. Dixit. The arts of smooth pasting. *In Fundamentals of Pure and Applied Economics (Lesourne and H. Sonnensche in, Eds.), Vol. 55*, 1993.

[21] M. U. Dothan. *Prices in Financial Markets.* Oxford University Press, New York, 1990.

[22] D. Duffie. *Futures Markets.* Prentice Hall, 1989.

[23] E. Fama. The behavior of stock prices. *Journal of Business 38*, pages 34–105, 1965.

[24] E. Fama. Efficient capital markets II. *Journal of Finance 26*, pages 1575–1617, 1991.

[25] J. Harrison and D. Kreps. Martingales and arbitrage in multiperiod securities markets. *Journal of Economic Theory 20*, pages 381–408, 1979.

[26] R. Jarrow and G. Oldfield. Forward contracts and futures contracts. *Journal of Financial Economics 9*, pages 373–382, 1981.

[27] R. A. Jarrow and A. Rudd. *Option Pricing.* Irwin, 1983.

[28] S. Karlin and H. Taylor. *A First Course in Stochastic Process.* Academic Press, New York, 1975.

[29] M. Knoll. Put call parity and the law. *Working Paper 94-12, USC Law School Working Paper Series.*

[30] O. Mangasarin. *Nonlinear Programming.* SIAM, Philadelphia, 1994.

[31] S. Mayhew. Implied volatility. *Financial Analysts Journal*, pages 8–20, July-August 1995.

[32] R. Merton. The relationship between put and call prices: Comment. *Journal of Finance 28*, pages 183–184, 1973.

[33] R. Merton. Theory of rational option pricing. *Bell Journal of Economics and Management Science 4*, pages 141–183, 1973.

[34] N. N. Neftci. *An Introduction to the Mathematics of Financial Derivatives.* Academic Press Inc., 1996.

[35] B. Oksendal. *Stochastic Differential Equations: An Introduction with Applications.* Springer-Verlag, 1995.

[36] E. Z. Prisman. Valuation of risky assets in arbitrage-free economies with frictions. *Journal of Finance 41*, pages 544–557, 1986.

[37] R. Rebonato. *Interest-Rate Options Models.* John Wiley and Sons, Ltd., England, 1998.

[38] R. Rendleman and B. Bartter. Two state option pricing. *Journal of Finance 34*, pages 1092–1110, 1979.

[39] S. Ross. A simple approach to the valuation of risky streams. *Journal of Business 51*, pages 453–475, 1978.

[40] S. Ross. *Stochastic Process.* John Wiley and Sons, Inc., New York, 1983.

[41] S. A. Ross. *Return, Risk and Arbitrage.* In Risk and Return in Finance (Friend, I. and J. Bicksler, Eds.). Ballinger, Cambridge, MA, 1976.

[42] M. Rubinstein. The valuation of an uncertain income stream and the pricing of options. *Bell Journal of Economics 7*, pages 407–425, 1976.

[43] P. Samulson. *Gibbs in Economics.* In Proceedings of the Gibbs Symposium (G. Caldi and G. D. Mostow Eds.). American Mathematical Society, 1990.

[44] S. M. Schaefer. Measuring a tax-specific term structure of interest rates in the market for British government securities. *The Economic Journal 91*, pages 415–438, 1981.

[45] H. Stoll. The relationship between put and call option prices. *Journal of Finance 24*, pages 801–824, 1969.

[46] M. Stutzer. A simple nonparametric approach to derivative security valuation. *Journal of Finance 51*, pages 1633–1652, 1996.

[47] P. Willmott, J. Dewynne, and S. Howision. *Option Pricing: Mathematical Models and Computation.* Oxford Financial Press, 1993.

Index

AllWays, 515
AmerCall, 675
AmerOpt, 675, 676, 679, 685, 687, 691, 698, 700, 731
AmerPut, 567, 568, 675, 676
Approximating functions, 384
Arbitrage, 1
  in the debt market, 94
Arbitrage opportunities, 325, 368, 375, 397, 403, 411, 412, 424, 425, 428, 430, 431, 439, 440, 443, 447, 451, 481, 500, 512, 530, 600, 638, 640, 641, 697, 707, 720
Arbitrage portfolio, 7–11, 30, 32, 42, 95–101, 166, 345, 347, 348, 372, 403, 407, 488, 503, 640
Arbitrage pricing
  debt markets, 91
  equity markets, 47
Arbitrage profit, 7, 9, 10, 32, 42, 495, 641
Arbitrage strategies, 144
Assignation, 171

Backwardation, 66
BDT, *see* Black, Derman and Toy model
BDTree, 727, 728
Bernoulli random variable, 593, 596
Binomial distribution, 520, 558, 561, 562, 565, 596, 731
  cumulative distribution function, 561
  density function, 561, 562
Binomial model, 47, 87, 505–508, 510–513, 519–522, 524, 528–530, 537, 539, 546, 554, 555, 557–559, 561, 564, 565, 567, 568, 571, 574–577, 579, 585, 587, 588, 590–592, 594, 599, 601, 633, 634, 636, 638, 639, 644, 646, 647, 649, 674–676, 681, 684, 685, 688, 691, 717, 730–732, 736
  multiperiod, 505, 510, 513, 559, 571
  n-period, 51, 513, 514, 524, 530, 600
  one-period, 50, 51, 505, 506, 513, 521, 522, 524, 528–530, 546, 554, 559, 571, 575, 576
  the effect of dividend, 568
  two-period, 506
Binomial tree, 507, 508, 511, 514, 515, 519, 521, 529, 550, 554, 555, 568, 570, 593, 679, 706, 714, 715, 717, 718, 722, 727, 730, 732
  non-recombining, 571
  recombining, 507, 508, 714

Binomvsln, 561
BinTree, 506, 508, 676, 677, 731
Black's model, 705, 706, 710, 712–714, 729, 730, 733
Black, Derman and Toy model, 706, 714, 715, 717, 727, 729, 730, 733, 736
Black-Scholes differential equation, 632, 642, 645, 650
Black-Scholes pricing formula, 228, 259, 323, 325, 339, 349, 370, 496, 497, 510, 557, 558, 565, 566, 570, 579, 590, 591, 593, 596, 597, 599, 600, 602, 632, 683, 684, 705–707, 710, 730, 736
  call option, 228, 229, 271, 565
  connection to Delta, Gamma and Theta, 370
Bond, 2, 15–17, 24, 28, 48–50, 52–54, 59, 60, 64, 65, 75, 78, 92–98, 103, 105, 111, 112, 327, 328, 330, 332, 337, 342, 343, 347, 353, 373, 377, 378, 380–384, 393, 398–403, 406–412, 418, 423–430, 432, 449, 450, 500
  coupon payments of, 92
  coupon rate, 92
  face value of, 92, 402
  floater, *see* variable rate bond
  maturity of, 91
  principal of, 92
  semi-annual coupon, 99
  variable rate, 399
Bond market, 29, 91, 93–95, 97, 99, 110–114, 373, 374, 397, 404–407, 412, 413, 423, 449, 450, 455, 475, 736
Bond option, *see* Options
Bond-rating agencies, 93
Brownian motion, 196, 557, 585, 586, 588, 612, 616, 622, 628–631, 643, 651, 652, 665, 716
Bs, 278, 314, 497, 679, 709
BstCall, 229, 230, 277, 312
BstPut, 230, 314, 315
Buy and hold strategy, 400

Call, 119, 122, 126, 127, 135, 143
Callable bonds, 93
CallPriceGap, 697
CallPricePL, 702, 703
Canadian Derivatives Clearing Corporation, 170
Cash, 4, 9, 12, 14, 15, 17, 126
Certainty equivalent, 25–28, 31
CheckNA, 7–9, 30, 32, 36, 48, 82, 83
Complete market, 7, 13, 14, 32, 33, 35, 37, 55, 97, 111, 112, 406, 506
COmswap, 473, 475
ConApp, 385, 389, 391
Contango, 66
Contingent cash flow, *see* Contingent claim, 21, 25–27, 29, 30, 34, 35, 37, 48, 58
Contingent claim, 3–5, 13, 20, 21, 24, 25, 27, 28, 36, 57, 62, 77, 116, 185, 190, 326, 514, 516–518, 530, 531, 565, 735–737
Continuous compounding, 108
Continuously compounded rate of return, 194, 196, 259, 599, 603, 604, 611, 616, 622, 631, 632, 674
  expected value of, 208

stochastic process of, 203
variance of, 203
Convenience yield, 67
Corporate bonds, 432
Cost-of-carry model, 65, 70, 427, 430, 431, 433, 472, 475
Covered call, *see* Options
Credit risk, 454, 481, 482

Debt market, 91, *see* Bond market, 94, 99, 103
Default risk, 403
Delivery date, 417
Delivery price, 417–419, 422, 429
Delta, 271, 281–286, 288, 290, 299, 312, 325–329, 332–334, 338, 339, 341–345, 347, 348, 352, 370, 371, 498, 524, 555, 652, 694, 731
comparison to Theta, 284
of a call option, 281
sign of, 317
of a call option on a dividend-paying stock, 312
sign of, 317
of a deep in-the-money option, 288
of a deep out-of-the-money option, 288
of a Digital option, 694
of a portfolio, 326, 328, 332, 334, 371
of a put option, 314
sign of, 318
of a put option on a dividend-paying stock, 315
sign of, 318
Delta hedging, 284, 288, 325, 326, 339, 344, 345, 348, 370, 731
of a call option, 339
Delta neutral portfolio, 326, 327, 338, 341–343, 370, 652
Derivative securities, xv–xviii, xx, xxi, xxvii, 117, 524, 532, 600–602, 621, 622, 628, 632–641, 643, 645, 647–650, 652, 732, 735, 737
path dependent, 510, 732
underlying security, 117, 524, 732
Deterministic term structure (DTS), 439–441, 445, 447, 449, 451
Differential equation, 110, 601, 602, 616, 621, 622, 628–630, 632, 633, 638, 639, 641, 642, 644, 646–650, 652, 736
initial condition of, 110, 605
numerical solution of, 647, 648
Diffusion coefficient, 620, 629, 630, 636, 641
Discount bond, 92
Discount factor function, 384, *see* Discount factors, 385, 386, 391, 402–404
approximating functions of, 411
continuous approximation of, 385
Discount factors, 17, 22, 24, 43, 52, 53, 91, 103–108, 110–112, 114, 373, 374, 376, 378–380, 382–387, 389, 391, 393, 398, 399, 402–404, 407, 411, 412, 414, 415, 417–419, 422, 424–426, 428, 439–441, 443, 449, 450, 454–460, 462, 463, 465, 467, 471, 472, 474–478, 481, 482, 519, 708, 720, 722, 724, 727

approximation of, 411
continuous approximation of, 385
estimation of, 384
monotonicity of, 106
Dividend, 485, 492, 493, 495–503, 568, 570, 571
Dividend yield, 255, 414, 423, 427, 430, 570, 633, 645, 651, 674, 676, 679, 680, 684, 707
Drift parameter, 716
Dynamic programming, 448

Elementary cash flow, 14–20, 126, 184
Eqbs, 497
EQswap, 477
Equity market, xx, 95, 96, 373, 376, 378, 736
erf, 225–228, 690, 691
relation with Normalcdf, 225
Eurodollar, 432, 435
Evolution of the term structure, 714, 715, 717, 722, 727, 729
Ex-dividend date, 495–501, 503
Exchange rate, 67–70, 72, 73, 80, 81, 84, 88, 462, 463, 466–471, 483
forward, 463, 466–468, 470, 483
spot, 88, 462, 463, 466, 467, 469, 470
Exercise price, 271, 486, 490, 492–494, 498
Expectation theory, 409, *see* Unbiased expectation theory
Expected rate of return, 28

Farkas' Lemma, 114
Fixed income securities, *see* Bond
ForVal, 418
Forward agreement, 396
Forward contract, 61–67, 70, 79, 86–88, 137, 371, 413–415, 417–419, 422–432, 435, 436, 438, 441, 443, 444, 447, 449, 463, 464, 467, 472, 478, 479, 731, 736
on exchange rate, 67, 68
Forward market, 62, 68
Forward price, 61, 65, 87, 88, 396, 413–415, 417, 419, 422–427, 429–431, 433, 438, 441, 443, 447–451
Forward rate, 393–396, 398–400, 403, 404, 408, 409, 425, 450, 457, 711–713, 722
Forward rate agreement, 403, 414, 432, 450, 457, 482
FRA, *see* Forward rate agreement
FRA rate, 432
Free boundary, 568, 648
Frictionless market, 2, *see* Perfect market
Futures contract, 61, 62, 66, 67, 371, 413, 435–438, 441, 443–445, 447, 448, 731, 736
Futures price, 61, 65–67, 436–438, 443, 447, 448, 451, 711
FxFlswap, 455, 457, 459, 460
FXswap, 464, 465, 469–471, 482, 483

Gamma, 271, 288, 290, 292, 299, 312, 334, 351, 352, 370, 555, 652
of a call option, 288
sign of, 317
of a call option on a dividend-

paying stock, 312
sign of, 317
of a put option, 314
sign of, 318
of a put option on a dividend-paying stock, 315
sign of, 318
Gamma neutral portfolio, 370, 652
Geometric Brownian motion, 622, 630, 631, 643, 652
Gmbm, 204
Government bonds, 93, 432, 450

Hedge portfolio, 326–329, 333, 334, 337–341, 344, 345, 348, 350–353, 357, 371
Hedged portfolio, 324, 326, 330, 332, 344, 345, 347, 350, 352, 353, 357, 359, 361, 364, 366, 368, 369, 656
optimized, 325, 364, 366
self-financing, 327–329, 350
Hedging, 60, 125, 136, 271, 323–326, 332, 337–340, 344, 345, 347, 348, 351, 364, 369–372, 694, 731
Delta hedging, 284
Delta neutral portfolio, 326
hedge ratio, 134, 286

Implied volatility, 259, 268
estimation of, 261
ImpliedVol, 263, 264
Incomplete market, 13, 14, 34–37, 103, 399, 412
arbitrage bounds, 35–37
Independent increments, 610, 611, 669, 670
Instantaneous increments, 715
Interest rate, 94
continuously compounded, 373
semi-annually compounded, 375, 377, 391, 405
Interest rate derivatives, 704, 714, 715, 722, 729, 730, 736
Interest rate options, *see* Options
Interest rate parity, 70
Interest rate process, 673, 704–706, 714, 730
Internal rate of return, 380, *see* Yield
Intrinsic value, 172
Inverted market, 66
Ito integral, 620, 621
Ito process, xviii, 622, 623, 628, 629, 632, 667
Ito's Lemma, 621–624, 628–630, 632, 634, 640, 641, 643, 650–652, 666, 667, 671, 674, 715, 736
Ito's lemma, 602
ItosLemma, 628, 630–632, 634, 635, 715
Ivtry, 268

Last trading day, 172
Law of one price, 1, 11–14, 18, 19, 32, 58
LEAPS, 677
LIBOR, 432–435, 450, 453, 457, 458, 482
Linear independence, 33
Liquidating operation, 172
Liquidity preference theory, 409, 410
Liquidity premium, 410
Lnrn, 211, 212, 216, 256, 257
Lognormal distribution, 211, 213, 555, 558, 561, 562, 565, 585, 590, 599, 680, 683,

684, 690, 706, 707, 713–715
cumulative distribution function, 561, 562
density function, 211, 213, 561, 564
risk-neutral, 212, 555, 561, 564, 690
Long position, 2, 3, 5, 11, 398, 401, 407, 415, 422, 424, 428, 429, 436–438, 441, 443, 444, 447, 494, 500, 641

Margin requirements, 2, 172
Market behavior, 389
Market price of risk, 736
Market segmentation theory, 410
Marking to market, 436
Markovian process, 574, 578
Markovian property, 574, 576
Martingale, 557, 577, 583, 590, 592, 631
Mean-reversion, 716
Minimum cost portfolio, 6
Money market account, 332, 342
Municipal bonds, 432

Naked position, 134
Narbit, 47–49, 62, 77, 82, 103, 104, 113
and NarbitB, 98
NarbitB, 98–100, 103, 106–108, 113, 374, 375, 378, 379, 381, 382, 384, 385, 387, 397, 404, 412, 418, 423, 424, 427, 428, 439, 455, 456, 459, 465, 467, 474, 475, 708, 727
Negative cost portfolio, 3, 6, 96
newerf, 228, 690, 691
No-Arbitrage Condition, xx, 1, 5, 7, 8, 11–14, 18, 20, 26, 30–32, 35, 42–45, 47, 48, 54, 77, 82, 83, 87, 88, 91, 93, 97, 99–107, 113, 183, 324, 373, 378, 382, 403, 405, 406, 512, 513, 529, 554, 557, 571, 575–577, 579, 582, 585, 590, 592, 599, 600, 633, 714, 717, 720, 722, 735, 736
Binomial model, 512, 577
theorem, 512
bond market, *see* debt market
debt market, 113
definition, 102, 113, 114
definition, 5, 12
geometric exposition, 42
one-period Binomial model, 52, 529
theorem, 27, 512
Normal distribution, 204, 206, 209, 559, 585, 596, 597, 715
cumulative distribution function, 199, 228, 597, 683, 690, 706
density function, 196
animation, 197
bell curve, 195, 197
Normal random variable, 715, 732
Normalcdf, 199, 211, 225, 227
relation with erf, 225
Normalpdf, 196, 199
Notional quantity, 472

Open interest, 173
Options, xvi, 78, 115, 116, 120, 555, 556, 599, 600, 632, 639, 645, 673–677, 679–682, 684, 685, 688–690, 694–696,

705, 706, 709, 710, 712, 729–732
American, 117, 171, 485–487, 490, 494, 496, 501, 502, 532, 543, 545, 546, 548, 549, 552, 554, 555, 567, 644, 645, 673, 674, 679, 732
arbitrage bounds, 736
early exercise, 487, 502
on a dividend-paying stock, 674
American call, 486–488, 490, 492, 493, 495, 500–503
arbitrage bounds, 486, 487
early exercise, 487, 500, 501, 503
on a dividend-paying stock, 495, 500, 501, 674
American put, 488–493, 501–503, 567, 568
arbitrage bounds, 488–490, 501–503
early exercise, 501–503
on a dividend-paying stock, 501–503
arbitrage bounds, 167
as you like it, 732
at-the-money, 171, 263
Bet, *see* Binary
Binary, 159, 161, 552, 689, 694
bond, 705, 706
call, 58, 143, 171–173, 185, 190, 294, 326–328, 330, 332, 334, 337, 339, 340, 342–344, 348, 350–352, 371, 673
payoff from long position, 118
payoff from short position, 121
cap, 710, 712, 713, 733
caplets, 710, 711, 733
cash on delivery, *see* paylater
chooser, 732
class of, 171
collar, 712
combination, 171
covered call, 133
covered writing, 172
currency, 679, 680, 730
Digital, 689–692, 694, 695, 697, 698, 700, 731
American, 691, 692, 731
European, 691
Digital put, 694
European, 117, 172, 225, 230, 485–487, 491, 492, 496, 503, 524, 532, 547–549, 552, 564, 673, 679, 732, 736
arbitrage bounds, 503
on a dividend-paying stock, 674
valuation, 564, 565
European call, 486–489, 493, 495, 556, 564–566, 632, 639
arbitrage bounds, 486, 487
definition, 117
on a dividend-paying stock, 496
European put, 488–491, 493
arbitrage bounds, 488–490, 502
exchange, 732
exercise price, 172, 486, 490, 492–494, 498
exercise price of, 167
exotic, 673, 688, 694, 730, 736
expiration date, 117, *see* maturity date
expiry cycle, 172
expiry date of, 172

floor, 710, 712
Gap, 695–698, 700
American, 698
call, 697
European, 697, 698
European call, 698
put, 698, 700
hedging, 125
in-the-money, 172, 494, 496
interest rate, 705, 706, 709, 710
long position, 120, 493, 494
lookback, 732
maturity date, 117
naked write, 173
on a dividend-paying stock, 673, 674
on foreign currency, 673
on forward contract, 731
on futures, 677, 730, 731
on indexes, 677, 730
out-of-the-money, 173, 494
path dependent, 556, 593
paylater, 700, 702, 703
paylater call, 700, 703
payoff from, 700
paylater put, 703
plain vanilla, 673, 677, 688–690, 692, 694, 695, 697, 698, 731, 736
portfolio of, 127, 132, 134, 155
bear spread, 171
bull spread, 171
calendar spread, 244–246
different underlying securities, 251
frown, 141
purchased straddle, 136
smile, 141
straddle, 240
premium, 117, 173
protective put, 130
put, 143, 171–173, 230, 673
payoff from long position, 122
payoff from short position, 124
risk reduction, 130
series of, 173
short position, 120, 493, 494
straddle, 173
struddle, 171
tax consequences, 556
to exchange one asset for another, 732
to obtain minimum between two assets, 732
trading strategies, 125
writer of, 120, 123, 173

Partial differential equation, 602, 638, 644, 645, 650
Pay, 251
Payoff diagrams, 141
PayoffCost, 250, 251
Perfect market, 2
PlotForPyf, 415
Portfolio
arbitrage, 30, 32, 345
self-financing, *see* Self-financing portfolio
Premium bond, 92
Present value, 376, 380, 382–384, 412, 454, 458, 462, 464, 471, 472, 475, 476, 478, 481, 488, 494, 499, 500
Price fixing assets, 61
Price process, 557–559, 561, 571, 573–575, 577–579, 585, 592, 601, 602, 673, 704, 715
instantaneous increment of, 602, 605, 608, 622, 629, 644,

656
Lognormal, 558, 585
Martingale property of, 557, 577, 592
relative, 577, 592
Pricing bounds, 143, 146
Pricing by replication, *see* Valuation by replication
Primary securities, 24
Probability distribution, 25
Protective put, *see* Options
Put, 122, 126, 127, 135
Put-call parity, 141, 148, 230, 233, 238, 485, 492, 493, 502, 503, 731
for American options, 502, 503
on a dividend-paying stock, 502, 503

Relative pricing, 141, 143, 147
Replicating portfolio, 55–57, 63–66, 70, 71, 74, 75, 521–523, 531–534, 537, 540, 543, 636, 639, 645
rho, 271, 299, 300, 312, 372
of a call option, 298
sign of, 317
of a call option on a dividend-paying stock, 312
sign of, 317
of a put option, 314
sign of, 318
of a put option on a dividend-paying stock, 315
sign of, 318
Risk averse investor, 26
Risk loving investor, 26
Risk-free
asset, *see* security, 54, 105, 327, 345, 350, 633, 635, 642
security, 2, 24, 31, 50, 60, 64, 93
Risk-free interest rate, *see* Risk-free rate
Risk-free rate, 2, 24, 47–50, 57, 60, 63, 68–70, 73, 76–78, 81, 85, 87–89, 101, 109, 110, 167, 271, 324, 325, 328, 340, 349, 353, 368, 374, 380, 400, 401, 414, 425, 431, 433–435, 443, 444, 447, 450, 476, 482, 490, 497, 498, 511, 521–523, 529, 533, 556, 561, 565, 574, 577, 591, 594, 599–601, 603, 604, 635, 640, 641, 643, 646, 650, 651, 676, 679–681, 684, 690, 691, 697, 702
continuously compounded, 167, 332, 449, 482, 529, 530, 561, 567
Risk-free rate of return, 28, *see* Expected rate of return
Risk-neutral investor, 26, 28, 194, 203
Risk-neutral probability, xvii, 22, 26, 28, 48, 54, 55, 57, 84–87, 183, 191, 203, 208, 209, 215, 216, 520, 521, 529, 530, 555, 559, 562, 564, 570–572, 574, 583, 590, 595, 599, 603, 631, 642–644, 646, 651, 674, 677, 688, 694, 715, 722, 729
Roll over strategy, 394

Secondary market, 92
Self-financing portfolio, 3, 5, 63, 327–330, 344, 345, 350, 352, 368, 372, 397, 652

definition, 3, 13
Sensitivity measures, xvii, 271, 312, 323, 325, 341, 347, 348, 350, 351, 353, 498
Delta, 281, 282, 347, 498, 524, 555
difference between Delta and Theta, 284
Gamma, 288, 292, 334, 555
of a portfolio, 325, 348
rho, 298, 299, 372
signs of, 317
Theta, 272, 358
Vega, 293, 296, 366
Short position, 2, 5, 8, 11, 12, 28, 30, 96, 398, 401, 404, 417, 422, 428–431, 433, 434, 436–438, 441, 444, 447, 494, 500
Short rate, 714, 715, 718, 720–724, 726–730, 733
evolution of, 714, 715, 722–724, 726–730, 733
volatility of, 727, 730
estimation, 724
Short sale, 2
Simudif, 607, 608, 610–612, 614
Spot curve, 373, 377, 380
Spot market, 62, 68, 89
Spot price, 414, 415, 417–419, 422, 424–426, 430, 436, 437, 443, 472–474, 479
Spot rate, 374–378, 380–384, 389, 391, 394, 404–406, 408–410, 680, 681, 704, 712
States of nature, 1, 2, 4, 6, 20, 24, 26, 27, 29–34, 47, 49, 50, 53, 57, 59, 70, 72, 82, 86, 324, 326, 347, 371, 510–514, 519, 521, 522, 599, 600, 632–634, 636, 639, 649, 736
continuum of, 453, 599, 736
States' prices, 23, *see* Stochastic discount factors, Valuation operator, Risk-neutral density
Static strategy, 400
Stochastic differential equation, 619, 622, 629–631, 641, 650, 651, 716
solution of, 631
Stochastic discount factor function, 215
Stochastic discount factors, xvii, 18, 23, 27–29, 37, 42, 48, 53, 54, 91, 93, 106, 107, 183, 184, 187, 189–191, 193, 212, 373, 403, 502, 510, 512–514, 518–522, 528, 530, 531, 540, 547, 570, 599, 642
risk-adjusted, 23
Stochastic evolution, 605, 608, 615
Straight bonds, 94
Stripay, 154–156, 159–161, 168, 169, 250
SumAbsDiv, 387, 418, 708
Swap, 72, 371, 449, 453–464, 467–479, 481–483, 600, 713, 733, 736
asset, *see* equity
buy down, 458, 482
buy up, 458, 482
commodity, 472, 473, 475, 479
currency, 72, 73, 83, 84, 453, 461–465, 468–470, 482, 483
deferred, 470, 483
definition, 72
equity, 74, 85, 432, 453, 472,

474, 475, 477, 481
fixed-for-float, 453, 456, 457, 475, 478, 481, 482, 712, 713
foreign exchange, *see* currency
notional principal of, 74, 453–455, 457, 459, 460, 475, 477, 712
par, 458
Swaptions, 706, 712, 713, 733
Synthetic loan, 152
Synthetic put, 152
Synthetic security, 152

Taylor series, 278, 283, 288, 290, 297, 601, 623–625
Term structure of interest rates, 373, 374, 377, 378, 383, 387, 389, 395, 403, 406, 408, 414, 415, 423, 427, 432, 438, 439, 448, 450, 457, 458, 468, 472, 477, 481–483, 712, 714, 720, 736
continuous approximation of, 391
downward sloping, 410
estimation of, 373, 383, 386, 389, 411
evolution of, 714, 715
flat, 383, 406, 410
smoothing of, 373, 383–385
continuous compounding, 389
upward sloping, 410
Term structure theories, 408
Theta, 271, 277, 288, 299, 312, 358, 370, 652
comparison to Delta, 284
of a call option, 272
sign of, 317
of a call option on a dividend-paying stock, 312
sign of, 317–319
of a put option, 314
sign of, 318, 320
of a put option on a dividend-paying stock, 315
sign of, 318, 320, 321
Theta neutral portfolio, 370, 652
Time value of money, 23, 91, 488, 490, 498, 499, 603
Time-value-of-money, xv
Trading strategies, 234
Transaction costs, 2
Treasury bills, 432

Unbiased expectation theory, 408, 409

Valbin, 565, 566, 675, 676
Valuation
by arbitrage, 47, 393, 403, 503, 600, 601, 635, 638, 640, 645, 650, 673, 735–737
by replication, 37, 54–56, 60, 63, 70, 74, 75, 77, 81, 85, 87, 373, 393, 431, 433, 450, 475, 500, 501, 503, 510, 519, 521, 522, 528, 530, 540, 555, 599, 600, 633, 635, 637, 642, 644, 650, 673
in an incomplete market, 35, 37
numerical, 528, 529
risk-neutral, 56, 57, 438, 521, 555, 594, 644, 650
via discount factors, 373, 393, 398, 400, 450, 519
via stochastic discount factors,

54–56, 510, 519, 521, 530, 531, 555
Valuation operator, 37, *see* Stochastic dicsount factors, Discount factors, Risk-neutral probability, 38, 42, 45
Valucash, 35, 77, 78
Value at risk (VAR), 737
Vdis, 20, 27, 47–49, 62, 63, 80, 82, 83, 85, 86, 374, 376
  and NarbitB, 107, 108
Vega, 271, 296, 297, 312, 366
  of a call option, 293
    sign of, 317
  of a call option on a dividend-paying stock, 312
    sign of, 317
  of a put option, 314
    sign of, 318
  of a put option on a dividend-paying stock, 315
    sign of, 318
  sign of, 295
Vega2, 297
Volatility, 173, 202, 259, 275, 339, 349, 351, 357, 361, 366, 367, 614, 620, 629, 679, 681, 704, 705, 714, 715, 717, 724, 727, 728, 730
  estimation of, 259, 261
    trial and error, 268
Volatility smile, 263

Wiener process, 196

Yield, 373, 380, 382, 383, 406, 408, 410
Yield curve, 373, 377, 380, 408–410
Yield to maturity, *see* Yield

Zero cost portfolio, 96
Zero sum game, 130
Zero-coupon bond, 97, 377–381, 383, 405, 406, 424, 432, 449, 467, 511, 591, 706, 720, 723, 724, 726–728, 733
Zero-coupon curve, 373, 377, 378

Waterloo Maple Inc.
License Agreement

IMPORTANT NOTICE: READ CAREFULLY BEFORE OPENING THE SEALED PACKAGE

THE ENCLOSED SOFTWARE IS PROVIDED TO YOU UNDER THE FOLLOWING LICENSE AGREEMENT. THIS LICENSE AGREEMENT IS A LEGAL AGREEMENT BETWEEN YOU (EITHER AN INDIVIDUAL OR A SINGLE ENTITY) AS END USER, AND WATERLOO MAPLE INC. ("WMI"), AND DEFINES WHAT YOU MAY DO WITH THE SOFTWARE, AND WHAT LIMITATIONS EXIST ON WARRANTIES AND REMEDIES RELATED TO THE SOFTWARE. BY OPENING THE SEALED PACKAGE, YOU BECOME A PARTY TO, AND AGREE TO BE BOUND BY ALL THE TERMS AND CONDITIONS OF, THIS LICENSE AGREEMENT. IF, AFTER READING THIS LICENSE AGREEMENT, YOU DO NOT ACCEPT OR AGREE TO THE TERMS AND CONDITIONS CONTAINED IN THIS LICENSE AGREEMENT, DO NOT OPEN THE SEALED PACKAGE. YOU MAY, WITHIN 15 DAYS OF PURCHASE, RETURN THE ENTIRE UNOPENED PACKAGE AND ALL OTHER ITEMS CONTAINED IN THIS BOX (INCLUDING THE PACKAGING), TOGETHER WITH YOUR RECEIPT, TO THE PLACE WHERE YOU OBTAINED THEM FOR A FULL REFUND OF THE AMOUNT PAID FOR THE PACKAGE.

LICENSE AGREEMENT

OWNERSHIP OF SOFTWARE.
- WMI owns and retains all right, title and interest, including all copyrights and other intellectual property rights, in MAPLE (which includes computer software and associated media (collectively, the "**Software**") as well as related printed materials, including the learning guide and/or programming guide (the "**Printed Materials**") contained in this package.
- WMI does not sell any rights in the Software, but rather grants the right to use the Software by means of a software license.
- WMI reserves all rights with respect to the Software and Printed Materials not expressly granted by this License Agreement.
- You own only the magnetic or other physical media on which the Software is recorded or fixed in this package.
- The Software and Printed Materials are protected by copyright laws and international copyright treaties, as well as other intellectual property laws and treaties.

GRANT OF LICENSE. In consideration of payment of the license fee, which is part of the price you paid for this package, and your agreement to abide by the terms and conditions of this License Agreement, WMI, as licensor, grants to you a non-exclusive, revocable, personal, non-transferable license and right to use this copy of the Software.

YOU MAY:
- *Install and use* this copy of the Software on a single computer at a single location, or physically transfer the Software from one computer to another (including permanently and completely deleting the Software from the first computer), provided the Software is used only by a single user on a single computer at one time. If you wish to use the Software for multiple users, you will need to purchase an additional copy of the Software for each user, and ensure that the Software is used only for the contracted number of concurrent users.
- *Copy* one copy of the Software solely for backup purposes (if the Software is not copy-protected) provided you reproduce and include the copyright notice on the backup copy and use the backup copy solely for archival purposes.

YOU MAY NOT:
- *Use, reproduce, transmit, modify, adapt or translate* the Software or Printed Materials, in whole or in part, to others, except as otherwise permitted by this License Agreement.
- *Reverse engineer, decompile, disassemble,* or create derivative works based on the Software.
- *Use* the Software in any manner whatsoever with the result that access to the Software may be obtained through the Internet including, without limitation, any web page, other than as may be specifically negotiated, in writing with WMI.
- *Rent, lease, license, transfer, assign, sell* or otherwise provide access to the Software or Printed Materials, in whole or in part, on a temporary or permanent basis, except as otherwise permitted by this License Agreement.
- *Alter, remove or cover* proprietary notices in or on the Software, Printed Materials or storage media.
- *Use* the Software in any unlawful manner whatsoever.

TERM. This License Agreement commences upon your acceptance by opening the sealed package, and is effective until terminated. WMI may terminate this License Agreement without notice to you if you fail to comply with any provision of this License Agreement. Upon termination, you agree to immediately cease using the Software and Printed Materials and to return all copies of the Software and Printed Materials in your possession to WMI within five days of termination. The Ownership of Software and Limited Warranty and Liability sections of this License Agreement shall continue in force after any termination.

WARRANTY. WMI warrants to you as original end user, that the storage media on which the Software is recorded is free from defects in materials and workmanship under normal use and service for a period of ninety (90) days from the date of delivery to you as evidenced by a copy of the invoice. If such a defect exists, return the entire package, including packaging, postage prepaid with a copy of the invoice to WMI at the address below, and WMI, at its option, shall either (a) return the purchase price or (b) replace the media. If failure of the media has resulted from accident, abuse, or misapplication, WMI shall have no responsibility whatsoever to refund the purchase price or replace the media. In the event of replacement of the media, the replacement media will be warranted for the remainder of the original warranty period or thirty (30) days, whichever is the longer. This remedy is your exclusive remedy for a breach of this warranty, and WMI's entire liability and only warranty made with respect to the Software and Printed Materials. THIS LIMITED WARRANTY GIVES YOU SPECIFIC LEGAL RIGHTS, AND YOU MAY HAVE OTHER RIGHTS WHICH VARY FROM JURISDICTION TO JURISDICTION.

LIMITED WARRANTY AND LIABILITY. OTHER THAN AS OUTLINED ABOVE AND TO THE MAXIMUM EXTENT PERMITTED BY APPLICABLE LAWS, THE SOFTWARE AND PRINTED MATERIALS ARE PROVIDED "AS IS" WITHOUT ANY WARRANTY OR CONDITION OF ANY KIND, EITHER EXPRESS OR IMPLIED, STATUTORY OR OTHERWISE, INCLUDING BUT NOT LIMITED TO THE IMPLIED WARRANTIES OR CONDITIONS OF MERCHANTABILITY AND FITNESS FOR A PARTICULAR PURPOSE. NO ORAL OR WRITTEN INFORMATION OR ADVICE GIVEN BY WMI, ITS DEALERS, DISTRIBUTORS, AGENTS OR EMPLOYEES (COLLECTIVELY, "**AGENTS**") SHALL CREATE A WARRANTY OR IN ANY WAY INCREASE THE SCOPE OF THIS WARRANTY. YOU ASSUME THE ENTIRE RISK AS TO THE USE AND PERFORMANCE OF THE SOFTWARE OR PRINTED MATERIALS IN TERMS OF CORRECTNESS, ACCURACY, RELIABILITY, CURRENTNESS, OR OTHERWISE. IN NO EVENT SHALL WMI, ITS AGENTS OR ANYONE ELSE WHO HAS BEEN INVOLVED IN THE CREATION, PRODUCTION OR DELIVERY OF THE SOFTWARE AND/OR PRINTED MATERIALS BE LIABLE TO YOU OR ANY OTHER PERSON FOR ANY DIRECT, INDIRECT, SPECIAL, CONSEQUENTIAL OR INCIDENTAL DAMAGES (INCLUDING WITHOUT LIMITATION, DAMAGES FOR LOSS OF REVENUES OR PROFITS, BUSINESS INTERRUPTION, LOSS OF BUSINESS INFORMATION, AND THE LIKE) ARISING OUT OF THE USE OR INABILITY TO USE THE SOFTWARE EVEN IF WMI OR ITS AGENTS HAVE BEEN ADVISED OF THE POSSIBILITY OF SUCH DAMAGE OR CLAIM, OR IT IS FORESEEABLE. BECAUSE SOME JURISDICTIONS DO NOT ALLOW THE EXCLUSION OR LIMITATION OF LIABILITY FOR CONSEQUENTIAL OR INCIDENTAL DAMAGES, THE ABOVE LIMITATION MAY NOT APPLY TO YOU. WMI'S MAXIMUM AGGREGATE LIABILITY TO YOU SHALL NOT EXCEED THE AMOUNT PAID BY YOU FOR THE SOFTWARE AND PRINTED MATERIALS. THE LIMITATIONS OF THIS SECTION SHALL APPLY WHETHER OR NOT THE ALLEGED BREACH OR DEFAULT IS A BREACH OF A FUNDAMENTAL CONDITION OR TERM. SOME JURISDICTIONS DO NOT ALLOW LIMITATIONS ON DURATION OF AN IMPLIED WARRANTY, SO THE ABOVE LIMITATION MAY NOT APPLY TO YOU.

ACKNOWLEDGEMENT. You acknowledge that you have read this License Agreement and limited warranty, understood them, and agree to be bound by their terms and conditions. You also agree that this License Agreement is the complete and exclusive agreement between you and WMI, and supersedes all prior agreements, representations and any other communications, oral or written, between you and WMI relating to the subject matter of the License Agreement including, without limitation, any warranties with respect to the Software and Printed Materials. This License Agreement may only be amended by written agreement of both parties.

EXPORT CONTROLS. The Software and Printed Materials is subject at all times to all applicable export control laws and regulations in force from time to time. You agree that you shall not make any disposition of the Software and Printed Materials purchased or licensed from WMI.

GOVERNING LAW. This License Agreement is governed by the laws of the Province of Ontario, Canada and, if the Software and Printed Materials were acquired within Canada, each of the parties hereto irrevocably attorns to the exclusive jurisdiction of the courts of the Province of Ontario without regard to conflicts of laws principles. If the Software and Printed Material were acquired outside Canada, each of the parties hereto irrevocably attorns to the non-exclusive jurisdiction of the courts of the Province of Ontario, provided that the Licensee agrees that any claim or action brought by the Licensee shall be commenced in the courts of the Province of Ontario. The parties agree that the *United Nations Convention on Contracts for the International Sale of Goods* does not apply to this License Agreement.

Should you have any questions concerning this License Agreement, or if you desire to contact WMI for any reason, please contact in writing:

Waterloo Maple Inc.
Customer Care Department
57 Erb Street W.
Waterloo, Ontario, Canada
N2L 6C2